AF903155

Advances in Experimental Medicine and Biology

Volume 1192

Series Editor

John D. Lambris, University of Pennsylvania, Philadelphia, PA, USA
Wim Crusio, CNRS University of Bordeaux UMR 5287, Institut de Neurosciences Cognitives et Intégratives d'Aquitaine, Pessac Cedex, France
Nima Rezaei, Children's Medical Center Hospital, Tehran University of Medical Sciences, Tehran, Iran

Advances in Experimental Medicine and Biology presents multidisciplinary and dynamic findings in the broad fields of experimental medicine and biology. The wide variety in topics it presents offers readers multiple perspectives on a variety of disciplines including neuroscience, microbiology, immunology, biochemistry, biomedical engineering and cancer research.

Advances in Experimental Medicine and Biology has been publishing exceptional works in the field for over 30 years and is indexed in Medline, Scopus, EMBASE, BIOSIS, Biological Abstracts, CSA, Biological Sciences and Living Resources (ASFA-1), and Biological Sciences. The series also provides scientists with up to date information on emerging topics and techniques. 2018 Impact Factor: 2.126.

More information about this series at http://www.springer.com/series/5584

Yong-Ku Kim
Editor

Frontiers in Psychiatry

Artificial Intelligence, Precision Medicine, and Other Paradigm Shifts

Springer

Editor
Yong-Ku Kim
College of Medicine
Korea University
Seoul, Gyeonggi-do, Korea (Republic of)

ISSN 0065-2598 ISSN 2214-8019 (electronic)
Advances in Experimental Medicine and Biology
ISBN 978-981-32-9720-3 ISBN 978-981-32-9721-0 (eBook)
https://doi.org/10.1007/978-981-32-9721-0

© Springer Nature Singapore Pte Ltd. 2019
This work is subject to copyright. All rights are reserved by the Publisher, whether the whole or part of the material is concerned, specifically the rights of translation, reprinting, reuse of illustrations, recitation, broadcasting, reproduction on microfilms or in any other physical way, and transmission or information storage and retrieval, electronic adaptation, computer software, or by similar or dissimilar methodology now known or hereafter developed.
The use of general descriptive names, registered names, trademarks, service marks, etc. in this publication does not imply, even in the absence of a specific statement, that such names are exempt from the relevant protective laws and regulations and therefore free for general use.
The publisher, the authors and the editors are safe to assume that the advice and information in this book are believed to be true and accurate at the date of publication. Neither the publisher nor the authors or the editors give a warranty, expressed or implied, with respect to the material contained herein or for any errors or omissions that may have been made. The publisher remains neutral with regard to jurisdictional claims in published maps and institutional affiliations.

This Springer imprint is published by the registered company Springer Nature Singapore Pte Ltd.
The registered company address is: 152 Beach Road, #21-01/04 Gateway East, Singapore 189721, Singapore

Preface

This book reviews recent advances and paradigm shifts in psychiatric clinical practice and research. Extraordinary discoveries that triggered paradigm shifts in medicine have been often ignored, denied, or underestimated at the time of initial discovery. Thus, the aim of this book, predicting ongoing or upcoming paradigm shifts in psychiatry, is challenging compared to the explanation of past paradigm shifts. However, we can infer the necessary characteristics from experience with paradigm shifts in medicine. This book therefore intends to classify recent research advances that are likely to be considered paradigm shifts in the future. This is the realistic and prudent goal of this article. Inventions that can show what we could not see before can lead to a paradigm shift.

This book reviews all the important aspects of the challenges and strategies of future psychiatry, covering issues, i.e., artificial intelligence, big data, precision medicine, and other paradigm shifts. In the first part, we introduce big data and discovery science, the shift of defining psychiatric disorders from a categorical approach into a dimensional approach, and genetic markers in psychiatry. Advances in information technology can lead to a paradigm shift that allows us to collect scattered, fragmented data and discover hidden meanings in stored data. To overcome the limitations of the categorical approach, psychiatrists have considered adopting a dimensional approach. However, their efforts were upset in the DSM-5 revision process. Genetic markers can be used for early prediction, detecting the risk of developing psychiatric disorders, novel subtypes of the diseases, and a tailored, personalized approach to therapy. Small vessel disease is strongly associated with a variety of neuropsychiatric syndromes in late-life, including late-life depression. Advanced imaging techniques can help identify mechanisms relating small vessel disease to depression and may help identify targets for prevention and treatment.

The second part addresses the applications of artificial intelligence (AI), machine learning, and imaging connectomics and neuroimaging-based biomarkers in psychiatry. Although the past few years have witnessed an increase in the use of AI in the medical practice, its role in psychiatry remains a complex and unanswered question. We provide the current state of knowledge of AI's use in the diagnosis, prediction, and treatment of psychiatric disorders, and examine the challenges and limitations of this approach in medical practice. Machine learning-based precision psychiatry for pretreatment prediction may become a reality in patient care following prospective large clinical trials to validate clinical factors and relevant biomarkers. Neuroimaging-based personalized medicine is emerging to characterize brain disorders and their evolution at the patient level. We present classic methods used to infer large-scale brain connectivity based on functional MRI. The future of neuroimaging for clinical practice is bright, especially for its potential to fill the current gap of predictive, personalized, and preventative psychiatry.

In the third part, we focus on precision medicine in psychiatry. Biomarker-guided tailored therapy is a promising future treatment of psychiatric disorders, and the personalized or precision medicine approach to psychiatric disorders is a very active avenue of investigation. We propose the necessity to shift the focus from population-based intervention to individual-based intervention in the design of clinical research. Women and children's psychiatry needs new paradigms to challenge researches and treatments. Smart health systems, through the brain mechanisms of smartphones and Internet game addiction, provide great opportunities for practicing precision medicine in behavioral addictions.

In the fourth part, we propose ongoing paradigm shifts in animal research, treatments, early identification of psychiatric disorders, and new research. We discuss challenges and future perspectives for modeling neuropsychiatric disorders using human-induced pluripotent stem cell-derived neurons. Rigorous efforts have been made for early detection and early intervention in psychiatry. We explore the current high-risk studies for various mental illnesses and aim to provide future direction for early detection studies in psychiatry. Multiple aspects of diseases and adapted animal models lead to difficulties for translational research in psychiatry. Animal research in psychiatry will provide good translational value.

The fifth part introduces important concepts for theoretical psychiatry, a developmental approach for defining psychopathology, the theory of constructed emotion, resilience, inflammation, and a frontier for suicide. Theoretical psychiatry will give all psychiatrists a common language, build bridges over academic gaps, and creatively export insights across disciplinary borders. Developmental psychopathology is a relevant paradigm of psychopathological research encompassing the study of the development of psychiatric disorders by means of a life course perspective. We provide a perspective on the complex formation of emotion and its operational use in neuroscience. The field of emotion research is wide and highly dynamic; yet it provides a crucial element in investigating psychiatric disorders and can be used both in basic science and in clinical practice. Resilience is the human capacity to adapt swiftly and successfully to stress and to revert to a positive state afterward. We posit the construct of resilience as a psychopathological construct for mental disorders. Psychiatric disorders are too multifactorial to be defined as a primarily inflammatory disorder, and an increased inflammatory response is not specific to mental disorders. It is possible to use inflammatory markers as depression biomarkers in subtypes of depression, which can serve as a basis for developing medications to treat the disorder.

In the final part, novel interventions for treatments in psychiatric disorders have been proposed. Intervention modifying the gut microbiome, computer (i.e., web, app)-based therapeutics to overcome low accessibility of psychotherapy, novel treatment related to the telomere–telomerase system, and neuromodulation augmenting cognitive control of emotion will be legitimate treatment options for various psychiatric disorders in the near future.

The book, written by leading experts from across the world, will be of value to all who seek a better understanding of ongoing or upcoming paradigm shift in psychiatry. I sincerely thank all of the authors for their valuable time that was spent preparing manuscripts.

Seoul, South Korea

Yong-Ku Kim
Editor

Contents

Challenges and Strategies

Big Data and Discovery Sciences in Psychiatry 3
Kyoung-Sae Na, Changsu Han and Yong-Ku Kim

Classification of Psychiatric Disorders 17
Yong-Ku Kim and Seon-Cheol Park

Genetic Markers of Alzheimer's Disease 27
Matea Nikolac Perkovic and Nela Pivac

Genetic Markers in Psychiatry 53
Gordana Nedic Erjavec, Dubravka Svob Strac, Lucija Tudor,
Marcela Konjevod, Marina Sagud and Nela Pivac

Neuroimaging of Small Vessel Disease in Late-Life Depression 95
Nadim S. Farhat, Robert Theiss, Tales Santini,
Tamer S. Ibrahim and Howard J. Aizenstein

Advances in Artificial Intelligence Technologies

Artificial Intelligence in Psychiatry 119
Marc Fakhoury

Machine Learning in Neural Networks 127
Eugene Lin and Shih-Jen Tsai

Imaging Connectomics and the Understanding of Brain Diseases 139
Andrea Insabato, Gustavo Deco and Matthieu Gilson

Development of Neuroimaging-Based Biomarkers in Psychiatry 159
Orkun Aydin, Pinar Unal Aydin and Ayla Arslan

Research Methods for Precision Psychiatry

Biomarker-Guided Tailored Therapy 199
Jessica Lydiard and Charles B. Nemeroff

Women's Psychiatry 225
Georgia Balta, Christina Dalla and Nikolaos Kokras

Understanding Mood Disorders in Children 251
Ho-Jun Lee, Seung-Hyun Kim and Moon-Soo Lee

Smart Healthcare Systems and Precision Medicine 263
Soo-Hyun Paik and Dai-Jin Kim

Ongoing Paradigm Shifts

Animal Research in Psychiatry 283
Michel Bourin

Modeling Psychiatric Diseases with Induced Pluripotent Stem Cells 297
Eline van Hugte and Nael Nadif Kasri

Psychopharmacology and Psychotherapy Research 313
Kiwon Kim and Hong Jin Jeon

An Integrated Bio-psycho-social Approach to Psychiatric Disorders 331
Seon-Cheol Park and Yong-Ku Kim

Early Identification of Psychiatric Disorders 341
Tae Young Lee, Minah Kim and Jun Soo Kwon

New Theoretical Frameworks for Research

Theoretical Psychiatry as a Link Between Academic and Clinical Psychiatry 355
Miro Jakovljevic and Ivan Jakovljevic

A Transdisciplinary Integrative Approach for Precision Psychiatry 399
Miro Jakovljevic and Ivan Jakovljevic

Applications of Developmental Psychopathology 429
Gonzalo Salazar de Pablo, Julio David Vaquerizo Serrano, Sandra Gómez Vallejo, Javier Sánchez Cerezo and Carmen Moreno Ruiz

Concepts and Dysfunctions of Emotion in Neuropsychiatric Research 453
Zumrut Duygu Sen, Lejla Colic, Vanessa Kasties and Martin Walter

Resilience as a Psychopathological Construct for Psychiatric Disorders 479
Amresh Shrivastava, Avinash De Sousa and Pragya Lodha

Role of Inflammation in Psychiatric Disorders 491
Sang Won Jeon, Ho-kyoung Yoon and Yong-Ku Kim

The Frontiers of Suicide 503
Sheikh Shoib and Yong-Ku Kim

Novel Interventions

Interventions Addressing the Telomere-Telomerase System 521
Ather Muneer

Neuromodulation and Cognitive Control of Emotion 545
Meysam Amidfar, Young-Hoon Ko and Yong-Ku Kim

Psychobiotics 565
Alper Evrensel, Barış Önen Ünsalver and Mehmet Emin Ceylan

Digital Interventions for Mental Disorders: Key Features, Efficacy, and Potential for Artificial Intelligence Applications 583
David Daniel Ebert, Mathias Harrer, Jennifer Apolinário-Hagen and Harald Baumeister

Cognitive Behavioral Therapy for Insomnia in the Digital Age 629
Seung-Gul Kang and Yong-Ku Kim

Challenges and Strategies

Big Data and Discovery Sciences in Psychiatry

Kyoung-Sae Na, Changsu Han and Yong-Ku Kim

Abstract

The modern society is a so-called era of big data. Whereas nearly everybody recognizes the "era of big data", no one can exactly define how big the data is a "big data". The reason for the ambiguity of the term big data mainly arises from the widespread of using that term. Along the widespread application of the digital technology in the everyday life, a large amount of data is generated every second in relation with every human behavior (i.e., measuring body movements through sensors, texts sent and received via social networking services). In addition, nonhuman data such as weather and Global Positioning System signals has been cumulated and analyzed in perspectives of big data (Kan et al. in Int J Environ Res Public Health 15(4), 2018 [1]). The big data has also influenced the medical science, which includes the field of psychiatry (Monteith et al. in Int J Bipolar Disord 3(1):21, 2015 [2]). In this chapter, we first introduce the definition of the term "big data". Then, we discuss researches which apply big data to solve problems in the clinical practice of psychiatry.

Keywords

Big data · Psychiatry · Electronic health records · Suicide · Delirium

K.-S. Na
Department of Psychiatry, Gachon University Gil Medical Center, Incheon, Republic of Korea
e-mail: ksna13@gmail.com

C. Han · Y.-K. Kim (✉)
Department of Psychiatry, College of Medicine, Korea University, Seoul, Republic of Korea
e-mail: yongku@korea.ac.kr; yongku@korea.edu

C. Han
e-mail: hancs@korea.ac.kr

© Springer Nature Singapore Pte Ltd. 2019

Y.-K. Kim (ed.), *Frontiers in Psychiatry*, Advances in Experimental Medicine and Biology 1192, https://doi.org/10.1007/978-981-32-9721-0_1

Definition

There is no single definition on the big data. Historically, a lot of researchers or institutions have defined and used the term big data. As expected, the most commonly used concept is volume or amount, which essentially refers the adjective of "big". In perspective of visualization, one of initial definition on the big data only referred the amount ("A dataset that is too big to fit on screen") [3]. Another researcher defined big data in similar perspectives as following: "Data that exceeds the processing capacity of conventional database systems" [4]. On the other hand, some companies defined big data in relation with industrial values as followings: "Big data is a term that describes the large volume of data—both structured and unstructured—that inundates a business on a day-to-day basis. But it's not the amount of data that's important" [5]. As shown in the above definition, if big data has no values, the current omnipresent use of the term big data would not be possible. Big data is present to be explored and finally unfolded [6], so that the hidden information inside the data could be utilized in the real world. Now the big data is one of the most important recourses for major industrial sections. Most of the global companies such as Netflix primarily exploit big data and improve their productivity [7]. The movie recommendation system based on the big data with machine learning of Netflix made the company grow fast.

However, big data cannot be directly used by end user's objectives. Rather, there are such hard jobs to make the big data available. The process is usually called to "data wrangling" or "preprocessing". The reason for the hard process is not solely due to the large amount of data. Rather, other inevitable property of big data makes it hard to be used easily. Nowadays, the term "big data" has inevitably included those nature of complexity.

Key Concept of Big Data: The *n* "Vs"

To explain the property of big data, the concept of "4 Vs" is widely used. The principal components of the 4 Vs were Volume, Velocity, Variety, and Veracity. Whereas the "volume" is one of the undeniable key features of big data, lack of the other Vs hinders utilizing the large amount of volume practically. Among the remained three Vs, the issues in terms of veracity make big data mistrustful. Recently, 5 Vs (4 Vs + Value) [8] and even 10 Vs (5 Vs + Validity, Venue, Vocabulary, Vagueness, and Variability) were proposed as key concepts for big data [9]. Whatever we define the number and the contents of the Vs in the big data, the only one emphasized by everyone is that the volume alone cannot create any values from the big data.

Volume

The size of data related to healthcare has grown. As the amount of data increase substantially, the measurement for the data has also increased (Table 1) [10].

Typical hospitals with 500 beds have approximately more 50 petabytes [11]. According to that resource, the total amount of digital healthcare data is estimated to be 153 exabytes, and it is expected to reach 2314 exabytes by 2020. The amount of digital data was 0.02 billion gigabyte in 1987, which had increased by more than 27,000 times in 2007 (276.12 billion gigabytes) [12]. Now the zettabyte has been frequently told. In 2015, Google processed 24 petabytes of data per day. In that period, a total amount of data processed in the Duke Heart Center was less than 30 terabyte per year [13]. Indeed, the velocity of generating data has become so fast recently. By 2013, it was reported that 90 of the world's data generated over the last two years (2011–2012) [14].

The progress of big data is not separable from the development of web-based information production. By 2016, the global Internet traffic exceeded one zettabyte (10^{21}). The current period is thus called "Zettabyte Era" [15].

Other Vs

The velocity of big data is close associated with real-time analysis. Buying and selling products occur every second. The fraud preventing system by using big data is common in most credit card companies [16]. In the field of healthcare, efforts to collect and analyze signals generated from the human body in a real time manner have been developed.

There are numerous cases that the problems in the veracity made the results from the big data wrong and useless. One of the most famous things would be the failure of the "Google Flu". The failure of the "Google Flu" evoked concerns for the validity of the big data in the fields of healthcare from multidisciplinary perspectives [17–20]. Among the various factors that interrupted the prediction of Google Flu, insufficient veracity was considered the main issue. As Google Flu is mainly relied on the searching terms in the Google website, if the searching terms

Table 1 Measures of data size

Bytes	1
Kilobytes	10^3 bytes
Megabytes	10^6 bytes
Gigabytes	10^9 bytes
Terabytes	10^{12} bytes
Petabytes	10^{15} bytes
Exabytes	10^{18} bytes
Zettabytes	10^{21} bytes
Yottabytes	10^{14} bytes

are wrong, then the results should be inevitably wrong. Unfortunately, the searching terms are not entered by medical professionals, but by laypeople. As a lot of warning for the flu by media, the laypeople searched "flu" and its related terms more and more, and the Google Flu misunderstood the rising flu and its related terms as the widespread flu. Eventually, the Google Flu overestimated the prevalence of flu two times more than the Center for Disease Control and Prevention. Since 2011 August, Google Flu overestimated the prevalence of flu more than actual events during 100 weeks out of a total of 108 weeks. The failure of Google Flu indeed emphasizes that good data is more important than big data as titled in the Harvard Business Review [18]. Particularly, medical science and healthcare are based on the professional knowledge and require more accuracy than other areas of sciences. However, big data and data-driven analysis have too many possibilities to improve the current state of medical technology and delivery so cannot be ignored due to the veracity problems.

Another form of healthcare big data is claimed data. For example, every citizen receives the National Health Insurance Service (NHIS) in the Republic of Korea [21]. The healthcare insurance data are freely available for qualified researchers. However, there are two critical limitations. First, the available contents of the data are so small. Basic information in the dataset of the NHIS are age group, gender, ICD code for the diagnoses, department of the medicine, type of vising to the hospital (i.e., outpatient, inpatient, emergency), date of vising to the hospital, prescriptions and related information (i.e., medication, prescribed length of the medication, number of the medication, amount of the medication, costs), and medical examinations (i.e., MRI, CT, ultrasonography, blood laboratories). Such information can be useful for analyzing cost-effectiveness research. However, there are no results of the laboratory examinations. Additionally, there are no clinical information judged by doctors. In the field of psychiatry, the use of the clinician-rated scales and interviews are particularly important. What makes it worse is the inaccurate diagnoses of the claimed data. To be approved by the Health Insurance and Review Assessment, there are cases in which psychiatric disorders are diagnosed to patients for claiming. Hence, although the size is big, veracity could not be met to solve causal relationships with time series data. The criticism for the issues in the claim data has been increasingly raised [22]. Those critics urge to construct electronic health records (EHR), which will be discussed later in this chapter.

On the other hand, the increasing speed of digital data is so fast; it may be not so meaningful to estimate the amount of data at a certain point. Medical science and healthcare is no exception from the big data. There is quite a bit of effort in the medical science, including the field of psychiatry. A lot of issues of big data have been discussed. And increasing results of data from big data has been reported in the fields of medical science and healthcare. In this section, we discuss essential features of big data and implication of big data in the prevention, diagnosis, and treatment of psychiatric conditions including suicide. The mobile mental health is one of the widely investigated and disseminated platforms where the big data and psychiatry generate synergistic effects [23].

Big Data for Identifying Mechanisms of Psychiatric Disorders

Big data can be used in two ways in the field of psychiatry, mechanism, and pragmatism. Omics is possibly one of the most widely investigated research fields for identifying etiological mechanisms of human diseases. Genomic and proteomic are the most widely known omics. Whereas traditional researches have focused on the genomics and proteomics, recent investigators more widen their scope to the metagenomic and metabolomics [24]. The two major catalytic factors for the widespread enthusiasm on the omics are technological development and the quantitative accumulation of the data, such as the big data [25]. Based on the integration of various omics, scientists move a step closer to identifying neurobiological mechanisms. However, the steps seem to be far away from the destination in a psychiatric world.

The biggest difference in diagnosing psychiatric and other medical diseases is the basis on which the diagnosis is made. Among healthcare, psychiatry has distinct positions in several perspectives. First, unlike physical diseases based on the measurable biological pathology, psychiatric disorders are still relied on the operational definition of diagnosis [26]. As the diagnosis and evaluation of the psychiatric disorder is primarily focused on the patients' own expression, behaviors reported by caregivers, and psychiatrists' evaluation, the phenomenological symptoms are still exceptionally important than other medical diseases. Psychopathology and phenomenological approach have been traditionally an important part of psychiatry for a long time [27, 28].

Although numerous neurobiological markers of psychiatric diseases have been researched and identified, most of those findings have not been embedded in the diagnostic criteria. The only biological diagnostic markers included in the Diagnostic and Statistical Manual of Mental Disorders, Fifth Edition (DSM-5) would be a hypocretin deficiency in narcolepsy [26]. Major psychiatric disorders such as schizophrenia, bipolar disorder, and depression are still diagnosed by the constellation of symptoms and the impairment in the major areas of psychosocial function. Given the lack of decisive neurobiological markers and reliance on the phenomenological symptoms, results from the omics big data such as genomics and proteomics have not been translated into the real-world clinical practices.

If the diagnosis is made based on the nonbiological, nonmeasurable basis then the results obtained from the omics research will not be psychiatric disease-specific nor elucidating etiological mechanisms of the disease. This is a fundamental limitation which is commonly applied to most psychiatric disorders. Nowadays, it is not unusual to say that major psychiatric disorders such as schizophrenia, major depressive disorder, and bipolar disorder are heterogeneous within their own diagnostic entities [29, 30]. The distance between psychiatric diagnosis and the omics approach is unavoidable because the two systems originated from different origins. The psychiatric diagnosis is based on the descriptive psychopathology [31].

The psychiatric diagnosis made on the descriptive psychopathology contributed to the high reliability of the psychiatric disorders with the aid of the structural interview tools such as the MINI and CIDI [32, 33]. However, there are two critical dark sides of the descriptive psychiatric diagnosis. First, the descriptive diagnosis does not reflect psychodynamic concepts which have been traditionally investigated. Second, the diagnostic criteria have no relationships with etiological mechanisms obtained in the omics levels. The dissociation between the diagnosis and the omics inevitably raise a circular problem from the causes to the consequences (Fig. 1). Let's assume that you begin to study the association between genomics and proteomics of schizophrenia. First, you should recruit patients with schizophrenia according to the diagnostic criteria. The patients diagnosed with schizophrenia might have common phenomenological characteristics such as delusion, hallucination, and thought problems which require operational criteria of the duration of symptoms and functional impairment. Some patients might have strong paranoid ideation and impulsivity with commanding auditory hallucination. On the other hand, some patients might be prominently withdrawn from social activity and have cognitive impairments. Regardless of the clinical subtypes of the schizophrenia, they are all the same schizophrenia according to the diagnostic criteria. Once you recruited patients, then you can measure various neurobiological factors such as neuroimaging, neuropsychological, genomics, and proteomics. Finally, the characteristics specific to the schizophrenia is obtained and reported. How much do the results reflect neurobiological etiology of schizophrenia? No one can certainly answer the question. With consideration on the diagnostic issues, efforts such as research diagnostic criteria (RDC) have been conducted, but tangible results have not yet been produced.

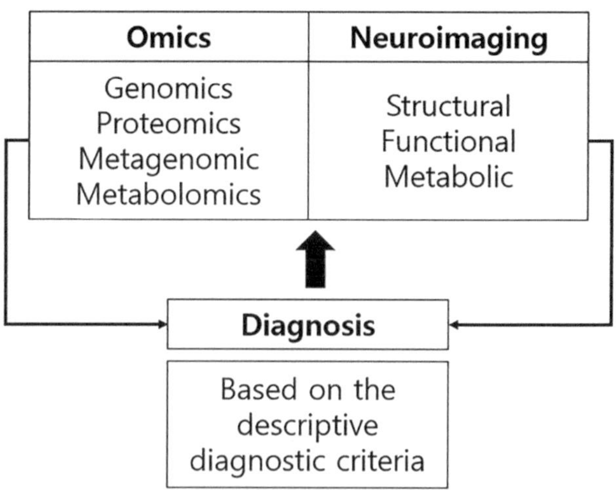

Fig. 1 Limitations of big data for identifying mechanisms of psychiatric disorders

Big Data for Pragmatism

The unavailability of biological markers accompanied by the importance of phenomenological evaluation naturally led to the detailed records on the patients' visit to the hospital. As the progress of the information and computer sciences, handwritten paper records now mostly became electronic health records (EHR) which is saved in and retrieved from the database. Although the accuracy and generalization of the analysis results still under debated and to be much further improved, the EHR may be one of the rapidly progressive edges of the big data approach in the psychiatry.

Some behavioral problems have been considered major psychiatric conditions. Suicide is one the representative one. Unlike several diseases that can be biologically and genetically determined, suicide occurs from very complicated complexes, which hinder from early detection and prevention. In addition, the occurrence rate of suicide is very low, which is measured by 100,000 population. For example, the suicide rate of the Republic of Korea, which have ranked first among the Organization for Economic Co-operation and Development (OECD) countries for the last 12 years, is 25.8 in 2015 [34]. One of the alternative approaches is to utilize suicide attempters, who are considered the most proximal surrogates for the suicide victims. However, not all the major components of suicide show similar patterns between suicide attempters and victims. For example, suicide methods are quite different between suicide victims and attempters. Among suicide victims, hanging is the most commonly used, which was followed by jumping from high places and poisoning. On the other hand, wrist cutting and poisoning [35]. If we plan to directly investigate the cases of the suicide victims, then we should search for their records from the health institutions.

The EHR has become one of the major resources in the field of healthcare big data. Unlike the Google search terms, the EHR is made by clinical experts who interpret and judge the patients' status. Hence, the quality of the veracity can be secured when appropriately utilized. There are discrepancies among countries which permit the range of using the EHR. For example, in case of Denmark, every citizen is registered in the healthcare, welfare, education, and employment data base [36]. With following-up from birth to the death, a huge amount of healthcare information can be analyzed with other major social domains.

When a huge amount of EHR with detailed clinical information is available, then the big data can contribute to the psychiatric management in the clinical settings. In the following section, we primarily focus on suicide and delirium as representative examples of the EHR-utilizing researches. The power of the EHR is more strengthen when it is combined with machine learning. Machine learning for the well-defined data of the EHR can lead to the data-driven analysis and decision. For example, a recent study suggested a personalized follow-up strategy for patients with MDD [37].

Suicide

Suicide is one of the most widely investigated psychiatric conditions in terms of big data and machine learning. Suicide has several distinct characteristics. First, suicide is still not a psychiatric disorder. There is no need for operative diagnostic criteria for suicide. It is a simple phenomenon. However, the simple definition of suicide does not guarantee its homogeneity. Although all suicide is the same phenomenologically, the underlying reasons for why they did are widely various. Although a lot of neurobiological etiological factors have been identified and reported [38, 39], no one clearly explains why someone kills themselves. Given the lack of common neurobiological factors which drives one to death, various and large amount of linguistic information are needed to understand suicide of each person. To date, psychological autopsy studies have identified risk factors for suicide based on suicide victims [40]. This may be one of the primary reasons why big data analysis based on the text analysis with EHR has been widely conducted for suicide.

A research group conducted a study to predict suicide behaviors, which include both suicide and nonfatal results of suicide attempt [41]. They analyzed records of a total of 8,980,954 person-year between 1998 and 2012. Bayesian analysis, which is one of the machine learning models, was used to predict the outcomes. Variables include laboratory findings, clinical symptoms, and sociodemographic characteristics. The predictive performance of the suicide behaviors was 0.33–0.45 of sensitivity and 0.90–0.95 of specificity, respectively. That performance profile can be useful to alert clinicians to prepare possible risk of suicide behaviors; however, managing a lot of risk groups can be burdensome due to the low sensitivity.

Another study predicted accidental death and suicide after medical or surgical discharge from general hospitals by using natural language processing [42]. They analyzed discharge notes narratively written. The study was conducted between 2005 and 2013 based on the overall 845,417 discharges with 458,053 individuals. Unlike other studies which mainly focused on or included the suicide attempt as a primary outcome measure, that study focused on the suicide death as compared to other types of death. Natural language processing revealed that positive valence in the discharge note was less associated with suicide.

On the other hand, one study built a machine learning model to predict suicide for 10 years [43]. That study analyzed suicide risk retrospectively by using data of the NHIS in the Republic of Korea. Based on the 1,016,583 subjects randomly selected from the 10-years of the NHIS records, 819,951 subjects were finally included. The number of suicide victims was 2546, whereas non-suicide death was 817,405. In the original set, AUC for predicting suicide was 0.688 (Cox regression), 0.576 (SVM), and 0.632 (deep neural network). However, when oversampling by increasing the number of suicide death, the AUC was 0.688 (Cox regression), 0.687 (SVM), and 0.683 (deep neural network). As shown in results, that study suggested that machine learning methodology was not superior over the traditional statistical approach for predicting suicide. One of the main reasons for the low predictability of machine learning models was the extremely imbalanced sample. As the prevalence of suicide is usually vey row, classifiers tend to put cases into the non-suicide category to increase the overall accuracy. Those problems of the imbalance can be treated by directly modifying the number of samples. Up-sampling and down-sampling are simple and frequently used methods. Those manipulations of the sample structures have a possibility to distort the original characteristics of the samples, and the results from the model would not be useful for real-world situation. To deal with such limitations, more delicate up-sampling methods such as the Synthetic Minority Oversampling Technique are increasingly used [44].

Facebook may be one of the practical companies which uses big data for suicide prevention. Using the platform and the machine learning-based evidence, Facebook sends users who show high suicide rescue resources such as 24-hour hotline [45].

Delirium

Delirium is one of the common neuropsychiatric conditions which threatens patients under unstable physical conditions. The prevalence of delirium is various, which ranges from 10 to 56% [46]. Some report that delirium is prevalent in nearly 50% of the elderly [47]. Delirium is one of the representative burdensome psychiatric conditions in several perspectives. First, delirium exerts harmful influences on the prognosis such as the mortality and cognitive function even if patients recovered from delirium [48]. Second, diagnosing delirium requires a considerable time and efforts due to its fluctuating course. On the contrary to dementia, delirium is characterized by abrupt onset and fluctuating diurnal variations in the symptoms. Finally, due to the above factors of delirium, the economic burden associated with delirium is substantially high. For example, approximately $38 billion to $152 billion were estimated to be associated with delirium in the United States [47].

Hence, measuring once a day usually cannot provide sufficient information on the diagnosis of delirium. Rather, at least three to four assessments per day is required to comprehensively evaluate changes.

Due to the difficulty and bothersome of the diagnosis of delirium, the needs for the automated process of recognition of delirium are high.

Given the difficulty in the diagnosis and harmful effects on the prognosis, it is particularly important to identify risk groups and prevent delirium. However, there are several common risk factors for the delirium [49], the accuracy of the prediction of delirium is relatively low so it is hardly applied to the clinical practice.

The combination of machine learning and big data can support the diagnosis and management of delirium in the clinical routines. Recent machine learning studies have shown promising results in diagnosing delirium by utilizing HER of a lot of health institutes. One study analyzed a total of 11,752 patients aged over 65 among 118 Veteran Affair (VA) medical centers [50]. The AUC values at admission ranged from 0.83 to 0.91. By conveniently predicting delirium at admission, it can be possible to promptly focus on the risk groups and recover them to the healthy state. In that study, as expected, cognitive function was suggested the most important variables for predicting delirium.

On the other hand, another study which similarly used machine learning with EHR data predicted delirium in a quite different approach [51]. That study predicted delirium among hospitalized patients without known cognitive impairment. They predicted delirium based only variables from the EHR recorded 24 h within admission. They used a total of 18,223 admissions. A total number of variables initially selected for machine learning were 796. The predictive performances were AUC of 0.855, specificity of 0.90, sensitivity of 0.60, positive predictive value of 0.23, and negative predictive value of 0.98 in the gradient boosting machine. Such predictive performance can be utilized for roughly classifying patients who will not be developed to delirium. However, it is apparently limited for classifying risk groups for delirium. Given the primary objective of screening test for all subjects is to do not omit possible risk groups, the usefulness of the predictive model can be questionable. On the other hand, distinct strength of such model is the automated process which requires the least additional efforts. As the model relies on the records already obtained in the EHR, users (doctors, nurses, and other health practitioners) can just check the results. Hence, although the current performance is insufficient for the direct use in the clinical setting, such approach will be more developed in the near future.

Considerable Issues

There are several considerations regarding pragmatic applying big data to the psychiatry. First, in case of unstructured text analysis, the algorithm in one language could not be generalized to others. The EHR is increasingly used; however, there are several methodological issues which should be further addressed. Results from the neurobiological studies in one language can be easily replicated and utilized in other research groups regardless of their language. However, analysis algorithm from one language, most commonly English, will not be able to be applied in other languages.

Second, to effectively use EHR, psychiatrists should pay more attention to their records. No matter how the large volume of the EHR can well be analyzed, there would be a limited availability when it was initially produced insufficiently. Third, the patients' factor should be considered. Although in the same country, not all patients belong to the same sociocultural context. One possible alternative for the heterogeneity is to generate interaction terms, or feature cross, so that sociodemographic variable-dependent verbal expression can be effectively analyzed. For example, in the society of Northeast Asia, somatization is frequently reported by the female elderly depressed [52, 53]. It is referred to as a "Hwa-Byung" in Korean, a cultural psychiatric condition. On the other hand, younger females in the same countries of Northeast Asia usually do not complaint depression in such way. The youngers may tend to express their emotional problems in a more direct way rather than feeling somatic symptoms. In such cases, individual input of the somatic complaints and sociodemographic context would be a noise rather than an explanatory feature. Given the domain knowledge for the sociocultural context and psychiatric conditions, such variables can be transformed to the interaction terms such as somatization among the female elderly, somatization among the male elderly, and somatization among the youth. If necessary, further division according to gender can be also conducted among the youth. This is only one case, and there might be more situations in which preprocessing should be carefully conducted in advance to the application of the predictive model. This is the reason why the domain knowledge is so important, and the availability of the so-called data-driven analysis is limited in psychiatry.

Future big data studies should be based on the more comprehensive and unified dataset than the current one. To obtain valid results, it is important to recruit valid materials in the initial stages, but not the analytical stage. The data wrangling and preprocessing do not more impact on the results than the data-building stage. Given the sold fundamental of the valid dataset, predictive algorithms with machine learning can produce useful and applicable products which can be applied to solve the real-world problems of psychiatry.

References

1. Kan Z, Tang L, Kwan MP, Zhang X. Estimating vehicle fuel consumption and emissions using GPS big data. Int J Environ Res Public Health. 2018;15(4).
2. Monteith S, Glenn T, Geddes J, Bauer M. Big data are coming to psychiatry: a general introduction. Int J Bipolar Disord. 2015;3(1):21.
3. Shneiderman B, editor. Extreme visualization: squeezing a billion records into a million pixels. In: Proceedings of the 2008 ACM SIGMOD international conference on management of data. ACM SIGMOD international conference on management of data. Canada, Vancouver: ACM; 2008.
4. Dumbill E. Making sense of big data. Big Data. 2013;1(1):1–2.
5. SAS Institute Inc. Big data: what it is and why it matters; 2018. https://www.sas.com/en_us/insights/big-data/what-is-big-data.html#dmhistory.
6. Wolfe PJ. Making sense of big data. Proc Natl Acad Sci USA. 2013;110(45):18031–2.
7. Westcott K. Netflix's data-driven strategy strengthens claim for 'Best Original Content' in 2018. Forbes; 2018. Updated 28 May 2018. https://www.forbes.com/sites/kristinwestcott-grant/2018/05/28/netflixs-data-driven-strategy-strengthens-lead-for-best-original-content-in-2018/#83ea93d3a94e.
8. Jain A. The 5 Vs of big data. IBM; 2016. https://www.ibm.com/blogs/watson-health/the-5-vs-of-big-data/.
9. Borne K. Top 10 list—the Vs of big data: data science central; 2014. https://www.datasciencecentral.com/profiles/blogs/top-10-list-the-v-s-of-big-data.
10. Sharpened Productions. What units of measurement are used for data storage? Sharpened Productions; 2018. Updated 13 Oct 2012. https://techterms.com/help/data_storage_units_of_measurement
11. IDC. The digital universe: driving data growth in healthcare. IDC; 2014. https://www.emc.com/analyst-report/digital-universe-healthcare-vertical-report-ar.pdf.
12. Hilbert M, Lopez P. The world's technological capacity to store, communicate, and compute information. Science. 2011;332(6025):60–5.
13. Tcheng JE. So just what is "Big Data"? In: 64th annual scientific session & expo. San Diego, CA, USA; 2015.

14. SINTEF. Big Data, for better or worse: 90% of world's data generated over last two years. ScienceDaily; 2013. https://www.sciencedaily.com/releases/2013/05/130522085217.htm.
15. Cisco. The Zettabyte Era: trends and analysis; 2017. Updated 7 July 2017. https://www.cisco.com/c/en/us/solutions/collateral/service-provider/visual-networking-index-vni/vni-hyperconnectivity-wp.html.
16. Hassell J. Using big data for financial fraud prevention. Hortonworks; 2018. https://hortonworks.com/article/using-big-data-for-financial-fraud-prevention/.
17. Salzberg S. Why Google Flu is a failure. Forbes; 2014. https://www.forbes.com/sites/stevensalzberg/2014/03/23/why-google-flu-is-a-failure/#67b43a6d5535.
18. Fung K. Google Flu Trends' failure shows good data > big data. Harvard Business Review; 2014. https://hbr.org/2014/03/google-flu-trends-failure-shows-good-data-big-data.
19. Lazer D, Kennedy R, King G, Vespignani A. Big data. The parable of Google Flu: traps in big data analysis. Science. 2014;343(6176):1203–5.
20. Butler D. When Google got flu wrong. Nature. 2013;494(7436):155–6.
21. NHIS. NIH Program. NHIS; 2010. http://www.nhis.or.kr/static/html/wbd/g/a/wbdga0401.html.
22. Kim J-A, Yoon S, Kim L-Y, Kim D-S. Towards actualizing the value potential of Korea health insurance review and assessment (HIRA) data as a resource for health research: strengths, limitations, applications, and strategies for optimal use of HIRA data. J Korean Med Sci. 2017;32(5):718–28.
23. Torous J, Staples P, Onnela JP. Realizing the potential of mobile mental health: new methods for new data in psychiatry. Curr Psychiatry Rep. 2015;17(8):602.
24. Turnbaugh PJ, Gordon JI. An invitation to the marriage of metagenomics and metabolomics. Cell. 2008;134(5):708–13.
25. Hasin Y, Seldin M, Lusis A. Multi-omics approaches to disease. Genome Biol. 2017;18(1):83.
26. American Psychiatric Association. Diagnostic and statistical manual of mental disorders. 5th ed. Arlington, VA: American Psychiatric Publishing; 2013.
27. Goldberg D. Psychopathology and classification in psychiatry. Soc Psychiatry Psychiatr Epidemiol. 2015;50(1):1–5.
28. Hoff P. Do social psychiatry and neurosciences need psychopathology-and if yes, what for? Int Rev Psychiatry. 2008;20(6):515–20.
29. Fanous AH, Kendler KS. Genetic heterogeneity, modifier genes, and quantitative phenotypes in psychiatric illness: searching for a framework. Mol Psychiatry. 2005;10(1):6–13.
30. Hodgson K, McGuffin P, Lewis CM. Advancing psychiatric genetics through dissecting heterogeneity. Hum Mol Genet. 2017;26(R2):R160–5.
31. Hafner H. Descriptive psychopathology, phenomenology, and the legacy of Karl Jaspers. Dialogues Clin Neurosci. 2015;17(1):19–29.
32. Sheehan DV, Lecrubier Y, Sheehan KH, Amorim P, Janavs J, Weiller E, et al. The mini-international neuropsychiatric interview (M.I.N.I.): the development and validation of a structured diagnostic psychiatric interview for DSM-IV and ICD-10. J Clin Psychiatry. 1998;59 Suppl 20:22–33;quiz 4–57.
33. Cooper L, Peters L, Andrews G. Validity of the composite international diagnostic interview (CIDI) psychosis module in a psychiatric setting. J Psychiatr Res. 1998;32(6):361–8.
34. OECD Data. Suicide rates 2018. OECD; 2018. https://data.oecd.org/healthstat/suicide-rates.htm
35. Han DG, Kang SG, Cho SJ, Cho SE, Na KS. Suicide methods according to age and sex: an analysis of data of 239,565 suicide victims in the Republic of Korea from 1991 to 2015. J Nerv Ment Dis. 2018;206(10):770–5.
36. Pedersen CB. The danish civil registration system. Scand J Public Health. 2011;39(7 Suppl):22–5.
37. Lin Y, Huang S, Simon GE, Liu S. Data-based decision rules to personalize depression follow-up. Sci Rep. 2018;8(1):5064.

38. Malhi GS, Das P, Outhred T, Irwin L, Morris G, Hamilton A, et al. Understanding suicide: Focusing on its mechanisms through a lithium lens. J Affect Disord. 2018;241:338–47.
39. Dwivedi Y. MicroRNAs in depression and suicide: recent insights and future perspectives. J Affect Disord. 2018;240:146–54.
40. Nock MK, Dempsey CL, Aliaga PA, Brent DA, Heeringa SG, Kessler RC, et al. Psychological autopsy study comparing suicide decedents, suicide ideators, and propensity score matched controls: results from the study to assess risk and resilience in service members (Army STARRS). Psychol Med. 2017;47(15):2663–74.
41. Barak-Corren Y, Castro VM, Javitt S, Hoffnagle AG, Dai Y, Perlis RH, et al. Predicting suicidal behavior from longitudinal electronic health records. Am J Psychiatry. 2017;174(2):154–62.
42. McCoy TH Jr, Castro VM, Roberson AM, Snapper LA, Perlis RH. Improving prediction of suicide and accidental death after discharge from general hospitals with natural language processing. JAMA Psychiatry. 2016;73(10):1064–71.
43. Choi SB, Lee W, Yoon JH, Won JU, Kim DW. Ten-year prediction of suicide death using Cox regression and machine learning in a nationwide retrospective cohort study in South Korea. J Affect Disord. 2018;231:8–14.
44. Blagus R, Lusa L. SMOTE for high-dimensional class-imbalanced data. BMC Bioinform. 2013;14:106.
45. Inc. F. How Facebook AI helps suicide prevention. Facebook Inc.; 2018. https://newsroom.fb.com/news/2018/09/inside-feed-suicide-prevention-and-ai/.
46. Inouye SK, Westendorp RG, Saczynski JS. Delirium in elderly people. Lancet. 2014;383(9920):911–22.
47. Leslie DL, Marcantonio ER, Zhang Y, Leo-Summers L, Inouye SK. One-year health care costs associated with delirium in the elderly population. Arch Intern Med. 2008;168(1):27–32.
48. Furukawa TA. Prognosis of delirium. Evid Based Ment Health. 2017;20(2):53–4.
49. Ahmed S, Leurent B, Sampson EL. Risk factors for incident delirium among older people in acute hospital medical units: a systematic review and meta-analysis. Age Ageing. 2014;43(3):326–33.
50. Halladay CW, Sillner A, Rudolph JL. Performance of electronic prediction rules for prevalent delirium at hospital admission. JAMA Netw Open. 2018;1(4):e181405.
51. Wong A, Young AT, Liang AS, Gonzales R, Douglas VC, Hadley D. Development and validation of an electronic health record-based machine learning model to estimate delirium risk in newly hospitalized patients without known cognitive impairment. JAMA Netw Open. 2018;1(4):e181018.
52. Pang KY. Hwabyung: the construction of a Korean popular illness among Korean elderly immigrant women in the United States. Cult Med Psychiatry. 1990;14(4):495–512.
53. Tonai S, Maezawa M, Kamei M, Satoh T, Fukui T. Illness behavior of housewives in a rural area in Japan: a health diary study. Cult Med Psychiatry. 1989;13(4):405–17.

Classification of Psychiatric Disorders

Yong-Ku Kim and Seon-Cheol Park

Abstract

Because of the poor link between psychiatric diagnosis and neurobiological findings, it is difficult to classify mental disorders. The changes made to psychiatric diagnostic systems over the years can be understood in terms of "practical conservatism." The Diagnostic and Statistical Manual of Mental Disorders (DSM)-I and DSM-II were theoretically supported by the psychoanalytic and psychodynamic approach. Subsequently, psychiatric diagnoses of this kind were opposed by the anti-psychiatry movement, as well as by the findings of the Rosenhan experiment. Thus, the DSM-III revolution contained more empiricism, aligning psychiatry with biomedicine. Psychiatric diagnoses are classified and defined in terms of Kraepelinian dualism, using a categorical approach. The empirical trend was continued in the DSM-IV. To overcome the limitations of current psychiatric diagnostic systems and integrate fundamental genetic, neurobiological, behavioral, environmental, and experimental components into psychiatry, the Research Domain Criteria (RDoC) were established. To overcome the limitations of the categorical approach, psychiatrists have considered adopting a dimensional approach. However, their efforts were frustrated in the DSM-5 revision process. Thus, the DSM-5 is characterized by the rearrangement of psychiatric diagnoses, the partial adoption of a dimensional approach,

Y.-K. Kim
Department of Psychiatry, College of Medicine, Korea University Ansan Hospital, 516, Gojan-dong, 425-707 Ansan, Gyeonggi Province, Republic of Korea
e-mail: yongku@korea.edu

S.-C. Park (✉)
Department of Psychiatry, Inje University Haeundae Paik Hospital, 875, Haeun-daero, Haeundae-gu, 48108 Busan, Republic of Korea
e-mail: cogito-ergo-sum@hanmail.net

© Springer Nature Singapore Pte Ltd. 2019

Y.-K. Kim (ed.), *Frontiers in Psychiatry*, Advances in Experimental Medicine and Biology 1192, https://doi.org/10.1007/978-981-32-9721-0_2

the introduction of new diagnoses, and harmonization with the International Classification of Diseases.

Keywords
Categorical approach · Dimensional approach · Psychiatric diagnosis · DSM · Research domain criteria (RDoC)

Introduction

It has been said that "the Catholic Church changes its pope more often than the American Psychiatric Association (APA) publishes a new version of the Diagnostic and Statistical Manual of Mental Disorders (DSM) [1]." This approach has been named "DSM conservative pragmatism" with regard to changes in the classification of psychiatric disorders [2]. In this parlance, the term "conservative" denotes resistance to the change—not in a partisan political sense, but rather in general [3], while "pragmatism" implies that changes to the DSM are primarily based on a pragmatic attitude and scientific sense.

One study proposed that DSM conservative pragmatism is influenced by the "medical–industrial complex" of mental health: the psychiatry equivalent of the military–industrial complex. Furthermore, it may be that researchers use the vague term "DSM conservative pragmatism" because psychiatric taxonomy is inherently obscure and arbitrary. Indeed, in the USA, psychiatric taxonomy can be influenced by mentally ill patients themselves, the pharmaceutical industry, the health-care system, politics, the National Institute for Mental Health (NIMH), popular demand, and academic medical centers [2].

Thus, psychiatric taxonomy can change, irrespective of neuroscientific research findings. In psychiatric diagnosis classifications, a "paradigm shift"—from a categorical approach to a dimensional approach—was proposed during the DSM-5 revision process. However, the suggestion was furiously resisted [4, 5]. The present chapter reviews and discusses the brief history, theoretical background, present condition, and controversies of the psychiatric classifications.

Brief History of the DSM

The DSM-I and DSM-II were published by the APA in 1952 and 1968, respectively, based mainly on psychoanalytic and psychodynamic perspectives. Whereas the DSM-I followed the Meyerian tradition for naming psychiatric disorders (e.g., schizophrenic reaction), the DSM-II adopted the name of the disease entity into the psychiatric classification (e.g., schizophrenia). However, many researchers criticized the inconsistency and arbitrariness of psychiatric diagnosis in the DSM-II [6–8].

Most importantly in this regard, the findings of the Rosenhan experiment were published in *Science*, whereby lengthy admissions to psychiatric hospitals occurred because the researchers falsely attributed psychosis symptoms to the patients [9]. In addition, dehumanizing depictions of the profession were popularized by an anti-psychiatry movement [10, 11], and there was a bitter and public debate on the diagnosis of homosexuality as a mental illness [12].

To overcome the limitations of the DSM-II, the "DSM-III revolution" saw an increase in empiricism and biological psychiatry in 1980. The principle of "Real, Recognizable, Unitary, and Stable" (RRUS), which formed the basis of the Kraepelinian nosological model, was revitalized by the DSM-III task force [13, 14]. Furthermore, to increase the reliability of diagnoses, Robert Spitzer oversaw a change in the psychiatric taxonomic system from a fluid, psychoanalytical understanding of mental illness to the standardized nosology of fixed disease categories. To align psychiatry more firmly with scientific medicine, the two Kraepelinian psychoses (schizophrenia and manic-depressive illness) took a central position in the DSM-III. In particular, based on the approach used by Sydenham, these psychiatric diagnoses were defined in terms of any one feature (i.e., etiology, symptomatology, or course) that co-occurred across patients [15]. Consequently, the DSM-III changed psychiatric diagnosis from a bottom-up process based on detailed life history, painstaking examination of mental states, and coordination from third-party informants to a top-down process based on manual symptom checklists [16]. Developing upon the empirical trends of the DSM-III, the DSM-IV was published by Allen Frances et al. in 1994.

In accordance with this increase of the empirical trends, operational diagnostic systems began to be used in psychiatric nosology. These were characterized by a descriptive approach, explicit inclusion—and exclusion—criteria, algorithms for the criteriology of each diagnosis, a nominalist understanding of psychiatric diagnoses, etiological neutrality, and a focus on severity, with a quantitative approach [17]. However, such rigid categorization of disease entities was criticized as "atheoretical theory." Most notably, the DSM categories no longer reflected discrete entities, and genetic variance could not be detected using the binary logic of the DSM-III (presence or absence of a mental disorder). Psychiatric diagnoses that were defined using an operational system were even regarded as Wittgensteinian language games, in that they were connected by family resemblance and had no essential or core connecting phenomenon [18, 19]. At that time, Steven Hyman, then-director of the NIMH, stated the limitations of the DSM as follows: "The problem is that DSM has been launched into under-researched waters, and this has been accepted in unquestioning way." In addition, using the DSM-IV, the category of "Not-Otherwise-Specified (NOS)" was often used by psychiatrists, and many patients were diagnosed with a high degree of comorbidity [1, 20].

As a reaction against aligning psychiatry with biomedicine, the eclectic, multidisciplinary biopsychosocial approach was adopted. However, it was criticized because it has boundless psychiatry and justifies the use of psychotropic drugs to treat symptoms [21–23]. Kendler and Zachar [24] responded to this nosological insecurity as follows: "Avoid planned obsolescence, study history to dampen the swings, be realistic about what has been accomplished, study traits as well as disorders, don't make small changes in diagnostic criteria, and become phenomenologists."

Andreasen [25] later claimed that the rigorous categorical approach of the DSMs had resulted in the unintentional "death of phenomenology." In DSM-III the greater reliability and validity of psychiatric diagnoses was traded off against the decline in careful clinical evaluation targeted at social context and the problems of individual patients. Parnas [26] suggested that the concepts of hebephrenia concept and schizophrenia Gestalt had disappeared when the European diagnostic criteria for schizophrenia were replaced by the DSM-IV. Consequently, phenomenological approaches were abandoned (disunity of consciousness, intrapsychic ataxia, autism, anthropological disproportion, crisis of "common sense," and changed modality of being) in favor of a return to pre-Kraepelinian stage of schizophrenia diagnosis.

Moreover, the DSM-III had used either incoherence or the marked loosening of associations as the diagnostic criteria for schizophrenia, because Grisinger, Kraepelin, and Bleuler had proposed Inkohärenz (incoherence), Zerfahrenheit (derailment), and associative loosening had been proposed as the prototypical concepts of formal thought disorder. However, with the increasing tendency toward empirical trends, the description of formal thought disorder was redefined as disorganized speech in the DSM-IV [27–29].

Frustration of the Paradigm Shift

To overcome the limitations of DSMs and create a framework for research into genomics- and neuroscience-based pathophysiology, the Research Domain Criteria (RDoC) project was launched by the NIMH in 2010. In this regard, the NIMH aimed to (1) identify the fundamental behavioral components that may span multiple disorders in clinical and basic sciences, (2) determine the full range of variation among these fundamental components—from normality to abnormality, (3) develop reliable and valid measures of these fundamental components, and (4) integrate the fundamental genetic, neurobiological, behavioral, environmental, and experimental components of psychiatric disorders [30].

The RDoC framework is based on three assumptions: (1) Psychiatric disorders can be regarded as brain disorders; (2) The tools of neuroscience, including electrophysiology, functional neuroimaging, and new methods for quantifying in vivo connections, can identify dysfunctions in neural circuits; (3) Data from genetics and clinical neuroscience can be used to identify biosignatures that accompany clinical symptoms and signs. Thus, the RDoC framework can be described as a matrix form, with rows that define the broad domains of functions (negative valence domain, positive valence systems, cognitive systems, systems for social process, and arousal/modulatory systems) and columns that define the different levels of analysis (genes, molecules, cells, circuits, physiology, behavior, self-report, and paradigms). With regard to the "neural circuits" column, the levels of analysis were dichotomized into the upper considerations of clinical relevance and lower considerations of genetic and molecular factors. The RDoC framework is expected to influence clinical practice. For instance, a brain-derived neurotrophic factor polymorphism can be used to identify patients with anxiety disorders who will not respond to behavior therapy, and a copy number variant can be used to define a form of psychosis with high remission rates [31, 32].

Under the partial influence of the RDoC framework, a paradigm shift was proposed whereby mental disorders differ from normality in quantity only, rather than quality. To circumvent the limitations of the current paradigm, encourage a research agenda that transcends current ways of thinking, and devise an etiologically and pathophysiologically based diagnostic system, the DSM-5 revision process proposed moving the DSM diagnostic system beyond a categorical and descriptive approach. To this end, the DSM-5 task force and working group aimed to incorporate well-replicated discoveries into the manual, revising the boundaries and structures of current diagnoses and grappling with a scientific base. Thus, the various proposals of the DSM-5, including the definition of diagnostic criteria, drastically differed from those of the DSM-IV, and each diagnosis adopted a new, dimensional approach [1, 20, 33].

In particular, the deconstruction of the two Kraepelinian psychoses was one of the most important issues in the revision of the psychosis diagnostic criteria. Dutta et al. proposed that the combined quantitative values of the positive, negative, disorganization, manic, and depressive domains could be used to define schizophrenia, schizoaffective disorder, and bipolar disorder, abandoning the Kraepelinian dualism [34]. However, the radical implications of such dimensionality were furiously resisted by defenders of neo-Kraepelinianism. The inertia of changes to the diagnostic criteria was resisted by Robert Spitzer and Allen Frances because of a lack of transparency, poor attention to the clinical utility of the changes, the introduction of new, problematic diagnoses without sufficient scientific justification, the exacerbation of false positives, and the introduction of untested and unproven dimensional assessment tools. Ultimately, the derailment of the DSM-5 was developed [4].

The categorical and dimensional approaches have finally been combined in the DSM-5. David Kupfer (chair of the DSM-5 task force) and Darrel Regier (vice-chair of the DSM-5 task force) [5] stated the following: "As we gradually build on our knowledge of mental disorders, we begin bridging the gap between what lies behind us (presumed etiologies based on phenomenology) and what we hope lies ahead (identifiable pathophysiological etiologies)." After the DSM-5 field trials were performed in large academic medical centers (April 2010 to December 2011) and routine clinical practices were subsequently established [35–37], the DSM-5 was published in 2013. The contents of the DSM-5 were rearranged according to age and sex, with the categorical diagnoses being preserved. In this way, a dimensional approach was partially adopted—the use of assessment scale severity, the introduction of newly defined diagnoses, including attenuated psychosis syndrome and persistent complex bereavement disorder, harmonization with the International Classification of Diseases (ICD), etc. [38].

Definition of a "Mental Disorder"

From the viewpoint of Karl Jaspers, classifying psychiatric disorders is like "drawing a line where none exists," and each psychiatric diagnosis can only be viewed as provisional. Conversely, the psychiatric diagnostic classifications of the DSM and ICD are based on disease essentialism, which assumes that observable symptoms or signs can be defined precisely and that psychiatric diagnoses can be identified clearly. Each psychiatric diagnosis is supported by neurobiological underpinnings in the categorical approach. However, most psychiatric disorders are not supported by essential biological findings, so psychiatric disorders cannot be classified in terms of natural boundaries, and boundlessness can therefore be considered a distinctive characteristic of psychiatric nosology and taxonomy, in contrast with medicine [8, 17, 39, 40]. In the definition of mental disorder, natural realities conflict with human constructs, which are only possible in theory. In other words, all psychopathological constructs are based on natural realities encountered in clinical practice [41].

The definition of mental disorders remains in conflict with the epistemological game proposed by Allen Frances [42]. Using the metaphor of baseball, mental disorders can be defined differently, as follows: "There are balls and there are strikes and I call them as they are (Robert Spitzer)," "There are balls and there are strikes and I call them as I see them (Allen Frances)," and "There are no balls and there are no strikes until I call them (Tom Szasz)." Furthermore, Frances [42] points out that the DSM is often misused to diagnose conditions at the intersection of normality and criminality. Since the dimensional approach, which ensures early detection and prevention, has been strengthened in the DSMs, the range of psychiatric diagnosis tends to encroach on the boundaries of normality, overdiagnosis is a problem in clinical psychiatry. Ultimately, it has been suggested that the concepts of mental disorders have no validity [43]. For example, Paris [44] stated the following: "The DSM definition of mental disorder neither successfully defines mental disorder nor provides a safe clinical space for psychiatry to exercise its authority." Hence, with the influence of pharmaceutical marketing, it has been speculated that psychotropic drug use in people at the healthier end of the mental illness continuum should be increased [44].

Conclusion

The categorical approach was used in the DSM-III and DSM-IV to overcome the limitations of the psychoanalytical and psychodynamic trends of the DSM-I and DSM-II. This approach defines psychiatric diagnoses on the basis of disease essentialism. However, the diagnostic classifications are not supported by essential neurobiological processes. The RDoC framework is based on the conceptualization of psychiatric disorders as brain disorders, whereby neural circuit dysfunctions and methods of quantifying connections can be identified, and biosignatures augment the clinical symptoms and signs of clinical management. Thus, RDoC research can be understood as a matrix, with rows defining broad domains of functions, and columns defining the different analysis levels. Partly consistent with the RDoC framework, the overarching ambition of the dimensional approach was discussed and abandoned in the DSM-5 revision process. Ultimately, the categorical approach was combined with the dimensional approach in the DSM-5, so the contents of the manual rearrange psychiatric diagnoses, adopt a dimensional approach, introduce new diagnoses, and harmonize with the ICD.

References

1. Adam D. Mental health: on the spectrum. Nature. 2013;496(7446):416–28.
2. Sadler JZ. Considering the economy of DSM alternatives. In: Paris J, Phillips J, editors. Making the DSM-5: concepts and controversies. New York: Spinger; 2013. p. 21–38.
3. Pincus H, Frances A, Davis WW, First M, Widiger T. DSM-IV and new diagnostic categories: holding the line on proliferation. Am J Psychiatry. 1992;149(1):112–7.
4. Whooley O, Horwitz AV. The paradox of professional success: grand ambition, furious resistance, and the derailment of the DSM-5 revision process. In: Paris J, Phillips J, editors. Making the DSM-5: concepts and controversies. New York: Springer; 2013. p. 75–92.
5. Kupfer DJ, Regier DA. Neuroscience, clinical evidence, and the future of psychiatric classification in DSM-5. Am J Psychiatry. 2011;168(7):672–4.
6. Shorter E. The history of DSM. In: Paris J, Phillips J, editors. Making the DSM-5: concepts and controversies. New York: Spinger; 2013. p. 3–20.
7. Ghaemi SN. The perils of open-mindedness: Adolf Meyer's psychobiology. In: Ghaemi SN, editor. The rise and fall of the biopsychosocial model: reconciling art & science in psychiatry. Baltimore: The Johns Hopkins University Press; 2010, p. 3–11.
8. Park S-C. Karl Jaspers' general psychopathology (Allgemeine Psychopathologie) and its implication for the current psychiatry. Psychiatry Investig. 2019;16(2):99–108.
9. Rosenhan DL. On being sane in insane places. Science. 1973;179(4070):250–8.
10. Goffman E. Asylums: essays on the social situations of mental patients and other inmates. Piscataway, NJ: Aldine Transaction; 2007.
11. Kesey K. One flew over the cuckoo's nest. New York: Penguin Classics; 2002.
12. Bayer R. Homosexality and American psychiatry: the politics of diagnosis. Princeton, NJ: Princeton University Press; 1987.
13. Engstrom EJ, Kendler KS. Emil Kraepelin: icon and reality. Am J Psychiatry 2015;172(2):1190–6
14. Kendler KS, Engstrom EJ. Criticisms of Kraepelin's psychiatric nosology: 1896-1927. Am J Psychiatry. 2018;175(4):316–26.
15. Kinghorn W. The Biopolitics of Defining "Mental Disorder.". In: Paris J, Phillips J, editors. Making the DSM-5: concepts and controversies. New York: Spinger; 2013. p. 47–61.
16. McHugh PR, Slavney PR. Mental illness—comprehensive evaluation or checklist? N Engl J Med. 2012;366(20):1853–5.
17. Sass H, Volpe U. Karl Jaspers' hierarchical principle and current psychiatric classification. In: Stanghellini G, Fuchs T, editors. One century of Karl Jaspers' general psychopathology. Oxford: Oxford University Press; 2013. p. 185–207.
18. Rosenman S, Nasti J. Psychiatric diagnoses are not mental processes: Wittgenstein on conceptual confusion. Aust N Z J Psychiatry. 2012;46(11):1046–52.
19. Park S-C, Kim JM, Jun TY, Lee MS, Kim JB, Yim HW, Park YC. How many different symptom combinations fulfill the diagnostic criteria for major depressive disorder? Results from the CRESCEND study. Nord J Psychiatry. 2017;71(3):217–22.

20. Regier DA, Narrow WE, Kuhl EA, Kupfer DJ. The conceptual development of DSM-V. Am J Psychiatry. 2009;166(6):645–50.
21. Engel GL. The need for a new medical model: a challenge for biomedicine. Science. 1977;196(4286):126–36.
22. Engel GL. The clinical application of the biopsychosocial model. Am J Psychiatry. 1980;137(5):535–44.
23. Park S-C, Kim Y-K. More integrated biopsychosocial approach to psychiatric disorders. In: Kim Y-K, editor. Research methods and interventions in psychiatry: artificial intelligence, precision medicine, and other paradigm shifts. New York: Springer; 2019 (in press).
24. Kendler KS, Zachar P. The incredible insecurity of psychiatric nosology. In: Kendler KS, Parnas J, editors. Philosophical issues in psychiatry: explanation, phenomenology, and nosology. Baltimore: The Johns Hopkins University Press; 2008. p. 368–82.
25. Andreasen NC. DSM and the death of phenomenology in America: an example of unintended consequences. Schizophr Bull. 2007;33(1):108–12.
26. Parnas J. DSM-IV and the founding prototype of schizophrenia: are we regressing to a Pre-Krapelinian nosology? In: Kendler KS, Parnas J, editors. Philosophical issues in psychiatry II: nosology. Oxford: Oxford University Press; 2012. p. 237–59.
27. Park S-C, Jang EY, Lee KU, Lee K, Lee HY, Choi J. Reliability and validity of the korean version of the scale for the assessment of thought, language, and communication. Compr Psychiatry. 2015;61:122–30.
28. Park S-C, Jang EY, Lee KU, Lee JG, Lee HY, Choi J. Psychometric properties of the korean version of the clinical language rating scale (CLANG). Clin Psychopharmacol Neurosci. 2016;15(1):49–56.
29. Park YC, Lee MS, Si TM, Chiu HFK., Kanba S, Chong M-Y, Tripathi A, Udomratn P, Chee KY, Tanra AJ, Rabbani G, Javel A, Kathiarachchi S, Myint WA, Cuoung TV, Sim K, Yang SY, Sartorius N, Tan CH, Shinfuku N, Park S-C. Psychotropic drug-prescribing correlates of disorganized speech in Asians with Schizophrenia: the REAP-AP study. Saudi Pharm J. 2019;27(2):246–253.
30. Cubert BN, Insel TR. Toward the future of psychiatric diagnosis: the seven pillars of RDoC. BMC Med. 2013;11:126.
31. Miller G. Beyond DSM: seeking a brain-based classification of mental illness. Science. 2010;327(5872):1437.
32. Insel T, Cuthbert B, Garvey M, Heissen R, Pine DS, Quinn K, Sanislow C, Wang P. Research domain criteria (RDoC): toward a new classification framework for research on mental disorders. Am J Psychiatry. 2010;167(7):748–51.
33. Kupfer DJ, First MB, Regier DA, editors. A research agenda for DSM-V. Washington, DC: American Psychiatric Association; 2002.
34. Dutta R, Greene T, Addington J, McKenzie K, Phillips M, Murray RM. Biological, life course, and cross-cultural studies all point toward the value of dimensional and developmental rating in the classification of psychosis. In: Tamminga CA, Sirovatka PJ, Regier DA, van OS J, editors. Deconstructing psychosis: refining the research agenda for DSM-V. Washington, DC: American Psychiatric Association; 2010. p. 11–44.
35. Clarke DE, Narrow WE, Regier DA, Kuramoto SJ, Kupfer DJ, Kuhl EA, Greiner L, Kraemer HC. DSM-5 field trials in the United States and Canada, part I: study design, sampling strategy, implementation, and analytic approaches. Am J Psychiatry. 2013;170(1):43–58.
36. Regier DA, Narrow WE, Clarke DE, Kraemer HC, Kuramoto SJ, Kuhl EA, Kupfer DJ. DSM-5 field trials in the United States and Canada, part II: test-retest reliability of selected categorical diagnoses. Am J Psychiatry. 2013;170(1):59–70.
37. Narrow WE, Clarke DE, Kuramoto SJ, Kraemer HC, Kupfer DJ, Greiner L, Regier DA. DSM-5 field trials in the United States and Canada, part III: development of reliability testing of a cross-cutting symptom assessment for DSM-5. Am J Psychiatry. 2013;170(1):71–82.
38. American Psychiatric Association. Diagnostic and statistical manual of mental disorders. 5th ed. Washington, D.C.: American Psychiatric Association; 2013.

39. Song JY. Introduction to psychopathology. Seoul: Jipmoondang Publishing Co.; 2012.
40. Mckinejad K, Sharifi V. Wittgenstein's phiolosophy and a dimensional approach to the classification of mental disorder: a preliminary scheme. Psychopathology. 2006;39:126–9.
41. Musalek M, Larach-Walters V, Lepine JP, Millet B, Gaebel W. WFSBP task force on nosology and psychopathology. Psychopathology in the 21st century. World J Biol Psychiatry. 2010;1(7):844–51.
42. Frances A. DSM in philosophyland: curiouser and curiouser. In: Paris J, Phillips J, editors. Making the DSM-5: concepts and controversies. New York: Spinger; 2013. p. 95–103.
43. Pierre JM. Overdiagnosis, underdiagnosis, synthesis: a dialectic for psychiatry and the DSM. In: Paris J, Phillips J, editors. Making the DSM-5: concepts and controversies. New York: Spinger; 2013. p. 105–24.
44. Paris J. The ideology behind DSM-5. In: Paris J, Phillips J, editors. Making the DSM-5: concepts and controversies. New York: Spinger; 2013. p. 39–44.

Genetic Markers of Alzheimer's Disease

Matea Nikolac Perkovic and Nela Pivac

Abstract

Alzheimer's disease is a complex and heterogeneous, severe neurodegenerative disorder and the predominant form of dementia, characterized by cognitive disturbances, behavioral and psychotic symptoms, progressive cognitive decline, disorientation, behavioral changes, and death. Genetic background of Alzheimer's disease differs between early-onset familial Alzheimer's disease, other cases of early-onset Alzheimer's disease, and late-onset Alzheimer's disease. Rare cases of early-onset familial Alzheimer's diseases are caused by high-penetrant mutations in genes coding for amyloid precursor protein, presenilin 1, and presenilin 2. Late-onset Alzheimer's disease is multifactorial and associated with many different genetic risk loci (>20), with the apolipoprotein E ε4 allele being a major genetic risk factor for late-onset Alzheimer's disease. Genetic and genomic studies offer insight into many additional genetic risk loci involved in the genetically complex nature of late-onset Alzheimer's disease. This review highlights the contributions of individual loci to the pathogenesis of Alzheimer's disease and suggests that their exact contribution is still not clear. Therefore, the use of genetic markers of Alzheimer's disease, for monitoring development, time course, treatment response, and prognosis of Alzheimer's disease, is still far away from the clinical application, because the contribution of genetic variations to the relative risk of developing Alzheimer's disease is limited. In the light of prediction and prevention of Alzheimer's disease, a novel approach could be found

M. Nikolac Perkovic · N. Pivac (✉)
Division of Molecular Medicine, Rudjer Boskovic Institute, Bijenicka 54, Zagreb 10000, Croatia
e-mail: npivac@irb.hr

© Springer Nature Singapore Pte Ltd. 2019

Y.-K. Kim (ed.), *Frontiers in Psychiatry*, Advances in Experimental Medicine and Biology 1192, https://doi.org/10.1007/978-981-32-9721-0_3

in the form of additive genetic risk scores, which combine additive effects of numerous susceptibility loci.

Keywords
Alzheimer's disease · Genetics · GWAS · Late-onset · Markers

Introduction

Alzheimer's Disease

Alzheimer's disease is a complex and heterogeneous brain disorder that can be classified, according to its stages, into dementia in Alzheimer's disease (F.00) and Alzheimer's disease (G.30), according to ICD-10. Namely, it is a severe neurodegenerative disease and the predominant form of dementia (50–75%), but when behavioral and psychotic symptoms of dementia (BPSD) develop during the course of Alzheimer's disease, it has to be treated as a severe mental, i.e., psychiatric disorder. These neuropsychiatric symptoms include depression, apathy, anxiety, irritability, agitation, euphoria, hallucinations, disinhibition, aberrant motor behavior, elation, delusions, and sleep or appetite changes; and they can occur in the early as well as in the middle and late stages of Alzheimer's disease [1].

The first sign of dementia in Alzheimer's disease is the gradual worsening of the ability to remember new information. However, during the course of Alzheimer's disease, multiple cognitive domains are disrupted [2, 3]. The cognitive disturbances affect universal domains such as attention, working memory, executive function, procedural learning and memory, speed of processing, fear-extinction learning and semantic memory, and some higher domains that include episodic memory, social cognition, theory of mind, verbal learning, memory, and language (i.e., use and understanding) [3, 4]. Alzheimer's disease is a slow, irreversible, progressive, complex, and lethal disorder, which represents a major health problem and fatal global epidemic worldwide [3]. It is characterized by progressive cognitive decline, disorientation, behavioral changes, and death. A latency phase of the Alzheimer's disease is without clinical symptoms although the pathophysiological processes are active [2]. The clear etiology of Alzheimer's disease is still unknown. However, the main risk factors are older age, genetic predisposition (especially the apolipoprotein E (ApoE) ε4 genotype), gender (female predominance), and presence of the mild cognitive impairment, but there are also modifiable factors such as cardiovascular risk factors, hypertension, diabetes, obesity, smoking, and high cholesterol levels [3]. Insulin signaling dysfunction and brain glucose metabolism disturbances are hallmarks of Alzheimer's disease, and therefore recently Alzheimer's disease was suggested to be considered as type 3 diabetes [5].

Genetic Background of Alzheimer's Disease

Alzheimer's disease can be divided into autosomal dominant Alzheimer's disease (or early-onset familial Alzheimer's disease), other cases of early-onset

Alzheimer's disease, and late-onset Alzheimer's disease [6]. Genetic background of Alzheimer's disease differs between early-onset familial AD, other cases of early-onset Alzheimer's disease, and late-onset Alzheimer's disease. Early-onset familial Alzheimer's disease, with a prevalence less than 1%, is caused by high-penetrant mutations in genes coding for amyloid precursor protein (APP), presenilin 1 (PSEN1), and presenilin 2 (PSEN2). Late-onset Alzheimer's disease is multifactorial and associated with many different genetic risk loci (>20), with the ApoE ε4 allele being a major genetic risk factor for late-onset Alzheimer's disease. Genome-wide association studies (GWAS) offered insight into many additional genetic risk loci involved in the genetically complex nature of late-onset Alzheimer's disease. This review focuses on the recent data from comprehensive meta-analysis and GWAS. However, it should be highlighted that the exact contributions of individual loci to the pathogenesis of Alzheimer's disease still remain unclear to date.

Early-Onset Familial Alzheimer's Disease

The discovery of the association between mutations in *APP*, presenilin *PSEN1* and *PSEN2* genes and the development of early-onset familial Alzheimer's disease provided knowledge about the molecular mechanisms underlying the Alzheimer's disease pathogenesis.

Amyloid Precursor Protein

The enzymatic cleavage of APP can lead to the formation of amyloid β-peptide (Aβ), which can be 38 to 43 amino acids long. Cleavage of APP by α- and γ-secretases results in the generation of nonpathogenic peptides, secreted form of APP (sAPPα) and C-terminal fragments. This pathway is known as nonamyloidogenic or constitutive pathway. Amyloidogenic pathway involves the proteolysis of APP by β- and γ-secretase, resulting in the formation of sAPPα, C-terminal fragments, and Aβ. We differentiate two main forms of Aβ, Aβ1–40, and Aβ1–42. Amyloid plaques are most commonly formed from more amyloidogenic Aβ1–42 form.

According to Alzheimer Disease and Frontotemporal Dementia Mutation Database (http://www.molgen.ua.ac.be/ADmutations/) and Alzforum (https://www.alzforum.org/mutations/app), there are around 35 different *APP* mutations that have been associated with Alzheimer's disease pathogenesis. These mutations include *APP* gene locus duplications and different point mutations in coding region of *APP* gene, resulting in an amino acid substitution. Duplication of the whole gene/locus lead to elevated levels of APP and Aβ, and increase the ratio of Aβ1–42 to Aβ1–40. Missense mutations can have different effects, depending on their position. If these mutations cause amino acid substitution near the β-proteolytic cleavage site (N-terminal of Aβ), they usually lead to increased β-secretase cleavage, increased total Aβ production, and increased aggregation and fibril formation (Table 1). Missense mutations in the Aβ sequence in general increase Aβ aggregation and fibril formation (Table 1). If the missense mutation is near the C-terminal of Aβ, then it will increase the relative production of Aβ1–42, compared to Aβ1–40 (Table 1).

Presenilin 1 and Presenilin 2

PSEN1 and PSEN2 are two homologous multi-transmembrane proteins that share around 67% of the sequence [7], and they represent the catalytic core of γ-secretase complex. These proteins are also involved in the cleavage of some other proteins, like cadherins, low-density lipoprotein receptor (LDLR)-related proteins, Notch-1 and ErbB4 [8–11]. At the cell level, presenilins can be found in the nuclear membrane, endoplasmic reticulum, the trans-Golgi network and at the plasma membrane. They are also widely expressed throughout the organism.

Mutations in *PSEN1* and *PSEN2* genes are the most frequent known cause of early-onset familial Alzheimer's disease, with emphasis on *PSEN1* gene. Mutations in these two genes usually cause an impairment in γ-secretase activity and lead to an increase in the ratio between Aβ1–42 and Aβ1–40, as a consequence of Aβ1–42 overproduction or Aβ1–40 underproduction, or as a combination of both (Table 1). *PSEN1* mutations were associated with the earliest disease onset ages, with an average age of onset around 43 years, from 25 until 65 years of age [12]. In *APP* mutation carriers the disease starts on average 8.4 years earlier (age of onset between 35 and 65 years of age), and in *PSEN2* mutation carriers on average 14.2 years earlier, with a much older age of onset (between 45 and 70 years of age) [12].

Late-Onset Alzheimer's Disease

Most of the genes that have been associated with late-onset Alzheimer's disease, detected through different candidate genes studies and GWAS, are involved in cholesterol and lipid metabolism (genes coding for ApoE (*APOE*), sortilin-related receptor-1 (*SORL1*), ATP-binding cassette subfamily A member 7 (*ABCA7*), and clusterin (*CLU*)), immune system and inflammation (genes coding for complement C3b/C4b receptor 1 (*CR1*), CD33 antigen, membrane-spanning 4-domains, subfamily A member (*MS4A*), triggering receptor expressed on myeloid cells 2 (*TREM2*), member of the major histocompatibility complex class II HLA-DRB5/HLA-DRB1, and a SH2-containing inositol 5-phosphatase 1 (*INPP5D*)), and/or endosome cycling (genes coding for bridging integrator protein-1 (*BIN1*), CD2-associated protein (*CD2AP*), phosphatidylinositol binding clathrin assembly protein (*PICALM*), ephrin type-A receptor 1 (*EPHA1*)). However, there are also studies that implicate some other genes, whose function is not so well known and described, with Alzheimer's disease pathology, like genes coding for thioredoxin domain-containing protein 3 (*NME8*), CUGBP Elav-like family member 1 (*CELF1*), cas scaffolding protein family member 4 (*CASS4*), protein-tyrosine kinase 2-beta (*PTK2B*), zinc finger CW-type PWWP domain protein 1 (*ZCWPW1*), fermitin family homolog 2 (*FERMT2*), sodium/potassium/calcium exchanger 4 (*SLC24A4*), and Ras and Rab interactor 3 (*RIN3*) [13–19]. In this

Table 1 Genes, chromosomes, pathways, and polymorphisms associated with Alzheimer's disease

Gene	Chr	Pathway	Mutations/Polymorphisms		References
			ID	Potential effect	
APP	21q21	APP processing	dupAPP[ALZ254], dupAPP[EXT187], dupAPP [3], dupAPP[EXT144], dupAPP[EXT145], dupAPP [2], dupAPP [1], dupAPP[PED3281], dupAPP[VI], dupAPP[EXT054], dupAPP[EXT279], dupAPP [4], dupAPP[EXT298], dupAPP[Swedish], dupAPP [5], dupAPP[PED2945], dupAPP[BRB]	Increased production of Aβ	[57–61]
			rs63751263, rs63750445, rs63750064, Glu682Lys, rs63751039, ΔGlu693	N-terminal of Aβ peptide (β-secretase site, increased total Aβ production)	[62–69]
			rs63750973, rs63750734, rs63750868, rs63750399, Ile716Phe, Ile716Thr, Ile716Met, rs63750264, rs63750264, rs63750264, rs63751122	Mutations in Aβ sequence (increased aggregation and fibril formation)	[57, 61, 70–114]
			rs63750151	C-terminal of Aβ peptide(γ-secretase site, increased Aβ1-42 production)	[115]
PSEN1	14q24.3	APP processing	rs63749824, rs63749967, rs63751144, rs63750831, rs63750601, rs63750852, Thr99Ala, rs63750325, Phe105Val, Phe105Cys, rs63750321, Arg108Gln, rs63749962, rs63750450, rs63750730, rs63750730, rs63750550, rs63749805, rs63750800, Glu120Gly, rs63751272, rs63750378, His131Arg, Leu134Arg, rs63750353, rs41345849, rs63751106, rs63750522, Met139Ile, Ile143Val, rs63750322, rs63750004, rs63751071, rs63750306, rs63750306, rs63750391, rs63750907, Leu150Pro, rs63751441, rs63751292, Tyr159Phe, rs63749885, His163Pro, rs63751010, Leu166Val, rs63751458, rs63750265, Ile168Thr, rs63750963, Leu173Phe, rs63751144, rs63751025, rs63749911, rs63749806, rs63750155, rs63750311, Ile202Phe, rs63750569, rs63750082, rs63749880, rs63750053, Ser212Tyr, rs63750861, rs63750861, rs63751309, His214Asp, rs63751003, Gly217Arg, rs63749987, rs63750761, Leu219Arg, rs63750009, rs63751072, rs63749961, rs63749970, Ser230Ile, Ser230Arg, rs63749836, rs63750799, rs63751287, rs63751024, rs63751479, rs63751130, Leu235Arg, rs63750858, Ile238Met, rs63750888, rs63750526, Leu248Pro, Leu248Arg, rs63751163, rs63751320, rs63751420, rs63750248, rs63750543, rs63751102, rs63751229, rs63750779, rs63750886, rs63750772, Glu273Gly, rs63750284, Ala275Val, rs63750231, Leu282Phe, rs63750050, rs63751139, rs63751235, Arg352Cys, rs63751164, rs63751174, Ser355Ala, rs63750941, rs63751051, rs63750323, rs63750646, rs63749860, Phe386Leu, rs63750883, rs63751066, rs63751416, rs63750929, Ala396Thr, rs63751254, Ile408Thr, rs63750227, rs6[illegible], rs63751316, rs63750802, rs63751032, rs63751223, rs63750083, rs63750528, rs63750341, rs63750001, rs63749925, rs63750249, Ile439Ser	Increased ratio of Aβ42 to Aβ40	[61, 75, 76, 82, 87, 94–96, 104–106, 110, 114, 116–224]
			Intron 4 (InsTAC), rs63750879, Ser169del, Leu174del	Increased ratio of Aβ42 to Aβ40	[94–96, 106, 205, 225–227]

(continued)

Table 1 (continued)

Gene	Chr	Pathway	Mutations/Polymorphisms		References
			ID	Potential effect	
PSEN2	1q31–q42	APP processing	Gly34Ser, rs150400387, Pro69Ala, Arg71Trp, rs63749851, Glu126Lys, rs63750197, Val139Met, Asn141Tyr, rs63750215, Ile235Phe, rs200670135, Leu238Pro, rs63749884, Ala258Val, Pro348Leu, Val393Met, Thr421Met, rs63750666, rs63750110, rs63750812, Lys161Arg, Ser175Cys, Val214Leu, rs63750880	Increased ratio of Aβ42 to Aβ40	[7, 61, 62, 71, 82, 86, 87, 106, 107, 110, 114, 144, 184, 186, 188, 189, 204, 211, 220, 228–244]
APOE	19q13.2	Lipid metabolism	rs429358 + rs7412	Less efficient clearance of soluble Aβ, amyloid plaques and/or neurofibrillary tangles (ApoE4)	[56, 245–251]
SORL1	11q23/24	Endocytosis Lipid metabolism	rs12285364, rs668387, rs3781835, rs117260922, rs143571823, rs11218343, rs2298813	Mutations possibly leading to SORL1 underexpression and overexpression of Aβ	[14, 34, 35, 252]
ABCA7	9p13.3	Lipid metabolism Immune response	rs3752246, rs3764650, rs115550680	Altered lipid homeostasis and/or immune response and/or the clearance of amyloid plaques	[37, 39, 47, 253]
CLU	8p21–p12	Cholesterol metabolism and immune response	rs11136000, rs1532278, rs9331888, rs867230, rs9331908, rs7982, p.T445_D447del	Alterations in CLU expressions and subcellular localization, modulating amyloid deposition	[29–32, 151, 254]
CR1	1q32	Immune response	rs3818361, rs6656401, rs6701713, rs1408077, LCR1-CNV	Alter the binding activity of CR1 and affect the CR1-mediated clearance of Aβ from blood	[14, 30, 37, 45, 46, 255]
CD33	19q13.3	Immune response	rs3865444, rs3826656, rs12459419	Possible effect on splicing efficiency or CR1 expression affecting Aβ clearance and microglia-mediated neuroinflammatory pathways	[37, 256, 257]

(continued)

Table 1 (continued)

Gene	Chr	Pathway	Mutations/Polymorphisms		References
			ID	Potential effect	
MS4A family	11q12.2	Immune response	rs610932, rs670139, rs4938933, rs667897	Alterations in MS4A expression affect Aβ generation, tau phosphorylation, and apoptosis by regulation of calcium homeostasis	[37, 47, 53, 258]
TREM2	6p21.1	Immune response	rs75932628, rs6916710, rs6922617	Decreased affinity of TREM2 for its ligands could affect the clearance of Aβ or lead to systemic inflammatory response	[92, 259–261]
BIN1	2q14.3	Endocytosis Synapse function	rs744373, rs7561528	Increased BIN1 expression modulates Tau pathology	[29, 30, 40, 41]
PICALM	11q14	Endocytosis Synapse function	rs561655a, rs3851179, rs541458	Increased PICALM expression facilitates Aβ and tau clearance	[14, 29, 42, 262, 263]
CD2AP	6p12	Endocytosis Synapse function	rs9296559 rs9349407	Reduced CD2AP expression associated with increased neuritic plaque pathology	[37, 47, 55, 264]
EPHA1	7q34	Immune response	rs11771145, rs11767557	Regulate EPHA1 gene expression and interferes with the pathological alteration in AD	[37, 47, 265]

chapter, some of the most interesting genes found to be associated with late-onset Alzheimer's disease will be described, with their possible involvement in certain biological pathways and mechanisms that might be relevant for Alzheimer's disease pathology.

Apolipoprotein E

One of the major risk loci for late-onset Alzheimer's disease is the ε4 allele of *APOE* gene, gene coding for the main apolipoprotein in the central nervous system. This glycoprotein plays an important role in lipid transport, and it has an undeniable role in growth, repair, reorganization, and maintenance of neurons. ApoE facilitates the cellular uptake of lipoproteins by binding to the members of LDLR family, or it takes part in the activation of signaling pathways involved in modulating lipid homeostasis [20]. Two amino acid substitutions at the positions 112 and 158 lead to three possible ApoE isoforms, ApoE2, ApoE3, ApoE4, which are encoded by three common alleles (ε2, ε3, ε4). The ε4 allele has been associated with Alzheimer's disease, and it is considered as a most important risk factor in the case of late-onset Alzheimer's disease (Table 1). The carriers of *APOE* ε4 allele have an earlier age of onset of Alzheimer's disease, and they also tend to have more pronounced accumulation of neurofibrillary tangles and amyloid plaques [21]. However, *APOE* ε2 allele was associated with reduced risk of developing Alzheimer's disease, with reduced accumulation of neurofibrillary tangles and amyloid plaques [22, 23], but also with significantly larger regional cortical thicknesses and volumes in subjects with cognitive impairment or Alzheimer's disease [24]. The amino acid substitution at the position 158 (arginine to cysteine) impairs the binding of ApoE2 to LDLR and its ability to promote clearance of TG-rich lipoprotein remnant particles. ApoE4 is characterized by an amino acid substitution at the position 112 (cysteine to arginine) that affects the stability of the N-terminal domain helix bundle and C-terminal domain, resulting in enhanced lipid-binding ability of ApoE4 [20] and less efficient clearance of soluble Aβ, amyloid plaques and/or neurofibrillary tangles [25].

Clusterin

Clusterin is a highly glycosylated cell-aggregating factor that is involved in different processes, including complement inhibition, inflammation, apoptosis, and lipid transport [26]. As a chaperone, it could be involved in the amyloid aggregation and pathogenesis of Alzheimer's disease [27]. Evidence suggests that clusterin forms complexes with Aβ in cerebrospinal fluid that are able to cross the brain–blood barrier [28]. Few GWAS studies suggested clusterin as a potential biomarker of Alzheimer's disease [29, 30]. A single nucleotide polymorphism (SNP) in the *CLU* gene was suggested to be associated with Alzheimer's disease pathology by affecting alternative splicing of *CLU* [31]. Other rare non-synonymous single nucleotide variations have also been identified, along with an in-frame 9-bp deletion, that could possibly/probably disturb clusterin structure and function

(Table 1). Findings summarized in Table 1 include mutations that could affect the β-chain domain of clusterin or are positioned in the intron sequence with high regulatory potential [32].

Sortilin-Related Receptor-1

Sortilin-related receptor-1 (SORL1) is considered a member of low-density lipoprotein receptor family and a member of the vacuolar protein sorting 10 (Vps10) family of receptors. There are indications that SORL1 could be involved in APP processing and trafficking, and that it could be responsible for directing Aβ toward lysosomes [33]. However, as a member of LDLR family and an ApoE receptor, SORL1 also plays a role in lipid metabolism. SORL1 was first suggested as a potential risk factor for late-onset Alzheimer's disease by Rogaeva and colleagues [34], and this was later confirmed by other more comprehensive studies [14, 35]. One of the possibilities is that the mutations in *SORL1* gene (Table 1) affect *SORL1* expression and BDNF-induced APP processing [36].

ATP-Binding Cassette Subfamily A Member 7

ATP-binding cassette subfamily A member 7 (ABCA7) belongs to a family of ABC transporters that are responsible for transporting various molecules across cellular membranes. The exact function of ABCA7 still unknown, but there are indications that this protein could play a role in lipid homeostasis and the immune system. Therefore, the mutations in *ABCA7* gene could contribute to Alzheimer's disease development by affecting its interaction with ApoE and lipid metabolism and/or by modulating the immune response and the clearance of amyloid plaques. ABCA7 was associated with Alzheimer's disease in 2011 in a large-scale GWAS analysis [37]. The reported mutations in *ABCA7* gene (Table 1) mostly lead to alterations in gene expression [38], but some rare loss-of-function mutations have also been reported [39].

Bridging Integrator Protein-1

Bridging integrator protein-1 (BIN1) is an amphiphysin involved in caspase-independent cell death pathways and clathrin-mediated endocytic pathway [29, 30]. Few GWAS have identified mutations in *BIN1* gene (Table 1) associated with Alzheimer's disease diagnosis [29, 30, 40]. The study by Chapuis and colleagues [41] suggested that increased *BIN1* gene expression in dementia patients mediates Alzheimer's disease risk by modulating tau pathology.

Phosphatidylinositol Binding Clathrin Assembly Protein

Phosphatidylinositol binding clathrin assembly protein (PICALM) is, similarly to BIN1, involved in clathrin-mediated endocytic pathway. Certain SNPs in *PICALM* gene (Table 1) were found to be associated with the risk of developing Alzheimer's disease. There are even indications that this protein is involved in the internalization of APP and Aβ production [42], Aβ and tau clearance [43, 44].

Complement C3b/C4b Receptor 1

Complement C3b/C4b receptor 1 (CR1) is a glycoprotein belonging to the receptors of complement activation (RCA) family. CR1 regulates complement activation, but it is also participating in innate immune responses. It is expressed by many cell types, including erythrocytes, leukocytes, and dendritic cells. Different SNPs and an intragenic functional copy-number variation [45, 46] in *CR1* gene (Table 1) have been associated with increased risk of developing Alzheimer's disease. Mentioned copy-number variation in *CR1* gene results in two different CR1 protein isoforms (CR1-F and CR1-S), differentiating in the number of C3b/C4b, and cofactor activity binding sites [45, 46], but the exact mechanism of the association with Alzheimer's disease pathology is not known.

CD33

CD33 is a cell surface receptor that mediates cell–cell interaction. This member of the sialic acid-binding receptor family transmembrane proteins is an important mediator of cell growth and survival, and one of key players in clathrin-independent endocytic pathway and innate and adaptive immune system functions [37, 47]. There is evidence of a positive correlation between the expression of CD33 in microglial cells, amyloid plaque burden and decline in cognitive functions [48, 49]. Different GWAS found a significant association between certain SNPs (Table 1) and late-onset Alzheimer's disease. One of these SNPs, rs3865444, was associated with the modifications in CD33 level and amyloid pathology [49, 50].

Membrane-Spanning 4-Domains Subfamily A Gene Cluster

Membrane-spanning 4-domains subfamily A gene cluster (MS4A*)* gene products are transmembrane proteins with at least four transmembrane domains. The genes belonging to this cluster family are not very well characterized, but they might play a role in immunity and intracellular protein trafficking in microglia [51, 52]. There are indications that genes within the *MS4A* gene cluster regulate soluble triggering receptor expressed on myeloid cells 2 (sTREM2) levels, linking this gene cluster family with Alzheimer's disease pathogenesis [52]. Three members of MS4A family (MS4A4A, MS4A4E, MS4A6E) have been linked to Alzheimer's disease by GWAS (Table 1), more precisely, SNPs rs670139 (*MS4A4E*), rs4938933 and rs1562990 (region between *MS4A4E* and *MS4A4A*), and rs610932 and rs983392 (*MS4A6A*) [53].

Triggering Receptor Expressed on Myeloid Cells 2

Triggering receptor expressed on myeloid cells 2 (*TREM2)* gene product is an important part of transmembrane receptor-signaling complex that is very abundant on the cell surface of microglia where it plays an important role in downregulation of inflammation, microglial survival and activation, and phagocytosis [54]. There is evidence of high involvement of TREM2 in Alzheimer's disease pathology.

Different *TREM2* variants (Table 1) have been associated with Alzheimer's disease by few meta-analysis and GWAS. These variants include missense mutation rs75932628, rs6916710, rs6922617 (Table 1). *TREM2* mutations have been linked to extensive brain atrophy in Alzheimer's disease patients and with other neuropathological phenotypes characteristic for Alzheimer's disease [52].

CD2-Associated Protein

CD2-associated protein (CD2AP) is a cytoplasmic protein involved in cytoskeletal structure regulation, receptor-mediated endocytosis, intracellular trafficking, cytokinesis, cell adhesion, and apoptosis. Few GWAS pointed to *CD2AP* SNPs (rs9296559, rs9349407) as loci possibly associated with late-onset Alzheimer's disease (Table 1). SNP rs9349407 could be associated with increased neuritic plaque burden in patients with diagnosed Alzheimer's disease [55].

Ephrin Type-A Receptor 1

Ephrin type-A receptor 1 (EPHA1) is a tyrosine kinase family member important during the nervous system development and during the formation of synapse. It binds to membrane-bound ephrin-A family ligands leading to bidirectional signaling between two adjacent cells, directing cell adhesion and migration. EPHA1 could potentially have a role in the microglial immune response in Alzheimer's disease. Two EPHA1 gene variants (rs11771145 and rs11767557) were associated with Alzheimer's disease risk (Table 1).

Conclusion

The knowledge about the genetic background of early-onset familial Alzheimer's disease allowed detection of mutations in *APP*, *PSEN1*, and *PSEN2* as a predictive/diagnostic screening, but only for these rare cases of autosomal dominant Alzheimer's disease. In the case of late-onset Alzheimer's disease, there is still much of the heritability that remains unexplained, even though, with the help from GWAS, there are now more than 20 different identified loci that have been associated with late-onset complex Alzheimer's disease. In the case of *APOE*, the evidence from an extensive meta-analysis shows that around 75% of individuals that carry one *APOE* ε4 allele never develop Alzheimer's disease, and around 50% of individuals diagnosed with Alzheimer's disease are not the carriers of this risk allele [56]. Therefore, the use of genetic variations identified by GWAS, for more effective diagnosis of Alzheimer's disease, is still far away from the clinical application, because the contribution of these variations to the relative risk of developing Alzheimer's disease is limited. In the light of prediction and prevention of Alzheimer's disease, there is much more we can expect from the additive genetic risk scores, which combine additive effects of numerous susceptibility loci.

References

1. Cerejeira J, Lagarto L, Mukaetova-Ladinska EB. Behavioral and psychological symptoms of dementia. Front Neurol. 2012;3:73.
2. Reitz C (2016) Toward precision medicine in Alzheimer's disease. Ann Transl Med. 2016;4(6):107.
3. Hampel H, O'Bryant S, Durrleman S, Younesi E, Rojkova K, Escott-Price V, et al. A precision medicine Initiative for Alzheimer's disease: the road ahead to biomarker-guided integrative disease modeling. Climacteric. 2017;20(2):107–18.
4. Nikolac Perkovic M, Svob Strac D, Tudor L, Konjevod M, Nedic Erjavec G, Pivac N. Catechol-O-methyltransferase, cognition and Alzheimer's disease. Curr Alzheimer Res. 2018;15(5):408–19.
5. Duarte A, Santos MS, Oliveira CR. Moreira P. Brain insulin signalling, glucose metabolism and females' reproductive aging: a dangerous triad in Alzheimer's disease. Neuropharmacology. 2018;136(Pt B):223–42.
6. Van Cauwenberghe C, Van Broeckhoven C, Sleegers K. The genetic landscape of Alzheimer disease: clinical implications and perspectives. Genet Med. 2015;18(5):421–30.
7. Levy-Lahad E, Wasco W, Poorkaj P, Romano DM, Oshima J, Pettingell WH, et al. Candidate gene for the chromosome 1 familial Alzheimer's disease locus. Science. 1995;269(5226):973–7.
8. Kopan R, Goate A. A common enzyme connects Notch signaling and Alzheimer's disease. Genes Dev. 2000;14(22):2799–806.
9. Marambaud P, Shioi J, Serban G, Georgakopoulos A, Sarner S, Nagy V, et al. A presenilin-1/γ-secretase cleavage releases the E-cadherin intracellular domain and regulates disassembly of adherens junctions. EMBO J. 2002;21(8):1948–56.
10. Lleó A, Waldron E, von Arnim CA, Herl L, Tangredi MM, Peltan ID, et al. Low density lipoprotein receptor-related protein (LRP) interacts with presenilin 1 and is a competitive substrate of the amyloid precursor protein (APP) for gamma-secretase. J Biol Chem. 2005;280:27303–9.
11. Vidal GA, Naresh A, Marrero L, Jones FE. Presenilin-dependent gamma-secretase processing regulates multiple ERBB4/HER4 activities. J Biol Chem. 2005;280:19777–83.
12. Dai MH, Zheng H, Zeng LD, Zhang Y. The genes associated with early-onset Alzheimer's disease. Oncotarget. 2017;9(19):15132–43.
13. Larsson M, Duffy DL, Zhu G, Liu JZ, Macgregor S, McRae AF, et al. GWAS findings for human iris patterns: associations with variants in genes that influence normal neuronal pattern development. Am J Hum Genet. 2011;89(2):334–43.
14. Lambert JC, Ibrahim-Verbaas CA, Harold D, Naj AC, Sims R, Bellenguez C, et al. Meta-analysis of 74,046 individuals identifies 11 new susceptibility loci for Alzheimer's disease. Nat Genet. 2013;45(12):1452–8.
15. Liu Y, Yu JT, Wang HF, Hao XK, Yang YF, Jiang T, et al. Association between NME8 locus polymorphism and cognitive decline, cerebrospinal fluid and neuroimaging biomarkers in Alzheimer's disease. PLoS ONE. 2014;9:e114777.
16. Ruiz A, Heilmann S, Becker T, Hernández I, Wagner H, Thelen M, et al. Follow-up of loci from the International Genomics of Alzheimer's Disease Project identifies TRIP4 as a novel susceptibility gene. Transl Psychiatry. 2014;4:e358.
17. Rosenthal SL, Barmada MM, Wang X, Demirci FY, Kamboh MI. Connecting the dots: potential of data integration to identify regulatory SNPs in late-onset Alzheimer's disease GWAS findings. PLoS ONE. 2014;9(4):e95152.
18. Allen M, Kachadoorian M, Carrasquillo MM, Karhade A, Manly L, Burgess JD, et al. Late-onset Alzheimer disease risk variants mark brain regulatory loci. Neurol Genet. 2015;1(2):e15.
19. Jiao B, Liu X, Zhou L. Polygenic analysis of late-onset Alzheimer's disease from mainland China. PLoS ONE. 2015;10:e0144898.

20. Phillips MC. Apolipoprotein E isoforms and lipoprotein metabolism. IUBMB Life. 2014;66(9):616–23.
21. Lai MK, Tsang SW, Garcia-Alloza M, Minger SL, Nicoll JA, Esiri MM, et al. Selective effects of the APOE ε4 allele on presynaptic cholinergic markers in the neocortex of Alzheimer's disease. Neurobiol Dis. 2006;22(3):555–61.
22. Corder EH, Saunders AM, Risch NJ, Strittmatter WJ, Schmechel DE, Gaskell PC Jr, et al. Protective effect of apolipoprotein E type 2 allele for late onset Alzheimer disease. Nat Genet. 1994;7(2):180–4.
23. Tiraboschi P, Hansen LA, Masliah E, Alford M, Thal LJ, Corey-Bloom J. Impact of APOE genotype on neuropathologic and neurochemical markers of Alzheimer disease. Neurology. 2004;62(11):1977–83.
24. Liu Y, Paajanen T, Westman E, Zhang Y, Wahlund LO, Simmons A, et al. APOE ε2 allele is associated with larger regional cortical thicknesses and volumes. Dement Geriatr Cogn Disord. 2010;30(3):229–37.
25. Deane R, Sagare A, Hamm K, Parisi M, Lane S, Finn MB, et al. ApoE isoform-specific disruption of amyloid beta peptide clearance from mouse brain. J Clin Invest. 2008;118(12):4002–13.
26. Matukumalli SR, Tangirala R, Rao CM. Clusterin: full-length protein and one of its chains show opposing effects on cellular lipid accumulation. Sci Reports. 2017;7:41235.
27. Li X, Ma Y, Wei X, Li Y, Wu H, Zhuang J, Zhao Z. Clusterin in Alzheimer's disease: a player in the biological behaviour of amyloid-beta. Neurosci Bull. 2014;30(1):162–8.
28. Zlokovic BV. Cerebrovascular transport of Alzheimer's amyloid beta and apolipoproteins J and E: possible anti-amyloidogenic role of the blood-brain barrier. Life Sci. 1996;59:1483–97.
29. Harold D, Abraham R, Hollingworth P, Sims R, Gerrish A, Hamshere ML, et al. Genome-wide association study identifies variants at CLU and PICALM associated with Alzheimer's disease. Nat Genet. 2009;41(10):1088–93.
30. Lambert JC, Heath S, Even G, Campion D, Sleegers K, Hiltunen M, et al. Genome-wide association study identifies variants at CLU and CR29 associated with Alzheimer's disease. Nat Genet. 2009;41(10):1094–9.
31. Szymanski M, Wang R, Bassett SS, Avramopoulos D. Alzheimer's risk variants in the clusterin gene are associated with alternative splicing. Transl Psychiatry. 2011;1(7):e18.
32. Bettens K, Brouwers N, Engelborghs S, Lambert JC, Rogaeva E, Vandenberghe R, et al. Both common variations and rare non-synonymous substitutions and small insertion/deletions in CLU are associated with increased Alzheimer risk. Mol Neurodegener. 2012;16(7):3.
33. Offe K, Dodson SE, Shoemaker JT, Fritz JJ, Gearing M, Levey AI, et al. The lipoprotein receptor LR11 regulates amyloid beta production and amyloid precursor protein traffic in endosomal compartments. J Neurosci. 2006;26(5):1596–603.
34. Rogaeva E, Meng Y, Lee JH, Gu Y, Kawarai T, Zou F, et al. The neuronal sortilin related receptor SORL1 is genetically associated with Alzheimer disease. Nat Genet. 2007;39(2):168–77.
35. Vardarajan BN, Zhang Y, Lee JH, Cheng R, Bohm C, Ghani M, et al. Coding mutations in SORL1 and Alzheimer disease. Ann Neurol. 2015;77:215–27.
36. Young JE, Boulanger-Weill J, Williams DA, Woodruff G, Buen F, Revilla AC, et al. Elucidating molecular phenotypes caused by the SORL1 Alzheimer's disease genetic risk factor using human induced pluripotent stem cells. Cell Stem Cell. 2015;16(4):373–85.
37. Hollingworth P, Harold D, Sims R, Gerrish A, Lambert JC, Carrasquillo MM, et al. Common variants at ABCA7, MS4A6A/MS4A4E, EPHA1, CD33 and CD2AP are associated with Alzheimer's disease. Nat Genet. 2011;43(5):429–35.
38. Vasquez JB, Fardo DW, Estus S. ABCA7 expression is associated with Alzheimer's disease polymorphism and disease status. Neurosci Lett. 2013;556:58–62.
39. Steinberg S, Stefansson H, Jonsson T, Johannsdottir H, Ingason A, Helgason H, et al. Loss-of-function variants in ABCA7 confer risk of Alzheimer's disease. Nat Genet. 2015;47(5):445–7.

40. Seshadri S, Fitzpatrick AL, Ikram MA, DeStefano AL, Gudnason V, Boada M, et al. Genome-wide analysis of genetic loci associated with Alzheimer disease. JAMA. 2010;303(18):1832–40.
41. Chapuis J, Hansmannel F, Gistelinck M, Mounier A, Van Cauwenberghe C, Kolen KV, et al. Increased expression of BIN1 mediates Alzheimer genetic risk by modulating tau pathology. Mol Psychiatry. 2013;18(11):1225–34.
42. Xiao Q, Gil SC, Yan P, Wang Y, Han S, Gonzales E, et al. Role of phosphatidylinositol clathrin assembly lymphoid-myeloid leukemia (PICALM) in intracellular amyloid precursor protein (APP) processing and amyloid plaque pathogenesis. J Biol Chem. 2012;287:21279–89.
43. Zhao Z, Sagare AP, Ma Q, Halliday MR, Kong P, Kisler K, et al. Central role for PICALM in amyloid-β blood-brain barrier transcytosis and clearance. Nat Neurosci. 2015;18:978–87.
44. Moreau K, Fleming A, Imarisio S, Lopez Ramirez A, Mercer JL, Jimenez-Sanchez M, et al. PICALM modulates autophagy activity and tau accumulation. Nat Commun. 2014;5:4998.
45. Brouwers N, Van Cauwenberghe C, Engelborghs S, Lambert JC, Bettens K, Le Bastard N, et al. Alzheimer risk associated with a copy number variation in the complement receptor 1 increasing C3b/C4b binding sites. Mol Psychiatry. 2012;17(2):223–33.
46. Hazrati LN, Van Cauwenberghe C, Brooks PL, Brouwers N, Ghani M, Sato C, et al. Genetic association of CR45 with Alzheimer's disease: a tentative disease mechanism. Neurobiol Aging. 2012;33(12):2949.e5–12.
47. Naj AC, Jun G, Beecham GW, Wang LS, Vardarajan BN, Buros J, et al. Common variants at MS4A4/MS4A6E, CD2AP, CD33 and EPHA1 are associated with late-onset Alzheimer's disease. Nat Genet. 2011;43(5):436–41.
48. Karch CM, Jeng AT, Nowotny P, Cady J, Cruchaga C, Goate AM. Expression of novel Alzheimer's disease risk genes in control and Alzheimer's disease brains. PLoS ONE. 2012;7:e50976.
49. Griciuc A, Serrano-Pozo A, Parrado AR, Lesinski AN, Asselin CN, Mullin K, et al. Alzheimer's disease risk gene CD33 inhibits microglial uptake of amyloid beta. Neuron. 2013;78(4):631–43.
50. Bradshaw EM, Chibnik LB, Keenan BT, Ottoboni L, Raj T, Tang A, et al. CD33 Alzheimer's disease locus: altered monocyte function and amyloid biology. Nat Neurosci. 2013;16(7):848–50.
51. Zuccolo J, Deng L, Unruh TL, Sanyal R, Bau JA, Storek J, et al. Expression of MS4A and TMEM176 genes in human B lymphocytes. Front Immunol. 2013;4:195.
52. Deming Y, Filipello F, Cignarella F, Cantoni C, Hsu S, Mikesell R, et al. The MS4A gene cluster is a key regulator of soluble TREM2 and Alzheimer disease risk. BioRxiv. 2018. https://doi.org/10.1101/352179.
53. Antunez C, Boada M, González-Pérez A, Gayán J, Ramírez-Lorca R, Marín J, et al. The membrane-spanning 4-domains, subfamily A (MS4A) gene cluster contains a common variant associated with Alzheimer's disease. Genome Med. 2011;3(5):33.
54. Ulrich JD, Ulland TK, Colonna M, Holtzman DM. Elucidating the Role of TREM2 in Alzheimer's disease. Neuron. 2017;94:237–48.
55. Shulman JM, Chen K, Keenan BT, Chibnik LB, Fleisher A, Thiyyagura P, et al. Genetic susceptibility for Alzheimer disease neuritic plaque pathology. JAMA Neurol. 2013;70(9):1150–7.
56. Farrer LA, Cupples LA, Haines JL, Hyman B, Kukull WA, Mayeux R, et al. Effects of age, sex, and ethnicity on the association between apolipoprotein E genotype and Alzheimer disease. A meta-analysis. APOE and Alzheimer disease meta analysis consortium. JAMA. 1997;278(16):1349–56.
57. Hooli BV, Mohapatra G, Mattheisen M, Parrado AR, Roehr JT, Shen Y, et al. Role of common and rare APP DNA sequence variants in Alzheimer disease. Neurology. 2012;78(16):1250–7.

58. Kasuga K, Shimohata T, Nishimura A, Shiga A, Mizuguchi T, Tokunaga J, et al. Identification of independent APP locus duplication in Japanese patients with early-onset Alzheimer disease. J Neurol Neurosurg Psychiatry. 2009;80(9):1050–2.
59. McNaughton D, Knight W, Guerreiro R, Ryan N, Lowe J, Poulter M, et al. Duplication of amyloid precursor protein (APP), but not prion protein (PRNP) gene is a significant cause of early onset dementia in a large UK series. Neurobiol Aging. 2012;33(2):426.e13–21.
60. Thonberg H, Fallström M, Björkström J, Schoumans J, Nennesmo I, Graff C. Mutation screening of patients with Alzheimer disease identifies APP locus duplication in a Swedish patient. BMC Res Notes. 2011;4:476.
61. Wallon D, Rousseau S, Rovelet-Lecrux A, Quillard-Muraine M, Guyant-Maréchal L, Martinaud O, et al. The French series of autosomal dominant early onset Alzheimer's disease cases: mutation spectrum and cerebrospinal fluid biomarkers. J Alzheimers Dis. 2012;30(4):847–56.
62. Brouwers N, Sleegers K, Van Broeckhoven C. Molecular genetics of Alzheimer's disease: an update. Ann Med. 2008;40(8):562–83.
63. Kamino K, Orr HT, Payami H, Wijsman EM, Alonso ME, Pulst SM, et al. Linkage and mutational analysis of familial Alzheimer disease kindreds for the APP gene region. Am J Hum Genet. 1992;51(5):998–1014.
64. Mullan M, Crawford F, Axelman K, Houlden H, Lilius L, Winblad B, et al. A pathogenic mutation for probable Alzheimer's disease in the APP gene at the N-terminus of beta-amyloid. Nat Genet. 1992;1(5):345–7.
65. Nilsberth C, Westlind-Danielsson A, Eckman CB, Forsell C, Axelman K, Luthman J, et al. The Arctic APP mutation (E693G) causes Alzheimer's disease through a novel mechanism: increased amyloid β protofibril formation and decreased amyloid β levels in plasma and conditioned media. Neurobiol Aging. 2000;21 Suppl 1:S58.
66. Nilsberth C, Westlind-Danielsson A, Eckman CB, Condron MM, Axelman K, Forsell C, et al. The 'Arctic' APP mutation (E693G) causes Alzheimer's disease by enhanced Abeta protofibril formation. Nat Neurosci. 2001;4(9):887–93.
67. Tomiyama T, Nagata T, Shimada H, Teraoka R, Fukushima A, Kanemitsu H, et al. A new amyloid beta variant favoring oligomerization in Alzheimer's-type dementia. Ann Neurol. 2008;63:377–87.
68. Wakutani Y, Watanabe K, Adachi Y, Wada Isoe K, Urakami K, Ninomiya H, et al. Novel amyloid precursor protein gene missense mutation (D678N) in probable familial Alzheimer's disease. J Neurol Neurosurg Psychiatry. 2004;75(7):1039–42.
69. Wakutani Y. Gene symbol: APP. Disease: Familial Alzheimer's disease. Hum Genet. 2005;117(2–3):299.
70. Ancolio K, Dumanchin C, Barelli H, Warter JM, Brice A, Campion D, et al. Unusual phenotypic alteration of beta amyloid precursor protein (betaAPP) maturation by a new Val-715 → Met betaAPP-770 mutation responsible for probable early onset Alzheimer's disease. Proc Natl Acad Sci USA. 1999;96(7):4119–24.
71. Blauwendraat C, Wilke C, Jansen IE, Schulte C, Simón-Sánchez J, Metzger FG, et al. Pilot whole-exome sequencing of a German early-onset Alzheimer's disease cohort reveals a substantial frequency of PSEN2 variants. Neurobiol Aging. 2016;37(208):e11–7.
72. Brooks WS, Martins RN, De Voecht J, Nicholson GA, Schofield PR, Kwok JB, et al. A mutation in codon 717 of the amyloid precursor protein gene in an Australian family with Alzheimer's disease. Neurosc Lett. 1995;199(3):183–6.
73. Brouwers N, Sleegers K, Engelborghs S, Bogaerts V, Serneels S, Kamali K, et al. Genetic risk and transcriptional variability of amyloid precursor protein in Alzheimer's disease. Brain. 2006;129(Pt 11):2984–91.
74. Campion D, Brice A, Hannequin D, Charbonnier F, Dubois B, Martin C, et al. No founder effect in three novel Alzheimer's disease families with APP 717 Val → Ile mutation. Clerget-darpoux. French Alzheimer's Disease Study Group. J Med Genet. 1996;33(8):661–4.

75. Campion D, Dumanchin C, Hannequin D, Dubois B, Belliard S, Puel M, et al. Early-onset autosomal dominant Alzheimer disease: prevalence, genetic heterogeneity, and mutation spectrum. Am J Hum Genet. 1999;65(3):664–70.
76. Campion D, Flaman JM, Brice A, Hannequin D, Dubois B, Martin C, et al. Mutations of the presenilin I gene in families with early-onset Alzheimer's disease. Hum Mol Genet. 1995;4(12):2373–7.
77. Clarimón J, Guerreiro R, Lleó A, Guardia C, Blesa F, Gómez-Isla T, et al. Genetic screening in a large cohort of early-onset Alzheimer's disease patients from Spain: novel mutations in the amyloid precursor protein and presenilines. Alzheimer Dement. 2008;4 Supp 2:T583.
78. Cruts M, Dermaut B, Kumar-Singh S, Rademakers R, Van den Broeck M, Stögbauer F, et al. Novel German APP V715A mutation associated with presenile Alzheimer's disease. Neurobiol Aging. 2002;23(1S):S327.
79. Cruts M, Dermaut B, Rademakers R, Van den Broeck M, Stogbauer F, Van Broeckhoven C. Novel APP mutation V715A associated with presenile Alzheimer's disease in a German family. J Neurol. 2003;250(11):1374–5.
80. De Jonghe C, Kumar-Singh S, Cruts M, Kleinert R, Vanderstichele H, Vanmechelen EJM, et al. Unusual Aβ amyloid deposition in Alzheimer's disease due tu an APP T714I mutation at the γ42-secretase site. Neurobiol Aging. 2000;21(1):S200.
81. De Jonghe C, Esselens C, Kumar-Singh S, Craessaerts K, Serneels S, et al. Pathogenic APP mutations near the gamma-secretase cleavage site differentially affect Abeta secretion and APP C-terminal fragment stability. Hum Mol Genet. 2001;10(16):1665–71.
82. Dobricic V, Stefanova E, Jankovic M, Gurunlian N, Novakovic I, Hardy J, et al. Genetic testing in familial and young-onset Alzheimer's disease: mutation spectrum in a Serbian cohort. Neurobiol Aging. 2012;3(1481):e7–12.
83. Eckman CB, Mehta ND, Crook R, Perez-tur J, Prihar G, Pfeiffer E, et al. A new pathogenic mutation in the APP gene (I716V) increases the relative proportion of A beta 42(43). Hum Mol Genet. 1997;6(12):2087–9.
84. Edwards-Lee T, Ringman JM, Chung J, Werner J, Morgan A, St George Hyslop P, et al. An African American family with early-onset Alzheimer disease and an APP (T714I) mutation. Neurology. 2005;64(2):377–9.
85. Fidani L, Rooke K, Chartier-Harlin MC, Hughes D, Tanzi R, Mullan M, et al. Screening for mutations in the open reading frame and promoter of the beta-amyloid precursor protein gene in familial Alzheimer's disease: identification of a further family with APP717 Val → Ile. Hum Mol Genet. 1992;1(3):165–8.
86. Finckh U, Muller-Thomsen T, Mann U, Eggers C, Marksteiner J, Meins W, et al. High prevalence of pathogenic mutations in patients with early-onset dementia detected by sequence analyses of four different genes. Am J Hum Genet. 2000;66(1):110–7.
87. Finckh U, Kuschel C, Anagnostouli M, Patsouris E, Pantes GV, Gatzonis S, et al. Novel mutations and repeated findings of mutations in familial Alzheimer disease. Neurogenetics. 2005;6(2):85–9.
88. Ghetti B, Hake AM, Murrell JR, Epperson F, Farlow MR, Vidal R, Spina S. Familial Alzheimer disease associated with the V717L amyloid precursor protein gene mutation: Neuropathological characterization. Alzheimers Dement. 2008;4 Suppl 4:T585.
89. Goate A, Chartier-Harlin MC, Mullan M, Brown J, Crawford F, Fidani L, et al. Segregation of a missense mutation in the amyloid precursor protein gene with familial Alzheimer's disease. Nature. 1991;349(6311):704–6.
90. Godbolt AK, Beck JA, Collinge JC, Cipolotti L, Fox NC, Rossor MN. A second family with familial AD and the V717L APP mutation has a later age at onset. Neurology. 2006;66(4):611–2.
91. Guardia-Laguarta C, Pera M, Clarimón J, Molinuevo JL, Sánchez-Valle R, Lladó A, et al. Clinical, neuropathologic, and biochemical profile of the amyloid precursor protein I716F mutation. J Neuropathol Exp Neurol. 2010;69(1):53–9.
92. Guerreiro R, Wojtas A, Bras J, Carrasquillo M, Rogaeva E, Majounie E, et al. TREM2 variants in Alzheimer's disease. N Engl J Med. 2013;368(2):117–27.

93. Hardy J, Mullan M, Chartier-Harlin M-C, Brown J, Goate A, Rossor M, Collinge J, et al. Molecular classification of Alzheimer's disease. Lancet. 1991;337(8753):1342–3.
94. Janssen JC, Beck JA, Campbell TA, Dickinson A, Fox NC, Harvey RJ, et al. Early onset familial Alzheimer's disease: mutation frequency in 31 families. Neurobiol Aging. 2002;23(1):S311.
95. Janssen JC, Beck JA, Campbell TA, Dickinson A, Fox NC, Harvey RJ, et al. Early onset familial Alzheimer's disease: mutation frequency in 31 families. Neurology. 2003;60(2):235–9.
96. Jiao B, Tang B, Liu X, Xu J, Wang Y, Zhou L, et al. Mutational analysis in early-onset familial Alzheimer's disease in Mainland China. Neurobiol Aging. 2014;35(8):1957.e1–6.
97. Kumar-Singh S, De Jonghe C, Cruts M, Kleinert R, Wang R, Mercken M, et al. Nonfibrillar diffuse amyloid deposition due to a gamma(42)-secretase site mutation points to an essential role for N-truncated abeta(42) in Alzheimer's disease. Hum Mol Genet. 2000;9(18):2589–98.
98. Kwok JBJ, Li Q-X, Hallupp M, Milward L, Whyte S, Schofield PR. Novel familial early-onset Alzheimer's disease mutation (Leu723Pro) in amyloid precursor protein (APP) gene increases production of 42(43) amino-acid isoform of amyloid beta peptide. Neurobiol Aging. 1998;19(4):S91.
99. Kwok JB, Li QX, Hallupp M, Whyte S, Ames D, Beyreuther K, Masters CL, Schofield PR. Novel Leu723Pro amyloid precursor protein mutation increases amyloid beta42(43) peptide levels and induces apoptosis. Ann Neurol. 2000;47(2):249–53.
100. Matsumura Y, Kitamura E, Miyoshi K, Yamamoto Y, Furuyama J, Sugihara T. Japanese siblings with missense mutation (717Val → Ile) in amyloid precursor protein of early-onset Alzheimer's disease. Neurology. 1996;46(6):1721–3.
101. Murrell J, Farlow M, Ghetti B, Benson MD. A mutation in the amyloid precursor protein associated with hereditary Alzheimer's disease. Science. 1991;254(5028):97–9.
102. Murrell JR, Hake AM, Quaid KA, Farlow MR, Ghetti B. Early-onset Alzheimer disease caused by a new mutation (V717L) in the amyloid precursor protein gene. Arch Neurol. 2000;57:885–7.
103. Naruse S, Igarashi S, Kobayashi H, Aoki K, Inuzuka T, Kaneko K, et al. Mis-sense mutation Val → Ile in exon 17 of amyloid precursor protein gene in Japanese familial Alzheimer's disease. Lancet. 1991;337(8747):978–9.
104. Park HK, Na DL, Lee JH, Kim JW, Ki CS. Identification of PSEN1 and APP gene mutations in Korean patients with early-onset Alzheimer's disease. J Korean Med Sci. 2008;23(2):213–7.
105. Raux G, Guyant-Marechal L, Martin C, Bou J, Penet C, Brice A, et al. Molecular diagnosis of autosomal dominant early onset Alzheimer's disease: an update. J Med Genet. 2005;42(10):793–5.
106. Sassi C, Guerreiro R, Gibbs R, Ding J, Lupton MK, Troakes C, et al. Exome sequencing identifies 2 novel presenilin 1 mutations (p.L166V and p.S230R) in British early-onset Alzheimer's disease. Neurobiol Aging. 2014;35(10):2422.e13–6.
107. Sassi C, Guerreiro R, Gibbs R, Ding J, Lupton MK, Troakes C, et al. Investigating the role of rare coding variability in Mendelian dementia genes (APP, PSEN1, PSEN2, GRN, MAPT, and PRNP) in late-onset Alzheimer's disease. Neurobiol Aging. 2014;35(12):2881.e1–6.
108. Sorbi S, Nacmias B, Forleo P, Piacentini S, Amaducci L, Provinciali L. APP717 and Alzheimer's disease in Italy. Nat Genet. 1993;4(1):10.
109. Sorbi S, Nacmias B, Forleo P, Piacentini S, Latorraca S, Amaducci L. Epistatic effect of APP717 mutation and apolipoprotein E genotype in familial Alzheimer's disease. Ann Neurol. 1995;38(1):124–7.
110. Tedde A, Nacmias B, Ciantelli M, Forleo P, Cellini E, Bagnoli S, et al. Identification of new presenilin gene mutations in early-onset familial Alzheimer disease. Arch Neurol. 2003;60(11):1541–4.

111. Terreni L, Fogliarino S, Franceschi, Forloni G. Novel pathogenic mutation in an Italian patient with familial Alzheimer's disease detected in APP gene. Neurobiol Aging. 2002;23(1):S319.
112. Yoshioka K, Miki T, Katsuya T, Ogihara T, Sakaki Y. The 717Val → Ile substitution in amyloid precursor protein is associated with familial Alzheimer's disease regardless of ethnic groups. Biochem Biophys Res Commun. 1991;178(3):1141–6.
113. Yoshizawa T, Komatsuzaki Y, Iwamoto H, Mizusawa H, Kanazawa I. Screening of the missense mutation producing the 717Val → Ile substitution in the amyloid precursor protein in Japanese familial and sporadic Alzheimer's disease. J Neurol Sci. 1993;117(1–2):12–5.
114. Zekanowski C, Styczynska M, Peplonska B, Gabryelewicz T, Religa D, Ilkowski J, et al. Mutations in presenilin 1, presenilin 2 and amyloid precursor protein genes in patients with early-onset Alzheimer's disease in Poland. Exp Neurol. 2003;184(2):991–6.
115. Theuns J, Marjaux E, Vandenbulcke M, Van Laere K, Kumar-Singh S, Bormans G, et al. Alzheimer dementia caused by a novel mutation located in the APP C-terminal intracytosolic fragment. Hum Mutat. 2006;27(9):888–96.
116. Alberici A, Bonato C, Borroni B, Cotelli M, Mattioli F, Binetti G, et al. Dementia, delusions and seizures: storage disease or genetic AD. Eur J Neurol. 2007;14(9):1057–9.
117. Aldudo J, Bullido MJ, Arbizu T, Oliva R, Valdivieso F. Identification of a novel mutation (Leu282Arg) of the human presenilin 1 gene in Alzheimer's disease. Neurosci Lett. 1998;240(3):174–6.
118. Aldudo J, Bullido MJ, Valdivieso F. DGGE method for the mutational analysis of the coding and proximal promoter regions of the Alzheimer's disease presenilin-1 gene: two novel mutations. Hum Mutat. 1999;14(5):433–9.
119. Antonell A, Balasa M, Oliva R, Lladó A, Bosch B, Fabregat N, et al. A novel PSEN1 gene mutation (L235R) associated with familial early-onset Alzheimer's disease. Neurosci Lett. 2011;496(1):40–2.
120. Aoki M, Abe K, Oda N, Ikeda M, Tsuda T, Kanai M, et al. A presenilin-1 mutation in a Japanese family with Alzheimer's disease and distinctive abnormalities on cranial MRI. Neurology. 1997;48(4):1118–20.
121. Arai N, Kishino A, Takahashi Y, Morita D, Nakamura K, Yokoyama T, et al. Familial cases presenting very early onset autosomal dominant Alzheimer's disease with I143T in presenilin-1 gene: implication for genotype-phenotype correlation. Neurogenetics. 2008;9(1):65–7.
122. Arango D, Cruts M, Torres O, Backhovens H, Serrano ML, Villareal E, et al. Systematic genetic study of Alzheimer disease in Latin America: mutation frequencies of the amyloid beta precursor protein and presenilin genes in Colombia. Am J Med Genet. 2001;103(2):138–43.
123. Athan ES, Williamson J, Ciappa A, Santana V, Romas SN, Lee JH, et al. A founder mutation in presenilin 1 causing early-onset Alzheimer disease in unrelated Caribbean Hispanic families. JAMA. 2001;286(18):2256–63.
124. Axelman K, Basun H, Lannfelt L. Wide range of disease onset in a family with Alzheimer disease and a His163Tyr mutation in the presenilin-1 gene. Arch Neurol. 1998;55(5):698–702.
125. Batelli S, Albani D, Prato F, Polito L, Franceschi M, Gavazzi A, Forloni G. Early-onset Alzheimer disease in an Italian family with presenilin-1 double mutation E318G and G394V. Alzheimer Dis Assoc Disord. 2008;22(2):184–7.
126. Boteva K, Vitek M, Mitsuda H, de Silva H, Xu PT, Small G, Gilbert JR. Mutation analysis of presenillin 1 gene in Alzheimer's disease. Lancet. 1996;347(8994):130–1.
127. Brouwers N, Sleegers K, Theuns J, Engelborghs S, Bogaerts V, Serneels S, et al. Contribution of dementia genes to Alzheimer's disease in Belgium. Alzheimer Dement. 2006;2(3):S191.
128. Cervenakova L, Brown P, Sandoval F, Foncin J-F, Garruto R, Polinsky RJ, et al. Identification of presenilin-1 gene point mutations in early-onset Alzheimer's disease families. Am J Hum Genet. 1996;59(Suppl):A252.

129. Church A, Prescott J, Lillis S, Rees J, Chance P, Williamson K, Morris HR. A novel presenilin 1 mutation, I202F occurring at a previously predicted pathogenic site causing autosomal dominant Alzheimer's disease. Neurobiol Aging. 2011;32(556):e1–2.
130. Clarimón J, Guerreiro R, Lleó A, Guardia C, Blesa F, Gómez-Isla T, et al. Genetic screening in a large cohort of early-onset Alzheimer's disease patients from Spain: novel mutations in the amyloid precursor protein and presenilines. Alzheimers Dement. 2008;4(Supp 2):T583.
131. Clark RF, Hutton M, Fuldner RA, Froelich S, Karran E, Talbot C, et al. The structure of the presenilin 1 (S182) gene and identification of six novel mutations in early onset AD families. Nat Genet. 1995;11(2):219–22.
132. Coleman P, Kurlan R, Crook R, Werner J, Hardy J. A new presenilin Alzheimer's disease case confirms the helical alignment of pathogenic mutations in transmembrane domain 5. Neurosci Lett. 2004;364(3):139–40.
133. Crook R, Ellis R, Shanks M, Thal LJ, Perez-Tur J, Baker M, et al. Early-onset Alzheimer's disease with a presenilin-1 mutation at the site corresponding to the Volga German presenilin-2. Ann Neurol. 1997;42(1):124–8.
134. Cruts M, Backhovens H, Wang SY, Gassen GV, Theuns J, De Jonghe CD, Wehnert A, De Voecht J, De Winter G, Cras P, Bruyland M, Datson N, Weissenbach J, den Dunnen JT, Martin J-J, Hendriks L, Van Broeckhoven C. Molecular genetic analysis of familial early-onset Alzheimer's disease linked to chromosome 14q24.3. Hum Mol Genet. 1995;4(12):2363–72.
135. Cruts M, van Duijn CM, Backhovens H, Van den Broeck M, Wehnert A, Serneels S, et al. Estimation of the genetic contribution of presenilin-1 and -2 mutations in a population-based study of presenile Alzheimer disease. Hum Mol Genet. 1998;7:43–51.
136. Doran M, Larner AJ. Familial Alzheimer's disease due to presenilin-1 Y115C mutation. J Neurol. 2006;253 Suppl 2:ii91,2006.
137. Dowjat K, Kuchna I, Wisniewski K, Wisniewski T, Wegiel J. Another highly pathogenic Alzheimer presenilin-1 mutation in codon 117: genotype-phenotype comparison of P117S and P117L mutations. Neurobiol Aging. 2002;23(1S):S219.
138. Dowjat WK, Kuchna I, Wisniewski T, Wegiel J. A novel highly pathogenic Alzheimer presenilin-1 mutation in codon 117 (Pro117Ser): comparison of clinical, neuropathological and cell culture phenotypes of Pro117Leu and Pro117Ser mutations. J Alzheimers Dis. 2004;6(1):31–43.
139. Dumanchin C, Brice A, Campion D, Hannequin D, Martin C, Moreau V, et al. De novo presenilin 1 mutations are rare in clinically sporadic, early onset Alzheimer's disease cases. French Alzheimer's Disease Study Group. J Med Genet. 1998;35(8):672–3.
140. Edwards-Lee T, Wen J, Bell J, Hardy J, Chung J, Momeni P. A presenilin-1 mutation (T245P) in transmembrane domain 6 causes early onset Alzheimer's disease. Neurosci Lett. 2006;398(3):251–2.
141. Ezquerra M, Carnero C, Blesa R, Oliva R. A novel presenilin 1 mutation (Leu166Arg) associated with early onset Alzheimer disease. Arch Neurol. 2000;57(4):485–8.
142. Fang B, Jia L, Jia J. Chinese Presenilin-1 V97L mutation enhanced Abeta42 levels in SH-SY5Y neuroblastoma cells. Neurosci Lett. 2006;406(1–2):33–7.
143. Fang BY, Jia JP. The effect of two newly Chinese presenilin-1 mutations on the sensitivity to trophic factor withdrawal in human neuroblastoma cells. Zhonghua Yi Xue Za Zhi. 2007;87(5):336–40.
144. Finckh U, Alberici A, Antoniazzi M, Benussi L, Fedi V, Giannini C, Gal A, Nitsch RM, Binetti G. Variable expression of familial Alzheimer disease associated with presenilin 2 mutation M239I. Neurology. 2000;54(10):2006–8.
145. Forsell C, Froelich S, Axelman K, Vestling M, Cowburn RF, Lilius L, et al. A novel pathogenic mutation (Leu262Phe) found in the presenilin 1 gene in early-onset Alzheimer's disease. Neurosci Lett. 1997;234(1):3–6.
146. Gallo M, Marcello N, Curcio SA, Colao R, Geracitano S, Bernardi L, et al. A novel pathogenic PSEN1 mutation in a family with Alzheimer's disease: phenotypical and neuropathological features. J Alzheimers Dis. 2011;25(3):425–31.

147. Gliebus G, Rosso A, Lippa CF. Progranulin and beta-amyloid distribution: a case report of the brain from preclinical PS-1 mutation carrier. Am J Alzheimers Dis Other Demen. 2009;24(6):456–60.
148. Golan M, Lipczynska-Lojkowska W, Krzysko KA, Styczynska M, Luczywek E, Filipek S, et al. Two novel mutations in presenilin 1 (PSEN1) gene connected with atypical familial early-onset Alzheimer's disease (EOAD). Alzheimer's and Parkinson's diseases: insights, progress and perspectives. In: 7th international conference AD/PD 2005 book of abstracts: 24, 2005.
149. Goldman JS, Reed B, Gearhart R, Kramer JH, Miller BL. Very early-onset familial Alzheimer's disease: a novel presenilin 1 mutation. Int J Geriatr Psychiatry. 2002;17(7):649–51.
150. Gómez-Tortosa E, Barquero S, Barón M, Gil-Neciga E, Castellanos F, Zurdo M, et al. Clinical-genetic correlations in familial Alzheimer's disease caused by presenilin 1 mutations. J Alzheimers Dis. 2010;19(3):873–84.
151. Guerreiro RJ, Baquero M, Blesa R, Boada M, Brás JM, Bullido MJ, et al. Genetic screening of Alzheimer's disease genes in Iberian and African samples yields novel mutations in presenilins and APP. Neurobiol Aging. 2010;31(5):725–31.
152. Gustafson L, Brun A, Englund E, Hagnell O, Nilsson K, Stensmyr M, et al. A 50-year perspective of a family with chromosome-14-linked Alzheimer's disease. Hum Genet. 1998;102(3):253–7.
153. Hamaguchi T, Morinaga A, Tsukie T, Kuwano R, Yamada M. A novel presenilin 1 mutation (L282F) in familial Alzheimer's disease. J Neurol. 2009;256(9):1575–7.
154. Harvey RJ, Ellison D, Hardy J, Hutton M, Roques PK, Collinge J, Fox NC, Rossor MN. Chromosome 14 familial Alzheimer's disease: the clinical and neuropathological characteristics of a family with a leucine → serine (L250S) substitution at codon 250 of the presenilin 1 gene. J Neurol Neurosurg Psychiatry. 1998;64(1):44–9.
155. Heckmann J, de Viliers C, Rutherfoord S, Ramesar R, Morris C, Low R, Kalaria R. Novel presenilin 1 mutation with profound neurofibrillary pathology in an indigenous South African family with early-onset Alzheimer's disease. Brain. 2004;127(Pt 1):133–42.
156. Higuchi S, Yoshino A, Matsui T, Matsushita S, Satoh A, Iimura T, et al. A novel PS1 Mutation (W165G) in a Japanese family with early-onset Alzheimer's disease. Alzheimers Rep. 2000;3(4):227–31.
157. Houlden H, Crook R, Dolan RJ, McLaughlin J, Revesz T, Hardy J. A novel presenilin mutation (M233V) causing very early onset Alzheimer's disease with Lewy bodies. Neurosci Lett. 2001;313(1–2):93–5.
158. Hutton M, Busfield F, Wragg M, Crook R, Perez-Tur J, Clark RF, et al. Complete analysis of the presenilin 1 gene in early onset Alzheimer's disease. NeuroReport. 1996;7(3):801–5.
159. Ikeda M, Sharma V, Sumi SM, Rogaeva EA, Poorkaj P, Sherrington R, et al. The clinical phenotype of two missense mutations in the presenilin I gene in Japanese patients. Ann Neurol. 1996;40(6):912–7.
160. Ikeda M, Yonemura K, Kakuda S, Tashiro Y, Fujita Y, Takai E, et al. Cerebrospinal fluid levels of phosphorylated tau and Aß1-38/Aß1-40/Aß1-42 in Alzheimer's disease with PS1 mutations. Amyloid. 2013;20(2):107–12.
161. Ikeuchi T, Kaneko H, Miyashita A, Nozaki H, Kasuga K, Tsukie T, et al. Mutational analysis in early-onset familial dementia in the Japanese population. The role of PSEN1 and MAPT R406 W mutations. Dement Geriatr Cogn Disord. 2008;26(1):43–9.
162. Jacquier M, Arango D, Torres O, Cruts M, Serrano M, Matallana M, et al. Presenilin mutations in a Colombian familial and sporadic AD sample. Neurobiol Aging. 2000;21(1):S176.
163. Janssen JC, Lantos PL, Fox NC, Harvey RJ, Beck J, Dickinson A, et al. Autopsy-confirmed familial early-onset Alzheimer disease caused by the l153V presenilin 1 mutation. Arch Neurol. 2001;58(6):953–8.
164. Jia J, Xu E, Shao Y, Jia J, Sun Y, Li D. One novel presenilin-1 gene mutation in a Chinese pedigree of familial Alzheimer's disease. J Alzheimers Dis. 2005;7(2):119–24.

165. Jiang HY, Li GD, Dai SX, Bi R, Zhang DF, Li ZF, et al. Identification of PSEN1 mutations p.M233L and p.R352C in Han Chinese families with early-onset familial Alzheimer's disease. Neurobiol Aging. 2015;36(3):1602.e3–6.
166. Jin SC, Pastor P, Cooper B, Cervantes S, Benitez BA, Razquin C, Goate A, et al. Pooled-DNA sequencing identifies novel causative variants in PSEN1, GRN and MAPT in a clinical early-onset and familial Alzheimer's disease Ibero-American cohort. Alzheimers Res Ther. 2012;4(4):34.
167. Jorgensen P, Bus C, Pallisgaard N, Bryder M, Jorgensen AL. Familial Alzheimer's disease co-segregates with a Met146Ile substitution in presenilin-1. Clin Genet. 1996;50(5):281–6.
168. Kamimura K, Tanahashi H, Yamanaka H, Takahashi K, Asada T, Tabira T. Familial Alzheimer's disease genes in Japanese. J Neurol Sci. 1998;160(1):76–81.
169. Kamino K, Sato S, Sakaki Y, Yoshiiwa A, Nishiwaki Y, Takeda M, et al. Three different mutations of presenilin 1 gene in early-onset Alzheimer's disease families. Neurosci Lett. 1996;208(3):195–8.
170. Kasuga K, Ohno T, Ishihara T, Miyashita A, Kuwano R, Onodera O, et al. Depression and psychiatric symptoms preceding onset of dementia in a family with early-onset Alzheimer disease with a novel PSEN1 mutation. J Neurol. 2009;256(8):1351–3.
171. Kauwe JS, Jacquart S, Chakraverty S, Wang J, Mayo K, Fagan AM, et al. Extreme cerebrospinal fluid amyloid beta levels identify family with late-onset Alzheimer's disease presenilin 1 mutation. Ann Neurol. 2007;61(5):446–53.
172. Kerchner GA, Holbrook K. Novel presenilin-1 Y159F sequence variant associated with early-onset Alzheimer's disease. Neurosci Lett. 2012;531(2):142–4.
173. Kim HJ, Kim HY, Ki CS, Kim SH. Presenilin 1 gene mutation (M139I) in a patient with an early-onset Alzheimer's disease: clinical characteristics and genetic identification. Neurol Sci. 2010;31(6):781–3.
174. Kim J, Bagyinszky E, Chang YH, Choe G, Choi BO, An SS, Kim S. A novel PSEN1 H163P mutation in a patient with early-onset Alzheimer's disease: clinical, neuroimaging, and neuropathological findings. Neurosci Lett. 2012;530(2):109–14.
175. Klunemann HH, Rogaeva E, Neumann M, Kretzschmar HA, Kandel M, Toulina A, et al. Novel PS1 mutation in a Bavarian kindred with familial Alzheimer disease. Alzheimer Dis Assoc Disord. 2004;18(4):256–8.
176. Knight WD, Kennedy J, Mead S, Rossor MN, Beck J, Collinge J, Mummery C. A novel presenilin 1 deletion (p.L166del) associated with early onset familial Alzheimer's disease. Eur J Neurol. 2007;14(7):829–31.
177. Kowalska A, Wender M, Florczak J, Pruchnik-Wolinska D, Modestowicz R, Szczech J, et al. Molecular genetics of Alzheimer's disease: presenilin 1 gene analysis in a cohort of patients from the Poznan region. J Appl Genet. 2003;44(2):231–4.
178. Kowalska A, Pruchnik-Wolinska D, Florczak J, Modestowicz R, Szczech J, Kozubski W, et al. Genetic study of familial cases of Alzheimer's disease. Acta Biochim Pol. 2004;51(1):245–52.
179. Kowalska A, Pruchnik-Wolinska D, Florczak J, Szczech J, Kozubski W, Rossa G, Wender M. Presenilin 1 mutations in Polish families with early-onset Alzheimer's disease. Folia Neuropathol. 2004;42(1):9–14.
180. Kwok JB, Taddei K, Hallupp M, Fisher C, Brooks WS, Broe GA, et al. Two novel (M233T and R278T) presenilin-1 mutations in early-onset Alzheimer's disease pedigrees and preliminary evidence for association of presenilin-1 mutations with a novel phenotype. NeuroReport. 1997;8(6):1537–42.
181. Kwok JB, Halliday GM, Brooks WS, Dolios G, Laudon H, Murayama O, et al. Presenilin-1 mutation L271V results in altered exon 8 splicing and Alzheimer's disease with non-cored plaques and no neuritic dystrophy. J Biol Chem. 2003;278(9):6748–54.
182. La Bella V, Liguori M, Cittadella R, Settipani N, Piccoli T, Manna I, Quattrone A, Piccoli F. A novel mutation (Thr116Ile) in the presenilin 1 gene in a patient with early-onset Alzheimer's disease. Eur J Neurol. 2004;11(8):521–4.

183. Lee P, Medina L, Ringman JM. The Thr354Ile substitution in PSEN1: disease-causing mutation or polymorphism? Neurology. 2006;66(12):1955–6.
184. Lee JH, Kahn A, Cheng R, Reitz C, Vardarajan B, Lantigua R, Medrano M, et al. Disease-related mutations among Caribbean Hispanics with familial dementia. Mol Genet Genomic Med. 2014;2(5):430–7.
185. Lendon CL, Martinez A, Behrens IM, Kosik KS, Madrigal L, Norton J, et al. E280A PS-1 mutation causes Alzheimer's disease but age of onset is not modified by ApoE alleles. Hum Mutat. 1997;10(3):186–95.
186. Lindquist S, Schwartz M, Batbayli M, Waldemar G, Nielsen J. Genetic testing in familial AD and FTD: Mutation and phenotype spectrum in a Danish cohort. Clin Genet. 2009;76(2):205–9.
187. Lladó A, Sánchez-Valle R, Rey MJ, Mercadal P, Almenar C, López-Villegas D, Fortea J, Molinuevo JL. New mutation in the PSEN1 (E120G) gene associated with early onset Alzheimer's disease. Neurologia. 2010;25(1):13–6.
188. Lleo A, Blesa R, Queralt R, Ezquerra M, Molinuevo JL, Pena-Casanova J, Rojo A, Oliva R. Frequency of mutations in the presenilin and amyloid precursor protein genes in early-onset Alzheimer disease in Spain. Arch Neurol. 2002;59(11):1759–63.
189. Lohmann E, Guerreiro RJ, Erginel-Unaltuna N, Gurunlian N, Bilgic B, Gurvit H, et al. Identification of PSEN1 and PSEN2 gene mutations and variants in Turkish dementia patients. Neurobiol Aging. 2012;33(8):1850.e17–27.
190. Luedecke D, Becktepe JS, Lehmbeck JT, Finckh U, Yamamoto R, Jahn H, Boelmans K. A novel presenilin 1 mutation (Ala275Val) as cause of early-onset familial Alzheimer disease. Neurosci Lett. 2014;566:115–9.
191. Miklossy J, Taddei K, Suva D, Verdile G, Fonte J, Fisher C, et al. Two novel presenilin-1 mutations (Y256S and Q222H) are associated with early-onset Alzheimer's disease. Neurobiol Aging. 2003;24(5):655–62.
192. Miravalle L, Murrell JR, Takao M, Glazier B, Piccardo P, Vidal R, Ghetti B. Genetic mutations associated with presenile dementia. Neurobiol Aging. 2002;23(1):S322.
193. Moehlmann T, Winkler E, Xia X, Edbauer D, Murrell J, Capell A, et al. Presenilin-1 mutations of leucine 166 equally affect the generation of the Notch and APP intracellular domains independent of their effect on Abeta 42 production. Proc Natl Acad Sci USA. 2002;99(12):8025–30.
194. Morelli L, Prat MI, Levy E, Mangone CA, Castano EM. Presenilin 1 Met146Leu variant due to an A → T transversion in an early-onset familial Alzheimer's disease pedigree from Argentina. Clin Genet. 1998;53(6):469–73.
195. Norton JB, Cairns NJ, Chakraverty S, Wang J, Levitch D, Galvin JE, Goate A. Presenilin1 G217R mutation linked to Alzheimer disease with cotton wool plaques. Neurology. 2009;73(6):480–2.
196. Palmer MS, Beck JA, Campbell TA, Humphries CB, Roques PK, Fox NC, Harvey R, Rossor MN, Collinge J. Pathogenic presenilin 1 mutations (P436S & I143F) in early-onset Alzheimer's disease in the UK. Hum Mutat. 1999;13(3):256.
197. Pantieri R, Pardini M, Cecconi M, Dagna-Bricarelli F, Vitali A, Piccini A, et al. A novel presenilin 1 L166H mutation in a pseudo-sporadic case of early-onset Alzheimer's disease. Neurol Sci. 2005;26(5):349–50.
198. Poorkaj P, Sharma V, Anderson L, Nemens E, Alonso ME, Orr H, et al. Missense mutations in the chromosome 14 familial Alzheimer's disease presenilin 1 gene. Hum Mutat. 1998;11(3):216–21.
199. Portet F, Dauvilliers Y, Campion D, Raux G, Hauw JJ, Lyon-Caen O, et al. Very early onset AD with a de novo mutation in the presenilin 1 gene (Met 233 Leu). Neurology. 2003;61(8):1136–7.
200. Queralt R, Ezquerra M, Castellvi M, Lleo A, Blesa R, Oliva R. Detection of the presenilin 1 gene mutation (M139T) in early-onset familial Alzheimer disease in Spain. Neurosci Lett. 2001;299(3):239–41.

201. Ramirez-Duenas MG, Rogaeva EA, Leal CA, Lin C, Ramirez-Casillas GA, Hernandez-Romo JA, et al. A novel Leu171Pro mutation in presenilin-1 gene in a Mexican family with early onset Alzheimer disease. Ann Genet. 1998;41(3):149–53.
202. Reznik-Wolf H, Treves TA, Davidson M, Aharon-Peretz J, St George Hyslop PH, Chapman J, et al. A novel mutation of presenilin 1 in familial Alzheimer's disease in Israel detected by denaturing gradient gel electrophoresis. Hum Genet. 1996;98(6):700–2.
203. Ringman JM, Gylys KH, Medina LD, Fox M, Kepe V, Flores DL, et al. Biochemical, neuropathological, and neuroimaging characteristics of early-onset Alzheimer's disease due to a novel PSEN1 mutation. Neurosci Lett. 2011;487(3):287–92.
204. Rogaev EI, Sherrington R, Rogaeva EA, Levesque G, Ikeda M, Liang Y, et al. Familial Alzheimer's disease in kindreds with missense mutations in a gene on chromosome 1 related to the Alzheimer's disease type 3 gene. Nature. 1995;376(6543):775–8.
205. Rogaeva EA, Fafel KC, Song YQ, Medeiros H, Sato C, Liang Y, et al. Screening for PS1 mutations in a referral-based series of AD cases: 21 novel mutations. Neurology. 2001;57(4):621–5.
206. Romero I, Jorgensen P, Bolwig G, Fraser PE, Rogaeva E, Mann D, et al. A presenilin-1 Thr116Asn substitution in a family with early-onset Alzheimer's disease. NeuroReport. 1999;10(11):2255–60.
207. Rossor MN, Fox NC, Beck J, Campbell TC, Collinge J. Incomplete penetrance of familial Alzheimer's disease in a pedigree with a novel presenilin-1 gene mutation. Lancet. 1996;347(9014):1560.
208. Sherrington R, Rogaev EI, Liang Y, Rogaeva EA, Levesque G, Ikeda M, et al. Cloning of a gene bearing missense mutations in early-onset familial Alzheimer's disease. Nature. 1995;375(6534):754–60.
209. Shrimpton AE, Schelper RL, Linke RP, Hardy J, Crook R, Dickson DW, et al. A presenilin 1 mutation (L420R) in a family with early onset Alzheimer disease, seizures and cotton wool plaques, but not spastic paraparesis. Neuropathology. 2007;27(3):228–32.
210. Smith MJ, Gardner RJ, Knight MA, Forrest SM, Beyreuther K, Storey E, et al. Early-onset Alzheimer's disease caused by a novel mutation at codon 219 of the presenilin-1 gene. NeuroReport. 1999;10(3):503–7.
211. Sorbi S, Tedde A, Nacmias B, Ciantelli M, Caffarra P, Ghidoni E, et al. Novel presenilin 1 and presenilin 2 mutations in early onset Alzheimer's disease families. Neurobiol Aging. 2002;23(1):S312.
212. Sugiyama N, Suzuki K, Matsumura T, Kawanishi C, Onishi H, Yamada Y, et al. A novel missense mutation (G209R) in exon 8 of the presenilin 1 gene in a Japanese family with presenile familial Alzheimer's disease. Hum Mutat. 1999;14(1):90.
213. Tanahashi H, Kawakatsu S, Kaneko M, Yamanaka H, Takahashi K, Tabira T. Sequence analysis of presenilin-1 gene mutation in Japanese Alzheimer's disease patients. Neurosci Lett. 1996;218(2):139–41.
214. Tedde A, Bartoli A, Piaccri I, Ferrara S, Bagnoli S, Serio A, et al. Novel presenilin 1 mutation (Ile408Thr) in an Italian family with late-onset Alzheimer's disease. Neurosci Lett. 2016;610:150–3.
215. Terreni L, Valeria C, Calella AM, Gavazzi A, Alberoni M, Grimaldi LM, et al. A novel missense mutation (L219F) in exon 8 of the presenilin 1 gene in an Italian family with presenile familial Alzheimer's disease. Neurobiol Aging. 2000;21(1):176–7.
216. Tiedt HO, Lueschow A, Winter P, Muller U. Previously not recognized deletion in presenilin-1 (p.Leu174del.) in a patient with early-onset familial Alzheimer's disease. Neurosci Lett. 2013;544:115–8.
217. Ting SK, Benzinger T, Kepe V, Fagan A, Coppola G, Porter V, et al. A novel PSEN1 mutation (I238M) associated with early-onset Alzheimer's disease in an African-American woman. J Alzheimers Dis. 2014;40(2):271–5.
218. Wasco W, Pettingell WP, Jondro PD, Schmidt SD, Gurubhagavatula S, Rodes L, et al. Familial Alzheimer's chromosome 14 mutations. Nat Med. 1995;1(9):848.

219. Wisniewski T, Dowjat WK, Buxbaum JD, Khorkova O, Efthimiopoulos S, Kulczycki J, et al. A novel Polish presenilin-1 mutation (P117L) is associated with familial Alzheimer's disease and leads to death as early as the age of 28 years. NeuroReport. 1998;9(2):217–21.
220. Yagi R, Miyamoto R, Morino H, Izumi Y, Kuramochi M, Kurashige T, et al. Detecting gene mutations in Japanese Alzheimer's patients by semiconductor sequencing. Neurobiol Aging. 2014;35(7):1780.e1–5.
221. Yasuda M, Maeda K, Ikejiri Y, Kawamata T, Kuroda S, Tanaka C. A novel missense mutation in the presenilin-1 gene in a familial Alzheimer's disease pedigree with abundant amyloid angiopathy. Neurosci Lett. 1997;232(1):29–32.
222. Yasuda M, Maeda K, Hashimoto M, Yamashita H, Ikejiri Y, Bird TD, et al. A pedigree with a novel presenilin 1 mutation at a residue that is not conserved in presenilin 2. Arch Neurol. 1999;56(1):65–9.
223. Yasuda M, Maeda S, Kawamata T, Tamaoka A, Yamamoto Y, Kuroda S, et al. Novel presenilin-1 mutation with widespread cortical amyloid deposition but limited cerebral amyloid angiopathy. J Neurol Neurosurg Psychiatry. 2000;68(2):220–3.
224. Zekanowski C, Golan MP, Krzysko KA, Lipczynska-Lojkowska W, Filipek S, Kowalska A, et al. Two novel presenilin 1 gene mutations connected with frontotemporal dementia-like clinical phenotype: genetic and bioinformatic assessment. Exp Neurol. 2006;200(1):82–8.
225. De Jonghe C, Cruts M, Rogaeva EA, Tysoe C, Singleton A, Vanderstichele H, et al. Aberrant splicing in the presenilin-1 intron 4 mutation causes presenile Alzheimer's disease by increased Abeta42 secretion. Hum Mol Genet. 1999;8(3):1539–40.
226. Guo J, Wei J, Liao S, Wang L, Jiang H, Tang B. A novel presenilin 1 mutation (Ser169del) in a Chinese family with early-onset Alzheimer's disease. Neurosci Lett. 2010;468(1):34–7.
227. Tysoe C, Whittaker J, Xuereb J, Cairns NJ, Cruts M, Van Broeckhoven C, et al. A presenilin-1 truncating mutation is present in two cases with autopsy-confirmed early-onset Alzheimer disease. Am J Hum Genet. 1998;62(1):70–6.
228. Bernardi L, Tomaino C, Anfossi M, Gallo M, Geracitano S, Puccio G, et al. Late onset familial Alzheimer's disease: novel presenilin 2 mutation and PS1 E318G polymorphism. J Neurol. 2008;255(4):604–6.
229. Beyer K, Lao JI, Fernandández-Novoa L, Cacabelos R. Identification of a novel mutation (V148I) in the TM2 domain of the presenilin 2 gene in a patient with late-onset Alzheimer disease. Neurob Aging. 1998;19(4):S87.
230. Ertekin-Taner N, Younkin LH, Yager DM, Parfitt F, Baker MC, Asthana S, et al. Plasma amyloid beta protein is elevated in late-onset Alzheimer disease families. Neurology. 2008;70(8):596–606.
231. Ezquerra M, Lleo A, Castellvi M, Queralt R, Santacruz P, Pastor P, et al. A novel mutation in the PSEN2 gene (T430 M) associated with variable expression in a family with early-onset Alzheimer disease. Arch Neurol. 2003;60(8):1149–51.
232. Gallo M, Tomaino C, Bernardi L, Maletta R, Anfossi M, Geracitano S, et al. PS1 polymporphism and a novel PS2 mutation in a patient with late-onset familial Alzheimer's disease. Alzheimers Dement. 2008;4(4):Suppl:T585.
233. Guerreiro RJ, Beck J, Gibbs JR, Santana I, Rossor MN, Schott JM, et al. Genetic variability in CLU and its association with Alzheimer's disease. PLoS ONE. 2010;5(3):e9510.
234. Lao JI, Beyer K, Fernandez-Novoa L, Cacabelos R. A novel mutation in the predicted TM2 domain of the presenilin 2 gene in a Spanish patient with late-onset Alzheimer's disease. Neurogenetics. 1998;1(4):293–6.
235. Li D, Parks SB, Kushner JD, Nauman D, Burgess D, Ludwigsen S, et al. Mutations of presenilin genes in dilated cardiomyopathy and heart failure. Am J Hum Genet. 2006;79(6):1030–9.
236. Lindquist SG, Hasholt L, Bahl JM, Heegaard NH, Andersen BB, Nørremølle A, et al. A novel presenilin 2 mutation (V393 M) in early-onset dementia with profound language impairment. Eur J Neurol. 2008;15(10):1135–9.

237. Lleo A, Blesa R, Gendre J, Castellvi M, Pastor P, Queralt R, Oliva R. A novel presenilin 2 gene mutation (D439A) in a patient with early-onset Alzheimer's disease. Neurology. 2001;57(10):1926–8.
238. Müller U, Winter P, Bolender C, Nolte D. Previously unrecognized missense mutation E126 K of PSEN2 segregates with early onset Alzheimer's disease in a family. J Alzheimers Dis. 2014;42(1):109–13.
239. Niu F, Yu S, Zhang Z, Yi X, Ye L, Tang W, Qiu C, Wen H, Sun Y, Gao J, Guo Y. Novel mutation in the PSEN2 gene (N141Y) associated with early-onset autosomal dominant Alzheimer's disease in a Chinese Han family. Neurobiol Aging. 2014;35(10):2420.e1–5.
240. Piscopo P, Talarico G, Crestini A, Gasparini M, Malvezzi-Campeggi L, Piacentini E, et al. A novel mutation in the predicted TMIII domain of the PSEN2 gene in an Italian pedigree with atypical Alzheimer's disease. J Alzheimers Dis. 2010;20(1):43–7.
241. Piscopo P, Talarico G, Spadoni O, Malvezzi-Campeggi L, Crestini A, Gasparini M, et al. A novel Italian presenilin 2 mutation (S175Y). Alzheimers Dement. 2008;4(4):Suppl 2:T595.
242. Sleegers K, Roks G, Theuns J, Aulchenko YS, Rademakers R, Cruts M, et al. Familial clustering and genetic risk for dementia in a genetically isolated Dutch population. Brain. 2004;127(Pt7):1641–9.
243. Tomaino C, Bernardi L, Anfossi M, Costanzo A, Ferrise F, Gallo M, et al. Presenilin 2 Ser130Leu mutation in a case of late-onset "sporadic" Alzheimer's disease. J Neurol. 2007;254(3):391–3.
244. Youn YC, Bagyinszky E, Kim H, Choi BO, An SS, Kim S. Probable novel PSEN2 Val214Leu mutation in Alzheimer's disease supported by structural prediction. BMC Neurol. 2014;14:105.
245. Corder EH, Saunders AM, Strittmatter WJ, Schmechel DE, Gaskell PC, Small GW, Roses AD, Haines JL, Pericak-Vance MA. Gene dose of apolipoprotein E type 4 allele and the risk of Alzheimer's disease in late onset families. Science. 1993;13;261(5123):921–3.
246. Poirier J, Davignon J, Bouthillier D, Kogan S, Bertrand P, Gauthier S. Apolipoprotein E polymorphism and Alzheimer's disease. Lancet. 1993;342(8873):697–9.
247. Rebeck GW, Reiter JS, Strickland DK, Hyman BT. Apolipoprotein E in sporadic Alzheimer's disease: allelic variation and receptor interactions. Neuron. 1993;11(4):575–80.
248. Roses AD. Apolipoprotein E alleles as risk factors in Alzheimer's disease. Annu Rev Med. 1996;47:387–400
249. Saunders AM, Strittmatter WJ, Schmechel D, George-Hyslop PH, Pericak-Vance MA, Joo SH, et al. Association of apolipoprotein E allele epsilon 4 with late-onset familial and sporadic Alzheimer's disease. Neurology. 1993;43(8):1467–72.
250. Saunders AM. Apolipoprotein E and Alzheimer disease: an update on genetic and functional analyses. J Neuropathol Exp Neurol. 2000;59(9):751–8.
251. Strittmatter WJ, Saunders AM, Schmechel D, Pericak-Vance M, Enghild J, Salvesen GS, Roses AD. Apolipoprotein E: high-avidity binding to β-amyloid and increased frequency of type 4 allele in late-onset familial Alzheimer disease. Proc Natl Acad Sci USA. 1993;90(5):1977–81.
252. Cuenco KT, Lunetta KL, Baldwin CT, McKee AC, Guo J, Cupples LA, et al. Association of distinct variants in SORL1 with cerebrovascular and neurodegenerative changes related to Alzheimer disease. Arch Neurol. 2008;65(12):1640–8.
253. Holton P, Ryten M, Nalls M, Trabzuni D, Weale ME, Hernandez D, et al. Initial assessment of the pathogenic mechanisms of the recently identified Alzheimer risk Loci. Ann Hum Genet. 2013;77(2):85–105.
254. Tycko B, Feng L, Nguyen L, Francis A, Hays A, Chung WY, et al. Polymorphisms in the human apolipoproteinJ/clusterin gene: ethnic variation and distribution in Alzheimer's disease. Hum Genet. 1996;98(4):430–6.
255. Kok EH, Luoto T, Haikonen S, Goebeler S, Haapasalo H, Karhunen PJ. CLU, CR255 and PICALM genes associate with Alzheimer's-related senile plaques. Alzheimers Res Ther. 2011;3(2):12.

256. Bertram L, Lange C, Mullin K, Parkinson M, Hsiao M, Hogan MF, et al. Genome-wide association analysis reveals putative Alzheimer's disease susceptibility loci in addition to APOE. Am J Hum Genet. 2008;83(5):623–32.
257. Malik M, Simpson JF, Parikh I, Wilfred BR, Fardo DW, Nelson PT, Estus S. CD33 Alzheimer's risk-altering polymorphism, CD33 expression, and exon 2 splicing. J Neurosci. 2013;33(33):13320–5.
258. Lacher SE, Alazizi A, Wang X, Bell DA, Pique-Regi R, Luca F, Slattery M. A hypermorphic antioxidant response element is associated with increased MS4A6A expression and Alzheimer's disease. Redox Biol. 2018;14:686–93.
259. Cruchaga C, Kauwe JS, Harari O, Jin SC, Cai Y, Karch CM, et al. GWAS of cerebrospinal fluid tau levels identifies risk variants for Alzheimer's disease. Neuron. 2013;78(2):256–68.
260. Jin SC, Benitez BA, Karch CM, Cooper B, Skorupa T, Carrell D, et al. Coding variants in TREM2 increase risk for Alzheimer's disease. Hum Mol Genet. 2014;23(21):5838–46.
261. Jonsson T, Stefansson H, Steinberg S, Jonsdottir I, Jonsson PV, Snaedal J, et al. Variant of TREM2 associated with the risk of Alzheimer's disease. N Engl J Med. 2013;368(2):107–16.
262. Schjeide BM, Schnack C, Lambert JC, Lill CM, Kirchheiner J, Tumani H, et al. The role of clusterin, complement receptor 1, and phosphatidylinositol binding clathrin assembly protein in Alzheimer disease risk and cerebrospinal fluid biomarker levels. Arch Gen Psychiatry. 2011;68(2):207–13.
263. Morgen K, Ramirez A, Frölich L, Tost H, Plichta MM, Kölsch H, et al. Genetic interaction of PICALM and APOE is associated with brain atrophy and cognitive impairment in Alzheimer's disease. Alzheimers Dement. 2014;10(5 Suppl):S269–76.
264. Chen H, Wu G, Jiang Y, Feng R, Liao M, Zhang L, et al. Analyzing 54,936 samples supports the association between CD2AP rs9349407 polymorphism and Alzheimer's disease susceptibility. Mol Neurobiol. 2015;52(1):1–7.
265. Wang HF, Tan L, Hao XK, Jiang T, Tan MS, Liu Y, et al. Effect of EPHA1 genetic variation on cerebrospinal fluid and neuroimaging biomarkers in healthy, mild cognitive impairment and Alzheimer's disease cohorts. J Alzheimers Dis. 2015;44(1):115–23.

Genetic Markers in Psychiatry

Gordana Nedic Erjavec, Dubravka Svob Strac, Lucija Tudor, Marcela Konjevod, Marina Sagud and Nela Pivac

Abstract

Psychiatric disorders such as addiction (substance use and addictive disorders), depression, eating disorders, schizophrenia, and post-traumatic stress disorder (PTSD) are severe, complex, multifactorial mental disorders that carry a high social impact, enormous public health costs, and various comorbidities as well as premature morbidity. Their neurobiological foundation is still not clear. Therefore, it is difficult to uncover new set of genes and possible genetic markers of these disorders since the understanding of the molecular imbalance leading to these disorders is not complete. The integrative approach is needed which will combine genomics and epigenomics; evaluate epigenetic influence on genes and their influence on neuropeptides, neurotransmitters, and hormones; examine gene × gene and gene × environment interplay; and identify abnormalities contributing to development of these disorders. Therefore, novel genetic approaches based on systems biology focused on improvement of the identification of the biological underpinnings might offer genetic markers of addiction, depression, eating disorders, schizophrenia, and PTSD. These markers might be used for early prediction, detection of the risk to develop these

G. Nedic Erjavec · D. Svob Strac · L. Tudor · M. Konjevod · N. Pivac (✉)
Division of Molecular Medicine, Rudjer Boskovic Institute,
Bijenicka 54, HR-10000 Zagreb, Croatia
e-mail: npivac@irb.hr

M. Sagud
School of Medicine, University of Zagreb, Salata 2, HR-10000 Zagreb, Croatia

M. Sagud
Department of Psychiatry, University Hospital Centre Zagreb,
Kispaticeva 12, HR-10000 Zagreb, Croatia

© Springer Nature Singapore Pte Ltd. 2019

Y.-K. Kim (ed.), *Frontiers in Psychiatry*, Advances in Experimental Medicine and Biology 1192, https://doi.org/10.1007/978-981-32-9721-0_4

disorders, novel subtypes of the diseases and tailored, personalized approach to therapy.

Keywords
Addiction · Depression · Eating disorders · Schizophrenia · Genetics · Markers · PTSD

Introduction

Psychiatric disorders are among the leading causes of disability worldwide. The classifications of psychiatric disorders, even the most recent ones such as DSM 5 [1] and ICD-10 [2], are based on clusters of symptoms, described as clinical syndromes, which have mostly unknown etiology. In the twenty-first century, unlike in any other fields of medicine, there is still lack of reliable diagnostic tools in psychiatry. However, strong evidence from numerous studies indicates that psychiatric disorders have a genetic background. This chapter focuses on the genetic basis of different psychiatric conditions. Various methods have been used in the studies, which investigated the association between genetic factors and the development of psychiatric disorders. Numerous family, twin, and adoption studies were carried out several decades ago and provided the first evidence of the genetic background of psychiatric disorders. They revealed more frequent occurrence of psychiatric diseases in the families of affected members than in control population. This familial aggregation might be due to the shared genes which are involved in the development of the disease (nature) or shared environment (nurture). Twin studies offer an advantage over even the most sophisticated molecular genetic studies, in terms of capturing all inherited genetic effects, and thus offering the best available measure of heritability [3]. Findings obtained from family, twin, and adoption studies provided strong evidence for the heritability of major psychiatric disorders. Heritability estimates the proportion of individual variation in particular trait, which is explained by inherited factors, while the rest of variation is explained by non-inherited factors [4]. Most psychiatric disorders are moderately to highly heritable [5]. Among psychiatric disorders, bipolar disorder is one of the most heritable medical disorders, with the heritability between 59 and 87% [6, 7]. Similarly, heritability estimates for schizophrenia are around 80% [8]. Some studies suggested that earlier onset of the disease is associated with increased familiar risk [6], while others did not confirm this finding [9]. Major depressive disorder (MDD) is generally considered to have moderate heritability [4]. However, recent study reported the heritability of MDD to be as high as 67% [9]. On the other hand, anxiety disorders have lower heritability rate, such as 43% for panic disorder and 32% for generalized anxiety disorder [10], while heritability estimates for obsessive-compulsive disorder (OCD) ranged from 27 to 65% [11]. For eating disorders, such as anorexia nervosa and bulimia nervosa, heritability rates were reported to be 28 to 58% and 54 to 83%, respectively [12]. Wide variations in aforementioned heritability estimates across studies might arise from differences in methodology, such as type of population, diagnostic criteria, different endophenotypes, issues of comorbidity, and statistical power. Modern methods include the field of molecular genetics. Although the development of all psychiatric disorders is influenced by both environmental and genetic factors, it is considered that the onset of such complex disorders is influenced by only small number of genes with a small effect. Namely, thousands of variants impact the risk for psychiatric disorders. Almost all psychiatric disorders are associated with potentially thousands of genes, each contributing with a very small effect, as reflected from the odds ratios mostly being between 1.01 and 1.2. Risk loci are located in both coding and noncoding portions of the genes. The human genome contains millions of loci that are commonly polymorphic. At the most polymorphic loci, hundreds of alleles may be found. They might impact gene expression and, consequently, increase the risk for the development of particular disorder. Association studies are commonly used in psychiatry, including case-control and family-based designs. The former includes large samples of cases and controls and is the only possible method in the case of diseases with a late onset such as dementias, for which parents of affected individuals are no longer available. However, in case-control studies, both cases and controls need to be carefully matched demographically, given the substantial differences in allele frequencies in different populations [13, 14]. Those candidate genes are coding different structures (i.e., receptors, transporters, metabolizing enzymes, and ion channels) in dopaminergic, adrenergic, serotonergic, glutamatergic, GABAergic, cannabinoid and opioid systems, neurotrophins, the hypothalamic–pituitary–adrenal (HPA) axis, and proteins involved in neuronal and synaptic functioning. In addition, association studies are typically addressing only few markers in candidate genes, which are hypothesized to be related to a particular disorder. Those limitations are overcome in genome-wide association studies (GWAS), including tens of thousands of samples and millions of different polymorphisms. GWAS including tens of thousands of participants confirmed genetic heritability and genetic correlation estimates for PTSD [3]. These studies are necessary for locating genetic effects in highly polygenic conditions such as psychiatric disorders. Majority of GWAS were focused on genetics of only one disorder. In psychiatry, however, the comorbidity is more a rule than an exception. Consequently, the cross-disorder genomics field has been recently introduced [5]. The large molecular study of cross-disorder genetics [15] reported the highest genetic correlations of common single-nucleotide polymorphisms (SNP) between schizophrenia and bipolar disorder, the prototypes of the two most severe and devastating psychiatric disorders. This finding is not completely unexpected, given that two disorders share several similarities, such as the onset in early adulthood, psychotic features, and the chronic course. Correlations of MDD were also significant with schizophrenia, bipolar disorder, and attention deficit hyperactivity disorder (ADHD) [5]. Moreover, a high proportion of co-heritable SNPs was detected across schizophrenia and bipolar disorder, unlike for any other pair of psychiatric disorders [5]. A recent study, which included more than 100,000 subjects, confirmed the presence of shared genetic substrate between schizophrenia and bipolar disorder, but also emphasized disease-specific genetic substrate [16]. More recently, a conceptual shift in molecular genomic studies occurred. Dimensional approach, rather than classical diagnostic criteria, holds promise for future phenotypic research [5].

The aim of this chapter is to summarize the knowledge about the role of genetic factors in the development of psychiatric disorders. For scientists and clinicians, those findings would not only improve our understanding of the biology of psychiatric disorders, but are also expected to eventually help in predicting the development of certain disorders and thus, apply appropriate preventive strategies. Moreover, it would help tailoring treatment to the individual patient, in order to maximize the chance of the most favorable outcome.

Genetics of Addiction

Substance Use and Addictive Disorders

Addiction is a chronic psychiatric disorder characterized mostly by the compulsive use of a certain drug or activity which strongly and directly activate the brain reward system leading to neglect of other activities resulting in repercussions on the individuals, their families, and society [17]. According to DSM-5 [1] criteria, addictions are grouped in substance use and addictive disorders. Substance use disorders are addictions related to different types of substances including, for example, alcohol or stimulant use disorder, while gambling disorder is the only addictive disorder included in DSM-5 as a diagnosable condition for now [1]. Some people are more prone to become addictive which depends on several factors, including intrinsic factors (genetic background, gender, age, medical history including other addictions or mental disorders), environmental factors (availability of addictive substance or activity, peer influences, social support, childhood adversity, socioeconomic status) and the nature of the addictive substance or activity (psychoactive properties, mode of use). The impact of these factors may be different through different stages of addiction. Usually, the first exposure is mostly influenced by peers and family environment, while further development of addiction depends more on genetic factors and psychopathology [18].

Genetic Background of Addiction

There are family, adoption, and twin studies emphasizing the importance of genetic background in the development of addictions [19–23]. Despite the fact that there are evidences indicating a role of genetic influences in the development of addiction, a sole identification of crucial genes and loci moderating vulnerability is a big challenge due to the genetic complexity of addictive disorders. During this search, scientists usually choose either candidate gene studies or genome-wide association studies (GWAS) approach.

One of the first GWAS [24] seeking to define specific genes included in the etiology of opioid use disorder (OUD) found no significant associations of 10,000

SNPs between relatively small groups of heroin-dependent patients and controls of European descent. After increasing sample size and including also patients of African ancestry, the same group of scientists found only one (rs10494334) out of 100,000 SNPs significantly associated with heroin addiction, but only in Europeans [24, 25]. A big GWAS of OUD [26] performed on more than 12,000 subjects of either European-American or African-American ancestry implicated the strongest association of opioid dependence with genes encoding potassium voltage-gated channel subunits such as KCNC1, KCNG2, and KCNA4 genes. A study conducted on 4 cohorts [27] identified a variant on chromosome 15, rs12442183, near RGMA (repulsive guidance molecule family member A), associated with opioid dependence. Additionally, they showed on 10 brain tissues derived from 134 healthy human brains from UK Biobank that the risk T allele is correlated with higher expression of a specific RGMA transcript variant in frontal cortex [27]. Considering the GWAS results in other addictions, the strongest association was found [28] between nicotine addiction and CHRNA5-CHRNA3-CHRNB4 gene cluster encoding for α5, α3, and β4 subunits of nicotinic acetylcholine receptors (nAChRs). On the other hand, alcohol dependence GWAS reported mostly results not reaching the genome-wide significance [29]. The most significant marker associated with alcoholism was rs6943555 in autism susceptibility candidate 2 gene (AUTS2) [30]. It seems that the GWAS results for addiction thus far defined quite small number of genes that could be a consequence of a relatively small sample size (<10,000) or cross-country heterogeneity.

In contrast to the non-hypothesis driven GWAS approach, the candidate gene approach focuses on discovering potential associations between specific disorder or phenotype and prespecified genes that could be involved in certain mechanism of disease development. One of the candidate gene in case of OUD is opioid receptor mu 1 gene (OPRM1) encoding for mu-opioid receptor that becomes activated by endogenous peptides, opioid analgesics or illegal substances. Maybe the most studied SNP in OPRM1 gene is rs1799971. Since this polymorphism changes amino acid sequence of the protein, one could assume that it could also alter its expression or function. This hypothesis was confirmed in several studies [31, 32], but association studies of this SNP in OUD provided opposite findings, suggesting that this association either exists [33] or does not exist [34]. Other SNP (rs3778150) found in OPRM1 gene that does not alter the amino acid sequence was significantly associated with opioid dependence in a population of subjects from the USA that were of either African or European ancestry [35]. The same effect was found in a replication study in a population of European ancestry, while, on the other hand, this effect was not significant in a replication study with subjects of African ancestry, suggesting that, as in most genetics studies, the whole genetic background could play a crucial role. Besides mu-opioid receptor, an important role in reward system has the delta-opioid receptor encoded by opioid receptor delta 1 (OPRD1) gene. Most studied variants of OPRD1 gene were rs2236861, rs2236857 and rs3766951. Findings of studies dealing with an association of those OPRD1 variants with OUD are again opposite, a phenomenon that could be explained by the different ancestry of included subjects [36–39].

Not only opioid but also dopamine receptors have an important role in the normal function of human brain reward system. The rs1079597 (Taq1B) variant of dopamine receptor D2 (*DRD2*) gene was found to be in high linkage disequilibrium with the rs1800497 (Taq1A) variant of a gene encoding a serine/threonine protein kinase (*ANKK1*) [40]. For both variants an association with opioid dependence in European [41] as well as in Han Chinese [42] subjects was found. Although not directly involved in the award system, the brain-derived neurotrophic factor (BDNF) is, through its role in a neuronal growth and differentiation, a gene candidate in OUD. The G allele of rs6265 (Val66Mat) BDNF variant was found to be associated with increased risk of opioid dependence in the samples from central China [43] and Taiwan [44], but, also, the lack of association between BDNF rs6265 and alcohol dependence or alcohol-related endophenotypes was reported in Caucasians with alcohol dependence [45]. In the case of alcohol dependence, probably the most important candidate genes are alcohol-metabolizing genes, alcohol dehydrogenase 1B (*ADH1B*) and aldehyde dehydrogenase 2 (*ALDH2*). The rs1229984 (His48Arg) variant in *ADH1B* gene directly affects catalytic efficiency of an enzyme. Namely, His48/His48 homozygotes have increased ADH1B activity resulting in higher rates rate of oxidation of ethanol to acetaldehyde [46]. On the other hand, the Lys487 allele according to rs671 (Glu487Lys) variant of *ALDH2* gene reduces the enzyme activity causing the same result, accumulation of acetaldehyde [46]. Subsequently, acetaldehyde accumulation leads to heightened alcohol-induced responses (flushing, nausea, headache, and tachycardia) which in the end makes the abovementioned alleles protective against alcohol dependence [47, 48]. Considering the role of monoamines in the modulation of emotionality, cognition, and reward, it is not surprising that some of the genes encoding for monoamines are also among candidate genes for alcohol dependence. For example, the activity of dopamine metabolizing enzyme catechol-O-methyltransferase (COMT) depends on rs4680 SNP in *COMT* gene. There are studies reporting no associations [49] between this polymorphism and addiction, but also indicating the Val158 allele as a risk allele in methamphetamine, nicotine, alcohol, and polysubstance addiction [50–53]. Opposite results were found in patients with late onset alcoholism from Finland, where increased risk was associated with the Met158 allele of the *COMT* rs4680 [54].

Since such complex phenotypes as addictions arise as a result of different interacting biochemical pathways influenced by the products of numerous genes, scientists often choose a study approach in which one phenotype would be better understood by clarifying the effects of gene variants in the context of specific biochemical pathways. Also, unless there is a direct impact of a particular gene on a substance metabolism, genetic influence on the development of addiction will be more evident if environmental exposure is also considered. Besides, one should take into account possible gene $\times$ gene interactions that could have additive or masking effect. After all, it is evident that the search for genetic markers of addiction is quite demanding, but promising and important field of science.

Genetics of Eating Disorders

Eating Disorders

Classical eating disorders include anorexia nervosa (AN), bulimia nervosa (BN), and binge eating disorder (BED), and they are gender-dependent, affecting mainly female population. These are severe mental disorders that carry a high social impact, an enormous public health cost, premature mortality in young people, and increased disability-adjusted life years [55]. According to the newest classification of mental disorders (DSM 5) [1], the chapter on feeding and eating disorders includes also pica, rumination disorder avoidant/restrictive food intake disorder (ARFID), other specified feeding or eating disorder (OSFED), and unspecified feeding or eating disorder (UFED). A better understanding of the neurobiological and genetic underpinning of eating disorders might improve the treatment, since novel knowledge might point to novel therapeutic targets, and identify validated and sensitive biomarkers [56]. Namely, besides psychological interventions, only two drugs are approved by the FDA for eating disorders, fluoxetine for bulimia nervosa and lisdexamfetamine for BED [57].

Genetic Background of Eating Disorders

Genetic background of eating disorders is confirmed by the family, twin, and adoption studies showing aggregations in families and estimated high heritability, between 40 and 65% [55, 57–59]. As in all complex psychiatric disorders, which are complicated entities, the neurobiological foundation of eating disorders is not clearly identified. Different neurotransmitters and their receptors (serotonin, dopamine, opioids), neuropeptides (BDNF) and hormones (neuropeptide Y, agouti-related peptide/AgRP/, alpha-melanocyte-stimulating hormone, orexins, proopiomelanocortin (POMC), CRH, cocaine- and amphetamine-regulated transcript (CART), oxytocin, ghrelin, leptin), other appetite regulators (glucagon-like peptide 1, cholecystokinin, peptide tyrosine tyrosine (PYY), anandamide, or N-arachidonoylethanolamine (AEA)), and many others are altered in eating disorders [55, 57, 58]. Serotonergic system has a role in regulation of mood and impulse control, satiety, anxious and obsessional behavior, appetite, and body weight, dopaminergic system affects feeding and addictive behavior via its reward system, opioid system via its receptors also modulate reward sensitivity and food intake, ghrelin, neuropeptide Y and AgRP possess appetite-stimulating effects and promote feeding and increase body mass index while POMC and CART have the opposite, appetite-suppressing effects; melanocortin 4 receptor (MC4R) is associated with obesity; BDNF moderates energy metabolism and regulates mood, feeding behavior, and weight [60]; oxytocin inhibits appetite; leptin promotes satiety and it is an important regulator of energy metabolism and balance;

cholecystokinin induces satiation, glucagon-like peptide 1 inhibits food intake [55, 57, 58].

Therefore, the genetic architecture of eating disorders is complex and polygenic, making the search for the risk/protective genetic markers even more complicated. Several approaches are used to study genetic background of all psychiatric disorders including eating disorders. These are quantitative and molecular genetic studies [57]. Quantitative studies include family, twin, and adoption studies that shed light on the aggregation in families and heritability [55]. Molecular genetic [57] studies might be hypothesis-free, such as linkage studies and GWAS, and hypothesis-based, such as candidate gene association studies. Linkage studies detect specific genomic regions associated with genes related to eating disorders (i.e., AN, BN, or BED), while candidate gene association studies compare the frequency of genotypes/alleles of the particular gene associated with eating disorders in affected subjects and controls. GWAS identify genetic variations of the whole genome in case/controls. However, characterization of genetic basis of eating disorders (AN, BN, or BED) has been particularly challenging, and a lot of discoveries were not replicated. GWAS are large-scale genomic studies and require very large samples; however due to the multiple testing corrections, rare genetic variants reach genome-wide significance [55]. To overcome these problems, meta-analyses were conducted, but they still failed to confirm GWAS significant results either in AN [61–63], BN [64], or general symptoms of eating disorders [63, 64].

Besides these methods, novel methods include copy number variation (changes/deletions or duplications/in the number of copies of a genomic region), high-throughput sequencing studies, transcriptomic and epigenetic studies [65], linkage disequilibrium score regression, genetic correlations, polygenic risk scores, gene-wide analysis and pathway analysis, and rare genetic variations [57]. All these novel methods might lead to precision medicine approach to eating disorders [57]. With the aim to evaluate which genetic variations contribute to a person's risk for development of particular eating disorder, a psychiatric consortium and working groups for eating disorders (especially AN) were established, resulting in large sets of data [57, 65]. However, consortiums for BN and BED and more confirmed GWAS data are still scarce.

As recently reviewed [55], few loci, genes, and their polymorphisms have been reported to be associated with eating disorders (AN, BN, BED), using association studies and GWAS, but few of them were confirmed with meta-analyses [58]. These are the genes related to potassium calcium-activated channel (KCNN3) and orexin receptor (HCRTR1) [66]; serotonin receptor type 2A (HTR2A) [59, 67–71], serotonin receptor type 1D (HTR1D) [62, 72–74], serotonin receptor type 1B (HT1B) [75]; serotonin transporter (5-HTT) [59, 76, 77], tryptophan hydroxylase (TPH2) [78]; OPRD1 [73, 79], dopamine receptor type 2/ankyrin repeat and kinase domain containing 1 (DRD2/ANKK1), COMT [71, 80–82]; and dopamine receptor type 4 (DRD4) [83, 84]; also genes for ghrelin [85, 86], ghrelin O-acyltransferase (GOAT) [87]; AgRP [85, 88]; estrogen receptor 1 (ESR1) [89]; type 1 cannabinoid receptor (CB1R) [90]; BDNF [91]; fat mass and obesity (FTO) [92]; calcium-activated potassium channel (SK3) [93]; sex determining region Y-Box 2 (SOX2) [63, 65, 94]; C-type lectin domain containing 5A (CLEC5A),

Table 1 Genes and polymorphisms associated with anorexia nervosa

Gene	Gene description	Polymorphism	References
Anorexia nervosa			
ORPD1	Delta-opioid receptor type 1	rs569356; rs521809; rs4654327	[72, 73]
5-HTR1D	Serotonin receptor type 1D	rs674386; rs856510	[72, 73]
5-HT2A	Serotonin receptor type 2A	rs6311	[67, 68, 84]
5-HTT	Serotonin transporter	*5-HTTLPR*	[76, 77]
TPH2	Tryptophan hydroxylase	rs1473473	[78]
DRD2	Dopaminergic receptor type 2	rs1799732; rs6275; rs1800497; rs6278	[72]
DRD4	*Dopamine receptor type 4*	rs1800955; DRD4 haplotypes	[83, 84]
COMT	Catechol-*O*-methyltransferase	rs4680	[80, 95]
GHRL	Ghrelin	rs696217	[85]
GOAT	Ghrelin O-acyl-transferase	rs10096097	[87]
AgRP	Agouti-related protein	rs5030980	[96]
BDNF	Brain-derived neurotrophic factor	rs6265	[91]
SK3	Calcium-activated potassium channel	CAG repeat	[93]

Table 2 Genes and polymorphisms associated with bulimia nervosa

Gene	Gene description	Polymorphism	References
Bulimia nervosa			
5-HT2A	Serotonin receptor type 2A	rs6311	[97, 98]
GHSR	Growth hormone secretagogue receptor	*rs*495225	[95]
ERβ	Estrogen receptor beta gene	rs4986938; rs928554	[99]
CB1R	Cannabinoid receptor type 1	rs1049353	[90]
FAAH	Fatty acid amide hydrolase	rs324420	[90]
FTO	Fat mass and obesity	rs9939609	[92]
BDNF	Brain derived neurotrophic factor	rs6265; rs56164415	[91]

peptidase S1 domain-containing protein (LOC136242), teashirt zinc finger homeobox 1 (TSHZ1), synaptotagmin like 5 (SYTL5) for AN, 5'-nucleotidase, cytosolic IB (NT5C1B) for BN and ATPase phospholipid transporting 8A2 (ATP8A2) for BED [64]; and RUN and FYVE domain-containing protein 1 (RUFY1), cyclin L1 (CCNL1), semaphorin 6D (SEMA6D), SHC adaptor protein 4 (SHC4), disks large-associated protein 1 (DLGAP1), serum deprivation-response protein (SDPR), and transcriptional repressor GATA binding 1 (TRPS1) [63]. Some of the polymorphisms significantly associated with AN (Table 1) or BN (Table 2) are shown [55, 58].

However, although some of these associations were suggestive, none of them reached genome-wide significance [55]. Namely, after correction for the multiple testing, none of these loci, genes, and polymorphisms were significantly associated with eating disorders, or were significant only at the trend level, or suggested that the existence of signal which was not significant [57, 58]. Genes related to serotonergic, dopaminergic, and opioid system were not confirmed, while some new genes with unknown biological function were suggested, but still not replicated in GWAS.

Until now, there are no significant genes that are associated with eating disorders [55, 57, 58]. The reasons for the inconsistent results in linkage, candidate gene association studies and especially GWAS lay in the fact that these studies need very narrowly defined groups (i.e., phenotypes, endophenotypes, and clinical subtypes) with particular eating disorders (AN, BN, or BED) and age and sex-matched control groups, a large number of subjects to increase statistical power and to detect association, and require the inclusion of patients without comorbid psychiatric disorders and symptoms, such as major depressive disorder, bipolar disorder, schizophrenia, alcohol and substance use disorders, personality disorder, obsessive-compulsive disorder, anxiety, emotional instability, suicidality, and obesity [55, 57–59, 100], which are frequent in eating disorders. On the other hand, some of these comorbid disorders (alcohol or substance use disorders) have been associated with eating disorders as they share similar disruptions of the control and reward systems [58, 69, 71]. A novel method, linkage disequilibrium score regression, merged GWAS data with genetic correlation between different phenotypes/traits and provided the first results showing an overlap between AN and schizophrenia, AN and major depressive disorder, and AN and bipolar disorder, suggesting a shared, common genetic cumulative risk [65, 100]. In addition, detection of the single gene is tempered by the small effect size of genes involved in eating disorders [57, 58].

Genetic Markers of Eating Disorders

To uncover new set of genes, and possible genetic markers, the integrative approach is needed which will combine genomics and epigenomics; evaluate epigenetic influence on genes and their influence on neuropeptides, neurotransmitters, and hormones; examine gene–environment interplay; and identify biological, but also psychological, and/or environmental abnormalities contributing to development of eating disorders. The aim is to improve the understanding of the molecular imbalance leading to eating disorders [55, 57, 58]. Therefore, novel genetic approaches, based on systems biology focused on improvement of the identification of the biological underpinnings of eating disorders, might offer genetic markers of eating disorders such as markers for early prediction, detection of risk to develop eating disorders, novel subtypes of the diseases, and tailored approach to therapy of particular eating disorders [55, 57, 58]. To achieve this goal, multidisciplinary teams and creation of new available databases are required.

Genetics of Post-traumatic Stress Disorder

Post-traumatic Stress Disorder

Although only a subset of individuals develops post-traumatic stress disorder (PTSD) (5–10%) [101] after witnessing or being exposed to the traumatic event, its debilitating nature is manifested in subsequent trauma reexperience, avoidance, numbing, negative cognitive and mood alternations, and hyperarousal symptoms (DSM-5) [1]. In addition to psychiatric, metabolic and cardiovascular comorbidities frequently present in individuals with PTSD [102] and lack of appropriate therapy, this disorder still represents a major health care challenge as well as social and financial burden. External factors such as type of trauma, early life adversity, sex, and age influence PTSD development and severity [101]. However, twin studies demonstrated substantial role of genetic inheritance in PTSD etiology, especially in women. It is estimated that PTSD heritability varies from 42% for general population and 72% in all-female subset of patients [103, 104] and is shared with other mental disorders, such as schizophrenia and major depression [104].

Candidate Gene Studies in PTSD

Previous strategies seeking to identify potential genetic markers of PTSD vulnerability/resilience have been mostly relying on understanding the neurobiological background of PTSD symptomatology and hence offering the genes potentially involved in those processes. This approach was mainly focused on genes involved in dopaminergic, serotonergic and opioid neurotransmitters systems, neurotrophins, and the HPA axis. Positive findings from candidate genes, as well as genome-wide studies, are presented in Table 3. Dopamine and serotonin are catecholamines involved in regulation of mood, memory processes, emotional reactivity, arousal, and attention [105], so commonly studied genes were ones encoding for their receptors and transporters.

The T allele of the SNP rs1800497 located in *DRD2* gene was more prevalent in PTSD subjects than in controls and was associated with higher alcohol intake [106–108], however, several negative results have also been reported [109, 110]. A study by Voisey et al. [111] did not confirm the association of the rs1800497 but showed a significant excess of the C allele of the SNP rs6277 in PTSD subjects, which is also located in the *DRD2* gene. The 3′ untranslated region (3′ UTR) of the dopamine active transporter 1 gene (*DAT1*) contains 40 bp variable tandem repeat (VNTR,) where lower number of repeats (9) showed higher risk of PTSD in contrast to higher number (10) of repeats [112–114]. However, a non-replication also exists [109]. The VNTR in dopamine receptor 4 gene (*DRD4*) did not predict the PTSD diagnosis, but seven and eight repeats were associated with more severe PTSD symptoms, especially avoidance [115].

Table 3 SNPs and VNTRs significantly associated with PTSD in candidate gene and GWA studies

Nearest gene	Gene description	SNP	References
Candidate gene association			
DRD2	Dopamine receptor 2DRD2/ ANKK1-Taq1A	rs1800497 rs6277	[106–108, 111]
DAT1	The dopamine active transporter 1 gene	VNTR (3′ UTR)	[112–114, 153]
DRD4	D4 dopamine receptor	VNTR (exon 3)	[115]
SLC6A4	Sodium-dependent serotonin transporter	rs4795541 rs25531	[121–125]
GABRA2	$GABA_A$ receptor subunit alpha 2	rs279836 rs279826 rs279871	[126]
COMT	Catechol-O-methyltransferase	rs4680	[127–129]
BDNF	Brain-derived neurotrophic factor	rs6265	[132, 133]
CNR1	Cannabinoid receptor 1	rs1049353	[136, 137]
ADCYAP1R1	Pituitary adenylate cyclase-activating polypeptide receptor	rs2267735	[138, 139]
FKBP5	FK506 binding Protein 51	rs3800373 rs9296158 rs1360780 rs9470080	[140–142]
GWAS			
RORA	RAR-related orphan receptor A	rs8042149	[145]
TLL1	Tolloid-like 1	rs6812849	[146]
LINC01090	Long noncoding RNA	rs10170218	[154]
ADCY8	ADCY8 adenylate cyclase 8	rs263232	[147]
PRTFDC1	Phosphoribosyl transferase domain containing 1	rs6482463	[148]
BC036345	Long noncoding RNA	rs717947	[155]
TBC1D2	TBC1 domain family member 2	rs7866350	[149]
ANKRD55 ZNF626	Ankyrin repeat domain 55 Zinc finger protein 626	rs159572 rs11085374	[150]
NLGN1 ZNRD1 AS1	Neuroligin 1 Long noncoding RNA	N/A N/A	[151]
OR11L1	Olfactory receptor family 11 subfamily L Member 1	rs6681483 rs6667389 rs10888255 rs10888257	[152]
KLHL1	Kelch-like protein 1	rs139558732	[144]

Serotonergic activity and concentration are regulated by 5-HT transporter, encoded by the *5HTT* or *SLC6A4* gene. The 5-HTTLPR polymorphism is a functional one, where the shorter variant (the S allele) leads to reduced 5-HTT expression and consequently lower serotonin reuptake. However, another polymorphism rs25531 (A>G substitution), in the same region, also strongly affects 5-HTT expression [116]. Therefore, this polymorphism (5HTTLPR) is considered triallelic. Studies have shown that reduced 5-HTT expression or the presence of the S allele [or long (L) allele carriers with G substitution] is associated with a potential risk for PTSD [117, 118]. Although some studies did not show any association of this polymorphism with PTSD [119, 120], its interaction with the environmental factors, such as number and severity of trauma, was reported to be related to PTSD symptoms [121–125]. These gene × environment (G×E) studies demonstrated potential effect of stress and external factors on genetic susceptibility to PTSD, possibly mediated via (de)methylation of DNA regions, as a way of epigenetic control of gene expression [122]. However, these results need to be replicated in larger cohorts.

The inhibitory GABAergic system, in addition to dopaminergic and serotonergic systems, is also involved in pathophysiology of anxiety and depression, which are common comorbidities in individuals with PTSD. The rs279836 T allele, rs279826 A allele and rs279871 A allele in the $GABA_A$ receptor subunit alpha 2 (*GABRA2*) were found to significantly interact with severity of childhood abuse to predict PTSD [126].

Another widely studied gene in psychiatric disorders is *COMT* which encodes for COMT enzyme included in degradation of catecholamines. Its functional G>A polymorphism (rs4680) leads to significantly diminished COMT activity in the brain and likely an excess of catecholamines (dopamine, noradrenaline) in the brain of A allele carriers, thereby affecting the neurocircuits of fear inhibition [105]. This SNP, rs4680 was reported to be significant predictor of PTSD in interaction with traumatic load [127–129] and also had an effect on cognitive functioning in PTSD patients [130].

Neurotrophin BDNF has numerous roles in central nervous system [131], such as regulation of the stress and fear response, neurogenesis, and synaptic plasticity. Since the fear response and lack of fear extinction are hallmark symptoms in PTSD, *BDNF* gene was studied as a candidate gene in PTSD. Functional G>A SNP at codon 66 (rs6265) results in G (valine) to A (methionine) substitution, leading to lower activity-dependent release of the BDNF in hippocampus and potentially inadequate stress response and memory consolidation. Zhang et al. [132] reported higher prevalence of the risk A allele of the rs6265 in veterans with PTSD, successfully integrating the environmental factors in the study. The presence of one or two A allele of the rs6265 was more frequently found in veterans with PTSD with psychotic features than in veterans without psychotic symptoms, or in veterans without PTSD [133].

The HPA axis activates noradrenergic system as a response to stress-induced release of corticotrophin-releasing hormone (CRH) and forms a complex feedback loop between glucocorticoid (GR) and endocannabinoid receptors (CNR1), along with regulating genes [134]. Dysregulation in any part of this system could mediate development and severity of PTSD symptoms. Besides its role in maintaining homeostasis of the HPA axis, endocannabinoid system has a role in memory processing, in particular, in the extinction of the fear memory [135]. The A allele of SNP rs1049353 of *CNR1* gene has been associated with risk for PTSD, and it was also found to interact with childhood physical maltreatment to increase severity of fear experience [136, 137].

A polymorphism (rs2267735) in *ADCYAP1R1* (gene that encodes for the receptor for pituitary adenylate cyclase-activating polypeptide—PACAP) was associated with PTSD only in women, but not in men [138], thereby suggesting potential role of estrogen in mechanism of regulation of this gene. Later, Mercer et al. [139] showed the allele-specific binding of estrogen within ADCYAP1R1, in which the risk (C) allele diminished binding of estrogen on estrogen response element in this gene, leading to the lower expression of *ADCYAP1R1* and increased risk for PTSD.

Various SNPs: rs9296158, rs3800373, 1360780, and rs9470080 in steroid receptor chaperone FK506 Binding Protein 51 gene (*FKBP5*), all in high linkage disequilibrium, showed association with PTSD in gene × environment fashion, and in particular, interaction with the childhood abuse [140, 141]. Klengel et al. [142] proposed selective demethylation of DNA in enhancer region of this gene as a response to glucocorticoids release induced by childhood trauma. Decreased methylation leads to the increased gene expression of FKBP5, a protein also involved in other pathologies such as violent behavior, depression, and suicide risk [143]. Although there are several successful candidate gene studies, especially ones which included environment as an important cofactor in predicting the diagnosis and severity of PTSD, with some of them even mechanistically well supported (for example, *FKBP5*), main limitation is that this type of studies cannot easily escape bias. Moreover, the proposed genes did not show a significant effect on PTSD in GWAS.

Genome-Wide Association Studies in PTSD

By the early 2000s, scientific and technological development of GWAS and next-generation sequencing (NGS) made analysis or very large number of SNPs and variable number tandem repeats (VNTRs) possible. These no-hypothesis-driven approaches, due to the numerous of analyzed genetic variations (minimum half a million per test), require large sample sizes (>10,000) and very stern statistical correction (p-value $< 5 \times 10^8$) to achieve appropriate power. Currently, the largest GWAS was done by Psychiatric Genomics Consortium (PGC)–PTSD which comprised of 5000 PTSD subjects and 15,000 controls (mostly trauma-exposed). This study did not provide any SNP in GWAS significant p-value, except for the rs139558732 in African-American subgroup [144] located close to Kelch-like protein 1 (*KLHL1*) gene, although this SNP failed to reach statistical significance in transethnic meta-analysis. This study also confirmed SNP-based heritability of PTSD, shared with other psychiatric disorders, primarily schizophrenia, but also bipolar disorder and major depression, and its higher impact on women.

Other findings in previous less powered GWAS highlighted SNPs in close vicinity of *RORA*, retinoid-related orphan receptor gene [145], zinc-dependent metalloprotease tolloid-like 1 or *TLL-1* [146], adenylate cyclase 8—*ADCY8* [147], a phosphoribosyl transferase domain containing 1—*PRTFDC1* [148], TBC1 domain family member 2—*TBC1D2* [149], ankyrin repeat domain 55—*ANKRD55* [150], neuroligin 1—*NLGN1* [151], and olfactory receptor family 11 subfamily L Member 1—*OR11L1* [152] as potential markers of PTSD (Table 3).

There is no straightforward connection of the mentioned genes determined with GWAS with systems explored in candidate gene studies. However, some of them are implicated in basic neurological processes. For example, changes in RORA levels could affect neuron response to oxidative stress and inflammation and was reported to interact with childhood trauma to predict PTSD [156] in addition to original GWAS study. Neuroligin 1 has an important role in synaptic formation, maturation, and maintenance, and is previously associated with autism and Alzheimer's disease. Its depletion in mouse models resulted in impaired fear memory [157]. The OR11L1 is a G protein-coupled receptor that initiates and transduces the sense of smell upon binding odorant molecules [158]. Other positive hits mostly had more systemic roles, for example, *ANKRD55* is a protein involved in autoimmune and inflammatory disorders [150] while *PRTFDC1* could act as a possible tumor-suppressor gene, although its role is not completely determined [159]. The TLL1 and KLHL1 (primarily expressed in brain tissue) have important roles in remodeling of the extracellular matrix [160] and actin fibers [161], respectively, while TBC1D2 affects cell–cell adhesion [162].

The main limitation of GWAS is that none of the mentioned genes was proven significant in an independent replication cohort. This could be due to the very strict statistical corrections, or complex gene × environment network with still undefined effect on PTSD diagnosis. In order to successfully identify genetic and therapeutic targets, the next step relies on founding the large datasets with detailed genetic and environmental information and later their confirmation in mechanistic animal models in addition to neuroimaging genetics in humans. Next iteration by PGC-PTSD will include even more subjects than previous one [144], with total number of participants reaching over 130,000, which could have necessary power to elucidate key genes involved in pathophysiology of PTSD [163].

Analysis of epigenetic mechanisms of gene regulation and focusing on gene expression could point to the molecular mechanisms underlying the PTSD symptomatology, whether analyzed in postmortem brain tissue, or more systemic, immunological response in peripheral blood. The latter could be easily available to reveal differential signature of PTSD but could also provide an insight into systemic changes reflecting or promoting brain changes [164] that follow the experience of the traumatic event.

Genetic Markers of PTSD

Genetic studies in PTSD revealed that genetic background of PTSD is shared with other psychiatric disorders, such as schizophrenia, bipolar disorder and major depression, and the importance of gender. These studies did not confirm any of the genes in the independent replication cohorts, presumably due to the complexity of the PTSD's underlying neurobiology, lack of statistical power, and inadequate sample sizes. Therefore, future studies should include large datasets of both genders and different ethnicities to find significant genetic markers of PTSD.

Genetics of Schizophrenia

Schizophrenia

Schizophrenia is a chronic, severe, debilitating psychiatric disorder with prevalence of 1% in the world population [8, 165]. Characteristic symptoms of schizophrenia can be divided into positive and negative clusters, which are characterized by delusions, hallucinations, cognitive impairment, lack of interest, and behavioral alterations [166]. Etiology, molecular mechanisms, and development of schizophrenia are still unclear; however, family and twin studies have shown that heritability is more than 80% compared with general population [165]. Schizophrenia is a complex, multifactorial disorder whereas the interaction between many suspected genes with the small size effect and environmental triggers are involved in its development [165, 167]. GWAS in schizophrenia discovered some potential genes that may represent the increased risk for schizophrenia. These genes are involved in many processes associated with neurogenesis, neurotransmission, and other intracellular processes [8], while some risk alleles are shared between schizophrenia and other neuropsychiatric disorders, like depression, bipolar disorder, and autism spectrum disorder [168].

Candidate Genes in Schizophrenia

Candidate genes in schizophrenia are the genes coding proteins involved in neurotransmission, differentiation, proliferation, motility of neurons and cell–cell adhesion. Among them, the first genes were the genes associated with dopaminergic system. COMT is an enzyme involved in catalysis of dopamine; therefore, it directly affects the dopamine level in the brain [169], especially in prefrontal cortex. It is encoded by the *COMT* gene, which is expressed in neural system. Due to the role of dopaminergic system in development of schizophrenia, *COMT* gene and its polymorphisms have been studied among patients with schizophrenia [169]. The SNP Val158Met (rs4680) in the *COMT* gene is a functional polymorphism that results in G (valine) to A (methionine) substitution at codon 108 for soluble COMT or at codon 158 for membrane-bound COMT [170, 171]. The A allele is associated with decreased catalysis of dopamine, better cognitive performance but also increased aggression and better response to antipsychotic drugs [169, 172], while the G allele, due to high dopamine metabolism in prefrontal cortex, is associated with reduced cognitive performance and represents higher risk for schizophrenia [169–171]. However, some studies did not find any association between *COMT* rs4680 and the risk for schizophrenia [171, 173, 174], and such conflicting data can be the result of different ethnic groups and different allele distribution among them [167, 175]. Another polymorphism in the *COMT* gene is rs4818 polymorphism, that has been studied among schizophrenia subjects [176] and it is often inherited together with rs4680 polymorphism in a haplotype block. This polymorphism results in leucine to leucine substitution and has a greater influence on dopamine metabolism in the brain than rs4680 polymorphism. Both *COMT* SNPs, rs4680 and rs4818, were associated with treatment resistance, since female, but not male patients with schizophrenia, carriers of the G alleles of the COMT rs4680 and rs4818, and G–G/G–G haplotype, had lower risk of treatment resistance [176].

Genes for dopamine receptors were also studied as risk genes for schizophrenia. There are five subtypes of dopamine receptors encoded with different genes (*DRD1–DRD5*). The D1 dopamine receptor encoded with D1-like receptor gene (*DRD1*) has a major role in cognitive processes and it is assumed to be associated with negative symptoms in schizophrenia [177]. The dopamine receptor gene *DRD2*, coding the D2 dopamine receptors, targets all antipsychotics acting as antagonists of DRD2 receptors [178]. The C allele of the C957T polymorphism (rs6277) in *DRD2* has been determined as a risk factor for schizophrenia [179]. Although the T allele of the rs6277 polymorphism is associated with poor mRNA stability, it is also linked with better cognitive performance [180]. In addition to polymorphisms in exon region, also polymorphisms in the intronic region can affect receptor function due to alternative splicing of exons. For example, rs1076560 polymorphism in intronic region of *DRD2* gene is associated with deficits in cognitive processes due to altered expression of DRD2. The TT genotype showed great association with schizophrenia risk in family studies, while carriers of the T allele had greater risk for schizophrenia compared with the T allele non-carriers [181]. Lack of association was reported for schizophrenia risk and rs1799732 and rs1800497 polymorphisms, while the rs1801028 polymorphism in *DRD2* gene represents a risk factor for schizophrenia [182]. Nonetheless, besides these polymorphisms, there are some synergistic effects with environmental factors, such as stress, that might affect *DRD2* expression [183]. In addition to *DRD1* and *DRD2* genes, it is assumed that rs6280 polymorphism in the *DRD3* gene that results in serine–glycine substitution is also associated with schizophrenia risk. However, for *DRD3* and *DRD4* genes, there were some associations and some studies failed to confirm association with schizophrenia risk [184]. Furthermore, downstream genes in dopaminergic signaling are also associated with schizophrenia risk due to changes that can affect signal transmission [183].

Another neurotransmitter system affected in schizophrenia is glutamatergic neurotransmission. Metabotropic glutamate receptor (GRM3 or mGluR3) is encoded with *GRM3* gene, and it is expressed in presynaptic neurons. Thus, it is assumed that the association between GRM3 gene and schizophrenia risk lays in the alterations of glutamatergic neurotransmission and excitatory amino acid transporter 2 (EAAT2) located in hippocampus and prefrontal cortex [185, 186]. Variation in this gene increases schizophrenia risk, while the A allele of the polymorphism 4 in intronic region was associated with much worse cognitive performance due to alterations in receptor's activity and impairments in protein synthesis. Therefore, the G allele is considered as a protective factor, though molecular mechanisms are still unknown [185]. In addition, multiple studies showed lack of association between alterations in this gene and risk of schizophrenia [186], due to the impact of this gene on cognitive performance even in healthy controls and different distribution of alleles among different ethnic groups [185].

Disrupted in schizophrenia 1 is a protein encoded with the conversed gene *DISC1*. DISC1 protein interacts with other cytoskeletal, centrosomal, and membrane proteins, therefore mediates various functions such as signal transmission, cell transport, and gene expression [187, 188]. Alterations in *DISC1* are associated with different psychiatric diseases. Various haplotypes in *DISC1* gene can be linked to different diseases, including schizophrenia and schizoaffective disorder. It is assumed that certain haplotypes on 5′ end of *DISC1* gene represent schizophrenia risk. Moreover, haplotype block 3 is associated with the risk of the disease. It contains allele that codes for Phe607 and mediates the interaction between DISC1 and other proteins. Furthermore, the T allele of rs6675281 polymorphism is associated with schizoaffective disorder, but also with bipolar disorder. Thus, different mechanisms are involved in DISC1 function, i.e., different mechanisms associated with *DISC1* gene and interaction between its product and other proteins can represent risk factors for schizophrenia development [187]. The Ser allele of the Ser704Cys polymorphism together with other SNPs that are in linkage disequilibrium in *DISC1* gene could represent risk factors for schizophrenia, due to alterations in hippocampus and cytoarchitectural impairments [188].

Neuregulin 1 (NRG1) is a member of family neuregulins that plays an important role in neuronal differentiation, proliferation, and plasticity, through activation of kinases or coding for epidermal growth factor domain. Its association with schizophrenia is still unclear. However, in more than 80% of studies, it is assumed that chromosome 8p is associated with schizophrenia risk. *NRG1* expression is altered among schizophrenia patients, confirming the assumption that this gene is involved in schizophrenia development. Several studies found a connection between the first two exons of *NRG1* gene with schizophrenia risk; however, these findings mostly reflect an association between schizophrenia risk and this gene in Caucasians. Therefore, ethnicity should be taken into account due to the different allelic distribution in various populations. Since schizophrenia is a complex disorder, the association between the *NRG1* gene with schizophrenia should be considered together with other potential gene candidates [189].

BDNF is the most abundant neurotrophin in adult human brain. It is one of the most important factors in neuron growth differentiation, proliferation, plasticity, and survival, but also in dopaminergic and serotonergic neurotransmission [190]. *BDNF* gene, which is located on 11th chromosome, codes for homonymous protein, and it is regulated by several promoters [191–193]. Multiple studies have been shown an association between schizophrenia risk and *BDNF* gene, due to the developmental theory of schizophrenia, pointing to alterations during brain development and alterations in dopaminergic system [191]. The functional polymorphism in *BDNF* gene is Val66Met (rs6265), characterized by G/A substitution resulting in alterations in transport and secretion of premature and mature BDNF [192]. Although there are differences in the allele distribution among various ethnic populations [13], the A allele is assumed to represent a risk factor for schizophrenia, due to its association with reduced hippocampal volume and alterations in its functioning [192]. In addition, the A allele is often associated with positive symptoms that occur in schizophrenia. However, many studies reported conflicting results [193]. Several studies found a lack of association between rs6265 polymorphism and schizophrenia, while some studies suggested a potential role of the G allele as a risk factor. These studies assume that the A allele can have a potential protective effect, while the G allele, which is mostly inherited from heterozigotic parents, represent a risk factor for schizophrenia. Namely, the GG homozygotes usually have more severe symptoms and more expressed negative symptoms compared to the A allele carriers. Therefore, further investigations are necessary to clarify the potential role of rs6265 polymorphism in *BDNF* gene as a risk factor in schizophrenia [193].

Genes and Schizophrenia Risk

Schizophrenia is a complex, multifactorial psychiatric disease in which development, environmental factors, and interactions with multiple genes have a synergistic effect. Therefore, various genes are involved in various processes associated with schizophrenia development, progress, and treatment response. Genes involved in schizophrenia risk encode for many proteins that play an important role in dopaminergic, serotonergic (*COMT*, activator of D-amino acid oxidase/*DAOA*/, protein phosphatase 1 regulatory inhibitor subunit 1B/*PPP1R1B*/, *DRD2*, *DRD4*) or glutamatergic neurotransmission (*GRM3*, *GRIA1*, *GRIN2A*), synaptic transmission (cholinergic receptor nicotinic alpha 7 subunit/*CHRNA7*/), differentiation, proliferation and motility of neurons (*DISC1*, *DISC2*, *ANK1*), cell–cell adhesion (*DISC1*, *NRG1*), organelle biogenesis (dystrobrevin beta/*DTNB1*/), processing of amino acids (methylenetetrahydrofolate reductase/*MTHFR*/), G protein signaling (regulator of G protein signaling 4/*RGS4*/), calcium transfer and signaling (calcium voltage-gated channel subunit alpha 1I/*CACNA1I*/, calcium voltage-gated channel auxiliary subunit beta 2/*CACNB2*/, calcium voltage-gated channel subunit alpha 1 C/*CACNA1C*/, protein phosphatase 3 catalytic subunit gamma/*PPP3CC*/), regulation of transcription (transcription factor 4/*TCF4*/, zinc finger protein 804A/*ZNF804A*/), immune system (major histocompatibility complex/*MHC*/gene family) metabolism of xenobiotics (cytochrome P450 family 2 subfamily D member 6/*CYP2D6*/) and many other intracellular and intercellular processes [8, 168, 194, 195]. Besides these gene variants, rare alleles represent also a schizophrenia risk [168].

Genetic Markers of Schizophrenia

The risk for development of schizophrenia mainly depends on the small size effect of the multiple genes involved in different systemic processes, together with environmental triggers and epigenetic regulation [8, 168, 194, 195].

Genetics of Depression

Depression

Depression is a highly prevalent neuropsychiatric disorder characterized by many heterogenous symptoms with varying severity. In spite of the extensive research, the complex underlying biological determinants of depression are still not clear, resulting in non-adequate treatment. In order to bring new insights into the genetic architecture of depression and offer novel potential targets for therapy, both candidate gene studies and GWAS were applied [196].

Family and twin research has provided strong evidence for the genetic background of depression. A meta-analysis of twin studies demonstrated that the heritability rate for depression is 37% [197], whereas other reports estimated that the depression heritability ranges from 25 to 45% for general population [198–200]. However, higher genetic influence has been suggested in certain depression subtypes, for instance, 48–72% in hospitalized depressed patients and even 72% for patients with severe, recurrent depression [201, 202]. Similar to many other neuropsychiatric diseases, a large number of genes expressed in the brain has been involved in the etiology of depression, with high population occurrence of common gene variants, as well as small contribution of each gene [203, 204]. Considering the polygenic features of depression, the investigation of various gene–gene interactions genes might represent a promising approach in depression research.

In addition to genetic factors, diverse environmental influences play an important role in the etiology of depression [196]. A study of Kendler et al. [205] demonstrated that genetic factors and life experiences contribute equally to depression risk. Moreover, it has been shown that stressors were 2.5 times more likely in depressed subjects in comparison to control individuals and that approximately 80% of depression cases reported stressful life events [206, 207]. However, as depression develops in only about 20% of subjects exposed to acute stress [208], vulnerability to stress and environmental factors may be dependent on the individual genetic background [209]. Therefore, the investigations of interaction between genes and environment might represent an important tool for the identification of novel candidates in depression.

Gene Candidate Studies in Depression

More than 100 candidate gene studies have been conducted in order to investigate potential associations with the development of depression [210], and a significant amount of evidence for the genetic influences in depression has been gathered from different sources [211–215]. Various genetic studies investigated polymorphisms in genes associated with serotonergic, adrenergic, and dopaminergic neurotransmission, including genes for dopamine receptors—*DRD3*, *DRD4*, and serotonin receptors—*HTR1A*, *HTR2A*, *HTR1B*, *HTR2C*; genes for serotonin, noradrenaline, and dopamine transporter—solute carrier family 6 member 4 (*SLC6A4*), solute carrier family 6 member 2 (*SLC6A2*), solute carrier family 6 member 3 (*SLC6A3*); genes for the enzymes monoamine oxidase A—*MAOA*, tyrosine hydroxylase—*TH*, tryptophan hydroxylase 1—*TPH1*, catechol-O-methyl transferase—*COMT*, and the piccolo presynaptic cytomatrix protein—*PCLO* [210]. In addition, polymorphisms in genes involved in regulation of the HPA axis, such as genes coding for glucocorticoid and mineralocorticoid receptors (*NR3C1 and NR3C2*), and CRH receptors (*CRHR1* and *CRHR2*); genes important for neurogenesis and neuroplasticity (*BDNF*), as well as genes involved in the inflammatory processes (including genes for interleukins—*IL1B* and *IL6*) and functioning of circadian system (brain and muscle Arnt-like protein-1/*BMAL1*/, circadian locomotor output cycles kaput/*CLOCK*/, *n*euronal PAS domain-containing protein 2/*NPAS2*/, *p*eriod circadian regulator 3/*PER3*/, cryptochrome circadian regulator 1/*CRY1*/, and timeless circadian regulator/*TIMELESS*/) have been studied [210]. The research also included various other genes, which are not linked with the general hypotheses of depression etiopathogenesis, with focus on genes for angiotensin-converting enzyme (*ACE*), apolipoprotein E (*APOE*), and methylenetetrahydrofolate reductase (*MTHFR*) [210].

However, a meta-analysis of the findings from 183 studies investigating depression, supported involvement of only several genetic polymorphisms, such as *APOE*, G protein subunit beta 3/*GNB3*/(C825T), *MTHFR* (C677T), solute carrier family 6 member 15 or sodium-dependent neutral amino acid transporter B(0)AT2/*SLC6A4*/(40 bp VNTR, *5HTTLPR*), and *SLC6A3* (44 bp Ins/Del) [216]. Subsequent meta-analyses demonstrated associations between depression and *5HTTLPR* [217, 218] and *MTHFR* C677T [219], while no associations were reported with *SLC6A2* T-182C and G1287A [220, 221], *HTR2A* rs6311 [222], *BDNF* rs6265 [223], and *CLOCK* [224] polymorphisms. Nevertheless, only *GNB3* and *APOE3* associations seem to be depression-specific [225]. As most of these candidate gene studies had low sample sizes and replication issues, analyses involving larger cohorts have been conducted; however, they have not confirmed observed associations [203, 226, 227]. Therefore, the research focus was shifted to the GWAS.

Genome-Wide Association Studies (GWAS)

GWAS have been increasingly applied during the past decade in order to investigate genetic loci associated with various complex traits. In an unconventional meta-analysis of GWAS data performed by the Psychiatric Genomics Consortium (PGC), none of the investigated polymorphisms reached a genome-wide significance level, demonstrating no consistent association with depression [228]. According to the review of Dunn et al. [229], the only genome-wide significant association with depression from 15 reported studies was the one with rs1545843 *SLC6A15* polymorphism, which was replicated at a nominally significant level in four studies [230]. The study conducted in 2015, within the CONVERGE Project including >9000 Chinese females found two loci significantly associated with depression: rs12415800 *SIRT1* polymorphism and rs35936514 of the phospholysine phosphohistidine inorganic pyrophosphate phosphatase (*LHPP*) polymorphism [231].

Since then, more GWAS findings on depression have been published [231–239]; however, only three replicate between different studies. The *PCLO* gene originally proposed as a risk gene for depression [240], was found to be significant in the study of Mbarek et al. [236] and replication study by Wray et al. [227]. The associations of rs12552 of the olfactomedin 4 (*OLFM4*) gene polymorphism, as well as different *NEGR1* gene polymorphisms, with depression have been demonstrated in two separate GWAS [234, 238]. The meta-analysis of GWAS performed by Wray et al. [238] identified 44 statistically significant independent loci, of which 30 were new and 14 were significant in prior studies of depression.

Regarding epigenetic modifications, the genome-wide analysis of DNA methylation profiles in depression demonstrated significant differences in methylation status in the number of frontal cortex regions between healthy subjects and depressed patients [241].

Moreover, in the GWAS of Wong et al. [242], 11 rare variants were associated with depression. Rare variants analyses revealed variations and mutations in different genes including lipase G, endothelial type (*LIPG*) gene [243], phospholysine phosphohistidine inorganic pyrophosphate phosphatase (*LHPP*) and carboxypeptidase X, M14 family member 2 (*CPXM2*) genes [244], syntaxin binding protein 5 (*STXBP5*), *r*egulating synaptic membrane exocytosis 1 (*RIMS1*), *c*atenin beta 1 (*CTNNB1*), Dmx like 2 (*DMXL2*), *s*ynapsin I (*SYN1*), *t*yrosine 3-monooxygenase/tryptophan 5-monooxygenase activation protein beta (*YWHAB*) and *t*yrosine 3-monooxygenase/tryptophan 5-monooxygenase activation protein eta (*YWHAH*) genes [245], slit guidance ligand 3 (*SLIT3*) gene [246–249], *SLC6A2* and *HTR1A* genes [250], and Huntingtin (*HTT*) gene [251] associated with depression and depressive symptoms.

Gene × Gene (G×G) and Gene × Environment (G×E) Interaction Studies

Different studies investigated the effects of various gene–gene (G×G) interactions on depression. Neff et al. [252] suggested possible interaction of 5HTTLPR polymorphism with an unknown gene on chromosome 4, whereas the study of

Gabriela Nielsen et al. [253] demonstrated interaction of *MTHFR* A1298C and *COMT* rs4680 polymorphisms. Interactions of polymorphisms within the *CRHR1* and arginine vasopressin receptor 1B (*AVPR1b*) genes may also affect depression [254], although some studies have not confirmed these findings [255]. Interactions between *CRHR1* gene polymorphisms and *BDNF* rs6265 polymorphism have been also reported [256]. In another study, depression was influenced by the interaction of polymorphisms in matrix metalloproteinase (*MMP*) genes and TIMP metallopeptidase inhibitor 2 (*TIMP-2*) gene [257]. The significant effect on depression was also observed for interaction between rs41423247 of the CCND1-Cyclin D1 (*BCL1*) gene and rs1044396 of the cholinergic receptor nicotinic alpha 4 subunit (*CHRNA4*) gene polymorphisms [258], as well as between trace amine-associated receptor 6 (*TAAR6*) and heat shock protein 70 (*HSP-70*) gene variations [259]. On the other hand, GWAS demonstrated no significant findings for pairwise G×G interactions, although several nominally significant interactions were found [260].

Considering research on gene and environment (G×E) interaction in depression, most studies investigated the interaction of specific genes with a variety of stress factors. A significant interaction between *SLC6A* gene polymorphism 5HTTLPR and stressful life events including childhood maltreatment has been demonstrated in depression [261]. This finding has been confirmed in several [262–264], but not in other meta-analyses [265–267]. The differences in the obtained results might be attributed to the various types of stress interacting with 5HTTLPR polymorphism in depressed subjects [268]. Similarly, the interaction of *MAOA* gene polymorphisms and childhood maltreatment, as well as difficulties during maternity, affected depression [202, 269, 270], although other studies observed no significant effects [269]. In addition, significant gene × environment effect on depression has been demonstrated between various stressful life events and polymorphisms in *BDNF* [271, 272], *COMT*, *SLC6A2*, *FKBP5* and *CRHR1* genes [269], gamma-aminobutyric acid type A receptor alpha 6 subunit (*GABRA6*) gene [273], *IL1B* and *IL-6* genes [274–277], galanin receptor 1 (*GALR1*) and galanin receptor 3 (*GALR3*) genes [278], *CNR1* gene [269, 279] and fatty acid amide hydrolase (*FAAH*) gene [280]. Studies have also assessed gene × environment effect on depression in a genome-wide scale. The study enrolling European subjects observed no significant effects of interaction between genes and childhood trauma [281]. The other GWAS in the Mexican American sample revealed 44 gene variants in interaction with stress and affecting depression [242]; however, only the association with PHD finger protein 21B (*PHF21B*) gene has been replicated in a European cohort. Dunn et al. [282] found in African Americans one GWAS significant gene × environment effect for polymorphism rs4652467 near centrosomal protein 350 (*CEP350*) gene, which could not be replicated. In Japanese subjects, marginally significant gene × environment effect between stress and rs10510057 polymorphism near the regulator of G protein signaling 10 (*RGS10*) gene was found [283].

Three-way gene × gene × environment interactions in depression were demonstrated for *BDNF* rs6265 and 5HTTLPR polymorphisms with childhood abuse [284] and family environment [285], suggesting the involvement of epigenetic regulating mechanisms triggered by stress [286]. *BDNF* rs6265 polymorphism also showed positive interaction with glycogen synthase kinase 3 beta (*GSK3B*) gene and recent life events in depression [287]. However, three-way environment × environment × gene interactions were also observed in depression, such as interaction of 5HTTLPR polymorphism with both recent life event and childhood abuse exposure [288]. The example of even higher order interactions is five-way interactions of *BDNF* rs6265 polymorphism with four different neurotrophic receptor tyrosine kinase 2 (*NTRK2*) gene polymorphisms [289].

Genetic Markers of Depression

A large number of candidate gene studies reported associations between various genes and depression; however, most of them have not been confirmed in replication studies and GWAS. The few significant genetic targets found in GWAS are related to mechanisms nonspecific for depression, such as neurogenesis, neuronal synapse, cell contact, and DNA transcription [196]. A possible reason for the observed contradictory results might be clinical and biological heterogeneity of depression, and identification of additional subgroups with more homogeneous phenotypes may help to better understand the genetic background of depression. In addition, the inconsistency between findings of various genetic studies might be due to the polygenic characteristics of depression with weak effect of individual genes and polymorphisms, but also to a prominent role of environmental factors such as stress, suggesting the involvement of epigenetic regulating mechanisms. Therefore, investigation of gene × gene and gene × environment interactions, as well as epigenetic influences, seem most promising approach for determination of additional candidates in depression pathophysiology and development of more efficient pharmacotherapy.

Conclusion

Given the moderate to high heritability of virtually all psychiatric disorders, modern technology was expected to unravel their genetic background. In the previous two decades, there was an explosion of genetic studies in psychiatry. In spite of faster-than-ever increasing amount of genomic data, crucial genes which moderate vulnerability were not identified in any psychiatric disorder, except for Alzheimer's dementia. During this intensive search, scientists used either candidate gene or GWAS. The candidate gene approach appears justified, given that polymorphisms in certain genes affect, among many others, the expression of receptors for neurotransmitters, catalytic efficiency of enzymes, or activity of transporters which are involved in the etiology of the disease. However, those studies frequently, and disappointedly, yielded opposite findings, even in the same disorder. After that, the research focus shifted to the GWAS. However, GWAS revealed no straightforward connection with candidate genes explored in association studies. Actually, majority of GWAS identified only one or two, out of tens of thousands of SNPs, to be significantly associated with the presence of particular disorder, and some of these genes have yet unknown biological function. Other GWAS did not reach genome-wide significance at all. The main reason for such a lack of robust findings, in any psychiatric disorder, is attributed to genetic complexity of psychiatric disorders and very strict statistical corrections. Accurate diagnosis, sufficient sample size (>10,000), careful consideration of comorbidity and reduction of cross-country heterogeneity are challenges for future GWAS.

The pathway to overcome sample heterogeneity might be to focus on narrowly defined group within a single diagnosis, and especially on endophenotype. In addition to genetic factors, diverse environmental influences contribute to the development of the disease. In spite of undisputable heritability of psychiatric disorders, the way that genetic material predisposes an individual to development of psychopathology is largely unknown. While genetic consortiums are established worldwide, other avenues of investigation hold promise. Those include studies of polygenic risk scores, epigenetic studies, studies of gene–gene and gene–environment interactions, gene × gene × environment interactions or environment × environment × gene interactions in a genome-wide scale, studies of cross-disorder genetics and imaging genetics.

From the clinical point of view, genetics in psychiatry is currently far away from its use in practice. While waiting for new data, psychiatrists rely on their own skills and knowledge: preventive measures in high-risk groups, early recognition and treatment of symptoms, destigmatization, and individualized approach to each patient, as much as possible.

References

1. American Psychiatric Association. Diagnostic and statistical manual of mental disorders. 5th ed. Washington: American Psychiatric Association; 2013.
2. World Health Organization. The ICD-10 classification of mental and behavioral disorders: clinical descriptions and diagnostic guidelines. Geneva: World Health Organization; 1992.
3. Duncan LE, Cooper BN, Shen H. Robust findings from 25 years of PSTD genetics research. Curr Psychiatry Rep. 2018;20(12):115.
4. McMahon FJ. Population-based estimates of heritability shed new light on clinical features of major depression. Am J Psychiatry. 2018;175(11):1058–60.
5. Docherty AR, Moscati AA, Fanous AH. Cross-disorder psychiatric genomics. Curr Behav Neurosci Rep. 2016;3(3):256–63.
6. Smoller JW, Finn CT. Family, twin, and adoption studies of bipolar disorder. Am J Med Genet C Semin Med Genet. 2003;123C(1):48–58.
7. Barnett JH, Smoller JW. The genetics of bipolar disorder. Neuroscience. 2009;164(1): 331–43.
8. Gejman PV, Sanders AR, Duan J. The role of genetics in the etiology of schizophrenia. Psychiatr Clin North Am. 2010;33(1):35–66.
9. Guffanti G, Gameroff MJ, Warner V, Talati A, Glatt CE, Wickramaratne P, Weissman MM. Heritability of major depressive and comorbid anxiety disorders in multi-generational families at high risk for depression. Am J Med Genet B Neuropsychiatr Genet. 2016;171(8):1072–9.

10. Hettema JM, Neale MC, Kendler KS. A review and meta-analysis of the genetic epidemiology of anxiety disorders. Am J Psychiatry. 2001;158(10):1568–78.
11. Browne HA, Gair SL, Scharf JM, Grice DE. Genetics of obsessive-compulsive disorder and related disorders. Psychiatr Clin North Am. 2014;37(3):319–35.
12. Thornton LM, Mazzeo SE, Bulik CM. The heritability of eating disorders: methods and current findings. Curr Top Behav Neurosci. 2011;6:141–56.
13. Pivac N, Kim B, Nedic G, Joo YH, Kozaric-Kovacic D, Hong JP, et al. Ethnic differences in brain-derived neurotrophic factor Val66Met polymorphism in Croatian and Korean healthy participants. Croat Med J. 2009;50(1):43–8.
14. Noskova T, Pivac N, Nedic G, Kazantseva A, Gaysina D, Faskhutdinova G, et al. Ethnic differences in the serotonin transporter polymorphism (5-HTTLPR) in several European populations. Progr Neuro-Psychopharmacol Biol Psychiatry. 2008;32(7):1735–9.
15. Sivakumaran S, Agakov F, Theodoratou E, Prendergast JG, Zgaga L, Manolio T, et al. Abundant pleiotropy in human complex diseases and traits. Am J Hum Genet. 2011;89(5):607–18.
16. Bipolar Disorder and Schizophrenia Working Group of the Psychiatric Genomics Consortium. Electronic address: douglas.ruderfer@vanderbilt.edu; Bipolar Disorder and Schizophrenia Working Group of the Psychiatric Genomics Consortium. Genomic dissection of bipolar disorder and schizophrenia, including 28 subphenotypes. Cell. 2018;173(7):1705–15.e16.
17. Koob GF. The neurobiology of addiction: a neuroadaptational view relevant for diagnosis. Addiction. 2006;101(Suppl 1):23–30.
18. Merikangas KR, Avenevoli S. Implications of genetic epidemiology for the prevention of substance use disorders. Addict Behav. 2000;25(6):807–20.
19. Tsuang MT, Lyons MJ, Eisen SA, Goldberg J, True W, Lin N, et al. Genetic influences on DSM-III-R drug abuse and dependence: a study of 3,372 twin pairs. Am J Med Genet. 1996;67(5):473–7.
20. Heath AC, Bucholz KK, Madden PA, Dinwiddie SH, Slutske WS, Bierut LJ, et al. Genetic and environmental contributions to alcohol dependence risk in a national twin sample: consistency of findings in women and men. Psychol Med. 1997;27(6):1381–96.
21. Bierut LJ, Dinwiddie SH, Begleiter H, Crowe RR, Hesselbrock V, Nurnberger JI Jr, et al. Familial transmission of substance dependence: alcohol, marijuana, cocaine, and habitual smoking: a report from the collaborative study on the genetics of alcoholism. Arch Gen Psychiatry. 1998;55(11):982–8.
22. Kendler KS, Karkowski LM, Neale MC, Prescott CA. Illicit psychoactive substance use, heavy use, abuse, and dependence in a US population-based sample of male twins. Arch Gen Psychiatry. 2000;57(3):261–9.
23. Li MD, Cheng R, Ma JZ, Swan GE. A meta-analysis of estimated genetic and environmental effects on smoking behavior in male and female adult twins. Addiction. 2003;98(1):23–31.
24. Nielsen DA, Ji F, Yuferov V, Ho A, Chen A, Levran O, et al. Genotype patterns that contribute to increased risk for or protection from developing heroin addiction. Mol Psychiatry. 2008;13(4):417–28.
25. Nielsen DA, Ji F, Yuferov V, Ho A, He C, Ott J, et al. Genome-wide association study identifies genes that may contribute to risk for developing heroin addiction. Psychiatr Genet. 2010;20(5):207–14.
26. Gelernter J, Kranzler HR, Sherva R, Koesterer R, Almasy L, Zhao H, et al. Genome-wide association study of opioid dependence: multiple associations mapped to calcium and potassium pathways. Biol Psychiatry. 2014;76(1):66–74.
27. Cheng Z, Zhou H, Sherva R, Farrer LA, Kranzler HR, Gelernter J. Genome-wide association study identifies a regulatory variant of RGMA associated with opioid dependence in European Americans. Biol Psychiatry. 2018;84(10):762–70.

28. Bierut LJ, Madden PA, Breslau N, Johnson EO, Hatsukami D, Pomerleau OF, et al. Novel genes identified in a high-density genome wide association study for nicotine dependence. Hum Mol Genet. 2007;16(1):24–35.
29. Bierut LJ, Agrawal A, Bucholz KK, Doheny KF, Laurie C, Pugh E, et al. A genome-wide association study of alcohol dependence. Proc Natl Acad Sci USA. 2010;107(11):5082–7.
30. Schumann G, Coin LJ, Lourdusamy A, Charoen P, Berger KH, Stacey D, et al. Genome-wide association and genetic functional studies identify autism susceptibility candidate 2 gene (AUTS2) in the regulation of alcohol consumption. Proc Natl Acad Sci USA. 2011;108(17):7119–24.
31. Kroslak T, Laforge KS, Gianotti RJ, Ho A, Nielsen DA, Kreek MJ. The single nucleotide polymorphism A118G alters functional properties of the human mu opioid receptor. J Neurochem. 2007;103(1):77–87.
32. Oertel BG, Kettner M, Scholich K, Renne C, Roskam B, Geisslinger G, et al. A common human micro-opioid receptor genetic variant diminishes the receptor signaling efficacy in brain regions processing the sensory information of pain. J Biol Chem. 2009;284(10):6530–5.
33. Kumar D, Chakraborty J, Das S. Epistatic effects between variants of kappa-opioid receptor gene and A118G of mu-opioid receptor gene increase susceptibility to addiction in Indian population. Prog Neuropsychopharmacol Biol Psychiatry. 2012;36(2):225–30.
34. Nikolov MA, Beltcheva O, Galabova A, Ljubenova A, Jankova E, Gergov G, et al. No evidence of association between 118A>G OPRM1 polymorphism and heroin dependence in a large Bulgarian case-control sample. Drug Alcohol Depend. 2011;117(1):62–5.
35. Hancock DB, Levy JL, Gaddis NC, Glasheen C, Saccone NL, Page GP, et al. Cis-Expression quantitative trait loci mapping reveals replicable associations with heroin addiction in OPRM1. Biol Psychiatry. 2015;78(7):474–84.
36. Levran O, Londono D, O'Hara K, Nielsen DA, Peles E, Rotrosen J, et al. Genetic susceptibility to heroin addiction: a candidate gene association study. Genes Brain Behav. 2008;7(7):720–9.
37. Beer B, Erb R, Pavlic M, Ulmer H, Giacomuzzi S, Riemer Y, et al. Association of polymorphisms in pharmacogenetic candidate genes (OPRD1, GAL, ABCB1, OPRM1) with opioid dependence in European population: a case-control study. PLoS ONE. 2013;8(9):e75359.
38. Nelson EC, Lynskey MT, Heath AC, Wray N, Agrawal A, Shand FL, et al. Association of OPRD1 polymorphisms with heroin dependence in a large case-control series. Addict Biol. 2014;19(1):111–21.
39. Randesi M, van den Brink W, Levran O, Blanken P, Butelman ER, Yuferov V, et al. Variants of opioid system genes are associated with non-dependent opioid use and heroin dependence. Drug Alcohol Depend. 2016;168:164–9.
40. Auton A, Brooks LD, Durbin RM, Garrison EP, Kang HM, Korbel JO, et al. A global reference for human genetic variation. Nature. 2015;526(7571):68–74
41. Vereczkei A, Demetrovics Z, Szekely A, Sarkozy P, Antal P, Szilagyi A, et al. Multivariate analysis of dopaminergic gene variants as risk factors of heroin dependence. PLoS ONE. 2013;8(6):e66592.
42. Tsou CC, Chou HW, Ho PS, Kuo SC, Chen CY, Huang CC, et al. DRD2 and ANKK1 genes associate with late-onset heroin dependence in men. World J Biol Psychiatry. 2017:1–11.
43. Jia W, Shi JG, Wu B, Ao L, Zhang R, Zhu YS. Polymorphisms of brain-derived neurotrophic factor associated with heroin dependence. Neurosci Lett. 2011;495(3):221–4.
44. Cheng CY, Hong CJ, Yu YW, Chen TJ, Wu HC, Tsai SJ. Brain-derived neurotrophic factor (Val66Met) genetic polymorphism is associated with substance abuse in males. Brain Res Mol Brain Res. 2005;140(1–2):86–90.
45. Nedic G, Perkovic MN, Sviglin KN, Muck-Seler D, Borovecki F, Pivac N. Brain-derived neurotrophic factor Val66Met polymorphism and alcohol-related phenotypes. Prog Neuropsychopharmacol Biol Psychiatry. 2013;40:193–8.

46. Edenberg HJ. The genetics of alcohol metabolism: role of alcohol dehydrogenase and aldehyde dehydrogenase variants. Alcohol Res Health. 2007;30(1):5–13.
47. Chen CC, Lu RB, Chen YC, Wang MF, Chang YC, Li TK, et al. Interaction between the functional polymorphisms of the alcohol-metabolism genes in protection against alcoholism. Am J Hum Genet. 1999;65(3):795–807.
48. Bierut LJ, Goate AM, Breslau N, Johnson EO, Bertelsen S, Fox L, et al. ADH1B is associated with alcohol dependence and alcohol consumption in populations of European and African ancestry. Mol Psychiatry. 2012;17(4):445–50.
49. Tammimaki AE, Mannisto PT. Are genetic variants of COMT associated with addiction? Pharmacogenet Genomics. 2010;20(12):717–41.
50. Vandenbergh DJ, Rodriguez LA, Miller IT, Uhl GR, Lachman HM. High-activity catechol-O-methyltransferase allele is more prevalent in polysubstance abusers. Am J Med Genet. 1997;74(4):439–42.
51. Jugurnauth SK, Chen CK, Barnes MR, Li T, Lin SK, Liu HC, et al. A COMT gene haplotype associated with methamphetamine abuse. Pharmacogenet Genomics. 2011;21(11):731–40.
52. Nedic G, Nikolac M, Sviglin KN, Muck-Seler D, Borovecki F, Pivac N. Association study of a functional catechol-O-methyltransferase (COMT) Val108/158Met polymorphism and suicide attempts in patients with alcohol dependence. Int J Neuropsychopharmacol. 2011;14(3):377–88.
53. Nedic Erjavec G, Nenadic Sviglin K, Nikolac Perkovic M, Muck-Seler D, Jovanovic T, Pivac N. Association of gene polymorphisms encoding dopaminergic system components and platelet MAO-B activity with alcohol dependence and alcohol dependence-related phenotypes. Prog Neuropsychopharmacol Biol Psychiatry. 2014;54:321–7.
54. Tiihonen J, Hallikainen T, Lachman H, Saito T, Volavka J, Kauhanen J, et al. Association between the functional variant of the catechol-O-methyltransferase (COMT) gene and type 1 alcoholism. Mol Psychiatry. 1999;4(3):286–9.
55. Cuesto G, Everaerts C, León LG, Acebes A. Molecular bases of anorexia nervosa, bulimia nervosa and binge eating disorder: shedding light on the darkness. J Neurogenet. 2017;31(4):266–87.
56. Lindvall Dahlgren C, Wisting l, Rø Ø. Feeding and eating disorders in the DSM-5 era: a systematic review of prevalence rates in non-clinical male and female samples. J Eat Disord. 2017;5:56.
57. Breithaupt L, Hübel C, Bulik CM. Updates on genome-wide association findings in eating disorders and future application to precision medicine. Curr Neuropharmacol. 2018;16(8):1102–10.
58. Trace SE, Baker JH, Penas-Lledo E, Bulik CM. The genetics of eating disorders. Annu Rev Clin Psychol. 2013;9:589–620.
59. Yilmaz Z, Hardaway JA, Bulik CM. Genetics and epigenetics of eating disorders. Adv Genomics Genet. 2015;5:131–50.
60. Svob Strac D, Nikolac Perkovic M, Nedic Erjavec G, Uzun S, Sagud M, Zivkovic M, et al. The role of brain-derived neurotrophic factor (BDNF) in cognition and treatment response. In: Bennet C, editor. Brain-derived neurotrophic factor (BDNF): therapeutic approaches, role in neuronal development and effects on cognitive health. New York: Nova Science Publishers Inc.; 2015 (Chapter 3). p. 67–146.
61. Nakabayashi K, Komaki G, Tajima A, Ando T, Ishikawa M, Nomoto J, et al. Identification of novel candidate loci for anorexia nervosa at 1q41 and 11q22 in Japanese by a genome-wide association analysis with microsatellite markers. J Hum Genet. 2009;54(9):531–7.
62. Wang K, Zhang H, Bloss CS, Duvvuri V, Kaye W, Schork NJ, et al. A genome-wide association study on common SNPs and rare CNVs in anorexia nervosa. Mol Psychiatry. 2011;16(9):949–59.
63. Boraska V, Franklin CS, Floyd JA, Thornton LM, Huckins LM, Southam L, et al. A genome-wide association study of anorexia nervosa. Mol Psychiatry. 2014;19(19):1085–94.

64. Wade TD, Gordon S, Medland S, Bulik CM, Heath AC, Montgomery GW, et al. Genetic variants associated with disordered eating. Int J Eat Disord. 2013;46(6):594–608.
65. Bulik CM, Kleiman SC, Yilmaz Z. Genetic epidemiology of eating disorders. Curr Opin Psychiatry. 2016;29(6):383–8.
66. Root TL, Szatkiewicz JP, Jonassaint CR, Thornton LM, Pinheiro AP, Strober M, et al. Association of candidate genes with phenotypic traits relevant to anorexia nervosa. Eur Eat Disord Rev. 2011;19(6):487–93.
67. Martaskova D, Slachtova L, Kemlink D, Zahorakova D, Papezova H. Polymorphisms in serotonin-related genes in anorexia nervosa. The first study in Czech population and meta analyses with previously performed studies. Folia Biol. (Praha) 2009;55(5):192–7.
68. Kipman A, Bruins-Slot L, Boni C, Hanoun N, Ades J, BlotP, et al. 5-HT2A gene promoter polymorphism as a modifying rather than a vulnerability factor in anorexia nervosa. Eur Psychiatry. 2002;17(4):227–29.
69. Gorwood P, Kipman A, Foulon C. The human genetics of anorexia nervosa. Eur J Pharmacol. 2003;480(1–3):163–70.
70. Koren R, Duncan AE, Munn-Chernoff MA, Bucholz KK, Lynskey MT, Heath AC, et al. Preliminary evidence for the role of HTR2A variants in binge eating in young women. Psychiatr Genet. 2014;24(1):28–33.
71. Munn-Chernoff MA, Baker JH. A primer on the genetics of comorbid eating disorders and substance use disorders. Eur Eat Disord Rev. 2016;24(2):91–100.
72. Bergen AW, van den Bree MBM, Yeager M, Welch R, Ganjei JK, Haque K, et al. Candidate genes for anorexia nervosa in the 1p33–36 linkage region: serotonin 1D and delta opioid receptor loci exhibit significant association to anorexia nervosa. Mol Psychiatry. 2003;8(4):397–406.
73. Brown KMO, Bujac SR, Mann ET, Campbell DA, Stubbins MJ, Blundell JE. Further evidence of association of OPRD1 & HTR1D polymorphisms with susceptibility to anorexia nervosa. Biol Psychiatry. 2007;61(3):367–73.
74. Kiezebrink K, Mann ET, Bujac SR, Stubbins MJ, Campbell DA, Blundell JE. Evidence of complex involvement of serotonergic genes with restrictive and binge purge subtypes of anorexia nervosa. World J Biol Psychiatry. 2010;11(6):824–33.
75. Hernández S, Camarena B, González L, Caballero A, Flores G, Aguilar A. A family-based association study of the HTR1B gene in eating disorders. Rev Bras Psiquiatr. 2016;38(3):239–42.
76. Lee Y, Lin PY. Association between serotonin transporter gene polymorphism and eating disorders: a meta-analytic study. Int J Eat Disord. 2010;43(6):498–504.
77. Calati R, De Ronchi D, Bellini M, Serretti A. The 5-HTTLPR polymorphism and eating disorders: a meta-analysis. Int J Eat Disord. 2011;4(3):191–9.
78. Slof-Op't Landt MC, Meulenbelt I, Bartels M, Suchiman E, Middeldorp CM, Houwing-Duistermaat JJ, et al. Association study in eating disorders: TPH2 associates with anorexia nervosa and self induced vomiting. Genes Brain Behav. 2011;10(2):236–43.
79. Grice DE, Halmi KA, Fichter MM, Strober M, Woodside DB, Treasure JT, et al. Evidence for a susceptibility gene for anorexia nervosa on chromosome 1. Am J Hum Genet. 2002;70(3):787–92.
80. Frisch A, Laufer N, Danziger Y, Michaelovsky E, Leor S, Carel C, et al. Association of anorexia nervosa with the high activity allele of the COMT gene: a family-based study in Israeli patients. Mol Psychiatry. 2001;6(2):243–5.
81. Mikolajczyk E, Smiarowska M, Grzywacz A, Samochowiec J. Association of eating disorders with catechol-O-methyltransferase gene functional polymorphism. Neuropsychobiology. 2006;54(1):82–6.
82. Davis C, Levitan RD, Yilmaz Z, Kaplan AS, Carter JC, Kennedy JL. Binge eating disorder and the dopamine D2 receptor: genotypes and sub-phenotypes. Prog Neuropsychopharmacol Biol Psychiatry. 2012;38(2):328–35.

83. Bachner-Melman R, Lerer E, Zohar AH, Kremer I, Elizur Y, Nemanov L, et al. Anorexia nervosa, perfectionism, and dopamine D4 receptor (DRD4). Am J Med Genet. B Neuropsychiatr Genet. 2007;144B (6):748–56.
84. Gervasini G, González LM, Gamero-Villarroel C, Mota-Zamorano S, Carrillo JA, Flores I, et al. Effect of dopamine receptor D4 (DRD4) haplotypes on general psychopathology in patients with eating disorders. Gene. 2018;654:43–8.
85. Dardennes RM, Zizzari P, Tolle V, Foulon C, Kipman A, Romo L, et al. Family trios analysis of common polymorphisms in the obestatin/ghrelin, BDNF and AGRP genes in patients with anorexia nervosa: association with subtype, body-mass index, severity and age of onset. Psychoneuroendocrinology. 2007;32(2):106–13.
86. Monteleone P, Tortorella A, Castaldo E, Di Filippo C, Maj M. The Leu72Met polymorphism of the ghrelin gene is significantly associated with binge eating disorder. Psychiatr Genet. 2007;17(1):13–6.
87. Muller TD, Tschöp MH, Jarick I, Ehrlich S, Scherag S, Herpertz-Dahlmann B, et al. Genetic variation of the ghrelin activator gene ghrelin O-acyltransferase (GOAT) is associated with anorexia nervosa. J Psychiatr Res. 2011;45(5):706–11.
88. Boraska V, Davis OS, Cherkas LF, Helder SG, Harris J, Krug I, et al. Genome-wide association analysis of eating disorder-related symptoms, behaviors, and personality traits. Am J Med Genet B Neuropsychtr Genet. 2012;159B(7):803–11.
89. Versini A, Ramoz N, Le Strat Y, Scherag S, Ehrlich S, Boni C, et al. Estrogen receptor 1 gene (ESR1) is associated with restrictive anorexia nervosa. Neuropsychopharmacology. 35(8):1818–25.
90. Monteleone P, Bifulco M, Di Filippo C, Gazzerro P, Canestrelli B, Monteleone F, et al. Association of CNR1 and FAAH endocannabinoid gene polymorphisms with anorexia nervosa and bulimia nervosa: evidence for synergistic effects. Genes Brain Behav. 2009;8(7):728–32.
91. Ribases M, Gratacòs M, Fernández-Aranda F, Bellodi L, Boni C, Anderluh M, et al. Association of BDNF with anorexia, bulimia and age of onset of weight loss in six European populations. Hum Mol Genet. 2004;13(12):1205–12.
92. Muller TD, Greene BH, Bellodi L, Cavallini MC, Cellini E, Di Bella D, et al. Fat mass and obesity-associated gene (FTO) in eating disorders: evidence for association of the rs9939609 obesity risk allele with bulimia nervosa and anorexia nervosa. Obes Facts. 2012;5(3):408–19.
93. Koronyo-Hamaoui M, Frisch A, Stein D, Denziger Y, Leor S, Michaelovsky E, et al. Dual contribution of NR2B subunit of NMDA receptor and Ca^{2+}-activated K^+ channel to genetic predisposition to anorexia nervosa. J Psychiatr Res. 2007;41(1–2):160–7.
94. Liu X, Bipolar Genome Study (BiGS), Kelsoe JR, Greenwood TA. A genome-wide association study of bipolar disorder with comorbid eating disorder replicates the SOX2-OT region. J Affect Disord. 2016;189:141–9.
95. Miyasaka K, Hosoya H, Sekime A, Ohta M, Amono H, Matsushita S, et al. Association of ghrelin receptor gene polymorphism with bulimia nervosa in a Japanese population. J Neural Transm. 2006;113(9):1279–85.
96. Vink T, Hinney A, van Elburg AA, van Goozen SH, Sandkuijl LA, Sinke RJ, et al. Association between an agouti-related protein gene polymorphism and anorexia nervosa. Mol. Psychiatry. 2001;6(3):325–8.
97. Nishiguchi N, Matsushita S, Suzuki K, Masanobu M, Shirakawa O, Higuchi S. Association between 5HT2A receptor gene promoter region polymorphism and eating disorders in Japanese patients. Biol Psychiatry. 2001;50(2):123–8.
98. Ricca V, Nacmias B, Cellini E, Di Bernardo M, Rotella CM, Sorbi S. 5-HT2A receptor gene polymorphism and eating disorders. Neurosci Lett. 2002;323(2):105–8.
99. Nilsson M, Naessen S, Dahlman I, Linden Hirschberg A, Gustafsson JA, Dahlman-Wright K. Association of estrogen receptor beta gene polymorphisms with bulimic disease in women. Mol Psychiatry. 2004;9(1):28–34.

100. Bulik-Sullivan B, Finucane H, Anttila V, Gusev A, Day F, ReproGen Consortium, et al. An atlas of genetic correlations across human diseases and traits. Nat Genet. 2015;47(11):1236–41.
101. Yehuda R, Hoge CW, McFarlane AC, Vermetten E, Lanius RA, Nievergelt CM, et al. Posttraumatic stress disorder. Nat Rev Dis Primers. 2015;1:15057.
102. Kessler RC. Posttraumatic stress disorder: the burden to the individual and to society. J Clin Psychiatry. 2000;61(Suppl 5):4–12.
103. Stein MB, Jang KJ, Taylor S, Vernon PA, Livesley WJ. Genetic and environmental influences on trauma exposure and posttraumatic stress disorder: a twin study. Am J Psychiatry. 2002;159(10):1675–81.
104. Sartor CE, Grant JD, Lynskey MT, McCutcheon VV, Waldron M, Statham DJ, et al. Common heritable contributions to low-risk trauma, high-risk trauma, posttraumatic stress disorder, and major depression. Arch Gen Psychiatry. 2012;69(3):293–9.
105. Amstadter AB, Nugent NR, Koenen KC. Genetics of PTSD: fear conditioning as a model for future research. Psychiatr Ann. 2009;39(6):358–67.
106. Comings DE, Comings BG, Muhleman D, Dietz G, Shahbahrami B, Tast D, et al. The dopamine D2 receptor locus as a modifying gene in neuropsychiatric disorders. JAMA. 1991;266(13):1793–800.
107. Comings DE, Muhleman D, Gysin R. Dopamine D2 receptor (DRD2) gene and susceptibility to posttraumatic stress disorder: a study and replication. Biol Psychiatry. 1996;40(5):368–72.
108. Young RM, Lawford BR, Noble EP, Kann B, Wilkie A, Ritchie T, et al. Harmful drinking in military veterans with post-traumatic stress disorder: association with the D2 dopamine receptor A1 allele. Alcohol Alcohol. 2002;37(5):451–6.
109. Bailey JN, Goenjian AK, Noble EP, Walling DP, Ritchie T, Goenjian HA. PTSD and dopaminergic genes, DRD2 and DAT, in multigenerational families exposed to the Spitak earthquake. Psychiatry Res. 2010;178(3):507–10.
110. Hemmings SM, Martin LI, Klopper M, van der Merwe L, Aitken L, de Wit E, et al. BDNF Val66Met and DRD2 Taq1A polymorphisms interact to influence PTSD symptom severity: a preliminary investigation in a South African population. Prog Neuropsychopharmacol Biol Psychiatry. 2013;40:273–80.
111. Voisey J, Swagell CD, Hughes IP, Morris CP, van Daal A, Noble EP, et al. The DRD2 gene 957C>T polymorphism is associated with posttraumatic stress disorder in war veterans. Depress Anxiety. 2009;26(1):28–33.
112. Drury SS, Theall KP, Keats BJB, Scheeringa M. The role of the dopamine transporter (DAT) in the development of PTSD in preschool children. J Trauma Stress. 2009;22(6):534–9.
113. Valente NLM, Vallada H, Cordeiro Q, Miguita K, Bressan RA, Andreoli SB, et al. Candidate-gene approach in posttraumatic stress disorder after urban violence: association analysis of the genes encoding serotonin transporter, dopamine transporter, and BDNF. J Mol Neurosci. 2011;44(1):59–67.
114. Li L, Bao Y, He S, Wang G, Guan Y, Ma D, et al. The association between genetic variants in the dopaminergic system and posttraumatic stress disorder: a meta-analysis. Medicine. 2016;95(11):e3074.
115. Dragan WL, Oniszczenko W. The association between dopamine D4 receptor exon III polymorphism and intensity of PTSD symptoms among flood survivors. Anxiety Stress Coping. 2009;22(5):483–95.
116. Nakamura M, Ueno S, Sano A, Tanabe H. The human serotonin transporter gene linked polymorphism (5-HTTLPR) shows ten novel allelic variants. Mol Psychiatry. 2000;5(1):32–8.
117. Lee HJ, Lee MS, Kang RH, Kim H, Kim SD, Kee BS, et al. Influence of the serotonin transporter promoter gene polymorphism on susceptibility to posttraumatic stress disorder. Depress Anxiety. 2005;21(3):135–9.

118. Wang Z, Baker DG, Harrer J, Hamner M, Price M, Amstadter A. The relationship between combat-related posttraumatic stress disorder and the 5-HTTLPR/rs25531 polymorphism. Depress Anxiety. 2011;28(12):1067–73.
119. Mellman TA, Alim T, Brown DD, Gorodetsky E, Buzas B, Lawson WB, et al. Serotonin polymorphisms and posttraumatic stress disorder in a trauma exposed African American population. Depress Anxiety. 2009;26(11):993–7.
120. Kovacic Petrovic Z, Nedic Erjavec G, Nikolac Perkovic M, Peraica T, Pivac N. No association between the serotonin transporter linked polymorphic region polymorphism and severity of posttraumatic stress disorder symptoms in combat veterans with or without comorbid depression. Psychiatry Res. 2016;244:376–81
121. Kilpatrick DG, Koenen KC, Ruggiero KJ, Acierno R, Galea S, Resnick HS, et al. The serotonin transporter genotype and social support and moderation of posttraumatic stress disorder and depression in hurricane-exposed adults. Am J Psychiatry. 2007;164(11):1693–9.
122. Koenen KC, Aiello AE, Bakshis E, Amstadter AB, Ruggiero KJ, Acierno R, et al. Modification of the association between serotonin transporter genotype and risk of posttraumatic stress disorder in adults by county-level social environment. Am J Epidemiol. 2009;169(6):704–11.
123. Xie P, Kranzler HR, Poling J, Stein MB, Anton RF, Brady K, et al. Interactive effect of stressful life events and the serotonin transporter 5-HTTLPR genotype on posttraumatic stress disorder diagnosis in 2 independent populations. Arch Gen Psychiatry. 2009;66(11):1201–9.
124. Gressier F, Calati R, Balestri M, Marsano A, Alberti S, Antypa N, et al. The 5-HTTLPR polymorphism and posttraumatic stress disorder: a meta-analysis. J Trauma Stress. 2013;26(6):645–53.
125. Liu Y, Garrett ME, Dennis MF, Green KT; VA Mid-Atlantic MIRECC Registry Workgroup, Ashley-Koch AE, et al. An examination of the association between 5-HTTLPR, combat exposure, and PTSD diagnosis among U.S. Veterans. PLoS ONE. 2015;10(3):e0119998.
126. Nelson EC, Agrawal A, Pergadia ML, Lynskey MT, Todorov AA, Wang JC, et al. Association of childhood trauma exposure and GABRA2 polymorphisms with risk of posttraumatic stress disorder in adults. Mol Psychiatry. 2009;14(3):234–5.
127. Kolassa I-T, Kolassa S, Ertl V, Papassotiropoulos A, De Quervain DJF. The risk of posttraumatic stress disorder after trauma depends on traumatic load and the catechol-O-methyltransferase Val158Met polymorphism. Biol Psychiatry. 2010;67(4):304–8.
128. Valente NLM, Vallada H, Cordeiro Q, Bressan RA, Andreoli SB, Mari JJ, et al. Catechol-O-methyltransferase (COMT) val158met polymorphism as a risk factor for PTSD after urban violence. J Mol Neurosci. 2011;43(3):516–23.
129. Clark R, DeYoung CG, Sponheim SR, Bender TL, Polusny MA, Erbes CR, et al. Predicting post-traumatic stress disorder in veterans: interaction of traumatic load with COMT gene variation. J Psychiatr Res. 2013;47(12):1849–56.
130. Havelka Mestrovic A, Tudor L, Nikolac Perkovic M, Nedic Erjavec G, Kovacic Petrovic Z, Svob Strac D, et al. Significant association between catechol-O-methyltransferase (COMT) Val158/108Met polymorphism and cognitive function in veterans with PTSD. Neurosci Lett. 2018;666:38–43.
131. Notaras M, Hill R, van den Buuse M. The BDNF gene Val66Met polymorphism as a modifier of psychiatric disorder susceptibility: progress and controversy. Mol Psychiatry. 2015;20(8):916–30.
132. Zhang L, Benedek DM, Fullerton CS, Forsten RD, Naifeh JA, Li XX, et al. PTSD risk is associated with BDNF Val66Met and BDNF overexpression. Mol Psychiatry. 2014;19(1):8–10.
133. Pivac N, Kozaric-Kovacic D, Grubisic-Ilic M, Nedic G, Rakos I, Nikolac M, et al. The association between brain-derived neurotrophic factor Val66Met variants and psychotic symptoms in posttraumatic stress disorder. World J Biol Psychiatry. 2012;13(4):306–11.

134. Raison CL, Miller AH. When not enough is too much: the role of insufficient glucocorticoid signaling in the pathophysiology of stress-related disorders. Am J Psychiatry. 2003;160(9):1554–65.
135. Lutz B. The endocannabinoid system and extinction learning. Mol Neurobiol. 2007;36(1):92–101.
136. Lu AT, Ogdie MN, Järvelin MR, Moilanen IK, Loo SK, McCracken JT, et al. Association of the cannabinoid receptor gene (CNR1) with ADHD and post-traumatic stress disorder. Am J Med Genet B Neuropsychiatr Genet. 2008;147B(8):1488–94.
137. Mota N, Sumner JA, Lowe SR, Neumeister A, Uddin M, Aiello AE, et al. The rs1049353 polymorphism in the CNR1 gene interacts with childhood abuse to predict posttraumatic threat symptoms. J Clin Psychiatry. 2015;76(12):1622–3.
138. Ressler KJ, Mercer KB, Bradley B, Jovanovic T, Mahan A, Kerley K, et al. Post-traumatic stress disorder is associated with PACAP and the PAC1 receptor. Nature. 2011;470(7335):492–7.
139. Mercer KB, Dias B, Shafer D, Maddox SA, Muller JG, Hu P, et al. Functional evaluation of a PTSD-associated genetic variant: estradiol regulation and ADCYAP1R1. Transl. Psychiatry. 2016;6(12):e978.
140. Binder EB, Bradley RG, Liu W, Epstein MP, Deveau TC, Mercer KB, et al. Association of FKBP5 polymorphisms and childhood abuse with risk of posttraumatic stress disorder symptoms in adults. JAMA. 2008;299(11):1291–305.
141. Xie P, Kranzler HR, Poling J, Stein MB, Anton RF, Farrer LA, et al. Interaction of FKBP5 with childhood adversity on risk for post-traumatic stress disorder. Neuropsychopharmacology. 2010;35(8):1684–92.
142. Klengel T, Mehta D, Anacker C, Rex-Haffner M, Pruessner JC, Pariante CM, et al. Allele-specific FKBP5 DNA demethylation mediates gene-childhood trauma interactions. Nat Neurosci. 2013;16(1):33–41.
143. Ramikie TS, Ressler KJ. Stress-related disorders, pituitary adenylate cyclase—activating peptide (PACAP)ergic system, and sex differences. Dialogues Clin Neurosci. 2016;18(4):403–13.
144. Duncan LE, Ratanatharathorn A, Aiello AE, Almli LM, Amstadter AB, Ashley-Koch AE, et al. Largest GWAS of PTSD (N = 20,070) yields genetic overlap with schizophrenia and sex differences in heritability. Mol Psychiatry. 2018;23(3):666–73.
145. Logue MW, Baldwin C, Guffanti G, Melista E, Wolf EJ, Reardon AF, et al. A genome-wide association study of post-traumatic stress disorder identifies the retinoid-related orphan receptor alpha (RORA) gene as a significant risk locus. Mol Psychiatry. 2013;18(8): 937–42.
146. Xie P, Kranzler HR, Yang C, Zhao H, Farrer LA, Gelernter J. Genome-wide association study identifies new susceptibility loci for posttraumatic stress disorder. Biol Psychiatry. 2013;74(9):656–63.
147. Wolf EJ, Rasmusson AM, Mitchell KS, Logue MW, Baldwin CT, Miller MW. A genome-wide association study of clinical symptoms of dissociation in a trauma-exposed sample. Depress Anxiety. 2014;31(4):352–60.
148. Nievergelt CM, Maihofer AX, Mustapic M, Yurgil KA, Schork NJ, Miller MW, et al. Genomic predictors of combat stress vulnerability and resilience in U.S. marines: a genome-wide association study across multiple ancestries implicates PRTFDC1 as a potential PTSD gene. Psychoneuroendocrinology. 2015;51:459–71.
149. Ashley-Koch AE, Garrett ME, Gibson J, Liu Y, Dennis MF, Kimbrel NA. Veterans Affairs Mid-Atlantic Mental Illness Research, Education, and Clinical Center Workgroup, Beckham JC, Hauser MA. Genome-wide association study of posttraumatic stress disorder in a cohort of Iraq-Afghanistan era veterans. J Affect Disord. 2015;184:225–34.
150. Stein MB, Chen C, Ursano RJ, Cai T, Gelernter J, Heeringa SG, et al. Army Study to Assess Risk and Resilience in Servicemembers (STARRS) Collaborators. Genome-wide association studies of posttraumatic stress disorder in 2 cohorts of us army soldiers. JAMA Psychiatry. 2016;73(7):695–704.

151. Kilaru V, Iyer SV, Almli LM, Stevens JS, Lori A, Jovanovic T, et al. Genome-wide gene-based analysis suggests an association between Neuroligin 1 (NLGN1) and post-traumatic stress disorder. Transl Psychiatry. 2016;6:e820.
152. Melroy-Greif WE, Wilhelmsen KC, Yehuda R, Ehlers CL. Genome-wide association study of post-traumatic stress disorder in two high-risk populations. Twin Res Hum Genet. 2017;20(3):197–207.
153. Segman RH, Cooper-Kazaz R, Macciardi F, Goltser T, Halfon Y, Dobroborski T, et al. Association between the dopamine transporter gene and posttraumatic stress disorder. Mol Psychiatry. 2002;7(8):903–7.
154. Guffanti G, Galea S, Yan L, Roberts AL, Solovieff N, Aiello AE, et al. Genome-wide association study implicates a novel RNA gene, the lincRNA AC068718.1, as a risk factor for posttraumatic stress disorder in women. Psychoneuroendocrinology. 2013;38(12): 3029–38.
155. Almli LM, Stevens JS, Smith AK, Kilaru V, Meng Q, Flory J, et al. A genome-wide identified risk variant for PTSD is a methylation quantitative trait locus and confers decreased cortical activation to fearful faces. Am J Med Genet B Neuropsychiatr Genet. 2015;168B(5):327–36.
156. Lowe SR, Meyers JL, Galea S, Aiello AE, Uddin M, Wildman DE, et al. RORA and post-traumatic stress trajectories: main effects and interactions with childhood physical abuse history. Brain Behav. 2015;5(4):e00323.
157. Sindi IA, Dodd PR. New insights into Alzheimer's disease pathogenesis: the involvement of neuroligins in synaptic malfunction. Neurodegener Dis Manag. 2015;5(2):137–45.
158. Mainland JD, Li YR, Zhou T, Liu WL, Matsunami H. Human olfactory receptor responses to odorants. Sci Data. 2015;2:150002.
159. Suzuki E, Imoto I, Pimkhaokham A, Nakagawa T, Kamata N, Kozaki KI, et al. PRTFDC1, a possible tumor-suppressor gene, is frequently silenced in oral squamous-cell carcinomas by aberrant promoter hypermethylation oncogene. 2007;26(57):7921–32.
160. Apte SS, Parks WC. Metalloproteinases: a parade of functions in matrix biology and an outlook for the future. Matrix Biol. 2015;44–46:1–6.
161. Nemes JP, Benzow KA, Moseley ML, Ranum LP, Koob MD. The SCA8 transcript is an antisense RNA to a brain-specific transcript encoding a novel actin-binding protein (KLHL1). Hum Mol Genet. 2000;9(10):1543–51.
162. Frasa MA, Maximiano FC, Smolarczyk K, Francis RE, Betson ME, Lozano E, et al. Armus is a Rac1 effector that inactivates Rab7 and regulates E-cadherin degradation. Curr Biol. 2010;20(3):198–208.
163. Nievergelt CM, Ashley-Koch AE, Dalvie S, Hauser HA, Morey RA, Smith AK, et al. Genomic approaches to posttraumatic stress disorder: the psychiatric genomic consortium initiative. Biol Psychiatry. 2018;83(10):831–9.
164. Andrews JA, Neises KD. Cells, biomarkers, and post-traumatic stress disorder: evidence for peripheral involvement in a central disease. J Neurochem. 2012;120(1):26–36.
165. Williams HJ, Owen MJ, O'Donovan MC. Schizophrenia genetics: new insights from new approaches. Br Med Bull. 2009;91:61–74.
166. Karayiorgou M, Gogos JA. Schizophrenia genetics: uncovering positional candidate genes. Eur J Hum Genet. 2006;14:512–9.
167. Gonzalez-Castro TB, Hernandez-Diaz Y, Juarez-Rojop IE, Lopez-Navarez ML, Tovilla-Zarate CA, Fresan A. The role of a catechol-O-methyltransferase (COMT) Val158Met genetic polymorphism in schizophrenia: a systematic review and updated meta-analysis on 32,816 subjects. NeuroMol Med. 2016;18(2):216–31.
168. Rees E, O'Donovan MC, Owen MJ. Genetics of schizophrenia. Curr Opin Behav Sci. 2015;2:8–14.
169. Bilder RM, Volavka J, Czbor P, Malhotra AK, Kennedy JL, Ni X, et al. Neurocognitive correlates of the COMT Val158Met polymorphism in chronic schizophrenia. Biol Psychiatry. 2002;52:701–7.

170. Egan MF, Goldberg TE, Kolachana BS, Callicott JH, Mazzanti CM, Straub RE, et al. Effect of COMT Val108/158met genotype on frontal lobe function and risk for schizophrenia. Proc Natl Acad Sci USA. 2001;98(12):6917–22.
171. Fan JB, Zhang CS, Gu NF, Li XW, Sun WW, Wang HY, et al. Catechol-O-methyltransferase gene Val/Met functional polymorphism and risk of schizophrenia: a large-scale association study plus meta-analysis. Biol Psychiatry. 2004;57:139–44.
172. Huang E, Zai CC, Lisoway A, Maciukiewicz M, Felsky D, Tiwari AK, et al. Catechol-O-methyltransferase Val158Met polymorphism and clinical response to antipsychotic treatment in schizophrenia and schizo-affective disorder patients: a meta-analysis. Int J Neuropsychopharmacol. 2016;19(5):1–12.
173. Glatt SJ, Faraone SV, Tsuang MT. Association between a functional catechol O-methyltransferase gene polymorphism and schizophrenia: meta-analysis of case-control and family-based studies. Am J Psychiatry. 2003;160:469–76.
174. Nikolac M, Sagud M, Nedic G, Nenadic Sviglin K, Mihaljevic Peles A, Uzun S, et al. The lack of association between catechol-O-methyl-transferase Val108/158Met polymorphism and smoking in schizophrenia and alcohol dependence. Letter to the Editor. Psychiatry Res. 2013;205:179–80.
175. Herken H, Emin Erdal M. Catechol-O-methyltransferase gene polymorphism in schizophrenia: evidence for association between symptomatology and prognosis. Psychiatr Genet. 2000;11:105–9.
176. Sagud M, Tudor L, Uzun S, Nikolac Perkovic M, Zivkovic M, Konjevod M, et al. Haplotypic and genotypic association of catecol-O-methyktransferase rs4680 and rs4818 polymorphisms and treatment resistance in schizophrenia. Front Pharmacol. 2018;9:1–14.
177. Huo R, Wei Z, Xiong Y, Jiang J, Liu Y, Yan Y, et al. Association of dopamine receptor D1 (DRD1) polymorphisms with risperidone treatment response in Chinese schizophrenia patients. Neurosci Lett. 2014;584:178–83.
178. Matsumoto J, Nagaoka A, Kunii Y, Miura I, Hino M, Niwa S-I, et al. Effects of the −141C insertion/deletion polymorphism in the dopamine D2 receptor gene on the dopamine system in the striatum in patients with schizophrenia. Psychiatry Res. 2018;264:116–8.
179. Smith CT, Dang LC, Buckholtz JW, Tetreault AM, Cowan RL, Kessler RM, et al. The impact of common dopamine D2 receptor gene polymorphisms on D2/3 receptor availability: C957T as a key determinant in putamen and ventral striatum. Transl Psychiatry. 2017;7:e1091.
180. Persson J, Stenfors C. Superior cognitive goal maintenance in carriers of genetic markers linked to reduced striatal D2 receptor density (C957T and DRD2/ANKK1-TaqIA). PLoS ONE. 2018;13(8):1–12.
181. Cohen OS, Weickert TW, Hess JL, Paish LM, McCoy SY, Rothmond DA, et al. A splicing-regulatory polymorphism in DRD2 disrupts ZRANB2 binding, impairs cognitive functioning and increases risk for schizophrenia in six Han Chinese samples. Mol Psychiatry 2015;1–8.
182. Yao J, Pan Y-Q, Ding M, Pang H, Wang B-J. Association between DRD2 (rs1799732 and rs1801028) and ANK1 (rs1800497) polymorphisms and schizophrenia: a meta-analysis. Am J Med Genet Part B. 2014;168B:1–13.
183. Howes OD, McCutcheon R, Owen MJ, Murray RM. The role of genes, stress, and dopamine in the development of schizophrenia. Biol Psychiatry. 2017;8:9–20.
184. Okuyama Y, Ishiguro H, Toru M, Arinami T. A genetic polymorphism in the promoter region of DRD4 associated with expression and schizophrenia. Biochem Biophys Res Commun. 1999;258:292–5.
185. Egan MF, Straub RE, Goldberg TE, Yakub I, Callicott JH, Hariri AR, et al. Variation in GRM3 affects cognition, prefrontal glutamate, and risk for schizophrenia. Proc Natl Acad Sci USA. 2004;101(34):12604–9.
186. Harrison PJ, Weinberger DR. Schizophrenia genes, gene expression, and neuropathology: on the matter of their convergence. Mol Psychiatry. 2005;10:40–68.

187. Hodkinson CA, Goldman D, Jaeger J, Persaud S, Kane JM, Lipsky RH, et al. Disrupted in schizophrenia 1 (DISC1): association with schizophrenia, schizoaffective disorder, and bipolar disorder. Am J Hum Genet. 2004;75:862–72.
188. Callicott JH, Straub RE, Pezawas L, Egan MF, Mattay VS, Hariri AR, et al. Variation in DISC1 affects hippocampal structure and function and increases risk for schizophrenia. Proc Natl Acad Sci USA. 2005;102(24):8627–32.
189. Harrison PJ, Law AJ. Neuregulin 1 and schizophrenia: genetics, gene expression, and neurobiology. Biol Psychiatry. 2006;60:132–40.
190. Huang E, Hettige NC. Zai G, Tomasi J, Huang J, Zai CC, et al. BDNF Val66Met polymorphism and clinical response to antipsychotic treatment in schizophrenia and schizoaffective disorder patients: a meta-analysis. Pharmacogenomics J. 2018. https://doi.org/10.1038/s41397-018-0041-5. [Epub ahead of print].
191. Hong CJ, Younger WY, Lin CH, Tsai SJ. An association study of a brain-derived neurotrophic factor Val66Met polymorphism and clozapine response of schizophrenic patients. Neurosci Lett. 2003;349:206–8.
192. Kheirollahi M, Kazemi E, Ashouri S. Brain-derived neurotrophic factor gene Val66Met polymorphism and risk of schizophrenia: a meta-analysis of case–control studies. Cell Mol Neurobiol. 2015;36(1):1–10.
193. Notaras M, Hill R, van den Buuse M. A role for the BDNF gene Val66Met polymorphism in schizophrenia? A comprehensive review. Neurosci Biobehav Rev. 2015;51:15–30.
194. Blasi G, Selvaggi P, Fazio L, Antonucci LA, Taurisano P, Masellis R, et al. Variation in dopamine D2 and serotonin 5HTt2A receptor genes is associated with working memory processing and response to treatment with antipsychotics. Neuropsychopharmacology. 2015;40(7):1600–8.
195. Kavanagh DH, Tansey KE, O'Donovan MC, Owen MJ. Schizophrenia genetics: emerging themes for a complex disorder. Mol Psychiatry. 2015;20(1):72–6.
196. Gonda X, Petschner P, Eszlari N, Baksa D, Edes A, Antal P, et al. Genetic variants in major depressive disorder: from pathophysiology to therapy. Pharmacol Ther. 2018; pii: S0163–7258(18):30156-6.
197. Sullivan PF, Neale MC, Kendler KS. Genetic epidemiology of major depression: review and meta-analysis. Am J Psychiatry. 2000;157(10):1552–62.
198. Kendler KS, Gatz M, Gardner CO, Pedersen NL. A Swedish national twin study of lifetime major depression. Am J Psychiatry. 2006;163(1):109–14.
199. Muñoz M, Pong-Wong R, Canela-Xandri O, Rawlik K, Haley CS, Tenesa A. Evaluating the contribution of genetics and familial shared environment to common disease using the UK Biobank. Nat Genet. 2016;48(9):980–3.
200. Wang K, Gaitsch H, Poon H, Cox NJ, Rzhetsky A. Classification of common human diseases derived from shared genetic and environmental determinants. Nat Genet. 2017;49(9):1319–25.
201. Sullivan PF, Daly MJ, O'Donovan M. Genetic architectures of psychiatric disorders: the emerging picture and its implications. Nat Rev Genet. 2012;13(8):537–51.
202. Uher R. Gene-environment interactions in severe mental illness. Front Psychiatry. 2014;5:48.
203. Flint J, Kendler KS. The genetics of major depression. Neuron. 2014;81(3):484–503.
204. Mullins N, Lewis CM. Genetics of Depression: progress at last. Curr Psychiatry Rep. 2017;19(8):43.
205. Kendler KS, Ohlsson H, Sundquist K, Sundquist J. Sources of parent-offspring resemblance for major depression in a national Swedish extended adoption study. JAMA Psychiatry. 2018;75(2):194–200.
206. Hammen C. Stress and depression. Annu Rev Clin Psychol. 2005;1:293–319.
207. Hammen C, Kim EY, Eberhart NK, Brennan PA. Chronic and acute stress and the prediction of major depression in women. Depress Anxiety. 2009;26(8):718–23.

208. Brown GW, Bifulco A, Harris TO. Life events, vulnerability and onset of depression: some refinements. Br J Psychiatry. 1987;150:30–42.
209. Kendler KS, Karkowski-Shuman L. Stressful life events and genetic liability to major depression: genetic control of exposure to the environment? Psychol Med. 1997;27(3):539–47.
210. Shadrina M, Bondarenko EA, Slominsky PA. Genetics factors in major depression disease. Front Psychiatry. 2018;9:334.
211. Kautzky A, Baldinger P, Souery D, Montgomery S, Mendlewicz J, Zohar J, et al. The combined effect of genetic polymorphisms and clinical parameters on treatment outcome in treatment-resistant depression. Eur Neuropsychopharmacol. 2015;25(4):441–53.
212. Network and Pathway Analysis Subgroup of Psychiatric Genomics Consortium. Psychiatric genome-wide association study analyses implicate neuronal, immune and histone pathways. Nat Neurosci. 2015;18(2):199–209.
213. Marx P, Antal P, Bolgar B, Bagdy G, Deakin B, Juhasz G. Comorbidities in the diseasome are more apparent than real: what Bayesian filtering reveals about the comorbidities of depression. PLoS Comput Biol. 2017;13(6):e1005487.
214. Purves KL, Coleman JRI, Rayner C, Hettema JM, DeckertJ, McIntosh AM, et al. The common genetic architecture of anxiety disorders. 2017; bioRxiv 203844.
215. Yu C, Baune BT, Licinio J, Wong ML. A novel strategy for clustering major depression individuals using whole-genome sequencing variant data. Sci Rep. 2017;7:44389.
216. López-León S, Janssens A, González-Zuloeta Ladd AM, Del-Favero J, Claes SJ, Oostra BA, et al. Meta-analyses of genetic studies on major depressive disorder. Mol Psychiatry. 2008;13(8):772–85.
217. Clarke H, Flint J, Attwood AS, Munafò MR. Association of the 5-HTTLPR genotype and unipolar depression: a meta-analysis. Psychol Med. 2010;40(11):1767–78.
218. Kiyohara C, Yoshimasu K. Association between major depressive disorder and a functional polymorphism of the 5-hydroxytryptamine (serotonin) transporter gene: a meta-analysis. Psychiatr Genet. 2010;20(2):49–58.
219. Wu YL, Ding XX, Sun YH, Yang HY, Chen J, Zhao X, et al. Association between MTHFR C677T polymorphism and depression: an updated meta-analysis of 26 studies. Prog Neuropsychopharmacol Biol Psychiatry. 2013;46:78–85.
220. Zhao X, Huang Y, Ma H, Jin Q, Wang Y, Zhu G. Association between major depressive disorder and the norepinephrine transporter polymorphisms T-182C and G1287A: a meta-analysis. J Affect Disord. 2013;150(1):23–8.
221. Zhou Y, Su H, Song J, Guo L, Sun Y. Association between norepinephrine transporter T-182C polymorphism and major depressive disorder: a meta-analysis. Neurosci Lett. 2014;561:64–8.
222. Jin C, Xu W, Yuan J, Wang G, Cheng Z. Meta-analysis of association between the -1438A/G (rs6311) polymorphism of the serotonin 2A receptor gene and major depressive disorder. Neurol Res. 2013;35(1):7–14.
223. Gyekis JP, Yu W, Dong S, Wang H, Qian J, Kota P, et al. No association of genetic variants in BDNF with major depression: a meta- and gene-based analysis. Am J Med Genet B Neuropsychiatr Genet. 2013;162B(1):61–70.
224. Kishi T, Yoshimura R, Fukuo Y, Kitajima T, Okochi T, Matsunaga S, et al. The CLOCK gene and mood disorders: a case-control study and meta-analysis. Chronobiol Int. 2011;28(9):825–33.
225. Gatt JM, Burton KL, Williams LM, Schofield PR. Specific and common genes implicated across major mental disorders: a review of meta-analysis studies. J Psychiatr Res. 2015;60:1–13.
226. Bosker FJ, Hartman CA, Nolte IM, Prins BP, Terpstra P, Posthuma D, et al. Poor replication of candidate genes for major depressive disorder using genome-wide association data. Mol Psychiatry. 2011;16(5):516–32.

227. Wray NR, Pergadia ML, Blackwood DH, Penninx BW, Gordon SD, Nyholt DR, et al. Genome-wide association study of major depressive disorder: new results, meta-analysis, and lessons learned. Mol Psychiatry. 2012;17(1):36–48.
228. Major Depressive Disorder Working Group of the Psychiatric GWAS Consortium, Ripke S, Wray NR, Lewis CM, Hamilton SP, Weissman MM, et al. A mega-analysis of genome-wide association studies for major depressive disorder. Mol Psychiatry. 2013;18(4):497–511.
229. Dunn EC, Brown RC, Dai Y, Rosand J, Nugent NR, Amstadter AB, et al. Genetic determinants of depression: recent findings and future directions. Harv Rev Psychiatry. 2015;23(1):1–18.
230. Kohli MA, Lucae S, Saemann PG, Schmidt MV, Demirkan A, Hek K, et al. The neuronal transporter gene SLC6A15 confers risk to major depression. Neuron. 2011;70(2):252–65.
231. Converge Consortium. Sparse whole-genome sequencing identifies two loci for major depressive disorder. Nature. 2015;523(7562):588–91.
232. Ware EB, Mukherjee B, Sun YV, Diez-Roux AV, Kardia SL, Smith JA. Comparative genome-wide association studies of a depressive symptom phenotype in a repeated measures setting by race/ethnicity in the multi-ethnic study of atherosclerosis. BMC Genet. 2015;16:118.
233. Huo YX, Huang L, Zhang DF, Yao YG, Fang YR, Zhang C, et al. Identification of SLC25A37 as a major depressive disorder risk gene. J Psychiatr Res. 2016;83:168–75.
234. Hyde CL, Nagle MW, Tian C, Chen X, Paciga SA, Wendland JR, et al. Identification of 15 genetic loci associated with risk of major depression in individuals of European descent. Nat Genet. 2016;48(9):1031–6.
235. Okbay ABaselmans BM, De Neve JE, Turley P, Nivard MG, Fontana MA, et al. Genetic variants associated with subjective well-being, depressive symptoms, and neuroticism identified through genome-wide analyses. Nat Genet. 2016;48(6):624–33.
236. Mbarek H, Milaneschi Y, Hottenga JJ, Ligthart L, de Geus EJC, Ehli EA, et al. Genome-wide significance for PCLO as a gene for major depressive disorder. Twin Res Hum Genet. 2017;20(4):267–70.
237. Power RA, Tansey KE, Buttenschøn HN, Cohen-Woods S, Bigdeli T, Hall LS, et al. Genome-wide association for major depression through age at onset stratification: major depressive disorder working group of the psychiatric genomics consortium. Biol Psychiatry. 2017;81(4):325–35.
238. Wray NR, Ripke S, Mattheisen M, Trzaskowski M, Byrne EM, Abdellaoui A, et al. Genome-wide association analyses identify 44 risk variants and refine the genetic architecture of major depression. Nat Genet. 2018;50(5):668–81.
239. Xiao X, Zheng F, Chang H, Ma Y, Yao YG, Luo XJ, et al. The gene encoding protocadherin 9 (PCDH9), a novel risk factor for major depressive disorder. Neuropsychopharmacology. 2018;43(5):1128–37.
240. Sullivan PF, de Geus EJ, Willemsen G, James MR, Smit JH, Zandbelt T, et al. Genome-wide association for major depressive disorder: a possible role for the presynaptic protein piccolo. Mol Psychiatry. 2009;14(4):359–75.
241. Sabunciyan S, Aryee MJ, Irizarry RA, Rongione M, Webster MJ, Kaufman WE, et al. Genome-wide DNA methylation scan in major depressive disorder. PLoS ONE. 2012;7(4):e34451.
242. Wong ML, Arcos-Burgos M, Liu S, Vélez JI, Yu C, Baune BT, et al. The PHF21B gene is associated with major depression and modulates the stress response. Mol Psychiatry. 2017;22(7):1015–25.
243. Amin N, Jovanova O, Adams HH, Dehghan A, Kavousi M, Vernooij MW, et al. Exome-sequencing in a large population-based study reveals a rare Asn396Ser variant in the LIPG gene associated with depressive symptoms. Mol Psychiatry. 2017;22(4):537–43.
244. Knowles EE, Kent JW Jr, McKay DR, Sprooten E, Mathias SR, Curran JE, et al. Genome-wide linkage on chromosome 10q26 for a dimensional scale of major depression. J Affect Disord. 2016;2016(191):123–31.

245. Pirooznia M, Wang T, Avramopoulos D, Potash JB, Zandi PP, Goes FS. High-throughput sequencing of the synaptome in major depressive disorder. Mol Psychiatry. 2016;21(5):650–5.
246. Glessner JT, Wang K, Sleiman PM, Zhang H, Kim CE, Flory JH, et al. Duplication of the SLIT3 locus on 5q35.1 predisposes to major depressive disorder. PLoS One. 2010;5(12):e15463.
247. Cukier HN, Dueker ND, Slifer SH, Lee JM, Whitehead PL, Lalanne E, et al. Exome sequencing of extended families with autism reveals genes shared across neurodevelopmental and neuropsychiatric disorders. Mol Autism. 2014;5(1):1.
248. Rucker JJ, Breen G, Pinto D, Pedroso I, Lewis CM, Cohen-Woods S, et al. Genome-wide association analysis of copy number variation in recurrent depressive disorder. Mol Psychiatry. 2013;18(2):183–9.
249. Rucker JJ, Tansey KE, Rivera M, Pinto D, Cohen-Woods S, Uher R, et al. Phenotypic association analyses with copy number variation in recurrent depressive disorder. Biol Psychiatry. 2016;79(4):329–36.
250. Haenisch B, Linsel K, Brüss M, Gilsbach R, Propping P, Nöthen MM, et al. Association of major depression with rare functional variants in norepinephrine transporter and serotonin1A receptor genes. Am J Med Genet B Neuropsychiatr Genet. 2009;150B(7):1013–6.
251. Perlis RH, Smoller JW, Mysore J, Sun M, Gillis T, Purcell S, et al. Prevalence of incompletely penetrant Huntington's disease alleles among individuals with major depressive disorder. Am J Psychiatry. 2010;167(5):574–9.
252. Neff CD, Abkevich V, Potter J, Riley R, Shattuck D, Katz DA. Evidence for epistasis between SLC6A4 and a chromosome 4 gene as risk factors in major depression. Am J Med Genet B Neuropsychiatr Genet. 2010;153B(1):321–2.
253. Gabriela Nielsen M, Congiu C, Bortolomasi M, Bonvicini C, Bignotti S, Abate M, et al. MTHFR: genetic variants, expression analysis and COMT interaction in major depressive disorder. J Affect Disord. 2015;183:179–86.
254. Szczepankiewicz A, Leszczyńska-Rodziewicz A, Pawlak J, Rajewska-Rager A, Wilkosc M, Zaremba D, et al. Epistatic interaction between CRHR1 and AVPR1b variants as a predictor of major depressive disorder. Psychiatr Genet. 2013;23(6):239–46.
255. Ben-Efraim YJ, Wasserman D, Wasserman J, Sokolowski M. Family-based study of AVPR1B association and interaction with stressful life events on depression and anxiety in suicide attempts. Neuropsychopharmacology. 2013;38(8):1504–11.
256. Xiao Z, Liu W, Gao K, Wan Q, Yang C, Wang H, et al. Interaction between CRHR1 and BDNF genes increases the risk of recurrent major depressive disorder in Chinese population. PLoS ONE. 2011;6(12):e28733.
257. Bobińska K, Szemraj J, Czarny P, Gałecki P. Role of MMP-2, MMP-7, MMP-9 and TIMP-2 in the development of recurrent depressive disorder. J Affect Disord. 2016;205:119–29.
258. Reuter M, Markett S, Melchers M, Montag C. Interaction of the cholinergic system and the hypothalamic-pituitary-adrenal axis as a risk factor for depression: evidence from a genetic association study. NeuroReport. 2012;23(12):717–20.
259. Pae CU, Drago A, Forlani M, Patkar AA, Serretti A. Investigation of an epistastic effect between a set of TAAR6 and HSP-70 genes variations and major mood disorders. Am J Med Genet B Neuropsychiatr Genet. 2010;153B(2):680–3.
260. Murk W, DeWan AT. Exhaustive genome-wide search for SNP-SNP interactions across 10 human diseases. G3 (Bethesda). 2016;6(7):2043–50.
261. Caspi A, Sugden K, Moffitt TE, Taylor A, Craig IW, Harrington H, et al. Influence of life stress on depression: moderation by a polymorphism in the 5-HTT gene. Science. 2003;301(5631):386–9.
262. Karg K, Burmeister M, Shedden K, Sen S. The serotonin transporter promoter variant (5-HTTLPR), stress, and depression meta-analysis revisited: evidence of genetic moderation. Arch Gen Psychiatry. 2011;68(5):444–54.

263. Sharpley CF, Palanisamy SK, Glyde NS, Dillingham PW, Agnew LL. An update on the interaction between the serotonin transporter promoter variant (5-HTTLPR), stress and depression, plus an exploration of non-confirming findings. Behav Brain Res. 2014;273:89–105.
264. Bleys D, Luyten P, Soenens B, Claes S. Gene-environment interactions between stress and 5-HTTLPR in depression: a meta-analytic update. J Affect Disord. 2018;226:339–45.
265. Munafò MR, Durrant C, Lewis G, Flint J. Gene × Environment interactions at the serotonin transporter locus. Biol Psychiatry. 2009;65(3):211–9.
266. Risch N, Herrell R, Lehner T, Liang KY, Eaves L, Hoh J, et al. Interaction between the serotonin transporter gene (5-HTTLPR), stressful life events, and risk of depression: a meta-analysis. JAMA. 2009;301(23):2462–71.
267. Culverhouse RC, Saccone NL, Horton AC, Ma Y, Anstey KJ, Banaschewski T, et al. Collaborative meta-analysis finds no evidence of a strong interaction between stress and 5-HTTLPR genotype contributing to the development of depression. Mol Psychiatry. 2018;23(1):133–42.
268. Gonda X, Eszlari N, Kovacs D, Anderson IM, Deakin JF, Juhasz G, et al. Financial difficulties but not other types of recent negative life events show strong interactions with 5-HTTLPR genotype in the development of depressive symptoms. Transl Psychiatry. 2016;6:e798.
269. Mandelli L, Serretti A. Gene environment interaction studies in depression and suicidal behavior: an update. Neurosci Biobehav Rev. 2013;37(10 Pt 1):2375–97.
270. Naoi M, Maruyama W, Shamoto-Nagai M. Type A monoamine oxidase and serotonin are coordinately involved in depressive disorders: from neurotransmitter imbalance to impaired neurogenesis. J Neural Transm (Vienna). 2018;125(1):53–66.
271. Hosang GM, Shiles C, Tansey KE, McGuffin P, Uher R. Interaction between stress and the BDNF Val66Met polymorphism in depression: a systematic review and meta-analysis. BMC Med. 2014;12:7.
272. Zhao M, Chen L, Yang J, Han D, Fang D, Qiu X, et al. BDNF Val66Met polymorphism, life stress and depression: a meta-analysis of gene-environment interaction. J Affect Disord. 2018;227:226–35.
273. Gonda X, Sarginson J, Eszlari N, Petschner P, Toth ZG, Baksa D, et al. A new stress sensor and risk factor for suicide: the T allele of the functional genetic variant in the GABRA6 gene. Sci Rep. 2017;7(1):12887.
274. Baumeister D, Akhtar R, Ciufolini S, Pariante CM, Mondelli V. Childhood trauma and adulthood inflammation: a meta-analysis of peripheral C-reactive protein, interleukin-6 and tumour necrosis factor-alpha. Mol Psychiatry. 2016;21(5):642–9.
275. Kovacs D, Eszlari N, Petschner P, Pap D, Vas S, Kovacs P, et al. Effects of IL1B single nucleotide polymorphisms on depressive and anxiety symptoms are determined by severity and type of life stress. Brain Behav Immun. 2016;56:96–104.
276. Kovacs D, Eszlari N, Petschner P, Pap D, Vas S, Kovacs P, et al. Interleukin-6 promoter polymorphism interacts with pain and life stress influencing depression phenotypes. J Neural Transm (Vienna). 2016;123(5):541–8.
277. Tartter M, Hammen C, Bower JE, Brennan PA, Cole S. Effects of chronic interpersonal stress exposure on depressive symptoms are moderated by genetic variation at IL6 and IL1beta in youth. Brain Behav Immun. 2015;46:104–11.
278. Juhasz G, Hullam G, Eszlari N, Gonda X, Antal P, Anderson IM, et al. Brain galanin system genes interact with life stresses in depression-related phenotypes. Proc Natl Acad Sci USA. 2014;111(16):E1666–73.
279. Juhasz G, Chase D, Pegg E, Downey D, Toth ZG, Stones K, et al. CNR1 gene is associated with high neuroticism and low agreeableness and interacts with recent negative life events to predict current depressive symptoms. Neuropsychopharmacology. 2009;34(8):2019–27.
280. Lazary J, Eszlari N, Juhasz G, Bagdy G. Genetically reduced FAAH activity may be a risk for the development of anxiety and depression in persons with repetitive childhood trauma. Eur Neuropsychopharmacol. 2016;26(6):1020–8.

281. Van der Auwera S, Peyrot WJ, Milaneschi Y, Hertel J, Baune B, Breen G, et al. Genome-wide gene-environment interaction in depression: a systematic evaluation of candidate genes: the childhood trauma working-group of PGC-MDD. Am J Med Genet B Neuropsychiatr Genet. 2018;177(1):40–9.
282. Dunn EC, Wiste A, Radmanesh F, Almli LM, Gogarten SM, Sofer T, et al. Genome-wide association study (GWAS) and genome-wide by enviroment interaction study (GWEIS) of depressive symptoms in African American and Hispanic/latina women. Depress Anxiety. 2016;33(4):265–80.
283. Otowa T, Kawamura Y, Tsutsumi A, Kawakami N, Kan C, Shimada T, et al. The first pilot genome-wide gene-environment study of depression in the japanese population. PLoS ONE. 2016;11(8):e0160823.
284. Grabe HJ, Schwahn C, Mahler J, Appel K, Schulz A, Spitzer C, et al. Genetic epistasis between the brain-derived neurotrophic factor Val66Met polymorphism and the 5-HTT promoter polymorphism moderates the susceptibility to depressive disorders after childhood abuse. Prog Neuropsychopharmacol Biol Psychiatry. 2012;36(2):264–70.
285. Dalton ED, Hammen CL, Najman JM, Brennan PA. Genetic susceptibility to family environment: BDNF Val66met and 5-HTTLPR influence depressive symptoms. J Fam Psychol. 2014;28(6):947–56.
286. Ignácio ZM, Réus GZ, Abelaira HM, Quevedo J. Epigenetic and epistatic interactions between serotonin transporter and brain-derived neurotrophic factor genetic polymorphism: insights in depression. Neuroscience. 2014;275:455–68.
287. Yang C, Xu Y, Sun N, Ren Y, Liu Z, Cao X, et al. The combined effects of the BDNF and GSK3B genes modulate the relationship between negative life events and major depressive disorder. Brain Res. 2010;1355:1–6.
288. Juhasz G, Gonda X, Hullam G, Eszlari N, Kovacs D, Lazary J, et al. Variability in the effect of 5-HTTLPR on depression in a large European population: the role of age, symptom profile, type and intensity of life stressors. PLoS ONE. 2015;10(3):e0116316.
289. Lin E, Hong CJ, Hwang JP, Liou YJ, Yang CH, Cheng D, et al. Gene-gene interactions of the brain-derived neurotrophic-factor and neurotrophic tyrosine kinase receptor 2 genes in geriatric depression. Rejuvenation Res. 2009;12(6):387–93.

Neuroimaging of Small Vessel Disease in Late-Life Depression

Nadim S. Farhat, Robert Theiss, Tales Santini, Tamer S. Ibrahim and Howard J. Aizenstein

Abstract

Cerebral small vessel disease is associated with late-life depression, cognitive impairment, executive dysfunction, distress, and loss of life for older adults. Late-life depression is becoming a substantial public health burden, and a considerable number of older adults presenting to primary care have significant clinical depression. Even though white matter hyperintensities are linked with small vessel disease, white matter hyperintensities are nonspecific to small vessel disease and can co-occur with other brain diseases. Advanced neuroimaging techniques at the ultrahigh field magnetic resonance imaging are enabling

N. S. Farhat · T. Santini · T. S. Ibrahim (✉) · H. J. Aizenstein (✉)
Department of Bioengineering, Swanson School of Engineering,
University of Pittsburgh, Pittsburgh, PA, USA
e-mail: tibrahim@pitt.edu

H. J. Aizenstein
e-mail: aizen@pitt.edu

R. Theiss
School of Medicine, University of Pittsburgh, Pittsburgh, PA, USA

T. S. Ibrahim · H. J. Aizenstein
Department of Psychiatry, School of Medicine,
University of Pittsburgh, Pittsburgh, PA, USA

T. S. Ibrahim
Department of Radiology, School of Medicine,
University of Pittsburgh, Pittsburgh, PA, USA

H. J. Aizenstein
Clinical and Translational Science Institute,
University of Pittsburgh, Pittsburgh, PA, USA

© Springer Nature Singapore Pte Ltd. 2019

Y.-K. Kim (ed.), *Frontiers in Psychiatry*, Advances in Experimental Medicine and Biology 1192, https://doi.org/10.1007/978-981-32-9721-0_5

improved characterization, identification of cerebral small vessel disease and are elucidating some of the mechanisms that associate small vessel disease with late-life depression.

Keywords
Cerebral small vessel disease · Late-life depression · Ultrahigh field · Magnetic resonance imaging · Cerebral microvascular mechanisms · Tic-tac-toe radiofrequency head coil

Introduction

Most older adults have neuroradiologic evidence of cerebral small vessel disease (SVD) [1]. Although many older adults carry the burden of SVD without clinical manifestations, population studies have clearly shown that a high burden of small vessel disease is associated with cognitive, physical, and mood dysfunction [2–4]. Specifically, small vessel disease has been linked to the thinning of the cerebral cortex, gray matter atrophy [5], greater cerebral blood flow decline [6, 7], increased white matter hyperintensity volume [8–11], and worsening cognition [8, 9, 11, 12]. It is also shown that small vessel disease is strongly associated with late-life depression (LLD) [13]. Depression is a leading cause of morbidity and mortality in the elderly [14–19]. Better understanding triggers for late-life depression like small vessel disease will facilitate both prevention strategies and treatment plans. However, the progress in understanding the role of small vessel disease and how it contributes to late-life depression has been stymied in part by a lack of specificity when it comes to the imaging and classification of white matter hyperintensities (WMH). White matter hyperintensities are currently identified using MRI imaging and are a hallmark of small vessel disease [13, 20–25]. White matter hyperintensities represent components of edema, gliosis, ischemia, and inflammation. There is an established longitudinal association between white matter hyperintensities and late-life depression; specifically, the progression of small vessel disease is based on imaging is associated with persistent depression [26, 27].

Furthermore, depression can cause many complications in the growing elderly population, and at times, it is poorly responsive to treatment [26, 27]. Characterizing the pathophysiologic subcomponents responsible for white matter hyperintensity progression in late-life depression through advanced neuroimaging techniques is of great importance, as it will allow for an understanding of how depression develops over time. The neuroimaging characterization of small vessel disease will, in turn, shape decision making about what to do about this growing public health problem moving forward [28].

Magnetic resonance imaging (MRI) neuroimaging plays an essential role in exploring and understanding small vessel disease (SVD) and its association with late-life depression (LLD). MRI is noninvasive, without ionizing radiation, and due to the low risk associated with this imaging technique, it is ideal for longitudinal studies. Lately, ultrahigh field (UHF $\geq$ 7 T) MRI has been extensively used in MRI neuroimaging research. This significant development in MRI coupled with the advances in computer hardware and medical imaging analysis paradigms are enhancing the visualization, analysis, and help in better understanding LLD and its association with SVD. Despite these recent advances in the available neuroimaging techniques for SVD and LLD, these techniques are mostly applied for research purposes and are under heavy development, and we expect that they will mature to be used in clinics in the next few years.

The potential clinical applications of these UHF MRI neuroimaging techniques cover diagnosis, individualized treatment options, and treatment monitoring. The MRI neuroimaging techniques available to investigate SVD and LLD are numerous and multimodal. Therefore, we will focus our discussion on three main topics: (1) introduction to UHF MRI, (2) the advances in the neuroimaging of small vessel disease structural imaging with ultrahigh field MRI, (3) novel methods in assessing microvascular mechanism in small vessel disease, and (4) neuroimaging connectomes for late-life depression and small vessel disease.

UHF MRI

Human MRI UHF is impacting the neuroimaging field with its higher image resolution, higher signal-to-noise ratio (SNR) and enhanced soft tissue contrast. 7 Tesla (T) MRI has reached a level of maturity that can meet the demand of clinical settings, being recently cleared by the Food and Drug Administration [29] for clinical use. At the time of writing this article, there are 70 human 7 T scanners installed worldwide [30]. These new MRI scanners are usually installed with optimized head coils designs [31], better gradients, and powerful computing engines. In this section, we will explain what the advantages of 7 T MRI are and discuss some of the main challenges of scanning at 7 T.

Benefits of Ultrahigh Field MR

There are three main benefits of increasing the magnetic field in MRI; a higher signal-to-noise ratio (SNR), which can be used to either increase spatial resolution or to perform faster scans with higher acceleration factors, increased susceptibility effects, which increases blood oxygenation level dependent (BOLD) contrast, and longer T1 and shorter T2/T2* relaxation times, which enhances contrast in certain applications.

MRI SNR increases with magnetic field strength [32]. Consequently, this increase in SNR means that scans to be collected at higher spatial resolution (more voxels per volume unit) without an increase in acquisition time. As an example: Table 1 shows that the MPRAGE at 7 T collects almost twice the amount of the voxels per volume with a comparable acquisition time of 3 T MPRAGE.

The other benefits are a higher contrast-to-noise ratio (CNR) and an enhanced T2* effect (relaxation time due to magnetic susceptibility effects) which grows with the magnetic field and leads to more significant variability of relaxation times between several tissues which can be exploited in differentiating between structures. Moreover, this higher CNR contrast enables more accurate quantification susceptibility maps (QSM) of the iron deposit in the brain [33].

Table 1 This table shows typical pulse sequences used for small vessel disease MRI imaging. The "Purpose" column describes the marker for SVD targeted by the sequence. The second column lists the typical resolutions used in clinics. The third column lists the typical resolutions in SVD research in 3 T MRI. The fourth column lists the typical resolutions in SVD research in 7 T MRI

Sequences	Purpose	Clinical brain MRI The thickness in-plane Resolution	3 T Research brain MRI The slice thickness and in-plane resolution	7 T research brain The slice thickness and in-plane resolution
T1-Weighted/MPRAGE	Brain Atrophy distinguishing between Lacunes and perivascular spaces	3–5 mm 1 mm × 1 mm	MPRAGE 1 mm Isotropic (5 min acquisition)	MPRAGE 0.75 isotropic (5 min acquisition) 0.55 mm isotropic (8 min acquisition)
Diffusion-weighted imaging (DWI)	Acute ischemic stroke	3–5 mm 2 mm × 2 mm	3–5 mm × 2 mm × 2 mm	1.5 mm isotropic
T2-Weighted	Brain structures perivascular spaces white matter hyperintensities	3–5 mm 2 mm × 2 mm	1 mm isotropic	0.6 mm isotropic
Fluid-attenuated inversion recovery (FLAIR)	White matter hyperintensities differentiate Lacunes from PvS	3–5 mm 1 mm × 1 mm	1.6 mm × 0.8 mm × 0.8 mm	1.5 × 0.75 mm × 0.75 mm
T2* GRE/susceptibility-weighted imaging	Microbleeds/hemorrhage	3–5 mm 1 mm × 1 mm	SWI 1 mm × 1 mm × 1 mm	SWI 0.75 mm × 0.35 mm × 0.35 mm
Magnetic resonance angiography	Mainly stenosis visualization for large vessels, such as middle cerebral artery and internal carotid	1 mm isotropic with contrast and reconstruction		0.38 mm isotropic no contrast

T1 relaxation time is longer with the magnetic field [34]; this effect is exploited to increase the CNR in time of flight (TOF) magnetic resonance angiography (MRA). This gain in CNR is utilized to detect smaller arteries [35]. Lastly, UHF has a higher sensitivity to blood oxygenation level dependent (BOLD) contrast [36]. The increased BOLD contrast has a significant effect on enhancing functional MRI spatial resolution, CNR, and temporal resolution of (fMRI) [37].

Radio Frequency Transmit Nonuniformity and Specific Absorption Rate

In MRI, the transmit radiofrequency (RF) coil produces the RF transverse magnetic field responsible for the spin excitation (B1+). The B1+ field distribution becomes more inhomogeneous with the increase of the magnetic field. The inhomogeneity of the B1+ field is an unwanted consequence that causes the appearance of inhomogeneities in the images or low contrast in certain parts of the brain, especially in high flip angle sequences like the fluid-attenuated inverse recovery (FLAIR) [38]. Parallel transmissions using several RF transmit coils and pulses sequences that are insensitive to the RF inhomogeneity (e.g., adiabatic pulses) [39] might alleviate the RF Tx field inhomogeneity. However, in certain instances, solving the RF transmit inhomogeneity issues might cause RF heating effects. Moreover, the B1+ field distribution changes with different head shapes (loads); therefore, these transmit fields change for each subject [40, 41].

Load Insensitive Head Coil Design and Full Head Coverage Parallel Transmission

The tic-tac-toe (TTT) RF system for 7 T head MRI is composed by a 16-channel transmit array (4 TTT elements located around the head) and a 32-channel receive insert, for optimal receiving performance. The TTT transmit array has been shown to produce a homogenous spin excitation at 7 T frequencies [42] and superior subject insensitiveness (low variations of the RF performance with different head shapes) [43]. It also presents lower global and local (SAR) when compared with other transmit coil designs [42, 43], increasing patient safety and minimizing the risks of temperature increase and tissue damage.

Future of UHF

Several centers are exploring neuroimaging at UHF MRI at 9.4, 10.5, 11.2, and 20 T [30]. The current state of the art in 7 T is to optimize sensitivity, spatiotemporal resolution, and encoding speed [44] and to improve B_0 homogeneity to address susceptibility artifacts [45].

Neuroimaging of Small Vessel Disease Structural Imaging with Ultrahigh Field MRI

The neuroimaging standards of SVD designate a set of distinct neuroimaging features in the brain. These features are acute small subcortical infarct, lacunes, white matter hyperintensities (WMH), enlarged perivascular spaces (PvS), cerebral microbleeds (CMBs), and atrophy [46]. Although these distinct SVD features were regarded as different types of tissue changes, there is increasing evidence that these features are associated and might be caused by the diffuse nature of the small vessel pathological changes. 7 T UHF has a huge impact on the imaging of PvS, CMB, lacunes, and cortical microinfarcts. Therefore, in the following section of the book chapter, we will focus our review on how 7 T UHF impacted PvS, CMB, lacunes, and cortical microinfarcts.

Lacunes and PvS at 7 T

Lacunes

Lacunes are small cavities filled with fluid and have a diameter between 3 and 15 mm. Lacunes often form as a result of acute small subcortical infarcts or hemorrhages of a nearby perforating arteriole. On MRI, their shapes are usually round or oval with a hyperintense rim. These lacunes cavities have hyperintense signals in T2 sequences similar to the cerebral spinal fluid signal. On the other hand, the lacunes signal is hypointense in T2-FLAIR. While the presence of some lacunes is mostly asymptomatic, large amounts of lacunes are correlated with dementia, cognitive impairment, gait disturbance, and increased risk of stroke [47–49].

Enlarged Perivascular Spaces (PvS)

Perivascular spaces are cerebrospinal fluid (CSF) filled cavities areas surrounding cerebral arteries or veins, and they extend from the subarachnoid space [50]. Perivascular spaces (PvS) are usually <5 mm and have an irregular shape [20]. Since PvS are filled with CSF, they are hyperintense in T2 and hypointense in T2-FLAIR. These spaces tend to enlarge with age. Despite being considered, as a normal aging phenomenon, PvS are increasingly thought to be associated with the occurrence of cerebral small vessel disease. Additionally, studies have shown that PvS presence could be a risk factor for antidepressant resistance [51].

UHF 7 T MRI Role in Distinguishing Between Lacunes and Perivascular Spaces

UHF 7 T MRI plays a major role in distinguishing between lacunes and PvS. At low resolution on imaging such as 1.5 and 3 T, PvS can look the same as lacunes. Both pathologies are filled with liquid, and therefore, they have hyperintense signal in T2. Consequently, larger PvS might have been confused with lacunes in several studies looking at small vessel changes [52]. The hyperintense rim surrounding the hypointense "hole" on FLAIR, which is thought to be a distinguishing feature of lacunes from PvS, can be absent in some cases. The colocalization of PvS with extensive WMH areas may mimic the hyperintense rim of lacunes, subsequently causing misidentification. The importance of distinguishing between lacunar infarct and PvS cannot be overstated [53], as they are the result of different cerebral pathologies and lead to distinct clinical outcomes. The superior capability of 7 T in detecting PvS has been noted in the literature and offers a solution to the current problem [54]. The resolving power of 7 T UHF can better detect the small vessels in the PvS, which leads to a better characterization of PvS when compared to lacunar infarcts [55]. When comparing 7 and 1.5 T (7 and 1.5 T) [56], 7 T was more sensitive to PvS than 1.5 and 3 T (Fig. 1).

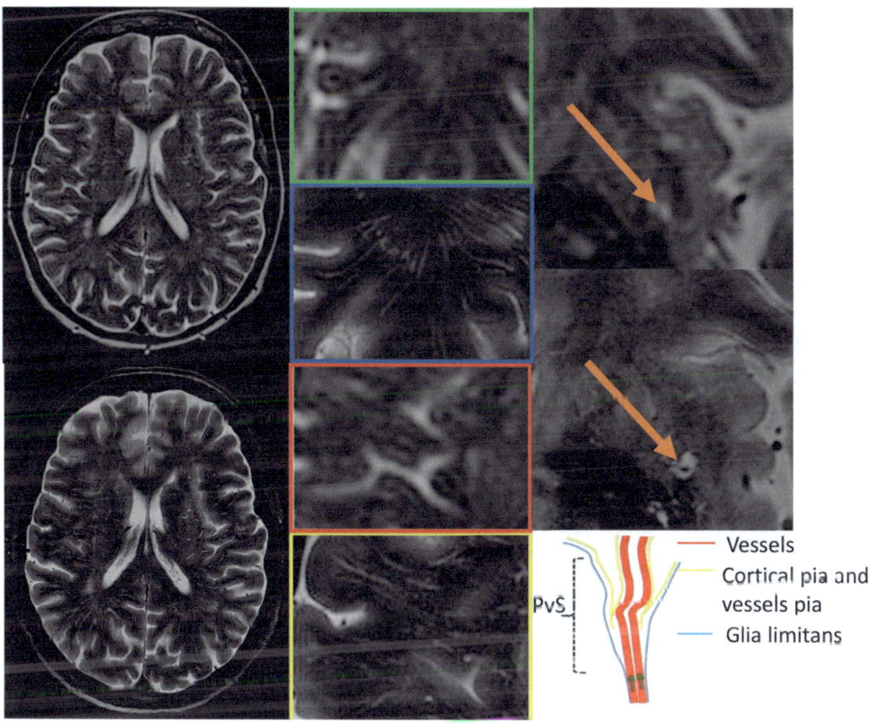

Fig. 1 This figure compares 7 T versus 3 T T2 turbo spin echo (TSE) resolution comparison of perivascular spaces appearances (PvS). The blue and yellow boxes show the PvS well-defined features at 7 T; the green and red boxes show PvS at 3 T. The top arrow shows PvS at 3 T; the bottom arrow shows PvS at 7 T. Note the black dot inside the PvS is a vessel and a distinctive feature of PvS that is most visible at 7 T. The bottom of the figure shows a diagram of a longitudinal view of PvS spaces and the morphology of perivascular spaces. PvS are delineated by basal membranes of glia, pia, and endothelium; the PvS depicts the space surrounding vessels penetrating into the parenchyma. These vessels penetrating the PvS are clearly seen in the 7 T image. The diagram is adapted from Ref. [130] under the Common Attribution 2.0 Generic License

Cerebral Microbleeds (CMB)

Cerebral microbleeds (CMBs) are small (typically 2–5 mm in diameter), ovoid hypointense signals on T2*. Susceptibility-based sequences are sensitive to microbleeds [46]. CMBs correlate with neurodegenerative brain diseases, including the small vessel disease seen in late-life depression [57, 58]. The increase in the magnetic field at 7 T makes CMBs more prominent than the surrounding tissue. Several studies have confirmed that 7 T is more reliable than 1.5 T at detecting CMBs, which lead researchers to investigate CMBs at 7 T as an imaging marker for SVD [59–61]. For example, in another study, the confidence in detecting microbleeds in vivo is slightly higher at 7 T compared to 3 T. However, increased susceptibility effects at 7 T also increases the number of artifacts on imaging. The susceptibility artifacts make detecting microbleeds near other structures challenging and will have to be addressed for better results in the future [62].

Cortical Microinfarcts

Cortical microinfarcts (CMI) were first observed in autopsy reports [63] in a mixed aging population. CMIs are found in the brains of non-demented elderly, Alzheimer disease, and vascular dementia patients [64]. There is an increased interest in studying CMI and their relations to small vessel disease (SVD). Cortical microinfarcts are sometimes visible in 3 T scanning [65], yet most of the time, they go undetected in the routine examination due to their size 0.2–2.9 mm [66]. While CMIs are relatively small, CMIs have an effect volume 12 times the volume of their core [67]. This affected volume causes disorganization in the adjacent axons, which might induce clinical deficits seen in the aging population [68]. Several studies reported that CMIs are more visible and easier to detect in both ex vivo and in vivo [69, 70] at 7 T MRI. There have even been new guidelines established for the detection and characterization of CMIs following these findings, which illustrates the potential of 7 T in helping to pick up small changes in the microvasculature and influence care [71].

Neuroimaging of Microvascular Mechanism in Small Vessel Disease

In the section above, "Neuroimaging of small vessel disease structural imaging with ultrahigh field MRI," we had discussed the neuroimaging methods investigating SVD that assess parenchymal and cortical structural damage, these methods in addition to WHM evaluation reflect damage in the later stage of the small vessel disease. With the technical advancement, the SVD research is heading toward detecting changes that happen earlier than overt disease on structural imaging. In addition to being a diffuse disease and a whole-brain disease [72], there is an accumulation of evidence that suggests the intrinsic aspect of small vessel disease.

The changes appear to start at the endothelial level which leads later on to disorders in the cerebral microvasculature, to fibrinoid necrosis, and then to perivascular tissue injury [73]; this new understanding of the pathophysiology of small vessel disease created a need to detect features at an earlier stage of SVD that are associated with the microvessels health. Subsequently, the following section will review the advancement of neuroimaging of small vessel disease in terms of cerebral microvascular changes (venules and arterioles), tissue pulsatility, and blood–brain barrier.

Small Arteries Imaging

MRA and TOF visualize large cerebral arteries such as middle cerebral artery (MCA) and the circle of Willis. Even with the administration of contrast, clinical MRI scanners are not able to visualize accurately perforating arteries at the lower magnetic field (1.5 and 3 T) [46, 74]. The visualization of perforating arteries and the consequences of their pathology can elucidate some of the SVD mechanism [75]. Ultrahigh field (7 T and more) has a higher contrast-to-noise ratio (CNR) and signal-to-noise ratio (SNR) clinical scanners. The imaging at UHF enables the visualization of some of the intracranial perforating arteries without the administration of contrast; several studies reported 0.3–0.2 mm^3 resolution for MRA [35]. Recent studies reported the visualization of lenticulostriate arteries in the basal ganglia (BPA) and the thalamus (TPA) [76, 77]. Although great progress has been made in the visualization of small arteries, MRA is still limited in the visualization of smaller arteries that are distal to the circle of Willis. The state of art research is to increase the contrast of 7 T UHF to detect small arteries either by the use of extrinsic contrast agents [78] or the use of blood flow as an intrinsic contrast agent [79].

Small Veins Imaging

An early cerebral microvascular study observed that arterioles become more tortuous in some of SVD earlier states [80]. The visualization of the venous side of the cerebral microvascular pathology was not possible without contrast; therefore, the alternations of the venules due to small vessel disease were rarely investigated and thus poorly understood [81]. The increase in susceptibility (T2*) at the Ultrahigh field (7 T) MR imaging enables the visualization of venous microcirculation [82, 83]. SWI enhances the contrast of deoxyhemoglobin compared to the brain parenchyma, therefore, enables the visualization venules without a contrast agent. This contrast enhancement leads to the development of methods to quantify venules in multiple sclerosis [84] sickle cell anemia, [85] CADASIL [86] and Alzheimer's disease (AD) [87]. In addition to arterioles pathology, venules pathology such occluded venules may be a contributing factor to small vessel disease [88] and venules tortuosity is a potential indicator of venule pathology, and the tortuosity

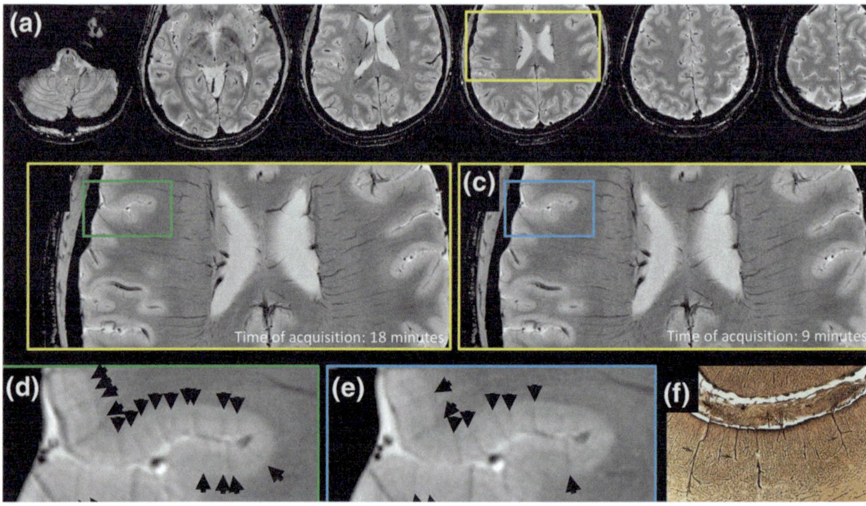

Fig. 2 High resolution in vivo gradient echo (GRE) whole-brain acquisitions at 7 T, showing details of the brain structure and a comparison with histology. The images were acquired with the tic-tac-toe RF system [42, 43] and with the following GRE sequence parameters: TR = 1960, TE = 14, in-plane resolution 0.21×0.21 mm^2, slice thickness of 1.5 mm, and 64 slices (114.9 mm) in total (20% gap between slices). In (**a**), selected slices showing the whole-brain acquisition; in (**b**), zoomed image showing the medullary draining veins throughout the deep white matter acquired without acceleration (18 min acquisitions); in (**c**), similar zoomed images acquired with acceleration factor (GRAPPA) 2 (9 min acquisition); in (**d**), the zoomed image in the cortical region, showing cortical microvessels (black arrows), without acceleration in the acquisition; in (**e**), similar slice showing the cortical regions and cortical microvessels (black arrows), acquired with GRAPPA 2; in (**f**), figure adapted from Ref. [130], showing similar vasculature structure in the histopathology images of the cortical region. The received profile of the MRI images (**a**–**e**) was corrected using the SPM12 software package

computation is done at 7 T without contrast [81]. More tortuous venules are more common with aging and mostly distributed near regions of white matter hyperintensities [89, 90]. The venules tortuosity is believed to be the result of the inflammation changes and is associated with white matter hyperintensities in the brain parenchyma and could predate radiologically overt white matter hyperintensities [81] (Fig. 2).

Arterial Pulsatility Index

In the past few years, there is an increasing amount of research papers that are showing an association between intracranial pulsatility and cerebral vessel disease. The increase in the intracranial pulsatility could be a contributing mechanism in small vessel disease [91]. Arterial pulsatility index is shown higher in more severe small vessel disease cases [92]. Most of these studies were done using transcranial Doppler Ultrasound and lower field MRI and investigated larger arteries such as the middle cerebral arteries [93]. Small perforating arteries have small diameters <300 μm [94, 95] and they are challenging to visualize in the current techniques let alone measuring their flow and pulsatility. Building on the relative increase in the T2* contrast of the arterial blood, cardiac gated 7 T MRI 2D phase-contrast method can measure the pulsatility and blood flow in small artery with good precision and repeatability [96]. This method measured a higher pulsatility in patients with lacunar infarct and intracerebral hemorrhages on both the basal ganglia and semioval centers [97]. Most of the studies in the MRI cerebral artery pulsatility are cross-sectional and have a limited number of patients. They need to be adjusted to several confounding factors such as age. Nonetheless, with the increase in centers using 7 T and development of automatic methods to segment arteries and estimating pulsatility index, small arterioles pulsatility index could help in unraveling poorly understood mechanisms in small vessel disease.

Cerebral Reactivity

Cerebral reactivity measurements might estimate the small vessel disease burden directly. The reactivity measures the response of microvasculature to the administration of acetazolamide and by manipulations of the air breathed by the subject (CO_2 challenge) breath-holding, and neuronal activation (high-frequency visual tasks). A reduced reactivity is a sign of exhausted autoregulation and impaired adaptation of microvasculature which can be a consequence of the loss of smooth muscle cells, vessel walls stiffness [98], or poor functioning of the pericytes [99].

As a consequence of the impaired autoregulation ischemia of the brain is more frequent [100]. The MRI methods to estimate cerebral reactivity are BOLD, arterial spin labels, phase-contrast MRI. BOLD MRI [101], which measures the increase of blood oxygenation and the neurovascular coupling, is the most popular method used in estimating the cerebral reactivity measures [102]. Though the interpretation of BOLD signal alterations and their relation to cerebral reactivity can be complex, it is the most widely used method for cerebral reactivity assessment. In a recent paper, the authors indicated that spontaneous low-frequency fluctuations (0.02–0.04 Hz) without gaseous challenge are a measure of cerebral reactivity [103]. As mentioned above, the increase in the magnetic field increases the BOLD signal contrast. Therefore, the cerebral reactivity measurement using BOLD signal is expected to have higher sensitivity at 7 T [104].

Blood–Brain Barrier

The integrity of the blood–brain barrier (BBB) is imaged with dynamic contrast-enhanced (DCE) MRI [105], this has demonstrated BBB breakdown associated with tumor infiltration. DCE MRI has recently been explored as a marker of blood–brain barrier breakdown in small vessel disease as well as in conditions such as diabetes and Alzheimer's disease. BBB acts as a regulator for material transfer to and out of the cerebrum, Also BBB maintains homeostasis of the brain cell environment. In addition to the breakdown of the BBB with tumor infiltration, the BBB may leak more with advancing age [76] and is leakier in white matter hyperintensities in small vessel disease [106]. In a recent case report using ultrahigh field MRI, it was observed that microbleeds were formed in the areas of BBB disruption [107].

Mechanical Assessment of Cerebral Tissue

Age and disease alter brain tissue mechanical properties [108, 109]. Elastography methods are imaging-based techniques that can estimate the elastic property of biological tissue [110]. The elastography techniques can use the endogenous small vessel pulsatility to estimate the displacement (<1 mm) in the brain tissue [111] or apply exogenous force to estimate stiffness. Magnetic resonance elastography (MRE) uses an exogenous force to estimate brain tissue stiffness. This exogenous force induces mechanical waves that travel perpendicular to the direction of the applied force. The wave speeds are measured using magnetic resonance elastography (MRE). MRE can be sensitive to tissue alterations [112]. Thus, MRE has the potential to assess tissue alterations in SVD at an early stage. The MRE is finding its way into the UHF field, and it is expected to be added as a routine research sequence [113].

Neuroimaging Connectomes for Late-Life Depression and Small Vessel Disease

Connectomes are defined as the structural and functional mapping of neural connections between different brain regions to understand the organization of the human brain [114]. This functional and structural organization or networks can be analyzed using MRI and has benefited from the advances of the ultrahigh field imaging [30].

Structural Networks

Diffusion-weighted imaging (DWI) may detect the weakened white matter integrity on a microstructure scale, in normal-appearing white matter [115]. Therefore, it can be used as an indicator of small vessel disease. Diffusion tensor imaging (DTI) and graph theory [116, 117] measures the white matter integrity of the whole-brain structural networks. In SVD, metrics such as path length (connecting two brain regions) and efficiency are used to investigate how the burden of SVD affects the brain. For example, an increase in path length between two brain regions means the information is routed through the long way. Thus, DTI-based connectomes can be used as an indicator of the severity of the disease [118].

Functional Networks

Neuronal activity at rest (under no cognitive command) induces spontaneous fluctuation throughout the brain; this fluctuation can be temporally correlated in different brain regions without direct fiber connectivity. This synchronized activity is called resting states; it was observed in BOLD fMRI [119]. Resting states fMRI (rs-fMRI) is analyzed using distinct computational methods (seed-based correlations [120], independent component analysis [121] and graph theory [122]) to extract resting brain networks. The resting brain networks provide information about the intrinsic organization of the brain networks, and the alteration of these networks can be indicative of a pathogenic brain process that affects mental health like major depression disorder [123], schizophrenia [124, 125] and cognition like attention-deficit/hyperactivity disorder [126, 127]. Similar to the structural connectomes, functional connectomes shows a disrupted organization of the functional brain networks in late-life depression [128] and small vessel disease [129].

Connectomes in the Clinics

The evidence is growing that the connectomes are a potential clinical tool that can map and characterize brain networks in healthy aging population as opposed to an aging population with small vessel disease. However, a considerable effort is required to validate the reliability, reproducibility, specificity, and sensitivity of SVD connectomes, if they are going to be used as a clinical tool and marker for the small vessel diseases.

Conclusion

In this chapter, we have highlighted how UHF MRI can enhance the imaging of small vessel disease (SVD). The advances rely on improvements in the basic MR coil design, which allow for uniform high SNR. We reviewed how the UHF MRI allows for improved identification of small microbleeds and microinfarcts, as well as improved characterization of the small arteries and veins. Small vessel disease is strongly associated with a variety of neuropsychiatric syndromes in late-life, including late-life depression. Advanced imaging techniques, such as those described in this chapter, can help identify mechanisms relating SVD to depression. Moreover, they may help identify targets for prevention and treatment.

References

1. De Leeuw FE, De Groot JC, Achten E, Oudkerk M, Ramos LMP, Heijboer R, et al. Prevalence of cerebral white matter lesions in elderly people: a population based magnetic resonance imaging study. The Rotterdam scan study. J Neurol Neurosurg Psychiatry. 2001;70(1):9–14.
2. Kuo H-K, Lipsitz LA. Cerebral white matter changes and geriatric syndromes: is there a link? J Gerontol Ser A Biol Sci Med Sci. 2004;59(8):M818–26. http://www.ncbi.nlm.nih.gov/pubmed/15345732.
3. Herrmann LL, Le Masurier M, Ebmeier KP. White matter hyperintensities in late life depression: a systematic review. J Neurol Neurosurg Psychiatry. 2007;79(6):619–24. http://jnnp.bmj.com/cgi/doi/10.1136/jnnp.2007.124651.
4. Silbert LC, Nelson C, Howieson DB, Moore MM, Kaye JA. Impact of white matter hyperintensity volume progression on rate of cognitive and motor decline. Neurology. 2008;71(2):108–13. http://www.neurology.org/cgi/doi/10.1212/01.wnl.0000316799.86917.37.
5. Steffens DC, Tupler LA, Ranga K, Krishnan R. Magnetic resonance imaging signal hypointensity and iron content of putamen nuclei in elderly depressed patients. Psychiatry Res Neuroimaging. 1998;83(2):95–103. https://linkinghub.elsevier.com/retrieve/pii/S0925492798000328.
6. Sabayan B, Westendorp RG, Grond J van der, Stott DJ, Sattar N, van Osch MJP, et al. Markers of endothelial dysfunction and cerebral blood flow in older adults. Neurobiol Aging. 2014;35(2):373–7. https://linkinghub.elsevier.com/retrieve/pii/S0197458013003540.
7. Beason-Held LL, Moghekar A, Zonderman AB, Kraut MA, Resnick SM. Longitudinal changes in cerebral blood flow in the older hypertensive brain. Stroke. 2007;38(6):1766–73. https://www.ahajournals.org/doi/10.1161/STROKEAHA.106.477109.
8. Debette S, Seshadri S, Beiser A, Au R, Himali JJ, Palumbo C, et al. Midlife vascular risk factor exposure accelerates structural brain aging and cognitive decline. Neurology. 2011;77(5):461–8. http://www.neurology.org/cgi/doi/10.1212/WNL.0b013e318227b227.
9. Jefferson AL, Himali JJ, Beiser AS, Au R, Massaro JM, Seshadri S, et al. Cardiac index is associated with brain aging. Circulation. 2010;122(7):690–7. https://www.ahajournals.org/doi/10.1161/CIRCULATIONAHA.109.905091.
10. Vidal J-S, Sigurdsson S, Jonsdottir MK, Eiriksdottir G, Thorgeirsson G, Kjartansson O, et al. Coronary artery calcium, brain function and structure. Stroke. 2010;41(5):891–7. https://www.ahajournals.org/doi/10.1161/STROKEAHA.110.579581.
11. Swan GE, DeCarli C, Miller BL, Reed T, Wolf PA, Jack LM, et al. Association of midlife blood pressure to late-life cognitive decline and brain morphology. Neurology. 1998;51(4):986–93. http://www.neurology.org/cgi/doi/10.1212/WNL.51.4.986.
12. Strassburger TL, Lee H-C, Daly EM, Szczepanik J, Krasuski JS, Mentis MJ, et al. Interactive effects of age and hypertension on volumes of brain structures. Stroke. 1997;28(7):1410–7. https://www.ahajournals.org/doi/10.1161/01.STR.28.7.1410.
13. Taylor WD, Aizenstein HJ, Alexopoulos GS. The vascular depression hypothesis: mechanisms linking vascular disease with depression. Mol Psychiatry. 2013;18(9):963–74. http://www.nature.com/articles/mp201320.
14. Blazer DG. Depression in late life: review and commentary. J Gerontol A Biol Sci Med Sci. 2003;58(3):249–65. http://www.ncbi.nlm.nih.gov/pubmed/12634292.
15. Alexopoulos GS. Depression in the elderly. Lancet. 2005;365(9475):1961–70. https://linkinghub.elsevier.com/retrieve/pii/S0140673605666652.
16. Gallegos-Carrillo K, García-Peña C, Mudgal J, Romero X, Durán-Arenas L, Salmerón J. Role of depressive symptoms and comorbid chronic disease on health-related quality of life among community-dwelling older adults. J Psychosom Res. 2009;66(2):127–35. https://linkinghub.elsevier.com/retrieve/pii/S0022399908003632.
17. Lebowitz BD, Pearson JL, Schneider LS, Reynolds CF, Alexopoulos GS, Bruce ML, et al. Diagnosis and treatment of depression in late life. Consensus statement update. JAMA. 1997;278(14):1186–90. http://www.ncbi.nlm.nih.gov/pubmed/9326481.

18. Vu NQ, Aizenstein HJ. Depression in the elderly. Curr Opin Neurol. 2013;26(6):656–61. http://content.wkhealth.com/linkback/openurl?sid=WKPTLP:landingpage&an=00019052-201312000-00011.
19. Suzuki A, Kondo T, Mihara K, Yasui-Furukori N, Otani K, Furukori H, et al. Association between Taq I A dopamine D2 receptor polymorphism and therapeutic response to bromperidol: a preliminary report. Eur Arch Psychiatry Clin Neurosci. 2001;251(2):57–9. http://link.springer.com/10.1007/s004060170053.
20. Sheline YI, Pieper CF, Barch DM, Welsh-Boehmer K, McKinstry RC, MacFall JR, et al. Support for the vascular depression hypothesis in late-life depression. Arch Gen Psychiatry. 2010;67(3):277. http://archpsyc.jamanetwork.com/article.aspx?doi=10.1001/archgenpsychiatry.2009.204.
21. Sheline YI, Price JL, Vaishnavi SN, Mintun MA, Barch DM, Epstein AA, et al. Regional white matter hyperintensity burden in automated segmentation distinguishes late-life depressed subjects from comparison subjects matched for vascular risk factors. Am J Psychiatry. 2008;165(4):524–32. http://psychiatryonline.org/doi/abs/10.1176/appi.ajp.2007.07010175.
22. Gouw AA, van der Flier WM, Fazekas F, van Straaten ECW, Pantoni L, Poggesi A, et al. Progression of white matter hyperintensities and incidence of new lacunes over a 3-year period. Stroke. 2008;39(5):1414–20. https://www.ahajournals.org/doi/10.1161/STROKEAHA.107.498535.
23. Quarantelli M, Lanzillo R, Del Vecchio W, Mollica C, Prinster A, Iadicicco L, et al. Modifications of brain tissue volumes in facioscapulohumeral dystrophy. Neuroimage. 2006;32(3):1237–42. https://linkinghub.elsevier.com/retrieve/pii/S105381190600526X.
24. Tupler LA, Krishnan KRR, McDonald WM, Dombeck CB, D'Souza S, Steffens DC. Anatomic location and laterality of MRI signal hyperintensities in late-life depression. J Psychosom Res. 2002;53(2):665–76. https://linkinghub.elsevier.com/retrieve/pii/S0022399902004257.
25. Nebes RD, Vora IJ, Meltzer CC, Fukui MB, Williams RL, Kamboh MI, et al. Relationship of deep white matter hyperintensities and apolipoprotein e genotype to depressive symptoms in older adults without clinical depression. Am J Psychiatry. 2001;158(6):878–84. http://psychiatryonline.org/doi/abs/10.1176/appi.ajp.158.6.878.
26. Taylor WD, Steffens DC, MacFall JR, McQuoid DR, Payne ME, Provenzale JM, et al. White matter hyperintensity progression and late-life depression outcomes. Arch Gen Psychiatry. 2003;60(11):1090. http://archpsyc.jamanetwork.com/article.aspx?doi=10.1001/archpsyc.60.11.1090.
27. Khalaf A, Edelman K, Tudorascu D, Andreescu C, Reynolds CF, Aizenstein H. White matter hyperintensity accumulation during treatment of late-life depression. Neuropsychopharmacology. 2015;40(13):3027–35. http://www.nature.com/articles/npp2015158.
28. Aizenstein HJ, Khalaf A, Walker SE, Andreescu C. Magnetic resonance imaging predictors of treatment response in late-life depression. J Geriatr Psychiatry Neurol. 2014;27(1):24–32.
29. R.A.Ochs. FDA approval of Magnetom Terra. 2017. https://www.accessdata.fda.gov/cdrh docs/pdf17/K170840.pdf.
30. Polimeni JR, Uludağ K. Neuroimaging with ultra-high field MRI: present and future. Neuroimage. 2018;168(February):1–6.
31. Ibrahim T, Hue Y, Gilbert R, Boada F. Tic Tac Toe: highly-coupled, load insensitive Tx/Rx array and a quadrature coil without lumped capacitors. Proceedings of the 16th scientific meeting, international society for magnetic resonance in medicine, Toronto, January 2008. p. 438. https://cds.ismrm.org/ismrm-2008/files/00438.pdf.
32. Pohmann R, Speck O, Scheffler K. Signal-to-noise ratio and MR tissue parameters in human brain imaging at 3, 7, and 9.4 tesla using current receive coil arrays. Magn Reson Med. 2016;75(2):801–9. http://doi.wiley.com/10.1002/mrm.25677.
33. Peters AM, Brookes MJ, Hoogenraad FG, Gowland PA, Francis ST, Morris PG, et al. T2* measurements in human brain at 1.5, 3 and 7 T. Magn Reson Imaging. 2007;25(6):748–53. https://linkinghub.elsevier.com/retrieve/pii/S0730725X07001701.

34. Rooney WD, Johnson G, Li X, Cohen ER, Kim S-G, Ugurbil K, et al. Magnetic field and tissue dependencies of human brain longitudinal $1H_2O$ relaxation in vivo. Magn Reson Med. 2007;57(2):308–18. http://doi.wiley.com/10.1002/mrm.21122.
35. Park C-A, Kang C-K, Kim Y-B, Cho Z-H. Advances in MR angiography with 7 T MRI: From microvascular imaging to functional angiography. Neuroimage. 2018;168(January 2017):269–78. https://doi.org/10.1016/j.neuroimage.2017.01.019.
36. van der Zwaag W, Francis S, Head K, Peters A, Gowland P, Morris P, et al. fMRI at 1.5, 3 and 7 T: characterising BOLD signal changes. Neuroimage. 2009;47(4):1425–34. http://dx.doi.org/10.1016/j.neuroimage.2009.05.015.
37. Yacoub E, Shmuel A, Pfeuffer J, Van De Moortele P-F, Adriany G, Andersen P, et al. Imaging brain function in humans at 7 Tesla. Magn Reson Med. 2001;45(4):588–94. http://doi.wiley.com/10.1002/mrm.1080.
38. Zwanenburg JJM, Hendrikse J, Visser F, Takahara T, Luijten PR. Fluid attenuated inversion recovery (FLAIR) MRI at 7.0 Tesla: comparison with 1.5 and 3.0 Tesla. Eur Radiol. 2010;20(4):915–22.
39. van Kalleveen IML, Koning W, Boer VO, Luijten PR, Zwanenburg JJM, Klomp DWJ. Adiabatic turbo spin echo in human applications at 7 T. Magn Reson Med. 2012;68(2):580–7. http://doi.wiley.com/10.1002/mrm.23264.
40. Ibrahim TS, Abduljalil AM, Baertlein BA, Jin JM, Chen J, Prock T, et al. B1 field homogeneity and SAR calculations for the birdcage coil. Phys Med Biol. 2001;46:609–19.
41. Ibrahim TS, Abduljalil AM, Baertlein BA, Lee R, Robitaille P-M-L. Analysis of B 1 field profiles and SAR values for multi-strut transverse electromagnetic RF coils in high field MRI applications. Phys Med Biol. 2001;46(10):2545–55. http://stacks.iop.org/0031-9155/46/i=10/a=303?key=crossref.47bbda7bd3ed5b0905ee30465694aa42.
42. Santini T, Zhao Y, Wood S, Krishnamurthy N, Kim J, Farhat N, et al. In-vivo and numerical analysis of the eigenmodes produced by a multi-level Tic-Tac-Toe head transmit array for 7 Tesla MRI. Lundberg P, editor. PLoS One. 2018;13(11):e0206127. http://dx.plos.org/10.1371/journal.pone.0206127.
43. Krishnamurthy N, Santini T, Wood S, Kim J, Zhao T, Aizenstein HJ, et al. Computational and experimental evaluation of the Tic-Tac-Toe RF coil for 7 Tesla MRI. Orzada S, editor. PLoS One. 2019;14(1):e0209663. http://dx.plos.org/10.1371/journal.pone.0209663.
44. Winkler SA, Schmitt F, Landes H, de Bever J, Wade T, Alejski A, et al. Gradient and shim technologies for ultra high field MRI. Neuroimage. 2018;168:59–70. https://linkinghub.elsevier.com/retrieve/pii/S1053811916306498.
45. Stockmann JP, Wald LL. In vivo B 0 field shimming methods for MRI at 7 T. Neuroimage. 2018;168:71–87. https://linkinghub.elsevier.com/retrieve/pii/S1053811917304822.
46. Wardlaw JM, Smith EE, Biessels GJ, Cordonnier C, Fazekas F, Frayne R, et al. Neuroimaging standards for research into small vessel disease and its contribution to ageing and neurodegeneration. Lancet Neurol. 2013;12:822–38.
47. Vermeer SE, Longstreth WT, Koudstaal PJ. Silent brain infarcts: a systematic review. Lancet Neurol. 2007;6(7):611–9. https://linkinghub.elsevier.com/retrieve/pii/S1474442207701709.
48. Snowdon DA. Brain infarction and the clinical expression of alzheimer disease-reply. J Am Med Assoc. 1997;278(2):114. http://jama.jamanetwork.com/article.aspx?doi=10.1001/jama.1997.03550020046024.
49. Vermeer SE, Prins ND, den Heijer T, Hofman A, Koudstaal PJ, Breteler MMB. Silent brain infarcts and the risk of dementia and cognitive decline. N Engl J Med. 2003;348(13):1215–22. http://www.nejm.org/doi/abs/10.1056/NEJMoa022066.
50. Schrag M, McAuley G, Pomakian J, Jiffry A, Tung S, Mueller C, et al. Correlation of hypointensities in susceptibility-weighted images to tissue histology in dementia patients with cerebral amyloid angiopathy: a postmortem MRI study. Acta Neuropathol. 2010;119(3):291–302. http://link.springer.com/10.1007/s00401-009-0615-z.

51. Patankar TF, Baldwin R, Mitra D, Jeffries S, Sutcliffe C, Burns A, et al. Virchow–Robin space dilatation may predict resistance to antidepressant monotherapy in elderly patients with depression. J Affect Disord. 2007;97(1–3):265–70. https://linkinghub.elsevier.com/retrieve/pii/S0165032706002916.
52. Potter GM, Marlborough FJ, Wardlaw JM. Wide variation in definition, detection, and description of lacunar lesions on imaging. Stroke. 2011;42(2):359–66. https://www.ahajournals.org/doi/10.1161/STROKEAHA.110.594754.
53. Benjamin P, Trippier S, Lawrence AJ, Lambert C, Zeestraten E, Williams OA, et al. Lacunar infarcts, but not perivascular spaces, are predictors of cognitive decline in cerebral small-vessel disease. Stroke. 2018;49(3):586–93. https://www.ahajournals.org/doi/10.1161/STROKEAHA.117.017526.
54. Madai VI, von Samson-Himmelstjerna FC, Bauer M, Stengl KL, Mutke MA, Tovar-Martinez E, et al. Ultrahigh-field MRI in human ischemic Stroke—a 7 Tesla study. Herholz K, editor. PLoS One. 2012;7(5):e37631. http://dx.plos.org/10.1371/journal.pone.0037631.
55. Zong X, Park SH, Shen D, Lin W. Visualization of perivascular spaces in the human brain at 7 T: sequence optimization and morphology characterization. Neuroimage. 2016;125:895–902. https://linkinghub.elsevier.com/retrieve/pii/S1053811915010022.
56. Bouvy WH, Zwanenburg JJM, Reinink R, Wisse LEM, Luijten PR, Kappelle LJ, et al. Perivascular spaces on 7 Tesla brain MRI are related to markers of small vessel disease but not to age or cardiovascular risk factors. J Cereb Blood Flow Metab. 2016;36(10):1708–17. http://journals.sagepub.com/doi/10.1177/0271678X16648970.
57. De Cocker LJ, Lindenholz A, Zwanenburg JJ, van der Kolk AG, Zwartbol M, Luijten PR, et al. Clinical vascular imaging in the brain at 7 T. Neuroimage. 2018;168(November 2016):452–8. https://doi.org/10.1016/j.neuroimage.2016.11.044.
58. Shoamanesh A, Preis SR, Beiser AS, Vasan RS, Benjamin EJ, Kase CS, et al. Inflammatory biomarkers, cerebral microbleeds, and small vessel disease: Framingham heart study. Neurology. 2015;84(8):825–32. http://www.neurology.org/cgi/doi/10.1212/WNL.0000000000001279.
59. Theysohn JM, Kraff O, Maderwald S, Barth M, Ladd SC, Forsting M, et al. 7 tesla MRI of microbleeds and white matter lesions as seen in vascular dementia. J Magn Reson Imaging. 2011;33(4):782–91. http://www.ncbi.nlm.nih.gov/pubmed/21448941
60. Stehling C, Wersching H, Kloska SP, Kirchhof P, Ring J, Nassenstein I, et al. Detection of asymptomatic cerebral microbleeds. Acad Radiol. 2008;15(7):895–900. https://linkinghub.elsevier.com/retrieve/pii/S1076633208000469.
61. Nandigam RNK, Viswanathan A, Delgado P, Skehan ME, Smith EE, Rosand J, et al. MR imaging detection of cerebral microbleeds: effect of susceptibility-weighted imaging, section thickness, and field strength. Am J Neuroradiol. 2009;30(2):338–43. http://www.ajnr.org/lookup/doi/10.3174/ajnr.A1355.
62. de Bresser J, Brundel M, Conijn MM, van Dillen JJ, Geerlings MI, Viergever MA, et al. Visual cerebral microbleed detection on 7 T MR imaging: reliability and effects of image processing. Am J Neuroradiol. 2013;34(6):E61–4. http://www.ajnr.org/lookup/doi/10.3174/ajnr.A2960.
63. White L, Petrovitch H, Hardman J, Nelson J, Davis DG, Ross GW, et al. Cerebrovascular pathology and dementia in autopsied honolulu-asia aging study participants. Ann N Y Acad Sci. 2002;977(1):9–23. http://doi.wiley.com/10.1111/j.1749-6632.2002.tb04794.x.
64. Brundel M, de Bresser J, van Dillen JJ, Kappelle LJ, Biessels GJ. Cerebral microinfarcts: a systematic review of neuropathological studies. J Cereb Blood Flow Metab. 2012;32(3):425–36. http://journals.sagepub.com/doi/10.1038/jcbfm.2011.200.
65. Wardlaw JM, Smith EE, Biessels GJ, Cordonnier C, Fazekas F, Frayne R, et al. Neuroimaging standards for research into small vessel disease and its contribution to ageing and neurodegeneration. Lancet Neurol. 2013;12(8):822–38. https://linkinghub.elsevier.com/retrieve/pii/S1474442213701248.

66. Smith EE, Schneider JA, Wardlaw JM, Greenberg SM. Cerebral microinfarcts: the invisible lesions. Lancet Neurol. 2012;11(3):272–82. https://linkinghub.elsevier.com/retrieve/pii/S1474442211703076.
67. Summers PM, Hartmann DA, Hui ES, Nie X, Deardorff RL, McKinnon ET, et al. Functional deficits induced by cortical microinfarcts. J Cereb Blood Flow Metab. 2017;37(11):3599–614. http://journals.sagepub.com/doi/10.1177/0271678X16685573.
68. Coban H, Tung S, Yoo B, Vinters H V., Hinman JD. Molecular disorganization of axons adjacent to human cortical microinfarcts. Front Neurol. 2017;8. http://journal.frontiersin.org/article/10.3389/fneur.2017.00405/full.
69. van Veluw SJ, Zwanenburg JJM, Engelen-Lee J, Spliet WGM, Hendrikse J, Luijten PR, et al. In vivo detection of cerebral cortical microinfarcts with high-resolution 7 T MRI. J Cereb Blood Flow Metab. 2013;33(3):322–9. http://journals.sagepub.com/doi/10.1038/jcbfm.2012.196.
70. van Veluw SJ, Zwanenburg JJ, Rozemuller AJ, Luijten PR, Spliet WG, Biessels GJ. The spectrum of MR detectable cortical microinfarcts: a classification study with 7-Tesla postmortem MRI and histopathology. J Cereb Blood Flow Metab. 2015;35(4):676–83. http://journals.sagepub.com/doi/10.1038/jcbfm.2014.258.
71. van Veluw SJ, Shih AY, Smith EE, Chen C, Schneider JA, Wardlaw JM, et al. Detection, risk factors, and functional consequences of cerebral microinfarcts. Lancet Neurol. 2017;16(9):730–40. http://dx.doi.org/10.1016/S1474-4422(17)30196-5.
72. Shi Y, Wardlaw JM. Update on cerebral small vessel disease: a dynamic whole-brain disease. BMJ. 2016;1(3):83–92. http://svn.bmj.com/cgi/doi/10.1136/svn-2016-000035.
73. Wardlaw JM, Smith C, Dichgans M. Mechanisms of sporadic cerebral small vessel disease: insights from neuroimaging. Lancet Neurol. 2013;12(5):483–97. https://linkinghub.elsevier.com/retrieve/pii/S1474442213700607.
74. Heverhagen JT, Bourekas E, Sammet S, Knopp M V., Schmalbrock P. Time-of-flight magnetic resonance angiography at 7 Tesla. Invest Radiol. 2008;43(8):568–73. https://insights.ovid.com/crossref?an=00004424-200808000-00004.
75. Pantoni L. Cerebral small vessel disease: from pathogenesis and clinical characteristics to therapeutic challenges. Lancet Neurol. 2010;9(7):689–701. https://linkinghub.elsevier.com/retrieve/pii/S1474442210701046.
76. Cho Z-H, Kang C-K, Han J-Y, Kim S-H, Kim K-N, Hong S-M, et al. Observation of the lenticulostriate arteries in the human brain in vivo using 7.0T MR Angiography. Stroke. 2008;39(5):1604–6. https://www.ahajournals.org/doi/10.1161/STROKEAHA.107.508002.
77. Kang C-K, Park C-A, Kim K-N, Hong S-M, Park C-W, Kim Y-B, et al. Non-invasive visualization of basilar artery perforators with 7 T MR angiography. J Magn Reson Imaging. 2010;32(3):544–50. http://doi.wiley.com/10.1002/jmri.22250.
78. Harteveld AA, De Cocker LJL, Dieleman N, van der Kolk AG, Zwanenburg JJM, Robe PA, et al. High-resolution postcontrast time-of-flight MR angiography of intracranial perforators at 7.0 Tesla. Paul F, editor. PLoS One. 2015;10(3):e0121051. https://dx.plos.org/10.1371/journal.pone.0121051.
79. Cho Z-H, Kang C-K, Park C-A, Hong S-M, Kim S-H, Oh S-T, et al. Microvascular functional MR angiography with ultra-high-field 7 t MRI: comparison with BOLD fMRI. Int J Imaging Syst Technol. 2012;22(1):18–22. http://doi.wiley.com/10.1002/ima.22008.
80. Moody DM, Brown WR, Challa VR, Ghazi-Birry HS, Reboussin DM. Cerebral microvascular alterations in aging, leukoaraiosis, and Alzheimer's disease. Ann N Y Acad Sci. 1997;826(1 Cerebrovascul):103–16. http://doi.wiley.com/10.1111/j.1749-6632.1997.tb48464.x.
81. Shaaban CE, Aizenstein HJ, Jorgensen DR, MacCloud RL, Meckes NA, Erickson KI, et al. In vivo imaging of venous side cerebral small-vessel disease in older adults: an MRI method at 7 T. Am J Neuroradiol. 2017;38(10):1923–8. http://www.ajnr.org/lookup/doi/10.3174/ajnr.A5327.
82. Mittal S, Wu Z, Neelavalli J, Haacke EM. Susceptibility-weighted imaging: technical aspects and clinical applications, Part 2. Am J Neuroradiol. 2009;30(2):232–52. http://www.ajnr.org/lookup/doi/10.3174/ajnr.A1461.

83. Haacke EM, Mittal S, Wu Z, Neelavalli J, Cheng Y-CN. Susceptibility-weighted imaging: technical aspects and clinical applications, Part 1. Am J Neuroradiol. 2009;30(1):19–30. http://www.ajnr.org/lookup/doi/10.3174/ajnr.A1400.
84. Sinnecker T, Bozin I, Dörr J, Pfueller CF, Harms L, Niendorf T, et al. Periventricular venous density in multiple sclerosis is inversely associated with T2 lesion count: a 7 Tesla MRI study. Mult Scler J. 2013;19(3):316–25. http://journals.sagepub.com/doi/10.1177/1352458512451941.
85. Novelli EM, Elizabeth Sarles C, Jay Aizenstein H, Ibrahim TS, Butters MA, Connelly Ritter A, et al. Brain venular pattern by 7 T MRI correlates with memory and haemoglobin in sickle cell anaemia. Psychiatry Res Neuroimaging. 2015;233(1):18–22. https://linkinghub.elsevier.com/retrieve/pii/S0925492715000931.
86. De Guio F, Vignaud A, Ropele S, Duering M, Duchesnay E, Chabriat H, et al. Loss of venous integrity in cerebral small vessel disease. Stroke. 2014;45(7):2124–6. https://www.ahajournals.org/doi/10.1161/STROKEAHA.114.005726.
87. Kuijf HJ, Bouvy WH, Zwanenburg JJM, Razoux Schultz TB, Viergever MA, Vincken KL, et al. Quantification of deep medullary veins at 7 T brain MRI. Eur Radiol. 2016;26(10):3412–8. http://link.springer.com/10.1007/s00330-016-4220-y.
88. Hartmann DA, Hyacinth HI, Liao F-F, Shih AY. Does pathology of small venules contribute to cerebral microinfarcts and dementia? J Neurochem. 2018;144(5):517–26. http://doi.wiley.com/10.1111/jnc.14228.
89. Fazekas F, Kleinert R, Offenbacher H, Schmidt R, Kleinert G, Payer F, et al. Pathologic correlates of incidental MRI white matter signal hyperintensities. Neurology. 1993;43(9):1683–9. http://www.ncbi.nlm.nih.gov/pubmed/8414012.
90. Moody DM, Brown WR, Challa VR, Anderson RL. Periventricular venous collagenosis: association with leukoaraiosis. Radiology. 1995;194(2):469–76. http://pubs.rsna.org/doi/10.1148/radiology.194.2.7824728.
91. Webb AJS, Simoni M, Mazzucco S, Kuker W, Schulz U, Rothwell PM. Increased cerebral arterial pulsatility in patients with leukoaraiosis. Stroke. 2012;43(10):2631–6. https://www.ahajournals.org/doi/10.1161/STROKEAHA.112.655837.
92. Shi Y, Thrippleton MJ, Marshall I, Wardlaw JM. Intracranial pulsatility in patients with cerebral small vessel disease: a systematic review. Clin Sci. 2018;132(1):157–71. http://www.clinsci.org/cgi/doi/10.1042/CS20171280.
93. Blair GW, Hernandez MV, Thrippleton MJ, Doubal FN, Wardlaw JM. Advanced neuroimaging of cerebral small vessel disease. Curr Treat Options Cardiovasc Med. 2017;19(7):56. http://link.springer.com/10.1007/s11936-017-0555-1.
94. Furuta A, Ishii N, Nishihara Y, Horie A. Medullary arteries in aging and dementia. Stroke. 1991;22(4):442–6. https://www.ahajournals.org/doi/10.1161/01.STR.22.4.442.
95. Miao Q, Paloneva T, Tuominen S, Pöyhönen M, Tuisku S, Viitanen M, et al. Fibrosis and stenosis of the long penetrating cerebral arteries: the cause of the white matter pathology in cerebral autosomal dominant arteriopathy with subcortical infarcts and leukoencephalopathy. Brain Pathol. 2006;14(4):358–64. http://doi.wiley.com/10.1111/j.1750-3639.2004.tb00078.x.
96. Geurts L, Biessels GJ, Luijten P, Zwanenburg J. Better and faster velocity pulsatility assessment in cerebral white matter perforating arteries with 7 T quantitative flow MRI through improved slice profile, acquisition scheme, and postprocessing. Magn Reson Med. 2018;79(3):1473–82. http://doi.wiley.com/10.1002/mrm.26821.
97. Geurts LJ, Zwanenburg JJM, Klijn CJM, Luijten PR, Biessels GJ. Higher pulsatility in cerebral perforating arteries in patients with small vessel disease related stroke, a 7 T MRI study. Stroke. 2019;50(1):62–8. https://www.ahajournals.org/doi/10.1161/STROKEAHA.118.022516.
98. Zwanenburg JJM, van Osch MJP. Targeting cerebral small vessel disease with MRI. Stroke. 2017;48(11):3175–82. https://www.ahajournals.org/doi/10.1161/STROKEAHA.117.016996.
99. Hall CN, Reynell C, Gesslein B, Hamilton NB, Mishra A, Sutherland BA, et al. Capillary pericytes regulate cerebral blood flow in health and disease. Nature. 2014;508(7494):55–60. http://www.nature.com/articles/nature13165.

100. Toth P, Tarantini S, Csiszar A, Ungvari Z. Functional vascular contributions to cognitive impairment and dementia: mechanisms and consequences of cerebral autoregulatory dysfunction, endothelial impairment, and neurovascular uncoupling in aging. Am J Physiol Circ Physiol. 2017;312(1):H1–20. http://www.physiology.org/doi/10.1152/ajpheart.00581.2016.
101. Ogawa S, Lee T-M, Nayak AS, Glynn P. Oxygenation-sensitive contrast in magnetic resonance image of rodent brain at high magnetic fields. Magn Reson Med. 1990;14(1):68–78. http://doi.wiley.com/10.1002/mrm.1910140108.
102. Blair GW, Doubal FN, Thrippleton MJ, Marshall I, Wardlaw JM. Magnetic resonance imaging for assessment of cerebrovascular reactivity in cerebral small vessel disease: a systematic review. J Cereb Blood Flow Metab. 2016;36(5):833–41. http://journals.sagepub.com/doi/10.1177/0271678X16631756.
103. Liu P, Li Y, Pinho M, Park DC, Welch BG, Lu H. Cerebrovascular reactivity mapping without gas challenges. Neuroimage. 2017;146:320–6. https://linkinghub.elsevier.com/retrieve/pii/S1053811916306723.
104. Huber L, Ivanov D, Krieger SN, Streicher MN, Mildner T, Poser BA, et al. Slab-selective, BOLD-corrected VASO at 7 tesla provides measures of cerebral blood volume reactivity with high signal-to-noise ratio. Magn Reson Med. 2014;72(1):137–48. http://doi.wiley.com/10.1002/mrm.24916.
105. Heye AK, Culling RD, Valdés Hernández MDC, Thrippleton MJ, Wardlaw JM. Assessment of blood–brain barrier disruption using dynamic contrast-enhanced MRI: a systematic review. NeuroImage Clin. 2014;6:262–74. https://linkinghub.elsevier.com/retrieve/pii/S2213158214001387.
106. Wardlaw JM, Makin SJ, Valdés Hernández MC, Armitage PA, Heye AK, Chappell FM, et al. Blood-brain barrier failure as a core mechanism in cerebral small vessel disease and dementia: evidence from a cohort study. Alzheimer's Dement. 2017;13(6):634–43. https://linkinghub.elsevier.com/retrieve/pii/S1552526016300401.
107. van Nieuwenhuizen KM, Hendrikse J, Klijn CJM. New microbleed after blood–brain barrier leakage in intracerebral haemorrhage. BMJ Case Rep. 2017;bcr-2016-218794. http://casereports.bmj.com/lookup/doi/10.1136/bcr-2016-218794.
108. Hiscox L V., Johnson CL, Barnhill E, McGarry MDJ, Huston J, van Beek EJR, et al. Magnetic resonance elastography (MRE) of the human brain: technique, findings and clinical applications. Phys Med Biol. 2016;61(24):R401–37. http://stacks.iop.org/0031-9155/61/i=24/a=R401?key=crossref.ecc79a2b0aeac0f13673535f9385f27a.
109. Sack I, Beierbach B, Wuerfel J, Klatt D, Hamhaber U, Papazoglou S, et al. The impact of aging and gender on brain viscoelasticity. Neuroimage. 2009;46(3):652–7. https://linkinghub.elsevier.com/retrieve/pii/S1053811909002237.
110. Sarvazyan AP, Skovoroda AR, Emelianov SY, Fowlkes JB, Pipe JG, Adler RS, et al. Biophysical bases of elasticity imaging. In 1995. p. 223–40. http://link.springer.com/10.1007/978-1-4615-1943-0_23.
111. Desmidt T, Brizard B, Dujardin P-A, Ternifi R, Réméniéras J-P, Patat F, et al. Brain tissue pulsatility is increased in midlife depression: a comparative study using ultrasound tissue pulsatility imaging. Neuropsychopharmacology. 2017;42(13):2575–82. http://www.nature.com/articles/npp2017113.
112. Johnson CL, Telzer EH. Magnetic resonance elastography for examining developmental changes in the mechanical properties of the brain. Dev Cogn Neurosci. 2018;33:176–81. https://linkinghub.elsevier.com/retrieve/pii/S1878929317300373.
113. Fehlner A, Hirsch S, Weygandt M, Christophel T, Barnhill E, Kadobianskyi M, et al. Increasing the spatial resolution and sensitivity of magnetic resonance elastography by correcting for subject motion and susceptibility-induced image distortions. J Magn Reson Imaging. 2017;46(1):134–41. http://doi.wiley.com/10.1002/jmri.25516.
114. Toga AW, Clark KA, Thompson PM, Shattuck DW, Van Horn JD. Mapping the human connectome. Neurosurgery. 2012;71(1):1–5. https://academic.oup.com/neurosurgery/article-lookup/doi/10.1227/NEU.0b013e318258e9ff.

115. Schmidt R, Ropele S, Ferro J, Madureira S, Verdelho A, Petrovic K, et al. Diffusion-weighted imaging and cognition in the leukoariosis and disability in the elderly study. Stroke. 2010;41(5). https://www.ahajournals.org/doi/10.1161/STROKEAHA.109.576629.
116. Tuladhar AM, van Dijk E, Zwiers MP, van Norden AGW, de Laat KF, Shumskaya E, et al. Structural network connectivity and cognition in cerebral small vessel disease. Hum Brain Mapp. 2016;37(1):300–10. http://doi.wiley.com/10.1002/hbm.23032.
117. Tuladhar AM, Lawrence A, Norris DG, Barrick TR, Markus HS, de Leeuw F-E. Disruption of rich club organisation in cerebral small vessel disease. Hum Brain Mapp. 2017;38(4):1751–66. http://doi.wiley.com/10.1002/hbm.23479.
118. Lawrence AJ, Tozer DJ, Stamatakis EA, Markus HS. A comparison of functional and tractography based networks in cerebral small vessel disease. NeuroImage Clin. 2018;18(February):425–32. https://doi.org/10.1016/j.nicl.2018.02.013.
119. Biswal B, Zerrin Yetkin F, Haughton VM, Hyde JS. Functional connectivity in the motor cortex of resting human brain using echo-planar mri. Magn Reson Med. 1995;34(4):537–41. http://doi.wiley.com/10.1002/mrm.1910340409.
120. Cole. Advances and pitfalls in the analysis and interpretation of resting-state FMRI data. Front Syst Neurosci. 2010; http://journal.frontiersin.org/article/10.3389/fnsys.2010.00008/abstract.
121. Smith SM, Fox PT, Miller KL, Glahn DC, Fox PM, Mackay CE, et al. Correspondence of the brain's functional architecture during activation and rest. Proc Natl Acad Sci. 2009;106(31):13040–5. http://www.pnas.org/cgi/doi/10.1073/pnas.0905267106.
122. Fornito A, Zalesky A, Bullmore ET. Fundamentals of brain network analysis. Elsevier; 2016. https://linkinghub.elsevier.com/retrieve/pii/C2012006036X.
123. Craddock RC, Holtzheimer PE, Hu XP, Mayberg HS. Disease state prediction from resting state functional connectivity. Magn Reson Med. 2009;62(6):1619–28. http://doi.wiley.com/10.1002/mrm.22159.
124. Venkataraman A, Kubicki M, Westin C-F, Golland P. Robust feature selection in resting-state fMRI connectivity based on population studies. In: 2010 IEEE computer society conference on computer vision and pattern recognition—workshops. IEEE; 2010. p. 63–70. http://ieeexplore.ieee.org/document/5543446/.
125. Yu Y, Shen H, Zhang H, Zeng L-L, Xue Z, Hu D. Functional connectivity-based signatures of schizophrenia revealed by multiclass pattern analysis of resting-state fMRI from schizophrenic patients and their healthy siblings. Biomed Eng Online. 2013;12(1):10. http://biomedical-engineering-online.biomedcentral.com/articles/10.1186/1475-925X-12-10.
126. Eloyan A, Muschelli J, Nebel MB, Liu H, Han F, Zhao T, et al. Automated diagnoses of attention deficit hyperactive disorder using magnetic resonance imaging. Front Syst Neurosci. 2012;6. http://journal.frontiersin.org/article/10.3389/fnsys.2012.00061/abstract.
127. Dai D, Wang J, Hua J, He H. Classification of ADHD children through multimodal magnetic resonance imaging. Front Syst Neurosci. 2012;6 http://journal.frontiersin.org/article/10.3389/fnsys.2012.00063/abstract.
128. Li W, Douglas Ward B, Liu X, Chen G, Jones JL, Antuono PG, et al. Disrupted small world topology and modular organisation of functional networks in late-life depression with and without amnestic mild cognitive impairment. J Neurol Neurosurg Psychiatry. 2015;86(10):1097–105. http://jnnp.bmj.com/lookup/doi/10.1136/jnnp-2014-309180.
129. Reijmer YD, Fotiadis P, Piantoni G, Boulouis G, Kelly KE, Gurol ME, et al. Small vessel disease and cognitive impairment: the relevance of central network connections. Hum Brain Mapp. 2016;37(7):2446–54. http://doi.wiley.com/10.1002/hbm.23186
130. Brinker T, Stopa E, Morrison J, Klinge P. A new look at cerebrospinal fluid circulation. Fluids Barriers CNS. 2014;11(1):10. http://fluidsbarrierscns.biomedcentral.com/articles/10.1186/2045-8118-11-10.
131. Moody DM, Bell MA, Challa VR. Features of the cerebral vascular pattern that predict vulnerability to perfusion or oxygenation deficiency: an anatomic study. Am J Neuroradiol. 1990;11:431–9. https://cds.ismrm.org/ismrm-2008/files/00438.pdf.

Advances in Artificial Intelligence Technologies

Artificial Intelligence in Psychiatry

Marc Fakhoury

Abstract

Scientific findings over the past few decades have shaped our understanding of the underlying neurobiology associated with psychiatric illnesses. However, despite significant advances in research, there is widespread disappointment with the overall pace of progress in detecting and treating psychiatric disorders. Current approaches for the diagnosis of psychiatric disorders largely rely on physician-patient questionnaires that are most of the time inaccurate and ineffective in providing a reliable assessment of symptoms. These limitations can, however, be overcome by applying artificial intelligence (AI) to electronic medical database and health records. AI in psychiatry is a general term that implies the use of computerized techniques and algorithms for the diagnosis, prevention, and treatment of mental illnesses. Although the past few years have witnessed an increase in the use of AI in the medical practice, its role in psychiatry remains a complex and unanswered question. This chapter provides the current state of knowledge of AI's use in the diagnosis, prediction, and treatment of psychiatric disorders, and examines the challenges and limitations of this approach in the medical practise.

Keywords

Artificial intelligence · Diagnosis · Language processing · Machine learning · Psychiatric disorders

M. Fakhoury (✉)
Faculty of Medicine, Department of Neuroscience,
Université de Montréal, Montreal, QC H3C 3J7, Canada
e-mail: marc.fakhoury@umontreal.ca

© Springer Nature Singapore Pte Ltd. 2019

Y.-K. Kim (ed.), *Frontiers in Psychiatry*, Advances in Experimental Medicine and Biology 1192, https://doi.org/10.1007/978-981-32-9721-0_6

Introduction

Psychiatric disorders are one of the leading causes of disability worldwide, affecting individuals from an early age [1]. They represent a major burden to the affected individuals in terms of years of life lost to disability or death, but also to the society in terms of the ensuing health care costs [2]. Some of the most frequently observed mental health disorders include mood, anxiety, and substance use disorder [3]. It is estimated that roughly 30% of adults will experience one of these mental health disorders across their lifetime [3]. There is therefore an urgent need to detect signs and symptoms of psychiatric disorders at an early stage as this could provide an opportunity to more effectively manage the symptoms.

The development of the latest edition of the Diagnostic and Statistical Manual of Mental Disorders (DSM-5), which was made available in 2013, has provided clinicians with a new approach to diagnose psychiatric disorders based on dimensions rather than categories [4]. However, despite introducing new and improved sets of diagnostic criteria, the DSM-5 suffers from relatively low reliability in clinical settings because of its inability to consistently identifying false positives and distinguishing risk from disorder [5]. The field of psychiatry also lacks reliable biomarkers that could effectively demarcate disease state from normal state, though advances in genetics and neuroimaging are slowly paving the way toward improved diagnosis [6].

For decades, even centuries, the field of psychiatry has largely relied on patient self-reports and clinical observations, often leading to subjective and inaccurate evaluations. In our new era of technological innovations, developments in artificial intelligence (AI) could provide a promising opportunity for improving, and even revolutionizing, interventions with psychiatric disorders [7, 8]. AI in psychiatry is a broad term that involves the use of advanced computerized techniques such as automated language processing and machine learning algorithms for assessing a patient's mental state beyond what could be measured with self-reports and clinical observations. Although still in its infancy, AI has already revolutionized mental health care and has profoundly influenced the way clinicians detect, predict, and treat psychiatric disorders [7, 9, 10].

This chapter provides an overview of recent evidence that support the use of AI in the diagnosis, prediction, and treatment of psychiatric disorders (which is summarized in Fig. 1), and highlights the existing challenges and limitations associated with this approach.

AI in the Diagnosis of Psychiatric Disorders

Although recent years have witnessed substantial progress in our understanding of the neurobiology underlying psychiatric disorders, the capacity of clinicians to diagnose these disorders has been hampered by the lack of objective and reliable clinical measures. One promising approach in the diagnosis of psychiatric disorders is the use of computerized techniques. For instance, latent semantic analysis

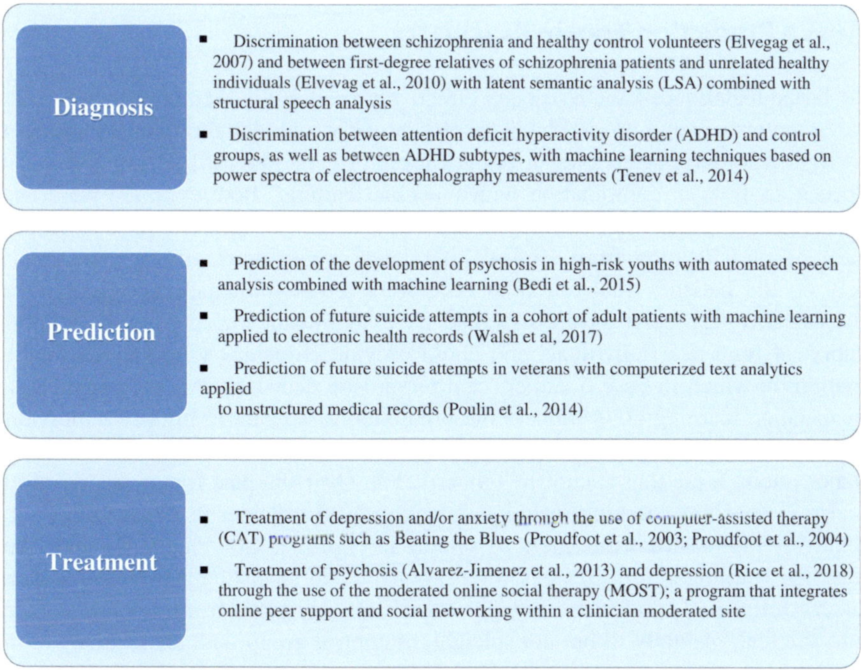

Fig. 1 Applications of AI in psychiatry. This figure shows a non-exhaustive list of evidence in support of the use of AI in the diagnosis, prediction, and treatment of psychiatric symptoms

(LSA), an automated high-dimensional tool for the analysis of speech transcripts, has been successfully employed to aid clinicians in the diagnosis of psychiatric disorders [9, 11, 12]. LSA is a computational technique in natural language processing for concept-based text analysis [13]. It acquires a representation of semantic knowledge based on the analysis of a wide array of words of natural discourse, and by unraveling the relationship between words and passage meanings [11]. The fundamental principle for developing a model of meaning is that words that are used in similar contexts tend to be more semantically related to each other compared to words that are used in different contexts [11, 14]. Using this approach, Elvevag et al. [11] analyzed the speech transcripts of patients with schizophrenia and healthy control volunteers, and by doing so, were successfully able to determine whether a particular discourse belonged to a patient or control group. LSA combined with structural speech analysis has also been successfully used to detect subtle deviations between first-degree relatives of schizophrenia patients and unrelated healthy individuals [12] More recently, machine learning techniques have been used to improve the discrimination between healthy individuals and adults with attention deficit hyperactivity disorder (ADHD) [15]. The use of AI could therefore have a huge impact on the health care system as it can be used to complement human clinical ratings in neuropsychological disorders, thereby reducing the number of false-negative and false-positive diagnoses.

AI as a Prediction Tool in Psychiatry

AI-based techniques have also been effectively used in the prediction of psychiatric symptoms including psychosis, which broadly includes the manifestations of thought disorders, behavioral disorganization, or catatonia [16]. Using automated speech analysis in combination with machine learning, Bedi et al. [9] were able to accurately predict the development of psychosis in high-risk youths, outperforming classification from clinical interviews, where much of the assessments rely on the patient's motivation to accurately report his experience. Enhancing the capacity to predict psychosis could have significant impacts for the identification of high-risk individuals and could provide clinicians with valuable information on which to base treatment and prognostic decisions. Another application of machine learning algorithms is the prediction of suicide in high-risk individuals. Accounting to roughly 800 thousand deaths worldwide every year, suicide is a major public issue that cannot be ignored [17]. Over the past few years, developments in machine learning techniques have proved efficient in determining with relatively high success the intent of suicide in high-risk individuals. For instance, machine learning algorithms based on linguistic and acoustic characteristics were successfully used to classify a cohort of subjects recruited from medical centers into suicidal, mentally ill but not suicidal, or control group with an accuracy of up to 85% [18]. More recently, Walsh et al. [19] were able to accurately predict future suicide attempts in a cohort of adult patients with a history of self-injury by applying machine learning to electronic health records. Results have been more than 80% accurate in predicting whether someone will make a suicide attempt within the next two years, and 92% accurate in predicting whether someone will make a suicide attempt within the following 7 days [19]. Last but not least, computerized text analytics applied to unstructured medical records predicted the risk of suicide in veterans with more than 65% accuracy, thereby allowing clinicians to better screen seemingly healthy individuals and to evaluate their risk for attempting suicide in the future [10].

AI-Assisted Therapy in Psychiatry

The majority of early intervention programs to individuals experiencing symptoms of mental health disorders provide services that do not last a long time, often resulting in poor outcomes and loss of benefits after their termination [7]. Computer-assisted therapy (CAT) could offer exciting prospects in this regard by delivering some aspects of psychotherapy or behavioral treatment [20]. CAT typically consists of programs made up of videos and questionnaires that are delivered to the patient through a computerized platform to help him cope with his symptoms. For instance, Beating the Blues, a computerized-assisted therapy recommended by the National Institute for Health and Clinical Excellence, was proven effective in reducing symptoms of depression and/or anxiety in randomized controlled trials [21–23].

CAT could also be delivered via the Internet, thus allowing a higher degree of interactivity between the patient and the program [20]. This approach is referred to as e-therapy. Existing e-therapies range from programs that provide support to the user through chat rooms to those with a high level of clinician involvement, where communication with the user is accomplished through e-mail correspondence [20, 24]. Considering that Internet is intricately merged into our daily lives, e-therapies could be an effective way to provide support for individuals suffering from mental health disorders [7, 25]. For instance, the moderated online social therapy (MOST) project, an Internet-based intervention, has been specifically designed to aid individuals with a mental health disorder through an online social therapy system [7, 26, 27]. MOST uniquely integrates online peer support and social networking within a clinician moderated site [7, 28]. To date, the MOST model has been effectively implemented in studies that included young people recovering from psychosis [29] or depression [30], though more work is needed to address the long term suitability of this model.

Pros and Cons of AI-Based Interventions

AI-based interventions offer several advantages to the user. One of the advantages of using computerized platforms is that it dissipates the stigma associated with revealing symptoms of mental disorders to the clinician [31]. In addition, AI-based interventions are usually cost-effective [32] and offer significant advantages to individuals who have limited mobility due to their symptoms [20]. Research over the past few years has also shown that CAT is quite effective in enhancing supportive relationships, decreasing isolation, and increasing self-disclosure [33–35]. Notwithstanding their numerous advantages, interventions that are based on AI techniques carry risks and limitations. For instance, treatment plans that rely on computerized techniques do not implement full psychiatric evaluations and do not display the emphatic concern and emotional awareness of physicians. As a result, individuals who only rely on AI-based interventions are often discouraged to pursue treatment [20]. Another major concern is that the majority of the studies on AI-based interventions have been conducted by their developers who want to demonstrate the efficacy of their product with personal financial stake in the outcome [20]. Finally, the implementation of AI-based treatments may face several ethical concerns regarding patient safety and privacy and could lead to a rise in unemployment in the field of psychiatry.

Conclusion

In summary, findings collected over the past few years support the utility of AI-based interventions in the diagnosis, prediction, and treatment of psychiatric disorders. Computerized techniques such as automated speech language analysis could provide the groundwork for future developments of reliable clinical tests for psychiatry, and may revolutionize the way clinicians treat psychiatric disorders. Innovations such as CAT and MOST have already proved effective in ameliorating symptoms of depression, anxiety and/or psychosis through online peer support, and could present an opportunity to facilitate the delivery of tailored therapy content while facilitating privacy and autonomy. Future work should focus on evaluating the effectiveness of AI-based therapy in large controlled trials and in ascertaining its advantage compared to the traditional clinical delivered therapy. These investigations will help bring a new dimension to the psychotherapeutic process and will play a crucial role in enhancing the quality of life of individuals who suffer from mental health disorders.

Conflict of Interest The author declares that he has no conflict of interest.

References

1. Kessler RC, Amminger GP, Aguilar-Gaxiola S, Alonso J, Lee S, Ustun TB. Age of onset of mental disorders: a review of recent literature. Curr Opin Psychiatry. 2007;20(4):359–64.
2. Trautmann S, Rehm J, Wittchen HU. The economic costs of mental disorders: do our societies react appropriately to the burden of mental disorders? EMBO Rep. 2016;17(9):1245–9.
3. Steel Z, Marnane C, Iranpour C, Chey T, Jackson JW, Patel V, et al. The global prevalence of common mental disorders: a systematic review and meta-analysis 1980–2013. Int J Epidemiol. 2014;43(2):476–93.
4. Stein DJ, Lund C, Nesse RM. Classification systems in psychiatry: diagnosis and global mental health in the era of DSM-5 and ICD-11. Curr Opin Psychiatry. 2013;26(5):493–7.
5. Wakefield JC. Diagnostic issues and controversies in DSM-5: return of the false positives problem. Annu Rev Clin Psychol. 2016;12:105–32.
6. Krystal JH, State MW. Psychiatric disorders: diagnosis to therapy. Cell. 2014;157(1):201–14.
7. D'Alfonso S, Santesteban-Echarri O, Rice S, Wadley G, Lederman R, Miles C, et al. Artificial intelligence-assisted online social therapy for youth mental health. Front Psychol. 2017;8:796.
8. Bzdok D, Meyer-Lindenberg A. Machine learning for precision psychiatry: opportunities and challenges. Biol Psychiatry Cogn Neurosci Neuroimaging. 2018;3(3):223–30.
9. Bedi G, Carrillo F, Cecchi GA, Slezak DF, Sigman M, Mota NB, et al. Automated analysis of free speech predicts psychosis onset in high-risk youths. NPJ Schizophr. 2015;1:15030.
10. Poulin C, Shiner B, Thompson P, Vepstas L, Young-Xu Y, Goertzel B, et al. Predicting the risk of suicide by analyzing the text of clinical notes. PLoS ONE. 2014;9(1):e85733.
11. Elvevag B, Foltz PW, Weinberger DR, Goldberg TE. Quantifying incoherence in speech: an automated methodology and novel application to schizophrenia. Schizophr Res. 2007;93(1–3):304–16.
12. Elvevag B, Foltz PW, Rosenstein M, Lynn ED. An automated method to analyze language use in patients with schizophrenia and their first-degree relatives. J Neurolinguistics. 2010;23(3):270–84.
13. Landauer T, Dumais S. A solution to Plato's problem: the latent semantic analysis theory of the acquisition, induction, and representation of knowledge. Psychol Rev. 1997;104:211–40.
14. Landauer TK, Foltz PW, Laham D. An introduction to latent semantic analysis. Discourse Process. 1998;25(2–3):259–84.
15. Tenev A, Markovska-Simoska S, Kocarev L, Pop-Jordanov J, Muller A, Candrian G. Machine learning approach for classification of ADHD adults. Int J Psychophysiol. 2014;93(1):162–6.

16. Stevens JR, Prince JB, Prager LM, Stern TA. Psychotic disorders in children and adolescents: a primer on contemporary evaluation and management. Prim Care Companion CNS Disord. 2014;16(2).
17. Varnik P. Suicide in the world. Int J Environ Res Public Health. 2012;9(3):760–71.
18. Pestian JP, Sorter M, Connolly B, Bretonnel Cohen K, McCullumsmith C, Gee JT, et al. A machine Learning approach to identifying the thought markers of suicidal subjects: a prospective multicenter Trial. Suicide Life Threat Behav. 2017;47(1):112–21.
19. Walsh CG, Ribeiro JD, Franklin JC. Predicting risk of suicide attempts over time through machine learning. Clin Psychol Sci. 2017;5(3):457–69.
20. Carroll KM, Rounsaville BJ. Computer-assisted therapy in psychiatry: be brave-it's a new world. Curr Psychiatry Rep. 2010;12(5):426–32.
21. Szanton SL, Thorpe RJ Jr, Gitlin LN. Beat the blues decreases depression in financially strained older African-American adults. Am J Geriatr Psychiatry. 2014;22(7):692–7.
22. Proudfoot J, Goldberg D, Mann A, Everitt B, Marks I, Gray JA. Computerized, interactive, multimedia cognitive-behavioural program for anxiety and depression in general practice. Psychol Med. 2003;33(2):217–27.
23. Proudfoot J, Ryden C, Everitt B, Shapiro DA, Goldberg D, Mann A, et al. Clinical efficacy of computerised cognitive-behavioural therapy for anxiety and depression in primary care: randomised controlled trial. Br J Psychiatry. 2004;185:46–54.
24. Postel MG, de Haan HA, De Jong CA. E-therapy for mental health problems: a systematic review. Telemed J E Health. 2008;14(7):707–14.
25. Kumar V, Sattar Y, Bseiso A, Khan S, Rutkofsky IH. The effectiveness of internet-based cognitive behavioral therapy in treatment of psychiatric disorders. Cureus. 2017;9(8):e1626.
26. Gleeson JF, Alvarez-Jimenez M, Lederman R. Moderated online social therapy for recovery from early psychosis. Psychiatr Serv. 2012;63(7):719.
27. Gleeson J, Lederman R, Koval P, Wadley G, Bendall S, Cotton S, et al. Moderated online social therapy: a model for reducing stress in carers of young people diagnosed with mental health disorders. Front Psychol. 2017;8:485.
28. Lederman R, Wadley G, Gleeson JF, Bendall S, Alvarez-Jimenez M. Moderated online social therapy: designing and evaluating. ACM Trans Comput Interact. 2014;2:1–27.
29. Alvarez-Jimenez M, Bendall S, Lederman R, Wadley G, Chinnery G, Vargas S, et al. On the HORYZON: moderated online social therapy for long-term recovery in first episode psychosis. Schizophr Res. 2013;143(1):143–9.
30. Rice S, Gleeson J, Davey C, Hetrick S, Parker A, Lederman R, et al. Moderated online social therapy for depression relapse prevention in young people: pilot study of a 'next generation' online intervention. Early Interv Psychiatry. 2018;12(4):613–25.
31. Houston TK, Cooper LA, Ford DE. Internet support groups for depression: a 1-year prospective cohort study. Am J Psychiatry. 2002;159(12):2062–8.
32. McCrone P, Knapp M, Proudfoot J, Ryden C, Cavanagh K, Shapiro DA, et al. Cost-effectiveness of computerised cognitive-behavioural therapy for anxiety and depression in primary care: randomised controlled trial. Br J Psychiatry. 2004;185:55–62.
33. O'Keeffe GS, Clarke-Pearson K, Council on C, Media. The impact of social media on children, adolescents, and families. Pediatrics. 2011;127(4):800–4.
34. Dennis CL. Peer support within a health care context: a concept analysis. Int J Nurs Stud. 2003;40(3):321–32.
35. Weisband S, Kiesler S. Self-disclosure on computer forms: meta-analysis and implications. In: CHI 96 electronic proceedings. https://dl.acm.org/citation.cfm?id=238387

Machine Learning in Neural Networks

Eugene Lin and Shih-Jen Tsai

Abstract

Evidence now suggests that precision psychiatry is becoming a cornerstone of medical practices by providing the patient of psychiatric disorders with the right medication at the right dose at the right time. In light of recent advances in neuroimaging and multi-omics, more and more biomarkers associated with psychiatric diseases and treatment responses are being discovered in precision psychiatry applications by leveraging machine learning and neural network approaches. In this article, we focus on the most recent developments for research in precision psychiatry using machine learning, deep learning, and neural network algorithms, together with neuroimaging and multi-omics data. First, we describe different machine learning approaches that are employed to assess prediction for diagnosis, prognosis, and treatment in various precision psychiatry studies. We also survey probable biomarkers that have been identified to be involved in psychiatric diseases and treatment responses.

E. Lin
Department of Electrical & Computer Engineering, University of Washington, Seattle, WA 98195, USA

Department of Biostatistics, University of Washington, Seattle, WA 98195, USA

Graduate Institute of Biomedical Sciences, China Medical University, Taichung, Taiwan

S.-J. Tsai (✉)
Department of Psychiatry, Taipei Veterans General Hospital, No. 201, Shih Pai Road, Sec. 2, 11217 Taipei, Taiwan
e-mail: tsai610913@gmail.com

Division of Psychiatry, National Yang-Ming University, Taipei, Taiwan

Institute of Brain Science, National Yang-Ming University, Taipei, Taiwan

© Springer Nature Singapore Pte Ltd. 2019

Y.-K. Kim (ed.), *Frontiers in Psychiatry*, Advances in Experimental Medicine and Biology 1192, https://doi.org/10.1007/978-981-32-9721-0_7

Furthermore, we summarize the limitations with respect to the mentioned precision psychiatry studies. Finally, we address a discussion of future directions and challenges.

Keywords
Artificial intelligence · Biomarker · Genomics · Multi-omics · Neural networks · Neuroimaging · Precision medicine

Introduction

The interdisciplinary fields of precision psychiatry, machine learning, neural network algorithms, and neuroimaging had been making good progress in recent years [1–3]. The objective of a machine learning method is to enable a data-driven algorithm that can generally learn from data of the past or present and leverage the learned knowledge to make a predictive decision for an unknown future event or for any unknown data in the future [4–6]. In the general terms, the roadmap for a machine learning method is comprised of three steps where we build the model from initial inputs in the first step, evaluate and tune the model in the second step, and then utilize the model for making a predictive decision in the third step [4–6]. In the field of precision psychiatry, machine learning approaches integrate multiple data types such as neuroimaging and multi-omics data by using state-of-the-art statistical and data mining algorithms that can automatically learn to perceive complicated patterns based on empirical datasets [1–3]. To address the pressing challenges precision psychiatry faces today, there is a tremendous need for the development of machine learning software frameworks that can achieve clinical predictions of a given categorical or quantitative phenotype using next-generation neuroimaging and multi-omics data [1–3].

Precision psychiatry, an emerging field of medicine, is growing into a cornerstone of medical practices with prospects of the customization of healthcare for patients with psychiatric disorders, which means that medical practices, decisions, and treatments are tailored to individual patients [7]. More precisely, entire patient populations are subdivided into groups by biomarkers such as neuroimaging and multi-omics data; thereby, medications might be adapted personally to each individual patient with relevant or comparable genetic and imaging characteristics [7]. To date, there are more and more accumulating biomarkers that could affect clinical drug response and disease prognosis for treatment of patients with psychiatric disorders [8]. Furthermore, it has long been acclaimed that selected single nucleotide polymorphisms (SNPs) and gene expression profiles could be used as biomarkers to influence clinical treatment response and adverse drug reactions for antidepressants in patients with major depressive disorder (MDD) [9, 10].

Recent advances in machine learning, especially deep learning, have pointed out its potential and ability to learn and recognize nonlinear and complicated hierarchical patterns based on enormous large-scale empirical datasets [11–15]. Due to new approaches such as the deployment of general-purpose computing on graphics processing units, deep learning has achieved state-of-the-art performances on a wide variety of applications such as precision psychiatry [2, 12–15]. In general, the goal of deep learning is to construct a machine learning algorithm to facilitate a hierarchical representation of the data by using multiple layers of abstraction such as neural networks [12–16]. In other words, a deep learning algorithm for classification applications such as medical diagnosis in precision psychiatry is a procedure for choosing the best hypothesis using a neural network with multiple layers, instead of using a neural network with only single layer [12–16].

With the advent of technology in neuroimaging and multi-omics sciences, novel diagnostic tools as well as new drugs are exhibiting high growth potentials to address the needs of precision psychiatry for treatment and therapeutic interventions [17]. The use of biomarkers based on machine learning approaches has played a vital role in precision medicine in psychiatry [17]. Recently, there were a number of key emerging diagnostic studies for various diseases and treatments of significance for psychiatry with consideration of machine learning methods [2]. To that end, it would be greatly fascinating to create machine learning models that are able to forecast the probable outcome of disease status and drug treatment for patients with psychiatric disorders [2, 17]. In addressing this need, machine learning approaches might provide invaluable tools to accomplish the promise of precision psychiatry by tailoring treatment based on individual biomarkers [2, 17]. In this article, we present various precision psychiatry studies for assessment of disease status and drug treatment with consideration of machine learning, deep learning, and neural network approaches. In addition, we summarize the limitations in these studies and provide a discussion of future directions and challenges.

Method

Literature Search and Analyses

In this review, we present relevant studies on precision psychiatry and machine learning applications after a comprehensive search of the electronic PubMed database (2015–present). Key words in the search included "machine learning," "deep learning," "psychiatry," "neuroimaging," "precision psychiatry," and "neural network." Furthermore, we employed a manual search procedure of bibliographical cross-referencing. We manually screened the obtained articles and aimed at identifying original papers and reviews with a particular focus on precision psychiatry and machine learning applications. While this article is by no means a comprehensive review of all potential studies reported in the literature, we merely pinpointed various examples for machine learning methods in precision psychiatry.

Machine Learning and Neural Network Applications

Here we describe selective studies that focus on four main arenas including diagnosis prediction, prognosis prediction, treatment prediction, and the detection of potential biomarkers in the context of machine learning and neural network methods in psychiatry. Clinical or biological suggestions from these four main categories could be a decision support aide for future prognosis and optimal treatments in translational psychiatry [18].

While this summary does not provide the entire set of relevant studies reported in the literature, it nonetheless provides a synthesis of those that can markedly influence public and population health-oriented applications in psychiatry and machine learning in the near to midterm future.

Diagnosis Prediction

In recent years, there has been a growing trend in combining machine learning techniques and structural and functional neuroimaging to provide new insights into brain disorders such as Alzheimer's disease, autism spectrum disorder, and schizophrenia [19]. In particular, we focus on emerging big data methodologies such as deep learning in the following selective studies.

In order to distinguish mild to severe sporadic Alzheimer's disease from normal aging, Kloppel et al. pioneered a project to utilize a machine learning method, support vector machines (SVMs), using structural magnetic resonance imaging (MRI) and achieved 89% accuracy in their model [20]. In their two-step procedure, SVMs learned the differences between patients with Alzheimer's disease and healthy controls in the first step, and then the framework was tested on a new brain scan in the second step [20].

In a recent study, Ju et al. proposed a deep learning approach, which consists of auto-encoders and a softmax regression layer, to predict the early diagnosis of Alzheimer's disease [21]. In general, the auto-encoder is an encoding architecture, which consists of a neural network with the input layer representing the MRI data, multiple hidden layers representing nonlinear transformations from the previous layer, and the output layer representing the reconstructed MRI instances. Compared to widely used single-kernel SVM (accuracy = 84.40%) and multi-kernel SVM (accuracy = 86.42%), their proposed auto-encoder model had a better accuracy (87.76%). Their work highlights that deep learning approaches have an advantage over traditional machine learning methods to predict and prevent Alzheimer's disease at an early stage [21].

With a belief network-based algorithm, Ortiz et al. employed a deep learning architecture that integrates gray matter images from brain areas and automated anatomical labeling data for identifying Alzheimer's disease with MRI data [22]. Their model was composed of an ensemble of two deep belief networks with four different voting schemes [22]. The analysis results demonstrated that the proposed deep belief network method provided good performances for differentiating images between healthy controls and Alzheimer's disease individuals (accuracy = 90%) as well as differentiating images between mild cognitive impairment and Alzheimer's disease (accuracy = 84%) [22].

In machine learning, a deep belief network is a class of deep neural networks which consist of multiple layers of latent variables and can be viewed as a composition of auto-encoders [23]. Deep belief networks have also been applied to discriminate young children with autism spectrum disorders using functional MRI [23]. In addition, Pinaya et al. utilized a deep belief network model to characterize differences between patients with schizophrenia and healthy controls (accuracy = 73.6%) using MRI data, and the performance was better than the SVM model (accuracy = 68.1%) [24].

Prognosis Prediction

In order to identify course trajectories of MDD, Schmaal et al. proposed a machine learning framework that integrates structural and functional MRI using Gaussian process classifiers [25]. Gaussian process classifiers are one type of multivariate pattern recognition methods, which are similar to SVMs and provide the advantage of predictive probabilities of class membership [25]. Schmaal et al. employed Gaussian process classifiers to evaluate three MDD trajectories (including chronic, gradual improving, and fast remission patients) using prognostic value of MRI and clinical data (such as baseline severity, duration, and comorbidity) [25]. Their analysis showed that their machine learning framework can discriminate chronic patients from remitted patients up to 73% accuracy.

To predict health status and better inform clinical decision making, Miotto et al. proposed a deep learning framework which constructs a general-purpose patient representation from electronic health records (EHRs) to facilitate clinical predictive modeling [26]. Specifically, the deep learning framework uses a multilayer neural network, which is a three-layer stack of de-noising auto-encoders with sigmoid activation functions to derive hierarchical regularities and dependencies in the dataset of about 700,000 patients [26]. Then, random forest classifiers were implemented to evaluate the probability that patients might develop a certain disease given their current clinical status [26]. In the proposed deep learning framework, de-noising auto-encoders were trained to reconstruct the input from a noisy version of the original data to avoid overfitting [26]. The findings indicated that schizophrenia and attention deficit hyperactivity disorder could be forecasted with high accuracy (area under the receiver operating characteristic curve (AUC) = 0.85) [26]. Moreover, Miotto et al. compared the proposed deep learning framework with well-known conventional machine learning algorithms, including principal component analysis, *k*-means clustering, Gaussian mixture model, and independent component analysis [26]. The performance metrics of the proposed deep learning framework were superior to those obtained by the conventional machine learning algorithms [26]. The main strength of the proposed approach is that preprocessing EHR data with the deep learning framework can help provide more effective predictions because of nonlinear transformations [26].

Treatment Prediction

The use of machine learning in terms of predicting treatment response in psychiatric drugs is still in its infancy as scant human studies have investigated methods to build prediction models for estimating treatment response. We focus on antidepressant treatment response in this section.

In order to predict patient-specific possible antidepressant treatment outcomes in MDD, Lin et al. carried out a deep learning prediction algorithm and leveraged it to the integrated datasets from several data types (including genetic data such as SNPs, demographic data such as age, sex, and marital status, and clinical data such as baseline Hamilton Rating Scale for Depression score, depressive episodes, and suicide attempt status of MDD patients) [27]. First, they conducted a genome-wide association study (GWAS) to pinpoint potentially significant SNPs of antidepressant treatment response and remission in a hypothesis-free manner [27]. Their deep learning prediction algorithm is called the multilayer feedforward neural network (MFNN) approach, which adapts the back-propagation algorithm. MFNN was employed to calculate the predicted complex relationship between antidepressant treatment response and biomarkers [27]. An advantage of their approach is that these predictive MFNN methods possess the benefits of nonlinear models, fault tolerance, real-time processing, and integrated systems [27]. In their analysis results, Lin et al. identified the MFNN model with three hidden layers (AUC = 0.81; sensitivity = 0.77; specificity = 0.66) for remission and the MFNN model with two hidden layers (AUC = 0.82; sensitivity = 0.75; specificity = 0.69) for antidepressant treatment response [27].

There are several studies that utilized traditional machine learning methods to predict antidepressant treatment response. A study by Kautzky et al. reported that a machine learning prediction model pinpointed 25% of responders correctly for treatment outcome by using clinical and genetic information [28]. Their model was based on the random forests algorithm, which is an ensemble learning method and is constructed as a multitude of decision trees to perform classification tasks [28]. Particularly, Kautzky et al. identified several potential biomarkers including a clinical variable called melancholia as well as three SNPs such as brain-derived neurotrophic factor (*BDNF*) rs6265, 5-hydroxytryptamine receptor 2A (*HTR2A*) rs6313, and protein phosphatase 3 catalytic subunit gamma (*PPP3CC*) rs7430 [28].

Moreover, by leveraging information such as age, structural imaging, and mini-mental status examination scores, the subsequent study by Patel et al. showed that a machine learning model predicted treatment response with 89% accuracy by using an alternating decision tree model, which generalizes decision trees and is related to boosting for primarily reducing bias and variance [29].

Furthermore, another study by Chekroud et al. demonstrated that a machine learning model estimated clinical remission by using 25 variables with 59% accuracy [30]. First, the top 25 predictors were identified by the elastic net, and then a tree-based ensemble method, a gradient boosting machine, was used to combine several weakly predictive models (typically decision trees) to form a final ensemble model [30]. In particular, Chekroud et al. identified the top three potential biomarkers of non-remission such as baseline depression severity, feeling restless during the past 7 days, and reduced energy level during the past 7 days [30]. On the other hand, the top three potential biomarkers of remission were total years of education, currently being employed, and loss of insight into one's depressive condition [30]. One advantage of their approach is that their model and predictors were externally validated by other independent cohorts [30].

Iniesta et al. also implicated that a machine learning model based on clinical and demographical characteristics can forecast response with clinically meaningful accuracy by using regularized regression models (AUC = 0.72) [31]. They utilized elastic net, an application of regularized regression models, which are general linear models with penalties to provide variable selection from a large number of variables while avoiding overfitting [32].

Finally, a recent study by Maciukiewicz et al. suggested that a machine learning model based on SNPs can predict treatment response with 52% accuracy by using both SVM-based and decision trees-based models [33]. They first conducted a GWAS study to search for genetic susceptibility loci of antidepressant treatment response in a hypothesis-free manner [33]. Then, they performed least absolute shrinkage and selection operator (LASSO) regression to identify potentially significant predictors such as rs2036270 and rs7037011 SNPs [33]. In order to enhance the prediction accuracy, LASSO performs both variable selection and regularization [34].

Detection of Potential Biomarkers

In order to carry out a novel risk factor called Alzheimer's disease pattern similarity scores, Casanova et al. implemented a high-dimensional machine learning framework which is a regularized logistic regression method with an elastic net to simultaneously accomplish data integration and prediction for Alzheimer's disease [35]. A regularized logistic regression method with the elastic net, also called the adaptive elastic net, has been successfully applied in high-dimensional data, where the adaptive elastic net uses elastic net estimates as the initial weight in the model [32, 36]. By using MRI data, their approach is based on a coordinate-wise descent technique [37] and is able to discriminate patients with Alzheimer's disease from healthy controls [35]. Casanova et al. revealed that Alzheimer's disease pattern similarity scores had strong associations with age, cognitive function, and cognitive status and may be used as an Alzheimer's disease risk factor [35].

Limitations

The findings as discussed in the previous sections should be interpreted by taking into account some limitations of these studies in psychiatry and machine learning applications. One limitation of the aforementioned studies is that the small size of the sample does not allow for drawing of definite conclusions due to the possibility of overfitting during the training process of machine learning and deep learning algorithms [38]. Second, it is important to replicate their results by comparing comparable data from an independent cohort [2]. However, most of these studies did not provide replication studies because more and larger datasets may not be available to facilitate subsequent studies. Furthermore, these findings may not be generalizable. An open challenge is that the most useful findings will generalize across various testing conditions, broader populations, different ethnic groups, numerous sites, and diverse real-life clinical settings [39].

In addition, variability in data quality such as missing data may make prior selection of predictive variables or driver biomarkers essential; for example, preselecting SNPs from the GWAS data [40]. However, we speculate that this preselection can be expected to impact the overall structure and implementation in the final machine learning model. In future work, large prospective clinical trials are necessary in order to answer whether the relevant biomarkers are reproducibly associated with disease status and drug response in machine learning studies.

While in this article we selected a few studies to exemplify relevant machine learning and deep learning algorithms in the neurobiology of psychiatric disorders, it should be pointed out that psychiatry research could further benefit from combining the advanced machine learning algorithms with multi-omics techniques [17]. Several existing biobanks have been established for multi-omics studies, including the COMBINE Biobank [41], the Korea Healthy Twin Study [42], LifeGene [43], and the Taiwan Biobank [44–51].

Finally, it should be noted that in order to demonstrate the robustness of a single biomarker or even a set of biomarkers, future studies in precision psychiatry and machine learning should assess the overlap or lack of overlap between biomarkers in diverse machine learning analyses by using tools such as Venn diagrams [52]. Current evidence indicates that various machine learning analysis strategies may often yield different biomarkers, even when they were applied to the same type of disease [53].

Perspectives

As suggested by the aforementioned studies, precision psychiatry promises to offer new therapeutic and diagnostic techniques for accurate diagnosis, prognosis, and treatment in a disease-specific and patient-specific manner [7, 8, 54, 55]. In the context of multi-omics and neuroimaging-driven approaches, it is of major importance that potential clinical applications involving machine learning for prediction of drug responses or disease might provide the appropriate solutions to global primary care, and thereby it would have been up-taken by users and governments. Moreover, it is hypothesized that the next generation of psychiatric therapies for disease treatment thus ought to take into consideration the interplay among neuroimaging and multi-omics data. The latest advances in single-cell sequencing and data-intensive life sciences will certainly trigger more novel machine learning software tools for population health in the next decade

[56]. Furthermore, it will also increasingly generate application-oriented solutions toward the field of public health in light of the pressing needs of precision psychiatry for innovative diagnostics [57]. Over the next few years, machine learning-based precision psychiatry for the pretreatment prediction may become a reality in patient care after prospective large clinical trials to validate clinical factors and the relevant biomarkers [58].

Acknowledgements This work was supported by grant MOST 107-2634-F-075-002 from Taiwan Ministry of Science and Technology, and grant V105D17-002-MY2-2 from the Taipei Veterans General Hospital. We thank Emily Ting for English editing.

References

1. Lin E, Lane HY. Machine learning and systems genomics approaches for multi-omics data. Biomark Res. 2017;5:2.
2. Bzdok D, Meyer-Lindenberg A. Machine learning for precision psychiatry: opportunities and challenges. Biol Psychiatry Cogn Neurosci Neuroimaging. 2018;3(3):223–30.
3. Davatzikos C. Machine learning in neuroimaging: progress and challenges. Neuroimage. 2018.
4. Iniesta R, Stahl D, McGuffin P. Machine learning, statistical learning and the future of biological research in psychiatry. Psychol Med. 2016;46(12):2455–65.
5. Lane HY, Tsai GE, Lin E. Assessing gene–gene interactions in pharmacogenomics. Mol Diagn Ther. 2012;16(1):15–27.
6. Lin E, Hwang Y, Liang KH, Chen EY. Pattern-recognition techniques with haplotype analysis in pharmacogenomics. Pharmacogenomics. 2007;8(1):75–83.
7. Torres EB, Isenhower RW, Nguyen J, Whyatt C, Nurnberger JI, Jose JV, et al. Toward precision psychiatry: statistical platform for the personalized characterization of natural behaviors. Front Neurol. 2016;7:8.
8. Gandal MJ, Leppa V, Won H, Parikshak NN, Geschwind DH. The road to precision psychiatry: translating genetics into disease mechanisms. Nat Neurosci. 2016;19(11):1397–407.
9. Lin E, Tsai SJ. Genome-wide microarray analysis of gene expression profiling in major depression and antidepressant therapy. Prog Neuropsychopharmacol Biol Psychiatry. 2016;64:334–40.
10. Lin E, Lane HY. Genome-wide association studies in pharmacogenomics of antidepressants. Pharmacogenomics. 2015;16(5):555–66.
11. Litjens G, Kooi T, Bejnordi BE, Setio AAA, Ciompi F, Ghafoorian M, et al. A survey on deep learning in medical image analysis. Med Image Anal. 2017;42:60–88.
12. Ching T, Himmelstein DS, Beaulieu-Jones BK, Kalinin AA, Do BT, Way GP, et al. Opportunities and obstacles for deep learning in biology and medicine. J R Soc Interface. 2018;15(141).
13. Dana D, Gadhiya SV, St Surin LG, Li D, Naaz F, Ali Q, et al. Deep learning in drug discovery and medicine; scratching the surface. Molecules. 2018;23(9).
14. Grapov D, Fahrmann J, Wanichthanarak K, Khoomrung S. Rise of deep learning for genomic, proteomic, and metabolomic data integration in precision medicine. OMICS. 2018.
15. Zhang S, Hosseini Bamakan SM, Qu Q, Li S. Learning for personalised medicine: a comprehensive review from deep learning perspective. IEEE Rev Biomed Eng. 2018.
16. Hulshoff Pol H, Bullmore E. Neural networks in psychiatry. Eur Neuropsychopharmacol. 2013;23(1):1–6.
17. Fernandes BS, Williams LM, Steiner J, Leboyer M, Carvalho AF, Berk M. The new field of 'precision psychiatry'. BMC Med. 2017;15(1):80.
18. Dwyer DB, Falkai P, Koutsouleris N. Machine learning approaches for clinical psychology and psychiatry. Annu Rev Clin Psychol. 2018;14:91–118.

19. Arbabshirani MR, Plis S, Sui J, Calhoun VD. Single subject prediction of brain disorders in neuroimaging: promises and pitfalls. Neuroimage. 2017;145(Pt B):137–65.
20. Kloppel S, Stonnington CM, Chu C, Draganski B, Scahill RI, Rohrer JD, et al. Automatic classification of MR scans in Alzheimer's disease. Brain. 2008;131(Pt 3):681–9.
21. Ju R, Hu C, Zhou P, Li Q. Early diagnosis of Alzheimer's disease based on resting-state brain networks and deep learning. IEEE/ACM Trans Comput Biol Bioinform. 2017.
22. Ortiz A, Munilla J, Gorriz JM, Ramirez J. Ensembles of deep learning architectures for the early diagnosis of the Alzheimer's disease. Int J Neural Syst. 2016;26(7):1650025.
23. Akhavan Aghdam M, Sharifi A, Pedram MM. Combination of rs-fMRI and sMRI data to discriminate autism spectrum disorders in young children using deep belief network. J Digit Imaging. 2018.
24. Pinaya WH, Gadelha A, Doyle OM, Noto C, Zugman A, Cordeiro Q, et al. Using deep belief network modelling to characterize differences in brain morphometry in schizophrenia. Sci Rep. 2016;6:38897.
25. Schmaal L, Marquand AF, Rhebergen D, van Tol MJ, Ruhe HG, van der Wee NJ, et al. Predicting the naturalistic course of major depressive disorder using clinical and multimodal neuroimaging information: a multivariate pattern recognition study. Biol Psychiatry. 2015;78(4):278–86.
26. Miotto R, Li L, Kidd BA, Dudley JT. Deep patient: an unsupervised representation to predict the future of patients from the electronic health records. Sci Rep. 2016;6:26094.
27. Lin E, Kuo PH, Liu YL, Yu YW, Yang AC, Tsai SJ. A deep learning approach for predicting antidepressant response in major depression using clinical and genetic biomarkers. Front Psychiatry. 2018;9:290.
28. Kautzky A, Baldinger P, Souery D, Montgomery S, Mendlewicz J, Zohar J, et al. The combined effect of genetic polymorphisms and clinical parameters on treatment outcome in treatment-resistant depression. Eur Neuropsychopharmacol. 2015;25(4):441–53.
29. Patel MJ, Andreescu C, Price JC, Edelman KL, Reynolds CF 3rd, Aizenstein HJ. Machine learning approaches for integrating clinical and imaging features in late-life depression classification and response prediction. Int J Geriatr Psychiatry. 2015;30(10):1056–67.
30. Chekroud AM, Zotti RJ, Shehzad Z, Gueorguieva R, Johnson MK, Trivedi MH, et al. Cross-trial prediction of treatment outcome in depression: a machine learning approach. Lancet Psychiatry. 2016;3(3):243–50.
31. Iniesta R, Malki K, Maier W, Rietschel M, Mors O, Hauser J, et al. Combining clinical variables to optimize prediction of antidepressant treatment outcomes. J Psychiatr Res. 2016;78:94–102.
32. Zou H, Zhang HH. On the adaptive elastic-net with a diverging number of parameters. Ann Stat. 2009;37(4):1733–51.
33. Maciukiewicz M, Marshe VS, Hauschild AC, Foster JA, Rotzinger S, Kennedy JL, et al. GWAS-based machine learning approach to predict duloxetine response in major depressive disorder. J Psychiatr Res. 2018;99:62–8.
34. Tibshirani R. The lasso method for variable selection in the Cox model. Stat Med. 1997;16(4):385–95.
35. Casanova R, Barnard RT, Gaussoin SA, Saldana S, Hayden KM, Manson JE, et al. Using high-dimensional machine learning methods to estimate an anatomical risk factor for Alzheimer's disease across imaging databases. Neuroimage. 2018;183:401–11.
36. Algamal ZY, Lee MH. Regularized logistic regression with adjusted adaptive elastic net for gene selection in high dimensional cancer classification. Comput Biol Med. 2015;67:136–45.
37. Friedman J, Hastie T, Tibshirani R. Regularization paths for generalized linear models via coordinate descent. J Stat Softw. 2010;33(1):1–22.
38. Walter M, Alizadeh S, Jamalabadi H, Lueken U, Dannlowski U, Walter H, et al. Translational machine learning for psychiatric neuroimaging. Prog Neuropsychopharmacol Biol Psychiatry. 2018.
39. Woo CW, Chang LJ, Lindquist MA, Wager TD. Building better biomarkers: brain models in translational neuroimaging. Nat Neurosci. 2017;20(3):365–77.

40. Moreno-De-Luca D, Ross ME, Ross DA. Leveraging the power of genetics to bring precision medicine to psychiatry: too little of a good thing? Biol Psychiatry. 2018;83(8):e45–6.
41. Folkersen L, Brynedal B, Diaz-Gallo LM, Ramskold D, Shchetynsky K, Westerlind H, et al. Integration of known DNA, RNA and protein biomarkers provides prediction of anti-TNF response in rheumatoid arthritis: results from the COMBINE study. Mol Med. 2016;22:322–8.
42. Gombojav B, Song YM, Lee K, Yang S, Kho M, Hwang YC, et al. The healthy twin study, Korea updates: resources for omics and genome epidemiology studies. Twin Res Hum Genet. 2013;16(1):241–5.
43. Qvarfordt M, Anderson M, Svartengren M. Quality and learning aspects of the first 9000 spirometries of the Life Gene study. NPJ Prim Care Respir Med. 2018;28(1):6.
44. Chen CH, Yang JH, Chiang CWK, Hsiung CN, Wu PE, Chang LC, et al. Population structure of Han Chinese in the modern Taiwanese population based on 10,000 participants in the Taiwan Biobank project. Hum Mol Genet. 2016;25(24):5321–31.
45. Lin JC, Fan CT, Liao CC, Chen YS. Taiwan Biobank: making cross-database convergence possible in the big data era. Gigascience. 2018;7(1):1–4.
46. Lin E, Kuo PH, Liu YL, Yang AC, Kao CF, Tsai SJ. Association and interaction of APOA5, BUD13, CETP, LIPA and health-related behavior with metabolic syndrome in a Taiwanese population. Sci Rep. 2016;6:36830.
47. Lin E, Kuo PH, Liu YL, Yang AC, Kao CF, Tsai SJ. Effects of circadian clock genes and environmental factors on cognitive aging in old adults in a Taiwanese population. Oncotarget. 2017;8(15):24088–98.
48. Lin E, Kuo PH, Liu YL, Yang AC, Kao CF, Tsai SJ. Effects of circadian clock genes and health-related behavior on metabolic syndrome in a Taiwanese population: evidence from association and interaction analysis. PLoS ONE. 2017;12(3):e0173861.
49. Lin E, Tsai SJ, Kuo PH, Liu YL, Yang AC, Kao CF, et al. The rs1277306 variant of the REST gene confers susceptibility to cognitive aging in an elderly Taiwanese population. Dement Geriatr Cogn Disord. 2017;43(3–4):119–27.
50. Lin E, Tsai SJ, Kuo PH, Liu YL, Yang AC, Kao CF, et al. The ADAMTS9 gene is associated with cognitive aging in the elderly in a Taiwanese population. PLoS ONE. 2017;12(2):e0172440.
51. Lin E, Yang AC, Tsai SJ. Association between metabolic syndrome and cognitive function in old adults in a Taiwanese population. Taiwanese J Psychiatry. 2017;31(3):232–40.
52. Ben-Hamo R, Efroni S. Biomarker robustness reveals the PDGF network as driving disease outcome in ovarian cancer patients in multiple studies. BMC Syst Biol. 2012;6:3.
53. Tebani A, Afonso C, Marret S, Bekri S. Omics-based strategies in precision medicine: toward a paradigm shift in inborn errors of metabolism investigations. Int J Mol Sci. 2016;17(9).
54. Dalvie S, Koen N, McGregor N, O'Connell K, Warnich L, Ramesar R, et al. Toward a global roadmap for precision medicine in psychiatry: challenges and opportunities. OMICS 2016;20(10):557–64.
55. Williams LM. Precision psychiatry: a neural circuit taxonomy for depression and anxiety Lancet Psychiatry. 2016;3(5):472–80.
56. Hu Y, Huse T, Li HP, Prabhakar S, Kitano H, Ng SK, et al. A machine learning approach for the identification of key markers involved in brain development from single-cell transcriptomic data. BMC Genom. 2016;17(Suppl 13):1025.
57. Serretti A. The present and future of precision medicine in psychiatry: focus on clinical psychopharmacology of antidepressants. Clin Psychopharmacol Neurosci. 2018;16(1):1–6.
58. Stein MB, Smoller JW. Precision psychiatry-will genomic medicine lead the way? JAMA Psychiatry. 2018;75(7):663–4.

Imaging Connectomics and the Understanding of Brain Diseases

Andrea Insabato, Gustavo Deco and Matthieu Gilson

Abstract

Neuroimaging-based personalized medicine is emerging to characterize brain disorders and their evolution at the patient level. In this chapter, we present the most classic methods used to infer large-scale brain connectivity based on functional MRI. We adopt a modeling perspective where every connectivity measure is linked to a specific model that allows to interpret the connectivity estimate. This perspective allows to analyze the quality of retrieved connectivity profiles in terms of modeling error and estimation error. In the first part of the chapter, we present undirected functional connectivity (Pearson's correlation and MI) and effective connectivity (partial correlation), as well as directed effective connectivity (VAR, MOU, Granger causality, DCM). In addition, some of these measures correspond to fully connected graphs (Pearson's correlation) while others to sparse ones (MOU, DCM), where the sparsity can come from the integration of functional and structural data. In the second part, we claim that machine learning tools are better suited than null-hypothesis testing to link the estimated connectomes with diagnosis and prognosis of

A. Insabato (✉)
Institut de Neurosciences de la Timone, UMR CNRS 7289, Aix Marseille University, 13385 Marseille, France
e-mail: andrea.insabato@upf.edu

G. Deco
Institució Catalana de la Recerca i Estudis Avanats (ICREA), Universitat Pompeu Fabra, Barcelona, Spain
e-mail: gustavo.deco@upf.edu

M. Gilson
Theoretical and Computational Neuroscience, Center for Brain and Cognition, Universitat Pompeu Fabra, C\Ramon Trias Fargas, 25–27, 08005 Barcelona, Spain
e-mail: mattieu.gilson@upf.edu

© Springer Nature Singapore Pte Ltd. 2019

Y.-K. Kim (ed.), *Frontiers in Psychiatry*, Advances in Experimental Medicine and Biology 1192, https://doi.org/10.1007/978-981-32-9721-0_8

neuropsychiatric diseases. Finally, we propose that linear models and features selection are preferable to more complex and nonlinear tools (when prediction performance is on a par) for building interpretable algorithms to predict clinical variables.

Keywords
Model-based connectivity · Functional connectivity · Effective connectivity · Interpretable machine learning · Whole-brain modeling

Neuroimaging-based personalized medicine is emerging to characterize brain disorders and their evolution at the patient level. Here we describe current approaches used to investigate brain diseases from functional magnetic resonance imaging (fMRI) data. In particular, we will introduce in a systematic way the most classical methods used to characterize the correlation patterns of fMRI activity between brain areas (regions of interest, ROIs), which is referred to in the general sense as the functional connectome. We then discuss the relation of each type of FC to a specific generative model of brain activity. Finally, we discuss how to link FC to clinically relevant variables (using ML) with the aim of providing an early diagnosis and complementing therapeutic strategies using fMRI-based biomarkers [1–3].

All code used in this chapter is available from a public Git repository on BitBucket: https://bitbucket.org/ainsabato/imaging_connectomics. The code is written in Python 3 and uses, beyond the typical suite of libraries for scientific computing (Scipy, Numpy, Matplotlib): NetworkX library to deal with networks, Scikit Learn for machine learning routines, Pandas and Seaborn for some graphical representations. All figures can be reproduced running the Jupyter notebook imaging_connectomics_chapter.ipynb and may easily modified by varying parameters to further explore the topics presented in the chapter.

Functional Connectivity Measures

In the literature, three families of connectomes have emerged:

- **Structural connectivity (SC)**: It measures the density or probability of anatomical pathways that connect two ROIs, mostly via the white matter [4], as measured by DTI and reconstructed with tractographic algorithms. This led to the definition of the human connectome at the whole-brain level [5, 6].
- **Functional connectivity (FC)**: Pairwise interactions between the activity of ROIs as measured by BOLD fMRI signal. The interaction is usually measured by Pearson's correlation or synchrony [7, 8] but also other measures have been tried like mutual information [9] or instantaneous phases [10, 11].
- **Effective connectivity (EC)**: Coming from electrophysiology [12], the concept has been brought to neuroimaging and further developed in the context of the dynamic causal model (DCM) [13–15]. Note that it is also still used close to its original formulation when measuring stimulation-driven responses in

neuroimaging [16]. Here we broadly use the concept as the underlying connectivity in a generative model, i.e., the graph that determines the connections between variables in the model (see section "Model-Based Versus Model-Free"). Roughly speaking the idea is that the observed time series (and as a consequence their correlation matrix) are generated by a non-observed interaction pattern, the EC, and the goal is to identify this EC.

While the interpretation of SC is not straightforward and deserves a dedicated treatment we will focus here only on connectomes based on functional data (FC and EC[1]).

Model-Based Versus Model-Free

A recurrent discussion about connectivity measures is model-free versus model-based approaches. Our viewpoint is that, even though measures can be calculated without explicitly assuming a model, their *interpretation is always related to a model*. For instance, in the simple bivariate case, Pearson's correlation coefficient between two time series can be directly evaluated from data and is thus considered to be model-free. However, Anscombe [18] famously showed that when the underlying assumptions are violated, the interpretation of the correlation coefficient becomes fragile.

In Fig. 1, we reproduce the Anscombe quartet, where three datasets violate the assumptions of Pearson's correlation coefficient. In all four panels, the coefficient is the same but only the top left panel shows the distribution of data that we would expect if we are told that $\rho = 0.816$. To interpret the correlation coefficient, e.g., to compare the respective coefficients for several groups of data points, we need to take into account the underlying assumptions, i.e., the statistical model. This same argument applies to all connectivity measures. In the next section, we discuss the specific assumptions underlying each model.

Before moving to the different models, further clarification is needed about causal interpretation of functional connectivity measures. FC (as well as EC) is in general a measure of dependence between brain regions and, while it is well known that correlation does not imply causation, it is common to claim a causal interpretation[2] for EC or when a model is involved. Given the broad definition of EC that we are using, this claim is not correct in general. Minimal requirements for an interpretation of EC in terms of causality are directionality and sparseness

---

[1]Some EC models use SC information in addition to functional data to constrain the estimated connectivity, e.g., [17]. In general, SC is used as a mask to retain only those anatomical connections that have probability higher than a given threshold. So, SC provides a kind of empirical prior upon the existence of connections, thereby making the estimated functional connectome sparse. However, the strength, or weight, of links is estimated using functional data.

[2]In the context of EC **statistic** causality is usually involved, where causes all contribute to the effect but none is individually necessary nor sufficient. See https://en.wikipedia.org/wiki/Causality for more information about different types of causality.

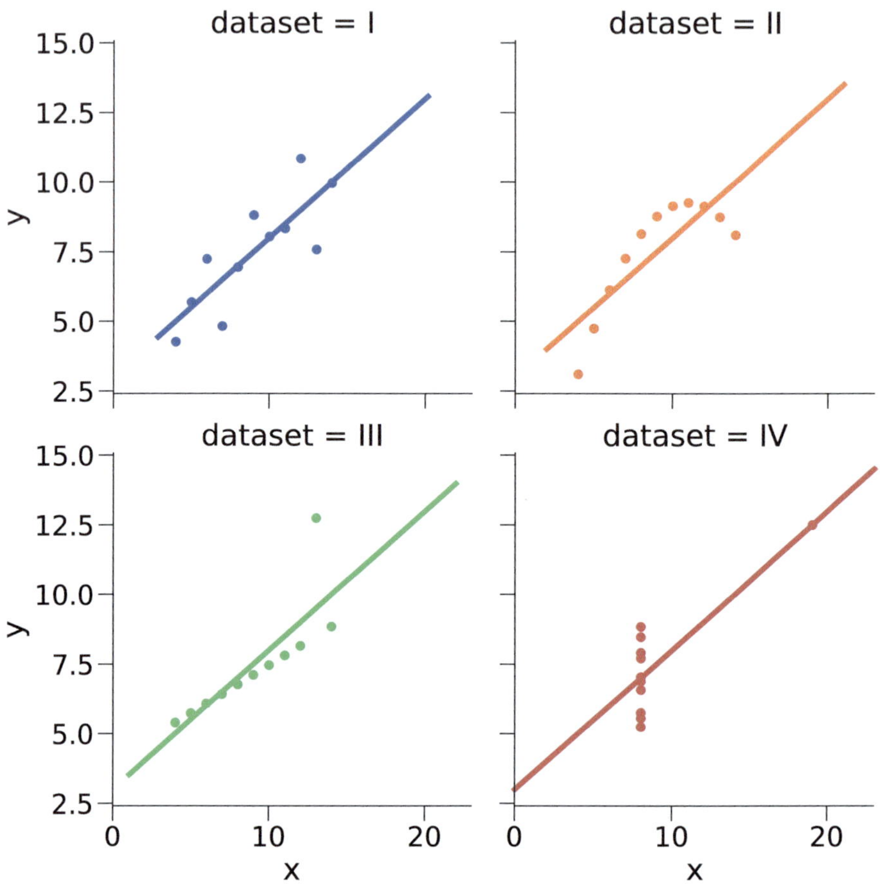

Fig. 1 Anscombe's quartet. The data points in each dataset have the same means, variances, Pearson's correlation, and linear regression coefficients but show very different relationships. Dataset I has Gaussian variables with linear dependence. Dataset II shows nonlinear dependence between y and x. Dataset III has a perfect linear relation but the estimates are off because of the presence of one outlier in the dependent variable. Dataset IV shows no relation between x and y but an apparently strong regression line due to one outlier in both the dependent and independent variables

of the connectivity [which are not necessarily fulfilled by every model, as described in section "Stationary Measures (Whole-Session Approach)"]. Without these requirements, it's impossible to identify different alternative configurations of the causal graph, as shown in Fig. 2. In the next section we'll see some methods to estimate sparse directed connectivity but in general, directionality is usually assessed by considering that only previous values of a variable can exert an effect on another variable. Sparseness allows for the exclusion of some spurious connections and is achieved in different ways.

As a final remark, focusing on the model behind a measure also allows for a better characterization of the fitting of the method (especially when the method

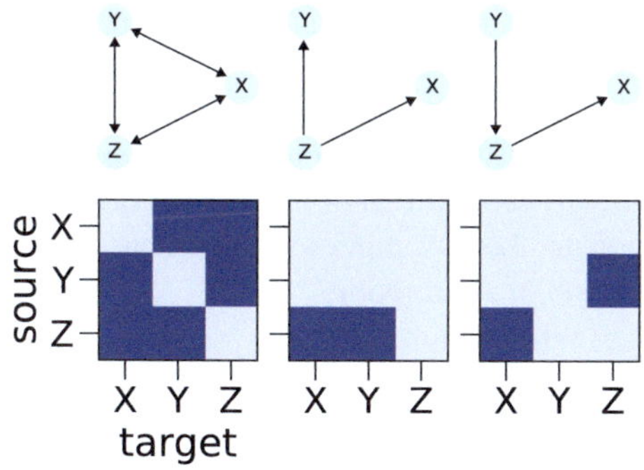

Fig. 2 Three different graphs and their associated adjacency matrices. Left: Fully connected graph. Center: *Z* is the common input to both *X* and *Y*. Right: Chain from *Y* to *X* through *Z*. A connectivity measure unable to estimate a sparse matrix cannot distinguish the three cases: all estimates would look like the one on the left. A nondirected measure would be unable to distinguish between the chain and the common input: both estimated matrices would be symmetric

fails and we have to fix it). If we consider that every measure is based on a model then the error of this model in predicting the data is the sum of the modeling error and estimation error. For example, dataset II of Anscombe quartet shows a clear modeling error: the relation between x and y is nonlinear while the model is linear. Dataset III and IV instead show an estimation error: data is actually well described by the model but an outlier influenced the estimation of the regression line; robust regression [19] would have solved the problem here. The use of a limited amount of data is what produces the estimation error even when the modeling error is zero. In practice, when the number of samples (time points of BOLD/fMRI) is comparable to the number of parameters (or, even worst, lower) the estimation error may be high for all measures.[3] Often multiple estimation procedures are available for the same model and they can imply substantial differences in the estimation error. Once we are sure that we have the best possible estimation for our model, we can consider if a different model could better explain the data. In the next section, we comment on different estimation methods for some model.

Stationary Measures (Whole-Session Approach)

All models described in this section assume stationarity, i.e., they generate time series using a dynamic system whose parameters do not change over time. An fMRI scanning session of resting state might be seen as stationary since

[3]This should be taken into consideration when choosing a brain atlas to define ROIs. Indeed, while an atlas with more ROIs will have a better spatial resolution, this also increases (in general quadratically) the number of links to be estimated. So while a larger atlas might produce a better model, the total error might be actually lower for a smaller atlas.

the external conditions do not vary during the session.[4] On the other hand, this assumption might be critical in settings such as behavioral tasks where cognitive functions are dynamically involved in the production of behavior. The issue of the timescale at which the brain dynamics can be assumed to be stationary is an open question. We will see more about this in section "Dynamic Connectivity Measures". In addition to the assumption of stationarity, each model described below entails a specific set of assumptions.

Pearson's Correlation and Covariance Matrix

Pearson's correlation is commonly used to characterize the functional relationships between brain areas. As an extension of what we have seen above with the Anscombe quartet for the bivariate case, in the multivariate case the correlation matrix is only meaningful when a multivariate Gaussian noise (MGN) model, i.e., an undirected graphical model where the variables follow a multivariate normal distribution, is an adequate representation of the data. Indeed, the MGN model is fully characterized by its covariance matrix, where each diagonal term determines a node's input variance and off-diagonal terms determine the strength of interactions. The MGN model entails the following assumptions (1) the dynamics of each node in the network has no memory (i.e., the past activity of one node has no effect on the same node or in other words the activity is i.i.d.), (2) the influence of one node over the others is linear, (3) interactions are symmetric, and (4) each node receives a noisy independent Gaussian input (i.e., inputs are uncorrelated). The FC estimated by Pearson's correlation thus corresponds to a nondirected fully connected graph. An estimation of covariance matrix (and of correlation matrix) can be easily obtained with a simple inner product of centered (demeaned) time series. As we noted above, for a fixed number of ROIs N, as the number of time points T decreases the noise in the estimation increases and when $T \leq N$ the resulting matrix is singular. In such cases, a good alternative is a regularized estimator [20, 21], see also Partial Correlation section.

Partial Correlation and Precision Matrix

The covariance matrix is an observable of the MGN model, so it's a type of FC. If we look at the connectivity graph of an MGN model, we have an instance of EC. The inverse covariance matrix, also called precision matrix, represents the graph of an MGN model (which is a special case of Markov random field, or undirected graphical model). This EC is undirected and a missing link (zero in the precision matrix) indicates independence between the two nodes conditioned upon all other nodes in the network. The partial correlation matrix is a normalized version of the precision matrix. The estimation of the precision matrix can be done by simple algebraic inversion of covariance matrix (provided the latter is not singular). However under the assumptions of this model, it makes sense to estimate a sparse

[4]However for long sessions, this assumption might be more questionable since internal conditions, like sleepiness, attention, body temperature, might change.

precision matrix. Sparsity can be imposed after inverting by using a null-hypothesis testing (Fisher test) to set to zero those entries of the empirical precision matrix whose value is not significantly different than zero. Alternatively, a regularized estimator, like the Graphical Lasso [22] can be used. In the case where few observations (time points) are available compared to the number of nodes, the Graphical Lasso is preferable since it provides a more robust estimation. Note that a Lasso-regularized estimate of the covariance matrix can be obtained by first estimating the precision matrix with Graphical Lasso and then inverting the result.

Mutual Information

The MGN model (and in turn Pearson's correlation and partial correlation) entails the assumption of linear dependence between nodes. This means that if two nodes are associated by a nonlinear relationship these methods are not going to detect this link or its strength will be wrongly estimated. A more general approach, able to detect both linear and nonlinear dependencies, is based on mutual information (MI). MI measures the amount of information (not only linear) between two brain areas; in other words, how much the uncertainty about the activity of one brain area is reduced when the activity of another area is known [23].

MI has the advantage of being able to detect nonlinear relationships but the estimated connections are symmetric, i.e., $I(X, Y)=I(Y, X)$, and therefore does not allow a causal interpretation for brain connectivity.

While dependence between ROIs is probably nonlinear, the system might still be reasonably well approximated by a linear system. Indeed it has been shown that MI capture only a small amount (about 5%) of information beyond Pearson's correlation [9, 24]. In general, nonlinear methods have more parameters than linear ones and need more samples to attain a given estimation error. Hence, even when data are generated by a nonlinear process, a linear approximation might still provide good results.

Vector Autoregression Model

The simplest model that enables the estimation of directed functional connectivity is the vector autoregression model (VAR). This model considers not only the spatial structure of the data as MGN model but also the temporal structure. The autocorrelation of the model does not decay to zero for time lags larger than 0, as for the MGN, i.e., the past activity influences the state of the system. The activity state of each node is described by the following difference equation:

$$x_t = \sum_{p=1}^{P} A_p x_{t-p} + e_t$$

where x_t is the vector representing the activity of the N ROIs at time t, each A_p is an $N \times N$ matrix of interactions and e_t is a vector of error terms (with positive definite covariance matrix). The number P of interaction matrices determines the order of the VAR model. If we consider an order 1 VAR model, matrix A corresponds to the EC of this model.

The matrix A is not required to be symmetric, and the estimation of directionality in the network comes from extracting information from spatio*temporal* covariance structure. Compared to the MGN model, assumptions 1 and 3 are dropped while assumption 4 is sometimes modified, allowing correlated Gaussian input. As a consequence, the VAR allows for the estimation of an asymmetric EC matrix that corresponds to a directed and weighted connectivity graph. Note that the VAR model assumes linear dependencies between nodes, so it is unable to detect nonlinear relationships.

VAR model has been used to estimate the connectivity pattern underlying fMRI BOLD signal [25, 26] also relaxing the assumption of stationarity [27] (see section "Dynamic Connectivity Measures").

Multivariate Ornstein–Uhlenbeck

The multivariate Ornstein–Uhlenbeck model (MOU) is a type of continuous-time autoregressive model. Each node in the network obeys the differential equation:

$$\frac{dx_i(t)}{dt} = \frac{-x_i(t)}{\tau_i} + \sum_{j \neq i} C_{ij} x_j + \sigma_i \frac{dB_i}{dt}$$

where τ_i is the time constant of the node, C_{ij} is the connectivity from node j to node i, and σ_i is the standard deviation of the Gaussian noise. The matrix C is the EC matrix of the MOU model.

MOU assumes the same hypothesis as the VAR model and as such is able to retrieve a directed EC. Recently, a method has been developed that efficiently estimates the parameters of the mOU [17]. The method is based on a natural gradient descent that extracts information from the spatiotemporal covariance of the signal. In addition, this estimation method incorporates DTI information (when available) to constrain the topology of EC (i.e., only tuning anatomically existing connections). The natural gradient method proved to outperform moment's method and Bayesian MAP estimation [28]. The sparseness that comes from retaining only nonzero or strong DTI values allows to reduce the number of parameters, which enhances the estimation robustness. This approach has recently been shown to improve the identification of subjects as compared to covariance-based FC [29].

Granger Causality

The idea behind Granger causality (GC), originally proposed in econometrics [30], is that a cause cannot come after the effect (for a detailed description see http://www.scholarpedia.org/article/Granger_causality). In the context of brain connectivity, if the activity of a brain region X influences (in a causal sense) the activity of another region Y, the former should improve the prediction of the latter region. Of course, the improvement cannot consider all the available information of the universe, so GC is usually tied to a model where all relevant variables are included. This concept of causality can be easily developed in the context of VAR models [31]. In practice, considering an order-1 VAR model, if the entry A_{ij} of the interaction matrix has value zero, region j has no (Granger) causal effect on region i (this is easily extended to higher order VAR). When VAR parameters are estimated from empirical time series there are no exact zeros due to estimation noise, so null-hypothesis testing (e.g., via a Wald test) is used to enforce sparseness in A. Essentially brain connectivity based on GC is still an EC under the VAR model, where the estimation procedure is extended to filter out non-causal connections.

Recently, covariance-based Granger causality has been proposed as an alternative to estimate EC in large networks [32].

Dynamic Causal Model

Dynamic causal model (DCM), initially proposed by Friston [13, 14], is now a classic for the estimation of EC. Initially, DCM was developed to infer directed coupling between few regions in task conditions and to test hypotheses about the interaction of coupling and experimental manipulations (e.g., sensory input, attentive state, etc.). Given the nonlinearity of the model and the complexity of the estimation procedure, the initial formulation was limited to small networks. More recently the model and its estimation procedure have been extended to account for resting-state recordings with many brain areas involved [33].

DCM is based on a state-space model, where the neural activity is considered a hidden (i.e., non-observed) variable and BOLD signal is represented by a hemodynamic response function acting on the neural activity. The neural activity is modeled by a deterministic bilinear differential equation. The model includes input variables that have a double effect: they allow directly influence the activity of a node and to modulate the connectivity between two nodes.

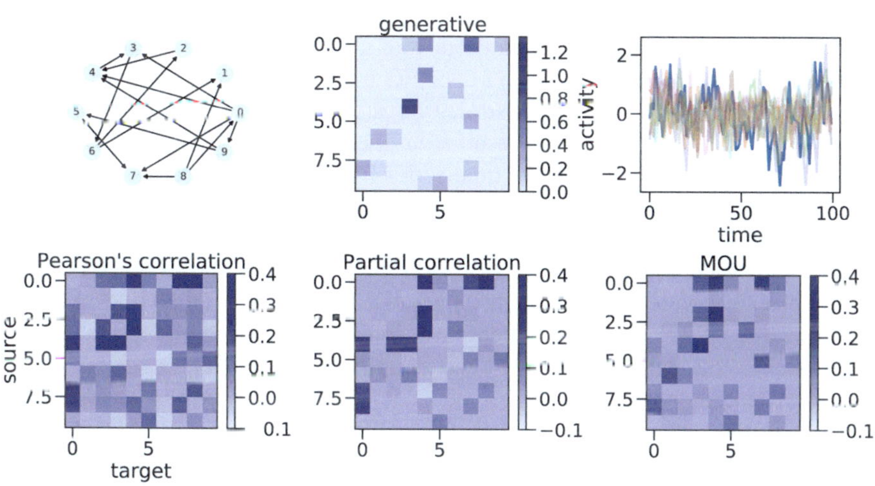

Fig. 3 Synthetic dataset generated by simulating an MOU and estimation of the connectivity matrix from the simulated time series using different measures. Upper row: graph (left) and connectivity matrix (center) of the generative model. Right: example of simulated time series (one node is highlighted for clarity). Bottom row: estimation using Pearson's correlation (left), partial correlation (center), and MOU (right). Partial correlation is sparser than Pearson's correlation and both are symmetric. MOU estimates are sparser and asymmetric as the generative matrix. Limited amount of time sample produces noisy estimates, even in the absence of modeling error (right)

Note that even though the parameters of the model are not time-varying, the modulation of the activity induced by the stimulus will make the generated time series nonstationary if the stimulus changes with time [14, 34].

A key feature of DCM is its hierarchical structure. While previous models directly model BOLD time series, DCM first models the hidden neural activity and then the BOLD signal: $u \rightarrow x = f(u) \rightarrow y = h(x)$. Alternatively, one can say that other models, like VAR or MOU, consider $h(x)$ to be the identity function.

While the introduction of the hemodynamic response function $h(x)$ makes the model more realistic, the precise relationship between neural activity and BOLD signals (related to the energy consumption of brain cells) is still under debate. In addition, it is not obvious that the incorporation of more realistic mechanisms in a model leads to better biomarkers for clinical prediction, as the optimization procedure becomes more complex and putatively noisier (Fig. 3).

Dynamic Connectivity Measures

All the connectivity measures presented in the previous section assume stationarity of BOLD time series (except DCM). Given the intrinsic dynamic nature of brain signals, there is increasing interest in the estimation of time-resolved connectivity measures. Indeed, the so-called dynamic FC [35] has been found to be informative about cognitive conditions [36].

The simplest approach to develop a time-resolved estimation of brain connectivity is to use a sliding window [37–39], which consist of the following steps: (1) a chunk of the BOLD signal is extracted from a time window of length l, shorter than the total signal length, (2) the FC or EC is estimated from this shorter time series, (3) the window is slid by a stride of s and the process is repeated until the end of the signal is reached. We recall from the previous discussion that models parameters estimated from shorter time series have a higher variability, i.e., their value is less reliable. Indeed even for stationary signals, the estimated parameters will fluctuate as the time window is slid over the signal and these fluctuations will be stronger as the window length l is made shorter [40]. Ideally, l and s should be set in order to capture the fluctuations of connectivity due to external (e.g., sensory input) or internal (e.g., urgency, attention) changes. In practice, the range of possible values for l and s is limited by the sampling rate and BOLD signal length. Imagine, for example, that the true time scale of FC modulation in a given condition is in the range of 20 s: the typical sampling frequency of BOLD signal is 0.5 Hz, so that we would have 10 time samples in each window; many brain atlas has about 100 ROIs, so we should estimate, for example, a 100×100 covariance matrix from 10 samples; the resulting matrix will be very noisy and singular. So we couldn't trust the values of connectivity estimated under these conditions.[5]

[5]Regularization, discussed in previous section, might help to stabilize the estimates but it's likely giving a very sparse, if not diagonal, estimate.

Alternative methods to study the dynamics of FC at shorter time scales have been explored using instantaneous phases obtained using the Hilbert transform on the BOLD signals [11]. As also noted above DCM is able to generate nonstationary time series, if the stimulus is not static, even when the parameters do not vary with time and has also been used to characterize variation in parameters over successive time windows [41]. However, it remains to be shown whether these methods provide a robust description for cognition or neuropathologies, for instance, with respect to session-to-session variability.

The work on FC and EC for nonstationary time series has mainly focused on the cognitive domain. For resting-state fMRI and clinical applications is not clear what is the time scale of connectivity modulation; however, the dynamic connectivity estimation might extract information relevant also to clinical purposes. Indeed, changes in functional and effective connectivity over different time scales may be informative about neuropsychiatric disorders.

Linking Functional Brain Connectivity with Clinical Conditions

In the previous section, we have seen different methods to estimate functional or effective connectivity from BOLD time series. In this section, we address some aspects related to the use of connectomes in clinical studies. First, we present some questions connectomes can shed light on, and then we give some arguments to favor machine learning (ML) over null-hypothesis testing for clinical purposes. Finally, we discuss some general problems in ML, how to address them, and how to build interpretable algorithms.

Questions

The functional connectomes presented in section "Functional Connectivity Measures" are potentially informative about clinically relevant variables. Some general questions that might be addressed with FC or EC are:

1. Can FC/EC be used to **diagnose** disease *X*? Can we predict the clinically relevant **score** *S* using FC/EC?
2. Can we predict the **evolution** of *X* using FC/EC?
3. Which **subsets** of links in FC/EC are more relevant to diagnose *X*?
4. Is disorder *X* homogeneous, or are there multiple **unknown subgroups** defined by FC/EC with potential clinical relevance?

Type-1 questions are interesting for their possible application to develop computer-aided diagnosis for neuropsychiatric disorders. For example, the ADHD-200 Global Competition asked the participant to develop imaging-based diagnosis able to distinguish between typically developing, ADHD primarily inattentive type and ADHD combined type. This type of question can focus on discrete categories, as in the ADHD-200 competition, defined by standard diagnostic practice or on the prediction of a pseudo-continuous score. For example, predicting ADHD rating scale instead of the diagnostic group. In both cases, connectome-based methods would constitute a real paradigm shift in clinical practice if the diagnosis could be made earlier than with traditional methods, which usually leads to more efficient therapies [42, 43]. Type-2 questions are more prognostic. For example, predicting the conversion from mild cognitive impairment to Alzheimer's disease. This is probably the field of application where there is more room for improvement of traditional methods [44]. While type-2 and type-3 questions are directly linked to clinical practice, type-3 questions focus more on understanding the neural underpinnings of diseases. Oftentimes conventional diagnostic methods are already highly accurate and probably cheaper than fMRI-based methods. However, investigating the influence of a disorder on brain functional connections can help to better characterize it. In particular, although many disorders are not localized to specific parts of the brain, only a few of the thousands of connections in FC/EC are likely to be involved in the diagnosis. The identification of this subset of links is key to gain a better understanding of the pathological conditions. Type-4 questions are more exploratory: Mental disorders are complex and often present heterogeneous symptoms and behavior. Brain connectivity can then be used to discover subgroups of patients with different characteristics. A pioneering work in this direction was that of Brodersen and colleagues, who found three subgroups of schizophrenic patients based on DCM EC and confirmed that these groups had different negative symptom severity [45]. A similar approach has been taken to distinguish different types of depression [46, 47] for a discussion about reproducibility of the results).

Statistics and Machine Learning

Questions of type 1, 2, and 3 can usually be reduced to the problem of whether a given measurement (brain connectivity) is different under two or several groups (diagnostic groups, either in the present or in the future).

Traditionally, this problem has been addressed by null-hypothesis testing (NHT) with techniques such as t-test, ANOVA, etc. Here we point out some issues inherent to NHT, in particular when applied to connectome data. We allege that ML tools are better suited to address questions about linking brain connectivity with clinical conditions.

A common problem of NHT is the prevalent focus on p-values and the relative neglect of effect size. For example, when looking at the difference in tau protein concentration between Alzheimer patients and controls, not only the probability of that effect (its p-value) is important but also its magnitude. An important concern about p-values is that every effect, even the tiniest, can be regarded as significant with a large enough sample size [48], as shown in Fig. 4. The emergence of large datasets in many disciplines is highlighting the relevance of this concern.

A related problem with p-values comes when a statistical test is repeated several times, for example, to test if there is a difference between a treatment and control group for each connection in an FC. This process leads to the well-known problem of multiple hypotheses testing. In NHT the p-value corresponds to the probability of a false positive result, or type I error. If we consider as significant all results where $p<\alpha$, over m hypotheses the probability of making one or more false discoveries is $1-(1-\alpha)^m$ (for $m=100$ hypotheses and usual $\alpha=0.05$, the probability of at least one false discovery is more than 0.99). In order to solve this problem, several correction procedures have been proposed. Some procedures, like Bonferroni, are more conservative: they accept a higher type II error rate (i.e., the probability of missing a true discovery) in order to keep type I error rate low. Some other methods, like false discovery rate controlling procedures, accept a higher type I error rate in order to make more discoveries.

Finally, statistical tests rely on underlying hypotheses (e.g., specific distribution of data, independence between samples, sidedness of the test, etc.), that, if not fulfilled, yield misleading results [48].

The application of NHT to functional or effective connectivity is further complicated by two features of connectome data: high dimensionality and noise distribution.

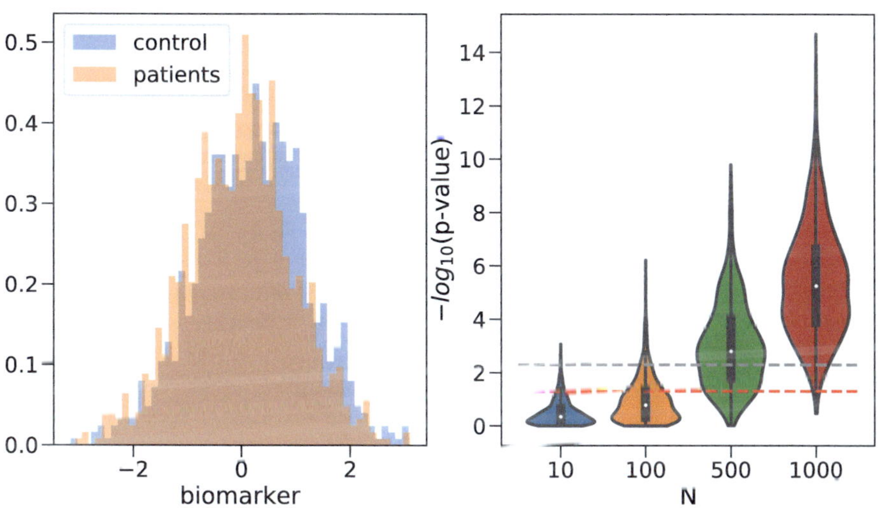

Fig. 4 Simulated data from two groups: control and patients. Data are simulated drawing samples from two Gaussian with unitary variance and a difference in means of 0.2. Left: Histogram with 1000 samples for each group; the difference is perceptible but the overlap is considerable. Right: Distribution of $\log_{10}$ (p-value) for a different number of samples; transformation is used to make clear the whole range of values. Typical threshold of 0.05 and 0.005 is shown in red and gray, respectively. For small sample size (10, 100), the effect is almost never detected, while for 1000 samples, it is almost always detected

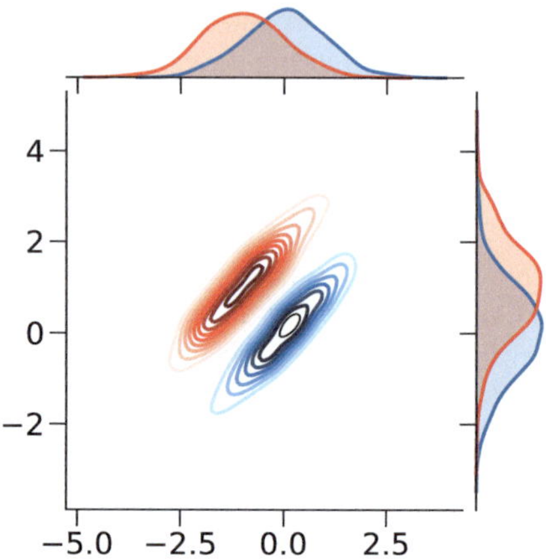

Fig. 5 Simulated data from a bivariate process in two conditions (red and blue). The two dimensions are correlated. The two conditions only differ by a small shift in the mean of the noise for each dimension. The marginal distribution of each dimension shows a considerable overlap between the conditions while the distribution in the bivariate space shows a perfect separation due to the correlation of the dimensions

FC or EC are composed of a large number of connections. For example, Pearson's correlation matrix of 100 brain regions has 4950 entries. So, each connectome can be considered as a sample point in a space with very high dimensionality. The problem of testing for differences in this space with thousands of dimensions can be decomposed in principle into many univariate tests, one for each dimension. Once the appropriate multiple tests controlling procedure is applied, there is nothing wrong in principle with this approach, in particular if the dimensions of the data are independent. However, univariate analysis can miss some information if there is correlation between dimensions. To illustrate this point we show, in Fig. 5, the situation where two groups, defined in a bivariate space, largely overlap in single dimensions but are perfectly separated in the 2D space.

In addition to the high dimensionality, connectome data presents a mixture of variability from different sources: session-to-session variability, subject-specific, and condition-related information in the simplest cases where additional confounders (like drug use) are not relevant. For example, the FC estimated from the same control participant over different scanning sessions present certain variability; the FC of different control participants will also be variable; finally, the addition of multiple groups (e.g., Alzheimer's disease and mild cognitive impairment) provides another source of variability. Sometimes even more noise sources are present, for example, if data acquired with different scanners even from same brand and model. Ideally, we would like the variability due to the clinical groups to be large compared with the other sources of variability but demixing the noise is a complex task.

ML tools nicely deal with high-dimensional data, hence do not incur in limitations of univariate methods and problems of multiple testing. ML techniques consist of building models for prediction of discrete (classifiers) or continuous variables (regression). So, they address more directly diagnostic and prognostic questions.[6] ML models have free parameters that need to be estimated, e.g., intercept and slope of a simple regression line. Once the parameters are estimated, they are kept fixed and the model can make predictions for each sample. In general, we are more interested in making predictions for new samples that have not been used in the estimating procedure. In this process, there is no test of null-hypothesis involved; hence, problems related to p-value are not an issue. ML models also have underlying assumptions that may affect the results. However, divergence from assumptions can be easily detected from poor predictive performance and other models with different assumptions can be tried.

Interpretable ML Algorithms

As we noted above, each connectome is a sample in a space where the number of dimensions scales approximately with the squared number of ROIs defined in the brain atlas. So, if we estimate one connectome from each scanning session (we might estimate more if we use dynamic connectivity estimation explained in section "Dynamic Connectivity Measures"), there will be as many samples as sessions (in some cases a subject might have multiple scans). It is common in ML and statistics to arrange data in the so-called design matrix, where each sample lies on one row and each dimension (also called features) stays in one column. So, for typical **big** clinical datasets, the design matrix will have about 1000 rows (1000 subjects, one scan each) and about 4000 columns (one per link considering an atlas of about 100 ROIs, although the trend is to use atlases with finer spatial resolution, in general much larger than 100 ROIs). This is then called a wide matrix.

Wide matrices are a problem in ML because they lead in general to poor predictions for new data. To understand why, let's imagine a simple classification scenario in one dimension: we want to find the point on our dimension that separates all samples in two classes; if there are only two samples, the separating point, the so-called boundary, always exists. If we move to samples in two dimensions, we ought to find a separating line: if there are no more than three samples we can always find that line (unless the three samples lie on the same line, but in that case, their actual dimensionality is one). With three-dimensional samples, we can always perfectly separate four samples. So, in general, n samples are separable in an $(n-1)$-dimensional space. While this might look like a positive situation, problems come when we want to make predictions for new samples. New samples will usually be close to those used in estimating the boundary but with some variability. This variability will produce errors in the prediction. This is a simple example of overfitting in the context of high-dimensional data. In general, overfitting

[6]Type 4 questions are naturally approached in ML as clustering problems (you can learn more about clustering in Ref. [49]).

occurs when the model is too complex compared to the number of samples. Here the complexity comes from the high number of dimensions.

The first step to fight overfitting is detecting it. To this aim, a subset of the samples, the so-called validation set, is kept separate from the data used to estimate the model's parameters, the so-called training set. This allows for the estimation of the prediction accuracy on "unseen" data. Since overfitting can be compensated by increasing the number of samples it seems a bad idea to decrease it by keeping some data for validation. So, there is a trade-off between a good estimation of validation accuracy and using many samples to control overfitting. Techniques to split the data in training and validation set are called cross-validation. The choice of the right cross-validation scheme is crucial for neuroimaging datasets where the number of samples is not very big, since this can lead to confirmation bias and non-reproducible results [50, 51]. Recently, random resampling has been proposed as a good procedure for neuroimaging, where a randomly chosen 20% of the samples is used for validation and the split is repeated 50 times [52]. In order to avoid biases, all preprocessing steps that involve the estimation of parameters need to be taken into account for cross-validation in order to avoid leaking of information from the validation set to the training set: for example, if a functional atlas is used to define ROIs (you can read more about best practices in these cases in Ref. [45]).

Since overfitting is produced by the high number of dimensions, a common approach to control overfitting is *dimensionality reduction*: a family of techniques to represent the data using fewer dimensions. The most common technique in this family is principal component analysis (PCA) that allows to represent the data in a new space where each dimension (or principal component) is a linear combination of all original dimensions and the principal components are ordered according to how much variance of the data they contain. Components with very low variance can then be discarded without altering too much the information contained in the dataset. Below we discuss also another dimensionality technique.

Prediction accuracy is not the only outcome of a computational pipeline for clinical applications. The ability to interpret the model, to understand how it makes such predictions, is equally important in order to gain a better understanding of the studied disorder and to avoid the lack of confidence associated to black-box solutions. The first step in the interpretation of models for high-dimensional data is to identify the features (brain connections here) that are most relevant in a pathological condition. Then this reduced connectome can be further studied in term of the involved brain structures or characterized using graph theory tools.

The identification of relevant features is a hard task in complex models such as deep neural networks or random forests that are very popular for their performance [42, 53, 54]. Simple linear models, in contrast, are easy to interpret, since their prediction is usually a linear combination of the features, such that each brain connection can be directly assigned a relevance weight by the coefficient of the linear combination. In addition, Lasso regularization[7] can be applied to enforce

[7]Similar to that used in the context of graphical models (see Partial Correlation in section "Stationary Measures (Whole-Session Approach)").

the use of only few features. Nonlinear models can sometimes have better accuracy than linear ones and when there is a big gap, we probably want to choose the model that gives the best predictions, especially for diagnostic or prognostic purpose. However, given the high dimensionality of neuroimaging data, linear models sometimes perform good enough and are therefore preferable for their amenability to interpretation. The identification of relevant brain connections can also be integrated into the dimensionality reduction phase. For this purpose approaches such as PCA are usually not helpful since all features participate in each principal component. The subfamily of techniques called *feature selection* [49] is a better choice in order to obtain an interpretable model. These techniques discard features in the original space or rank features [55–57] in terms of prediction accuracy, such that the best subset can easily be chosen in terms of prediction accuracy. Recursive feature elimination [55], one of these techniques has been recently used to identify the subnetworks related to subject specific and cognitive-state information [29].

In summary, in this chapter, we have seen that all connectivity measures based on BOLD fMRI are based on a model (those that come from the connectivity matrix parameter of the model are called EC) and that quality of fit is given by the sum of modeling error and estimation error. Estimation error can be quite large when the number of brain region is comparable to the number of time samples. We have seen some undirected FC (Pearson's correlation and MI) and EC (partial correlation), as well as directed EC (VAR, MOU, Granger causality, DCM). Some measures are fully connected graphs (Pearson's correlation), others are sparse (MOU, DCM) and the sparsity can come from the combination of functional and DTI data. We have also seen that a simple approach to go beyond the stationarity assumption is the use of sliding window estimation. Finally, we have seen that ML is better suited than null-hypothesis testing to link the estimated connectomes with diagnosis and prognosis of neuropsychiatric diseases and that linear models and features selection are good tools for building interpretable algorithm to predict clinical variables.

References

1. Devanand DP, Pradhaban G, Liu X, Khandji A, De Santi S, Segal S, et al. Hippocampal and entorhinal atrophy in mild cognitive impairment: prediction of Alzheimer disease. Neurology. 2007;68(11):828–36.
2. Greicius M. Resting-state functional connectivity in neuropsychiatric disorders. Curr Opin Neurol. 2008;21(4):424.
3. Hohenfeld C, Werner CJ, Reetz K. Resting-state connectivity in neurodegenerative disorders: is there potential for an imaging biomarker? NeuroImage Clin. 2018;18:849–70.
4. Schmahmann JD, Pandya DN. Cerebral white matter–historical evolution of facts and notions concerning the organization of the fiber pathways of the brain. J Hist Neurosci. 2007;16(3):237–67.
5. Sporns O, Tononi G, Kötter R. The human connectome: a structural description of the human brain. PLoS Comput Biol. 2005;1(4):e42.
6. Hagmann P, Cammoun L, Gigandet X, Meuli R, Honey CJ, Wedeen VJ, et al. Mapping the structural core of human cerebral cortex. PLoS Biol. 2008;6(7):e159.

7. Gillebert CR, Mantini D. Functional connectivity in the normal and injured brain. Neurosci Rev J Bringing Neurobiol Neurol Psychiatry. 2013;19(5):509–22.
8. Buckner RL. Human functional connectivity: new tools, unresolved questions. Proc Natl Acad Sci U S A. 2010;107(24):10769–70.
9. Hlinka J, Palus M, Vejmelka M, Mantini D, Corbetta M. Functional connectivity in resting-state fMRI: is linear correlation sufficient? NeuroImage. 2011;54(3):2218–25.
10. Cabral J, Hugues E, Sporns O, Deco G. Role of local network oscillations in resting-state functional connectivity. NeuroImage. 2011;57(1):130–9.
11. Cabral J, Vidaurre D, Marques P, Magalhães R, Silva Moreira P, Miguel Soares J, et al. Cognitive performance in healthy older adults relates to spontaneous switching between states of functional connectivity during rest. Sci Rep. 2017;7(1):5135.
12. Aertsen AM, Gerstein GL, Habib MK, Palm G. Dynamics of neuronal firing correlation: modulation of "effective connectivity". J Neurophysiol. 1989;61(5):900–17.
13. Friston KJ. Functional and effective connectivity: a review. Brain Connect. 2011;1(1):13–36.
14. Friston KJ, Harrison L, Penny W. Dynamic causal modelling. NeuroImage. 2003;19(4):1273–302.
15. Valdes-Sosa PA, Roebroeck A, Daunizeau J, Friston K. Effective connectivity: influence, causality and biophysical modeling. NeuroImage. 2011;58(2):339–61.
16. Lafleur L-P, Tremblay S, Whittingstall K, Lepage J-F. Assessment of effective connectivity and plasticity with dual-coil transcranial magnetic stimulation. Brain Stimulat. 2016;9(3):347–55.
17. Gilson M, Moreno-Bote R, Ponce-Alvarez A, Ritter P, Deco G. Estimation of directed effective connectivity from fMRI functional connectivity hints at asymmetries of cortical connectome. PLoS Comput Biol. 2016;12(3):e1004762.
18. Anscombe FJ. Graphs in statistical analysis. Am Stat. 1973;27(1):17–21.
19. Rousseeuw PJ, Leroy AM. Robust regression and outlier detection. Wiley; 1987 [cited 2018 Dec 31]. (Wiley series in probability and mathematical statistics. Applied probability and statistics). http://gen.lib.rus.ec/book/index.php?md5=DDF4DC5F86EDCE42F8A9588034 40B025.
20. Ledoit O, Wolf M. Nonlinear shrinkage estimation of large-dimensional covariance matrices. Ann Stat. 2012;40(2):1024–60.
21. Bartz D, Müller K-R. Covariance shrinkage for autocorrelated data. In: Ghahramani Z, Welling M, Cortes C, Lawrence ND, Weinberger KQ, editors. Advances in neural information processing systems 27. Curran Associates, Inc.; 2014 [cited 2018 Dec 31]. p. 1592–1600. http://papers.nips.cc/paper/5399-covariance-shrinkage-for-autocorrelated-data.pdf.
22. Friedman J, Hastie T, Tibshirani R. Sparse inverse covariance estimation with the graphical lasso. Biostatistics. 2008;9(3):432–41.
23. MacKay DJC. Information theory, inference and learning algorithms. Cambridge University Press; 2003. 644 p.
24. Hartman D, Hlinka J, Paluš M, Mantini D, Corbetta M. The role of nonlinearity in computing graph-theoretical properties of resting-state functional magnetic resonance imaging brain networks. Chaos. 2011 [cited 2018 Oct 24];21(1). https://www.ncbi.nlm.nih.gov/pmc/articles/PMC4108645/.
25. Hyvärinen A, Zhang K, Shimizu S, Hoyer PO. Estimation of a structural vector autoregression model using non-gaussianity. J Mach Learn Res. 2010;11(May):1709–31.
26. Gates KM, Molenaar PCM, Hillary FG, Ram N, Rovine MJ. Automatic search for fMRI connectivity mapping: an alternative to granger causality testing using formal equivalences among SEM path modeling, VAR, and unified SEM. NeuroImage. 2010;50(3):1118–25.
27. Goebel R, Roebroeck A, Kim D-S, Formisano E. Investigating directed cortical interactions in time-resolved fMRI data using vector autoregressive modeling and Granger causality mapping. Magn Reson Imaging. 2003;21(10):1251–61.

28. Insabato A, Cunningham JP, Gilson M. Bayesian estimation for large scale multivariate Ornstein-Uhlenbeck model of brain connectivity. ArXiv180510050 Cs Stat. 2018 May 25 [cited 2018 May 30]. http://arxiv.org/abs/1805.10050.
29. Pallarés V, Insabato A, Sanjuán A, Kühn S, Mantini D, Deco G, et al. Extracting orthogonal subject- and condition-specific signatures from fMRI data using whole-brain effective connectivity. NeuroImage. 2018;1(178):238–54.
30. Granger CWJ. Investigating causal relations by econometric models and cross-spectral methods. Econometrica. 1969;37(3):424–38.
31. Lütkepohl H. New introduction to multiple time series analysis. Berlin, Heidelberg: Springer; 2005 [cited 2018 Dec 31]. www.springer.com/us/book/9783540401728.
32. Brovelli A, Chicharro D, Badier J-M, Wang H, Jirsa V. Characterization of cortical networks and corticocortical functional connectivity mediating arbitrary visuomotor mapping. J Neurosci. 2015;35(37):12643–58.
33. Frässle S, Lomakina EI, Kasper L, Manjaly ZM, Leff A, Pruessmann KP, et al. A generative model of whole-brain effective connectivity. NeuroImage. 2018;1(179):505–29.
34. Li B, Wang X, Yao S, Hu D, Friston K. Task-dependent modulation of effective connectivity within the default mode network. Front Psychol. 2012;3:206.
35. Bassett DS, Wymbs NF, Porter MA, Mucha PJ, Carlson JM, Grafton ST. Dynamic reconfiguration of human brain networks during learning. Proc Natl Acad Sci USA. 2011;108(18):7641–6.
36. Mitra A, Snyder AZ, Tagliazucchi E, Laufs H, Raichle ME. Propagated infra-slow intrinsic brain activity reorganizes across wake and slow wave sleep. eLife. 2015;4:e10781.
37. Park H-J, Friston K. Structural and functional brain networks: from connections to cognition. Science. 2013;342(6158):1238411.
38. Gonzalez-Castillo J, Hoy CW, Handwerker DA, Robinson ME, Buchanan LC, Saad ZS, et al. Tracking ongoing cognition in individuals using brief, whole-brain functional connectivity patterns. Proc Natl Acad Sci. 2015;112(28):8762–7.
39. Brovelli A, Badier J-M, Bonini F, Bartolomei F, Coulon O, Auzias G. Dynamic reconfiguration of visuomotor-related functional connectivity networks. J Neurosci. 2016;15:1672–16.
40. Hindriks R, Adhikari MH, Murayama Y, Ganzetti M, Mantini D, Logothetis NK, et al. Can sliding-window correlations reveal dynamic functional connectivity in resting state fMRI? NeuroImage. 2016;15(127):242–56.
41. Park H-J, Friston KJ, Pae C, Park B, Razi A. Dynamic effective connectivity in resting state fMRI. NeuroImage. 2018;180(Pt B):594–608.
42. Rahim M, Thirion B, Bzdok D, Buvat I, Varoquaux G. Joint prediction of multiple scores captures better individual traits from brain images. NeuroImage. 2017;1(158):145–54.
43. Meskaldji D-E, Preti MG, Bolton TA, Montandon M-L, Rodriguez C, Morgenthaler S, et al. Prediction of long-term memory scores in MCI based on resting-state fMRI. NeuroImage Clin. 2016;12:785–95.
44. Gao S, Calhoun VD, Sui J. Machine learning in major depression: from classification to treatment outcome prediction. CNS Neurosci Ther. 2018;24(11):1037–52.
45. Brodersen KH, Deserno L, Schlagenhauf F, Lin Z, Penny WD, Buhmann JM, et al. Dissecting psychiatric spectrum disorders by generative embedding. NeuroImage Clin. 2013;16(4):98–111.
46. Drysdale AT, Grosenick L, Downar J, Dunlop K, Mansouri F, Meng Y, et al. Resting-state connectivity biomarkers define neurophysiological subtypes of depression. Nat Med. 2017;23(1):28–38.
47. Dinga R, Schmaal L, Penninx B, Tol MJ van, Veltman D, Velzen L van, et al. Evaluating the evidence for biotypes of depression: attempted replication of Drysdale et al. 2017. bioRxiv. 2018 Sep 14;416321.
48. Wagenmakers E-J. A practical solution to the pervasive problems ofp values. Psychon Bull Rev. 2007;14(5):779–804.

49. Friedman J, Hastie T, Tibshirani R. The elements of statistical learning. Vol. 1. In: Springer series in statistics New York, NY, USA; 2001.
50. Varoquaux G. Cross-validation failure: small sample sizes lead to large error bars. NeuroImage. 2017 [cited 2018 Feb 18]. http://www.sciencedirect.com/science/article/pii/S1053811917305311.
51. Arbabshirani MR, Plis S, Sui J, Calhoun VD. Single subject prediction of brain disorders in neuroimaging: promises and pitfalls. NeuroImage. 2017;15(145):137–65.
52. Varoquaux G, Raamana PR, Engemann DA, Hoyos-Idrobo A, Schwartz Y, Thirion B. Assessing and tuning brain decoders: cross-validation, caveats, and guidelines. NeuroImage. 2017;15(145):166–79.
53. Sarica A, Cerasa A, Quattrone A. Random forest algorithm for the classification of neuroimaging data in Alzheimer's disease: a systematic review. Front Aging Neurosci. 2017 [cited 2018 Feb 23];9. https://www.frontiersin.org/articles/10.3389/fnagi.2017.00329/full.
54. Vieira S, Pinaya WHL, Mechelli A. Using deep learning to investigate the neuroimaging correlates of psychiatric and neurological disorders: methods and applications. Neurosci Biobehav Rev. 2017;1(74):58–75.
55. Guyon I, Weston J, Barnhill S, Vapnik V. Gene selection for cancer classification using support vector machines. Mach Learn. 2002;46(1):389–422.
56. Roffo G, Melzi S, Cristani M. Infinite feature selection. In: 2015 IEEE international conference on computer vision (ICCV); 2015. p. 4202–10.
57. Roffo G, Melzi S, Castellani U, Vinciarelli A. Infinite latent feature selection: a probabilistic latent graph-based ranking approach. In 2017 [cited 2018 Nov 13]. p. 1398–406. http://openaccess.thecvf.com/content_iccv_2017/html/Roffo_Infinite_Latent_Feature_ICCV_2017_paper.html.

Development of Neuroimaging-Based Biomarkers in Psychiatry

Orkun Aydin, Pinar Unal Aydin and Ayla Arslan

Abstract

This chapter presents an overview of accumulating neuroimaging data with emphasis on translational potential. The subject will be described in the context of three disease states, i.e., schizophrenia, bipolar disorder, and major depressive disorder, and for three clinical goals, i.e., disease risk assessment, subtyping, and treatment decision.

Keywords

Connectomics · Neuroimaging · Amygdala · Hippocampus · Prefrontal cortex · White matter · Gray matter · Functional magnetic resonance imaging · Magnetic resonance imaging · Positron emission tomography · Diffusion tensor imaging · Voxel-based morphometry · Machine learning · Biomarker · Schizophrenia · Depression · Diagnosis · Prognosis · Predictive · Personalized · Bipolar disorder

Abbreviations

AD	Alzheimer's disease
ADHD	Attention deficit hyperactivity disorder

O. Aydin · P. Unal Aydin
Psychology Program, Faculty of Arts and Social Sciences, International University of Sarajevo, Hrasnička cesta br. 15, 71210 Sarajevo, Bosnia and Herzegovina

A. Arslan (✉)
Genetics and Bioengineering Program, Faculty of Engineering and Natural Sciences, International University of Sarajevo, Hrasnička cesta br. 15, 71210 Sarajevo, Bosnia and Herzegovina
e-mail: aarslan@ius.edu.ba

© Springer Nature Singapore Pte Ltd. 2019

Y.-K. Kim (ed.), *Frontiers in Psychiatry*, Advances in Experimental Medicine and Biology 1192, https://doi.org/10.1007/978-981-32-9721-0_9

ACC	Anterior cingulate cortex
BD	Bipolar disorder
BOLD	Blood oxygenation level dependent
CS	Chronic schizophrenia
CT	Computerized tomography
DALYs	Disability-adjusted life years
DSM	Diagnostics and Statistics Manual
DMN	Default mode network
DLPFC	Dorsolateral prefrontal cortex
DTI	Diffusion tensor imaging
EPI	Echo-planar imaging
FA	Fractional anisotropy
FCS	Functional connectivity strength
FES	First episode schizophrenia
fMRI	Functional magnetic resonance imaging
ICD	International Classification of Diseases
MCI	Mild cognitive impairment
MRI	Magnetic resonance imaging
MDD	Major depression disorder
mPFC	Medial prefrontal cortex
NCD	Neurocognitive disorder
OCD	Obsessive-compulsive disorder
P4	Predictive, preventative, personalized, and participatory
PET	Positron emission tomography
RDoC	Research domain criteria
SAD	Social anxiety disorder
SZ	Schizophrenia
SPECT	Single-photon emission computed tomography
VBM	Voxel-based morphometry
YLDs	Years lived with disability

Introduction

One of the greatest challenges in psychiatry is to identify biomarkers that aid the diagnosis, prognosis, and treatment of a given pathological condition. Moreover, a repertoire of biomarkers is desired for predictive, preventative, and personalized treatment. This is particularly important as the so-called P4 (predictive, preventative, personalized and participatory)—medicine is the direction of modern medical practices [1]. This direction has been fueled by recent progress in various disciplines including, molecular biology, omics technology, information technology, medical technology. On the other hand, psychiatry does not seem to benefit from this progress as desired [2–4]. This is partially due to the complexity of gene–environment interactions involved in the etiology of mental disorders [5–9] and partially due to the complexity of psychiatric phenotypes, i.e., "phenotype bottleneck" [3]. Nevertheless, a continuous effort made to address these problems leading to the emergence of several important progress including the development of Research domain criteria [RDoC) [10] and (neuroimaging) endophenotypes [11] to reduce the phenotype bottleneck, for example.

The concept of the endophenotype, which was introduced already in 1960s [12] was first adapted to psychiatry in the 1970s as a strategy to facilitate gene discovery [13]. Thus, Gottesman's original endophenotype concept was formulated as measurable heritable characteristics that "*co-segregate with a psychiatric illness, yet be present even when the disease is not (i.e. state independent), and be found in non-affected family members at a higher rate than in the population.*" Since then, this concept with some modifications and perhaps with some overestimation has been increasingly applied to psychiatric research [14].

Methodologies that assist the characterization of endophenotypes commonly involve metabolic, anatomical, electrophysiological or neural activity measurements. Regarding this, one approach is based on the measurement of blood oxygenation level dependent (BOLD) contrast of blood vessels. The first demonstration of this approach, which was reported in rat brain by Ogawa et al. [17], was followed by its application on human brain [15–17]. These, together with the advancements such as echo-planar imaging (EPI) [18] have led to the emergence of structural magnetic resonance imaging (MRI), and functional magnetic resonance imaging (fMRI), mapping of whole brain with significant spatial and temporal resolution. In the meanwhile, other neuroimaging approaches such as diffusion tensor imaging (DTI), computerized tomography (CT), spectroscopy, and positron emission tomography (PET) have been increasingly used integratively leading to the multimodal neuroimaging. Among these, the last two (spectroscopy and PET) correspond to molecular neuroimaging, which refers to the visualization, characterization, and measurement of biological processes at the molecular and cellular level in vivo [19, 20]. The principle of the molecular neuroimaging systems relies on a probe interacting with the molecular or cellular substrate and producing a signal which is amplified to form an image [20].

The Current State of Neuroimaging in the Neuropsychiatric Clinic

Despite the accumulation of such neuroimaging studies in the literature, translation into psychiatric clinical practice was not fruitful and studies mostly retained in the theoretical level [21–29]. This is well reflected by several reviews screening neuroimaging studies as potential biomarker for suicidality [30], Huntington's disease [31], bipolar disorder [32], for example. Thus, the current neuroimaging biomarkers in clinical practice mostly involve their application in the diagnosis of neurocognitive disorders (NCD). Table 1 shows the current practice of neuroimaging markers based on international diagnostic guidelines. Below a summary of these conditions is presented:

Table 1 Neuroimaging biomarkers for brain diseases/disorders currently used in the clinical practice. Each disease condition is represented as DSM-5 and ICD-10 categories. (CT: Computer tomography; DSM: Diagnostics and Statistics Manual; fMRI; ICD: International Classification of Diseases MRI: SPECT: PET: Positron emission tomography)

Condition	Modality	Criteria	Biomarker	Reference
Major or mild frontotemporal neurocognitive disorder (DSM-5) Frontotemporal dementia (ICD-10)	CT MRI fMRI	Diagnostic criteria	Disproportionate frontal and/or temporal lobe involvement from neuroimaging	DSM-5 ICD-10 G31.0
Major or mild neurocognitive disorder with Lewy bodies (DSM-5) dementia with Lewy bodies (ICD-10)	SPECT PET	Diagnostic criteria	Reduced dopamine transporter uptake in basal ganglia	DSM-5 ICD-10 G31.83
Major or mild vascular neurocognitive disorder (DSM-5) Vascular dementia (ICD-10)	CT MRI	Diagnostic criteria	Significant parenchymal injury attributed to cerebrovascular disease	DSM-5 ICD-10 F01
Major or mild neurocognitive disorder due to traumatic brain injury (DSM-5) Dementia due to head trauma (ICD-10)	CT MRI	Diagnostic criteria	Neuroimaging demonstrating injury	DSM-5 ICD-10 F02.8X, S09.8X
Substance/medication-induced major or mild neurocognitive disorder (DSM-5) Dementia in other diseases classified elsewhere (ICD-10)	MRI DTI MR SPECT	Helpful in diagnosis	Structural changes in white matter and brain cortex	DSM-5 ICD-10 F10.26 F02, F10–F19
Major or mild neurocognitive disorder due to HIV infection (DSM-5) Dementia in other diseases classified elsewhere (ICD-10)	MRI DTI	Helpful in diagnosis	Structural changes in white matter and brain cortex	DSM-5 ICD-10 F02, B20
Major or mild neurocognitive disorder due to prion disease (DSM-5) Dementia in other diseases classified elsewhere (ICD-10)	MRI with DWI/ FLAIR	Diagnostic criteria	Multifocal gray matter hyperintensities in subcortical and cortical regions	DSM-5 ICD-10 F02, A81.9
Major or mild neurocognitive disorder due to Parkinson's disease (DSM-5) Dementia associated with Parkinson's disease (ICD-10)	MRI-DaT	Helpful in diagnosis	Structural changes in brain and dopamine transporters	DSM-5 ICD-10 G20, G31.83
Major or mild neurocognitive disorder due to Huntington's disease (DSM-5) Dementia due to Huntington's disease (ICD-10)	MRI	Helpful in diagnosis	Structural changes in basal ganglia	DSM-5 ICD-10 G10

(i) Major or Mild Frontotemporal NCD: The neuroimaging is essential to discriminate probable and possible frontotemporal neurocognitive disorder (Probable frontotemporal neurocognitive disorder is diagnosed if evidence of disproportionate frontal and/or temporal lobe involvement from neuroimaging is present. Distinctive atrophy or reduced activity, hypoperfusion, and/or cortical hypometabolism in frontotemporal regions can be shown on structural or functional imaging.

(ii) Major or Mild NCD with Lewy Bodies: The neuroimaging is beneficial in diagnosing probable and possible neurocognitive disorder with Lewy bodies. One of the indicative biomarkers is low striatal dopamine transporter uptake on single-photon emission computed tomography (SPECT) or positron emission tomography (PET) scan. This biomarker can be used in diagnosing possible neurocognitive disorder with Lewy bodies regardless of core clinical features. Moreover, this indicative biomarker can be used in probable neurocognitive disorder with Lewy bodies if core clinical features are present. Also, several supportive biomarkers are identified: relative preservation of medial temporal structures on computed tomography (CT)/magnetic resonance imaging (MRI) brain scan; reduced striatal dopamine transporter uptake on SPECT/PET scan; generalized low uptake on SPECT/PET perfusion scan with reduced occipital activity and/or the cingulate island sign on FDG-PET imaging.

(iii) Major or Mild Vascular NCD: The neuroimaging-supported evidence such as significant parenchymal injury attributed to cerebrovascular disease is sufficient to diagnose probable vascular neurocognitive disorder in addition to clinical criteria. If clinical criteria are met but neuroimaging is not available and the temporal relationship of the neurocognitive syndrome with one or more cerebrovascular events is not established, then the diagnosis will be possible vascular neurocognitive disorder. For mild vascular NCD, history of a single stroke or extensive white matter disease is generally sufficient. For major vascular NCD, two or more strokes, a strategically placed stroke, or a combination of white matter disease and one or more lacunes are generally necessary.

(iv) Major or Mild NCD Due to Traumatic Brain Injury: Neuroimaging-based injury demonstration is essential for diagnosis.

(v) Substance/Medication-Induced Major or Mild NCD: The neuroimaging outputs aid in diagnosing but they are not mandatory on diagnosis. That is, the cortical thinning, white matter loss, and enlargement of sulci and ventricles (MRI), damage to specific white matter tracts (DTI), reduction in N-acetylaspartate, and increase in markers of inflammation (e.g., myoinositol) or white matter injury (e.g., choline) (MRSpectroscopy) can be seen in chronic alcohol abuse cases. The hyperintensities suggestive of microhemorrhages or larger areas of infarction (MRI) may occur in methamphetamine use disorder.

(vi) Major or Mild NCD Due to HIV Infection: The neuroimaging can be used to ensure diagnosis. MRI may present reduction in total brain volume, cortical thinning, reduction in white matter volume, and patchy areas of abnormal white matter (hyperintensities); moreover, damage to specific white matter tracts can be encountered in DTI.

(vii) Major or Mild NCD Due to Prion Disease: The one of the characteristic biomarkers that is necessary for diagnosing is multifocal gray matter hyperintensities in subcortical and cortical regions on magnetic resonance imaging with DWl (diffusion-weighted imaging) or FLAIR (fluid-attenuated inversion recovery).

(viii) Major or Mild NCD Due to Parkinson's Disease: Neuroimaging is not obligatory in diagnosis but structural neuroimaging and dopamine transporter scans, such as DaT scans, may differentiate Lewy body-related dementias (e.g., Alzheimer's disease) from Lewy body-related dementias (Parkinson's and dementia with Lewy bodies) and can sometimes be helpful in the evaluation of this disorder.

(ix) Major or Mild NCD due to Huntington's Disease: The MRI can show volume loss in the basal ganglia, particularly the caudate nucleus and putamen; in addition, this reduction progresses over the course of illness, but it is not a diagnostic criterion.

On the other hand, biomarkers are urgently required in psychiatric practice [33]. The global burden of psychiatric disease appears to be more than expected: data show that the global burden of mental disorders accounts for 32.4% of years lived with disability (YLDs) and 13.0% of disability-adjusted life years (DALYs) [34]. In addition, the burden due to misdiagnosis and subsequent mistreatment is not uncommon [35].

Attempts to utilize neuroimaging for psychiatric conditions mostly involve studies which have a focus of comparison between states of disease and health. For example, a recent survey shows that 75% of the 615 neuroimaging studies are this type of translational studies [36]. This is well reflected by studies describing the neural bases of schizophrenia (SZ). For example, alterations of brain volume were described as a comparison of SZ patients and healthy controls: Total volume of frontal lobe bilaterally reduced in schizophrenia patients [37, 38] in addition to total reductions in brain volume for white and gray matter [37, 38] as well as regional gray matter changes including frontal lobe, thalamus, temporal lobe [39–42] basal ganglia [43], amygdala and hippocampus [37, 44–46], intracranial volume reduction and extracranial volume increase [47]; reduced volume, density or thickness of the cingulate cortex [48–52], the increased volume of cavum septum pellucidum [53]; bilaterally reduced volumes frontal lobe [37, 43]; superior temporal gyri volume reductions [37, 54]; bilateral volume reductions in parahippocampal [55], total ventricular volume increase [38, 46].

In one study conducted with over 2000 SZ patients and healthy controls that were recruited from 15 centers, it was noted that the volumes of subcortical areas were different between patients and controls. They ranked the brain regions that sustain the difference according to the effect sizes. Due to those results, patients with schizophrenia had smaller hippocampus, amygdala, thalamus, accumbens, and intracranial volumes, respectively. Moreover, patient group had larger pallidum and lateral ventricle volumes, respectively [56]. In a recent study, the hippocampal volume of high-risk subjects who subsequently developed psychotic disorders reduced. Moreover, volume reduction among converters was localized to the CA1 and subiculum subregions and was most prominent in the anterior left hippocampus [57]. The study results which belong to assessment of the cortical thickness among 1985 patients with schizophrenia revealed that left, medial orbitofrontal cortex thickness was related with negative symptom severity [58], despite previous studies with contradictive results [25].

Nevertheless, the studies described so far, being very important for fundamental understanding of neurobiological bases of brain disorders, may not offer a better solution than the current diagnostic guidelines for clinical applications [36]. For example for SZ, there is a need for objective measurable and reliable criteria for treatment decision. Moreover, objective predictors for the treatment response of a particular pharmacotherapy for individual patients are required. If a patient fails to respond to a specific treatment adequately, how long the treatment should be maintained before substitution with another treatment or considering psychosocial interventions? Addressing these problems, suitable psychiatric biomarkers would ideally serve for several purposes: Risk assessment and prediction, differential diagnosis, and disease subtyping, predicting treatment outcome [36].

Thus, what is really needed in clinics is something that would refine the current diagnostic guidelines which have a space for risk assessment for disease prevention, response prediction for treatment decision and diagnosis for disease subtyping or early detection. This can only be achieved by objective biological markers which may be addressed by clinically significant neuroimaging and/or other biological measures. In this chapter, we will provide a description of neuroimaging-based strategies aiming to address this need for three psychiatric conditions: schizophrenia (SZ), bipolar disorder (BD), and major depressive disorder (MDD).

The Potential Neuroimaging Markers for SZ

Accumulating neuroimaging studies characterize brain structure/function changes that can aid risk assessment in SZ. Moreover, brain structure/function changes due to illness duration and treatment in SZ can aid treatment decision, while some other neuroimaging data may help characterize the disease differentially. Nevertheless, replication studies are required [59] and the field is still immature [24]. Long story short, the data for potential biomarkers for SZ require further investigation, as there are inconsistencies which may be due to the lack of replication studies with increased and standardized sample sizes [60]. Below we will present these candidate biomarkers of neuroimaging for risk assessment in SZ, potential biomarkers for treatment response, for prognosis, disease classification, and subtyping.

Risk Assessment in SZ

Determination of at risk mental state (ARMS) or early diagnosis of the psychosis may be critical to postpone or prevent the disease onset. Accumulating studies address the prediction of ARMS by several neuroimaging measures such as total or regional brain volumes, thickness, and densities as well as structural and functional connectivity alterations. Functional alterations and risk determination is especially the result of functional MRI studies. For example, functional connectivity studies link the psychosis risk with aberrant connectivity of brain regions such as the temporal cortex [61], inferior frontal gyri [62], and thalamus [63]. Also, combination of genetic findings with functional neuroimaging sheds light on the association of polygenic risk scores with specific neural activity patterns in regions of interest. For example, one study generated a psychosis related polygenic risk score profiles for 1841 healthy adolescents and found an association with polygenic risk scores and the activity of ventral striatum [64]. Nevertheless, the field is still immature [65].

Analysis of chronic SZ (CS) and first episode schizophrenia (FES) may be important for early detection. Regarding this, the mid-sagittal area of corpus callosum was reduced in both CS and FES [66]. The gray matter volume in postcentral gyrus of the parietal lobe reduced in CS and FES [48, 52]. The reduction of gray matter density in the anterior cingulate cortex was reported in CS and FES [51, 52]. Bilateral volume reductions in parahippocampal area were shown in CS [55]. Besides, the voxel-based morphometry (VBM) studies in CS patients showed similar results related with white matter reductions in particular brain regions such as frontal lobe including DLPFC when compared with healthy controls [52]. Later studies found that temporal lobe and corpus callosum volume reductions might be observed in both FES and CS [42].

One way to test the initial utility of the above findings is the cumulative assessment of volume abnormalities for selective regions. Several studies have conducted a synthesis of accumulating data to determine their potential for the assessment of ARMS individuals. Accordingly, gray matter abnormalities [67], prefrontal cortex, hippocampal, and amygdala abnormalities are suggested as biomarker for risk assessment [68]. Indeed, most consistency found across the studies for gray matter abnormalities and alteration of prefrontal cortex and hippocampal volume besides to functional studies which describe the hyperactivity of PFC during executive control relevant tasks [69]. In line with these, other review analysis and longitudinal studies report progressive structural abnormalities that can predict the onset of psychosis differentially [70–74].

Unfortunately, the search of associated genetic factors such as polygenic risk scores that may underlie the so far mentioned structural brain alterations was not very fruitful [65, 75, 76] but this may derive from the methodological reasons [77]. Nevertheless, in combination with other measures [78] such as electrophysiology [79], inflammation and oxidative stress [80] the accuracy may increase, which would suggest the possibility of early intervention in SZ [81, 82]. Moreover, advancements in the analytical field are likely contributing to this goal. For example, the first study using multivariate neuroanatomical pattern analysis used whole-brain neuroanatomical alterations that may classify the ARMS [83]. Support vector machine (SVM) is a machine learning methodology used for multivariate pattern analysis, which may be useful for identification of individuals with ARMS [84]. Besides to SVM, Deep Learning (DL) is another machine learning methodology gaining momentum in recent years to analyze neuroimaging data [85].

Potential Biomarkers for Prognosis and Treatment Response

In general, neuroimaging seems to have good potential to predict treatment response and to monitor prognosis for better intervention. Several studies suggest that gray matter and white matter volumes are associated with SZ illness duration [51, 86–88], but more studies and replications are required. In general, greater severity induces more diffuse alterations in the brain [51], particularly the altered posterior gray matter volume might be observed [88]. In a recent study gray matter volume reductions in the left parahippocampal gyrus and white matter reductions of the superior temporal gyrus and right precuneus were found to be associated with positive symptoms in SZ [89]. Other gray matter alterations such as insula, superior temporal gyrus, and anterior cingulate cortex were negatively correlated with duration of illness in schizophrenia [86]. Also, decreased white matter volumes of the superior frontal gyrus, inferior temporal gyrus, and superior temporal gyrus were characterized patients with SZ [86]. One study with a very large sample of SZ patients (2028) revealed that volume increase in pallidum and putamen was related with the duration of illness, and deficits of hippocampus were associated with the proportion of medication-naive patients [56]. Nevertheless, the current research for prognostic biomarkers for SZ is still far from translational border.

It is essential to identify the impact of a specific drug on the structural and functional brain measures for prognostic interventions. Structural connectome, white matter integrity or white matter microstructure may be used to predict treatment response [90–92]. For example, poor treatment response to antipsychotics in the first episode of psychosis may be predicted by "Connectomic study of gyrification" [93, 94].

Region-specific functional and structural analysis may be essential to probe the specific improvements. Analysis of studies reporting the executive function related fronto-cingulo-parietal network shows activity dependent patterns matching with the most common transdiagnostic effects of pharmacological treatment as predictors [95]. Improvement in frontal cortex activation during a working memory task after substitution of risperidone in patients with SZ has been observed [96]. The effect of 6 weeks of amisulpride treatment may be assessed by normalization of white matter alterations especially alterations in frontal fasciculi integrity, which are associated with psychotic symptoms [97]. Regarding this, some studies suggest that clozapine is associated with distinctive structural and functional neuroimaging measures [98, 99] that are not shared with other antipsychotics [100] or neuroleptics [101]. Other studies suggest that measures of limbic circuitry determined by PET are suggested as the predictor for treatment response to first-generation antipsychotics flupentixol and haloperidol, and of the second-generation antipsychotics risperidone, aripiprazole, quetiapine, sertindole, ziprasidone, paliperidone, and olanzapine [102]. PET is also suggested for predicting extrapyramidal side effects from antipsychotic treatment [102]. Accumulating studies suggest dosage adjusted effects of risperidone and zuclopenthixol on cognitive functions in first episode drug-naive schizophrenic patients [103] or effect of risperidone on the volume of gray matter first episode drug-naive schizophrenic patients [104], effects of clozapine and risperidone in relation to cortical thinning in FES [105] or Clozapine association with progressive brain atrophy and cortical thinning in SZ [99]. In long-term antipsychotic treatment, if switching from typical to atypical antipsychotic medication was present, reduction in basal ganglia and thalamus volume was found [106].

On the other hand, the results of neuroimaging studies that evaluated patients undergoing initial antipsychotic treatment were inconsistent. In general, frontal and temporal lobe volume reduced in patients who took typical antipsychotics, thalamic volume reduction was observed under atypical antipsychotic treatment. Anterior cingulate cortex volume decreased in both typical and atypical antipsychotic treatment. Basal ganglia volume increased in patients who were treated with typical antipsychotics though results are not consistent [107–110]. Thus, the associations between treatment and changes of brain structure remain unclear because of the confounding effect of the illness severity, the alterations in the brain structure might reflect treatment effect [108, 109].

A recent survey of 98 studies shows various biological predictors of Clozapine response [111]. Among these, structural integrity and activity of prefrontal cortex and a lower ratio of the dopamine and serotonin metabolites in cerebrospinal fluid (CSF) have been suggested as predictor of clozapine response in patients of SZ [111]. Functional connectivity of the striatum with prefrontal and limbic regions may be a biomarker for prediction of treatment response in the first episode or chronic psychosis as well as to utilize the relation between longer duration of untreated psychosis and worsened response to treatment [112–114]. Structural data or functional connectivity assessed by machine learning method were reported as beneficial to predict treatment responses for interventions such as rTMS and Electroconvulsive therapy [115–117]. One question is whether treatment-resistant cases [118] can be predicted by neuroimaging, as about 30% of SZ patients show an inadequate response to antipsychotics. Recent literature suggests some distinct features of neuroimaging measures [119–121] compensating the required replications in the field [122]; thus, these findings such as decrease in gray matter could be potential biomarker for treatment-resistant SZ. However, there is still a need for further confirmation with larger sample sizes.

Neuroimaging for Disease Subtyping and Prediction

It is well known that clinical diagnostics guidelines such as DSM are reliable but they lack biological validity [3]. Thus, some studies in the literature focus on the comparison between biological dimensions in SZ with clinical diagnostic groups. Regarding this, one study used a range of biomarker panel that includes neuroimaging and analyzed the patients of SZ, schizoaffective disorder, and BD with psychosis ($N=711$), their first-degree relatives ($N=883$), and HCs ($N=278$). Data show that there may exist three psychosis biotypes which do not match with clinical diagnostic categories [123]. In line with this, Ivleva et al. [124] reported biotype classification based on whole-brain gray matter density and compared it with clinical diagnostics categories in a total of 1409 subjects of probands, their relatives, and healthy controls. Data suggest that biotypes caused better differentiation compared to the clinical diagnoses. Biotype1 was characterized by extensive and diffusely distributed loss of gray matter volumes with the largest effects in frontal, anterior/middle cingulate cortex, and temporal regions. Biotype 2 was characterized by moderate and rather localized reductions, with the largest impact on insula and frontotemporal regions. Biotype 3 was characterized by small reductions localized to anterior limbic regions [124]. Interestingly, one study used the functional connectivity patterns in terms of four dimensions of psychopathology—mood, psychosis, fear, and externalizing behavior. According to the results, authors suggested a link between the connectivity with mood and psychosis, crossing the clinical diagnostic categories. Moreover, sex differences were reported for connectivity related to mood and fear [125]. As the comorbidity is high among psychiatric conditions, this way of dimensional approach guided by neuroimaging might be a solution.

Other studies showed the beneficial effect of the fMRI in delineation of SZ border [83, 126]. The neural responses to verbal fluency task that observed in fMRI showed diagnostic specificity for SZ patients [127]. Structural MRI is assumed as feasible method to discriminate SZ patients from healthy controls [128, 129]. Diffusion-weighted MRI studies also supported the identification of illness with small effect size [130]. Another study supplemented by the fMRI data showed altered activations in prefrontal cortex, ventral cortex, and dorsolateral prefrontal cortex could predict the patients' distinct group belongings due to their different symptom severity and clinical symptomatology [131]. DTI studies revealed the white matter abnormalities particularly in the left superior longitudinal fasciculus in the early phase of SZ; moreover, the individuals exhibited global white matter abnormalities suffering from negative symptoms [132]. Data show that distinct subgroups of patients with SZ have different forms of white matter pathology but there is a need for filling the gap of large sampled studies. In a very recent study, the structural MRI results were combined with severity of symptoms, cognitive failure, and longitudinal symptom change parameters for discriminating the two different subtypes of schizophrenia patients. Subcortical and cortical volume reduction and the alterations in other defined parameters could predict the particular subtype group with high accuracy rates [133].

Indeed, gray matter volume may be one of the characteristics which aid the differentiation of SZ and BD. Regarding this, more prominent reductions in the gray matter volume in SZ compared to BD was observed in a total of 802 subjects: 243 SZ, 176 BD, 383 HC [134]. Other candidate biomarkers are decreased gray matter density in the vermis and tonsil of cerebellum for SZ with cognitive deficits [62], greater alteration of gray matter deficits in earlier age of disease onset [87]. Also, a recent meta-analysis evaluated the gray matter differences in non-affective and affective first episode psychosis [135]. Results show a shared gray matter alteration in the regions of frontotemporal and anterior cingulate cortex, while unique differences in gray matter alterations have been detected in amygdala for non-affective first episode psychosis and in hippocampus and insula for the affective first episode psychosis. Nevertheless, none of these studies are conclusive to reach clinical significance [135, 136]. In the future longitudinal studies with larger sample sizes are required.

Several studies used connectome-based biomarkers to discriminate schizophrenic patients from healthy controls or from other disease states [137–142]. Some multisite studies used functional connectivity to discriminate SZ patients with relatively high accuracy [138, 143, 144]. Functional connectivity-based features for classification of SZ and BP patients at the individual level have also been studied [141, 145, 146].

Use of neuroimaging data for single subject prediction to classify subjects into specific categories has been suggested as an alternative approach for differential diagnosis. Single subject prediction is based on classifying each subject into one of the groups and determining its accuracy. This is different than other studies [82, 147–150] which rely on the average value of alterations in a patient group compared to controls [137]. Single subject prediction assessed by functional connectivity has been used for classification of SZ and BP with more than 80% accuracy in some studies [141, 145, 146, 151]. Regarding this, one study reported that dynamic connectivity, time-resolved (scale of tens of seconds) analysis of functional connectivity was superior compared to static connectivity, which assumes the connectivity as an average of 5–10 min. Another study reported the single subject prediction to distinguish SZ from mood disorders [139]. Thus, candidate biomarkers derived from the analysis of dynamic functional connectivity will likely increase in the future for disease classification and subtyping [152, 153] despite its limitations [154]. A recent multisite study [155] developed a deep discriminant autoencoder network with sparsity (DANS) network for the diagnosis of SZ patients. By using the largest multisite functional connectivity data obtained from more than 1000 participants (474 SZ patients), the accuracies were more than 81% [155]. However, there are some caveats of these kinds of studies, such as the potential confounding variables, limited sample size, and lack of multimodal data integration.

The Potential Neuroimaging Markers for BD

BD corresponds to a range of affective disorders characterized by episodes of depression, and episodes of either mania and/or irritable mood. BD shares common characteristics with SZ and MDD [43, 156]. This is well reflected by high rates of delayed diagnosis, misdiagnosis, and the problems of comorbid conditions causing patients with BD to have high disability, unemployment, and mortality [157].

Potential Neuroimaging Biomarkers for Risk Assessment

A literature review reports that studies for the risk assessment for BD are mostly based on the evaluation of familial risk and biomarkers are far from translation to clinical settings [158]. On the other hand, empirical studies accumulate. According to meta-analysis of 65 fMRI studies which compare 1040 patients of BD with healthy controls, BD patients had decreased activity in inferior frontal cortex (during cognitive and emotional processing/in manic episode) and increased activity in limbic regions (during emotional processing) of the medial temporal lobe including parahippocampus, hippocampus, and amygdala as well as regions of basal ganglia [159]. Extending on these several lines of evidences focus on structural connectivity alterations. For example, some studies suggest WM integrity as a potential biomarker that maybe useful for risk identification of BD [160–162]. Regarding this, fronto-limbic dysconnectivity has been suggested as a neuroimaging biomarker to predict individuals at risk [162].

Differential Diagnosis and Disease Subtyping by Neuroimaging

There are several structural neuroimaging studies that were found to be notable in distinguishing BD from healthy participants and unipolar depression. The results for those studies that used MRI with VBM showed reduced gray matter volumes in the amygdala and hippocampus and in addition reductions in white matter volume within the cerebellum and hippocampus in patients with BD [163]. Gray matter and white matter alterations localized within cortical and subcortical structures are assumed as useful tools for discriminating BD patients and controls [164]. According to the whole-brain DTI assessments, fractional anisotropy (FA) was examined and decreased FA was showed in anterior thalamic region, superior longitudinal fasciculus, and corpus callosum which were the most implicated areas in bipolar disorders [165]. In addition to those common shared regions in bipolar patients, bipolar disorder type 1 patients had decreased FA in the right precuneus, else [165]. For the adolescent bipolar disorders, the left cingulum hippocampus FA increases after regular lithium treatment and the results of this study claimed that pretreatment FA of the left cingulum hippocampus white matter could predict the symptom severity [166]. Due to tractography results, the main affected areas were inferior longitudinal fasciculus and the inferior fronto-occipital fasciculus [167]. In a recent meta-analysis, the bipolar and major depressive disorders compared with healthy controls. The study results revealed that anterior cingulate cortex, bilateral insula, ventromedial and dorsomedial prefrontal cortex gray matter volume reductions were observed in both depressive and bipolar disorder patients.

Moreover, gray matter volume was decreased substantially in left hippocampus and right dorsolateral prefrontal cortex, in addition with temporal, cerebellar, and parietal regions among patients with MDD but not in BD [168].

Indeed, gray matter volume may be one of the characteristics which aid the differentiation of SZ and BD. Regarding this, more prominent reductions in the gray matter volume in SZ compared to BD was observed in a total of 802 subjects: 243 SZ, 176 BD, 383 HC [134]. In addition, an MRI study of 184 young patients (39 anxious, 20 BD, 52 MDD, 20 ADHD, and 53 HC) reported both shared and specific features in gray matter volumes [169]. This was especially significant dorsolateral prefrontal cortex (DLPFC) to differentiate between BD, anxiety, and HCs. Gray matter volume has also been suggested to differentiate between BD with and without psychotic features but results are not precisely clear [170]. One study reported that BD patients with psychotic features have gray matter volume abnormalities in right temporoparietal cortex compared to both HC and non-psychotic BD patients [171]. Nevertheless, recent studies suggest that the differences of brain phenotypes respective to disease states may not be as prominent as once thought before [172]. For example, white matter characterization measured by MRI based methods have been very successful to identify neurobiological substrates involved in the psychopathologies but this identification may not suggest a unique specificity to disease states [172, 173]. Perhaps this artifact arises from the studies with small sample size [174–176]. On the other hand, refinement of DTI and associated techniques may allow to capture the specific characteristics that are not accessible previously. For example, the neurite orientation dispersion density imaging (NODDI) assisted gray matter measurements [177] coupled with alternative analytical methods are on the rise [178, 179].

In a study of 537 subjects (SZ, 126; BD, 97; MDD, 126; and HCs, 188), resting-state functional connectivity has been compared. Interestingly, results show an increasing gradient alterations between disease categories (SZ>BD>MDD) [172]. In another study the SCZ>BD>MDD gradient was also detected in the degree of randomization, a degree of deviation from network circuitry in a subject of 512 individuals (121 with SZ, 100 with BD, 108 with MDD, and 183 HC) [125]. Amygdala-PFC connectivity has been studied by different groups to investigate its profile in different states. Some studies suggest that this circuitry may aid the differentiation between remitted-BD and depressed-BD subjects [180] and between MDD and BD [181], SZ and BD [182]. Nevertheless, it seems like the direction of future studies should be carefully considered in order to validate these findings. For example, a recent review of MRI studies for SZ and BD suggest the requirement for studies with large sample sizes and multimodal neuroimaging in medication-naïve first episode patients [183].

Predicting Treatment Outcome by Neuroimaging

In general, white matter microstructure and gray matter changes are suggested for probing illness severity and progression of BD. Tighe et al. [184] reviewed the structural neuroimaging studies for association with response to lithium treatment. Regarding this, hippocampal volume, gray matter volumes of total brain, prefrontal cortex, and left subgenual cortex, cerebral gray matter volume have been suggested as areas affected. Which of these areas might be a biomarker for treatment response?

One longitudinal brain imaging study suggests increased prefrontal gray matter [185] while another study suggests increase of lithium-induced gray matter volume as candidate biomarker for treatment response in BD [186]. Also, one study investigated if response to lithium treatment of children with BD can be predicted by changes in white matter microstructure [166]. By assessing the white matter in cortico-limbic tracts involved in emotion regulation authors suggest that there is a link between the response in lithium treatment and normalization of white matter microstructure in regions associated with emotion processing [166]. A recent study using a combination of fMRI and proton magnetic resonance spectroscopy (1 H-MRS) data developed a machine learning model called LITHium Intelligent Agent (LITHIA) to predict Lithium treatment response with an accuracy of more than 80 percent [187].

In conclusion, the field is still not mature. Implementation of multivariate analysis may trigger the identification of neural markers for clinical characterization. In this context, the harmonization of neuroimaging data together with other biological signatures seems to be essential. For example, in addition to distinct structural and functional brain markers, molecular factors such as neurotrophic factors, cytokines, and oxidative stress molecules may be important for BPD [32]. Future studies for longitudinal assessment of patients with different courses of the disease as well as correlation of neuroimaging measures with clinical, cognitive, and genetic marker are required for better understanding [188].

The Potential Neuroimaging Markers for MDD

The exact neurobiological mechanism underlying major depressive disorder (MDD) is not known; however, converging evidences suggest a complex interaction of genetic and environmental factors [140, 189]. Biomarkers in general and neuroimaging markers in particular are essential for differential diagnosis, risk assessment, and treatment prediction [21].

Neuroimaging studies reported structural and functional deficits which have most consistently been identified in the hippocampus [190], anterior [191] and posterior cingulate [192], orbitofrontal [193], lateral temporal and occipital cortices [194], and amygdala [195]. Along with the medial prefrontal cortex (mPFC) and the amygdala, the role of the anterior cingulate cortex (ACC) and the hippocampus have been shown to have functional abnormalities in MDD. fMRI studies have found elevated activity in the ACC in MDD patients, which is thought to be evidence for abnormal processing of environmental stimuli in MDD [196].

Cortical regions such as the ACC are thought to have a regulatory role over the limbic structures that process emotional stimuli [197]. A breakdown in this circuitry could potentially explain the development of depressive characteristics such as negative bias in interpersonal feedback and somatic complaints [198]. Indeed, a number of different research groups have found abnormal functional connectivity incorporating the ACC and limbic structures [197–200], suggesting its importance as a biomarker for MDD [201]. Other studies address the assessment of severity [117], or risk of suicidality [202] of major depression disorder.

When considering functional connectivity networks in MDD, the default mode network (DMN) [203] and the pre-frontal–amygdalar–pallidostriatal–mediothalamic mood-regulating circuit (MRC) are frequently cited as regions of abnormal synchrony [197]. In a direct resting-state fMRI comparison between depressed patients and healthy controls, the MD group was found to have decreased functional connectivity between the ACC and three structures: the amygdala, the pallidostriatum, and the medial thalamus, suggesting a decreased regulatory effect of the ACC over the mood-regulating limbic areas in MD [197]. Lui et al. extended the findings of Anand et al. [204–206] by showing bilaterally reduced functional connectivity within the prefrontal-limbic-thalamic areas, in particular in regions subserved by the left amygdala-ACC and the right insula–precuneus region, in depressed patients [182]. Are there any promising candidate biomarkers for disease subtyping or treatment guidance which could be assessed by neuroimaging? Below sections explore this possibility.

Risk Assessment and Prediction by Neuroimaging

Structural and functional variations in various brain regions have been associated with risk for MDD. Among these, reductions in the cortical thickness of the fusiform gyrus [207] reduced volume in DLPFC and hippocampus [208, 209], activation in amygdala and nucleus accumbens [210], and anterior cingulate cortex [211] have been suggested as predictive for the susceptibility of MDD.

Accumulating studies report functional and structural neural correlates of MDD compared to HC. In a recent study, the whole-brain analysis revealed reduced activation of MDD patients compared with HC primarily in the basal ganglia, in frontal and temporal areas, the cingulate gyrus spreading to the amygdala, the insula, and the hippocampus. Also, MD patients showed a reduced activation to fearful faces in the superior frontal gyrus and ACG, and middle cingulate gyrus compared with HC [212]. DTI studies consistently identified a reduction in anisotropy in subjects with MDD compared with control subjects [213]. Reductions in anisotropy were largely localized to the frontal and temporal lobes or white matter tracts and detected in both hemispheres. A large pooled effect size was detected in superior frontal white matter, which contains fibers of the DLPFC and ACC circuits [214]. A significant decrease in fractional anisotropy in the left hemisphere, including the anterior limb of the internal capsule and the inferior parietal portion of the superior longitudinal fasciculus has been detected in patients with MDD compared with healthy controls [213] Thus, findings support the theory that white matter abnormalities play a role in a disconnection syndrome between frontal and subcortical regions and may contribute as a risk factor for affective disorders. Within the DLPFC circuit, the anterior thalamic radiation connects the prefrontal cortex (PFC) to the thalamus through the anterior limb of the internal capsule [214]. Nevertheless, the utility of these cross-sectional studies for risk determination is dependent on specially designed longitudinal studies with selected subject groups.

Differential Diagnosis and Disease Subtyping

Regions important for individual diagnosis have been featured within the cortico-striato-pallido-thalamic loops, which include the medial and orbital prefrontal cortices, amygdala, hippocampus, medial thalamus, and striatum [215], and cortico-cortical circuits from the medial prefrontal cortex connecting the parahippocampus, posterior cingulate, and superior temporal cortices [216]. Meta-analyses have pointed to a reduction in hippocampal size in MDD patients, suggesting its possible use as a diagnostic neurobiomarker for MDD [217, 218]. One study reported that decreased gray matter density in the right subgenual anterior cingulate, superior temporal cortex, medial frontal gyrus, precuneus, hippocampus, and thalamus, as well as in the left inferior occipital cortex, parietal cortex, and cerebellum regions showed the highest contribution to the diagnosis of depression. As a diagnostic marker, the accuracy was 67.6% from whole-brain structural neuroanatomy. Yet the structural MRI results had shown limited potential for diagnosis [219].

One challenge for depression diagnosis derives from disease subtyping and several studies may help development of neuroimaging-based tools for this challenge. Although major neuroanatomical regions and neural networks are common in one diagnostic entity, prominently disturbed features differ between subtypes that should be determined by neuroimaging. Regarding this, melancholic depression, which is commonly accompanied by anhedonia and psychomotor retardation, is associated with the medical forebrain bundle in MDD [220]. On the other hand, rumination and autobiographical memories were commonly associated with abnormal connectivity in the default mode network [221]. One study reported a differential pattern in the hippocampal, amygdala, and subgenual prefrontal cortex volumes in melancholic versus psychotic depression [222]. Accumulating studies suggest that youth depression is characterized by alterations in the activity of ventromedial frontal regions, the anterior cingulate, and amygdala [223]. Indeed, amygdala connectivity has been found as differential in MDD in adults and youth according to a recent meta-analysis [224]. Also, connectivity within the amygdala circuitry (amygdala intrinsic connectivity) was suggested as a clinical marker for detection of early-onset depression compared to adult-onset depression [225].

An interesting study [226] reported four "biotypes" of depression, which were also associated with differences in treatment outcome [226, 227]. Drysdale et al. [226] suggest that a comparison of fMRI scans of more than 1100 patients of depression and HCs revealed that patients with depression can be divided into four "biotypes" according to connectivity in limbic and fronto-striatal networks and different clinical symptoms. For example, type 1 and type 4 are characterized as having most severe reduction in the fronto-amygdala connectivity, a circuitry involved in the regulation of fear and reappraisal of negative emotional stimuli. A decreased connectivity in anterior cingulate and orbitofrontal areas, involved in motivation and incentive-salience evaluation were other areas most severe in type 1 and 2. On the other hand, reward processing, adaptive motor control, and action initiation relevant connectivity, which involves the thalamic and fronto-striatal pathways, were especially prominent in type 3 and type 4, which are characterized by anhedonia and psychomotor retardation.

Although there are quite a bit of common alterations across the different disease states such as SZ, BD, and MDD, the degree of alteration in these common regions may help differentiate disease borders. For example, alterations in bilateral orbital frontal cortex (OFC), left OFC extending to putamen, visual, auditory, and motor cortices, right supplemental motor area, and bilateral thalami represented a gradient such that SZ>BD>MDD [172].

One of the problems associated with differential diagnosis is to distinguish between the patients of MDD and patients of BD during the first episode of depression [228]. Both of these disorders are associated with alterations in limbic regions. Cross-disorder classifications are particularly difficult especially when the disorders that need to be distinguished are similar to the case in BD and MDD. Accumulating studies investigate neural activity patterns. In an fMRI study, BD compared with MDD patients, significantly increased activation was found in the fearful condition primarily in the right frontal and parietal regions, and the cingulate gyrus [212]. Another fMRI study analyzed the amygdala activation to sad>neutral faces and found an increased responsiveness in MDD relative to BD. In contrast, responsiveness to happy>neutral faces showed the opposite pattern, with higher amygdala activity in BD than in MDD [229]. Other findings that may serve for differentiation is anterior cingulate gyrus activation [212], insula and amygdala resting-state functional connectivity in [230].

Based on cortical, thalamic, and striatal regions, is it possible to differentiate between BD and MDD? In this regard, the lack of studies with large sample sizes may be compensated by meta-analysis. For example, meta-analysis results show that increased activity in the thalamus and basal ganglia and decreased engagement of ventrolateral prefrontal cortical regions was associated with BD where as a hypoactivation of the sensorimotor cortices was seen in MDD [231]. Also, BD could be distinguished from MDD on the basis of multivariate patterns based on gray matter differences [163]. In line with this, a recent meta-analysis [232] reported the features of gray matter in MD and BD by VBM studies, composed of 4101 individuals with MDD and 2407 individuals with BD. This study suggests that there are distinct and common patterns of gray matter volumes when these two conditions were compared. Especially, volume reductions in bilateral insula, anterior cingulate cortex, and medial prefrontal cortex were common for both MDD and BD. On the other hand, the volumetric reductions in certain regions such as right dorsolateral prefrontal cortex and left hippocampus were prominent in MDD. Other recent meta-analysis showed that the rostral middle frontal cortex was thinner in BD compared to MDD [233] and here is a common pattern of pituitary gland enlargement moderated by age and sex in both BD and MDD [136].

Machine learning methodologies such as support vector machine is one way to integrate multivariate data and being increasingly used to identify neural biomarkers to predict remission to pharmacotherapy in late-life depression [234], predicting antidepressant response in major depression [235], besides to other utilizations such as differentiation of generalized anxiety disorder from MD [236]. Also, one study [219] using machine learning approach suggests that gray matter density allowed prediction of treatment response. Accordingly, remission was predicted by greater gray matter density in the right rostral anterior cingulate cortex, left posterior cingulate cortex, left middle frontal gyrus, and right occipital cortex.

Support vector machine (SVM) studies investigating MD using functional neuroimaging data have been conducted [237–240]. In the first of these, Fu and colleagues reported a significant pattern of discrimination between patients with MD and healthy controls (HC) based on the neural response as processing of sad facial expressions [240]. Similarly, Marquand et al. [239] were able to use subject responses to a variable load version of the verbal memory "N-back" task which allowed them to discriminate between depression patients and HCs. Although functional neuroimaging of working memory represents a statistically significant, it remained as clinically moderate for a diagnostic biomarker for depression. Despite these findings, for the differentiation of MDD and BPD large studies are still missing to a degree that allows definitive conclusions to be drawn.

Predicting Treatment Outcome of MDD by Neuroimaging

Selection of best matching treatment is essential to improve response rates in MD. Thus, predictive biomarkers guiding treatment selection is crucial. Many studies compare the effect of drug treatments. One study reported that antidepressant treatment increased resting-state functional connectivity between ACC and the subcortical structures such as the medial thalamus and the pallidostriatum in the depressed group, but not in the control group [205]. The pattern of hypoconnectivity between the amygdala and several cortical and subcortical structures changed and connectivity increased after the treatment. It has also been suggested that the amygdala connectivity becomes normalized after antidepressant treatment [241]. The changes in brain connectivity between two groups of depressed patients: one treated with venlafaxine and the other with mirtazapine are compared. After the treatment, the patients presented increases in connectivity between dmPFC and cerebellum, cingulate cortex, parietal cortex, and decreases in connectivity between the OFC and the right medial cingulate cortex, the middle temporal gyrus, the superior occipital gyrus, the right fusiform gyrus, and the inferior temporal gyrus. Also the same paper showed decreases in connectivity between left the OFC and the left superior parietal gyrus, the precuneus, and the postcentral gyrus, the left medial temporal gyrus, the cuneus, the calcarine fissure, and the angular gyrus [242].

In some studies, connectivity alterations were observed in DMN after the treatment [243–245]. Abbott et al. [246] found that patients with MDD present a hyperconnectivity pattern in the posterior areas of DMN, PFDLC, and dmCPF, which tends to recover after the treatment. Wu et al. [247] showed a mixed pattern of hypoconnectivity in the posterior areas of DMN, and hyperconnectivity in the most anterior areas of the same network at rest in depressed patients. In the same study, hypoconnectivity between PCC and sgACC recovered, but not PCC-dmPFC hyperconnectivity. In a study, depressed patients show smaller functional connectivity strength maps (FCS) than healthy subjects in both hippocampi, and the opposite pattern in the dmPFC cortexes of both hemispheres. Antidepressant treatment tended to reverse the pattern so that, after treatment, FCS increased between both hippocampi and it decreased between dmPFCs. Additionally, the changes in dmPFC connectivity correlated positively with the clinical improvement measures [62]. In a resting-state fMRI, patients with treatment-resistant depression (TRD) had increased right-thalamic fractional amplitude of low-frequency fluctuations (fALFF) values compared with patients without TRD. Also, patients with TRD showed increased fALFF values in the right inferior frontal gyrus (IFG), inferior parietal lobule (IPL) and vermis, compared with patients with non-TRD and HC subjects [248].

Nevertheless, to what extent the studies presented so far can guide the treatment selection is dependent on the studies specifically designed to test the predictive potential of relevant test paradigms associated with above circuitries. Regarding this one study assessed the potential of neuroimaging data to be used as predictors of antidepressant response. A positive emotion regulation task during fMRI has been studied in a case/control study. Results show that neural activation pattern may be useful to predict the responsiveness to Behavioral activation therapy for depression (BATD) which appear to be more effective for patients with greater disturbances in brain reward network function [249]. Another study used neural and performance predictors during a cognitive control task to predict treatment response. Haemodynamic response function modeling activation during Commission errors in the rostral and dorsal anterior cingulate, mid-cingulate, dorsomedial prefrontal cortex, and lateral orbital frontal cortex predicted treatment response [250]. Spies et al. [251] suggested the DMN activity as a measure of early antidepressant response. Among the most replicated findings are the hyperactivity in ACC and the reduction of hippocampal volume as well as activity in medial prefrontal cortices and partially amygdala, which may help determine the patients who benefit from specific treatment [66, 252]. For example, persistence of hyperactivity in ACC and greater loss of hippocampal volume are both associated with poor treatment response. As reviewed by Phillips et al. [252] the most consistent literature findings suggest key areas such as anterior cingulate and medial prefrontal cortices and partially amygdala, which may help determine the patients who benefit from psychotherapy such as cognitive behavior therapy (CBT) compared with pharmacotherapy. Thus, activity in these areas may help predict treatment outcome. Also, the response to serotonergic reuptake inhibitors (SRIs) and nonserotonergic antidepressants can be predicted by medial prefrontal cortex and pregenual anterior cingulate cortex. Moreover, in late-life depression lower fractional anisotropy predicts response to SRI treatment. Increased arterial spin labeling measures of cerebral blood flow in the ventral anterior cingulate cortex may be predictive for better response to SRI treatment [252]. One interesting progress might aid personalized treatment strategies. For example, insula resting-state activity may be a predictor of individualized assessment of patient's response to pharmacological or psychological treatment approaches [66].

Conclusion

Current literature shows that neuroimaging data increasingly accumulate but they are not mature yet for translation into clinical practice. This is well reflected with the inconsistencies across the studies and with the lack of replication studies, besides to small sample sizes, lack of sufficient longitudinal studies and the lack of multimodal studies [29, 253, 254]. On the other hand, these problems are likely to be solved in the light of recent progress such as data-sharing initiatives, multisite consortiums, and technical/methodological advancements.

Neuroimaging data-sharing initiatives come with the opportunity of increased scale and scope. For example, more than 8000 shared MRI data is available online [255]. Moreover, open data for animal neuroimaging will boost the research by allowing cross-species studies [256]. In addition to open data, initiatives like "The Consortium for Reliability and Reproducibility (CoRR)" will address the problems of test–retest reliability [257]. Besides, progress in neuroimaging technology and associated methodologies will likely contribute to better analysis of brain disorders by allowing the analysis of structural/functional networks differentially with better spatiotemporal resolution [258–264].

Also, progressive approaches may help overcome methodological limitations by allowing the harmonization of neuroimaging data together with other biological measures. One elaboration of this is the imaging genetics which combines the genetic variables with neuroimaging phenotypes [3]. Although the main focus of imaging genetics is to create fundamental knowledge about the neurobiology of brain disorders, it may be especially important for the disease risk assessment since it typically focuses on the analysis of genetic risk factors and their link with neuroimaging phenotypes in healthy individuals.

It is thought that multimodal imaging may help better predictive capacity [260, 265], and with the advancements in statistics, it will boost neuroimaging. In addition to "mass-univariate analysis" which is commonly used statistical model for the analysis of neuroimaging data processed by a voxel-by-voxel approach [36], multivariate analysis relies on the correlation/covariance of activation across brain regions [266, 267]. In this regard, support vector machine (SVM) is a machine learning-based classification algorithm for multivariate neuroimaging data analysis. Application of SVM to neuroimaging data can particularly facilitate the identification of biomarkers. Literature suggests that it has potential applications in disease subtyping and treatment prediction [86, 119].

In conclusion, the impact of accumulating neuroimaging studies on the clinical practice has been very limited. However in the light of current progress described so far, the future of neuroimaging for clinical practice is bright especially for its potential to fill the current gap of predictive, personalized, and preventative psychiatry.

References

1. Hood L, Friend SH. Predictive, personalized, preventive, participatory (P4) cancer medicine. Nat Rev Clin Oncol. 2011;8(3):184.
2. Hyman SE, Fenton WS. What are the right targets for psychopharmacology? Science. 2003;299(5605):350–1.
3. Arslan A. Genes, brains, and behavior: imaging genetics for neuropsychiatric disorders. J Neuropsychiatry Clin Neurosci. 2015;27(2):81–92.
4. Fernandes BS, Williams LM, Steiner J, Leboyer M, Carvalho AF, Berk M. The new field of 'precision psychiatry'. BMC Med BioMed Central. 2017;15(1):80.
5. Cataldo I, Azhari A, Lepri B, Esposito G. Oxytocin receptors (OXTR) and early parental care: an interaction that modulates psychiatric disorders. Res Dev Disabil. 2018;82:27–38.
6. Mascheretti S, Andreola C, Scaini S, Sulpizio S. Beyond genes: a systematic review of environmental risk factors in specific reading disorder. Res Dev Disabil. 2018.
7. Vieta E, Berk M, Schulze TG, Carvalho AF, Suppes T, Calabrese JR, et al. Bipolar disorders. Nat Rev Dis Prim. 2018 [cited 2018 Dec 24];4:18008. http://www.ncbi.nlm.nih.gov/pubmed/29516993.
8. Kim Y-K, Shin C. The microbiota-gut-brain axis in neuropsychiatric disorders: pathophysiological mechanisms and novel treatments. Curr Neuropharmacol. 2018;16(5):559–73.
9. Rogers J, Renoir T, Hannan AJ. Gene-environment interactions informing therapeutic approaches to cognitive and affective disorders. Neuropharmacology. 2019 [cited 2018 Dec 24];145(Pt A):37–48. http://www.ncbi.nlm.nih.gov/pubmed/29277490.
10. Insel TR. The NIMH research domain criteria (RDoC) project: precision medicine for psychiatry. Am J Psychiatry. 2014;171(4):395–7.
11. Arslan A. Mapping the Schizophrenia genes by neuroimaging: the opportunities and the challenges. Int J Mol Sci. 2018;19(1):219.
12. John B, Lewis KR. Chromosome variability and geographic distribution in insects. Science. 1966;152(3723):711–21.
13. Gottesman II, Shields J. Genetic theorizing and schizophrenia. Br J Psychiatry. 1973;122(566):15–30.
14. Flint J, Munafò MR. The endophenotype concept in psychiatric genetics. Psychol Med. 2007 [cited 2018 Dec 25];37(02):163. http://www.journals.cambridge.org/abstract_S0033291706008750.
15. Bandettini PA, Wong EC, Hinks RS, Tikofsky RS, Hyde JS. Time course EPI of human brain function during task activation. Magn Reson Med. 1992;25(2):390–7.
16. Kwong KK, Belliveau JW, Chesler DA, Goldberg IE, Weisskoff RM, Poncelet BP, et al. Dynamic magnetic resonance imaging of human brain activity during primary sensory stimulation. Proc Natl Acad Sci. 1992;89(12):5675–9.
17. Ogawa S, Lee TM, Kay AR, Tank DW. Brain magnetic resonance imaging with contrast dependent on blood oxygenation. Proc Natl Acad Sci USA. 1990 [cited 2018 Dec 24];87(24):9868–72. http://www.ncbi.nlm.nih.gov/pubmed/2124706.

18. Mansfield JR. Evaluation and refinement of a system and a method for the use of hyperspectral imaging for metabolic monitoring. 2003 [cited 2018 Dec 25]. https://apps.dtic.mil/docs/citations/ADA420083.
19. Mankoff DA. Molecular imaging as a tool for translating breast cancer science. Breast Cancer Res. 2008;10(1):S3.
20. Le Roux LG, Schellingerhout D. Molecular neuroimaging: the basics. In: Seminars in roentgenology. Elsevier; 2014. p. 225–33.
21. Arslan A. Application of neuroimaging in the diagnosis and treatment of depression. In: Understanding depression. Springer; 2018. p. 69–81.
22. Arslan A. Imaging genetics of schizophrenia in the post-GWAS era. Prog Neuro-Psychopharmacology Biol Psychiatry. 2018;80:155–65.
23. Xia L, Li S, Wang T, Guo Y, Meng L, Feng Y, et al. Spontaneous alterations of regional brain activity in patients with adult generalized anxiety disorder. Neuropsychiatr Dis Treat. 2017 [cited 2018 Dec 24];13:1957–65. http://www.ncbi.nlm.nih.gov/pubmed/28790831.
24. Fusar-Poli P, Meyer-Lindenberg A. Forty years of structural imaging in psychosis: promises and truth. Acta Psychiatr Scand. 2016;134(3):207–24.
25. Guo S, Iwabuchi S, Balain V, Feng J, Liddle P, Palaniyappan L. Cortical folding and the potential for prognostic neuroimaging in schizophrenia. Br J Psychiatry. 2015;207(5):458–9.
26. Hahn C, Lim HK, Lee CU. Neuroimaging findings in late-onset schizophrenia and bipolar disorder. J Geriatr Psychiatry Neurol. 2014;27(1):56–62.
27. Tamashiro JH, Zung S, Zanetti MV, de Castro CC, Vallada H, Busatto GF, et al. Increased rates of white matter hyperintensities in late-onset bipolar disorder. Bipolar Disord. 2008 [cited 2018 Dec 24];10(7):765–75. http://www.ncbi.nlm.nih.gov/pubmed/19032708.
28. Nakajima S, Takeuchi H, Plitman E, Fervaha G, Gerretsen P, Caravaggio F, et al. Neuroimaging findings in treatment-resistant schizophrenia: a systematic review: lack of neuroimaging correlates of treatment-resistant schizophrenia. Schizophr Res. 2015;164(1–3):164–75.
29. Ball TM, Stein MB, Ramsawh HJ, Campbell-Sills L, Paulus MP. Single-subject anxiety treatment outcome prediction using functional neuroimaging. Neuropsychopharmacology. 2014;39(5):1254.
30. Desmyter S, Van Heeringen C, Audenaert K. Structural and functional neuroimaging studies of the suicidal brain. Prog Neuro-Psychopharmacology Biol Psychiatry. 2011;35(4):796–808.
31. Pagano G, Niccolini F, Politis M. Current status of PET imaging in Huntington's disease. Eur J Nucl Med Mol Imaging. 2016 [cited 2018 Dec 24];43(6):1171–82. http://www.ncbi.nlm.nih.gov/pubmed/26899245.
32. Teixeira AL, Barbosa IG, Machado-Vieira R, Rizzo LB, Wieck A, Bauer ME. Novel biomarkers for bipolar disorder. Expert Opin Med Diagn. 2013 [cited 2018 Dec 24];7(2):147–59. http://www.ncbi.nlm.nih.gov/pubmed/23530885.
33. Won E, Kim Y-K. Neuroimaging in psychiatry: steps toward the clinical application of brain imaging in psychiatric disorders. Prog Neuro Psychopharmacology Biol Psychiatry. 2018 [cited 2018 Dec 24];80(Pt B):69–70. http://www.ncbi.nlm.nih.gov/pubmed/28962844.
34. Vigo D, Thornicroft G, Atun R. Estimating the true global burden of mental illness. The Lancet Psychiatry. 2016 [cited 2018 Dec 24];3(2):171–8. http://www.ncbi.nlm.nih.gov/pubmed/26851330.
35. Grudzinski AN. Considerations in the treatment of anxiety disorders: a pharmacoeconomic review. Expert Opin Pharmacother. 2001;2(10):1557–69.
36. Woo C-W, Chang LJ, Lindquist MA, Wager TD. Building better biomarkers: brain models in translational neuroimaging. Nat Neurosci. 2017 [cited 2018 Dec 24];20(3):365–77. http://www.ncbi.nlm.nih.gov/pubmed/28230847.

37. Davidson LL, Heinrichs RW. Quantification of frontal and temporal lobe brain-imaging findings in schizophrenia: a meta-analysis. Psychiatry Res Neuroimaging. 2003;122(2):69–87.
38. Wright IC, Rabe-Hesketh S, Woodruff PWR, David AS, Murray RM, Bullmore ET. Meta-analysis of regional brain volumes in schizophrenia. Am J Psychiatry. 2000;157(1):16–25.
39. Adriano F, Spoletini I, Caltagirone C, Spalletta G. Updated meta-analyses reveal thalamus volume reduction in patients with first-episode and chronic schizophrenia. Schizophr Res. 2010;123(1):1–14.
40. Konick LC, Friedman L. Meta-analysis of thalamic size in schizophrenia. Biol Psychiatry. 2001;49(1):28–38.
41. Hulshoff Pol HE, Kahn RS. What happens after the first episode? A review of progressive brain changes in chronically ill patients with schizophrenia. Schizophr Bull. 2008;34(2):354–66.
42. Olabi B, Ellison-Wright I, McIntosh AM, Wood SJ, Bullmore E, Lawrie SM. Are there progressive brain changes in schizophrenia? A meta-analysis of structural magnetic resonance imaging studies. Biol Psychiatry. 2011;70(1):88–96.
43. Ellison-Wright I, Bullmore E. Anatomy of bipolar disorder and schizophrenia: a meta-analysis. Schizophr Res. 2010;117(1):1–12.
44. Buehlmann E, Berger GE, Aston J, Gschwandtner U, Pflueger MO, Borgwardt SJ, et al. Hippocampus abnormalities in at risk mental states for psychosis? A cross-sectional high resolution region of interest magnetic resonance imaging study. J Psychiatr Res. 2010;44(7):447–53.
45. Borgwardt SJ, Riecher-Rössler A, Dazzan P, Chitnis X, Aston J, Drewe M, et al. Regional gray matter volume abnormalities in the at risk mental state. Biol Psychiatry. 2007;61(10):1148–56.
46. Vita A, De Peri L, Silenzi C, Dieci M. Brain morphology in first-episode schizophrenia: a meta-analysis of quantitative magnetic resonance imaging studies. Schizophr Res. 2006;82(1):75–88.
47. Woods BT, Ward KE, Johnson EH. Meta-analysis of the time-course of brain volume reduction in schizophrenia: implications for pathogenesis and early treatment. Schizophr Res. 2005;73(2–3):221–8.
48. Fornito A, Yung AR, Wood SJ, Phillips LJ, Nelson B, Cotton S, et al. Anatomic abnormalities of the anterior cingulate cortex before psychosis onset: an MRI study of ultra-high-risk individuals. Biol Psychiatry. 2008;64(9):758–65.
49. Röthlisberger M, Riecher-Rössler A, Aston J, Fusar-Poli P, Radü E-W, Borgwardt S. Cingulate volume abnormalities in emerging psychosis. Curr Pharm Des. 2012 [cited 2018 Dec 24];18(4):495–504. http://www.ncbi.nlm.nih.gov/pubmed/22239580.
50. Takayanagi Y, Kulason S, Sasabayashi D, Takahashi T, Katagiri N, Sakuma A, et al. Reduced thickness of the anterior cingulate cortex in individuals with an at-risk mental state who later develop psychosis. Schizophr Bull. 2017 [cited 2018 Dec 24];43(4):907–13. http://www.ncbi.nlm.nih.gov/pubmed/28338751.
51. Bora E, Fornito A, Radua J, Walterfang M, Seal M, Wood SJ, et al. Neuroanatomical abnormalities in schizophrenia: a multimodal voxelwise meta-analysis and meta-regression analysis. Schizophr Res. 2011;127(1–3):46–57.
52. Chan RCK, Di X, McAlonan GM, Gong Q. Brain anatomical abnormalities in high-risk individuals, first-episode, and chronic schizophrenia: an activation likelihood estimation meta-analysis of illness progression. Schizophr Bull. 2009;37(1):177–88.
53. Trzesniak C, Oliveira IR, Kempton MJ, Galvão-de Almeida A, Chagas MHN, Ferrari MCF, et al. Are cavum septum pellucidum abnormalities more common in schizophrenia spectrum disorders? A systematic review and meta-analysis. Schizophr Res. 2011;125(1):1–12.
54. Sun D, Stuart GW, Jenkinson M, Wood SJ, McGorry PD, Velakoulis D, et al. Brain surface contraction mapped in first-episode schizophrenia: a longitudinal magnetic resonance imaging study. Mol Psychiatry. 2009;14(10):976.

55. Glahn DC, Laird AR, Ellison-Wright I, Thelen SM, Robinson JL, Lancaster JL, et al. Meta-analysis of gray matter anomalies in schizophrenia: application of anatomic likelihood estimation and network analysis. Biol Psychiatry. 2008;64(9):774–81.
56. van Erp TGM, Hibar DP, Rasmussen JM, Glahn DC, Pearlson GD, Andreassen OA, et al. Subcortical brain volume abnormalities in 2028 individuals with schizophrenia and 2540 healthy controls via the ENIGMA consortium. Mol Psychiatry. 2016;21(4):547.
57. Lieberman JA, Girgis RR, Brucato G, Moore H, Provenzano F, Kegeles L, et al. Hippocampal dysfunction in the pathophysiology of schizophrenia: a selective review and hypothesis for early detection and intervention. Mol Psychiatry. 2018.
58. Walton E, Hibar DP, van Erp TGM, Potkin SG, Roiz-Santiañez R, Crespo-Facorro B, et al. Positive symptoms associate with cortical thinning in the superior temporal gyrus via the ENIGMA Schizophrenia consortium. Acta Psychiatr Scand. 2017;135(5):439–47.
59. Lee J, Chon M-W, Kim H, Rathi Y, Bouix S, Shenton ME, et al. Diagnostic value of structural and diffusion imaging measures in schizophrenia. NeuroImage Clin. 2018;18:467–74.
60. McGuire P, Dazzan P. Does neuroimaging have a role in predicting outcomes in psychosis? World Psychiatry. 2017;16(2):209–10.
61. Colibazzi T, Yang Z, Horga G, Yan C-G, Corcoran CM, Klahr K, et al. Aberrant temporal connectivity in persons at clinical high risk for psychosis. Biol Psychiatry Cogn Neurosci Neuroimaging. 2017;2(8):696–705.
62. Li T, Wang Q, Zhang J, Rolls ET, Yang W, Palaniyappan L, et al. Brain-wide analysis of functional connectivity in first-episode and chronic stages of schizophrenia. Schizophr Bull. 2017;43(2):436–48.
63. Anticevic A, Haut K, Murray JD, Repovs G, Yang GJ, Diehl C, et al. Association of thalamic dysconnectivity and conversion to psychosis in youth and young adults at elevated clinical risk. JAMA Psychiatry. 2015;72(9):882–91.
64. Lancaster TM, Linden DE, Tansey KE, Banaschewski T, Bokde ALW, Bromberg U, et al. Polygenic risk of psychosis and ventral striatal activation during reward processing in healthy adolescents. JAMA Psychiatry. 2016;73(8):852–61.
65. Reus LM, Shen X, Gibson J, Wigmore E, Ligthart L, Adams MJ, et al. Association of polygenic risk for major psychiatric illness with subcortical volumes and white matter integrity in UK Biobank. Sci Rep. 2017 [cited 2018 Dec 24];7(1):42140. http://www.ncbi.nlm.nih.gov/pubmed/28186152.
66. Arnone D, Wise T, Cleare A, Young A, Herane A. Diagnostic and therapeutic utility of neuroimaging in depression: an overview. Neuropsychiatr Dis Treat. 2014 [cited 2018 Dec 24];10:1509. http://www.dovepress.com/diagnostic-and-therapeutic-utility-of-neuroimaging-in-depression-an-ov-peer-reviewed-article-NDT.
67. Brent BK, Thermenos HW, Keshavan MS, Seidman LJ. Gray matter alterations in schizophrenia high-risk youth and early-onset schizophrenia: a review of structural MRI findings. Child Adolesc Psychiatr Clin. 2013;22(4):689–714.
68. Ganzola R, Maziade M, Duchesne S. Hippocampus and amygdala volumes in children and young adults at high-risk of schizophrenia: research synthesis. Schizophr Res. 2014;156(1):76–86.
69. Thermenos HW, Keshavan MS, Juelich RJ, Molokotos E, Whitfield-Gabrieli S, Brent BK, et al. A review of neuroimaging studies of young relatives of individuals with schizophrenia: a developmental perspective from schizotaxia to schizophrenia. Am J Med Genet Part B Neuropsychiatr Genet. 2013 [cited 2018 Dec 24];162(7):604–35. http://www.ncbi.nlm.nih.gov/pubmed/24132894.
70. Ordóñez AE, Luscher ZI, Gogtay N. Neuroimaging findings from childhood onset schizophrenia patients and their non-psychotic siblings. Schizophr Res. 2016 [cited 2018 Dec 24];173(3):124–31. http://www.ncbi.nlm.nih.gov/pubmed/25819937.
71. Arango C, Moreno C, Martínez S, Parellada M, Desco M, Moreno D, et al. Longitudinal brain changes in early-onset psychosis. Schizophr Bull. 2008;34(2):341–53.

72. Arango C, Rapado-Castro M, Reig S, Castro-Fornieles J, González-Pinto A, Otero S, et al. Progressive brain changes in children and adolescents with first-episode psychosis. Arch Gen Psychiatry. 2012;69(1):16–26.
73. Mechelli A, Riecher-Rössler A, Meisenzahl EM, Tognin S, Wood SJ, Borgwardt SJ, et al. Neuroanatomical abnormalities that predate the onset of psychosis: a multicenter study. Arch Gen Psychiatry. 2011;68(5):489–95.
74. Pantelis C, Velakoulis D, McGorry PD, Wood SJ, Suckling J, Phillips LJ, et al. Neuroanatomical abnormalities before and after onset of psychosis: a cross-sectional and longitudinal MRI comparison. Lancet. 2003 [cited 2018 Dec 24];361(9354):281–8. http://www.ncbi.nlm.nih.gov/pubmed/12559861.
75. Caseras X, Tansey KE, Foley S, Linden D. Association between genetic risk scoring for schizophrenia and bipolar disorder with regional subcortical volumes. Transl Psychiatry. 2015;5(12):e692.
76. Ranlund S, Rosa MJ, de Jong S, Cole JH, Kyriakopoulos M, Fu CHY, et al. Associations between polygenic risk scores for four psychiatric illnesses and brain structure using multivariate pattern recognition. NeuroImage Clin. 2018 [cited 2018 Dec 24];20:1026–36. http://www.ncbi.nlm.nih.gov/pubmed/30340201.
77. Ranlund S, Calafato S, Thygesen JH, Lin K, Cahn W, Crespo-Facorro B, et al. A polygenic risk score analysis of psychosis endophenotypes across brain functional, structural, and cognitive domains. Am J Med Genet Part B Neuropsychiatr Genet. 2018 [cited 2018 Dec 24];177(1):21–34. http://www.ncbi.nlm.nih.gov/pubmed/28851104.
78. Savill M, D'Ambrosio J, Cannon TD, Loewy RL. Psychosis risk screening in different populations using the Prodromal Questionnaire: a systematic review. Early Interv Psychiatry. 2018 [cited 2018 Dec 24];12(1):3–14. http://www.ncbi.nlm.nih.gov/pubmed/28782283.
79. Pantlin LN, Davalos D. Neurophysiology for detection of high risk for psychosis. Schizophr Res Treatment. 2016 [cited 2018 Dec 24];2016:1–5. http://www.ncbi.nlm.nih.gov/pubmed/27579180.
80. Perkins DO, Jeffries CD, Addington J, Bearden CE, Cadenhead KS, Cannon TD, et al. Towards a psychosis risk blood diagnostic for persons experiencing high-risk symptoms: preliminary results from the NAPLS project. Schizophr Bull. 2015 [cited 2018 Dec 24];41(2):419–28. http://www.ncbi.nlm.nih.gov/pubmed/25103207.
81. Sommer IE, Bearden CE, van Dellen E, Breetvelt EJ, Duijff SN, Maijer K, et al. Early interventions in risk groups for schizophrenia: what are we waiting for? NPJ Schizophr. 2016 [cited 2018 Dec 24];2(1):16003. http://www.ncbi.nlm.nih.gov/pubmed/27336054.
82. Chung Y, Cannon TD. Brain imaging during the transition from psychosis prodrome to schizophrenia. J Nerv Ment Dis. 2015;203(5):336.
83. Koutsouleris N, Meisenzahl EM, Davatzikos C, Bottlender R, Frodl T, Scheuerecker J, et al. Use of neuroanatomical pattern classification to identify subjects in at-risk mental states of psychosis and predict disease transition. Arch Gen Psychiatry. 2009;66(7):700–12.
84. Valli I, Marquand AF, Mechelli A, Raffin M, Allen P, Seal ML, et al. Identifying individuals at high risk of psychosis: predictive utility of support vector machine using structural and functional MRI data. Front Psychiatry. 2016 [cited 2018 Dec 24];7:52. http://www.ncbi.nlm.nih.gov/pubmed/27092086.
85. Vieira S, Pinaya WHL, Mechelli A. Using deep learning to investigate the neuroimaging correlates of psychiatric and neurological disorders: methods and applications. Neurosci Biobehav Rev. 2017 [cited 2018 Dec 24];74(Pt A):58–75. http://www.ncbi.nlm.nih.gov/pubmed/28087243.
86. Kim Y-K, Na K-S. Application of machine learning classification for structural brain MRI in mood disorders: critical review from a clinical perspective. Prog Neuro-Psychopharmacology Biol Psychiatry. 2018;80:71–80.
87. Torres LC, Burke E, Shin JH, Straub R, Tao R, Hyde T, et al. 264. Unique molecular correlates of schizophrenia and its genetic risk in the hippocampus compared to frontal cortex. Biol Psychiatry. 2018;83(9):S107.

88. Mitelman SA, Buchsbaum MS, Young DS, Haznedar MM, Hollander E, Shihabuddin L, et al. Increased white matter metabolic rates in autism spectrum disorder and schizophrenia. Brain Imaging Behav. 2018;12(5):1290–305.
89. Kim G-W, Kim Y-H, Jeong G-W. Whole brain volume changes and its correlation with clinical symptom severity in patients with schizophrenia: a DARTEL-based VBM study. PLoS One. 2017;12(5):e0177251.
90. Crossley NA, Marques TR, Taylor H, Chaddock C, Dell'Acqua F, Reinders AATS, et al. Connectomic correlates of response to treatment in first-episode psychosis. Brain. 2016;140(2):487–96.
91. Zeng B, Ardekani BA, Tang Y, Zhang T, Zhao S, Cui H, et al. Abnormal white matter microstructure in drug-naive first episode schizophrenia patients before and after eight weeks of antipsychotic treatment. Schizophr Res. 2016 [cited 2018 Dec 24];172(1–3):1–8. http://www.ncbi.nlm.nih.gov/pubmed/26852402.
92. Reis Marques T, Taylor H, Chaddock C, Dell'Acqua F, Handley R, Reinders AATS, et al. White matter integrity as a predictor of response to treatment in first episode psychosis. Brain. 2014 [cited 2018 Dec 24];137(1):172–82. http://www.ncbi.nlm.nih.gov/pubmed/24253201.
93. Palaniyappan L, Marques TR, Taylor H, Mondelli V, Reinders AATS, Bonaccorso S, et al. Globally efficient brain organization and treatment response in psychosis: a connectomic study of gyrification. Schizophr Bull. 2016 [cited 2018 Dec 24];42(6):1446–56. http://www.ncbi.nlm.nih.gov/pubmed/27352783.
94. Schmidt A, Borgwardt S. Third-generation neuroimaging: translating research into clinical utility. Front Psychiatry. 2016;7:170.
95. van Amelsvoort T, Hernaus D. Effect of pharmacological interventions on the fronto-cingulo-parietal cognitive control network in psychiatric disorders: a transdiagnostic systematic review of fMRI studies. Front Psychiatry. 2016 [cited 2018 Dec 24];7:82. http://journal.frontiersin.org/Article/10.3389/fpsyt.2016.00082/abstract.
96. Honey GD, Bullmore ET, Soni W, Varatheesan M, Williams SCR, Sharma T. Differences in frontal cortical activation by a working memory task after substitution of risperidone for typical antipsychotic drugs in patients with schizophrenia. Proc Natl Acad Sci. 1999;96(23):13432–7.
97. Ebdrup BH, Raghava JM, Nielsen MØ, Rostrup E, Glenthøj B. Frontal fasciculi and psychotic symptoms in antipsychotic-naive patients with schizophrenia before and after 6 weeks of selective dopamine D2/3 receptor blockade. J Psychiatry Neurosci JPN. 2016;41(2):133.
98. Molina V, Reig S, Sarramea F, Sanz J, Artaloytia JF, Luque R, et al. Anatomical and functional brain variables associated with clozapine response in treatment-resistant schizophrenia. Psychiatry Res Neuroimaging. 2003;124(3):153–61.
99. Ahmed M, Cannon DM, Scanlon C, Holleran L, Schmidt H, McFarland J, et al. Progressive brain atrophy and cortical thinning in schizophrenia after commencing clozapine treatment. Neuropsychopharmacology. 2015;40(10):2409.
100. Garcia GJ, Chagas MH, Silva CH, Machado-de-Sousa JP, Crippa JA, Hallak JE. Structural and functional neuroimaging findings associated with the use of clozapine in schizophrenia: a systematic review. Rev Bras Psiquiatr. 2015;37(1):71–9.
101. Pérez-Gómez M, Junqué C. Clozapine: neuropsychological and neuroimaging studies. Actas Esp Psiquiatr. [Cited 2018 Dec 24];27(5):341–6. http://www.ncbi.nlm.nih.gov/pubmed/10545667.
102. Vyas NS, Patel NH, Nijran KS, Al-Nahhas A, Puri BK. The use of PET imaging in studying cognition, genetics and pharmacotherapeutic interventions in schizophrenia. Expert Rev Neurother. 2011 [cited 2018 Dec 24];11(1):37–51. http://www.ncbi.nlm.nih.gov/pubmed/21158554.
103. Fagerlund B, Mackeprang T, Gade A, Hemmingsen R, Glenthøj BY. Effects of low-dose risperidone and low-dose zuclopenthixol on cognitive functions in first-episode drug-naive schizophrenic patients. CNS Spectr. 2004;9(5):364–74.

104. Massana G, Salgado-Pineda P, Junqué C, Pérez M, Baeza I, Pons A, et al. Volume changes in gray matter in first-episode neuroleptic-naive schizophrenic patients treated with risperidone. J Clin Psychopharmacol. 2005;25(2):111–7.
105. Molina V, Taboada D, Aragüés M, Hernández JA, Sanz-Fuentenebro J. Greater clinical and cognitive improvement with clozapine and risperidone associated with a thinner cortex at baseline in first-episode schizophrenia. Schizophr Res. 2014;158(1–3):223–9.
106. Ho B-C, Andreasen NC, Ziebell S, Pierson R, Magnotta V. Long-term antipsychotic treatment and brain volumes: a longitudinal study of first-episode schizophrenia. Arch Gen Psychiatry. 2011;68(2):128–37.
107. Moncrieff J, Leo J. A systematic review of the effects of antipsychotic drugs on brain volume. Psychol Med. 2010;40(9):1409–22.
108. Navari S, Dazzan P. Do antipsychotic drugs affect brain structure? A systematic and critical review of MRI findings. Psychol Med. 2009;39(11):1763–77.
109. Smieskova R, Fusar-Poli P, Allen P, Bendfeldt K, Stieglitz RD, Drewe J, et al. The effects of antipsychotics on the brain: what have we learnt from structural imaging of schizophrenia? A systematic review. Curr Pharm Des. 2009;15(22):2535–49.
110. Glenthoj A, Glenthoj BY, Mackeprang T, Pagsberg AK, Hemmingsen RP, Jernigan TL, et al. Basal ganglia volumes in drug-naive first-episode schizophrenia patients before and after short-term treatment with either a typical or an atypical antipsychotic drug. Psychiatry Res Neuroimaging. 2007;154(3):199–208.
111. Samanaite R, Gillespie A, Sendt K-V, McQueen G, MacCabe JH, Egerton A. Biological predictors of clozapine response: a systematic review. Front Psychiatry. 2018 [cited 2018 Dec 24];9:327. http://www.ncbi.nlm.nih.gov/pubmed/30093869.
112. Sarpal DK, Robinson DG, Lencz T, Argyelan M, Ikuta T, Karlsgodt K, et al. Antipsychotic treatment and functional connectivity of the striatum in first-episode schizophrenia. JAMA Psychiatry. 2015 [cited 2018 Dec 24];72(1):5. http://www.ncbi.nlm.nih.gov/pubmed/25372846.
113. Sarpal DK, Argyelan M, Robinson DG, Szeszko PR, Karlsgodt KH, John M, et al. Baseline striatal functional connectivity as a predictor of response to antipsychotic drug treatment. Am J Psychiatry. 2016 [cited 2018 Dec 24];173(1):69–77. http://www.ncbi.nlm.nih.gov/pubmed/26315980.
114. Sarpal DK, Robinson DG, Fales C, Lencz T, Argyelan M, Karlsgodt KH, et al. Relationship between duration of untreated psychosis and intrinsic corticostriatal connectivity in patients with early phase schizophrenia. Neuropsychopharmacology. 2017 [cited 2018 Dec 24];42(11):2214–21. http://www.ncbi.nlm.nih.gov/pubmed/28294137.
115. Tolmeijer E, Kumari V, Peters E, Williams SCR, Mason L. Using fMRI and machine learning to predict symptom improvement following cognitive behavioural therapy for psychosis. NeuroImage Clin. 2018 [cited 2018 Dec 24];20:1053–61. http://www.ncbi.nlm.nih.gov/pubmed/30343250.
116. Koutsouleris N, Wobrock T, Guse B, Langguth B, Landgrebe M, Eichhammer P, et al. Predicting response to repetitive transcranial magnetic stimulation in patients with schizophrenia using structural magnetic resonance imaging: a multisite machine learning analysis. Schizophr Bull. 2017.
117. Li J, Yang Y, Zhu Y, Zhou L, Han Y, Yin T, et al. Towards characterizing the regional cerebral perfusion in evaluating the severity of major depression disorder with SPECT/CT. BMC Psychiatry. 2018;18(1):70.
118. Gillespie AL, Samanaite R, Mill J, Egerton A, MacCabe JH. Is treatment-resistant schizophrenia categorically distinct from treatment-responsive schizophrenia? A systematic review. BMC Psychiatry. 2017;17(1):12.
119. Gao S, Calhoun VD, Sui J. Machine learning in major depression: From classification to treatment outcome prediction. CNS Neurosci Ther. 2018;24(11):1037–52.
120. Nucifora FC, Woznica E, Lee BJ, Cascella N, Sawa A. Treatment resistant schizophrenia: clinical, biological, and therapeutic perspectives. Neurobiol Dis. 2018 [cited 2018 Dec 24]. https://www.sciencedirect.com/science/article/pii/S0969996118304947?via%3Dihub.

121. Paul S, Sharfman N. Functional connectivity as a means to delineate differences between treatment-resistant and treatment-responsive schizophrenia. J Neurophysiol. 2016 [cited 2018 Dec 24];116(2):229–31. http://www.ncbi.nlm.nih.gov/pubmed/26763784.
122. Mouchlianitis E, McCutcheon R, Howes OD. Brain-imaging studies of treatment-resistant schizophrenia: a systematic review. The Lancet Psychiatry. 2016;3(5):451–63.
123. Clementz BA, Sweeney JA, Hamm JP, Ivleva EI, Ethridge LE, Pearlson GD, et al. Identification of distinct psychosis biotypes using brain-based biomarkers. Am J Psychiatry. 2015;173(4):373–84.
124. Ivleva EI, Clementz BA, Dutcher AM, Arnold SJM, Jeon-Slaughter H, Aslan S, et al. Brain structure biomarkers in the psychosis biotypes: findings from the bipolar-schizophrenia network for intermediate phenotypes. Biol Psychiatry. 2017;82(1):26–39.
125. Xia CH, Ma Z, Ciric R, Gu S, Betzel RF, Kaczkurkin AN, et al. Linked dimensions of psychopathology and connectivity in functional brain networks. Nat Commun. 2018 [cited 2018 Dec 24];9(1):3003. http://www.nature.com/articles/s41467-018-05317-y.
126. Honorio J, Tomasi D, Goldstein RZ, Leung H-C, Samaras D. Can a single brain region predict a disorder? IEEE Trans Med Imaging. 2012;31(11):2062–72.
127. Costafreda SG, Fu CHY, Picchioni M, Toulopoulou T, McDonald C, Kravariti E, et al. Pattern of neural responses to verbal fluency shows diagnostic specificity for schizophrenia and bipolar disorder. BMC Psychiatry. 2011;11(1):18.
128. Nieuwenhuis M, van Haren NEM, Pol HEH, Cahn W, Kahn RS, Schnack HG. Classification of schizophrenia patients and healthy controls from structural MRI scans in two large independent samples. Neuroimage. 2012;61(3):606–12.
129. Sabuncu MR, Konukoglu E, Initiative ADN. Clinical prediction from structural brain MRI scans: a large-scale empirical study. Neuroinformatics. 2015;13(1):31–46.
130. Zhu J, Zhuo C, Qin W, Wang D, Ma X, Zhou Y, et al. Performances of diffusion kurtosis imaging and diffusion tensor imaging in detecting white matter abnormality in schizophrenia. NeuroImage Clin. 2015;7:170–6.
131. Brodersen KH, Deserno L, Schlagenhauf F, Lin Z, Penny WD, Buhmann JM, et al. Dissecting psychiatric spectrum disorders by generative embedding. NeuroImage Clin. 2014;4:98–111.
132. Sun H, Lui S, Yao L, Deng W, Xiao Y, Zhang W, et al. Two patterns of white matter abnormalities in medication naive patients with first-episode schizophrenia revealed by diffusion tensor imaging and cluster analysis. JAMA Psychiatry. 2015;72(7):678–86.
133. Dwyer DB, Cabral C, Kambeitz-Ilankovic L, Sanfelici R, Kambeitz J, Calhoun V, et al. Brain subtyping enhances the neuroanatomical discrimination of schizophrenia. Schizophr Bull. 2018.
134. Maggioni E, Crespo-Facorro B, Nenadic I, Benedetti F, Gaser C, Sauer H, et al. Common and distinct structural features of schizophrenia and bipolar disorder: the European network on psychosis, affective disorders and cognitive trajectory (ENPACT) study. PLoS One. 2017;12(11):e0188000.
135. Calvo A, Delvecchio G, Altamura AC, Soares JC, Brambilla P. Gray matter volume differences between affective and non-affective first episode psychosis: a review of magnetic resonance imaging studies. J Affect Disord. 2018.
136. Delvecchio G, Altamura AC, Soares JC, Brambilla P. Pituitary gland in bipolar disorder and major depression: evidence from structural MRI studies: special section on "translational and neuroscience studies in affective disorders". J Affect Disord. 2017;218:446–50.
137. Arbabshirani MR, Plis S, Sui J, Calhoun VD. Single subject prediction of brain disorders in neuroimaging: promises and pitfalls. Neuroimage. 2017;145:137–65.
138. Cheng H, Newman S, Goñi J, Kent JS, Howell J, Bolbecker A, et al. Nodal centrality of functional network in the differentiation of schizophrenia. Schizophr Res. 2015;168(1–2):345–52.
139. Koutsouleris N, Meisenzahl EM, Borgwardt S, Riecher-Rössler A, Frodl T, Kambeitz J, et al. Individualized differential diagnosis of schizophrenia and mood disorders using neuroanatomical biomarkers. Brain. 2015;138(7):2059–73.

140. Kim YK. Molecular neurobiology of major depressive disorder. Prog Neuropsychopharmacol Biol Psychiatry. 2016;64:275.
141. Rashid B, Chen J, Rashid I, Damaraju E, Liu J, Miller R, et al. A framework for linking resting-state chronnectome/genome features in schizophrenia: a pilot study. Neuroimage. 2019;184:843–54.
142. Mikoláš P. Improving the diagnosis of first-episode schizophrenia from magnetic resonance imaging using machine learning. Univerzita Karlova, 3. lékařská fakulta; 2018.
143. Rozycki M, Satterthwaite TD, Koutsouleris N, Erus G, Doshi J, Wolf DH, et al. Multisite machine learning analysis provides a robust structural imaging signature of schizophrenia detectable across diverse patient populations and within individuals. Schizophr Bull. 2018 [cited 2018 Dec 24];44(5):1035–44. http://www.ncbi.nlm.nih.gov/pubmed/29186619.
144. Skåtun KC, Kaufmann T, Doan NT, Alnæs D, Córdova-Palomera A, Jönsson EG, et al. Consistent functional connectivity alterations in schizophrenia spectrum disorder: a multisite study. Schizophr Bull. 2017 [cited 2018 Dec 24];43(4):914–24. http://www.ncbi.nlm.nih.gov/pubmed/27872268.
145. Calhoun VD, Maciejewski PK, Pearlson GD, Kiehl KA. Temporal lobe and "default" hemodynamic brain modes discriminate between schizophrenia and bipolar disorder. Hum Brain Mapp. 2008;29(11):1265–75.
146. Arribas JI, Calhoun VD, Adali T. Automatic Bayesian classification of healthy controls, bipolar disorder, and schizophrenia using intrinsic connectivity maps from FMRI data. IEEE Trans Biomed Eng. 2010;57(12):2850–60.
147. Fusté M, Pauls A, Worker A, Reinders AATS, Simmons A, Williams SCR, et al. Brain structure in women at risk of postpartum psychosis: an MRI study. Transl Psychiatry. 2017;7(12):1286.
148. Kasai K, Shenton ME, Salisbury DF, Onitsuka T, Toner SK, Yurgelun-Todd D, et al. Differences and similarities in insular and temporal pole MRI gray matter volume abnormalities in first-episode schizophrenia and affective psychosis. Arch Gen Psychiatry. 2003;60(11):1069–77.
149. Kubicki M, Shenton ME, Salisbury DF, Hirayasu Y, Kasai K, Kikinis R, et al. Voxel-based morphometric analysis of gray matter in first episode schizophrenia. Neuroimage. 2002;17(4):1711–9.
150. Hirayasu Y, Shenton ME, Salisbury DF, Dickey CC, Fischer IA, Mazzoni P, et al. Lower left temporal lobe MRI volumes in patients with first-episode schizophrenia compared with psychotic patients with first-episode affective disorder and normal subjects. Am J Psychiatry. 1998;155(10):1384–91.
151. Ardekani BA, Tabesh A, Sevy S, Robinson DG, Bilder RM, Szeszko PR. Diffusion tensor imaging reliably differentiates patients with schizophrenia from healthy volunteers. Hum Brain Mapp. 2011;32(1):1–9.
152. Sakoğlu Ü, Pearlson GD, Kiehl KA, Wang YM, Michael AM, Calhoun VD. A method for evaluating dynamic functional network connectivity and task-modulation: application to schizophrenia. Magn Reson Mater Physics, Biol Med. 2010 [cited 2018 Dec 24];23(5–6):351–66. http://www.ncbi.nlm.nih.gov/pubmed/20162320.
153. Du Y, Fu Z, Calhoun VD. Classification and prediction of brain disorders using functional connectivity: promising but challenging. Front Neurosci. 2018;12.
154. Laumann TO, Snyder AZ, Mitra A, Gordon EM, Gratton C, Adeyemo B, et al. On the stability of BOLD fMRI correlations. Cereb Cortex. 2016;27(10):4719–32.
155. Zeng L-L, Wang H, Hu P, Yang B, Pu W, Shen H, et al. Multi-site diagnostic classification of schizophrenia using discriminant deep learning with functional connectivity MRI. EBioMedicine. 2018 [cited 2018 Dec 24];30:74–85. http://www.ncbi.nlm.nih.gov/pubmed/29622496.
156. Phillips ML, Kupfer DJ. Bipolar disorder diagnosis: challenges and future directions. Lancet. 2013 [cited 2018 Dec 24];381(9878):1663–71. http://www.ncbi.nlm.nih.gov/pubmed/23663952.

157. Leboyer M, Kupfer DJ. Bipolar disorder: new perspectives in health care and prevention. J Clin Psychiatry. 2010;71(12):1689.
158. Benarous X, Consoli A, Milhiet V, Cohen D. Early interventions for youths at high risk for bipolar disorder: a developmental approach. Eur Child Adolesc Psychiatry. 2016;25(3):217–33.
159. Chen C, Suckling J, Lennox BR, Ooi C, Bullmore ET. A quantitative meta-analysis of fMRI studies in bipolar disorder. Bipolar Disord. 2011;13(1):1–15.
160. Ganzola R, Nickson T, Bastin ME, Giles S, Macdonald A, Sussmann J, et al. Longitudinal differences in white matter integrity in youth at high familial risk for bipolar disorder. Bipolar Disord. 2017;19(3):158–67.
161. Roberts G, Perry A, Lord A, Frankland A, Leung V, Holmes-Preston E, et al. Structural dysconnectivity of key cognitive and emotional hubs in young people at high genetic risk for bipolar disorder. Mol Psychiatry. 2018 [cited 2018 Dec 24];23(2):413–21. http://www.ncbi.nlm.nih.gov/pubmed/27994220.
162. Jeganathan J, Perry A, Bassett DS, Roberts G, Mitchell PB, Breakspear M. Fronto-limbic dysconnectivity leads to impaired brain network controllability in young people with bipolar disorder and those at high genetic risk. NeuroImage Clin. 2018;19:71–81.
163. Redlich R, Almeida JR, Grotegerd D, Opel N, Kugel H, Heindel W, et al. Brain morphometric biomarkers distinguishing unipolar and bipolar depression: a voxel-based morphometry–pattern classification approach. JAMA Psychiatry. 2014;71(11):1222–30.
164. Rocha-Rego V, Jogia J, Marquand AF, Mourao-Miranda J, Simmons A, Frangou S. Examination of the predictive value of structural magnetic resonance scans in bipolar disorder: a pattern classification approach. Psychol Med. 2014;44(3):519–32.
165. Nortje G, Stein DJ, Radua J, Mataix-Cols D, Horn N. Systematic review and voxel-based meta-analysis of diffusion tensor imaging studies in bipolar disorder. J Affect Disord. 2013;150(2):192–200.
166. Kafantaris V, Spritzer L, Doshi V, Saito E, Szeszko PR. Changes in white matter microstructure predict lithium response in adolescents with bipolar disorder. Bipolar Disord. 2017;19(7):587–94.
167. Wakana S, Caprihan A, Panzenboeck MM, Fallon JH, Perry M, Gollub RL, et al. Reproducibility of quantitative tractography methods applied to cerebral white matter. Neuroimage. 2007;36(3):630–44.
168. Wise T, Radua J, Nortje G, Cleare AJ, Young AH, Arnone D. Voxel-based meta-analytical evidence of structural disconnectivity in major depression and bipolar disorder. Biol Psychiatry. 2016;79(4):293–302.
169. Gold AL, Brotman MA, Adleman NE, Lever SN, Steuber ER, Fromm SJ, et al. Comparing brain morphometry across multiple childhood psychiatric disorders. J Am Acad Child Adolesc Psychiatry. 2016;55(12):1027–37.
170. Wang X, Tian F, Wang S, Cheng B, Qiu L, He M, et al. Gray matter bases of psychotic features in adult bipolar disorder: a systematic review and voxel-based meta-analysis of neuroimaging studies. Hum Brain Mapp. 2018 [cited 2018 Dec 24];39(12):4707–23. http://www.ncbi.nlm.nih.gov/pubmed/30096212.
171. Altamura AC, Maggioni E, Dhanoa T, Ciappolino V, Paoli RA, Cremaschi L, et al. The impact of psychosis on brain anatomy in bipolar disorder: a structural MRI study. J Affect Disord. 2018;233:100–9.
172. Wei Y, Chang M, Womer FY, Zhou Q, Yin Z, Wei S, et al. Local functional connectivity alterations in schizophrenia, bipolar disorder, and major depressive disorder. J Affect Disord. 2018 [cited 2018 Dec 24];236:266–73. http://www.ncbi.nlm.nih.gov/pubmed/29751242.
173. Pasternak O, Kelly S, Sydnor VJ, Shenton ME. Advances in microstructural diffusion neuroimaging for psychiatric disorders. Neuroimage. 2018 [cited 2018 Dec 24];182:259–82. http://www.ncbi.nlm.nih.gov/pubmed/29729390.

174. Tréhout M, Leroux E, Delcroix N, Dollfus S. Relationships between corpus callosum and language lateralization in patients with schizophrenia and bipolar disorders. Bipolar Disord. 2017 [cited 2018 Dec 24];19(6):496–504. http://www.ncbi.nlm.nih.gov/pubmed/28834020.
175. Nenadić I, Hoof A, Dietzek M, Langbein K, Reichenbach JR, Sauer H, et al. Diffusion tensor imaging of cingulum bundle and corpus callosum in schizophrenia vs. bipolar disorder. Psychiatry Res Neuroimaging. 2017 [cited 2018 Dec 24];266:96–100. https://www.sciencedirect.com/science/article/pii/S0925492716303651?via%3Dihub.
176. Li J, Edmiston EK, Chen K, Tang Y, Ouyang X, Jiang Y, et al. A comparative diffusion tensor imaging study of corpus callosum subregion integrity in bipolar disorder and schizophrenia. Psychiatry Res Neuroimaging. 2014;221(1):58–62.
177. Nazeri A, Mulsant BH, Rajji TK, Levesque ML, Pipitone J, Stefanik L, et al. Gray matter neuritic microstructure deficits in schizophrenia and bipolar disorder. Biol Psychiatry. 2017 [cited 2018 Dec 24];82(10):726–36. http://www.ncbi.nlm.nih.gov/pubmed/28073491.
178. Parvathaneni P, Rogers BP, Huo Y, Schilling KG, Hainline AE, Anderson AW, et al. Gray matter surface based spatial statistics (GS-BSS) in diffusion microstructure. In: Medical image computing and computer-assisted intervention: MICCAI. International conference on medical image computing and computer-assisted intervention. 2017 [cited 2018 Dec 24]. p. 638–46. http://www.ncbi.nlm.nih.gov/pubmed/29226284.
179. Huo Y, Bao S, Parvathaneni P, Landman BA. Improved stability of whole brain surface parcellation with multi-atlas segmentation. In: Medical imaging 2018: image processing. International Society for Optics and Photonics; 2018. p. 1057438.
180. Perlman SB, Almeida JR, Kronhaus DM, Versace A, LaBarbara EJ, Klein CR, et al. Amygdala activity and prefrontal cortex-amygdala effective connectivity to emerging emotional faces distinguish remitted and depressed mood states in bipolar disorder. Bipolar Disord. 2012 [cited 2018 Dec 24];14(2):162–74. http://www.ncbi.nlm.nih.gov/pubmed/22420592.
181. Fournier JC, Chase HW, Almeida J, Phillips ML. Within-and between-session changes in neural activity during emotion processing in unipolar and bipolar depression. Biol Psychiatry Cogn Neurosci Neuroimaging. 2016;1(6):518–27.
182. Liu H, Tang Y, Womer F, Fan G, Lu T, Driesen N, et al. Differentiating patterns of amygdala-frontal functional connectivity in schizophrenia and bipolar disorder. Schizophr Bull. 2013;40(2):469–77.
183. Birur B, Kraguljac NV, Shelton RC, Lahti AC. Brain structure, function, and neurochemistry in schizophrenia and bipolar disorder—a systematic review of the magnetic resonance neuroimaging literature. NPJ Schizophr. 2017;3(1):15.
184. Tighe SK, Mahon PB, Potash JB. Predictors of lithium response in bipolar disorder. Ther Adv Chronic Dis. 2011 [cited 2018 Dec 24];2(3):209–26. http://www.ncbi.nlm.nih.gov/pubmed/23251751.
185. Moore GJ, Cortese BM, Glitz DA, Zajac-Benitez C, Quiroz JA, Uhde TW, et al. A longitudinal study of the effects of lithium treatment on prefrontal and subgenual prefrontal gray matter volume in treatment-responsive bipolar disorder patients. J Clin Psychiatry. 2009 [cited 2018 Dec 24];70(5):699–705. http://www.psychiatrist.com/abstracts/abstracts.asp?abstract=200905/050908.htm.
186. Lyoo IK, Dager SR, Kim JE, Yoon SJ, Friedman SD, Dunner DL, et al. Lithium-induced gray matter volume increase as a neural correlate of treatment response in bipolar disorder: a longitudinal brain imaging study. Neuropsychopharmacology. 2010;35(8):1743.
187. Fleck DE, Ernest N, Adler CM, Cohen K, Eliassen JC, Norris M, et al. Prediction of lithium response in first-episode mania using the lithium intelligent agent (LITHIA): pilot data and proof-of-concept. Bipolar Disord. 2017;19(4):259–72.
188. Dusi N, De Carlo V, Delvecchio G, Bellan M, Soares JC, Brambilla P. MRI features of clinical outcome in bipolar disorder: a selected review. J Affect Disord. 2018.
189. Fakhoury M. New insights into the neurobiological mechanisms of major depressive disorders. Gen Hosp Psychiatry. 2015;37(2):172–7.

190. Campbell S, Marriott M, Nahmias C, MacQueen GM. Lower hippocampal volume in patients suffering from depression: a meta-analysis. Am J Psychiatry. 2004;161(4):598–607.
191. Drevets WC, Price JL, Simpson Jr JR, Todd RD, Reich T, Vannier M, et al. Subgenual prefrontal cortex abnormalities in mood disorders. Nature. 1997;386(6627):824.
192. Caetano SC, Kaur S, Brambilla P, Nicoletti M, Hatch JP, Sassi RB, et al. Smaller cingulate volumes in unipolar depressed patients. Biol Psychiatry. 2006;59(8):702–6.
193. Bremner JD, Narayan M, Anderson ER, Staib LH, Miller HL, Charney DS. Hippocampal volume reduction in major depression. Am J Psychiatry. 2000;157(1):115–8.
194. Frodl T, Jäger M, Smajstrlova I, Born C, Bottlender R, Palladino T, et al. Effect of hippocampal and amygdala volumes on clinical outcomes in major depression: a 3-year prospective magnetic resonance imaging study. J Psychiatry Neurosci JPN. 2008;33(5):423.
195. Hamilton JP, Siemer M, Gotlib IH. Amygdala volume in major depressive disorder: a meta-analysis of magnetic resonance imaging studies. Mol Psychiatry. 2008;13(11):993.
196. Ebmeier K, Rose E, Steele D. Cognitive impairment and fMRI in major depression. Neurotox Res. 2006;10(2):87–92.
197. Anand A, Li Y, Wang Y, Wu J, Gao S, Bukhari L, et al. Activity and connectivity of brain mood regulating circuit in depression: a functional magnetic resonance study. Biol Psychiatry. 2005;57(10):1079–88.
198. Cullen KR, Gee DG, Klimes-Dougan B, Gabbay V, Hulvershorn L, Mueller BA, et al. A preliminary study of functional connectivity in comorbid adolescent depression. Neurosci Lett. 2009;460(3):227–31.
199. Liu Z, Xu C, Xu Y, Wang Y, Zhao B, Lv Y, et al. Decreased regional homogeneity in insula and cerebellum: a resting-state fMRI study in patients with major depression and subjects at high risk for major depression. Psychiatry Res Neuroimaging. 2010;182(3):211–5.
200. Horn DI, Yu C, Steiner J, Buchmann J, Kaufmann J, Osoba A, et al. Glutamatergic and resting-state functional connectivity correlates of severity in major depression–the role of pregenual anterior cingulate cortex and anterior insula. Front Syst Neurosci. 2010;4:33.
201. Wang L, Hermens DF, Hickie IB, Lagopoulos J. A systematic review of resting-state functional-MRI studies in major depression. J Affect Disord. 2012;142(1–3):6–12.
202. Zhang S, Chen J, Kuang L, Cao J, Zhang H, Ai M, et al. Association between abnormal default mode network activity and suicidality in depressed adolescents. BMC Psychiatry. 2016 [cited 2018 Dec 24];16(1):337. http://www.ncbi.nlm.nih.gov/pubmed/27688124.
203. Broyd SJ, Demanuele C, Debener S, Helps SK, James CJ, Sonuga-Barke EJS. Default-mode brain dysfunction in mental disorders: a systematic review. Neurosci Biobehav Rev. 2009;33(3):279–96.
204. Anand A, Li Y, Wang Y, Wu J, Gao S, Bukhari L, et al. Antidepressant effect on connectivity of the mood-regulating circuit: an FMRI study. Neuropsychopharmacology. 2005;30(7):1334.
205. Anand A, Li Y, Wang Y, Gardner K, Lowe MJ. Reciprocal effects of antidepressant treatment on activity and connectivity of the mood regulating circuit: an FMRI study. J Neuropsychiatry Clin Neurosci. 2007;19(3):274–82.
206. Anand A, Li Y, Wang Y, Lowe MJ, Dzemidzic M. Resting state corticolimbic connectivity abnormalities in unmedicated bipolar disorder and unipolar depression. Psychiatry Res Neuroimaging. 2009;171(3):189–98.
207. Foland-Ross LC, Gilbert BL, Joormann J, Gotlib IH. Neural markers of familial risk for depression: an investigation of cortical thickness abnormalities in healthy adolescent daughters of mothers with recurrent depression. J Abnorm Psychol. 2015;124(3):476.
208. Rao U, Chen L-A, Bidesi AS, Shad MU, Thomas MA, Hammen CL. Hippocampal changes associated with early-life adversity and vulnerability to depression. Biol Psychiatry. 2010 [cited 2018 Dec 24];67(4):357–64. http://www.ncbi.nlm.nih.gov/pubmed/20015483.
209. Amico F, Meisenzahl E, Koutsouleris N, Reiser M, Möller H-J, Frodl T. Structural MRI correlates for vulnerability and resilience to major depressive disorder. J Psychiatry Neurosci JPN. 2011;36(1):15.

210. Monk CS, Klein RG, Telzer EH, Schroth EA, Mannuzza S, Moulton JL, et al. Amygdala and nucleus accumbens activation to emotional facial expressions in children and adolescents at risk for major depression. Am J Psychiatry. 2008;165(1):90–8. https://doi.org/10.1176/appi.ajp.2007.06111917.
211. Mannie ZN, Filippini N, Williams C, Near J, Mackay CE, Cowen PJ. Structural and functional imaging of the hippocampus in young people at familial risk of depression. Psychol Med. 2014;44(14):2939–48.
212. Bürger C, Redlich R, Grotegerd D, Meinert S, Dohm K, Schneider I, et al. Differential abnormal pattern of anterior cingulate gyrus activation in unipolar and bipolar depression: an fMRI and pattern classification approach. Neuropsychopharmacology. 2017;42(7):1399.
213. Zou K, Huang X, Li T, Gong Q, Li Z, Ou-yang L, et al. Alterations of white matter integrity in adults with major depressive disorder: a magnetic resonance imaging study. J Psychiatry Neurosci JPN. 2008;33(6):525.
214. Sexton CE, Mackay CE, Ebmeier KP. A systematic review of diffusion tensor imaging studies in affective disorders. Biol Psychiatry. 2009;66(9):814–23.
215. Öngür D, Price JL. The organization of networks within the orbital and medial prefrontal cortex of rats, monkeys and humans. Cereb Cortex. 2000;10(3):206–19.
216. Kondo H, Saleem KS, Price JL. Differential connections of the perirhinal and parahippocampal cortex with the orbital and medial prefrontal networks in macaque monkeys. J Comp Neurol. 2005;493(4):479–509.
217. Cole J, Costafreda SG, McGuffin P, Fu CHY. Hippocampal atrophy in first episode depression: a meta-analysis of magnetic resonance imaging studies. J Affect Disord. 2011;134(1–3):483–7.
218. Kempton MJ, Salvador Z, Munafò MR, Geddes JR, Simmons A, Frangou S, et al. Structural neuroimaging studies in major depressive disorder: meta-analysis and comparison with bipolar disorder. Arch Gen Psychiatry. 2011;68(7):675–90.
219. Costafreda SG, Chu C, Ashburner J, Fu CHY. Prognostic and diagnostic potential of the structural neuroanatomy of depression. PLoS One. 2009;4(7):e6353.
220. Bracht T, Horn H, Strik W, Federspiel A, Schnell S, Höfle O, et al. White matter microstructure alterations of the medial forebrain bundle in melancholic depression. J Affect Disord. 2014;155:186–93.
221. Lois G, Wessa M. Differential association of default mode network connectivity and rumination in healthy individuals and remitted MDD patients. Soc Cogn Affect Neurosci. 2016;11(11):1792–801.
222. Vassilopoulou K, Papathanasiou M, Michopoulos I, Boufidou F, Oulis P, Kelekis N, et al. A magnetic resonance imaging study of hippocampal, amygdala and subgenual prefrontal cortex volumes in major depression subtypes: melancholic versus psychotic depression. J Affect Disord. 2013 [cited 2018 Dec 24];146(2):197–204. http://www.ncbi.nlm.nih.gov/pubmed/23021193.
223. Kerestes R, Davey CG, Stephanou K, Whittle S, Harrison BJ. Functional brain imaging studies of youth depression: a systematic review. NeuroImage Clin. 2014;4:209–31.
224. Tang S, Lu L, Zhang L, Hu X, Bu X, Li H, et al. Abnormal amygdala resting-state functional connectivity in adults and adolescents with major depressive disorder: a comparative meta-analysis. EBioMedicine. 2018 [cited 2018 Dec 24];36:436–45. http://www.ncbi.nlm.nih.gov/pubmed/30316866.
225. Clark DL, Konduru N, Kemp A, Bray S, Brown EC, Goodyear B, et al. The impact of age of onset on amygdala intrinsic connectivity in major depression. Neuropsychiatr Dis Treat. 2018;14:343.
226. Drysdale AT, Grosenick L, Downar J, Dunlop K, Mansouri F, Meng Y, et al. Resting-state connectivity biomarkers define neurophysiological subtypes of depression. Nat Med. 2017;23(1):28.
227. Williams LM. Defining biotypes for depression and anxiety based on large-scale circuit dysfunction: a theoretical review of the evidence and future directions for clinical translation. Depress Anxiety. 2017;34(1):9–24.

228. Dunlop BW, Mayberg HS. Neuroimaging advances for depression. In: Cerebrum: the Dana forum on brain science. Dana Foundation; 2017.
229. Grotegerd D, Stuhrmann A, Kugel H, Schmidt S, Redlich R, Zwanzger P, et al. Amygdala excitability to subliminally presented emotional faces distinguishes unipolar and bipolar depression: an fMRI and pattern classification study. Hum Brain Mapp. 2014;35(7):2995–3007.
230. Ambrosi E, Arciniegas DB, Madan A, Curtis KN, Patriquin MA, Jorge RE, et al. Insula and amygdala resting-state functional connectivity differentiate bipolar from unipolar depression. Acta Psychiatr Scand. 2017;136(1):129–39.
231. Delvecchio G, Fossati P, Boyer P, Brambilla P, Falkai P, Gruber O, et al. Common and distinct neural correlates of emotional processing in bipolar disorder and major depressive disorder: a voxel-based meta-analysis of functional magnetic resonance imaging studies. Eur Neuropsychopharmacol. 2012;22(2):100–13.
232. Wise T, Radua J, Via E, Cardoner N, Abe O, Adams TM, et al. Common and distinct patterns of grey-matter volume alteration in major depression and bipolar disorder: evidence from voxel-based meta-analysis. Mol Psychiatry. 2017;22(10):1455.
233. Niu M, Wang Y, Jia Y, Wang J, Zhong S, Lin J, et al. Common and specific abnormalities in cortical thickness in patients with major depressive and bipolar disorders. EBioMedicine. 2017;16:162–71.
234. Karim HT, Wang M, Andreescu C, Tudorascu D, Butters MA, Karp JF, et al. Acute trajectories of neural activation predict remission to pharmacotherapy in late-life depression. NeuroImage Clin. 2018.
235. Lin E, Kuo P-H, Liu Y-L, Yu YWY, Yang A, Tsai S-J. A deep learning approach for predicting antidepressant response in major depression using clinical and genetic biomarkers. Front Psychiatry. 2018;9:290.
236. Hilbert K, Lueken U, Muehlhan M, Beesdo-Baum K. Separating generalized anxiety disorder from major depression using clinical, hormonal, and structural MRI data: a multimodal machine learning study. Brain Behav. 2017;7(3):e00633.
237. Sacchet MD, Prasad G, Foland-Ross LC, Thompson PM, Gotlib IH. Support vector machine classification of major depressive disorder using diffusion-weighted neuroimaging and graph theory. Front Psychiatry. 2015 [cited 2018 Dec 24];6:21. http://www.ncbi.nlm.nih.gov/pubmed/25762941.
238. Orru G, Pettersson-Yeo W, Marquand AF, Sartori G, Mechelli A. Using support vector machine to identify imaging biomarkers of neurological and psychiatric disease: a critical review. Neurosci Biobehav Rev. 2012;36(4):1140–52.
239. Marquand AF, Mourão-Miranda J, Brammer MJ, Cleare AJ, Fu CHY. Neuroanatomy of verbal working memory as a diagnostic biomarker for depression. Neuroreport. 2008;19(15):1507–11.
240. Fu CHY, Mourao-Miranda J, Costafreda SG, Khanna A, Marquand AF, Williams SCR, et al. Pattern classification of sad facial processing: toward the development of neurobiological markers in depression. Biol Psychiatry. 2008;63(7):656–62.
241. Chen C-H, Suckling J, Ooi C, Fu CHY, Williams SCR, Walsh ND, et al. Functional coupling of the amygdala in depressed patients treated with antidepressant medication. Neuropsychopharmacology. 2008;33(8):1909.
242. Lisiecka D, Meisenzahl E, Scheuerecker J, Schoepf V, Whitty P, Chaney A, et al. Neural correlates of treatment outcome in major depression. Int J Neuropsychopharmacol. 2011;14(4):521–34.
243. Li B, Liu L, Friston KJ, Shen H, Wang L, Zeng L-L, et al. A treatment-resistant default mode subnetwork in major depression. Biol Psychiatry. 2013;74(1):48–54.
244. Posner J, Hellerstein DJ, Gat I, Mechling A, Klahr K, Wang Z, et al. Antidepressants normalize the default mode network in patients with dysthymia. JAMA Psychiatry. 2013;70(4):373–82.

245. Liston C, Chen AC, Zebley BD, Drysdale AT, Gordon R, Leuchter B, et al. Default mode network mechanisms of transcranial magnetic stimulation in depression. Biol Psychiatry. 2014;76(7):517–26.
246. Abbott CC, Lemke NT, Gopal S, Thoma RJ, Bustillo J, Calhoun VD, et al. Electroconvulsive therapy response in major depressive disorder: a pilot functional network connectivity resting state FMRI investigation. Front Psychiatry. 2013;4:10.
247. Wu F, Tang Y, Xu K, Kong L, Sun W, Wang F, et al. Whiter matter abnormalities in medication-naive subjects with a single short-duration episode of major depressive disorder. Psychiatry Res Neuroimaging. 2011;191(1):80–3.
248. Yamamura T, Okamoto Y, Okada G, Takaishi Y, Takamura M, Mantani A, et al. Association of thalamic hyperactivity with treatment-resistant depression and poor response in early treatment for major depression: a resting-state fMRI study using fractional amplitude of low-frequency fluctuations. Transl Psychiatry. 2017;6(3):e754.
249. Walsh EC, Eisenlohr-Moul TA, Minkel J, Bizzell J, Petty C, Crowther A, et al. Pretreatment brain connectivity during positive emotion upregulation predicts decreased anhedonia following behavioral activation therapy for depression. J Affect Disord. 2019 [cited 2018 Dec 24];243:188–92. http://www.ncbi.nlm.nih.gov/pubmed/30245249.
250. Crane NA, Jenkins LM, Bhaumik R, Dion C, Gowins JR, Mickey BJ, et al. Multidimensional prediction of treatment response to antidepressants with cognitive control and functional MRI. Brain. 2017;140(2):472–86.
251. Spies M, Kraus C, Geissberger N, Auer B, Klöbl M, Tik M, et al. Default mode network deactivation during emotion processing predicts early antidepressant response. Transl Psychiatry. 2017 [cited 2018 Dec 24];7(1):e1008. http://www.ncbi.nlm.nih.gov/pubmed/28117844.
252. Phillips ML, Chase HW, Sheline YI, Etkin A, Almeida JRC, Deckersbach T, et al. Identifying predictors, moderators, and mediators of antidepressant response in major depressive disorder: neuroimaging approaches. Am J Psychiatry. 2015 [cited 2018 Dec 24];172(2):124–38. http://www.ncbi.nlm.nih.gov/pubmed/25640931.
253. Fonseka TM, MacQueen GM, Kennedy SH. Neuroimaging biomarkers as predictors of treatment outcome in major depressive disorder. J Affect Disord. 2018;233:21–35.
254. Lener MS, Iosifescu DV. In pursuit of neuroimaging biomarkers to guide treatment selection in major depressive disorder: a review of the literature. Ann N Y Acad Sci. 2015;1344(1):50–65.
255. Poldrack RA, Gorgolewski KJ. Making big data open: data sharing in neuroimaging. Nat Neurosci. 2014 [cited 2018 Dec 24];17(11):1510–7. http://www.ncbi.nlm.nih.gov/pubmed/25349916.
256. Milham MP, Ai L, Koo B, Xu T, Amiez C, Balezeau F, et al. An open resource for nonhuman primate imaging. Neuron. 2018;100(1):61–74.
257. Zuo X-N, Anderson JS, Bellec P, Birn RM, Biswal BB, Blautzik J, et al. An open science resource for establishing reliability and reproducibility in functional connectomics. Sci Data. 2014 [cited 2018 Dec 24];1:140049. http://www.ncbi.nlm.nih.gov/pubmed/25977800.
258. Cui L-B, Liu L, Wang H-N, Wang L-X, Guo F, Xi Y-B, et al. Disease definition for schizophrenia by functional connectivity using radiomics strategy. Schizophr Bull. 2018.
259. Tamminga CA, Pearlson GD, Stan AD, Gibbons RD, Padmanabhan J, Keshavan M, et al. Strategies for advancing disease definition using biomarkers and genetics: the bipolar and schizophrenia network for intermediate phenotypes. Biol Psychiatry Cogn Neurosci Neuroimaging. 2017 [cited 2018 Dec 24];2(1):20–7. http://www.ncbi.nlm.nih.gov/pubmed/29560884.
260. Proudfoot M, Woolrich MW, Nobre AC, Turner MR. Magnetoencephalography. Pract Neurol. 2014 [cited 2018 Dec 24];14(5):336–43. http://www.ncbi.nlm.nih.gov/pubmed/24647614.
261. Brookes MJ, Woolrich M, Luckhoo H, Price D, Hale JR, Stephenson MC, et al. Investigating the electrophysiological basis of resting state networks using magnetoencephalography. Proc Natl Acad Sci. 2011. https://doi.org/10.1073/1112685108.

262. Beckmann CF, DeLuca M, Devlin JT, Smith SM. Investigations into resting-state connectivity using independent component analysis. Philos Trans R Soc London B Biol Sci. 2005;360(1457):1001–13.
263. Smith SM, Miller KL, Moeller S, Xu J, Auerbach EJ, Woolrich MW, et al. Temporally-independent functional modes of spontaneous brain activity. Proc Natl Acad Sci USA. 2012 [cited 2018 Dec 24];109(8):3131–6. http://www.ncbi.nlm.nih.gov/pubmed/22323591.
264. Luckhoo H, Hale JR, Stokes MG, Nobre AC, Morris PG, Brookes MJ, et al. Inferring task-related networks using independent component analysis in magnetoencephalography. Neuroimage. 2012;62(1):530–41.
265. Patel MJ, Andreescu C, Price JC, Edelman KL, Reynolds CF, Aizenstein HJ, et al. Machine learning approaches for integrating clinical and imaging features in late-life depression classification and response prediction. Int J Geriatr Psychiatry. 2015 [cited 2018 Dec 24];30(10):1056–67. http://www.ncbi.nlm.nih.gov/pubmed/25689482.
266. Guillaume B, Wang C, Poh J, Shen MJ, Ong ML, Tan PF, et al. Improving mass-univariate analysis of neuroimaging data by modelling important unknown covariates: application to epigenome-wide association studies. Neuroimage. 2018;173:57–71.
267. Habeck C, Stern Y, Initiative ADN. Multivariate data analysis for neuroimaging data: overview and application to Alzheimer's disease. Cell Biochem Biophys. 2010;58(2):53–67.

Research Methods for Precision Psychiatry

Biomarker-Guided Tailored Therapy

Jessica Lydiard and Charles B. Nemeroff

Abstract

Personalized medicine aims to integrate a number of characteristics such as genetic and epigenetic variations, other biomarkers, clinical symptoms, and environmental factors in order to predict susceptibility to disease, aid in diagnosis, and identify efficacious treatments with maximum likelihood of favorable response and minimal chance of adverse effects. The use of personalized medicine approaches in psychiatry is underdeveloped, but has a profound potential for improving prevention and treatment. There are a number of studies that have found promising associations between a variety of biomarkers and clinical response to psychopharmacological treatment in various psychiatric disorders. These biomarkers include neuroimaging, electrophysiology, peripheral serum, and plasma biomarkers, and variations in genomics, epigenetics, proteomics, and metabolomics. Ultimately, the best model for precision medicine in complex, multifactorial diseases such as psychiatric illnesses will likely involve integrated methodology that combines information from multiple sources including biologic, clinical, and environmental data. While much progress has been made in the development of valid biomarkers

J. Lydiard · C. B. Nemeroff (✉)
Department of Psychiatry and Behavioral Sciences, University of Miami Miller School of Medicine, Miami, FL 33133, USA
e-mail: cnemeroff@austin.utexas.edu

J. Lydiard
e-mail: jessica.lydiard@jhsmiami.org

C. B. Nemeroff
Department of Psychiatry, University of Texas Austin, Dell Medical School, Austin, TX, USA

© Springer Nature Singapore Pte Ltd. 2019

Y.-K. Kim (ed.), *Frontiers in Psychiatry*, Advances in Experimental Medicine and Biology 1192, https://doi.org/10.1007/978-981-32-9721-0_10

in psychiatric disorders, there is much work to be done in determining their clinical utility.

Keywords
Major depressive disorder · Bipolar disorder · Schizophrenia · Personalized medicine · Precision medicine · Epigenetics · Pharmacogenomics · Neuroimaging

Introduction

Personalized medicine aims to tailor individualized treatment for each patient based on a number of characteristics such as genetic and epigenetic variations, other biomarkers, clinical symptoms, and environmental factors. The ultimate goal is to predict susceptibility to disease, aid in diagnosis, and identify efficacious treatments with maximum likelihood of favorable response and minimal chance of adverse effects [1]. Over the past decade, the Food and Drug Administration has approved a number of genetically prescreened drugs, labeled as pharmacogenomics, leading to significant advances in patient care in fields such as oncology, rheumatology, pulmonary medicine, and gastroenterology [2, 3].

Psychiatric illness is a huge burden to global society and the leading cause of years lost to disability [4]. The use of personalized medicine approaches in psychiatry is underdeveloped, but has a profound potential for improving prevention and treatment. The number and diversity of available treatments for psychiatric disorders alone illustrate the difficulty of matching the best treatment for the individual patient. The mechanisms by which psychotropic drugs act are incompletely understood and each drug has side effects. When an ineffective drug is prescribed, time is wasted for the patient with persistent symptoms and often bothersome side effects. Traditionally, psychiatric diagnosis has been based on the mental health professional's interpretation of patient (and family members) reports of behavioral signs and symptoms, and objective clinical measures such as biomarkers have not yet been incorporated into diagnostic criteria or monitoring of treatment response [4]. The development of personalized medicine approaches in psychiatry will be realized by the identification of relevant disease biomarkers, which can provide an objective measure of biologic function (or dysfunction), response to treatment, and disease severity. Biomarkers under investigation for psychiatric disorders include peripheral serum or plasma markers, neuroimaging findings, electrophysiology, and variations in genomics, epigenetics, proteomics, and metabolomics.

In this chapter, evidence to support the use of biomarkers to guide treatment and prevention in psychiatric disorders will be reviewed. The evidence is most robust in the area of predicting response to treatment for mood disorders, so much of the chapter will focus on this topic. Schizophrenia and psychotic disorders will only be briefly discussed as well as future directions, including biomarker-guided prevention efforts.

Major Depressive Disorder

Overview of Treatment of Major Depressive Disorder

Major depressive disorder (MDD) is the most prevalent psychiatric disorder, estimated to affect more than 100 million people worldwide. Up to 40% of individuals suffering from MDD never seek treatment, and approximately half of suicide victims visited a primary care provider in the month prior to their death [5]. This devastating disease exhibits considerable genetic heritability [6]. There are currently approximately 30 FDA-approved medications for the treatment of MDD that are believed to modulate the availability or activity of different neurotransmitter systems in the brain. These include serotonin, norepinephrine, and dopamine [7]. There is no established paradigm for choosing psychotropic medications and treatment is largely chosen based on physician preference and factors such as family or personal history of good treatment response, affordability, and side effect profile. Remission rates after treatment with a single antidepressant are modest, estimated to be roughly 30–50% after 12 weeks. An additional 30–40% of patients fail to achieve adequate response after multiple medication trials over 1 year of treatment [8]. Another commonly used treatment modality for MDD is evidence-based psychotherapies (i.e., cognitive behavioral therapy (CBT) and interpersonal psychotherapy (IPT)), which have demonstrated clinical efficacy in multiple randomized controlled clinical trials [7]. There are also three FDA-approved somatic nonpharmacological treatments for MDD: electroconvulsive therapy (ECT), repetitive transcranial magnetic stimulation (rTMS), and vagus nerve stimulation (VNS) [7]. However, these modalities are not as commonly used and biomarker-based prediction of response has not been well studied.

Peripheral Serum Markers

Hypothalamic–Pituitary–Adrenal (HPA) Axis

HPA axis hyperactivity is a well-established finding in a significant portion of individuals suffering from MDD and has been replicated in dozens of studies over the last several decades. Patients with MDD have been found to have increased cerebrospinal fluid (CSF) concentrations of corticotropin-releasing hormone (CRH), as well as increased levels of cortisol in blood, urine, and CSF [9]. Elevated levels of cortisol are observed in up to 70% of patients with MDD after a dexamethasone suppression test (DST) [10], evidence of HPA axis hyperactivity. Several studies have shown that treatment with selective serotonin reuptake inhibitors (SSRIs) and clinical recovery leads to normalization of HPA axis hyperactivity [11]. The DST or combined CRH/DST challenge test may have clinical utility as a laboratory marker for treatment outcome. MDD patients who were initially non-suppressors who exhibit normalization of the DST response show clinical improvement after antidepressant treatment [12]. In multiple studies, persistent non-suppression in response to DST after antidepressant treatment predicts poorer treatment response [11, 12]. The major confound in these studies is the finding that early life trauma in the form of child abuse or neglect is associated with alterations in HPA axis activity [13].

Hypothalamic–Pituitary–Thyroid (HPT) Axis

There is a well-documented relationship between MDD and abnormalities in the hypothalamic–pituitary–thyroid (HPT) axis. This has been consistently replicated in clinical trials over the last 40 years [14, 15]. Patients with thyroid dysfunction are more likely to develop depressive symptoms, and abnormalities in thyroid function are observed in a subset of patients with MDD [15]. Plasma concentrations of thyroid hormones (T3, T4) are typically normal in depressed patients, however other abnormalities have been observed. Patients with MDD have been found to have elevated TRH concentrations in the CSF [15, 16]. Another well-replicated finding is the presence of higher levels of circulating antithyroid antibodies in the plasma of depressed patients compared to nondepressed controls [15–17]. Plasma thyroid-stimulating hormone (TSH) responses after intravenous administration of thyrotropin-releasing hormone (TRH) as part of the TRH stimulation test are blunted in approximately 25% and abnormally elevated in approximately 15% of patients with MDD [14]. Research studies examining the prognostic utility of the TRH stimulation test in clinical trials have yielded mixed findings [18, 19]. In a study of 95 depressed patients treated with either nortriptyline ($N=46$) or fluoxetine ($N=49$) for 6 weeks, significant differences were seen in the maximum TSH response (ΔmaxTSH) to a TRH stimulation test between patients treated with each medication; ΔmaxTSH concentrations decreased significantly in fluoxetine responders after treatment and increased significantly in nortriptyline responders. Additionally, a significant decrease in plasma concentrations of thyroxine (T4) and free thyroxine (FT4) was observed after treatment with both medications in responders, but not nonresponders [20]. Several other studies have shown a significant decrease in plasma T4 and/or FT4 concentrations after at least 4 weeks of antidepressant treatment with SSRIs or TCAs in treatment responders, but not in nonresponders [21–23] Although the majority of patients with MDD do not have clinically significant thyroid disease, existing literature indicates a relationship between MDD and subclinical thyroid dysfunction. Triiodothyronine (T3) has been definitively shown to augment the efficacy of several antidepressants and is commonly used as adjunctive therapy in clinical practice for treatment-resistant patients, many of whom have thyroid indices within normal range [15, 16, 24]. While the mechanisms of thyroid supplementation in initially nonresponsive patients receiving antidepressant treatment remains unclear, assessment of thyroid function in depressed patients may serve as a useful personalized medicine approach to treatment of MDD.

Inflammatory Biomarkers

The association between MDD and inflammation has been consistently replicated in a multitude of clinical studies. One meta-analysis revealed that elevated

C-reactive protein (CRP) levels are associated with increased risk of future MDD [25]. Another meta-analysis found that treatment with antidepressants resulted in a marginally significant decrease in CRP levels, and moreover higher baseline CRP concentrations were associated with a greater reduction in depressive symptoms [26]. In contrast, Chang et al. found higher baseline CRP levels predicted poorer response to 2 weeks of treatment with either venlafaxine or fluoxetine [27]. Several recent trials [28, 29] suggest that lower pretreatment levels of CRP (<0.1 mg/L) are associated with a good response to SSRIs while individuals with higher pretreatment CRP levels (≥0.1 mg/L) may require combination therapy or the use of agents that act on additional neurotransmitter systems. In a clinical trial designed to determine if tumor necrosis factor (TNF) antagonism would reduce depressive symptoms (measured by HAM-D scores) in patients with treatment-resistant depression and if higher baseline levels of inflammatory biomarkers would predict treatment response, subjects with moderate treatment-resistant depression (as determined by the Massachusetts General Hospital Staging method) were randomized to receive three infusions of infliximab (baseline, 2 and 6 weeks), a TNF antagonist, or placebo in a 12-week study. There was no significant difference in HAM-D change scores between the two treatment groups over the 12-week period. However, subjects with baseline CRP >5 mg/L treated with infliximab showed significantly greater treatment response (≥50% reduction in HAM-D score at any point during treatment) as compared to the placebo group. This suggests that although TNF antagonism did not have a significant effect in treatment-resistant depression overall, it may have efficacy in a subset of patients with elevated baseline inflammatory biomarkers [30].

Multiple studies have demonstrated elevated levels of pro-inflammatory cytokines in depressed patients compared to controls [31]. In a meta-analysis of studies investigating the effect of antidepressant treatment on inflammatory cytokines IL-1β, IL-6, and TNF-α, antidepressant treatment lowered levels of IL-1β and possibly IL-6, but had no significant effect on TNF-α. Further analysis of antidepressant classes revealed that treatment with SSRIs, in particular, significantly lowered levels of IL-6 and TNFα [32]. A separate meta-analysis confirmed that antidepressant drugs decreased levels of IL-6, and higher baseline IL-6 levels were associated with larger decreases in depressive symptoms [26]. Dahl et al. measured plasma levels of cytokines in depressed patients and healthy controls at baseline and after 12 weeks of antidepressant treatment. At baseline, MDD patients had significantly higher levels of IL-1β, IL-1 receptor antagonist (IL-1Ra), IL-5, IL-6, IL-7, IL-8, IL-10, granulocyte colony-stimulating factor (G-CSF), and interferon gamma (IFNγ) compared to healthy controls ($p = 0.01–0.047$). After 12 weeks of treatment with antidepressant medications and/or psychotherapy, plasma levels of seven cytokines (IL-1Ra, IL-6, IL-7, IL-8, IL-10, G-CSF, and IFNγ) had decreased significantly in patients meeting criteria for recovery and did not differ significantly from levels in healthy controls. Depressive symptoms were simultaneously significantly reduced, and cytokine levels did not normalize in MDD patients who did not meet criteria for recovery [31].

One study by Danese et al. in 1,000 subjects revealed that maltreated children exhibit a significant and graded increase in CRP 20 years later, independent of adult behavior, health, or recent life stressors; this effect was especially robust in depressed adults with a history of childhood maltreatment. In a separate study, the same group found significant elevations in CRP in 12-year-olds with depression and a history of maltreatment compared to depressed only, maltreated only, or age-matched controls. A study of 7,642 individuals in the UK reported that separation from parents in childhood was associated with a significant increase in CRP levels at age 44. Slopen et al. found that adverse events prior to age 8 predicted increased inflammatory markers, including IL-6 and CRP, at age 10 [13].

In summary, inflammatory markers are more likely to be elevated in depressed patients with a history of early life stressors (ELS) such as childhood abuse and/or neglect. Multiple studies have revealed an increase in inflammatory markers including CRP and IL-6 in patients who are depressed and have a history of exposure to ELS, especially in patients exposed to childhood abuse or maltreatment before age 8. Several studies have revealed persistently elevated levels up to 20 years later [13].

Protein S100B

Protein S100B is a neurotrophic protein that is used as a biomarker for glial alterations and neuroplasticity. Loss of neuroplasticity has been hypothesized to contribute to the pathogenesis of mood disorders [33] and several studies have found elevated serum levels of S100B in mood disorders, with higher levels seen in patients with MDD compared to bipolar disorder [34]. In a combined analysis of three studies in 46 MDD patients, Shroeter et al. found a positive correlation between clinical treatment response (measured by the HAM-D) and increases in serum S100B, suggesting that S100B could be a reliable indicator of treatment efficacy [34]. A separate study by Shroeter et al. [33] in 20 patients with a mood disorder and 12 age-matched healthy controls found that antidepressant treatment reduced serum S100B; the magnitude of the decrease was correlated with decreased severity of depressive symptoms (measured by HAM-D) ($\text{rho} < 0.01$); depression severity was positively correlated with serum S100B ($r(s) = 0.51$, $\text{rho} < 0.005$) [33]. Higher levels of serum S100B have been found to predict a better response to MDD treatment with various classes of antidepressants chosen by clinician based on symptomatology and anticipated side effects, after 4 or 6 weeks [35].

Neuroimaging

Neuroimaging studies show promise in identifying potential biomarkers to aid in the prediction of treatment outcomes for MDD. A 2013 systematic review of neural predictors of response to pharmacotherapy and psychotherapy in subjects with MDD who underwent neuroimaging (PET or structural MRI) prior to treatment initiation revealed a number of interesting findings [36]. The functional imaging meta-analysis included 20 studies from 15 independent samples. Across studies, increased baseline activity in the anterior cingulate and medial prefrontal cortices was predictive of a higher likelihood of symptom improvement and increased baseline activation in the insula and striatum was associated with a poor clinical response. For the amygdala, there was inconsistency with some reports of increased activity and some of decreased activity associated with a positive therapeutic outcome [36]. The structural imaging (MRI) meta-analysis included nine studies from six independent samples and found that a decrease in right hippocampal volume was a statistically significant predictor of lower likelihood of good treatment response [36]. Again a potential confound here is early life trauma, which is associated with reduced hippocampal volume [13].

Using PET, McGrath et al. investigated predictors of response to escitalopram or CBT in 38 subjects with MDD. Right anterior insula hypometabolism was associated with remission of symptoms after CBT and poor response to escitalopram, whereas insula hypermetabolism was associated with remission of symptoms after escitalopram and poor response to CBT [37]. In another PET study, McGrath investigated predictors of response to citalopram and/or CBT in a two-phase study. Nonresponders to both therapies had significantly higher subcallosal cingulate (SCC) metabolism compared to the remitters [38]. A study by Dunlop et al. [39] evaluated functional MRI resting-state functional connectivity in 122 treatment-naive patients with MDD who completed 12 weeks of treatment with either CBT or antidepressant medication (escitalopram or venlafaxine). The resting-state functional connectivity of the SCC with three brain regions (left anterior ventrolateral prefrontal cortex/insula, the dorsal midbrain, and the left ventromedial prefrontal cortex) was differentially associated with outcomes of remission and treatment failure to psychotherapeutic and psychopharmacological interventions. Greater positive summed functional connectivity with the SCC was associated with remission to CBT and treatment failure with medication. Negative or absent summed functional connectivity with the SCC was associated with remission to medication and treatment failure with CBT; resting-state functional connectivity patterns differentiating CBT and medication outcomes were consistent for both escitalopram and venlafaxine [39].

In a systematic review of 21 studies investigating the use of resting-state fMRI to predict antidepressant treatment response, increased functional connectivity between the frontal lobe and limbic system was associated with response to antidepressant treatment, potentially suggesting greater inhibitory control over neural circuitry involved in emotional processing. Connectivity in visual recognition circuits (including the lingual gyrus, middle occipital gyrus, fusiform gyrus, and cuneus) showed potential utility in several studies in distinguishing treatment-responsive from treatment-resistant MDD patients. Finally, subcallosal cingulate gyrus connectivity consistently predicted treatment response to rTMS and was also implicated in treatment response to antidepressants [40].

Electrophysiologic Biomarkers

A 2018 meta-analysis by Widge et al. including 76 articles reporting 81 biomarkers sought to quantify the reliability of quantitative EEG (QEEG) for response prediction in MDD. The analysis found that no specific QEEG biomarker or specific treatment showed greater predictive value than the all-studies estimate in a meta-regression, suggesting that QEEG is not a reliable clinical measure for predicting treatment response. However, funnel plot analysis suggested substantial publication bias such as lack of out-of-sample validation, underreporting of negative results, and insufficient direct replication of previous findings [41].

Genetic Predictors

Genetic heritability is thought to account for approximately 40–70% of the risk of developing MDD and also appears to play a role in treatment response [6]. A genome-wide association study (GWAS) evaluating the role of genetic variation in antidepressant treatment response found that common genetic variants accounted for 42% of individual differences in antidepressant treatment response [42]. Another GWAS study including a total of over 1500 subjects with MDD found associations with a few single nucleotide polymorphisms (SNPs) and antidepressant treatment response; however, no effect withstood correction for multiple testing. Of interest, 46 SNPs were in the same direction and significant before correction and they found a significant association of the number of response alleles (high versus low) to treatment response. The least favorable outcome was observed in patients with comorbid anxiety disorder in combination with a low number of response alleles [43].

Studies targeting specific SNPs involved in biologic pathways known to be involved in MDD have generally proven more promising. For example, genes involved in modulation of the HPA axis theoretically could impact the risk for developing MDD and treatment response [44]. O'Connell et al. found that a specific SNP of the corticotropin-releasing hormone-binding protein (rs28365143) was a positive predictor for treatment response to antidepressants, confirming a previous report from the Star D sample. Patients homozygous for the G allele of rs28365143 had greater remission, response rates, and symptom reduction in response to escitalopram and sertraline (SSRIs) compared to A allele carriers. There was no association observed between this genotype and treatment response to venlafaxine, an SNRI [44]. In a separate study that evaluated genetic variants in the corticotropin-releasing hormone-binding protein, African American and Hispanic carriers of the T allele of a specific SNP (rs10473984) had significantly poorer treatment response to citalopram (SSRI) [45]. Binder et al. identified SNPs in the FK506-binding protein 5 (FKBP5), a key component of the glucocorticoid receptor complex, that were associated with both a more rapid therapeutic response to antidepressant therapy and an increased recurrence of depressive episodes [46]. These SNPs were associated with increased intracellular FKBP5 protein and less HPA axis hyperactivity during depressive episodes [46].

Genes encoding for components of the serotonin (5-HT) system have also been studied. Serotonin reuptake, regulated by the serotonin transporter (5-HTT), is involved in the regulation of serotonergic activity and is thought to be a major site of action for many antidepressant medications. A polymorphism in 5-HTTLPR, the promoter region of 5-HTT, influences both risk for developing MDD and response to antidepressant therapy. A number of studies, primarily in Caucasians, have linked the 5-HTTLPR long allele (l/l) with greater therapeutic response and the short allele (s/s) with slower or poorer response to SSRIs [47–50]. However, conflicting studies show no significant relationship between these genotypes and antidepressant response [6, 51]. The short allele has also been associated with greater therapeutic response in several studies in Asian populations, elderly patients, and patients being simultaneously treated with lithium or pindolol [48, 52]. Several clinical trials investigating the role of genetic polymorphisms in serotonin receptors, 5HTR1A and 5HTR2A, have yielded inconsistent results [6, 51, 53, 54]. Genetic variation in the expression of the cytochrome P450 (CYP450) isoenzymes impacts the metabolism of many antidepressants and psychotropic medications, as well as other commonly prescribed medications. There are several commercially available combinatorial pharmacogenetic tests that classify phenotypes of hepatic CYP450 metabolism as poor, intermediate, extensive/normal, or ultrarapid metabolizers [55]. These could plausibly have clinical utility in guiding medication choice and improving treatment course. For example, when determining the appropriate starting dose for a medication, higher doses would be needed for ultrarapid metabolizers, whereas poor metabolizers would be more susceptible to adverse effects which would prompt a prescriber to start at a low dose or refrain from prescribing that medication [55]. They also claim to predict treatment efficacy, but the database is insufficient at this time to recommend their use. Indeed, there are few well-designed clinical trials investigating the efficacy and clinical applicability of these pharmacogenetic tests, and further research is clearly needed before this testing is integrated into routine clinical practice [55, 56].

Bipolar Affective Disorder

Overview of Treatment of Bipolar Disorder

Bipolar disorder (BD) is a heterogeneous disease with significant variability in clinical presentation. Lithium is currently the first-line treatment for BD and, based on current evidence, is the treatment most likely to achieve long-term relapse prevention [2]. However, lithium is associated with remission in only approximately 30% of patients. Additionally, it has a low therapeutic index, and many patients are non-adherent due to side effects such as tremor, weight gain, nausea, and sedation which likely contributes to suboptimal dosing and lower remission rates [57]. Other FDA approved treatments for mania or bipolar depression include lamotrigine, olanzapine, valproate, quetiapine, aripiprazole, and risperidone; however, it remains difficult to predict which drug will produce the best response in an individual patient [2]. In general, it is estimated that only 50% of patients with BD respond to monotherapy with a mood stabilizer, with an additional benefit of approximately 20% with addition of a second mood stabilizer

[58]. Because early intervention and providing an effective treatment early in the progression of the disease is important to long-term outcomes, better information about tailoring treatment for BD is critical [59, 60]. It is of paramount importance to identify biomarkers with predictive value for treatment response to enable early and efficacious intervention.

Peripheral Biomarkers

Inflammatory Biomarkers

Several studies have investigated the impact of treatment response on inflammatory biomarkers in BD during manic, depressive, and euthymic states. Kim et al. found that IL-6 levels were increased in acute mania and decreased after 6 weeks of treatment with lithium, valproate or a combination of both [61]. In a separate study, this group also found that levels of IL-12 decreased significantly in bipolar patients after 8 weeks of treatment with lithium or valproate [62]. Su et al. evaluated the relationship between lithium response and levels of interferon gamma (IFN-ɣ) and IL-10 and found no differences in patients medicated with lithium compared to unmedicated patients [63]. Guloksuz et al. found increased levels of TNF-a and IL-4 in euthymic bipolar patients treated with lithium ($N=15$) compared to unmedicated euthymic bipolar patients ($N=16$) and healthy controls ($N=16$); however, no significant differences were found in unmedicated euthymic bipolar patients compared to healthy controls [64]. Two studies evaluated the relationship between treatment response and soluble IL-2 (sIL-2R) and IL-6 (sIL-6R) receptors with inconsistent results. One study found that patients with rapid cycling bipolar disorder exhibited normalization of the sIL-2R and sIL-6R levels after 30 days of treatment with lithium [65] while another study found no significant change following treatment with valproate for 30 days [66]. Boufidou et al. investigated cytokine production in isolated lymphocytes in 40 euthymic BD patients on chronic lithium therapy and found a significant reduction in cells secreting IL-2, IL-6, IL-10, and IFN-ɣ in patients on lithium compared to healthy controls [67].

HPA Axis

A recent meta-analysis [68] of 41 case–control studies investigating HPA axis activity in 1069 BD patients and 1836 healthy controls showed that BD patients had higher basal cortisol than controls at all time points assessed, and cortisol levels measured over 12 or 24 hours were also significantly higher in bipolar patients compared to controls. Another meta-analysis comparing 19 case–control studies investigating morning cortisol levels revealed increased morning cortisol levels in BD patients compared to controls, with greater morning cortisol levels observed in non-manic BD outpatients [69]. As seen in MDD, patients with BD have frequently been found to be non-suppressors in the dexamethasone suppression test (DST) or DST/CRH combination test, particularly in the depressed or mixed state [68, 70]. These tests have been proposed as biomarkers to assess HPA axis response to treatment in mood disorders.

Thyroid

Thyroid dysfunction has been frequently observed in patients with BD, with one study indicating that BD patients are 2.55 times more likely to experience thyroid dysfunction compared to healthy controls [71]. Cole et al. studied thyroid function in a group of 65 BD patients treated with mood stabilizers (lithium ($N = 57$, 88%), divalproex, combination, or carbamazepine). Lower free thyroxine index (FTI) values and higher thyroid-stimulating hormone (TSH) levels were significantly associated with a poorer lithium response. Lower TSH combined with higher FTI was associated with a substantially more rapid remission of depression [72]. A 2017 systematic review of 11 studies investigating the relationship between BD and thyroid autoimmunity found an increased prevalence of circulating thyroid autoantibodies in depressed and mixed BD patients; however, no evidence was found to support a relationship between specific autoimmune thyroid diseases and BD. Findings from a study in twins with BD suggested that autoimmune thyroiditis is related to the genetic vulnerability to develop BD rather than the pathogenesis of the disease itself. These findings suggest that thyroid autoantibodies could have potential utility as a biomarker of vulnerability for BD [73].

Brain-Derived Neurotrophic Factor (BDNF)

Lithium may exert its therapeutic action, in part, by upregulating BDNF expression to achieve mood stabilization, suggested by the finding that serum BDNF levels positively correlate with lithium levels [74]. Lithium increases BDNF expression in cultured rodent neurons [75], and platelet BDNF mRNA increases after 8 weeks of medication treatment for BD (lithium, valproate, or atypical antipsychotic) [76]. In one study, chronic administration of mood stabilizers (carbamazepine and lamotrigine) in rats increased BDNF mRNA expression and protein concentrations in the frontal cortex [77]. Increased blood concentrations of BDNF have been reported following treatment with antidepressants or mood stabilizers in BD and other mood disorders [78]. A 16-week open trial of quetiapine XR for BD indicated that independent of clinical response to treatment, serum BDNF concentrations increased in patients with bipolar depression but decreased in manic and mixed patients [79]. There is conflicting data concerning changes in levels of BDNF associated with different medication treatments for BD with some investigators finding changes associated with specific medication treatments [80] and others with discordant findings [81–83]. Suwalska et al. reported that lithium treated patients as a group had lower BDNF levels compared to healthy controls, and moreover patients that did not respond to lithium had significantly lower BDNF levels compared with healthy control subjects [84]. Several studies have attempted to correlate peripheral BDNF levels to treatment response in acute mania. In one study, a significant increase in plasma BDNF levels was observed after 28 days of lithium monotherapy for acute mania, with 87% of responders showing an increase in BDNF levels after treatment compared to baseline [85].

Tramontina et al. found that BDNF levels were significantly decreased in acutely manic patients when compared to controls; after treatment, however, there was no longer a significant difference observed between the two groups. In the BD patients, a sharp increase in BDNF was observed after effective treatment of acute mania [86]. A systematic review and meta-regression analysis of 3 studies found that BDNF levels increase after treatment for acute mania [87].

Overall, the data suggest that BDNF may play an important role in the pathophysiology of BD and lithium responders may upregulate BDNF in response to treatment. BDNF also appears to be a potential marker for mood state, as it is generally lower during mood episodes in BD and often normalizes in euthymia. There have been conflicting findings, however, and future research should focus on BDNF levels over time, individual patient mood state, and treatment response. Whether peripheral BDNF concentrations represent an index of CNS BDNF activity remains unclear.

Neuroimaging

Although the investigation of imaging techniques to predict treatment response in BD is in its infancy, there are several promising findings. Using (15)O water PET to measure changes in regional cerebral blood flow during an induced sadness task, different activity patterns were demonstrated in BD lithium ($n=9$) and valproate ($N=9$) responders with both groups showing changes in premotor cortex, dorsal anterior cingulate, and anterior insula, and valproate responders showing a larger magnitude of change in most regions. Comparison of the change patterns revealed differences in the rostral anterior cingulate and the dorsolateral prefrontal cortex between lithium and valproate responders [83]. In a small study ($n=20$) using fMRI and proton magnetic resonance spectroscopy (H-MRS) to predict lithium treatment outcome in BD subjects, investigators applied a data algorithm that showed excellent predictive power in determining a good lithium response [88].

A recent systematic review of 60 studies investigating neuroimaging and behavioral predictors of treatment efficacy in unipolar and bipolar patients found that good response to pharmacotherapy for depression was predicted by lower baseline responsivity in limbic regions coupled with heightened medial and dorsal prefrontal responses to emotional stimuli. Alternatively, good treatment response to psychotherapy was predicted by heightened baseline limbic and ventral prefrontal reactivity to emotional stimuli [89].

Genetic Predictors

GWAS findings suggest a genetic component in the response to specific drugs in BD patients. Specifically, several studies suggest that those who respond well to lithium are a genetically unique subset of patients. One prospective GWAS (average follow-up of 12 years) in 247 individuals from 31 families (106 diagnosed with BD) with a history of good lithium response revealed strong evidence for linkage with a locus on chromosome 15q14 (ACTC, lod score = 3.46, locus-specific p-value = 0.000014). Further analyses of these results suggested that this locus may be associated with the underlying etiology of BD. A possible linkage was also observed for a marker on chromosome 7q11.2 (D7S1816, lod score = 2.68, locus-specific p-value = 0.00011), with further analyses suggesting that this locus could potentially be useful in predicting response to lithium treatment. [90]. Another study by the same group comparing 136 bipolar patients with good lithium response and 163 healthy controls revealed that one polymorphism in the PLCG1 gene was observed at a significantly higher frequency in the lithium-responder bipolar group compared to controls ($p = 0.033$). A follow-up study with a Norwegian population confirmed these findings [91]. PLCG1 gene codes for a gamma-1 isozyme of phospholipase C, an enzyme that plays an important role in the phosphoinositide cycle. This second messenger system is believed to be involved in the mechanism of action of lithium in mood stabilization [92].

A GWAS in 294 BD Type 1 patients of Han Chinese descent revealed a strong association between good treatment response to lithium and two SNPs located in the introns of glutamic acid decarboxylase–like protein 1 (GADL1): rs17026688 ($p = 5.50 \times 10^{-37}$) and rs17026651 ($i = 2.52 \times 10^{-37}$). These two SNPs had 93% sensitivity for predicting lithium response and differentiated between patients with a good and poor response [93]. These findings were not replicated in a follow-up study in an Indian population [94].

In a recent GWAS on 2586 BD patients, the International Consortium on Lithium Genetics [95] assessed response to lithium treatment. A polygenic score for schizophrenia (PGS) was constructed using estimates from 36, 989 schizophrenia patients and cross-trait analysis was performed. A high polygenic score for schizophrenia was inversely associated ($p < 0.05$) with a good lithium response [96]. This finding concurs with evidence that patients with BD with a family history of schizophrenia have a poor response to lithium [95].

Circadian rhythm dysfunction has been posited to be involved in the pathogenesis of BD. A recent prospective study following 170 BPD patients over a 27-year period investigated the influence of polymorphisms in the Rev-erb-alpha gene, an important component of the mammalian circadian rhythm cycle, on response to lithium treatment. Patients carrying the T allele for the rs2314339 SNP were 3.5 times more likely to show no improvement from lithium prophylaxis or to experience worsened symptoms with treatment [97].

Multiple studies have found associations between lithium response and polymorphisms in the serotonin transporter gene 5-HTTLPR. Serretti et al. found that individuals with the s/s variant of the gene showed a poorer lithium response compared to those with l/s or l/l variant [98]. A follow-up study replicated the finding of better response in the l/s variant but did not reveal a significantly poorer response in individuals with the s/s variant [2]. Another study found that the s/s genotype variant and s allele were significantly more frequent in patients who did not respond to lithium compared to patients with a partial or excellent response to lithium [99].

The Val158Met polymorphism in the COMT gene may have predictive value for treatment response to mood stabilizers. Lee et al. conducted a study in a Korean population with 144 manic BDI patients and 157 controls and found that the Met/Met genotype was more frequent in nonresponders to mood stabilizers (lithium, valproate, or carbamazepine) than in treatment responders. There were no significant differences between BD patients and controls [100].

There has been an association reported between treatment response to valproate and the 116C/G polymorphism in the promoter region of the X-box-binding protein 1 (XBP1). In a sample of 51 BD patients, the G allele was associated with a better response to valproate compared to the C allele. An association has also been observed between the 116C/G polymorphism and clinical response to lithium [58].

Schizophrenia and Schizoaffective Disorders

Overview of Treatment

Schizophrenia is a devastating disease with enormous variability in presentation and treatment response. It is widely believed to be a neurodevelopmental disorder with early alterations in neuronal migration leading to absent and aberrant connections that alter development of surrounding cerebral regions. Many of the causative factors involved in the etiology of schizophrenia likely exist at birth, but do not become apparent until late adolescence/early adulthood as a result of environmental and genetic factors. This makes identification of biomarkers in schizophrenia challenging, as some abnormalities may occur years before disease presentation and may no longer be detectable at the time of disease presentation [101]. Treatment of schizophrenia largely involves the use of antipsychotic medications. A meta-analysis of 65 trials involving 6493 patients showed that treatment response to antipsychotics was superior compared with placebo, evidenced by relapse and readmission rates less than half in the medication groups compared to placebo groups. With maintained medication adherence, relapse rates did not change after several years of treatment [102]. There are a number of antipsychotic drugs, but all block dopamine D2 receptors which are considered to be key to antipsychotic efficacy. Various antipsychotics have differing actions on other neurotransmitter receptors/systems and their side effect profiles differ accordingly. As with the other treatments discussed in this chapter, there is not an abundance of evidence to guide antipsychotic drug choice, and side effect profile and clinician preference weigh heavily in decision-making. As with other medications discussed, side effects can be troubling and, in some cases, long-lasting. Positive treatment response is associated with long-term functional improvement, and being able to efficiently choose the right drug for the right patient can have profound implications. With the exception of clozapine which is clearly effective in many patients who have failed treatment with other antipsychotics, all of the other agents have equal efficacy [2].

Peripheral Biomarkers

Inflammatory Markers

Inflammation has long been posited to play a role in the pathophysiology of schizophrenia, and several studies have revealed elevated inflammatory markers in schizophrenic patients. Multiple studies revealed that CRP concentrations were higher in schizophrenic patients when compared to healthy controls, with some studies demonstrating higher CRP levels associated with more severe symptoms as determined by the Positive and Negative Syndrome Scale (PANSS) [103].

A meta-analysis by Miller et al. included 40 studies investigating levels of various cytokines in acutely relapsed schizophrenic (AR) patients and/or patients with first episode psychosis (FEP). In AR patients, blood IL-10 levels were significantly decreased ($p \leq 0.006$) and IL-6, IL-8, TNF-α, IFN-γ, transforming growth factor-β (TGF-β), and IL-1RA levels were significantly increased when compared with control subjects ($p \leq 0.02$ for all). There were no significant differences in blood levels of IL-2 or soluble IL-2 receptor (sIL-2R) between AR and control subjects. The most replicated finding (5/6 studies) was significantly increased IL-6 in AR compared to control subjects. In first episode psychosis patients, blood IL-1β, IL-6, IL-12, IFN-γ, TNF-α, TGF-β, and sIL-2R levels were significantly increased compared to control subjects ($p \leq 0.003$ for all). There was no significant difference observed in blood IL-2 levels between FEP and control subjects. The most replicated finding was for TNF-α, which was increased in FEP subjects in all four studies analyzed [104]. The same group also investigated the effect of antipsychotic treatment on cytokine levels in 488 patients after an acute exacerbation of schizophrenia. After a mean of 53 days of treatment with antipsychotics (which were not standardized in seven (58%) of the studies included in the analysis), there was a significant decrease in IL-1β, IL-6, and TGF-β ($p \leq 0.005$ for all) and a significant increase in sIL-2R ($p = 0.04$) and IL-12 ($p = 0.02$) levels [104]. Another meta-analysis by Tourjman et al. that included 23 follow-up studies in 762 subjects found an increase of sIL-2 and decrease of IL-1β and IFN-γ levels after treatment with antipsychotics [105]. A longitudinal study in 68 FEP patients found that those who did not respond to 12 weeks of antipsychotic treatment had higher pretreatment IL-6 and IFN-γ levels compared to patients who did respond [106].

HPA Axis

HPA axis dysfunction may potentially play a role in the pathophysiology of schizophrenia, but studies have yielded conflicting results. Patients with schizophrenia have been found to exhibit elevated basal levels of cortisol [69] and non-suppression of cortisol in response to a DST compared to controls [103]. However, findings have been inconsistent, with studies showing both hyper- and hypoactivity of the HPA axis [107]. Some studies have shown decreases in cortisol after treatment with atypical antipsychotics (olanzapine, clozapine, quetiapine) and increases after treatment with typical antipsychotics (fluphenazine) [107].

BDNF

Peripheral BDNF levels could also have potential clinical utility in monitoring therapeutic response to antipsychotics in patients with schizophrenia and/or first-onset psychosis. Significantly decreased CSF BDNF concentrations have been reported, as well as in the hippocampus and prefrontal cortex of schizophrenic patients compared to controls [103]. A meta-analysis of 16 studies comparing plasma BDNF levels in schizophrenic patients versus healthy controls found a moderate reduction in schizophrenic patients [108]. Studies investigating the effect of antipsychotic treatment on peripheral BDNF levels in schizophrenia have yielded inconsistent results. In the previously mentioned meta-analysis, no significant difference in BDNF levels was observed between medicated and medication-naive schizophrenic patients. One study found a significant correlation between clozapine and BDNF levels, with the implication that clozapine may promote cognitive enhancement in schizophrenic patients [103]. Laboratory animal studies revealed that haloperidol and risperidone decreased BDNF mRNA expression and protein levels in multiple areas of the brain, including the hippocampus and prefrontal cortex. BDNF levels in the rat hippocampus normalized after switching from haloperidol treatment to olanzapine [103].

Neuroimaging

There have been relatively few studies of neuroimaging biomarkers in schizophrenia. Kapur et al. [109] studied the response to haloperidol in 21 patients with schizophrenia. The responders had significantly higher dopamine D2 receptor occupancy, as determined with raclopride and PET, after two weeks of treatment ($p<0.009$). Using it as a predictor with a 65% cutoff, D2 occupancy provided optimal separation: 80% of the responders were above it and 67% of nonresponders were below ($p=0.04$, Fisher's exact test). A study done in 51 adolescents and young adults at clinical high risk (CHR) for psychosis and 47 matched healthy controls used resting-state fMRI to systematically characterize functional connectivity (FC) and determine if abnormalities in FC during this period are associated with psychosis risk and severity of psychosis. The findings revealed between-group differences in whole-brain connectivity patterns of bilateral temporal regions, primarily affecting functional connections to the thalamus, which is consistent with well-established FC abnormalities observed in the thalamus and temporal regions of schizophrenic patients. In those individuals who went on to develop psychosis over the next 3.9 years ($N=12$), more severe positive symptoms were associated with greater FC abnormalities in the anterior cingulate and frontal cortex [110].

Genetic Predictors

Schizophrenia is thought to be one of the most heritable of the psychiatric disorders, with genetics contributing to 50–80% of risk. However, studies to date have demonstrated that schizophrenia is genetically complex with a large polygenic component [6]. In a GWAS study of 117 Chinese Han patients with schizophrenia, Reynolds et al. [111] investigated the dopamine 3 (D3), D2 and the 5-HT2C receptor promoter polymorphisms following 10 weeks of antipsychotic treatment with risperidone or chlorpromazine. The D3 receptor polymorphism was significantly associated with both improved symptoms and behavioral symptomatology on admission. The 5-HT2C receptor polymorphism was associated with treatment improvement but not baseline behavioral symptoms. Another study [112] of genetic variation underlying individual differences in response to a variety of antipsychotic medications in 738 subjects with schizophrenia found that two SNPs mediated the effect of ziprasidone on positive symptoms: rs17390445 on chromosome 4p15, with a q-value of slightly less than 0.05 ($p=9.8$ Å ~ 10), and rs11722719 with a q-value of less than 0.15 ($p=5.4$ Å ~ 10). SNP rs7968606 in the ANKS1B gene showed a q-value of 0.16 ($p=3.2$ Å ~ 10) for mediating the effect of olanzapine on negative symptoms. Two SNPs were found to mediate the effect of risperidone on negative symptoms: rs17727261 in the CNTNAP5 gene with a q-value of 0.13 ($p=5.4$ Å ~ 10) and rs17815774 in the TRPM1 gene, with a q-value of 0.4 ($p=3.3$ Å ~ 10) [112].

Zhang et al. [113] scrutinized a dopamine D2 receptor (DRD2) locus in a large-scale GWAS from the Psychiatric Genomics Consortium [114] to investigate whether the rs2514218 SNP could predict antipsychotic response in a cohort of 100 FEP patients, half treated with risperidone and half with aripiprazole for 12 weeks. Linear mixed model analysis showed that homozygotes for the risk (C) allele had a significantly greater reduction in positive symptoms during 12 weeks of treatment compared with the T allele carriers ($p=0.044$). Ikeda et al. [115] also showed that an SNP in DRD2 was a significant predictor of the response to risperidone along with an SNP in TaqIA and two SNPs in AKT1.

Multiple studies have found a significant association between the BDNF Val66Met polymorphism and antipsychotic treatment response, showing a higher frequency of the Val/Val homozygous genotype in patients with good clinical response to antipsychotic treatment (clozapine or olanzapine) compared to nonresponders. However, subsequent studies failed to confirm the association between the BDNF Val66Met polymorphism and treatment response to clozapine [103].

Electrophysiologic Biomarkers

Mismatch negativity (MMN) is an event-related potential response, passively evoked when a sequence of repetitive standard auditory stimuli are interrupted by deviations in pitch or duration and represents an automatic, preconscious process of detecting a mismatch between auditory sensory memory and deviant stimuli. In dozens of trials over the last 20 years, reduced MMN amplitude has consistently been found in schizophrenic patients compared to healthy controls [116–119]. A meta-analysis of 32 studies found that the effect sizes of MMN reduction were significantly correlated with duration of illness, suggesting that this may be a useful index of progression of neuropathologic changes in schizophrenia. In a study by Lee et al. of 25 patients with schizophrenia, 21 first-degree relatives and 29 healthy controls, MMN was a stronger predictor of functional outcomes in schizophrenia than neurocognition or theory of mind [117]. A similar study found that higher MMN activity in frontocentral regions of schizophrenic patients was correlated with better social perception, work, and independent living. [116]. While MMN is one of the most promising biomarkers for tracking response to therapeutic interventions in schizophrenia, it has yet to be used to study response to specific medication. [118]. Targeting early auditory perceptual processing in schizophrenic patients with MMN deficits in an effort to decrease the deficit has been hypothesized to improve cognitive and psychosocial functioning. Targeted cognitive training (TCT) uses neuroplasticity-based computerized cognitive tasks with the ultimate goal of inducing plastic changes within the neural substrates of low-level information processing, which in turn leads to improvement in higher order cognitive operations [120]. One study of 55 clinically stable schizophrenic patients found that subjects randomly assigned to 50 hours of TCT showed significant improvements in verbal working memory ($p<0.05$), verbal learning, verbal memory, and global cognition ($p<0.01$). [121]. Multiple clinical trials have demonstrated that changes in MMN are detectable in early stages of cognitive training, predict generalized improvements in higher order cognitive domains, and correspond to objective changes of cortical plasticity [118]. While these trials have shown promise at a group level, individual responses to TCT are variable and it is important to identify which patients are most likely to benefit as it is very resource and time-consuming [120]. Several trials have demonstrated that a larger baseline MMN predicted greater response to TCT, [101, 122] and in one smaller trial with 13 schizophrenic patients, it predicted greater response to a 3-month social skills training program [123]. This suggests that higher baseline MMN could be used to identify patients who are more likely to respond to TCT and social skills training.

Conclusions/Future Directions

Although the field of precision medicine for psychiatry is in its infancy, there are a number of studies that have found promising associations between a variety of biomarkers and clinical response to psychopharmacological treatment. These biomarkers include neuroimaging, electrophysiology, peripheral serum measures of HPA activity and inflammation, single nucleotide polymorphisms, and others.

Although some neuroimaging studies have shown promise in response prediction, the findings have not been overwhelmingly consistent. This fact, combined with the expense and limited accessibility to neuroimaging (especially of PET) in many front-line settings has limited the clinical utility of neuroimaging as a biomarker.

In recent years, DNA sequencing has become substantially more efficient and affordable, with the cost of DNA sequencing of a single human subject reduced from \$300 million in 2001 to \$1000 in 2014 [124]. Extensive GWAS investigation has revealed genetic variants linked with the risk for a number of psychiatric disorders. However, the contribution of individual gene variants appears to be small, and in many cases, follow-up studies have failed to consistently replicate the initial findings. One potential area for improvement is the expansion to include transcriptomics (the study of all expressed messenger RNA (mRNA)) and proteomics (study of expressed proteins). It is estimated that 98% of the human genome is not translated into protein, and multiple epigenetic changes (DNA methylation, histone modification, alternative splicing, RNA editing, and nontranscriptional gene silencing via microRNAs) occur as DNA is transcribed into RNA, which is subsequently translated into proteins. The ability to analyze all of these processes would immensely improve the knowledge base and could enhance the ability to make informed treatment decisions [6].

Another promising area is combinatorial pharmacogenetic testing, which has been attracting more attention in recent years with over 30 tools commercially available [55]. However, the clinical applicability of these tests is questionable as the companies that market them often do not report the specific genetic variants that are included in the tests and fail to disclose how the pharmacogenetic algorithms integrate and weigh important genetic variants [55]. In addition, the literature investigating the clinical efficacy is limited by small sample size and lack of scientific rigor with the majority of studies sponsored by the companies that market the tests or institutions with a commercial stake in the tests used [56]. As such, there is currently insufficient evidence to justify the widespread clinical use of these tests and further investigation is warranted with blinded, randomized controlled trials in larger samples.

Although biomarkers have promise in guiding future treatment of psychiatric disorders, one single biomarker is unlikely to definitively determine the most ideal treatment option but rather a combination of different biomarkers should be considered. The International Study to Predict Optimized Treatment in Depression (iSPOT-D) [125] is a large ongoing trial that aims to identify genetic, behavioral, and biological predictors of treatment response to commonly used antidepressants (escitalopram, venlafaxine-XR, sertraline) in order to ultimately develop a treatment model that incorporates a variety of predictors and moderators. These data are still being analyzed but have already provided an abundance of valuable data.

Biomarkers can have some predictive value with regard to treatment response, but it is important to consider the role of these biomarkers combined with behavioral and physiologic data captured in a naturalistic setting. Mobile health (mHealth) is a growing field that utilizes multiple tools and resources such as mobile and wireless communication devices to deliver and improve healthcare services, outcomes, and research. This technology aims to incorporate remotely acquired patient data (self-report mood scales, exposure to stressful environmental triggers, etc.) with biomarker data to enhance clinical outcomes. This field is rapidly expanding to also collect physiologic data using mobile biosensor devices to assess measures such as autonomic nervous system functioning, electrodermal activity, ECG, and breath alcohol or carbon monoxide levels [126].

Ultimately, the best model for precision medicine in complex, multifactorial diseases such as psychiatric illnesses will likely involve integrated methodology that combines information from multiple sources including biologic, clinical, and environmental data. While much progress has been made in the development of valid biomarkers in psychiatric disorders, there is much work to be done in determining their clinical utility.

Financial Disclosures Dr. Nemeroff has received grants or research support from NIH and the Stanley Medical Research Institute; he has served as a consultant for Bracket (Clintara), Dainippon Pharma, Fortress Biotech, Intra-Cellular Therapies, Janssen Research and Development, Magstim, Prismic Pharmaceuticals, Sumitomo Navitor Pharmaceuticals, Sunovion, Taisho Pharmaceutical, Takeda, TC MSO, and Xhale; he has served on scientific advisory boards for the American Foundation for Suicide Prevention (AFSP), the Anxiety Disorders Association of America (ADAA), Bracket (Clintara), the Brain and Behavior Research Foundation, the Laureate Institute for Brain Research, Skyland Trail, and Xhale and on directorial boards for ADAA, AFSP, and Gratitude America; he is a stockholder in AbbVie, Antares, BI Gen Holdings, Celgene, Corcept Therapeutics, OPKO Health, Seattle Genetics, and Xhale; he receives income or has equity of $10,000 or more from American Psychiatric Publishing, Bracket (Clintara), CME Outfitters, Intra-Cellular Therapies, Magstim, Takeda, and Xhale; and he holds patents on a method and devices for transdermal delivery of lithium (patent 6,375,990B1) and a method of assessing antidepressant drug therapy via transport inhibition of monoamine neurotransmitters by ex vivo assay (patent 7,148,027B2). Dr. Lydiard has no financial interests or relationships to disclose.

References

1. Gadad BS, Jha MK, Czysz A, et al. Peripheral biomarkers of major depression and antidepressant treatment response: current knowledge and future outlooks. J Affect Disord. 2018;233:3–14.
2. Stern S, Linker S, Vadodaria KC, Marchetto MC, Gage FH. Prediction of response to drug therapy in psychiatric disorders. Open Biol. 2018;8(5).
3. Garraway LA, Verweij J, Ballman KV. Precision oncology: an overview. J Clin Oncol. 2013;31(15):1803–5.
4. Biologically-inspired biomarkers for mental disorders. EBioMedicine. 2017;17:1–2.
5. Miller DB, O'Callaghan JP. Personalized medicine in major depressive disorder—opportunities and pitfalls. Metabolism. 2013;62(Suppl 1):S34–9.
6. Ozomaro U, Wahlestedt C, Nemeroff CB. Personalized medicine in psychiatry: problems and promises. BMC Med. 2013;11:132.
7. Nemeroff CB. The holy grail of psychiatry. Cerebrum. 2015;2015.
8. Petkova E, Ogden RT, Tarpey T, et al. Statistical analysis plan for stage 1 EMBARC (establishing moderators and biosignatures of antidepressant response for clinical care) study. Contemp Clin Trials Commun. 2017;6:22–30.
9. Nemeroff CB, Widerlov E, Bissette G, et al. Elevated concentrations of CSF corticotropin-releasing factor-like immunoreactivity in depressed patients. Science. 1984;226(4680):1342–4.
10. Perna GBR, Nemeroff CB. Precision psychiatry: personalized clinical approach to depression. In: Kim YK, editor. Understanding depression, vol. 1. Singapore: Springer; 2018. p. 245–62.
11. Binder EB, Kunzel HE, Nickel T, et al. HPA-axis regulation at in-patient admission is associated with antidepressant therapy outcome in male but not in female depressed patients. Psychoneuroendocrinology. 2009;34(1):99–109.
12. Greden JF, Gardner R, King D, Grunhaus L, Carroll BJ, Kronfol Z. Dexamethasone suppression tests in antidepressant treatment of melancholia. The process of normalization and test-retest reproducibility. Arch Gen Psychiatry. 1983;40(5):493–500.

13. Nemeroff CB. Paradise lost: the neurobiological and clinical consequences of child abuse and neglect. Neuron. 2016;89(5):892–909.
14. Nemeroff CB. Clinical significance of psychoneuroendocrinology in psychiatry: focus on the thyroid and adrenal. J Clin Psychiatry. 1989;50 Suppl:13–20; discussion 21–12.
15. Hage MP, Azar ST. The link between thyroid function and depression. J Thyroid Res. 2012;2012:590648.
16. Musselman DL, Nemeroff CB. Depression and endocrine disorders: focus on the thyroid and adrenal system. Br J Psychiatry Suppl. 1996;30:123–8.
17. Eller T, Metskula K, Talja I, Maron E, Uibo R, Vasar V. Thyroid autoimmunity and treatment response to escitalopram in major depression. Nord J Psychiatry. 2010;64(4):253–7.
18. Kirkegaard C, Norlem N, Lauridsen UB, Bjorum N. Prognostic value of thyrotropin-releasing hormone stimulation test in endogenous depression. Acta Psychiatr Scand. 1975;52(3):170–7.
19. Amsterdam JD, Fava M, Maislin G, Rosenbaum J, Hornig-Rohan M. TRH stimulation test as a predictor of acute and long-term antidepressant response in major depression. J Affect Disord. 1996;38(2–3):165–72.
20. Gendall KA, Joyce PR, Mulder RT, Luty SE. Thyroid indices and response to fluoxetine and nortriptyline in major depression. J Psychopharmacol. 2003;17(4):431–7.
21. Joffe RT, Singer W. The effect of tricyclic antidepressants on basal thyroid hormone levels in depressed patients. Pharmacopsychiatry. 1990;23(2):67–9.
22. Kusalic M, Engelsmann F, Bradwejn J. Thyroid functioning during treatment for depression. J Psychiatry Neurosci. 1993;18(5):260–3.
23. Brady KT, Anton RF. The thyroid axis and desipramine treatment in depression. Biol Psychiatry. 1989;25(6):703–9.
24. Iosifescu DV, Nierenberg AA, Mischoulon D, et al. An open study of triiodothyronine augmentation of selective serotonin reuptake inhibitors in treatment-resistant major depressive disorder. J Clin Psychiatry. 2005;66(8):1038–42.
25. Valkanova V, Ebmeier KP, Allan CL. CRP, IL-6 and depression: a systematic review and meta-analysis of longitudinal studies. J Affect Disord. 2013;150(3):736–44.
26. Hiles SA, Baker AL, de Malmanche T, Attia J. Interleukin-6, C-reactive protein and interleukin-10 after antidepressant treatment in people with depression: a meta-analysis. Psychol Med. 2012;42(10):2015–26.
27. Chang HH, Lee IH, Gean PW, et al. Treatment response and cognitive impairment in major depression: association with C-reactive protein. Brain Behav Immun. 2012;26(1):90–5.
28. Jha MK, Minhajuddin A, Gadad BS, et al. Can C-reactive protein inform antidepressant medication selection in depressed outpatients? Findings from the CO-MED trial. Psychoneuroendocrinology. 2017;78:105–13.
29. Miller AH, Trivedi MH, Jha MK. Is C-reactive protein ready for prime time in the selection of antidepressant medications? Psychoneuroendocrinology. 2017;84:206.
30. Raison CL, Rutherford RE, Woolwine BJ, et al. A randomized controlled trial of the tumor necrosis factor antagonist infliximab for treatment-resistant depression: the role of baseline inflammatory biomarkers. JAMA Psychiatry. 2013;70(1):31–41.
31. Dahl J, Ormstad H, Aass HC, et al. The plasma levels of various cytokines are increased during ongoing depression and are reduced to normal levels after recovery. Psychoneuroendocrinology. 2014;45:77–86.
32. Hannestad J, DellaGioia N, Bloch M. The effect of antidepressant medication treatment on serum levels of inflammatory cytokines: a meta-analysis. Neuropsychopharmacology. 2011;36(12):2452–9.
33. Schroeter ML, Abdul-Khaliq H, Diefenbacher A, Blasig IE. S100B is increased in mood disorders and may be reduced by antidepressive treatment. NeuroReport. 2002;13(13):1675–8.
34. Schroeter ML, Sacher J, Steiner J, Schoenknecht P, Mueller K. Serum S100B represents a new biomarker for mood disorders. Curr Drug Targets. 2013;14(11):1237–48.

35. Jang BS, Kim H, Lim SW, Jang KW, Kim DK. Serum S100B levels and major depressive disorder: its characteristics and role in antidepressant response. Psychiatry Investig. 2008;5(3):193–8.
36. Fu CH, Steiner H, Costafreda SG. Predictive neural biomarkers of clinical response in depression: a meta-analysis of functional and structural neuroimaging studies of pharmacological and psychological therapies. Neurobiol Dis. 2013;52:75–83.
37. McGrath CL, Kelley ME, Holtzheimer PE, et al. Toward a neuroimaging treatment selection biomarker for major depressive disorder. JAMA Psychiatry. 2013;70(8):821–9.
38. McGrath CL, Kelley ME, Dunlop BW, Holtzheimer PE 3rd, Craighead WE, Mayberg HS. Pretreatment brain states identify likely nonresponse to standard treatments for depression. Biol Psychiatry. 2014;76(7):527–35.
39. Dunlop BW, Rajendra JK, Craighead WE, et al. Functional connectivity of the subcallosal cingulate cortex and differential outcomes to treatment with cognitive-behavioral therapy or antidepressant medication for major depressive disorder. Am J Psychiatry. 2017;174(6):533–45.
40. Dichter GS, Gibbs D, Smoski MJ. A systematic review of relations between resting-state functional-MRI and treatment response in major depressive disorder. J Affect Disord. 2015;172:8–17.
41. Widge AS, Bilge MT, Montana R, Chang W, Rodriguez CI, Deckersbach T, Carpenter LL, Kalin NH, Nemeroff CB. Electroencephalographic biomarkers for treatment response prediction in major depressive illness: a meta-analysis. Am J Psychiatry. 2018 (In Press).
42. Tansey KE, Guipponi M, Hu X, et al. Contribution of common genetic variants to antidepressant response. Biol Psychiatry. 2013;73(7):679–82.
43. Ising M, Lucae S, Binder EB, et al. A genomewide association study points to multiple loci that predict antidepressant drug treatment outcome in depression. Arch Gen Psychiatry. 2009;66(9):966–75.
44. O'Connell CP, Goldstein-Piekarski AN, Nemeroff CB, et al. Antidepressant outcomes predicted by genetic variation in corticotropin-releasing hormone binding protein. Am J Psychiatry. 2017:appiajp201717020172.
45. Binder EB, Owens MJ, Liu W, et al. Association of polymorphisms in genes regulating the corticotropin-releasing factor system with antidepressant treatment response. Arch Gen Psychiatry. 2010;67(4):369–79.
46. Binder EB, Salyakina D, Lichtner P, et al. Polymorphisms in FKBP5 are associated with increased recurrence of depressive episodes and rapid response to antidepressant treatment. Nat Genet. 2004;36(12):1319–25.
47. Smits KM, Smits LJ, Peeters FP, et al. The influence of 5-HTTLPR and STin2 polymorphisms in the serotonin transporter gene on treatment effect of selective serotonin reuptake inhibitors in depressive patients. Psychiatr Genet. 2008;18(4):184–90.
48. Porcelli S, Drago A, Fabbri C, Gibiino S, Calati R, Serretti A. Pharmacogenetics of antidepressant response. J Psychiatry Neurosci. 2011;36(2):87–113.
49. Durham LK, Webb SM, Milos PM, Clary CM, Seymour AB. The serotonin transporter polymorphism, 5HTTLPR, is associated with a faster response time to sertraline in an elderly population with major depressive disorder. Psychopharmacology. 2004;174(4):525–9.
50. Arias B, Catalan R, Gasto C, Gutierrez B, Fananas L. 5-HTTLPR polymorphism of the serotonin transporter gene predicts non-remission in major depression patients treated with citalopram in a 12-weeks follow up study. J Clin Psychopharmacol. 2003;23(6):563–7.
51. Peters EJ, Slager SL, McGrath PJ, Knowles JA, Hamilton SP. Investigation of serotonin-related genes in antidepressant response. Mol Psychiatry. 2004;9(9):879–89.
52. Pollock BG, Ferrell RE, Mulsant BH, et al. Allelic variation in the serotonin transporter promoter affects onset of paroxetine treatment response in late-life depression. Neuropsychopharmacology. 2000;23(5):587–90.
53. Kato M, Fukuda T, Wakeno M, et al. Effect of 5-HT1A gene polymorphisms on antidepressant response in major depressive disorder. Am J Med Genet B Neuropsychiatr Genet. 2009;150B(1):115–23.

54. Illi A, Setala-Soikkeli E, Viikki M, et al. 5-HTR1A, 5-HTR2A, 5-HTR6, TPH1 and TPH2 polymorphisms and major depression. NeuroReport. 2009;20(12):1125–8.
55. Zeier Z, Carpenter LL, Kalin NH, et al. Clinical implementation of pharmacogenetic decision support tools for antidepressant drug prescribing. Am J Psychiatry. 2018:appiajp201817111282.
56. Zubenko GS, Sommer BR, Cohen BM. On the marketing and use of pharmacogenetic tests for psychiatric treatment. JAMA Psychiatry. 2018.
57. Oedegaard KJ, Alda M, Anand A, et al. The pharmacogenomics of bipolar disorder study (PGBD): identification of genes for lithium response in a prospective sample. BMC Psychiatry. 2016;16:129.
58. Geoffroy PA, Bellivier F, Leboyer M, Etain B. Can the response to mood stabilizers be predicted in bipolar disorder? Front Biosci (Elite Ed). 2014;6:120–38.
59. Berk M, Hallam KT, McGorry PD. The potential utility of a staging model as a course specifier: a bipolar disorder perspective. J Affect Disord. 2007;100(1–3):279–81.
60. Perry A, Tarrier N, Morriss R, McCarthy E, Limb K. Randomised controlled trial of efficacy of teaching patients with bipolar disorder to identify early symptoms of relapse and obtain treatment. BMJ. 1999;318(7177):149–53.
61. Kim YK, Jung HG, Myint AM, Kim H, Park SH. Imbalance between pro-inflammatory and anti-inflammatory cytokines in bipolar disorder. J Affect Disord. 2007;104(1–3):91–5.
62. Kim YK, Suh IB, Kim H, et al. The plasma levels of interleukin-12 in schizophrenia, major depression, and bipolar mania: effects of psychotropic drugs. Mol Psychiatry. 2002;7(10):1107–14.
63. Su KP, Leu SJ, Yang YY, Shen WW, Chou YM, Tsai SY. Reduced production of interferon-gamma but not interleukin-10 in bipolar mania and subsequent remission. J Affect Disord. 2002;71(1–3):205–9.
64. Guloksuz S, Cetin EA, Cetin T, Deniz G, Oral ET, Nutt DJ. Cytokine levels in euthymic bipolar patients. J Affect Disord. 2010;126(3):458–62.
65. Rapaport MH, Guylai L, Whybrow P. Immune parameters in rapid cycling bipolar patients before and after lithium treatment. J Psychiatr Res. 1999;33(4):335–40.
66. Maes M, Bosmans E, Calabrese J, Smith R, Meltzer HY. Interleukin-2 and interleukin-6 in schizophrenia and mania: effects of neuroleptics and mood stabilizers. J Psychiatr Res. 1995;29(2):141–52.
67. Boufidou F, Nikolaou C, Alevizos B, Liappas IA, Christodoulou GN. Cytokine production in bipolar affective disorder patients under lithium treatment. J Affect Disord. 2004;82(2):309–13.
68. Belvederi Murri M, Prestia D, Mondelli V, et al. The HPA axis in bipolar disorder: systematic review and meta-analysis. Psychoneuroendocrinology. 2016;63:327–42.
69. Girshkin L, Matheson SL, Shepherd AM, Green MJ. Morning cortisol levels in schizophrenia and bipolar disorder: a meta-analysis. Psychoneuroendocrinology. 2014;49:187–206.
70. Gonzalez SD, Williams AJ, Blacker CJ, et al. Putative biological predictors of treatment response in bipolar disorders. Personalized Medicine in Psychiatry. 2017;1–2:39–58.
71. Krishna VN, Thunga R, Unnikrishnan B, et al. Association between bipolar affective disorder and thyroid dysfunction. Asian J Psychiatr. 2013;6(1):42–5.
72. Cole DP, Thase ME, Mallinger AG, et al. Slower treatment response in bipolar depression predicted by lower pretreatment thyroid function. Am J Psychiatry. 2002;159(1):116–21.
73. Barbuti M, Carvalho AF, Kohler CA, et al. Thyroid autoimmunity in bipolar disorder: a systematic review. J Affect Disord. 2017;221:97–106.
74. Tunca Z, Ozerdem A, Ceylan D, et al. Alterations in BDNF (brain derived neurotrophic factor) and GDNF (glial cell line-derived neurotrophic factor) serum levels in bipolar disorder: the role of lithium. J Affect Disord. 2014;166:193–200.
75. Hashimoto R, Takei N, Shimazu K, Christ L, Lu B, Chuang DM. Lithium induces brain-derived neurotrophic factor and activates TrkB in rodent cortical neurons: an essential step for neuroprotection against glutamate excitotoxicity. Neuropharmacology. 2002;43(7):1173–9.

76. Pandey GN, Rizavi HS, Dwivedi Y, Pavuluri MN. Brain-derived neurotrophic factor gene expression in pediatric bipolar disorder: effects of treatment and clinical response. J Am Acad Child Adolesc Psychiatry. 2008;47(9):1077–85.
77. Chang YC, Rapoport SI, Rao JS. Chronic administration of mood stabilizers upregulates BDNF and bcl-2 expression levels in rat frontal cortex. Neurochem Res. 2009;34(3):536–41.
78. Gonzalez SD, Williams AJ, Blacker CJ, Voort JLV, Schak KM, Nemeroff CB, Widge AS, Tohen M. Putative biological predictors of treatment response in bipolar disorders. Pers Med Psychiatry. 2017;1–2:39–58.
79. Grande I, Kapczinski F, Stertz L, et al. Peripheral brain-derived neurotrophic factor changes along treatment with extended release quetiapine during acute mood episodes: an open-label trial in drug-free patients with bipolar disorder. J Psychiatr Res. 2012;46(11):1511–4.
80. Dias VV, Brissos S, Frey BN, Andreazza AC, Cardoso C, Kapczinski F. Cognitive function and serum levels of brain-derived neurotrophic factor in patients with bipolar disorder. Bipolar Disord. 2009;11(6):663–71.
81. Barbosa IG, Huguet RB, Mendonca VA, et al. Increased plasma levels of brain-derived neurotrophic factor in patients with long-term bipolar disorder. Neurosci Lett. 2010;475(2):95–8.
82. Fernandes BS, Gama CS, Kauer-Sant'Anna M, Lobato MI, Belmonte-de-Abreu P, Kapczinski F. Serum brain-derived neurotrophic factor in bipolar and unipolar depression: a potential adjunctive tool for differential diagnosis. J Psychiatr Res. 2009;43(15):1200–4.
83. Kruger S, Alda M, Young LT, Goldapple K, Parikh S, Mayberg HS. Risk and resilience markers in bipolar disorder: brain responses to emotional challenge in bipolar patients and their healthy siblings. Am J Psychiatry. 2006;163(2):257–64.
84. Suwalska A, Sobieska M, Rybakowski JK. Serum brain-derived neurotrophic factor in euthymic bipolar patients on prophylactic lithium therapy. Neuropsychobiology. 2010;62(4):229–34.
85. de Sousa RT, van de Bilt MT, Diniz BS, et al. Lithium increases plasma brain-derived neurotrophic factor in acute bipolar mania: a preliminary 4-week study. Neurosci Lett. 2011;494(1):54–6.
86. Tramontina JF, Andreazza AC, Kauer-Sant'anna M, et al. Brain-derived neurotrophic factor serum levels before and after treatment for acute mania. Neurosci Lett. 2009;452(2):111–3.
87. Fernandes BS, Gama CS, Cereser KM, et al. Brain-derived neurotrophic factor as a state-marker of mood episodes in bipolar disorders: a systematic review and meta-regression analysis. J Psychiatr Res. 2011;45(8):995–1004.
88. Fleck DE, Ernest N, Adler CM, et al. Prediction of lithium response in first-episode mania using the LITHium Intelligent Agent (LITHIA): pilot data and proof-of-concept. Bipolar Disord. 2017;19(4):259–72.
89. Seeberg I, Kjaerstad HL, Miskowiak KW. Neural and behavioral predictors of treatment efficacy on mood symptoms and cognition in mood disorders: a systematic review. Front Psychiatry. 2018;9(337).
90. Turecki G, Grof P, Grof E, et al. Mapping susceptibility genes for bipolar disorder: a pharmacogenetic approach based on excellent response to lithium. Mol Psychiatry. 2001;6(5):570–8.
91. Lovlie R, Berle JO, Stordal E, Steen VM. The phospholipase C-gamma1 gene (PLCG1) and lithium-responsive bipolar disorder: re-examination of an intronic dinucleotide repeat polymorphism. Psychiatr Genet. 2001;11(1):41–3.
92. Turecki G, Grof P, Cavazzoni P, et al. Evidence for a role of phospholipase C-gamma1 in the pathogenesis of bipolar disorder. Mol Psychiatry. 1998;3(6):534–8.
93. Chen CH, Lee CS, Lee MT, et al. Variant GADL1 and response to lithium therapy in bipolar I disorder. N Engl J Med. 2014;370(2):119–28.
94. Kotambail A, Mathur A, Bhat SM, Rai PS, Sharma PS, Satyamoorthy K. GADL1 gene polymorphisms and lithium response in bipolar I disorder: lack of association from an Indian population. Psychiatr Genet. 2015;25(1):39–40.

95. International Consortium on Lithium G, Amare AT, Schubert KO, et al. Association of polygenic score for schizophrenia and HLA antigen and inflammation genes with response to lithium in bipolar affective disorder: a genome-wide association study. JAMA Psychiatry. 2018;75(1):65–74.
96. Duffy A, Alda M, Milin R, Grof P. A consecutive series of treated affected offspring of parents with bipolar disorder: is response associated with the clinical profile? Can J Psychiatry. 2007;52(6):369–76.
97. Campos-de-Sousa S, Guindalini C, Tondo L, et al. Nuclear receptor rev-erb-{alpha} circadian gene variants and lithium carbonate prophylaxis in bipolar affective disorder. J Biol Rhythms. 2010;25(2):132–7.
98. Serretti A, Lilli R, Mandelli L, Lorenzi C, Smeraldi E. Serotonin transporter gene associated with lithium prophylaxis in mood disorders. Pharmacogenomics J. 2001;1(1):71–7.
99. Rybakowski JK, Suwalska A, Czerski PM, Dmitrzak-Weglarz M, Leszczynska-Rodziewicz A, Hauser J. Prophylactic effect of lithium in bipolar affective illness may be related to serotonin transporter genotype. Pharmacol Rep. 2005;57(1):124–7.
100. Lee HY, Kim YK. Catechol-O-methyltransferase Val158Met polymorphism affects therapeutic response to mood stabilizer in symptomatic manic patients. Psychiatry Res. 2010;175(1–2):63–6.
101. Light GA, Swerdlow NR. Neurophysiological biomarkers informing the clinical neuroscience of schizophrenia: mismatch negativity and prepulse inhibition of startle. Curr Top Behav Neurosci. 2014;21:293–314.
102. Leucht S, Tardy M, Komossa K, et al. Antipsychotic drugs versus placebo for relapse prevention in schizophrenia: a systematic review and meta-analysis. Lancet. 2012;379(9831):2063–71.
103. Perkovic MN, Erjavec GN, Strac DS, Uzun S, Kozumplik O, Pivac N. Theranostic biomarkers for schizophrenia. Int J Mol Sci. 2017;18(4).
104. Miller BJ, Buckley P, Seabolt W, Mellor A, Kirkpatrick B. Meta-analysis of cytokine alterations in schizophrenia: clinical status and antipsychotic effects. Biol Psychiatry. 2011;70(7):663–71.
105. Tourjman V, Kouassi E, Koue ME, et al. Antipsychotics' effects on blood levels of cytokines in schizophrenia: a meta-analysis. Schizophr Res. 2013;151(1–3):43–7.
106. Mondelli V, Ciufolini S, Belvederi Murri M, et al. Cortisol and inflammatory biomarkers predict poor treatment response in first episode psychosis. Schizophr Bull. 2015;41(5):1162–70.
107. Bradley AJ, Dinan TG. A systematic review of hypothalamic-pituitary-adrenal axis function in schizophrenia: implications for mortality. J Psychopharmacol. 2010;24(4 Suppl):91–118.
108. Green MJ, Matheson SL, Shepherd A, Weickert CS, Carr VJ. Brain-derived neurotrophic factor levels in schizophrenia: a systematic review with meta-analysis. Mol Psychiatry. 2011;16(9):960–72.
109. Kapur S, Zipursky R, Jones C, Remington G, Houle S. Relationship between dopamine D(2) occupancy, clinical response, and side effects: a double-blind PET study of first-episode schizophrenia. Am J Psychiatry. 2000;157(4):514–20.
110. Colibazzi T, Yang Z, Horga G, et al. Aberrant temporal connectivity in persons at clinical high risk for psychosis. Biol Psychiatry Cogn Neurosci Neuroimaging. 2017;2(8):696–705.
111. Reynolds GP, Yao Z, Zhang X, Sun J, Zhang Z. Pharmacogenetics of treatment in first-episode schizophrenia: D3 and 5-HT2C receptor polymorphisms separately associate with positive and negative symptom response. Eur Neuropsychopharmacol. 2005;15(2):143–51.
112. McClay JL, Adkins DE, Aberg K, et al. Genome-wide pharmacogenomic analysis of response to treatment with antipsychotics. Mol Psychiatry. 2011;16(1):76–85.
113. Zhang JP, Robinson DG, Gallego JA, et al. Association of a schizophrenia risk variant at the DRD2 locus with antipsychotic treatment response in first-episode psychosis. Schizophr Bull. 2015;41(6):1248–55.
114. Schizophrenia Working Group of the Psychiatric Genomics C. Biological insights from 108 schizophrenia-associated genetic loci. Nature. 2014;511(7510):421–427.

115. Ikeda M, Yamanouchi Y, Kinoshita Y, et al. Variants of dopamine and serotonin candidate genes as predictors of response to risperidone treatment in first-episode schizophrenia. Pharmacogenomics. 2008;9(10):1437–43.
116. Wynn JK, Sugar C, Horan WP, Kern R, Green MF. Mismatch negativity, social cognition, and functioning in schizophrenia patients. Biol Psychiatry. 2010;67(10):940–7.
117. Lee SH, Sung K, Lee KS, Moon E, Kim CG. Mismatch negativity is a stronger indicator of functional outcomes than neurocognition or theory of mind in patients with schizophrenia. Prog Neuropsychopharmacol Biol Psychiatry. 2014;48:213–9.
118. Light GA, Swerdlow NR. Future clinical uses of neurophysiological biomarkers to predict and monitor treatment response for schizophrenia. Ann N Y Acad Sci. 2015;1344:105–19.
119. Umbricht D, Krljes S. Mismatch negativity in schizophrenia: a meta-analysis. Schizophr Res. 2005;76(1):1–23.
120. Vinogradov S, Fisher M, de Villers-Sidani E. Cognitive training for impaired neural systems in neuropsychiatric illness. Neuropsychopharmacology. 2012;37(1):43–76.
121. Fisher M, Holland C, Merzenich MM, Vinogradov S. Using neuroplasticity-based auditory training to improve verbal memory in schizophrenia. Am J Psychiatry. 2009;166(7):805–11.
122. Menning H, Roberts LE, Pantev C. Plastic changes in the auditory cortex induced by intensive frequency discrimination training. NeuroReport. 2000;11(4):817–22.
123. Kawakubo Y, Kamio S, Nose T, et al. Phonetic mismatch negativity predicts social skills acquisition in schizophrenia. Psychiatry Res. 2007;152(2–3):261–5.
124. Perna G, Grassi M, Caldirola D, Nemeroff CB. The revolution of personalized psychiatry: will technology make it happen sooner? Psychol Med. 2018;48(5):705–13.
125. Williams LM, Rush AJ, Koslow SH, et al. International study to predict optimized treatment for depression (iSPOT-D), a randomized clinical trial: rationale and protocol. Trials. 2011;12:4.
126. Adams Z, McClure EA, Gray KM, Danielson CK, Treiber FA, Ruggiero KJ. Mobile devices for the remote acquisition of physiological and behavioral biomarkers in psychiatric clinical research. J Psychiatr Res. 2017;85:1–14.

Women's Psychiatry

Georgia Balta, Christina Dalla and Nikolaos Kokras

Abstract

Brain disorders and mental diseases, in particular, are common and considered as a top global health challenge for the twenty-first century. Interestingly, women suffer more frequently from mental disorders than men. Moreover, women may respond to psychotropic drugs differently than men, and, through their lifespan, they endure sex-orientated social stressors. In this chapter, we present how women may differ in the development and manifestation of mental health issues and how they differ from men in pharmacokinetics and pharmacodynamics. We discuss issues in clinical trials regarding women participation, issues in the use of psychotropic medications in pregnancy, and challenges that psychiatry faces as a result of the wider use of contraceptives, of childbearing at older age, and of menopause. Such issues, among others, demand further women-oriented psychiatric research that can improve the care for women during the course of their lives. Indeed, despite all these known sex differences, psychiatry for both men and women patients uses the same approach. Thereby, a modified paradigm for women's psychiatry, which takes into account all these differences, emerges as a necessity, and psychiatric research should take more vigorously into account sex differences.

Keywords

Sex differences · Women · Pharmacokinetics · Pharmacodynamics · Pregnancy · Contraceptives · Menopause

G. Balta · C. Dalla · N. Kokras (✉)
Department of Pharmacology, Medical School, National and Kapodistrian University of Athens, Mikras Asias 75, Goudi, 11527 Athens, Greece
e-mail: nkokras@med.uoa.gr

N. Kokras First Department of Psychiatry, Eginition Hospital, Medical School, National and Kapodistrian University of Athens, Athens, Greece

© Springer Nature Singapore Pte Ltd. 2019

Y.-K. Kim (ed.), *Frontiers in Psychiatry*, Advances in Experimental Medicine and Biology 1192, https://doi.org/10.1007/978-981-32-9721-0_11

Introduction

Brain disorders and mental diseases, in particular, are common and considered as a top global health challenge for the twenty-first century [1]. In the United States of America, mental disorders are recognized as the most costly category of diseases and one of the leading causes of disability [2, 3]. At the same time, it is estimated that 30% of women and 23% of men in the European Union suffer from neuropsychiatric disorders [1]. It cannot go unnoticed that women suffer more frequently from mental disorders than men. The World Health Organization recognizes sex as a crucial factor in determining the prevalence and severity of mental health disorders [4]. Women are more prone to suffer from certain diseases, they may respond to psychotropic drugs differently than men, and, through their lifespan, they endure sex-orientated social stressors. In fact, men and women exhibit a differential behavioral response to stress, with the traditional flight or fight response being mostly male. Actually, recent research suggests that different biological mechanisms may have a role in how both sexes respond to stress [5–7]. Many neural and behavioral functions are affected, including, among others, cognitive function, pain and opioid sensitivity, depressive symptoms such as irritability, insomnia, appetite, and general physical well-being [8, 9]. Despite all these known differences, the psychiatric approach for both men and women uses the same methodology. However, a modified paradigm for women in psychiatry, that takes into account all these differences, emerges as a necessity. In this chapter, we briefly present differences among genders concerning the fields of clinical psychopathology, biology, psychopharmacology, and social interactions that clinicians should bear in mind when treating women and investigators should take into consideration when performing research. Furthermore, we highlight key areas that can be addressed in order to meet the needs of women suffering from mental disorders, and we identify key areas for future research.

Women and Mental Health in the Social Context

It is generally accepted that society has a great impact on its members and their mental health. One of the clearest social divisions is quite often between sexes. Sex and gender define our life choices, influence access to resources, characterize the way we relate to others and define society's expectations, along with the ones toward ourselves. Consequently, sex and gender affect our way of thinking and acting and, therefore, our emotions. Even in modern societies that gender extends beyond the strict male/female distinction, sex and gender are important in far too many aspects of life, but mental health needs of LGBT individuals (which is another emerging and important aspect of modern psychiatry) is beyond the scope of this chapter.

Sex and gender differences in mental health can be explained, to some extent, by social, cultural, and economic status [10, 11]. Understanding gendered social practices that frequently impair women's (but sometimes also men's) mental health is of great importance when trying to understand and treat mental diseases. The industrial revolution has caused tremendous changes in the structure of modern society and, therefore, to the women's status. Nowadays women work outside the house and they earn their own money. This is of great importance, as they can be independent and satisfy their own needs. This independency extends in all fields of personal activity and life's choices. Women in developed, and in some developing countries, are now free to get educated, participate in the social and the political life, they have the liberty to choose whether to marry or stay single, whether to give birth to children, whether to get divorced or not. All these possibilities were usually not possible for women in the past centuries, as there were certain notions and traditions, concerning the structure of society and everyday life. In that given reality, several values and beliefs had been formed, which affirmed the basis of the so-called "masculine superiority". Given the change that has occurred nowadays, still both men and women try to adjust, find new roles, and determine themselves through different roles, but this process is not easy. There are certain values and attitudes ingrained in the social unconscious—the collective unconscious—[12], which still support thesis as: "women are inferior to men," or "women ought to get married, stay home, and raise their children," "women ought to obey their husbands". These "instructions of life," which are the most solid and the hardest to change, still exist in many parts of the world and exert their power, even if the reality has changed. Thus, in our time we are struggling to combine new gender roles, needs and possibilities, along with old beliefs, archetypes, and ideals. This situation is confusing, frustrates both men and women and can harm women's mental health.

Dominant conceptions of masculinity that associated men with characteristics of assertiveness, competitiveness, and independence [13–15] still have their place, even if these qualities are present in women, as well. Women have their own salaries, but on the other hand, they are underpaid, even in the same jobs that require the same training and experience with men. Women experience this difference as devaluation of their skills [16, 17]. Not rarely, income is also divided unequally within the family, with women receiving less and spending most of it on their children while several studies attest to the negative effects of low income on mental health [18]. Moreover, serving multiple roles, being a mother and a worker in a competitive economy, is quite demanding. In traditional societies, women are still regarded as responsible for childcare and housework, even if they work long hours, and equally contribute financially, as their partners [19–22]. These multiple tasks and requests have been proposed to raise women's levels of depressive and anxious symptoms [23–29] but when the partner contributes to these tasks, the level of women's symptoms of depression and anxiety resemble those among men, which are lower. In addition, high levels of distress may rise in women by unpredictable childcare situations [30–32], and usually, women feel more distress than men, when spending time away from their young children [33]. Because of these multiple social demands, women often experience intense internal conflicts and challenges which contribute to their higher distress in combining these roles [34].

Another social factor that interferes with women's psychology is the model of a lean female body. The rising prevalence of obesity in the past three decades is a factor that may be related to gender differences in mental health, as women are encouraged to maintain a lean body, in order to be acceptable and desirable among peers. These gendered expectations for an "appropriate" body weight may have an impact on women's mental health. Obesity has been found to be related to depression, among white women [35], well-educated persons, or those who are dieting [36]. Marijnissen et al. have found a connection between visceral fat and the risk of mood disorders in people 50–70 years of age, especially women [37]. This social pressure relating to obesity may lead women to further internalize pejorative feelings, thereby leading to mental health decline [38].

Furthermore, although work is rewarding for both men and women, there seems to be an overall cost for women: work outside the home detracts from time spent with family, which explains why married mothers have greater internalizing problems than married fathers [39]. On the other hand, the quality of the marriage is an important factor, as well, since there is a two-way dynamic process between marital conflicts and depressive symptoms [40]: "depressive symptoms lead to more marital conflict, which, in turn, lead to more depressive symptoms over time." In fact, marital status has long attracted the attention of sociologists because of the primacy of the bonds in marriage and family. In this context, it is important to bear in mind that women are more frequently the primary victims of domestic violence, a risk factor for mental health problems [11]. Despite those issues, married people, generally, manifest better mental health than non-married persons, but this could be due to selection processes in which persons with more mental health problems are less likely to be married [41]. However, due to the fact that women outlive men, most married women will experience widowhood, but the consequences of bereavement on mental health are quite varied [42]. When the spouse passes away, women may have to face numerous stressors, other than the loss of the partner per se. The need for relocation, assuming new roles, along with financial problems may emerge and put women in distress [43]. Additionally, prolonged caregiving for the spouse facing a serious illness appears to have a lasting negative impact on the survivor's mental health [44]. Interestingly enough, at the same time, long-term caregiving fosters anticipation of the loss, which in turn may be helpful for restorative activities, especially for persons of higher education [45].

Sex Differences in Psychopathology

Current diagnostic classification systems ignore the patient's gender, when proposing diagnostic criteria for mental illnesses. This is so, even though sex differences in mental disorders are well known. In particular, regarding mood disorders, women experience major depression at roughly twice the rate of males until late middle age, when they transition to menopause. Women seeking depression treatment report greater symptom severity, and more frequently gastrointestinal symptoms, hypersomnia, somatization, crying, anger, hostility, and interpersonal difficulties [46–48]. They are more likely than men to have a comorbid anxiety disorder, bulimia nervosa, or somatoform disorder and more frequently report past suicide attempts [49]. Men, on the other hand, are more likely to report comorbid alcohol and substance abuse [48]. Other studies support the observation that women are more likely to experience disturbances of sleep, appetite, and energy during depression, as well as more likely to experience "atypical" depression, with increased appetite, increased weight, and hypersomnia [50]. Atypical depression is also associated with younger age of onset, more comorbidity with social anxiety and specific phobias, as well as more severity, disability and suicide attempts [49]. In addition, 3–8% of women of reproductive age meet the criteria for premenstrual dysphoric disorder, while the percentage of reproductive women who experience premenstrual syndrome rises to 20% [51, 52].

On the other hand, bipolar disorder (BD), a complex disorder with extreme fluctuations in mood, from manic highs to depressive lows, affects men and women equally [53]. Although the lifetime incidence of BD is approximately 1:1 in men and women, men experience more frequently manic episodes and unipolar mania [54]. Bipolar disorder type-II, which is characterized by prolonged hypomania with frequent depressive episodes, and hypomania seem to have an increased prevalence in women, whereas men have significantly better general functioning [54]. Women with BD often report increased rates of rapid cycling, recurrent depressive polarity, and a depressive or mixed onset, while mania is more prevalent in men at the first onset of the disease [54–60]. Researchers have provided evidence supporting the use of antidepressants, benzodiazepines, electroconvulsive therapy and psychotherapy in women with BD [55, 61]. On the other hand, men are more often treated with lithium [61], but sex differences in the clinical response are not evident [58]. Women with BD, when treated with lithium, have increased rates of hypothyroidism and are at increased risk of migraine; rates of metabolic syndrome appear to be equal between sexes [54, 62].

Moreover, post-traumatic stress disorder is a psychiatric disorder that may emerge after experiencing a traumatic event involving threat to life or physical integrity [53]. It is characterized by intrusive memories and nightmares of the trauma, avoidance of trauma reminders, dysregulation in mood and cognition along with hyperarousal in the hypothalamic–pituitary axis [63]. There are several known risk factors for PTSD, including previous psychiatric or trauma history, family psychiatric history, age at trauma onset, and most importantly, the female gender [64–66]. Longitudinally, the sex difference on PTSD is highest between the ages of 21 and 25, with men and women reporting their highest rates of PTSD in their 40 s and 50 s, respectively [67]. Blanco et al. in 2018 after studying a sample of >34,000 participants, reported that lifetime prevalence for PTSD was found to be 9.48% in women and 5.97% in men, and that women experienced childhood maltreatment and assaultive violence more often than men [68]. An enlightening finding of this study was that the difference in reactivity to trauma explained more than 60% of the differential PTSD risk in women and men [68]. Women appear to respond better to PTSD treatment; previous studies reported a better treatment outcome and better maintenance after exposure, as well as behavioral therapies [69–71]. Wade et al. also performed a meta-analysis of 48 studies, which confirmed that women display a better treatment response to behavioral interventions, whereas sex differences in response to pharmacological treatment were not detected [72].

Regarding psychotic disorders, men take a greater burden as the incidence of schizophrenia is slightly more prevalent in men (1.4:1, male: female ratio) [49, 73, 74]. However, women experience a second peak of disease onset around 45–49 years, which could to be attributed to the decline of ovarian hormones due to menopause [74, 75]. Sex differences have also been reported concerning the disease progression and its prognosis, since men show less responsiveness to antipsychotic medication, and longer, more frequent hospitalizations [76, 77]. Thorup et al., after examining a sample of 578 patients with a first-episode psychosis, found males to have significantly higher levels of negative symptoms than females, at all times of follow-up. Poor compliance and substance abuse were more common among male patients and maybe this is another contributory factor. Women, on the other hand, were much more likely to live with children, to be compliant with medication and to maintain recovery at a 5-year follow-up [78]. Indeed, women have better social prognoses, including retainment of marriages, interpersonal relationships, and employment [74, 79, 80]. Men with psychotic disorders also exhibit lower self-esteem, self-dislike and mental clouding, greater substance abuse, social isolation, withdrawal, and death by suicide than women [81–86].

An important sex difference has been reported regarding suicidal behavior, with an overrepresentation of females in nonfatal suicidal behavior and a preponderance of males in completed suicide, also known as the "gender paradox of suicidal behavior" [87]. Specifically, males take their own lives at nearly four times the rate of females and comprise approximately 80% of all suicides, even though female suicide attempt rates are estimated to be three to four times higher than men's [88].

Women in Clinical Trials

Unfortunately, sex differences in psychiatry are not accompanied by sex-differentiated approaches, neither when clinical and preclinical research is conducted, nor when psychiatric drugs are prescribed. It is true that there is also relatively little knowledge on potential sex differences in the pharmacotherapy of mental diseases. A plausible explanation could lie on the fact that, until the 1990s, women were underrepresented in clinical trials for new drugs because the menstrual cycle was treated as a confounding factor, and women of childbearing age were excluded because of fear for potential harm to the fetus [89–91]. In 1990, the National Institutes of Health and the U.S. Food and Drug Administration (and later the

European Medicines Agency) enacted guidelines for the inclusion of women in clinical trials [92]. However, psychiatric research on sex differences is still hindered by tools that were not designed to detect sex differences. In fact, it is surprising that so little research has been devoted to the sex-dependent performance of widely used rating scales and that so little adaptation has been performed to correctly identify sex differences in the performance of psychiatric rating scales. For example, regarding one of the most widely used rating scales in depression studies, the Hamilton Depression Rating Scale, it is astonishing that so little research has been performed on its sex-dependent performance and its ability to detect the aforementioned differences [93, 94]. Similarly, the Beck's Depression scale has conflicting results about its ability to detect sex differences [95], and to the best of our knowledge, the Montgomery–Asberg Depression Rating Scale, which is also widely used in drug trials, has not been studied at all in relation to sex. Inclusion criteria in clinical trials often require specific rating scale scores, which are common for both sexes. In that case, it could be that men and women are recruited on the assumption that they suffer from the same severity, as they have the same score on the rating scale, whereas in reality severity of the disease may be different between the sexes and the scales are not correctly scoring it because they could be insensitive to sex-specific symptoms. Moreover, a rating scale insensitive to sex differences could attenuate the response to a specific treatment that might be more beneficial for men or for women. Therefore, it was recently pointed out that, if clinical researchers are using tools and designs that ignore, attenuate, and correct for potential sex differences, it might be the case that we are missing the discovery of sex-dependent or sex-specific treatments by not conducting sex-aware research [93, 96]. Turning a blind eye to sex differences in drug development may have consequences for the patients.

For many years, preclinical studies used mostly, or exclusively, one-sex (male) rodents [97]. This fact has created a knowledge void that unfortunately migrated into clinical studies. In this way, drugs are reaching with safety and efficacy profiles that contain very little information about potential sex differences. For example, pharmacovigilance revealed that severe side effects of zolpidem were more frequent in women than men [98]. Indeed, zolpidem was the first psychotropic drug the US Food and Drug Administration recommended to have a different dosing regimen in men versus women [99]. This could be the beginning of approaching differences in treatment efficacy and safety in men and women, and a broader use of treatment protocols separated by sex may help in this direction [100]. Allopregnanolone is another interesting example, highlighting the potential of sex-specific interventions in psychiatry. Allopregnanolone is the predominant metabolite of progesterone, and preclinical evidence provided proof that rapid changes in levels of allopregnanolone confer dramatic behavioral changes, and may trigger postpartum depression [101–103]. Recently, brexanolone was licensed as a novel therapeutic drug specifically for postpartum depression, following successful completion of clinical trials [104–106]. It is important to mention that conflicting results have been reported with regard to potential sex differences in drug response, and the clinical significance of many sex differences in pharmacokinetics and pharmacodynamics is unclear [107, 108]. On the other hand, a dearth of experimental and clinical evidence indicates that such differences actually do exist. Indeed, there is increasing evidence that sex is an important biological variable modulating efficacy of psychotropic medications and there are many arguments in favor of including both sexes in preclinical and clinical assessment of novel pharmaceuticals [109]. In any case, developing a branch of psychopharmacology focused on sex differences will lead us to structure the evidence and link preclinical researchers to clinical professionals, so these data are eventually ingrained in everyday medical knowledge and practice.

Sex Differences in Pharmacokinetics and Pharmacodynamics

Sex differences in pharmacokinetics and pharmacodynamics have been repeatedly studied [108, 110–116]. Acquiring strong data concerning sex differences in these areas would facilitate sex-oriented treatments involving sex-specific drugs and dose adjustments. Unfortunately, regarding pharmacokinetics, results so far are not always consistent. Generally, women exhibit both lower secretion of gastric acid and slower transit times from the stomach and intestine [117], which results in increased oral absorption and bioavailability. On the other hand, delayed intestinal transit in females delays absorption and drug peak levels. Hormonal variations due to the menstrual cycle affect gastrointestinal function, and estrogens may inhibit gastric acid and gastrointestinal mobility [117]. Moreover, women generally weigh less than men; they have a lower intravascular blood volume, lower organ blood flow, and less muscle [114]. Furthermore, women have a higher percentage of body fat [115]. This is important because drugs acting in the brain are highly lipophilic [118]. These parameters may all interfere with the higher volume of distribution in women [108]. Moreover, there are substantial differences, both among individuals and between sexes, in the metabolism of drugs, resulting from genetic polymorphisms, diseases, hormonal factors, concomitant medications, diet, and other epigenetic factors [119, 120]. The female hormonal milieu is thought to play a crucial role in this context. Similar antidepressants may be affected differentially by sex, depending on the affinity of each chemical molecule for P450 isozymes [121–124]. Gonadotropin hormone (GH), which controls directly the expression and activity of the CYP enzyme, has a different secretion profile in men and women [125]. Regarding Phase II drug metabolism, it is generally accepted that conjugation reactions are less efficient in women than in men [110, 111, 121]. Clearance of antidepressants depends on the rate of blood flow into the liver and kidney, and women are known to have a smaller liver and a lower organ blood flow rate than men. Renal glomerular filtration is consistently lower in women than in men [121]. Lately, the efflux transport proteins attract a lot of interest. The drug-efflux pump P-glycoprotein, which serves as barrier for the brain and for the gastrointestinal tract, helps to eliminate drugs from the brain was also found lower in women [114]. Furthermore, gonadal hormones have been shown to downregulate P-glycoprotein expression, but enhance drug absorption, via inhibition of P-glycoprotein function in the gastrointestinal tract [125]. In summary, drug pharmacokinetics are not similar in both sexes, and depending on the drug properties, more or less pronounced sex differences can be observed. As Kokras et al. noted [108], a key finding from the reviewed literature is that, generally, women, when dosed in a similar way as men, should be expected to have higher exposure to CNS-acting drugs. This may impact significantly in sex differences observed in treatment response and frequency of adverse effects.

Regarding sex differences in pharmacodynamics and despite significant research, no clear evidence has emerged on whether there are clinically significant sex-related differences. For example, many studies have not detected sex differences in the efficacy of antidepressants [126–135], but others have shown that males have a significantly greater therapeutic response, when treated with tricyclic antidepressants [46, 136, 137]. Numerous studies find premenopausal women to respond better than men to selective serotonin reuptake inhibitors [127, 136, 138–141]. In different studies, females under SSRI treatment had 15%, 23%, and 40% greater improvement than males [136, 139, 140], which constitutes a significant variability in the sex-differential response [112]. Men responded significantly better to imipramine than premenopausal females, but premenopausal women responded better to fluvoxamine than postmenopausal women and men [142, 143]. Berlanga and Flores-Ramos reported that women were more likely to respond to citalopram than to reboxetine, whereas men had similar response rates to both drugs [140]. Women have been reported to respond better to sertraline for the treatment of behavioral disturbances in Alzheimer's disease than do men [144]. With regard to adverse effects, females have a 1.5–1.7-fold greater risk of developing an adverse drug reaction than male patients [145]. Women are at risk for drug-induced long QT syndrome, with two-thirds of all cases of drug-induced torsades de pointes occurring in females [146, 147]. Women with dysthymia were reported to have a greater number of symptoms in discontinuation syndrome with paroxetine than did men [148]. Women were also more likely to develop hyperprolactinemia than men [149]. Moreover, premenopausal women are more likely to report nausea, as an adverse event, and men dyspepsia, sexual dysfunction, and urinary frequency [136].

Moreover, the effective dosage to elicit an adverse drug reaction is likely not the same for males and females. Studies have shown that male patients may require higher amounts of antipsychotic medication [150, 151]. On the other hand, women show greater improvement in psychotic symptoms and more severe adverse side effects with typical antipsychotic agents than do men [152, 153]. It is generally accepted that women are more susceptible to weight gain, diabetes, metabolic syndrome, and specific cardiovascular risks, when treated with antipsychotics compared with men [154, 155]. Atypical antipsychotics can raise prolactin levels, resulting in more pronounced symptoms in women than men, including galactorrhea, atrophic changes in the urethra and vaginal mucosa, reduced vaginal lubrication, dyspareunia, decreased libido, ovarian dysfunction, infertility, oligomenorrhea, and amenorrhea [156]. In recent years, epidemiological, clinical, and biological evidence has drawn the attention on the influence of sex and gender on Alzheimer's disease (AD) [157]. Nevertheless, not enough attention has been paid to their impact on treatment outcomes, and there is generally a considerable lack of data regarding sex differences in the efficacy, safety, and tolerability of medications for AD [157].

Another area of conflicting evidence regarding sex differences is the patient's adherence to treatment. Indeed, a significant parameter which determines the drug efficacy and does not include the "drug-human body" interaction per se is the adherence to treatment. Adherence is defined as compliance with dosage and regimen as prescribed for a duration considered sufficiently adequate for therapeutic response [112]. A historical cohort study of 310,994 individuals who filled antidepressant prescriptions during a 4-year period found adherence was significantly higher for males aged 20–40 years than for females of that age, but this relationship reversed later in life for those aged 50–70 years [158]. A historic cohort study of three Italian local health units of 88,755 patients with a prescription for antidepressants found that female sex was a predictor of better adherence [159]. On the other hand, a sample of 3684 patients with long-term prescription of antidepressants found compliance rates across sexes were similar, with 21.4% compliance for males and 22.4% compliance for females [160]. More clinical research comes as a necessity, in order to obtain more coherent results.

Contraceptives and Mental Health

Oral contraceptives (OCs) have become the most commonly prescribed medication in women of reproductive age [161]. It is estimated that 80% of sexually active women in the United States use hormonal contraception to prevent pregnancy, most typically OCs [162]. OCs contain synthetic analogues of estrogen and progesterone that centrally disrupt the hypothalamic–pituitary–ovarian axis, acting locally on reproductive organs and modifying the female hormonal milieu [163]. As a significant part of the population uses these drugs, necessity indicated that psychiatry should investigate their impact on women's mental health. Current research suggests that contraceptives are unlikely to cause or aggravate a mental illness. On the other hand, it has been found in human studies that the use of hormonal contraception seams to protect against depressive symptoms. For example, Toffol et al. in Finland found that hormonal contraceptive use reduced the levels of depressive symptoms [164]. Also, in 2013 in the United States, Keyes et al. after examining 6.654 sexually active non-pregnant women found that hormonal contraception may reduce levels of depressive symptoms among young women [165]. Cheslack-Postava et al. examined the association between contraceptive use (any current use, duration, and type) and Major Depressive Disorder, Generalized Anxiety Disorder and Panic Disorder in a nationally representative sample of 1105 women in the USA. They found evidence for an inverse association between reported contraceptives use and Panic Disorder [166]. Other researchers have reported consistent evidence of no association, positive or negative, between hormonal contraceptive use and depressive symptoms [167, 168]. As far as suicidality is concerned, some studies found no increase in suicide risk for contraceptive use [169, 170], while others argue in favor of an increase in violent deaths, including suicide, among oral contraceptive users compared with non- users [171]. Schaffir et al. reviewed studies from the past 30 years that focused on contraceptives and mood. They concluded that it is difficult to identify which contraceptives users are at risk for adverse mood effects and prompted clinicians to recognize that such effects are infrequent and contraceptives may be prescribed with confidence [172]. As for the cognitive function of women under contraceptives, Warren et al. in 2013, after reviewing 22 studies, found that verbal memory may be enhanced by contraceptive use and that visuospatial ability may be impacted either positively or negatively depending on the androgenicity of the progestin in the OC in question; results, which were consistent with other studies as well [163, 173, 174].

Childbearing at Older Age

During the last decades, there have been significant changes in the age women decide to give birth. Birth rates in developed countries are declining in younger and increase in older women [175]. Indeed, from 1970 to 2014, the average age of first-time mothers in the United States has increased from 21 to 26 years [176, 177] and the same trend is also observed in Europe [178, 179]. As maternal age has risen, gestation in women over 35 years old is considered to be a worldwide phenomenon [180]. Indeed, the percentage of first births among women over 35 years increased nearly eight times in the last 40 years [181], and 1 out of 12 first births are to women over 35 years old in comparison to just 1 out of 100 in 1970 [182, 183]. This new phenomenon has been attributed to several socioeconomic factors, as mentioned previously [184]. Nowadays, women are prompted not only to get educated but to obtain more advanced degrees, as well. These ambitions cause the lengthening of the educational period and require years spent in studying rather than in starting a family. These goals—women getting educated and financially independent—were not so common just a few decades ago in most societies. However, achieving academic and vocational objectives takes years to accomplish and coincides with the most fertile period of women's lives [185]. These changes pose challenges not only to obstetric medicine but also to psychiatry, as pregnancy and the perinatal period are periods of high risk for women's mental health and are nowadays observed in a different, from the point of aging, population. So, the new mother-model usually describes mothers who are older, more educated, more financially independent, and sometimes less available to their children than before. This phenomenon raises several questions about its impact on people's lives and mental health, which are largely unaddressed by modern psychiatry.

Use of Psychotropics During Pregnancy

It is known that dynamic changes in the female physiology during pregnancy affect several pharmacokinetic parameters [108, 186]. Pregnancy decelerates women's gastric emptying, as well as small bowel and colonic transit time of the drug [187]. The maternal plasma volume increases, the protein-binding capacity of the drugs undergoes changes as albumin levels are lowered, and the ratio of muscle to adipose tissue is decreased. All these changes may lead to lower peak plasma concentrations but the volume of distribution and the free fraction of drugs may increase [187–190]. As gestation progresses, the renal blood flow, the glomerular filtration rate, and the hepatic blood flow are increased, leading to high drug extraction ratios [187, 188, 190]. Sex hormones induce CYP3A4, CYP2A6, CYP2D6 and inhibit CYP1A2 and CYP2C19, thus significantly altering the metabolism of many psychotropics. Such changes require dose adjustments of drugs but this field is still not well researched. During pregnancy, fetal exposure depends on the drug dose, the maternal absorption, distribution and elimination, the placental transfer and the fetal distribution, and elimination of the drug [187, 190]. Most psychotropic drugs are highly lipophilic, cross the placental barrier, reach and distribute to the fetal compartment with varying transfer rates. Therefore, they can cause teratogenic effects, neonatal toxicity, withdrawal syndromes, or long-term neurobehavioral effects [188–191]. After delivery, the previous equilibrium between the fetus and its mother via the placenta does not apply to the neonate. For this reason, high concentrations of a particular drug in the fetus may result in significantly prolonged effects postnatally [190, 192]. For example, in pregnancy, increased metabolism reduces the levels of many antidepressants, such as sertraline, citalopram, fluoxetine, paroxetine, venlafaxine, and thus their overall availability [193]. However, some antidepressants have active metabolites. Fetal exposure to an active metabolite may be unpredictably increased, due to the increased metabolism of the parent drug in pregnancy. This holds true for sertraline, citalopram, fluoxetine, and Buproprion [187, 194, 195]. Moreover, after delivery, the rapid normalization of previously modified factors may lead to high drug maternal plasma concentrations, especially if the dose administered during pregnancy was increased and is not readily monitored and modified during the postpartum period [187].

The administration of an antidepressant with short or long elimination half-life will determine whether it is more likely to cause fetal withdrawal symptoms or toxicity [196]. Other notable examples include mood stabilizers (lithium, valproic acid, carbamazepine, lamotrigine), which cross the placenta [197]. Already in early stages of pregnancy, the lithium clearance increases, because of the increased renal function. In the postpartum period, lithium clearance rapidly returns to baseline values, so a dose adjustment is of significant importance [198, 199]. During pregnancy lamotrigine clearance increases and after delivery its clearance quickly reverts to the prepregnancy levels, but lamotrigine is extensively transferred to the fetus and after birth, the newborn clears the drug at a slow rate, thus careful dose adjustments are required [197, 200–202]. Finally, benzodiazepines pass through the placenta through diffusion, which is not constant across pregnancy. Indeed, in late pregnancy, changes in the placental circulation, size, permeability, and lipid content lead to an increased diffusion in comparison to early pregnancy. In addition, some benzodiazepines have long half-lives, and long-term administration may lead to drug accumulation in the maternal and fetal compartment. In fact, the fetal levels of benzodiazepines may be even higher than maternal levels, due to the rapid maternal drug distribution [203–205]. Regarding antipsychotics, olanzapine and haloperidol have the highest placental passage, risperidone and then quetiapine, with the latter having the lowest passage ratio of all [206–208].

Women and Menopause

The direct relationship between psychiatric symptoms and hormonal changes, such as estrogen decrease, has not been clearly understood yet, although significant research highlights the connection. Perhaps stress, educational level, ethnicity, socioeconomic factors, and partner status may influence and modulate the prevalence and clinical course of both menopause and psychiatric symptoms, thus making difficult to identify the exact mechanisms [209]. Interestingly enough, in a qualitative study of the transition to menopause, Winterich and Umberson [210] found that most women did not view menopause as a major event; rather, they identified other midlife events as more stressful and consequential. Indeed, Rossi [211] found that more than 60% of postmenopausal American women described menopause as "only relief." Thus, even though women do not necessarily regard menopause as a major negative life event, scholars in psychiatry, endocrinology, and related fields do find connections between the menopausal transition and mental health. In fact, in a recent review, Freeman [212] reported that most studies found the prevalence of depression to be higher in women undergoing the menopausal transition than during premenopause. Similarly, de Kruif et al. in a meta-analysis found that perimenopausal women are particularly vulnerable to develop depressive symptoms and to show higher symptom severity, when compared to the premenopausal women [213]. Weber et al. in their review also reported an increased vulnerability to cognitive decline and an increased risk for depression and this was also confirmed in longitudinal studies [214, 215]. As for possible explanations, risk factors for the development of depression during the menopause transition include the presence of vasomotor symptoms, a personal history of depression (particularly depression that is related to pregnancy or hormonal changes through the menstrual cycle), surgical menopause, adverse life events, and negative attitudes to menopause and aging [215]. Indeed, prior history of depression critically influences the risk for depression during menopausal transition. Women with a prior history of depression are far more likely to experience depression during the transition [212]. Moreover, vasomotor symptoms were positively related to depressive symptoms during menopausal transition, and several lines of research support the close link between somatic, psychological, and psychopathological symptoms of menopause [213, 216–219]. In summary, most studies give results that are in accordance with the thesis that menopausal women are vulnerable to depressive symptoms and a bio-psycho-sociocultural model of the processes, which might lead to a depressive disorder in midlife, might be the most plausible explanation [220]. Moreover, Llaneza et al. [209] claimed that bio-psycho-social and partner factors may have a significant influence on middle-aged women's sexuality and depressive disorders. Moreover, postmenopausal women form a subpopulation that is increasing during the last decades. This is because life expectancy increases globally and the mean life expectancy nowadays is over 75 years in developed countries [221]. Consequently, women will generally spend the last third of their lifetime in menopause. However, declined estrogens put women at risk, as is in the case of increased vulnerability to psychotic disorders after menopause [222]. Psychiatric research, so far, has given little attention to mental health issues in the, nowadays prolonged, period of life spent in menopause.

Conclusion: Toward a Sex-Aware Psychiatry

In everyday practice, mental health professionals may encounter sex-specific situations which affect exclusively women, as is the case with perinatal mental health and menopause. Nevertheless, evidence-based sex-specific interventions are scarcely available when addressing mental health problems that may affect both sexes. Indeed, there is guidance for the perinatal period [223] but evidence from translational research and limited clinical research that addresses women's psychiatry and does not refer to pregnancy or breastfeeding has yet to reach broad recognition and adoption. Even for perimenopause and postmenopausal women, psychiatric research is still struggling to provide solid evidence for best practice. As a result, it is rarely appreciated the variation of drug efficacy and safety according to sex, even when pharmacokinetic or pharmacodynamic sex differences are known [224, 225]. Increasing awareness is required to implement change toward a more sex-aware psychopharmacology and psychiatry [224]. Indeed, in the context of a new paradigm for women's psychiatry, there are multiple needs that should be covered. Some were described in this chapter, but without doubt more needs exist, especially when considering significant differences between developed and developing societies and economies. By adopting a more sex-aware psychiatric practice, both sexes can benefit by interventions better tailored to their specific needs. Clinical research should catch-up with emerging evidence form preclinical research in the field of sex differences [93], in order to develop evidence-based best practices regarding sex differences in psychiatry.

References

1. Wittchen HU, Jacobi F, Rehm J, Gustavsson A, Svensson M, Jonsson B, et al. The size and burden of mental disorders and other disorders of the brain in Europe 2010. Eur Neuropsychopharmacol. 2011;21(9):655–79.
2. Roehrig C. Mental disorders top the list of the most costly conditions in the United States: $201 Billion. Health Aff (Millwood). 2016;35(6):1130–5.
3. Theis KA, Roblin DW, Helmick CG, Luo R. Prevalence and causes of work disability among working-age U.S. adults, 2011–2013, NHIS. Disabil Health J. 2018;11(1):108–15.
4. WHO. Global Health Observatory Data 2015. https://www.who.int/gho/mental_health/suicide_rates_male_female/en/.

5. Li SH, Graham BM. Why are women so vulnerable to anxiety, trauma-related and stress-related disorders? The potential role of sex hormones. Lancet Psychiatry. 2017;4(1):73–82.
6. McEwen BS, Milner TA. Understanding the broad influence of sex hormones and sex differences in the brain. J Neurosci Res. 2017;95(1–2):24–39.
7. Taylor SE, Klein LC, Lewis BP, Gruenewald TL, Gurung RA, Updegraff JA. Biobehavioral responses to stress in females: tend-and-befriend, not fight-or-flight. Psychol Rev. 2000;107(3):411–29.
8. Kornstein SG, Young EA, Harvey AT, Wisniewski SR, Barkin JL, Thase ME, et al. The influence of menopause status and postmenopausal use of hormone therapy on presentation of major depression in women. Menopause. 2010;17(4):828–39.
9. Young E, Korszun A. Sex, trauma, stress hormones and depression. Mol Psychiatry. 2010;15(1):23–8.
10. Kuehner C. Why is depression more common among women than among men? Lancet Psychiatry. 2017;4(2):146–58.
11. World Health O. Women's mental health: an evidence based review. Geneva: World Health Organization; 2000.
12. Jung CG. The structure of the unconscious. Coll Wks. 1916;7:1916.
13. Rosenfield S, Lennon MC, White HR. The self and mental health: self-salience and the emergence of internalizing and externalizing problems. J Health Soc Behav. 2005;46(4):323–40.
14. Connell RW. Masculinities. Berkeley: University of California Press; 1995.
15. Flax J. Disputed subjects (RLE Feminist theory): essays on psychoanalysis, politics and philosophy. Routledge; 2012.
16. Cohen PN, Huffman ML. Individuals, jobs, and labor markets: the devaluation of women's work. Am Sociol Rev. 2003:443–63.
17. Kilbourne BS, England P, Farkas G, Beron K, Weir D. Returns to skill, compensating differentials, and gender bias: effects of occupational characteristics on the wages of white women and men. Am J Sociol. 1994;100(3):689–719.
18. Becker GS. Nobel lecture: the economic way of looking at behavior. J Polit Econ. 1993;101(3):385–409.
19. Greenstein TN. Economic dependence, gender, and the division of labor in the home: a replication and extension. J Marriage Fam. 2000;62(2):322–35.
20. Hochschild A, Machung A. The second shift: working families and the revolution at home. Penguin; 2012.
21. Lennon MC, Rosenfield S. Relative fairness and the division of housework: the importance of options. Am J Sociol. 1994;100(2):506–31.
22. Pleck JH. Working wives/working husbands. 1985.
23. Bird CE. Gender, household labor, and psychological distress: the impact of the amount and division of housework. J Health Soc Behav. 1999;40(1):32–45,
24. Krause N, Markides KS. Employment and psychological well-being in Mexican American women. J Health Soc Behav. 1985;26(1):15–26.
25. Lennon MC, Rosenfield S. Women and mental health: the interaction of job and family conditions. J Health Soc Behav. 1992;33(4):316–27.
26. Rosenfield S. The effects of women's employment: personal control and sex differences in mental health. J Health Soc Behav. 1989;30(1):77–91.
27. Roxburgh S. 'There just aren't enough hours in the day': the mental health consequences of time pressure. J Health Soc Behav. 2004;45(2):115–31.
28. Mirowsky J. Age and the sense of control. Soc Psychol Q. 1995:31–43.
29. Rosenfield S. Factors contributing to the subjective quality of life of the chronic mentally ill. J Health Soc Behav. 1992:299–315.
30. Ross CE, Mirowsky J. Child care and emotional adjustment to wives' employment. J Health Soc Behav. 1988:127–38.
31. Ross CE. Social causes of psychological distress. Routledge; 2017.
32. Ross CE, Mirowsky J. Households, employment, and the sense of control. Soc Psychol Q. 1992:217–35.

33. Milkie MA, Peltola P. Playing all the roles: gender and the work-family balancing act. J Marriage Fam. 1999:476–90.
34. Simon RW. Gender, multiple roles, role meaning, and mental health. J Health Soc Behav. 1995;36(2):182–94.
35. Schieman S, Plickert G. Functional limitations and changes in levels of depression among older adults: a multiple-hierarchy stratification perspective. J Gerontol B Psychol Sci Soc Sci. 2007;62(1):S36–42.
36. Ross CE. Overweight and depression. J Health Soc Behav. 1994;35(1):63–79.
37. Marijnissen RM, Bus BA, Holewijn S, Franke B, Purandare N, de Graaf J, et al. Depressive symptom clusters are differentially associated with general and visceral obesity. J Am Geriatr Soc. 2011;59(1):67–72.
38. Schafer MH, Ferraro KF. The stigma of obesity: does perceived weight discrimination affect identity and physical health? Soc Psychol Q. 2011;74(1):76–97.
39. Simon RW. The meanings individuals attach to role identities and their implications for mental health. J Health Soc Behav. 1997;38(3):256–74.
40. Choi H, Marks NF. Marital conflict, depressive symptoms, and functional impairment. J Marriage Fam. 2008;70(2):377–90.
41. Ferraro KF, Wilkinson LR. Age, aging, and mental health. Handbook of the sociology of mental health: Springer; 2013. p. 183–203.
42. Stroebe M, Schut H, Stroebe W. Health outcomes of bereavement. Lancet. 2007;370(9603):1960–73.
43. Stroebe M, Schut H. The dual process model of coping with bereavement: a decade on. Omega (Westport). 2010;61(4):273–89.
44. Richardson VE. Length of caregiving and well-being among older widowers: implications for the dual process model of bereavement. Omega (Westport). 2010;61(4):333–56.
45. Moller J, Bjorkenstam E, Ljung R, Yngwe MA. Widowhood and the risk of psychiatric care, psychotropic medication and all-cause mortality: a cohort study of 658,022 elderly people in Sweden. Aging Ment Health. 2011;15(2):259–66.
46. Frank E, Carpenter LL, Kupfer DJ. Sex differences in recurrent depression: are there any that are significant? Am J Psychiatry. 1988;145(1):41–5.
47. Marcus SM, Young EA, Kerber KB, Kornstein S, Farabaugh AH, Mitchell J, et al. Gender differences in depression: findings from the STAR*D study. J Affect Disord. 2005;87(2–3):141–50.
48. Marcus SM, Kerber KB, Rush AJ, Wisniewski SR, Nierenberg A, Balasubramani GK, et al. Sex differences in depression symptoms in treatment-seeking adults: confirmatory analyses from the Sequenced Treatment Alternatives to Relieve Depression study. Compr Psychiatry. 2008;49(3):238–46.
49. Altemus M, Sarvaiya N, Neill Epperson C. Sex differences in anxiety and depression clinical perspectives. Front Neuroendocrinol. 2014;35(3):320–30.
50. Blanco C, Vesga-Lopez O, Stewart JW, Liu SM, Grant BF, Hasin DS. Epidemiology of major depression with atypical features: results from the national epidemiologic survey on alcohol and related conditions (NESARC). J Clin Psychiatry. 2012;73(2):224–32.
51. Halbreich U, Borenstein J, Pearlstein T, Kahn LS. The prevalence, impairment, impact, and burden of premenstrual dysphoric disorder (PMS/PMDD). Psychoneuroendocrinology. 2003;28(Suppl 3):1–23.
52. Halbreich U, Backstrom T, Eriksson E, O'Brien S, Calil H, Ceskova E, et al. Clinical diagnostic criteria for premenstrual syndrome and guidelines for their quantification for research studies. Gynecol Endocrinol. 2007;23(3):123–30.
53. American Psychiatric A. Diagnostic and statistical manual of mental disorders (DSM-5®). American Psychiatric Association Publishing; 2013.
54. Diflorio A, Jones I. Is sex important? Gender differences in bipolar disorder. Int Rev Psychiatry. 2010;22(5):437–52.

55. Baldassano CF, Marangell LB, Gyulai L, Ghaemi SN, Joffe H, Kim DR, et al. Gender differences in bipolar disorder: retrospective data from the first 500 STEP-BD participants. Bipolar Disord. 2005;7(5):465–70.
56. Robb JC, Young LT, Cooke RG, Joffe RT. Gender differences in patients with bipolar disorder influence outcome in the medical outcomes survey (SF-20) subscale scores. J Affect Disord. 1998;49(3):189–93.
57. Kessing LV. Gender differences in the phenomenology of bipolar disorder. Bipolar Disord. 2004;6(5):421–5.
58. Viguera AC, Baldessarini RJ, Tondo L. Response to lithium maintenance treatment in bipolar disorders: comparison of women and men. Bipolar Disord. 2001;3(5):245–52.
59. Kawa I, Carter JD, Joyce PR, Doughty CJ, Frampton CM, Wells JE, et al. Gender differences in bipolar disorder: age of onset, course, comorbidity, and symptom presentation. Bipolar Disord. 2005;7(2):119–25.
60. Suppes T, Leverich GS, Keck PE, Nolen WA, Denicoff KD, Altshuler LL, et al. The Stanley foundation bipolar treatment outcome network. II. Demographics and illness characteristics of the first 261 patients. J Affect Disord. 2001;67(1–3):45–59.
61. Karanti A, Bobeck C, Osterman M, Kardell M, Tidemalm D, Runeson B, et al. Gender differences in the treatment of patients with bipolar disorder: a study of 7354 patients. J Affect Disord. 2015;174:303–9.
62. Saunders EF, Nazir R, Kamali M, Ryan KA, Evans S, Langenecker S, et al. Gender differences, clinical correlates, and longitudinal outcome of bipolar disorder with comorbid migraine. J Clin Psychiatry. 2014;75(5):512–9.
63. Pitman RK, Rasmusson AM, Koenen KC, Shin LM, Orr SP, Gilbertson MW, et al. Biological studies of post-traumatic stress disorder. Nat Rev Neurosci. 2012;13(11):769–87.
64. Christiansen DM, Hansen M. Accounting for sex differences in PTSD: a multi-variable mediation model. Eur J Psychotraumatol. 2015;6:26068.
65. Kessler RC, Aguilar-Gaxiola S, Alonso J, Benjet C, Bromet EJ, Cardoso G, et al. Trauma and PTSD in the WHO World Mental Health Surveys. Eur J Psychotraumatol. 2017;8(sup5):1353383.
66. Tolin DF, Foa EB. Sex differences in trauma and posttraumatic stress disorder: a quantitative review of 25 years of research. Psychol Bull. 2006;132(6):959–92.
67. Ditlevsen DN, Elklit A. The combined effect of gender and age on post traumatic stress disorder: do men and women show differences in the lifespan distribution of the disorder? Ann Gen Psychiatry. 2010;9:32.
68. Blanco C, Hoertel N, Wall MM, Franco S, Peyre H, Neria Y, et al. Toward understanding sex differences in the prevalence of posttraumatic stress disorder: results from the national epidemiologic survey on alcohol and related conditions. J Clin Psychiatry. 2018;79(2).
69. Felmingham KL, Bryant RA. Gender differences in the maintenance of response to cognitive behavior therapy for posttraumatic stress disorder. J Consult Clin Psychol. 2012;80(2):196–200.
70. Galovski TE, Blain LM, Chappuis C, Fletcher T. Sex differences in recovery from PTSD in male and female interpersonal assault survivors. Behav Res Ther. 2013;51(6):247–55.
71. Stenmark H, Guzey IC, Elbert T, Holen A. Gender and offender status predicting treatment success in refugees and asylum seekers with PTSD. Eur J Psychotraumatol. 2014;5.
72. Wade D, Varker T, Kartal D, Hetrick S, O'Donnell M, Forbes D. Gender difference in outcomes following trauma-focused interventions for posttraumatic stress disorder: systematic review and meta-analysis. Psychol Trauma. 2016;8(3):356–64.
73. McGrath J, Saha S, Welham J, El Saadi O, MacCauley C, Chant D. A systematic review of the incidence of schizophrenia: the distribution of rates and the influence of sex, urbanicity, migrant status and methodology. BMC Med. 2004;2:13.
74. Hafner H. Gender differences in schizophrenia. Psychoneuroendocrinology. 2003;28(Suppl 2):17–54.

75. Sun J, Walker AJ, Dean B, van den Buuse M, Gogos A. Progesterone: the neglected hormone in schizophrenia? A focus on progesterone-dopamine interactions. Psychoneuroendocrinology. 2016;74:126–40.
76. Szymanski S, Lieberman JA, Alvir JM, Mayerhoff D, Loebel A, Geisler S, et al. Gender differences in onset of illness, treatment response, course, and biologic indexes in first-episode schizophrenic patients. Am J Psychiatry. 1995;152(5):698–703.
77. Usall J, Ochoa S, Araya S, Marquez M, Group N. Gender differences and outcome in schizophrenia: a 2-year follow-up study in a large community sample. Eur Psychiatry. 2003;18(6):282–4.
78. Thorup A, Albert N, Bertelsen M, Petersen L, Jeppesen P, Le Quack P, et al. Gender differences in first-episode psychosis at 5-year follow-up-two different courses of disease? Results from the OPUS study at 5-year follow-up. Eur Psychiatry. 2014;29(1):44–51.
79. Vila-Rodriguez F, Ochoa S, Autonell J, Usall J, Haro JM. Complex interaction between symptoms, social factors, and gender in social functioning in a community-dwelling sample of schizophrenia. Psychiatr Q. 2011;82(4):261–74.
80. Cotton SM, Lambert M, Schimmelmann BG, Foley DL, Morley KI, McGorry PD, et al. Gender differences in premorbid, entry, treatment, and outcome characteristics in a treated epidemiological sample of 661 patients with first episode psychosis. Schizophr Res. 2009;114(1–3):17–24.
81. Koster A, Lajer M, Lindhardt A, Rosenbaum B. Gender differences in first episode psychosis. Soc Psychiatry Psychiatr Epidemiol. 2008;43(12):940–6.
82. Ochoa S, Usall J, Cobo J, Labad X, Kulkarni J. Gender differences in schizophrenia and first-episode psychosis: a comprehensive literature review. Schizophr Res Treatment. 2012;2012:916198.
83. Zhang XY, Chen DC, Xiu MH, Yang FD, Haile CN, Kosten TA, et al. Gender differences in never-medicated first-episode schizophrenia and medicated chronic schizophrenia patients. J Clin Psychiatry. 2012;73(7):1025–33.
84. Bekker MH, van Mens-Verhulst J. Anxiety disorders: sex differences in prevalence, degree, and background, but gender-neutral treatment. Gend Med. 2007;4 Suppl B:S178–93.
85. Eaton NR, Keyes KM, Krueger RF, Balsis S, Skodol AE, Markon KE, et al. An invariant dimensional liability model of gender differences in mental disorder prevalence: evidence from a national sample. J Abnorm Psychol. 2012;121(1):282–8.
86. Zetin M, Sklansky GJ, Cramer M. Sex differences in inpatients with major depression. J Clin Psychiatry. 1984;45(6):257–9.
87. Schrijvers DL, Bollen J, Sabbe BG. The gender paradox in suicidal behavior and its impact on the suicidal process. J Affect Disord. 2012;138(1–2):19–26.
88. Callanan VJ, Davis MS. Gender differences in suicide methods. Soc Psychiatry Psychiatr Epidemiol. 2012;47(6):857–69.
89. Hrdina P. Sex-related differences: do they matter? J Psychiatry Neurosci. 2000;25(4):319–20.
90. Simon V. Wanted: women in clinical trials. Science. 2005;308(5728):1517.
91. Uhl K, Parekh A, Kweder S. Females in clinical studies: where are we going? Clin Pharmacol Ther. 2007;81(4):600–2.
92. Merkatz RB, Temple R, Subel S, Feiden K, Kessler DA. Women in clinical trials of new drugs. A change in food and drug administration policy. The working group on women in clinical trials. N Engl J Med. 1993;329(4):292–6.
93. Kokras N, Dalla C. Preclinical sex differences in depression and antidepressant response: implications for clinical research. J Neurosci Res. 2017;95(1–2):731–6.
94. Bagby RM, Ryder AG, Schuller DR, Marshall MB. The Hamilton depression rating scale: has the gold standard become a lead weight? Am J Psychiatry. 2004;161(12):2163–77.
95. Beck AT, Steer RA, Carbin MG. Psychometric properties of the Beck depression inventory: twenty-five years of evaluation. Clin Psychol Rev. 1988;8(1):77–100.

96. Kokras N, Dalla C. Sex differences in animal models of psychiatric disorders. Br J Pharmacol. 2014;171(20):4595–619.
97. Beery AK, Zucker I. Sex bias in neuroscience and biomedical research. Neurosci Biobehav Rev. 2011;35(3):565–72.
98. Ferretti MT, Women's Brain P, Galea LAM. Improving pharmacological treatment in brain and mental health disorders: the need for gender and sex analyses. Front Neuroendocrinol. 2018;50:1–2.
99. FDA. FDA Drug Safety Communication: FDA approves new label changes and dosing for zolpidem products and a recommendation to avoid driving the day after using Ambien CR 2013. https://www.fda.gov/Drugs/DrugSafety/ucm352085.htm.
100. Mendrek A, Stip E. Sexual dimorphism in schizophrenia: is there a need for gender-based protocols? Expert Rev Neurother. 2011;11(7):951–9.
101. Majewska MD, Harrison NL, Schwartz RD, Barker JL, Paul SM. Steroid hormone metabolites are barbiturate-like modulators of the GABA receptor. Science. 1986;232(4753):1004–7.
102. Osborne LM, Gispen F, Sanyal A, Yenokyan G, Meilman S, Payne JL. Lower allopregnanolone during pregnancy predicts postpartum depression: an exploratory study. Psychoneuroendocrinology. 2017;79:116–21.
103. Luisi S, Petraglia F, Benedetto C, Nappi RE, Bernardi F, Fadalti M, et al. Serum allopregnanolone levels in pregnant women: changes during pregnancy, at delivery, and in hypertensive patients. J Clin Endocrinol Metab. 2000;85(7):2429–33.
104. Kanes SJ, Colquhoun H, Doherty J, Raines S, Hoffmann E, Rubinow DR, et al. Open-label, proof-of-concept study of brexanolone in the treatment of severe postpartum depression. Hum Psychopharmacol. 2017;32(2).
105. Meltzer-Brody S, Colquhoun H, Riesenberg R, Epperson CN, Deligiannidis KM, Rubinow DR, et al. Brexanolone injection in post-partum depression: two multicentre, double-blind, randomised, placebo-controlled, phase 3 trials. Lancet. 2018;392(10152):1058–70.
106. Kanes S, Colquhoun H, Gunduz-Bruce H, Raines S, Arnold R, Schacterle A, et al. Brexanolone (SAGE-547 injection) in post-partum depression: a randomised controlled trial. Lancet. 2017;390(10093):480–9.
107. Roux P, Deluol AM, Denis M, Bachoux I, Benlakhal N, Mayaud C, et al. The association of pneumocystosis and pulmonary toxoplasmosis in a female patient with HIV infection. Rev Pneumol Clin. 1989;45(1):42.
108. Kokras N, Dalla C, Papadopoulou-Daifoti Z. Sex differences in pharmacokinetics of antidepressants. Expert Opin Drug Metab Toxicol. 2011;7(2):213–26.
109. Williams AV, Trainor BC. The impact of sex as a biological variable in the search for novel antidepressants. Frontiers in neuroendocrinology. 2018.
110. Harris RZ, Benet LZ, Schwartz JB. Gender effects in pharmacokinetics and pharmacodynamics. Drugs. 1995;50(2):222–39.
111. Gleiter CH, Gundert-Remy U. Gender differences in pharmacokinetics. Eur J Drug Metab Pharmacokinet. 1996;21(2):123–8.
112. Sramek JJ, Murphy MF, Cutler NR. Sex differences in the psychopharmacological treatment of depression. Dialogues Clin Neurosci. 2016;18(4):447–57.
113. Schwartz JB. The influence of sex on pharmacokinetics. Clin Pharmacokinet. 2003;42(2):107–21.
114. Meibohm B, Beierle I, Derendorf H. How important are gender differences in pharmacokinetics? Clin Pharmacokinet. 2002;41(5):329–42.
115. Anderson GD. Sex and racial differences in pharmacological response: where is the evidence? Pharmacogenetics, pharmacokinetics, and pharmacodynamics. J Womens Health (Larchmt). 2005;14(1):19–29.
116. Anderson GD. Gender differences in pharmacological response. Int Rev Neurobiol. 2008;83:1–10.

117. Kashuba AD, Nafziger AN, Kearns GL, Leeder JS, Shirey CS, Gotschall R, et al. Quantification of intraindividual variability and the influence of menstrual cycle phase on CYP2D6 activity as measured by dextromethorphan phenotyping. Pharmacogenetics. 1998;8(5):403–10.
118. Nicolas JM, Espie P, Molimard M. Gender and interindividual variability in pharmacokinetics. Drug Metab Rev. 2009;41(3):408–21.
119. Waxman DJ, Holloway MG. Sex differences in the expression of hepatic drug metabolizing enzymes. Mol Pharmacol. 2009;76(2):215–28.
120. Pitychoutis PM, Zisaki A, Dalla C, Papadopoulou-Daifoti Z. Pharmacogenetic insights into depression and antidepressant response: does sex matter? Curr Pharm Des. 2010;16(20):2214–23.
121. Martin G, Gan TJ. Gender difference: hype or fact? Semin Anesth, Perioper Med Pain. 2000;19:76–82.
122. Sinues B, Fanlo A, Mayayo E, Carcas C, Vicente J, Arenaz I, et al. CYP2A6 activity in a healthy Spanish population: effect of age, sex, smoking, and oral contraceptives. Hum Exp Toxicol. 2008;27(5):367–72.
123. Higashi E, Fukami T, Itoh M, Kyo S, Inoue M, Yokoi T, et al. Human CYP2A6 is induced by estrogen via estrogen receptor. Drug Metab Dispos. 2007;35(10):1935–41.
124. Lamba V, Lamba J, Yasuda K, Strom S, Davila J, Hancock ML, et al. Hepatic CYP2B6 expression: gender and ethnic differences and relationship to CYP2B6 genotype and CAR (constitutive androstane receptor) expression. J Pharmacol Exp Ther. 2003;307(3):906–22.
125. Nicolson TJ, Mellor HR, Roberts RR. Gender differences in drug toxicity. Trends Pharmacol Sci. 2010;31(3):108–14.
126. Quitkin FM, Stewart JW, McGrath PJ, Taylor BP, Tisminetzky MS, Petkova E, et al. Are there differences between women's and men's antidepressant responses? Am J Psychiatry. 2002;159(11):1848–54.
127. Baca E, Garcia-Garcia M, Porras-Chavarino A. Gender differences in treatment response to sertraline versus imipramine in patients with nonmelancholic depressive disorders. Prog Neuropsychopharmacol Biol Psychiatry. 2004;28(1):57–65.
128. Himmelhoch JM, Thase ME, Mallinger AG, Houck P. Tranylcypromine versus imipramine in anergic bipolar depression. Am J Psychiatry. 1991;148(7):910–6.
129. Parker G, Parker K, Austin MP, Mitchell P, Brotchie H. Gender differences in response to differing antidepressant drug classes: two negative studies. Psychol Med. 2003;33(8):1473–7.
130. Wohlfarth T, Storosum JG, Elferink AJ, van Zwieten BJ, Fouwels A, van den Brink W. Response to tricyclic antidepressants: independent of gender? Am J Psychiatry. 2004;161(2):370–2.
131. Thiels C, Linden M, Grieger F, Leonard J. Gender differences in routine treatment of depressed outpatients with the selective serotonin reuptake inhibitor sertraline. Int Clin Psychopharmacol. 2005;20(1):1–7.
132. Pinto-Meza A, Usall J, Serrano-Blanco A, Suarez D, Haro JM. Gender differences in response to antidepressant treatment prescribed in primary care. Does menopause make a difference? J Affect Disord. 2006;93(1–3):53–60.
133. Kornstein SG, Clayton AH, Soares CN, Padmanabhan SK, Guico-Pabia CJ. Analysis by age and sex of efficacy data from placebo-controlled trials of desvenlafaxine in outpatients with major depressive disorder. J Clin Psychopharmacol. 2010;30(3):294–9.
134. Kornstein SG, Pedersen RD, Holland PJ, Nemeroff CB, Rothschild AJ, Thase ME, et al. Influence of sex and menopausal status on response, remission, and recurrence in patients with recurrent major depressive disorder treated with venlafaxine extended release or fluoxetine: analysis of data from the PREVENT study. J Clin Psychiatry. 2014;75(1):62–8.
135. Cuijpers P, Weitz E, Twisk J, Kuehner C, Cristea I, David D, et al. Gender as predictor and moderator of outcome in cognitive behavior therapy and pharmacotherapy for adult depression: an "individual patient data" meta-analysis. Depress Anxiety. 2014;31(11):941–51.

136. Kornstein SG, Schatzberg AF, Thase ME, Yonkers KA, McCullough JP, Keitner GI, et al. Gender differences in treatment response to sertraline versus imipramine in chronic depression. Am J Psychiatry. 2000;157(9):1445–52.
137. Raskin A. Age-sex differences in response to antidepressant drugs. J Nerv Ment Dis. 1974;159(2):120–30.
138. Haykal RF, Akiskal HS. The long-term outcome of dysthymia in private practice: clinical features, temperament, and the art of management. J Clin Psychiatry. 1999;60(8):508–18.
139. Khan A, Brodhead AE, Schwartz KA, Kolts RL, Brown WA. Sex differences in antidepressant response in recent antidepressant clinical trials. J Clin Psychopharmacol. 2005;25(4):318–24.
140. Berlanga C, Flores-Ramos M. Different gender response to serotonergic and noradrenergic antidepressants. A comparative study of the efficacy of citalopram and reboxetine. J Affect Disord. 2006;95(1–3):119–23.
141. Yang SJ, Kim SY, Stewart R, Kim JM, Shin IS, Jung SW, et al. Gender differences in 12-week antidepressant treatment outcomes for a naturalistic secondary care cohort: the CRESCEND study. Psychiatry Res. 2011;189(1):82–90.
142. Naito S, Sato K, Yoshida K, Higuchi H, Takahashi H, Kamata M, et al. Gender differences in the clinical effects of fluvoxamine and milnacipran in Japanese major depressive patients. Psychiatry Clin Neurosci. 2007;61(4):421–7.
143. Vermeiden M, van den Broek WW, Mulder PG, Birkenhager TK. Influence of gender and menopausal status on antidepressant treatment response in depressed inpatients. J Psychopharmacol. 2010;24(4):497–502.
144. Lanctôt KL, Herrmann N, van Reekum R, Eryavec G, Naranjo CA. Gender, aggression and serotonergic function are associated with response to sertraline for behavioral disturbances in Alzheimer's disease. Int J Geriatr Psychiatry. 2002;17(6):531–41.
145. Kando JC, Yonkers KA, Cole JO. Gender as a risk factor for adverse events to medications. Drugs. 1995;50(1):1–6.
146. Drici MD, Clement N. Is gender a risk factor for adverse drug reactions? The example of drug-induced long QT syndrome. Drug Saf. 2001;24(8):575–85.
147. Makkar RR, Fromm BS, Steinman RT, Meissner MD, Lehmann MH. Female gender as a risk factor for torsades de pointes associated with cardiovascular drugs. JAMA. 1993;270(21):2590–7.
148. Bogetto F, Bellino S, Revello RB, Patria L. Discontinuation syndrome in dysthymic patients treated with selective serotonin reuptake inhibitors: a clinical investigation. CNS Drugs. 2002;16(4):273–83.
149. Park YM. Serum prolactin levels in patients with major depressive disorder receiving selective serotonin-reuptake inhibitor monotherapy for 3 months: a prospective study. Psychiatry Investig. 2017;14(3):368–71.
150. Rabinowitz J, Werbeloff N, Caers I, Mandel FS, Stauffer V, Menard F, et al. Determinants of antipsychotic response in schizophrenia: implications for practice and future clinical trials. J Clin Psychiatry. 2014;75(4):e308–16.
151. Melkersson KI, Hulting AL, Rane AJ. Dose requirement and prolactin elevation of antipsychotics in male and female patients with schizophrenia or related psychoses. Br J Clin Pharmacol. 2001;51(4):317–24.
152. Chouinard G, Annable L. Pimozide in the treatment of newly admitted schizophrenic patients. Psychopharmacology. 1982;76(1):13–9.
153. Chouinard G, Annable L, Steinberg S. A controlled clinical trial of fluspirilene, a long acting injectable neuroleptic, in schizophrenic patients with acute exacerbation. J Clin Psychopharmacol. 1986;6(1):21–6.
154. Riordan HJ, Antonini P, Murphy MF. Atypical antipsychotics and metabolic syndrome in patients with schizophrenia: risk factors, monitoring, and healthcare implications. Am Health Drug Benefits. 2011;4(5):292–302.

155. Seeman MV. Secondary effects of antipsychotics: women at greater risk than men. Schizophr Bull. 2009;35(5):937–48.
156. Byerly M, Suppes T, Tran QV, Baker RA. Clinical implications of antipsychotic-induced hyperprolactinemia in patients with schizophrenia spectrum or bipolar spectrum disorders: recent developments and current perspectives. J Clin Psychopharmacol. 2007;27(6):639–61.
157. Canevelli M, Quarata F, Remiddi F, Lucchini F, Lacorte E, Vanacore N, et al. Sex and gender differences in the treatment of Alzheimer's disease: a systematic review of randomized controlled trials. Pharmacol Res. 2017;115:218–23.
158. Krivoy A, Balicer RD, Feldman B, Hoshen M, Zalsman G, Weizman A, et al. The impact of age and gender on adherence to antidepressants: a 4-year population-based cohort study. Psychopharmacology. 2015;232(18):3385–90.
159. Degli Esposti L, Piccinni C, Sangiorgi D, Fagiolini A, Buda S. Patterns of antidepressant use in Italy: therapy duration, adherence and switching. Clin Drug Investig. 2015;35(11):735–42.
160. Serna MC, Real J, Cruz I, Galvan L, Martin E. Monitoring patients on chronic treatment with antidepressants between 2003 and 2011: analysis of factors associated with compliance. BMC Public Health. 2015;15:1184.
161. Mazza D, Harrison C, Taft A, Brijnath B, Britt H, Hobbs M, et al. Current contraceptive management in Australian general practice: an analysis of BEACH data. Med J Aust. 2012;197(2):110–4.
162. Chandra A, Martinez GM, Mosher WD, Abma JC, Jones J. Fertility, family planning, and reproductive health of U.S. women: data from the 2002 national survey of family growth. Vital Health Stat 23. 2005(25):1–160.
163. Warren AM, Gurvich C, Worsley R, Kulkarni J. A systematic review of the impact of oral contraceptives on cognition. Contraception. 2014;90(2):111–6.
164. Toffol E, Heikinheimo O, Koponen P, Luoto R, Partonen T. Hormonal contraception and mental health: results of a population-based study. Hum Reprod. 2011;26(11):3085–93.
165. Keyes KM, Cheslack-Postava K, Westhoff C, Heim CM, Haloossim M, Walsh K, et al. Association of hormonal contraceptive use with reduced levels of depressive symptoms: a national study of sexually active women in the United States. Am J Epidemiol. 2013;178(9):1378–88.
166. Cheslack-Postava K, Keyes KM, Lowe SR, Koenen KC. Oral contraceptive use and psychiatric disorders in a nationally representative sample of women. Arch Womens Ment Health. 2015;18(1):103–11.
167. Duke JM, Sibbritt DW, Young AF. Is there an association between the use of oral contraception and depressive symptoms in young Australian women? Contraception. 2007;75(1):27–31.
168. Joffe H, Cohen LS, Harlow BL. Impact of oral contraceptive pill use on premenstrual mood: predictors of improvement and deterioration. Am J Obstet Gynecol. 2003;189(6):1523–30.
169. Colditz GA. Oral contraceptive use and mortality during 12 years of follow-up: the nurses' health study. Ann Intern Med. 1994;120(10):821–6.
170. Vessey M, Painter R, Yeates D. Mortality in relation to oral contraceptive use and cigarette smoking. Lancet. 2003;362(9379):185–91.
171. Hannaford PC, Iversen L, Macfarlane TV, Elliott AM, Angus V, Lee AJ. Mortality among contraceptive pill users: cohort evidence from Royal College of General Practitioners' Oral Contraception Study. BMJ. 2010;340:c927.
172. Schaffir J, Worly BL, Gur TL. Combined hormonal contraception and its effects on mood: a critical review. Eur J Contracept Reprod Health Care. 2016;21(5):347–55.
173. Mordecai KL, Rubin LH, Maki PM. Effects of menstrual cycle phase and oral contraceptive use on verbal memory. Horm Behav. 2008;54(2):286–93.
174. Wharton W, Hirshman E, Merritt P, Doyle L, Paris S, Gleason C. Oral contraceptives and androgenicity: influences on visuospatial task performance in younger individuals. Exp Clin Psychopharmacol. 2008;16(2):156–64.

175. Martin JA, Hamilton BE, Ventura SJ, Osterman MJ, Kirmeyer S, Mathews TJ, et al. Births: final data for 2009. Natl Vital Stat Rep. 2011;60(1):1–70.
176. Fritz R, Jindal S. Reproductive aging and elective fertility preservation. J Ovarian Res. 2018;11(1):66.
177. Mathews TJ, Hamilton BE. Mean age of mothers is on the rise: United States, 2000–2014. NCHS Data Brief. 2016;232:1–8.
178. Eurostat. Women in the EU gave birth to their first child at almost 29 years of age on average 2015. https://ec.europa.eu/eurostat/documents/2995521/6829228/3–13052015-CP-EN.pdf.
179. Eurostat. Fertility Statistics. 2018.
180. Huang L, Sauve R, Birkett N, Fergusson D, van Walraven C. Maternal age and risk of stillbirth: a systematic review. CMAJ. 2008;178(2):165–72.
181. Kenny LC, Lavender T, McNamee R, O'Neill SM, Mills T, Khashan AS. Advanced maternal age and adverse pregnancy outcome: evidence from a large contemporary cohort. PLoS ONE. 2013;8(2):e56583.
182. Mathews TJ, Hamilton BE. Mean age of mother, 1970–2000. Natl Vital Stat Rep. 2002;51(1):1–13.
183. Hamilton BE, Martin JA, Ventura SJ. Births: preliminary data for 2009. Natl Vital Stat Rep. 2010;59(3):1–19.
184. Group ECW. Social determinants of human reproduction. Hum Reprod. 2001;16(7):1518–26.
185. Sauer MV. Reproduction at an advanced maternal age and maternal health. Fertil Steril. 2015;103(5):1136–43.
186. Damoiseaux VA, Proost JH, Jiawan VC, Melgert BN. Sex differences in the pharmacokinetics of antidepressants: influence of female sex hormones and oral contraceptives. Clin Pharmacokinet. 2014;53(6):509–19.
187. Deligiannidis KM, Byatt N, Freeman MP. Pharmacotherapy for mood disorders in pregnancy: a review of pharmacokinetic changes and clinical recommendations for therapeutic drug monitoring. J Clin Psychopharmacol. 2014;34(2):244–55.
188. Chisolm MS, Payne JL. Management of psychotropic drugs during pregnancy. BMJ. 2016;532:h5918.
189. Eberhard-Gran M, Eskild A, Opjordsmoen S. Treating mood disorders during pregnancy: safety considerations. Drug Saf. 2005;28(8):695–706.
190. Goldberg HL. Psychotropic drugs in pregnancy and lactation. Int J Psychiatry Med. 1994;24(2):129–47.
191. DeVane CL, Stowe ZN, Donovan JL, Newport DJ, Pennell PB, Ritchie JC, et al. Therapeutic drug monitoring of psychoactive drugs during pregnancy in the genomic era: challenges and opportunities. J Psychopharmacol. 2006;20(4 Suppl):54–9.
192. Jain AE, Lacy T. Psychotropic drugs in pregnancy and lactation. J Psychiatr Pract. 2005;11(3):177–91.
193. Pariente G, Leibson T, Carls A, Adams-Webber T, Ito S, Koren G. Pregnancy-associated changes in pharmacokinetics: a systematic review. PLoS Med. 2016;13(11):e1002160.
194. Fokina VM, Xu M, Rytting E, Abdel-Rahman SZ, West H, Oncken C, et al. Pharmacokinetics of bupropion and its pharmacologically active metabolites in pregnancy. Drug Metab Dispos. 2016;44(11):1832–8.
195. Neuman G, Colantonio D, Delaney S, Szynkaruk M, Ito S. Bupropion and escitalopram during lactation. Ann Pharmacother. 2014;48(7):928–31.
196. Kieviet N, Dolman KM, Honig A. The use of psychotropic medication during pregnancy: how about the newborn? Neuropsychiatr Dis Treat. 2013;9:1257–66.
197. Iqbal MM, Gundlapalli SP, Ryan WG, Ryals T, Passman TE. Effects of antimanic mood-stabilizing drugs on fetuses, neonates, and nursing infants. South Med J. 2001;94(3):304–22.
198. Schou M, Amdisen A, Steenstrup OR. Lithium and pregnancy. II. Hazards to women given lithium during pregnancy and delivery. Br Med J. 1973;2(5859):137–8.

199. Iqbal MM, Sohhan T, Mahmud SZ. The effects of lithium, valproic acid, and carbamazepine during pregnancy and lactation. J Toxicol Clin Toxicol. 2001;39(4):381–92.
200. Grover S, Avasthi A. Mood stabilizers in pregnancy and lactation. Indian J Psychiatry. 2015;57(Suppl 2):S308–23.
201. de Haan GJ, Edelbroek P, Segers J, Engelsman M, Lindhout D, Devile-Notschaele M, et al. Gestation-induced changes in lamotrigine pharmacokinetics: a monotherapy study. Neurology. 2004;63(3):571–3.
202. Franco V, Mazzucchelli I, Gatti G, Specchio LM, La Neve A, Papantonio A, et al. Changes in lamotrigine pharmacokinetics during pregnancy and the puerperium. Ther Drug Monit. 2008;30(4):544–7.
203. Kanto JH. Use of benzodiazepines during pregnancy, labour and lactation, with particular reference to pharmacokinetic considerations. Drugs. 1982;23(5):354–80.
204. Gaillot J, Heusse D, Hougton GW, Marc Aurele J, Dreyfus JF. Pharmacokinetics and metabolism of zopiclone. Pharmacology. 1983;27(Suppl 2):76–91.
205. McElhatton PR. The effects of benzodiazepine use during pregnancy and lactation. Reprod Toxicol. 1994;8(6):461–75.
206. Babu GN, Desai G, Chandra PS. Antipsychotics in pregnancy and lactation. Indian J Psychiatry. 2015;57(Suppl 2):S303–7.
207. Klier CM, Mossaheb N, Saria A, Schloegelhofer M, Zernig G. Pharmacokinetics and elimination of quetiapine, venlafaxine, and trazodone during pregnancy and postpartum. J Clin Psychopharmacol. 2007;27(6):720–2.
208. Pinheiro EA, Wisner KL, Clark CT. Quetiapine dose adjustments in pregnant and postpartum women with bipolar disorder. J Clin Psychopharmacol. 2018;38(1):89–91.
209. Llaneza P, Garcia-Portilla MP, Llaneza-Suarez D, Armott B, Perez-Lopez FR. Depressive disorders and the menopause transition. Maturitas. 2012;71(2):120–30.
210. Winterich JA, Umberson D. How women experience menopause: the importance of social context. J Women Aging. 1999;11(4):57–73.
211. Rossi AS. The menopausal transition and aging processes. How healthy are we. 2004:153–201.
212. Freeman EW. Associations of depression with the transition to menopause. Menopause. 2010;17(4):823–7.
213. de Kruif M, Spijker AT, Molendijk ML. Depression during the perimenopause: a meta-analysis. J Affect Disord. 2016;206:174–80.
214. Weber MT, Maki PM, McDermott MP. Cognition and mood in perimenopause: a systematic review and meta-analysis. J Steroid Biochem Mol Biol. 2014;142:90–8.
215. Vivian-Taylor J, Hickey M. Menopause and depression: is there a link? Maturitas. 2014;79(2):142–6.
216. Vousoura E, Spyropoulou AC, Koundi KL, Tzavara C, Verdeli H, Paparrigopoulos T, et al. Vasomotor and depression symptoms may be associated with different sleep disturbance patterns in postmenopausal women. Menopause. 2015;22(10):1053–7.
217. Kokras N, Papadopoulos L, Zervas IM, Spyropoulou A, Stamatelopoulos K, Rizos D, et al. Psychological but not vasomotor symptoms are associated with temperament and character traits. Climacteric. 2014;17(4):500–9.
218. Chatzicharalampous C, Rizos D, Pliatsika P, Leonardou A, Hasiakos D, Zervas I, et al. Reproductive hormones and postpartum mood disturbances in Greek women. Gynecol Endocrinol. 2011;27(8):543–50.
219. Zervas IM, Lambrinoudaki I, Spyropoulou AC, Koundi KL, Voussoura E, Tzavara C, et al. Additive effect of depressed mood and vasomotor symptoms on postmenopausal insomnia. Menopause. 2009;16(4):837–42.
220. Judd FK, Hickey M, Bryant C. Depression and midlife: are we overpathologising the menopause? J Affect Disord. 2012;136(3):199–211.
221. Chetty R, Stepner M, Abraham S, Lin S, Scuderi B, Turner N, et al. The association between income and life expectancy in the United States, 2001–2014. JAMA. 2016;315(16):1750–66.

222. Gogos A, Ney LJ, Seymour N, Van Rheenen TE, Felmingham KL. Sex differences in schizophrenia, bipolar disorder, and post-traumatic stress disorder: are gonadal hormones the link? Br J Pharmacol. 2019.
223. National Collaborating Centre for Mental H, Britain G. Antenatal and postnatal mental health: clinical management and service guidance. National Institute for Health and Clinical Excellence; 2007.
224. Bolea-Alamanac B, Bailey SJ, Lovick TA, Scheele D, Valentino R. Female psychopharmacology matters! Towards a sex-specific psychopharmacology. J Psychopharmacol. 2018;32(2):125–33.
225. Legato MJ, Bilezikian JP. Principles of gender-specific medicine. Gulf Professional Publishing; 2004.

Understanding Mood Disorders in Children

Ho-Jun Lee, Seung-Hyun Kim and Moon-Soo Lee

Abstract

Mood disorders include all types of depression and bipolar disorder, and mood disorders are sometimes called affective disorders. We will discuss newly developing two issues in affective disorders in children and adolescents. Those are the new diagnostic challenges using neuroimaging techniques in affective disorders and the introduction of disruptive mood dysregulation disorder (DMDD). During the 1980s, mental health professionals began to recognize symptoms of mood disorders in children and adolescents, as well as adults. However, children and adolescents do not necessarily have or exhibit the same symptoms as adults. It is more difficult to diagnose mood disorders in children, especially because children are not always able to express how they feel. Child mental health professionals believe that mood disorders in children and adolescents remain one of the most underdiagnosed mental health problems. We are currently trying to introduce the new diagnostic technique—machine learning in children and adolescents with MDD. We will discuss the current progress in the clinical application of machine learning for MDD. After that, we would also discuss a new challenging diagnosis—DMDD. We are still suffering from a lack of evidence when trying to treat the patients with DMDD. In addition, there are

H.-J. Lee · M.-S. Lee (✉)
Department of Child and Adolescent Psychiatry,
Korea University Guro Hospital, 148,
Gurodong-ro, Guro-gu, Seoul 08308, Korea
e-mail: npboard@korea.ac.kr

S.-H. Kim Department of Psychiatry,
Korea University Guro Hospital, Seoul, Korea

© Springer Nature Singapore Pte Ltd. 2019

Y.-K. Kim (ed.), *Frontiers in Psychiatry*, Advances in Experimental Medicine and Biology 1192, https://doi.org/10.1007/978-981-32-9721-0_12

some debates about the diagnostic validity of DMDD. We will explain the current situation of DMDD studies and the future directions in the study of DMDD.

Keywords
Mood disorder · Bipolar disorder · Machine learning · Artificial intelligence · Disruptive mood dysregulation disorder · Treatment strategies

Machine Learning as a Study Technique for Major Depressive Disorder in Children and Adolescents

Significance of Study on Major Depressive Disorder in Child and Adolescent Population

Major depressive disorder (MDD) in childhood and adolescence is known as one of the most disabling psychiatric disorders worldwide, placing large burdens on patients and their families. MDD in adolescence is common. The annual incidence rate under the age of 13 is 1–2%, while over the age of 13 is 3–7%. In general, the ratio of occurrences between male and female appears approximately the same in prepubertal children, but MDD occurs more commonly in women after adolescence, where the gender ratio varied between 1:2 and 1:3 [1]. According to Diagnostic and Statistical Manual of Mental Disorders, fifth edition (DSM-5), a patient with MDD presents depressive mood, loss of pleasure (anhedonia) or appetite (anorexia), sleep deprivation, a slowing down and a reduction of movement, agitation, chronic fatigue, loss of energy, feelings of worthlessness or excessive guilty, diminished concentration, or suicidal thoughts. While the major symptoms of MDD are shared among children and adolescents, there is also a unique difference between these two groups and adult patient groups. Irritable mood as a symptom of MDD is more prevalent among children and adolescents than depressive mood, itself. This suggests that MDD rather than simple changes in weight in children and adolescents should be considered when the failure of expected weight gain occurs [2]. In addition, mood reactivity and fluctuation are more common than consistent depressed mood in child and adolescent patients with MDD. The child and adolescent patients complain of various and ambiguous somatic symptoms and these symptoms are often accompanied by other comorbid psychiatric disorders. Therefore, the complexity of symptoms places an obstacle in accurately diagnosing both child and adolescent patients with MDD. In particular, a major depressive episode as the first episode of bipolar disorder may appear, and during this period, there is a higher risk of developing manic switching after standard antidepressant medication is prescribed. This requires clinicians to be more cautious in diagnosing MDD. Suicidal thoughts or self-mutilation are often reported from children and adolescents with MDD. Furthermore, the relapse and recurrence rates are high which gives rise to a higher continuum in adulthood. However, child and adolescent patients are less responsive to the standard pharmacological treatment compared to adult patients. Pharmacological treatments, such as an antidepressant medication that can be used with consideration of age, are limited. As an example, the antidepressants currently approved by the FDA for adolescents are fluoxetine and escitalopram—this makes the treatment of pediatric depression more difficult. Therefore, the importance of an early, accurate diagnosis and proper treatment is emphasized in child and adolescent MDD [3].

Limitations of the Current Diagnostic System in MDD

Up to date, clinical assessment in MDD is operated based on the diagnostic classification systems DSM-5. The DSM-5 narrows down its scope to the major symptoms, which leads to functional impairments within a certain period of time. However, this symptom-based approach may result in diagnostic discrepancy among clinicians since MDD contains heterogeneous, multifaceted factors that have various biological and psychosocial etiologies. Furthermore, an early diagnosis with clinical signs requires professionals who have expertise in highly specialized mental health services such as psychiatrists and psychologists [4]. Therefore, using more objective and reliable diagnostic tools, such as neuroimaging, genetic, immunologic, and neuropsychological assessments, can improve identifying biomarkers. Identifying a biomarker is critical not only because it helps diagnose the disorder earlier, but it also increases the probability of identifying subtypes of MDD, containing heterogeneous features to predict treatment responses and allows the application of more specific treatments for individuals. Of the above techniques, non-invasive neuroimaging studies allow observation of brain structures and functions in patients with depression, which has received much attention from researchers in the field of psychiatry. Along with this, machine learning techniques have been adopted for accurate diagnosis and predict patient's response to a treatment. Classifiers found through pattern recognition enable psychiatrists to find the best fit model for a patient while considering individual differences. This technique goes beyond identifying the structural and functional differences across a healthy control and patients.

Neuroimaging Studies in MDD

Neuroimaging studies have two components in a macroscopic view: structural and functional scans. Various studies have been performed and reported employing the machine learning technique to predict the diagnosis of MDD. In structural brain imaging studies comparing MDD patients and healthy controls, two types of sequences are used. T1-weighted image is used to more accurately observe gray matter. This is possible because T1-weighted images contrast gray and white matter structures. More specifically, in T1-weighted images, gray matter is expressed with a darker color, while white matter is white. On the contrary, T2-weighted images express gray matter as white and white matter as gray. T1 images are acquired for a shorter period of time than T2 images. The primary focus using T1-weighted image was previously on volumetric comparisons in regions of interest, such as cortical thickness. However, scans of patients with some other psychiatric disorders, such as schizophrenia, autism spectrum disorder, and Alzheimer's disease, have shown that the differences in patients and healthy controls are not only from volumes but also in geometric characteristics (e.g., depth of a sulcus) [5]. Similarly, considering the various and complex etiologies of MDD, neuroanatomical abnormalities in MDD are likely to be accompanied by several changes in volume and shape rather than by simple abnormalities of the structure [6]. T2-weighted image focused on white matters' hyperintensity is the other imaging technique of structural study, which is used especially in late life depression since it demonstrates changes in white matter in ischemic or pre-ischemic periods. Diffusion tensor imaging (DTI) also has been used as a method of investigating structures. DTI visualizes and measures the diffusion of water in brain tissue and is a useful method for detecting white matter abnormalities. Meanwhile, functional studies can be performed using functional MRI (fMRI), and these have already been used to predict diagnosis of depression, which contain various methods such as task-related fMRI. Task-related fMRI is a method of observing the responses in the brain while patients perform a task, while resting state fMRI is a method to observe functional connectivity in brain during the task-free resting state. An analysis on studies using machine learning in major depression in 2018, estimated that there have been approximately 20 structural imaging studies, 20 task-related fMRI studies, 20 resting state fMRI studies, and 10 other DTI studies [7].

Development of Machine Learning Techniques and Application to Neuroimaging Study in Children and Adolescents with MDD

The artificial intelligence (AI) was first introduced by Dr. John McCarthy at Dartmouth University during the Dartmouth Summer Research Project Conference in 1956. With a classical view, AI only refers to a human-like intelligent being whereas AI nowadays achieves more than human-like cognition and perception. Machine learning is a specific approach AI adopts to find out algorithms of an underlying function and best fit statistical model. Machine learning means that a computer can learn how to perform a task through algorithms independently, without humans entering specific parameters for decision-making into the computer. In machine learning, humans are still involved in the process extracting features from given data for the machine to learn. Machine learning includes some methods to create a predictive model from data to accurately predict new data. There are three types of learning methods; supervised learning, semi-supervised learning, and unsupervised learning. In supervised learning, when data sets are given to a computer to label data, the label is also given classifiers. In unsupervised learning, the labels are not given. Laid between these two categories, semi-supervised learning refers to a learning technique when some data is given with interpretation and others are given data only. A goal of machine learning is to define the accurate characteristics of given data sets and define the variety of algorithmic techniques involved. One of the most commonly used techniques is the support vector machine technique in the supervised learning model and it is used for the classification and regression analysis of given data. Support vector machine technique is a type of multivariate pattern analysis technique which learns by computers alone on how to classify complex, high-dimensional data sets and generalizes and categorizes data sets that have not yet been labeled. In general, supportive vector machine includes two stages; the first stage is for the computer system to learn and train well-classified data labeled by a human, while the second stage is to adopt new data sets with this learned system and then reclassify the data sets [4, 8]. Apart from the support vector machine technique, other algorithms such as Gaussian process classifier, linear discriminant analysis, and decision tree areas are variously used as well.

While traditional machine learning techniques focus on the ability to process raw data, deep learning techniques rely on the ability to learn very complex functions by enabling simple expressions to be converted into more abstract expressions by forming nonlinear modules. Deep learning, which is a branch of machine learning, executes more improved performance in recent studies. Deep learning is an unsupervised learning model characterized by utilizing the given data itself as input data. In other words, it's not training the characteristics in the ways a human would, but learning the important features in the data itself through the machine, which is called end-to-end machine learning. Deep learning reduces the errors that result from human interventions in the process. Deep learning minimizes human intervention so that computers can analyze data and draw optimal conclusions without much of human knowledge, but it requires a lot of data, and high-quality data is important for accurate analysis. Thus, it's more important to understand how to obtain high-quality data for AI than simply which algorithm to choose.

The studies on machine learning for diagnosis of mental disorders have been conducted and shared for more than 10 years. Indicators generally used to assess these models are accuracy, sensitivity, and specificity. Indicators are highly accurate as they have consistently been reported from the late 60s to the mid-90s and that the sensitivity and specificity are each reported in the 70s to late 80s [9]. While existing studies demonstrated how excellent machine learning technique is throughout many different fields, there are some concerns in applying these machine learning theories to clinical practice. Models should be widely applicable, but complexities in individual patients hinder the generalization of data sets. Thus, machine learning has not yet been linked to clinical usage. Another main reason why the machine learning method in depression is not easy to apply is the heterogeneity in data acquisition such as MRI acquisition variables, and post processing methods. Consequently, there are challenges in generalizing the results from different experimental parameters.

Commonly, real-life data are high-dimensional (i.e., even each sample includes a large number of characteristics) and paradoxically each sample is limited in its size. Therefore, it's problematic to assume a generalizable learning model. The small sample size is a common problem that most studies of depression reported so far. Predictive models made from a single data set shows good results, but their reliability is significantly reduced in predictive models made from two or more different data sets. Recently, multi-cited studies have been conducted by many research institutes to compensate for these problems. Some problems, however, still existed because the utility of the data is somewhat depreciated due to the absence of criteria on data collection [8]. The study of depression through machine learning will be improved if standards on neuroimaging data collection in MDD are created and quality data is continuously accumulated through multi-center cooperation.

Future Directions in the Application of Machine Learning for Pediatric Mood Disorder

In summary, as a large number of patients experience a major depressive episode from childhood and adolescence, proper treatment targeting these periods is an important factor. The treatment during childhood and adolescence will also help determine the prognosis. Thus, longitudinal research on MDD will provide the basis of the treatment [10]. MDD studies through machine learning techniques will accelerate current research, especially on identifying biomarkers of depression. Along with it, pathogenesis will help understand the underlying mechanisms of MDD. Early, but the elaborate diagnosis will be possible, avoiding a symptom-oriented diagnosis.

Further, the improvement of diagnosis will enable psychiatrists to predict appropriate treatments for depression subtypes and their prognosis. In fact, the current treatment of depression takes one of several already proven treatments. For instance, in order to select an antidepressant, a psychiatrist waits to see patient's reactions to the antidepressant and modifies a prescription (e.g., adding different types of drugs, or nonpharmacological psychotherapy) or tries a new medication until the patient gets better. This whole process requires clinicians' experience and the duration of the treatment is expanded. It is expected that if treatment information on depression is further accumulated through machine learning, there is greater potential to improve prescriptions and the accuracy of treatment methods for each depression subtype. The end result of this improvement would mean that individual suffering would be reduced along with the incidence of treatment-resistant depression.

Disruptive Mood Dysregulation Disorder (DMDD)—A New Diagnostic Challenge

History of Disruptive Mood Dysregulation Disorder (DMDD) Since the Diagnosis of SMD

There is a diagnostic issue in mood dysregulation that the diagnosis is done within the context of pediatric bipolar spectrum disorder. In the past, pediatric onset mania was conceptualized as severe non-episodic irritability. The broader conceptualization of severe non-episodic irritability as a form of mania within pediatric bipolar spectrum disorder has caused the exponential (over 40-times) increase in bipolar disorder diagnosis for less than a decade [11]. Leibenluft et al. suggested severe mood dysregulation (SMD) criteria for patients who have a chronic, non-episodic illness that does not include the hallmark symptoms of mania, but shares the narrower phenotypes of bipolar disorder symptoms such as severe irritability and hyperarousal [12]. The key concerns in SMD studies were whether symptoms in SMD predicted the development of bipolar disorder. Chronic irritability in SMD has been known as a distinct construct when compared with episodic irritability. There were some longitudinal studies tracking the long-term consequences of SMD, and those studies showed that youth with SMD are at high risk for depressive disorders, but not for bipolar disorder. In contrast, bipolar disorder, not otherwise specified, requires identifiable, episodic mood cycles in children that have been associated with increased risk of narrow phenotype bipolar disorder. Accordingly, episodic irritability is more predictive of full-blown bipolar disorder than non-episodic, chronic irritability.

Irritability is a quite common reason for a referral to child and adolescent psychiatry department. However, there is no consensus for assessment tool for irritability as a gold standard measure yet [13]. Irritability can be understood as having a low threshold for experiencing anger in response to frustration. Irritability can fall into mood symptoms such as anxiety, depression, mania, and emotional dysregulation. It can also be classified as behavior symptoms such as hyperactivity, aggression, impulsiveness, suicidal behavior and attention deficit. It is a quite sensitive, but not specific symptom. When there is a lack of skills in managing irritability, irritability can be accompanied by temper outbursts. When evaluating the temper outbursts for the diagnosis of DMDD, the triggers, duration frequency, intensity of temper outbursts, the behavioral characteristics during the outbursts, methods to resolve the outbursts, the outbursts in different settings (home, school, and other), and the mood between outbursts should be extensively investigated.

Similarities and Differences of DMDD from SMD

The diagnosis of DMDD originated from SMD [12]. The definition of DMDD was branched out from the definition of a broad phenotype of juvenile mania. Conceptually, broad phenotype of juvenile mania was based on pediatric bipolar disorder. Regarding DMDD and SMD, however, significant changes were made in DSM-5 [2]. For example, the SMD was revised and renamed as DMDD with following changes: (1) The criterion of hyperarousal as defined by at least three of the following six symptoms (insomnia, agitation, distractibility, racing thoughts or flight of ideas, pressured speech, and intrusiveness) was removed in DMDD. (2) In the definition of abnormal mood, sadness was also removed. (3) The criterion of low intelligence (IQ<80) from the exclusionary criteria was eliminated. (4) The age of onset was lowered from 12 to 10 years old. With the modifications, many with SMD would not have DMDD diagnosis. Such differences are not trivial and could affect the clinical profiles of SMD and DMDD. In the Great Smoky Mountains Study, only 38.9% of youth with SMD also met the criteria for DMDD [14]. These results show that the application of SMD study results on DMDD patients has some essential limitations. As the "hyperarousal" criteria exist for SMD, but not for DMDD, treatments that are known to effectively decrease hyperarousal symptoms may not be more effective in DMDD than in SMD.

In case of comorbidity, despite the lack of direct comparison between the two clinical entities, DMDD most often co-occurs with depressive disorders and oppositional defiant disorder (ODD), but less with attention deficit hyperactivity disorder (ADHD) compared to SMD [15]. However, the rate of comorbidity between DMDD and externalizing disorders is high, especially between DMDD and ADHD.

Criticism for DMDD Diagnosis

The majority of youth with DMDD already meet criteria for another DSM disorder. It is quite uncommon to find patients who fall only in DMDD criteria. There is also a concern about the overlap between DMDD and ODD. DSM-5 specifically indicates that patients who meet the criteria for both DMDD and ODD should be given the diagnosis of DMDD, and not ODD. In the Longitudinal Assessment of Manic Symptoms study (known as LAMS study), more than half (58%) of the participants with ODD met DMDD criteria, while almost all youth with DMDD met the criteria for ODD, making it hard to differentiate DMDD from ODD. In addition, the patients with ODD who met DMDD criteria did not differ from the patients with ODD, but without DMDD, in terms of symptoms severity, comorbidity, or functional impairment [15].

For these reasons, there are some debates about the diagnostic validity of DMDD. Currently, some researchers think that DMDD may be a modifier of ODD, when considering that nearly all youth with DMDD meet the criteria for ODD and that irritability in the DMDD appears to be a distinct domain from the other symptoms of ODD [13]. These thoughts were reflected in the International Classification of Disease, 11th revision (ICD-11) and DMDD is now coded under ODD with the diagnostic code and name: 6C90.0Z, ODD with chronic irritability–anger, unspecified [16].

Treatment of DMDD

To date, there have been very limited DMDD treatments due to its novelty. The treatment of DMDD is composed of two main axes: pharmacological intervention and psychotherapeutic intervention.

Pharmacological Treatment

As severe, non-episodic irritability is conceptualized as a phenotype of unipolar depression, the risk for mania would be low. Non-episodic irritability also shows high comorbidity with other pediatric psychiatric disorders such as ADHD, anxiety disorder, and ODD. Antidepressants such as selective serotonin reuptake inhibitors can be a reasonable treatment option. When considering the high rate of comorbidity, we can also consider using the CNS stimulants and antimanic agents such as lithium and valproate. No pharmacological studies of youth with DMDD have been conclusive with results yet. Although symptoms of SMD appear quite similar to DMDD, it is not identical with DMDD and whether treatment options for SMD would work for DMDD management is unknown.

Various medications such as antidepressants, mood stabilizers, psychostimulants, antipsychotics, and alpha-2-agonists can be used for treatment for temper tantrum and chronic inter-outbursts irritability. However, clinical evidence of DMDD medication is insufficient. Therapists also need to pay attention to the possibility of polypharmacy in patients. There are high rates of comorbidities and medications and treatment should be given to target comorbid disorders. In addition, the use of medication selections for targeting each symptom without considering the overall diagnosis could contribute to high rates of polypharmacy. Accordingly, psychiatrists should avoid using polypharmacy in this young age group.

There are some completed and ongoing pharmacological trials in SMD and DMDD youths. Dickstein et al. led a placebo-controlled randomized trial in youths with SMD to test the benefit of lithium over placebo, but failed to show significant between-group (patients and control) differences [17]. Krieger et al. reported the results of risperidone on 21 youths with SMD in an open-label trial. They reported clinically significant improvements shown as the reduction of the Aberrant Behavior Checklist-irritability subscale score [18]. There are also other currently ongoing studies specifically targeted for DMDD patients rather than SMD [19]. The results from those studies will soon be available.

Psychotherapeutic Treatment

As irritability has nonspecific symptoms and occurs in many pediatric psychiatric disorders, it is necessary to evaluate contextual meanings of irritability rather than seeking the existence of non-episodic irritability for the diagnosis of DMDD. Evaluation of irritability and aggression in different settings, such as home, school, and extracurricular activities, and precipitating factors for the irritability and aggression are needed. The multidimensional, integrative approach including liaison with the school and involvement of all available community resources would be in demand for optimal treatment results. In addition, evidence-based parent training interventions and other psychotherapeutic interventions should be always considered in the treatment of DMDD. The parenting training for DMDD patients comes from evidence-based treatments for pediatric behavioral problems with managing antecedents for temper outbursts and training parents for emotion regulation coaching skills. Children patient group should learn how to identify effect, control negative influences, and solve the problems, sequentially.

Future Directions in the Study of DMDD

Non-episodic irritability is now conceptualized as a feature of unipolar depression rather than bipolar disorder. However, irritability itself is a rather nonspecific symptom and can occur in a wide range of psychopathologies. It is hard to assess irritability adequately and classify the more elaborate characteristics of irritability in young patients for clinical purposes. Therefore, a systematic assessment tool is required for diagnosing irritability in children and adolescents using their own words.

Currently, the results of pharmacological intervention of DMDD come from SMD, not DMDD. Recruiting more DMDD patient subjects is necessary. In addition, treatments have been primarily focused on the control of irritability or aggression. Accordingly, the only randomized, placebo-controlled trial of SMD is about the use of lithium over placebo. In addition, CNS stimulant has been administered as a treatment of DMDD. When psychiatrists focus on irritability or persistent aggression, they tend to use the combination of CNS stimulant and antipsychotics. However, psychiatrists should consider the fact that DMDD belongs to unipolar depressive disorder group in DSM-5 diagnosis criteria. Then, another related question raised is whether using an antidepressant would be safe. Evaluation of the clinical efficacy and long-term results of antidepressants in DMDD should be followed.

References

1. Alsaad AJ, Al Nasser Y. Depression in children. In: StatPearls, editor. Treasure Island (FL): StatPearls Publishing. StatPearls Publishing LLC; 2018
2. Diagnostic and statistical manual of mental disorders. 5th ed. Arlingtion, VA: American Psychiatric Association; 2013.
3. Foland-Ross LC, Sacchet MD, Prasad G, Gilbert B, Thompson PM, Gotlib IH. Cortical thickness predicts the first onset of major depression in adolescence. Int J Dev Neurosci. 2015;46:125–31.
4. Qiu L, Huang X, Zhang J, Wang Y, Kuang W, Li J, et al. Characterization of major depressive disorder using a multiparametric classification approach based on high resolution structural images. J Psychiatry Neurosci. 2014;39(2):78–86.
5. Klöppel S, Abdulkadir A, Jack CR, Koutsouleris N, Mourão-Miranda J, Vemuri P. Diagnostic neuroimaging across diseases. NeuroImage. 2012;61(2):457–63.
6. Patel MJ, Khalaf A, Aizenstein HJ. Studying depression using imaging and machine learning methods. Neuroimage Clin. 2016;10:115–23.
7. Gao S, Calhoun VD, Sui J. Machine learning in major depression: from classification to treatment outcome prediction. CNS Neurosci Ther. 2018;24(11):1037–52.

8. Mwangi B, Ebmeier KP, Matthews K, Steele JD. Multi-centre diagnostic classification of individual structural neuroimaging scans from patients with major depressive disorder. Brain. 2012;135(Pt 5):1508–21.
9. Wu MJ, Wu HE, Mwangi B, Sanches M, Selvaraj S, Zunta-Soares GB, et al. Prediction of pediatric unipolar depression using multiple neuromorphometric measurements: a pattern classification approach. J Psychiatr Res. 2015;62:84–91.
10. Lewinsohn PM, Rohde P, Seeley JR. Treatment of adolescent depression: frequency of services and impact on functioning in young adulthood. Depress Anxiety. 1998;7(1):47–52.
11. Moreno C, Laje G, Blanco C, Jiang H, Schmidt AB, Olfson M. National trends in the outpatient diagnosis and treatment of bipolar disorder in youth. Arch Gen Psychiatry. 2007;64(9):1032–9.
12. Leibenluft E, Charney DS, Towbin KE, Bhangoo RK, Pine DS. Defining clinical phenotypes of juvenile mania. Am J Psychiatry. 2003;160(3):430–7.
13. Baweja R, Mayes SD, Hameed U, Waxmonsky JG. Disruptive mood dysregulation disorder: current insights. Neuropsychiatr Dis Treat. 2016;12:2115–24.
14. Copeland WE, Angold A, Costello EJ, Egger H. Prevalence, comorbidity, and correlates of DSM-5 proposed disruptive mood dysregulation disorder. Am J Psychiatry. 2013;170(2):173–9.
15. Axelson D, Findling RL, Fristad MA, Kowatch RA, Youngstrom EA, Horwitz SM, et al. Examining the proposed disruptive mood dysregulation disorder diagnosis in children in the longitudinal assessment of manic symptoms study. J Clin Psychiatry. 2012;73(10):1342–50.
16. International classification of diseases 11th revision for mortality and morbidity statistics. World Health Organization; 2018. https://icd.who.int/browse11/l-m/en#/http%3a%2f%2fid.who.int%2ficd%2fentity%2f818792113.
17. Dickstein DP, Towbin KE, Van Der Veen JW, Rich BA, Brotman MA, Knopf L, et al. Randomized double-blind placebo-controlled trial of lithium in youths with severe mood dysregulation. J Child Adolesc Psychopharmacol. 2009;19(1):61–73.
18. Krieger FV, Pheula GF, Coelho R, Zeni T, Tramontina S, Zeni CP, et al. An open-label trial of risperidone in children and adolescents with severe mood dysregulation. J Child Adolesc Psychopharmacol. 2011;21(3):237–43.
19. Benarous X, Consoli A, Guile JM, Garny de La Riviere S, Cohen D, Olliac B. Evidence-based treatments for youths with severely dysregulated mood: a qualitative systematic review of trials for SMD and DMDD. Eur Child Adolesc Psychiatry. 2017;26(1):5–23.

Smart Healthcare Systems and Precision Medicine

Soo-Hyun Paik and Dai-Jin Kim

Abstract

This article gives an overview of the concept and brain mechanisms of Internet game and smartphone addiction and the applicability of precision medicine and smart healthcare system. Internet game and smartphone addiction are categorized as behavioral addictions, which share similar phenomenology and neurobiological underpinnings with substance addictions. Neuroimaging studies revealed the alteration in the functional activity and structure of individuals with Internet game and smartphone addiction, which also can be potent biomarkers. Precision medicine is defined as treatments targeted to the individual patients on the basis of genetic, biomarker, phenotypic or psychosocial characteristics. Recent advances in high-throughput technology and bioinformatics have enabled us to integrate these big data with behavioral data collected from smartphones or other wearable devices. Data collected via smart devices can be transferred to medical institute and integrated in order to diagnose current status precisely and to provide optimal intervention. The feedbacks of intervention are sent back to the medical provider via self-reports or objective measures

S.-H. Paik
Addiction Center, Keyo Hospital, Gyonggi-do, South Korea
e-mail: suehime@gmail.com

D.-J. Kim (✉)
Department of Psychiatry, Seoul St. Mary's Hospital,
The Catholic University of Korea, Seoul, South Korea
e-mail: kdj922@catholic.ac.kr

© Springer Nature Singapore Pte Ltd. 2019

Y.-K. Kim (ed.), *Frontiers in Psychiatry*, Advances in Experimental Medicine and Biology 1192, https://doi.org/10.1007/978-981-32-9721-0_13

to evaluate the appropriateness of the intervention. In conclusion, Internet game and smartphone addiction can be diagnosed precisely using high-throughput technology and optimally managed via smart healthcare system.

Keywords
Internet gaming disorder · Smartphone addiction · Neuroimaging · Precision medicine · Smart healthcare system

The Evolution and Conceptualization of Internet Game and Smartphone Addiction

Playing game has been a popular and pleasurable leisure activity which provides enjoyment and relaxation to everyday life. The penetration rate of gamers among the general population rose from 58% in 2013 to 66% in 2018 in the United States and from 57.2% in 2014 to 67.9% in 2015 in South Korea [1, 2]. The dark side of gaming has also been highlighted in psychological and psychiatric research area in the 1980s, approximately 10 years after the release of the first commercial video game in the early 1970s [3]. Soper and Miller reported "video game addiction", which is characterized as compulsive involvement, a lack of interest in other activities, association and friendship circles mainly with other video game addicts, and physical and mental symptoms when attempting to stop the behavior and was similar to other behavioral addictions [4]. After the 2000s, articles on online video games have been published extensively, especially in association with the introduction of the massive multiplayer online role-playing games (MMORPGs) such as World of Warcraft and Starcraft. MMORPGs are played by hundreds or thousands of players around the world simultaneously. Players interact and gather under a guild or other form of communities via the Internet. They participate in a variety of quests which increase in complexity, reward, and time involvement that typically operate on a random-ratio reinforcement schedule [5]. These features of MMORPGs satisfy the socialization need and provide a sense of achievement and competitiveness, which can be a strong drive to play online games. In addition, the virtual world is continued to exist even if players are offline, which makes it hard to stay offline. Some gamers feel it very hard to be offline and are very irritable and aggressive when game is interrupted by any reasons, which seems like "withdrawal". According to O'Brien and Holden's definition of "addiction", a phenomenon that manifests tolerance, withdrawal symptoms, and dependence, accompanied by social problems [6, 7], these features can be considered as cardinal features of behavioral addiction.

Nomenclature of this phenomenon was very heterogeneous for decades; video game addiction [4], Internet addiction [8], compulsive Internet use [9], problematic online gaming [10], mentioned only a few. Since most Internet overuse is due to excessive gaming, online game addiction is often considered as Internet addiction. In addition, the conceptualization of this phenomenon was heterogeneous among researchers. Some adopted definition from pathological gambling of the Diagnostic and Statistical Manual of Mental Disorders, 4th edition (DSM-IV) [6, 11], while others adopted from substance dependence criteria or combined these two criteria [12]. Accordingly, official bodies such as American Psychiatric Association and other scholars have suggested the need for the unification and consensus for the assessment of Internet game addiction, and finally, this conceptualization was listed on the Diagnostic and Statistical Manual of Mental Disorder, 5th edition (DSM-5) as “Internet gaming disorder (IGD)” on section III as condition for further study [13]. In 2018, gaming disorder was added to the addictive disorder section in the final draft of the 11th revision of the International Classification of Disease (ICD-11) by the World Health Organization [14]. Description of DSM-5 IGD and ICD-11 gaming disorder is given in Table 1. The inclusion of gaming disorder into ICD-11 can be a double-edged sword; if used wisely, it increases the capacity and treatment of disordered gamers, but there are also concerns that casual gamers can be stigmatized.

Recent spread of smartphone has explosively increased the number of smartphone gamers. In 2014, only 19.1% of gamers played only by personal computers (PCs) while 56.9% played only by their smartphones, and remaining 24% played by both devices [15]. In addition, the evolution of smartphone games has changed classical demographics and characteristics of online gamers. According to a Korean national survey conducted in 2015, 86.5% of the forties, 85.4% of the fifties, and 88.8% of female smartphone users reported they had experiences of

Table 1 Description of Internet gaming disorder and gaming disorder

Proposed criteria of Internet gaming disorder (DSM-5)	Description of gaming disorder (ICD-11)
Persistent and recurrent use of the Internet to engage in games, leading to clinically significant impairment or distress as indicated by 5 or more of the following in a 12 month period: 1. Preoccupation with Internet games 2. Withdrawal symptoms when Internet gaming is taken away 3. Tolerance: the need to spend increasing amounts of time engaged in Internet games 4. Unsuccessful attempts to control participation in Internet games 5. Loss of interests in previous hobbies and entertainment as a result of, and with the exception of, Internet games 6. Continued excessive use of Internet games despite knowledge of psychosocial problems 7. Has deceiving family members, therapists, or others regarding the amount of Internet gaming 8. Use of Internet games to escape or relieve negative moods 9. Has jeopardized or lost a significant relationship, job, or education or career opportunity because of participation in Internet games	Gaming disorder is characterized by a pattern of persistent or recurrent gaming behavior (“digital gaming” or “video gaming”), which may be online (i.e., over the Internet) or offline, manifested by: 1. Impaired control over gaming (e.g., onset, frequency, intensity, duration, termination, context) 2. Increasing priority given to gaming to the extent that gaming takes precedence over other life interests and daily activities 3. Continuation or escalation of gaming despite the occurrence of negative consequences The behavior pattern is of sufficient severity to result in significant impairment in personal, family, social, educational, occupational, or other important areas of functioning. The pattern of gaming behavior may be continuous or episodic and recurrent. The gaming behavior and other features are normally evident over a period of at least 12 months in order for a diagnosis to be assigned, although the required duration may be shortened if all diagnostic requirements are met and symptoms are severe

playing online games [15]. Thus, it is necessary to contemplate the role of smartphone on gaming disorder and the possibility of smartphone addiction as a separate diseases entity.

Despite the social networking advantages and work productivity enhanced by smartphone use, a growing number of literatures have reported the dark side of smartphone as well. Problematic smartphone use is associated with health problems such as dry eye and musculoskeletal deformity [16], hazardous situations such as texting while driving [17], accidents and injuries [18], decreased academic performance [19], and psychological problems such as anxiety, depression, and poor sleep patterns [20, 21]. Accordingly, a conceptualization that encompasses excessive and uncontrolled use of smartphone, loss of control over smartphone use, and negative consequences associated with these maladaptive behavioral patterns is considered among researchers and is nomenclated as "smartphone addiction" or "problematic mobile phone use (PMPU)" [22, 23]. In 2008, Billieux, van der Linden, and Rochart have devised Problematic Mobile Use Questionnaire (PMPUQ), targeting (1) dangerous/prohibited use, (2) financial problems, and (3) dependence-related symptoms [23]. In 2013, Kwon and colleagues firstly proposed Smartphone Addiction Scale (SAS), which is a 33-item, 6-point Likert scale based on Korean Internet addiction scale (K-scale) motivated by Young Internet Addiction Test (YIAT) [22]. SAS is composed of six factors, daily-life disturbance, positive anticipation, withdrawal, cyberspace-oriented relationship, overuse, and tolerance, PMPUQ, SAS and its short version (SAS-SV) are widely used to assess smartphone addiction around the world.

The Korean government also conducted annual national survey to assess at- and high-risk of "smartphone addiction". According to the national study, smartphone addiction is defined as (1) increased salience toward smartphone as a consequence of excessive smartphone use and (2) loss of control over smartphone use, and is composed of three domains: (1) self-control failure; (2) salience; and (3) serious negative consequences in physical, psychological and social aspects [24]. The prevalence of at- and high-risk of smartphone addiction in South Korea is dramatically rose from 8.4% in 2011 to 18.6% in 2017 (Fig. 1). The proportion of highly dependent smartphone users in European countries ranged from 1% in Poland to 3.9% in Belgium when assessed by PMPUQ [25].

The contents consumed by smartphones may influence smartphone addiction. According to Korean national survey, messengers and games were most frequently used smartphone contents, followed by searching news and listening music [24]. Interestingly, individuals with smartphone addiction and normal smartphone users

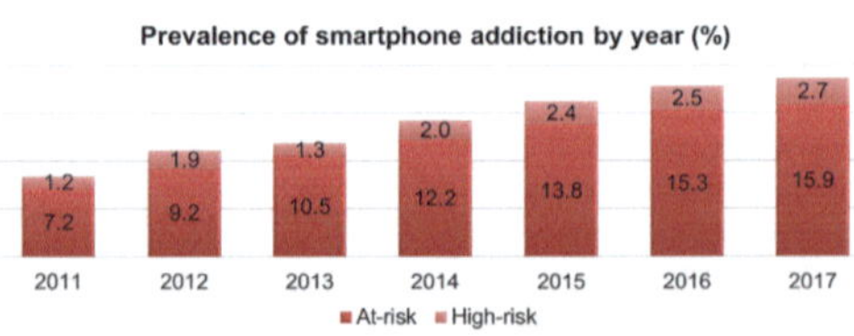

Fig. 1 Prevalence of smartphone addiction in South Korea by year (%), adopted from survey on smartphone overdependence 2017 [24]

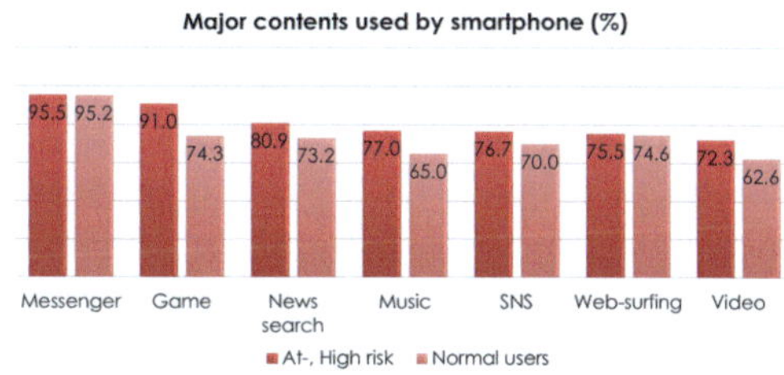

Fig. 2 Major contents used by smartphone (%), adopted from survey on smartphone overdependence 2017 [24]

were different in their game and music contents usage (Fig. 2). Fernandez and colleagues also reported that problematic mobile phone use was associated with engaging in social networking, playing video games, shopping and viewing TV shows, downloading contents, and messaging [25].

There still exist debates on admitting smartphone addiction as a separate disease entity and lack in scientific evidence related to diagnostic assessment and clinical course. Panova and Carbonell reported that they could not found sufficient support from the addictive perspective to confirm the disease entity of smartphone addiction at this time, and proposed to label as problematic or maladaptive smartphone use [26]. However, about half of the smartphone users felt it helpful to receive counseling service and preventive education on smartphone addiction and 40.3 and 37.2% of the smartphone users responded they would like to consult with counseling services and receive preventive education, respectively, if they have problems in the future [24]. Thus, more empirical studies with unyielding evidence are necessary to figure out neurobiological underpinnings and draw a consensus on the conceptualization of smartphone addiction.

Brain Mechanisms of Internet Gaming Disorder and Smartphone Addiction

Over the last few years, enormous articles have been published to elucidate the brain mechanisms of IGD. Neurocognitive studies have observed that individuals with IGD showed impulsivity, tendency of making risky decision and enhanced reactivity to gaming cues [27–29]. Recent advances in brain imaging techniques facilitated the elucidation of morbid brain. Neuroimaging research is a worldwide used noninvasive technique to study neurobiological correlates of psychiatric conditions and has a potential to fill the gap between molecular basis of psychiatric disorders and their clinical manifestations [30–32].

Task based functional magnetic resonance imaging (fMRI) studies have investigated neurobiological correlates associated with neurocognitive deficits observed in IGD. In cognitive control tasks (i.e., a Go/No-Go task), which measures impulsivity and response inhibition, IGD showed decreased activity of the supplementary motor area during no-go trials with no differences in behavioral performance [33]. Another study revealed that IGD showed hyperactivation of superior and middle frontal gyrus during no-go trials [34]. In a guessing task paradigm, which required participants to bet on different outcomes and assess brain response to win or loss condition to evaluate reward sensitivity, individuals with IGD showed hyperactivity on left orbitofrontal cortex (OFC) in win condition with hypoactivation on anterior cingulate cortex (ACC) in loss condition [35]. Another guessing task-related fMRI study showed hyperactivation in left superior frontal gyrus in win condition and hypoactivation in posterior cingulate cortex (PCC) in IGD compared to healthy control [36]. It seems that individuals with IGD are much more sensitive to rewards while less sensitive to loss. In cue-reactivity paradigms, which measures brain response to stimuli that are designed to induce carving for gaming, individuals with IGD showed hyperactivation of OFC, dorsolateral prefrontal cortex (DLPFC), medial frontal cortex (MFC), precuneus, parahippocampus, ACC, PCC, nucleus accumbens (NAc), and caudate nucleus while they saw gaming pictures [37, 38]. In addition, cue-induced activity of left DLPFC was decreased after 6 weeks of bupropion treatment in IGD [39]. Differences between IGD and healthy control were observed when processing swear words which express strong emotion and induce strong physiological responses; adolescents with IGD showed reduced activation in the right OFC which is related to cognitive control, and in the dorsal ACC, which was related to social rejection during swear word condition [40]. These findings suggested that individuals with IGD differed from healthy comparisons in the functioning of several brain regions involved in cognitive control, reward sensitivity and cur reactivity even when they were behaviorally intact.

Resting-state fMRI (rsfMRI) is a popular approach in clinical studies since the images can be obtained during 5–6 min of passive fMRI scan, measuring default status [30]. A systematic review of rsfMRI studies on IGD has summarized that the most relevant abnormalities were localized in the superior temporal gyrus (STG), which is responsible for the processing of visuo-auditory information, medial frontal regions (ACC, supplementary motor area), which is associated with self-control and impulsivity, limbic and parietal regions [41–43]. Another meta-analysis reported that individuals with IGD showed a significant activation in the bilateral medial frontal gyrus, left cingulate gyrus, left medial temporal gyrus, and fusiform gyrus, implicating the role of dysfunctional prefrontal lobe in the neurobiological mechanisms of IGD [44]. A rsfMRI study based on topological network revealed that brains of IGD showed higher global efficiency and lower local efficiency compared to healthy controls, suggesting shifting toward the random topological architecture, as exhibited in other pathological states [45]. These findings suggested that individuals with IGD shares some neurobiological features with substance addiction such as functional deficit in frontal and limbic regions, as well as possesses its unique features such as abnormalities in STG.

Structural MRI studies have demonstrated that IGD is associated with structural alteration in the brain. Decrease in gray matter volume of the bilateral ACC, precuneus, SMA, superior parietal cortex, left DLPFC, left insula and bilateral cerebellum, decrease in the gray matter density in the bilateral inferior frontal gyrus, cingulate gyrus, insula, right precuneus and hippocampus and in the white matter density in the inferior frontal gyrus, insula, amygdala, and anterior cingulate were observed among individuals with IGD [46, 47]. Adolescents with IGD showed increased cortical thickness in the left precentral cortex, precuneus, middle frontal cortex, inferior temporal, and middle temporal cortices while decreased in the left lateral OFC, insula, lingual gyrus, right postcentral gyrus, entorhinal cortex, and inferior parietal cortex [48]. Structural alteration in the prefrontal cortex, specifically left DLPFC, mediated the relationship between prolonged gaming and depressed mood [49]. These brain imaging parameters can be served as biomarkers to distinguish individuals with IGD from without. Park and colleagues searched for a link between symptom-based categorization and computation-based classification by using neuroanatomical biomarkers in the diagnosis of IGD and demonstrated the discrimination of disordered and non-gamers with accuracy exceeding 98% [50]. These findings are very fruitful in that structural brain alterations may serve as potential neurobiomarkers to diagnose IGD.

However, the majority of neuroimaging study sample involved male adolescents and only a small number of participants were included in each study. Thus, further studies with larger sample and prospective design and with application of high-throughput computational methods are necessary to elucidate more stable biomarkers of IGD.

Currently, the brain mechanisms of smartphone addiction have yet been studied extensively. Researches have been focused on whether smartphone use, especially media multitasking via smartphone, would interrupt important cognitive function such as attention and memory. Lee and colleagues reported that individuals with higher smartphone addiction scale scored significantly lower on self-regulatory learning and learning flow scales [51]. In an attention-demanding task, receiving mobile phone notification alone significantly disrupted performance even when participants did not directly interact with a mobile device during the task, similar in magnitude to active phone usage [52]. In a working memory task paradigm, frequent media multitaskers exhibited lower working memory performance and long-term memory functioning [53]. However, empirical researches regarding the impact of smartphones on cognition are still very limited though it seems that smartphones have the potential to affect a wide range of cognitive domains.

Some neuroimaging studies regarding smartphone addiction have been published recently. Frequent media multitaskers exhibited relatively increased activity in right frontal area, suggesting that increased daily multitasking leads to greater difficulty in recruiting cognitive control resources [54]. Hu and colleagues demonstrated that individuals with smartphone addiction assessed by the Mobile Phone Addiction Tendency Scale (MPATS) had significantly lower white matter integrity in the superior longitudinal fasciculus, superior corona radiate, internal and external capsule, sagittal stratum, fornix/stria terminalis, and midbrain structures [55]. A rsfMRI study demonstrated that adolescents with smartphone addiction assessed by the Smartphone Addiction Proneness Scale (SAPS) had significantly lower functional connectivity between the right OFC and NAc, and between the left OFC and midcingulate cortex (MCC) compared to healthy control, and the OFC-NAc connectivity was negatively correlated with withdrawal symptoms [56]. These findings implicated the alteration in the frontostriatal circuit in the smartphone addiction. A task-based fMRI study reported that excessive smartphone users showed neural deactivation in the DLPFC and dorsal ACC during the facial emotional processing task, which was aimed to assess social interaction [57]. This finding implicated that excessive smartphone users seemed to fail on cognitive control during emotional processing and provided an important neural basis for smartphone addiction. However, neuroimaging studies as well as neurocognitive studies on smartphone addiction are very scarce at this time and further studies are necessary to overcome inevitable debates on the conceptualization of smartphone addiction.

Precision Medicine in Addiction Psychiatry

Precision medicine is defined as treatments targeted to the needs of individual patients on the basis of genetic, biomarker, phenotypic, or psychosocial characteristics that distinguish a patient from other patients with similar clinical presentations [58]. This concept is familiar since clinicians have been tackled on personalizing or individualizing medical intervention which was attuned to patients' unique characteristics. Thanks to the brilliant advances in genetics, neuroimaging, bioinformatics, and other related technologies, the scope of precision medicine includes predicting susceptibility, labeling unique phenotypic profile, and offering personalized treatment options with maximal therapeutic effects and minimal adverse events. The scope of the precision medicine has extended from genetic disorders and cancers to psychiatric illnesses [59].

Despite advances in addiction psychiatry, there still reside major issues to be tackled. First, though addictive disorders are considered as a chronic mental disorder with relapsing nature, some individuals recover from them spontaneously without intervention. It seems that some important predicting factors of addictive disorders are hidden. Second, addictive disorders are frequently comorbid with other psychiatric and physical illness which have much influence on the course of addictive disorders. Third, though risk factors and vulnerabilities have been studied extensively, they are often recognized at very late stage of disease or retrospectively. Fourth is the lack of effective synchronization of the range of interventions to a patient's individual characteristics. Fifth, only limited effective treatment options are available. Is precision medicine an effective solution for these issues? The answer seems to be "Yes". Precision medicine already has provided some hints to these issues. In alcohol use disorder (AUD), the selection of anti-craving agents can be aided by the Mu-opioid receptor gene (OPRM1) polymorphism [60]. The combination of susceptible genetic polymorphisms of ADH1B*47Arg and ALDH2*Glu can predict the risk of AUD 91 times more than the combination of protective polymorphisms of ADH1B*47His and ALDH2*Lys [61]. Novel biomarkers that would differentiate patients with similar clinical presentations are vigorously identified via genomics, proteomics, and other multi-omics technologies [62–65]. Neuroimaging studies seem to have the potential to complement standard clinical measures and to optimize relapse prediction. Moeller and Paulus suggested that individuals who relapsed had enhanced activation to drug-related cues and rewards, weakened functional connectivity of corticolimbic and corticostriatal regions, and reduced gray and white matter volume and connectivity in prefrontal regions by doing extensive literature review [66]. The cue-reactive patterns of ventral striatum had a predictive value for the treatment response to naltrexone in AUD [67]. Electrical health records (EHR) can be used as a potential source for precision medicine since they have superiority in extracting phenomenological information. Three-item alcohol use disorder identification test consumption measure (AUDIT-C) extracted from longitudinal EHR data had a good validity in detecting harmful alcohol use [68]. The application of machine learning techniques can aid to select treatment option. Connor and colleagues demonstrated the predictive factors of abstinence among those who underwent 12-week cognitive behavioral therapy program by using Bayesian and decision-tree classifiers, two major machine learning techniques, 77% and 73%, respectively [69]. This poses an important clinical implication that matching patients' characteristics with optimal treatment option would increase the treatment efficacy.

It seems that precision medicine is useful when tackling to IGD and smartphone addiction. Not like to substance addiction, the brain and physicals of individuals with Internet game and smartphone addiction are not damaged by direct effect of substance. However, studies on behavioral addiction have demonstrated that addictive behaviors are associated with structural and functional alteration of brain. Behavioral data of Internet game and smartphone addiction, such as duration and frequency of gaming or smartphone use, the type of game or application, and other valuable information, is very important as the duration and amount of alcohol use is important for AUD. Under the permission of the users, behavioral data can be recorded from devices and transferred to the database, and the data can be a powerful source for the precision medicine in the Internet game and smartphone addiction.

Application of Smart Healthcare System to Internet Game and Smartphone Addiction

Smart healthcare refers to delivering ubiquitously available healthcare services using wired and wireless networking technology of the smart and digital technology [70]. Smart healthcare system aims to deliver quality care to patients while reducing the healthcare cost and providing "on-time" medical services. This service includes not only monitoring patients' condition but also offering primary healthcare to normal individuals.

Previously, several attempts have been made for the intervention of addictive disorders using smart healthcare. Computerized intervention targeting alcohol and tobacco use disorder was proven to be effective with relatively low cost and was suggested as a cost-effective and highly accessible means of treating uncomplicated substance use and related problems [71]. Text messaging tailored to smoking habits and barriers to quitting has successfully supported smoking cessation [72, 73]. Recent advances of smartphone technologies have enabled to support addictive disorders using smartphone applications. A smartphone application that helps smoking cessation with short messages and interactive tools, "the Real e Quit mobile application (REQ-Mobile)", was proven to be feasible for delivering cessation support [74]. The Addiction-Comprehensive Health Enhancement Support System (A-CHESS) is a smartphone application designed to improve continuing care or AUD by offering emotional and instrumental support at almost any time and place [75]. In particular, A-CHESS contained both static content (i.e., audio-guided relaxation) and interactive features, which collected real-time location data (i.e., a bar he or she used to frequent) and provided a timely notification of high-risk situation and an alert asking the patient if he or she wanted to be there. A-CHESS was successful in reducing risky drinking days [76]. These results suggest that smartphones would serve as a promising platform for delivering smart healthcare services for the Internet game and smartphone addiction as well.

Wearable devices also have been used in addictive disorders. Wearable biosensors that continuously monitor human physiological response to environmental exposure provide useful information on the behavior. The Secure Continuous Remote Alcohol Monitor (SCRAM) and the Giner WrisTAS, a bracelet worn around the ankle and wrist, respectively, can detect transdermal alcohol vapor electrochemically and estimate the drinking schedules [77–79]. These biosensors were proven to be useful in recovery program, especially contingency management of AUD [80, 81]. As mentioned above, behavioral data that is obtained from gaming devices and smartphones can be a source of therapeutic intervention for Internet game and smartphone addiction. Wearable devices such as smartwatch and activity tracker can gather physiological information such as pulse rate, breathing rate, and location information by global positioning systems (GIS) or geographic information systems (GIS). Thus, wearable device would be able to detect physiological signs of excessive involvement in online gaming and smartphone use, or track individuals in a high-risk location, and send alarming messages or intervene promptly. Further combination with Bluetooth techniques would facilitate automatic data transfer from wearable devices to or web database and offer more relevant management.

The "DETOX" project has launched since 2011 and set its goal to develop smart healthcare system through the investigation of the brain mechanisms of Internet game and smartphone addiction. The "Smart Healthcare Platform for Internet Addiction" is designed to collect data through mobile application and wearable devices, and to construct big database that is able to store not only behavioral data collected from smart device but also psychological data, neuroimaging, genomic data, and biosensor data. These data are sent to healthcare providers, and optimal medical feedback is sent back to each user and is analyzed to predict risk and detect abnormality, which harbor a huge value in healthcare. Behavioral intervention can be performed through this platform as well. Figure 3 shows the overview of the Smart Healthcare Platform for Internet Addiction.

A mobile application that can collect data on type of applications that the individual mostly access, usage time, amount of central processing unit (CPU) usage, network traffic, global positioning system (GPS) usage, total battery usage, and

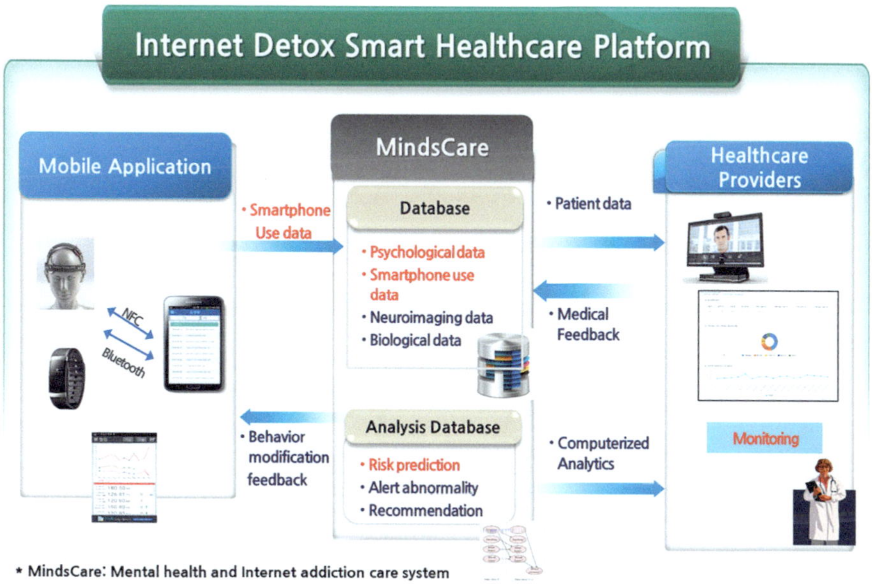

Fig. 3 Overview of the smart healthcare platform for Internet addiction

other valuable information was developed to unveil risk and protective factors of the Internet game and smartphone addiction. Data collected by this application is analyzed via artificial intelligence (AI)-based methods and to process and explore novel predictive factors. Some smartphone usage patterns that are effective to assess and predict smartphone addiction are derived from the usage data collected from smartphones via tensor factorization methods [82]. Six important predictors and patterns of the Internet game addiction were found via decision-tree model: gaming cost (50%), average weekday gaming time (23%), offline Internet gaming community meeting attendance (13%), average weekend and holiday gaming time (7%), marital status (4%), and self-perceptions of addiction to Internet game use (3%) [83].

Effective and optimal intervention for the Internet game and smartphone addiction using smart healthcare system is under development. Since Internet and smartphone are indispensable to our life, it is unrealistic to set a goal as "total abstinence" from the usage of the Internet and smartphones. The theoretical basis for this intervention is self-determination theory proposed by Ryan and Deci [84]. According to self-determination theory, people have three basic psychological needs: (1) competence: a self-perception of efficacy, (2) relatedness: need to experience connection with others, and (3) autonomy: sense that one's actions and experiences are volitional. This model is suitable for e-health system that provides the knowledge and skills to repair and new alternatives for social connection [85]. In the "Smart Healthcare Platform for Internet Addiction", the aim is to control excessive and uncontrolled smartphone usage by users themselves, improving sense of control and self-efficacy. Individuals who want to participate in this

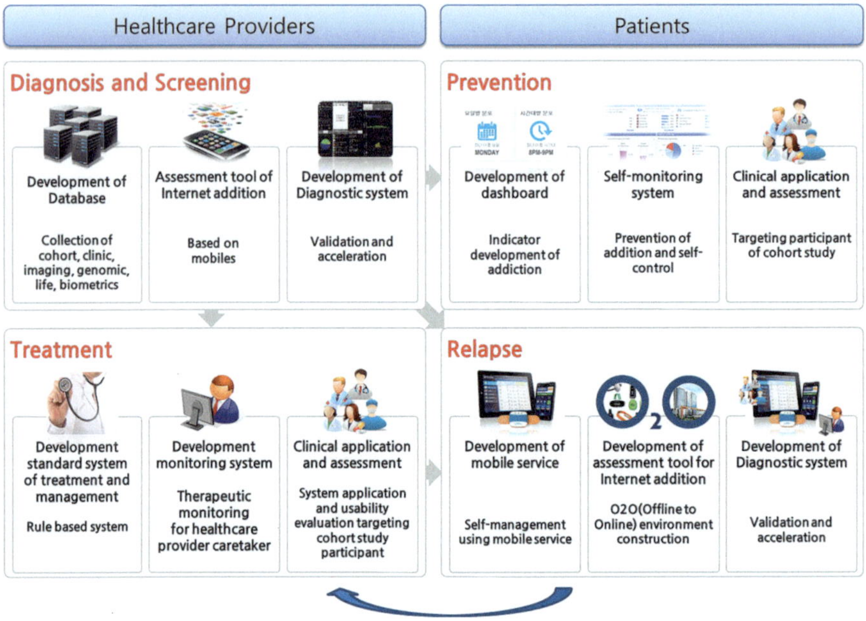

Fig. 4 Flow of smart healthcare system

program and allow transporting their smartphone usage data can view their usage patterns on the web or application and set a usage limitation. If their usage exceeds their limit, an alert or message is sent. Then the users may be able to decide whether they stop to use smartphones or games or not. Static contents that contain general information about Internet game and smartphone addiction, self-screening tools, and information on relaxation or alternative behaviors will be provided as well. Figure 4 is a flow of smart healthcare system for smartphone addiction. First, we devised a database and collected data from cohort, a tool to assess Internet and smartphone addiction by their mobile phones, and build a diagnostic system. These data are visualized for healthcare providers to overview the patients' characteristics and to design intervention methods. The visualized data is also available for the patients to monitor themselves. Longitudinal accumulation of data would make it possible to visualize the course of the Internet and smartphone addiction and assess the efficacy of detection and intervention methods. And relapse of the behavioral addiction can be tracked via the smart healthcare system.

Conclusion

Here we overviewed the evolution and conceptualization of Internet gaming disorder (IGD) and smartphone addiction. During the last few years, excessive online gaming and smartphone usage have brought social concerns and enormous articles have been published on these issues. Though debates exist, IGD is now listed on DSM-5 as a condition for further study and ICD-11. Numerous neuroimaging studies have been performed to elucidate the brain mechanisms of IGD and smartphone addiction. It seems that brains of IGD and smartphone addiction are different from those of normal gamers and users. Further studies with larger sample and prospective design would be necessary to validate previous findings. Then we overviewed the applicability of precision medicine in addiction psychiatry. It seems that many unsolved issues can be successfully tackled by precision medicine. Smart healthcare system would be one of the effective approaches that is tailored to individuals' need and offer optimized treatment options.

References

1. Penetration rate of gamers among the general population in the United States from 2013 to 2018. https://www.statista.com/statistics/748835/us-gamers-penetration-rate/.
2. Game users survey report 2016. Korea Creative Content Agency. 2016. http://www.kocca.kr/cop/bbs/view/B0000147/1831102.do?menuNo=200904.
3. D Griffiths M, J Kuss D, L King D. Video game addiction: past, present and future. Curr Psychiatry Rev. 2012;8(4):308–18.
4. Soper WB, Miller MJ. Junk-time junkies: an emerging addiction among students. Sch Couns. 1983;31(1):40–3.
5. Yee N. The demographics, motivations, and derived experiences of users of massively multi-user online graphical environments. Presence: Teleoperators Virtual Environ. 2006;15(3):309–29.
6. Holden C. 'Behavioral' addictions: do they exist? American Association for the Advancement of Science; 2001.
7. O'brien C. Addiction and dependence in DSM-V. Addiction. 2011;106(5):866–7.
8. Young KS. Internet addiction: The emergence of a new clinical disorder. Cyberpsychol & Behav. 1998;1(3):237–44.
9. Van Rooij AJ, Schoenmakers TM, Van de Eijnden RJ, Van de Mheen D. Compulsive internet use: the role of online gaming and other internet applications. J Adolesc Health. 2010;47(1):51–7.
10. Király O, Nagygyörgy K, Griffiths MD, Demetrovics Z. Problematic online gaming. Behav Addict: Elsevier; 2014. p. 61–97.
11. Lemmens JS, Valkenburg PM, Peter J. Development and validation of a game addiction scale for adolescents. Media Psychol. 2009;12(1):77–95.
12. Tejeiro Salguero RA, Morán RMB. Measuring problem video game playing in adolescents. Addiction. 2002;97(12):1601–6.
13. American Psychiatric Association. Diagnostic and statistical manual of mental disorders (DSM-5®). American Psychiatric Association Publishing; 2013
14. World Health Organization. ICD-11 for mortality and morbidity statistics DRAFT. 2018. https://icd.who.int/browse11/l-m/en#/http%3a%2f%2fid.who.int%2ficd%2fentity%2f1448597234.
15. Kwon M, Nam K, Seo B. A survey on internet overdependence; 2015. Ministry of Science: ICT, and Future Planning, Korea; 2016.
16. İNal EE, Demİrcİ k, Çetİntürk A, Akgönül M, Savaş S. Effects of smartphone overuse on hand function, pinch strength, and the median nerve. Muscle & Nerve. 2015;52(2):183–8.
17. Cazzulino F, Burke RV, Muller V, Arbogast H, Upperman JS. Cell phones and young drivers: a systematic review regarding the association between psychological factors and prevention. Traffic Inj Prev. 2014;15(3):234–42.

18. Kim H-J, Min J-Y, Kim H-J, Min K-B. Accident risk associated with smartphone addiction: a study on university students in Korea. J Behav Addict. 2017;6(4):699–707.
19. Hawi NS, Samaha M. To excel or not to excel: strong evidence on the adverse effect of smartphone addiction on academic performance. Comput Educ. 2016;98:81–9.
20. Boumosleh JM, Jaalouk D. Depression, anxiety, and smartphone addiction in university students-a cross sectional study. PLoS ONE. 2017;12(8):e0182239.
21. Demirci K, Akgönül M, Akpinar A. Relationship of smartphone use severity with sleep quality, depression, and anxiety in university students. J Behav Addict. 2015;4(2):85–92.
22. Kwon M, Lee J-Y, Won W-Y, Park J-W, Min J-A, Hahn C, et al. Development and validation of a smartphone addiction scale (SAS). PLoS ONE. 2013;8(2):e56936.
23. Billieux J, Van der Linden M, Rochat L. The role of impulsivity in actual and problematic use of the mobile phone. Appl Cogn Psychol: Off J Soc Appl Res Mem Cogn. 2008;22(9):1195–210.
24. Ministry of Science and ICT NISA. A survey on smartphone overdependence 2017. 2018.
25. Lopez-Fernandez O, Kuss DJ, Romo L, Morvan Y, Kern L, Graziani P, et al. Self-reported dependence on mobile phones in young adults: a European cross-cultural empirical survey. J Behav Addict. 2017;6(2):168–77.
26. Panova T, Carbonell X. Is smartphone addiction really an addiction? J Behav Addict. 2018:1–8.
27. Zhang Y, Ndasauka Y, Hou J, Chen J, Wang Y, Han L, et al. Cue-induced behavioral and neural changes among excessive internet gamers and possible application of cue exposure therapy to internet gaming disorder. Front Psychol. 2016;7:675.
28. Yao Y-W, Chen P-R, Li S, Wang L-J, Zhang J-T, Yip SW, et al. Decision-making for risky gains and losses among college students with Internet gaming disorder. PLoS ONE. 2015;10(1):e0116471.
29. Choi J-S, Park SM, Roh M-S, Lee J-Y, Park C-B, Hwang JY, et al. Dysfunctional inhibitory control and impulsivity in internet addiction. Psychiatry Res. 2014;215(2):424–8.
30. Walton E, Turner JA, Ehrlich S. Neuroimaging as a potential biomarker to optimize psychiatric research and treatment. Int Rev Psychiatry. 2013;25(5):619–31.
31. Bullmore E. The future of functional MRI in clinical medicine. Neuroimage. 2012;62(2):1267–71.
32. Mitterschiffthaler MT, Ettinger U, Mehta MA, Mataix-Cols D, Williams SC. Applications of functional magnetic resonance imaging in psychiatry. J Magn Reson Imaging: Off J Int Soc Magn Reson Med. 2006;23(6):851–61.
33. Chen CY, Huang MF, Yen JY, Chen CS, Liu GC, Yen CF, et al. Brain correlates of response inhibition in internet gaming disorder. Psychiatry Clin Neurosci. 2015;69(4):201–9.
34. Ding W-n, Sun J-h, Sun Y-w, Chen X, Zhou Y, Zhuang Z-g, et al. Trait impulsivity and impaired prefrontal impulse inhibition function in adolescents with internet gaming addiction revealed by a Go/No-Go fMRI study. Behav Brain Funct. 2014;10(1):20.
35. Dong G, Huang J, Du X. Enhanced reward sensitivity and decreased loss sensitivity in Internet addicts: an fMRI study during a guessing task. J Psychiatr Res. 2011;45(11):1525–9.
36. Dong G, Hu Y, Lin X. Reward/punishment sensitivities among internet addicts: Implications for their addictive behaviors. Prog Neuropsychopharmacol Biol Psychiatry. 2013;46:139–45.
37. Ko C-H, Liu G-C, Hsiao S, Yen J-Y, Yang M-J, Lin W-C, et al. Brain activities associated with gaming urge of online gaming addiction. J Psychiatr Res. 2009;43(7):739–47.
38. Ko CH, Liu GC, Yen JY, Chen CY, Yen CF, Chen CS. Brain correlates of craving for online gaming under cue exposure in subjects with Internet gaming addiction and in remitted subjects. Addict Biol. 2013;18(3):559–69.
39. Han DH, Hwang JW, Renshaw PF. Bupropion sustained release treatment decreases craving for video games and cue-induced brain activity in patients with Internet video game addiction. 2011.
40. Chun J, Choi J, Cho H, Lee S, Kim D. Dysfunction of the frontolimbic region during swear word processing in young adolescents with Internet gaming disorder. Transl Psychiatry. 2015;5(8):e624.

41. Sepede G, Tavino M, Santacroce R, Fiori F, Salerno RM, Di Giannantonio M. Functional magnetic resonance imaging of internet addiction in young adults. World J Radiol. 2016;8(2):210.
42. Kim H, Kim YK, Gwak AR, Lim J-A, Lee J-Y, Jung HY, et al. Resting-state regional homogeneity as a biological marker for patients with Internet gaming disorder: a comparison with patients with alcohol use disorder and healthy controls. Prog Neuropsychopharmacol Biol Psychiatry. 2015;60:104–11.
43. Dong G, Huang J, Du X. Alterations in regional homogeneity of resting-state brain activity in internet gaming addicts. Behav Brain Funct. 2012;8(1):41.
44. Meng Y, Deng W, Wang H, Guo W, Li T. The prefrontal dysfunction in individuals with Internet gaming disorder: a meta-analysis of functional magnetic resonance imaging studies. Addict Biol. 2015;20(4):799–808.
45. Park Ch, Chun JW, Cho H, Jung YC, Choi J, Kim DJ. Is the I nternet gaming-addicted brain close to be in a pathological state? Addict Biol. 2017;22(1):196–205.
46. Wang H, Jin C, Yuan K, Shakir TM, Mao C, Niu X, et al. The alteration of gray matter volume and cognitive control in adolescents with internet gaming disorder. Front Behav Neurosci. 2015;9:64.
47. Lin X, Dong G, Wang Q, Du X. Abnormal gray matter and white matter volume in 'Internet gaming addicts'. Addict Behav. 2015;40:137–43.
48. Yuan K, Cheng P, Dong T, Bi Y, Xing L, Yu D, et al. Cortical thickness abnormalities in late adolescence with online gaming addiction. PLoS ONE. 2013;8(1):e53055.
49. Choi J, Cho H, Kim J-Y, Jung DJ, Ahn KJ, Kang H-B, et al. Structural alterations in the prefrontal cortex mediate the relationship between internet gaming disorder and depressed mood. Sci Rep. 2017;7(1):1245.
50. Park C-h, Chun J-W, Cho H, Kim D-J. Discriminating pathological and non-pathological internet gamers using sparse neuroanatomical features. Front Psychiatry. 2018;9:291.
51. Lee J, Cho B, Kim Y, Noh J. Smartphone addiction in university students and its implication for learning. Emerging issues in smart learning. Springer; 2015. p. 297–305.
52. Stothart C, Mitchum A, Yehnert C. The attentional cost of receiving a cell phone notification. J Exp Psychol Hum Percept Perform. 2015;41(4):893.
53. Uncapher MR, Thieu MK, Wagner AD. Media multitasking and memory: differences in working memory and long-term memory. Psychon Bull Rev. 2016;23(2):483–90.
54. Moisala M, Salmela V, Hietajärvi L, Salo E, Carlson S, Salonen O, et al. Media multitasking is associated with distractibility and increased prefrontal activity in adolescents and young adults. NeuroImage. 2016;134:113–21.
55. Hu Y, Long X, Lyu H, Zhou Y, Chen J. Alterations in white matter integrity in young adults with smartphone dependence. Front Hum Neurosci. 2017;11:532.
56. Chun J-W, Choi J, Cho H, Choi M-R, Ahn K-J, Choi J-S, et al. Role of frontostriatal connectivity in adolescents with excessive smartphone use. Front Psychiatry. 2018;9:437.
57. Chun J-W, Choi J, Kim J-Y, Cho H, Ahn K J, Nam J-H, et al. Altered brain activity and the effect of personality traits in excessive smartphone use during facial emotion processing. Sci Rep. 2017;7(1):12156.
58. Jameson JL, Longo DL. Precision medicine—personalized, problematic, and promising. Obstet Gynecol Surv. 2015;70(10):612–4.
59. Insel TR. The NIMH research domain criteria (RDoC) project: precision medicine for psychiatry. Am J Psychiatry. 2014;171(4):395–7.
60. Chamorro AJ, Marcos M, Mirón-Canelo JA, Pastor I, González-Sarmiento R, Laso FJ. Association of µ-opioid receptor (OPRM1) gene polymorphism with response to naltrexone in alcohol dependence: a systematic review and meta-analysis. Addict Biol. 2012;17(3):505–12.
61. Kim D-J, Choi I-G, Park BL, Lee B-C, Ham B-J, Yoon S, et al. Major genetic components underlying alcoholism in Korean population. Hum Mol Genet. 2007;17(6):854–8.

62. Farris SP, Pietrzykowski AZ, Miles MF, O'Brien MA, Sanna PP, Zakhari S, et al. Applying the new genomics to alcohol dependence. Alcohol. 2015;49(8):825–36.
63. Agrawal A, Bierut LJ. Identifying genetic variation for alcohol dependence. Alcohol Res Curr Rev. 2012;34(3):274.
64. Gorini G, Harris RA, Mayfield RD. Proteomic approaches and identification of novel therapeutic targets for alcoholism. Neuropsychopharmacology. 2014;39(1):104.
65. Mostafa H, Amin AM, Teh C-H, Murugaiyah V, Arif NH, Ibrahim B. Metabolic phenotyping of urine for discriminating alcohol-dependent from social drinkers and alcohol-naive subjects. Drug Alcohol Depend. 2016;169:80–4.
66. Moeller SJ, Paulus MP. Toward biomarkers of the addicted human brain: Using neuroimaging to predict relapse and sustained abstinence in substance use disorder. Prog Neuropsychopharmacol Biol Psychiatry. 2018;80:143–54.
67. Schacht JP, Anton RF, Myrick H. Functional neuroimaging studies of alcohol cue reactivity: a quantitative meta-analysis and systematic review. Addict Biol. 2013;18(1):121–33.
68. Justice AC, McGinnis KA, Tate JP, Xu K, Becker WC, Zhao H, et al. Validating harmful alcohol use as a phenotype for genetic discovery using phosphatidylethanol and a polymorphism in ADH 1B. Alcohol: Clin Exp Res. 2017;41(5):998–1003.
69. Connor J, Symons M, Feeney G, Young RM, Wiles J. The application of machine learning techniques as an adjunct to clinical decision making in alcohol dependence treatment. Subst Use Misuse. 2007;42(14):2193–206.
70. Sivagami S, Revathy D, Nithyabharathi L. Smart health care system implemented using IoT. Int J Contemp Res Comput Sci Technol. 2016;2(3).
71. Rooke S, Thorsteinsson E, Karpin A, Copeland J, Allsop D. Computer-delivered interventions for alcohol and tobacco use: a meta-analysis. Addiction. 2010;105(8):1381–90.
72. Free C, Knight R, Robertson S, Whittaker R, Edwards P, Zhou W, et al. Smoking cessation support delivered via mobile phone text messaging (txt2stop): a single-blind, randomised trial. The Lancet. 2011;378(9785):49–55.
73. Rodgers A, Corbett T, Bramley D, Riddell T, Wills M, Lin R-B, et al. Do u smoke after txt? Results of a randomised trial of smoking cessation using mobile phone text messaging. Tob Control. 2005;14(4):255–61.
74. Buller DB, Borland R, Bettinghaus EP, Shane JH, Zimmerman DE. Randomized trial of a smartphone mobile application compared to text messaging to support smoking cessation. Telemed E-Health. 2014;20(3):206–14.
75. Gustafson DH, Shaw BR, Isham A, Baker T, Boyle MG, Levy M. Explicating an evidence-based, theoretically informed, mobile technology-based system to improve outcomes for people in recovery for alcohol dependence. Subst Use Misuse. 2011;46(1):96–111.
76. Gustafson DH, McTavish FM, Chih M-Y, Atwood AK, Johnson RA, Boyle MG, et al. A smartphone application to support recovery from alcoholism: a randomized clinical trial. JAMA Psychiatry. 2014;71(5):566–72.
77. Greenfield TK, Bond J, Kerr WC. Biomonitoring for improving alcohol consumption surveys: the new gold standard? Alcohol Res Curr Rev. 2014;36(1):39.
78. Sakai JT, Mikulich-Gilbertson SK, Long RJ, Crowley TJ. Validity of transdermal alcohol monitoring: fixed and self-regulated dosing. Alcohol: Clin Exp Res. 2006;30(1):26–33.
79. Swift R. Transdermal alcohol measurement for estimation of blood alcohol concentration. Alcohol: Clin Exp Res. 2000;24(4):422–3.
80. Barnett NP, Celio MA, Tidey JW, Murphy JG, Colby SM, Swift RM. A preliminary randomized controlled trial of contingency management for alcohol use reduction using a transdermal alcohol sensor. Addiction. 2017;112(6):1025–35.
81. Barnett NP, Tidey J, Murphy JG, Swift R, Colby SM. Contingency management for alcohol use reduction: a pilot study using a transdermal alcohol sensor. Drug Alcohol Depend. 2011;118(2–3):391–9.
82. Choi J, Rho MJ, Kim Y, Yook IH, Yu H, Kim D-J, et al. Smartphone dependence classification using tensor factorization. PLoS ONE. 2017;12(6):e0177629.

83. Rho MJ, Jeong J-E, Chun J-W, Cho H, Jung DJ, Choi IY, et al. Predictors and patterns of problematic Internet game use using a decision tree model. J Behav Addict. 2016;5(3):500–9.
84. Ryan RM, Deci EL. Self-determination theory and the facilitation of intrinsic motivation, social development, and well-being. Am Psychol. 2000;55(1):68.
85. Pingree S, Hawkins R, Baker T, DuBenske L, Roberts LJ, Gustafson DH. The value of theory for enhancing and understanding e-health interventions. Am J Prev Med. 2010;38(1):103–9.

Ongoing Paradigm Shifts

Animal Research in Psychiatry

Michel Bourin

Abstract

Animal research in psychiatry suffers from a poor translational value. It is the same for all other disciplines. Our purpose in this chapter is therefore to highlight all the parameters that can lead to a non-reproducibility of interlaboratory experiments as well as intralaboratory. This is to point out the experimental parameters that are likely to lead to bias. Parameters are essentially: breeding conditions, animal strains, housing, handling, illumination, weather conditions, age, and the actual experimental conditions. Controlling these parameters is not enough if there is no consensus of the scientific community to implement them in a standardized way. However, it is possible to improve the translational concept by taking stock of what has been operational without forgetting to standardize as much as possible the essential parameters of behavioral research. Now there are calls to take a different approach to animal experimentation, by asking not what was controlled in an experiment, but what was ignored. This new school of thinking has been termed "therioepistemology"; the study of how knowledge is gained from animal research. The focus is on what's been ignored in an animal data set, why it's been ignored, and how it affects the model or experiment.

Keywords

Age · Breeding · Experimental parameters · Illumination · Handling · Housing · Reproducibility · Strains · Therioepistemology

M. Bourin (✉)
Neurobiology of Anxiety and Mood Disorders, University of Nantes,
98, rue Joseph Blanchart, 44100 Nantes, France
e-mail: michel.bourin@univ-nantes.fr

© Springer Nature Singapore Pte Ltd. 2019

Y.-K. Kim (ed.), *Frontiers in Psychiatry*, Advances in Experimental Medicine and Biology 1192, https://doi.org/10.1007/978-981-32-9721-0_14

Introduction

Interindividual differences have always been an important factor of variability in biology, and their systematic study has often contributed to the clarification of mechanisms of action. In the field of behavioral pharmacology, the desire to act in a controlled manner on this source of variation has been at the origin of a new discipline, behavioral psychopharmacology.

The animal models in psychiatry after being developed have been shunned for a few years, in particular because the new animal research drugs have become rare. It is important to understand the pitfalls encountered in using such models.

"If there were no animals, human nature would be even more incomprehensible," wrote Buffon in the "Discourse on the Nature of Animals." This great naturalist only had in mind the biological nature of man and did not even suspect all that the study of animals has begun to reveal to us about the human psyche [1]. Animal psychology, by revealing to us the fundamental psychological processes common to man and animal, has become capable of enlightening us on certain obscure sides of the human mind. But such investigations are fraught with pitfalls, in particular:

The anthropomorphic illusion is to project on the animal our human ways of seeing.

The zoomorphic illusion can lead to the blind and abusive extrapolation of an animal study to human behaviors whose determinism is inextricable.

The tautological reasoning that can lead to the proof of a pathological animal behavior based on the fact that this behavior is sensitive to a therapy considered effective in the human disorder. In practice, modeling is therefore based on a necessarily reductionist compromise. Indeed, an animal model is an experimental simulation in which a simple system represents another system, complex and less immediately accessible to knowledge. Ideally, an animal model in psychopharmacology should be similar to the human psychiatric disorder both at:

Behaviors induced, i.e., CREATIVE VALIDITY or FACE VALIDITY,

The etiology and neurobiological mechanisms underlying the disorder, i.e., THEORETICAL VALIDITY or CONSTRUCT VALIDITY,

Only the treatment response that has been clinically effective, i.e., PREDICTIVE VALIDITY.

Theoretically, the ideal would be to take as a model, a test that meets all three criteria of validity, but, at present, there is no animal model of psychiatric disease, with both a predictive potential and explanatory. Indeed, the fundamental communication deficit between humans and animals makes the stakes of their behaviors difficult, even impossible to elucidate, and leads to the anthropomorphic drift which consists in depicting and labeling behaviors deemed abnormal or aberrant, using a cognitive and emotional jargon made for human disorders.

In addition, animal models are used to study a heterogeneous theory of the pathogenesis of the human disorder, but most often only explore a symptom, sometimes not very specific, of a complex syndrome, particularly in psychopharmacology.

Finally, some effective treatments on the model are not necessarily in human clinic.

However, behavioral models in psychopharmacology remain the best tools for studying psychiatric disorders. Indeed, if a model is sufficient to mimic a major symptom of a psychiatric disease, it might be interesting to explore the phylogenetic bases common to the disorder in humans and the adaptive response of the animal to aversive situations [2]. In addition, treatment may help to understand the disease. In a behavioral test, the experimenter proposes, in general, to measure certain parameters of the animal's response to a given experimental situation. It delineates the framework of possible studies. Take the case of the conditioned avoidance response which is the tool of choice of many psychopharmacologists; however, the parameters of the experiment should not be only those of the apparatus.

The purpose of this chapter is to identify different parameters to give animal models increased reliability and especially to increase their reproducibility from one research laboratory to another. After about 50 years of use of animal models in psychiatry, it is necessary to better understand why the search for new drugs stagnates. Indeed, it is necessary to control all the steps and all the environments in which the animals are placed. It is necessary to propose new tracks or new methods of experimentation, if we want to obtain reliable and reproducible models. Indeed, there are unfortunately examples where the modification of the experimental parameters has led the pharmaceutical industry to expensive development failures. The shift in thinking has developed over the past decade, as pharmaceutical companies noticed that every drug that fails in human trials worked on animals in the laboratory. That led to pharmaceutical companies disinvesting in internal animal research and development, and passing on the costs to startup and academic labs.

Breeding Conditions

In addition to the genetic fund, it is suggested other epigenetic factors such as intrauterine position. Indeed, the presence of several fetuses in the same uterus allows the sex hormones, male and female, secreted by some to spread to all others [3]. The relative position of male and female fetuses in the uterus in relation to the genus of the adjacent fetus may affect the degree of behavioral organization [4]. It is the prenatal exposure to different physiological titers of hormones that is able to modulate the process of early organization of the brain systems that will underlie motivated exploration behavior during adolescence. The age of weaning is another epigenetic factor that influences the future behavioral organization and the adult phenotype [5]. The dopaminergic system is maturing during the preweaning and postweaning periods [6], and the age of weaning can affect this system and the response to dopaminergic substances [7]. Multiple epidemiological studies in humans and the results of numerous experiments in animals show that individuals born to mothers under-nuanced during pregnancy or malnourished during the early stages of postnatal life have a high risk of developing obesity, diabetes, and cardiovascular diseases. Prenatal malnutrition also has deleterious effects on brain function in adults, resulting in cognitive deficits and responses to altered stressors. These observations lead to the hypothesis of "fetal programming" according to which aversive environmental stimuli during development, including undernutrition and stress, cause permanent physiological changes in the newborn that predispose him to developing newborns metabolic diseases and psychiatric disorders such as depression or anxiety in adulthood. However, the cellular and molecular mechanisms that underlie this pathological susceptibility remain poorly understood.

One team attempted to define the possible correlation between the process of neuronal plasticity and the metabolic and behavioral alterations induced by prenatal malnutrition [8]. Specifically, this team is investigating whether the proliferation and/or survival of newly created neurons in the hippocampus of adult animals (neurogenesis) are altered by undernutrition during development and gain an overall understanding of the malnutrition impact on the susceptibility of the CNS to develop diseases, from its molecular aspects to behavioral changes. Elucidation of the molecular mechanisms underlying long-term alterations induced by prenatal malnutrition can have important consequences in the establishment of preventive strategies, based on nutrition, brain damage associated with diabetes and aging (cognitive deficits, anxiety, depression, etc.), or even neurodegenerative disorders such as Alzheimer's disease.

In the same vein, it was discussed evidence supporting the idea that increased susceptibility to the expression of anxiety behaviors in humans, primates, and rodents may result from abnormal development [9]. There are differences in anxiety behaviors and anxiety levels among individuals that appear to persist throughout life [10, 11] and reflect fundamental differences in organization. Brain is influenced by individual genetic factors and the environment to which they have been subjected during their lifetime. Animal studies confirm the powerful effects of the quality of maternal care on emotional behavior and brain function throughout life, especially among primates. Indeed, the replacement of the mother in the rhesus monkey by an inanimate manikin induces long-term deficiencies in coping capacity and social interactions and is associated with an increased risk of developing behaviors related to anxiety such as grooming and swaying movements [12, 13]. In rats, maternal separation for several hours per day during the early postnatal period increases the frequency of anxiety behaviors and hormonal responsiveness to stress in adults [14].

The supposed molecular mechanisms are derived from studies on knockout (KO) subjects, particularly in mice. It is hypothesized that a mutation in the 5-HT1A receptor may increase anxiety behavior [15, 16] and suppresses the expression of this receptor until the age of 4 weeks is sufficient to produce these disorders in the adult. These findings underscore the importance of serotonin in establishing normal patterns of anxiety modulation during postnatal development. The crucial period for the expression of the KO mouse phenotype is between the third and fourth week during which synaptic genesis and dendritic growth are very important in the forebrain, especially in the CA1 region of the hippocampus. This role is crucial for the regulation of innate anxious and abnormal behavior in KO mice for the 5-HT1A receptor [17, 18].

These conditions can be related to those in humans for post-traumatic stress disorder (PTSD), which occurs in approximately 15% of individuals who experience or witness a traumatic experience such as rape, murder or a military fight. One of the most compelling findings regarding PTSD is a trend toward a decrease in the volume of the hippocampus, a medial temporal lobe structure of the brain responsible for associative memory [19]. It is easily damaged by stress hormones [20, 21], and it is thought that the decrease in its size in patients with PTSD is a direct consequence of the state of chronic stress induced by stress trauma [22, 23]. It is possible that the small size of hippocampus may increase an individual's susceptibility to environmental stress. Thus, the conditions for rearing and separating offspring from their mothers can be considered as a relevant model of PTSD. However, the human child is subject to multiple separations that are more or less traumatic, without inducing a PTSD. Thus, there is matter for reflection in order to build new research schemas concerning the origin of the PTSD.

Strain

The psychopharmacologist most often uses different strains by a behavioral trait to study the role of this behavior in the determinism of the effects of a drug or the specificity of the substance on the behavior studied. Such experiments, to be useful, presuppose the more intimate knowledge of the nature of the differences between strains. Strain differences are essential when practicing behavioral psychopharmacology. Even on spontaneous locomotor activity, there are differences between the results on strains [24]: in this study on the effects of antidepressants (Ads) in several strains of mice, classifies the controls of the four strains studied in three groups according to their spontaneous basal locomotor activity. Thus, the strain C57BL/6J has the most important locomotor activity of the four strains, the strain DBA/2J, the weakest and the SWISS and NMRI strains, an intermediate and neighboring activity. Indeed, while all other strains have a similar spontaneous locomotor activity, the NMRI strain seems more active. In comparison with laboratory strains such as BALB/c and C57/BL6, the wild strain Mus musculus exhibits lower spontaneous locomotor activity and greater avoidance for open spaces, but there are no differences between strains concerning motivated exploration. These wild mice have a particular behavioral risk assessment strategy. Indeed, they take extra precautions before entering a potentially hazardous area but explore all areas after assessing them as non risky [25].

In comparison with Swiss strain mice, male wild mice show greater exploratory activities and overall preference for open arms in elevated plus maze EPM [26]. In addition, during a new exposure to the test, the wild mice further explore the open arms and further evaluate the risk while mice of the Swiss strain take refuge in the closed arms. Wild mice can therefore be described as weakly reactive [27]. On the other hand, in a single session, wild-type female mice also show a preferential profile of open arms but exploratory and locomotor activities identical to females of the SWISS strain. The exploration patterns of wild-type mice have similarities with those of some strains whose behavior is usually studied, such as the BALB/c strain. However, it was added that wild mice exhibit unusual behaviors such as successful or unsuccessful jumps from EPM, frozen spontaneous freezing attitudes (significantly higher in females than males), and upper arms showing a profile not only low anxiety but also an important motivation for the exhaust and high responsiveness [26].

It should be noted here that we have also observed some of these "extreme" behaviors, unusual in the Swiss strain that we usually use, in non-consanguineous strains, including freezing in the strain BALB/c and jumps and exploration of closed arms with the DBA/2J and C57BL/6J strains, respectively, in the control animals but also under the effect of certain treatments. Unfortunately, we did not have the necessary equipment to measure the ethological elements; we could not exploit them but we realized the interest of this approach which not only reduces the false positive and negative results but also increases the sensitivity of the models to new anxiolytics.

It is difficult to compare the results of studies involving different strains in the EPM as we have seen in the literature synthesis given the many experimental schemes that exist. The possibility that the SWISS strain we have been using for a long time and with which we have gathered all the results we rely on could have "derived" has led us to compare several noninbred strains with each other. On the other hand, we sought to establish a direct comparison between strains known to have extreme behaviors (inbred strains) and more conventional strains (non-consanguineous strains) [27].

Thus, in front of any experiment involving the evaluation of the behavior of the animal, more specifically if one wishes to develop an anxiolytic, an antidepressant, a mood stabilizer, or an antipsychotic, it will be advisable to choose species and strain with precaution. Literature reports many examples in which it is impossible to reproduce a result, because the change of animal strain and/or animal breeding causes great disparities in results. Regarding the strains, it may be necessary to go further with neurotransmitter assays (5-HT, NA, DA) in different brain areas such as amygdala, hippocampus, cortex, and hypothalamus (Table 1).

In another paper of our team [28], we have shown the importance of variability in the response of antidepressants to forced swimming test (FST) or tail suspension test (TST) according to the strain of mouse used. The Swiss mice are the most sensitive to highlight the antidepressants with serotonergic or noradrenergic activity on the FST while the strain C57BL/6JRj is more sensitive to highlight the dopaminergic activity of antidepressants. On the other hand, on the TST all the antidepressants studied decreased the immobility in Swiss and C57BL/6JRj mice. These behavioral differences on various strains of mice are likely to refine the screening of new antidepressants for the pharmaceutical industry. On the other hand, the use of DBA/2 inbred mice may be limited, as an absence of antidepressant-like response was observed in the FST. The lack of sensitivity to antidepressant treatment in DBA/2 strains could be due to high DA, NA, and 5-HT whole brain concentrations [24]. We obtained similar results but with different strains on the tail suspension test (TST) [29] which allowed us to build the decision tree for antidepressant screening [28]. It is surprising to note

Table 1 Examples of different mice strains currently use in psychopharmacology

INBRED MICE	OUTBRED MICE
Aged C57BL/6JRj	NMRI
BALB/cAnNRj	SWISS
BALB/cByJRj	
BALB/cJRj	
C3H	
C57BL/6JRj	
C57BL/6NRj	
CBA	
DBA/1	
DBA/2	
FVB	
SJL	
129/Sv	

that the literature is teeming with contradictory results concerning the efficacy of the predictive tests of the antidepressant or anxiolytic activity only because the selected strain is not taken into account. There is a worsening success rate in human trials—currently, one in nine drugs entering human trials will succeed—combined with an explosion of interest in reproducibility. The choice of strains must become a reasoned phenomenon in terms of clinical applications and not only for the behavioral psychopharmacologist in order to obtain positive results.

Housing

This is one of the most difficult settings to control although regulated. In fact, animal housing premises must have ventilation or air treatment system that is appropriate to the requirements of the species being housed. Ventilation must provide clean air and must regulate temperature and humidity, reduce odors, toxic gas content, dust level, and the presence of pathogenic microorganisms. The air in the premises must be renewed frequently, avoiding harmful drafts. It is forbidden to smoke in the premises where there are animals. The temperature, relative humidity, pressure, and hourly renewal of the animal accommodation premises must be controlled to ensure that they are in good health. In rooms without windows, it is necessary to provide controlled artificial lighting to meet the biological and behavioral requirements of the animals. Animal housing and experimental rooms should be isolated from sources of high noise, ultrasound, and vibration to avoid animal behavior and physiological disturbances. In the context of the development of psychotropic products and especially anxiolytics, the conditions of accommodation are a crucial point. First of all, these are accommodations after the birth, which are difficult to control by the experimenter who bought them most often from animal suppliers, given the difficulty of conducting such farms in research laboratories.

Housing is important to prevent animals from getting sick, but also so that they are the least anxious possible, as they are accustomed to for at least a week at the places where the experimentation is carried out. Rodents used in behavioral psychopharmacology are stored in cages with small dimensions that are far from their natural habitat [30]. These living conditions are not optimal for their cerebral development and therefore their behavior. These are the constraints that must be understood in order to have an unbiased reading of rodent behaviors used in biomedical research. There is a need for researchers to better study the behavior of naive animals deprived of their freedom. For example, the observation of behavioral disorders resulting from detention in captivity often provides a wealth of unexpected information about the nature and formation of the behavioral chain in question.

Handling

Another problem related to experimentation that is often underestimated is that of handling. It has often been found within the same research laboratory that results involving behavior tests such as anxiety tests as well as depression tests may show different results from one day to the next. The external parameters were the same: temperature, illumination, and atmospheric pressure; only the experimenter was different. In looking at the problem it was found that the handling of the animals was noticeably different. Handling small animals such as rats can lead to several adverse effects [31]. It has been found within the same laboratory that some results that seemed aberrant could be due to a change of experimenter. This is clinically comparable to white coat syndrome. Animals may feel fear or discomfort depending on the intake of different experimenters. Methods have been developed to minimize the negative effects of handling; however, habituation methods are eminently variable and often ineffective.

Minimizing disturbance of study animals is a major consideration in ethological and ecological research design. It is certainly one of the most difficult parameters to control in a research team, a fortiori from one laboratory to another. Standardized methods are proposed in the literature but they are not controlled and therefore not used [32, 33].

Illumination

Illumination conditions play a fundamental role in behavioral research, especially for anxiety models. Unfortunately, in the experimental conditions, it is rarely taken into account. This is why the example of results obtained in our laboratory is developed in this chapter.

Thus in the light/dark paradigm (BW test), the witnesses from the other rooms that the housing room spend less time in the dark compartment without the motor component are affected. This behavior is observed to a similar extent to that of animals treated with diazepam. The aversive conditions of the test thus seem to favor, in the control animals, whatever the illumination conditions imposed before the test, a less anxious behavior. The anxiolytic efficacy of diazepam, demonstrated in comparison with the group treated with diazepam from the animal facility, also appears to be preserved despite the manipulation of the circadian rhythm except in animals subjected to conditions of permanent darkness. In animals from this room, diazepam shows an anxiety-like profile probably evidenced by the acute lighting conditions inflicted on animals affected by chronic lack of light, in comparison with witnesses who are on the contrary less anxious. Regarding the motor component of this observation, the test conditions could also exacerbate a disruption of masked locomotor activity in the actimetry test [34].

In the elevated plus maze (EPM), on the one hand, the manipulation of the circadian rhythm does not affect the anxiolytic effects of diazepam illustrated by the increase of the parameters in the open arms (inputs and time), on the other hand, this effect is more important for inputs than for time. Our results corroborate those that are conventionally observed with benzodiazepines for which this model has been widely validated [35]. In addition, it is interesting to note that diazepam significantly increases closed-arm inputs without affecting the time spent in these arms, regardless of the illumination conditions imposed on the mice before the test. This effect could illustrate the disinhibitory component of diazepam, more than a true anxiolytic effect.

The conditions of in the light/dark paradigm therefore seem to affect the behavior of the controls more than that of animals treated with diazepam, and light is the predominant aversive factor in this test, where the stress it causes in animals that have been deprived of them could modify the anxiolytic response to diazepam conventionally observed [36].

Weather Conditions

These are parameters never mentioned in the results of behavioral pharmacology. We found significant differences in our results as a function of atmospheric pressure. We often had to give up experimenting when there was a storm so the results were disturbed. Very few data able are available on the subject. Yet, it was observed that the thermoregulation of animals is disturbed under the influence in particular noradrenergic neurotransmission induces negative effects on performance [37]. It would be interesting to measure the central temperature of the animals in order to verify their presence and their thermoregulation is disturbed in case of sudden changes in atmospheric pressure.

Age

It is usual to choose animals for experimentation in a range of fairly narrow weight for mice 20 to 23 g by example. The animals are bought with a weight of 15–16 g, under their brain maturity, they are kept for a week in an air-conditioned room outside the experimentation room, so that they rest and that their brain becomes mature. Generally, these are 3 weeks old mice that have this weight of 15 g; however, it is possible that mouse suppliers change their diet. So 15 g mice can be born only for 15 days and already weigh 15 g but cerebral maturation is far from over. This is how we did for months, not able to get results with the elevated plus maze. Both controls and treated mice spent as much time in the open arms as in the closed arms. It was therefore impossible under these conditions to use this test. It took us several months and a visit to our supplier to understand that the food of mice had changed and that the 15 g mice were only 15 days old [38].

However when weight and age are controlled, it is possible to make an excellent translational research. This has been the case in studying antidepressants in mice of different ages. We were able to demonstrate that, as in humans, older mice responded better to tricyclic antidepressants than to specific serotonin inhibitors (SSRIs) [39, 40].

Experimental Conditions

These are still conditions that are only very rarely described in published articles. We have tried to contribute to improvement and regularity by reducing the phenomena of neophobia [41]. It is necessary to reduce the anxiety of the animals to have reproducible results. To do this, we must avoid cleaning the experimental equipment with products whose smell is repulsive to animals. In addition, before each manipulation, animals move and cause soiling to make the environment more familiar to the animals that will be used for the experiment itself [42].

Conclusion

Elucidating the historical and conceptual foundations of our contemporary vision of animal models is an important step in understanding what psychiatric pathology is. The elusive nature of this experience, as well as, the immense diversity of its physiological and behavioral manifestations, are interweaving with other forms of psychopathology and its existential dimension, for the Human. It could be an ideological quest if it were not primarily a major economic issue in developed countries that could benefit from active research activity in both the public and private sectors.

Nobody knows what an animal is. We spend our time with people we do not know, yet they are not transparent. What's different is the approach to how laboratory animals are treated, both in rearing and during experimentation, and how these factors are accounted for during analysis. This movement essentially calls for a shift from viewing animals as tools to seeing them as patients. A new discipline is born therioepistemology: the study of how knowledge is we feel that gained from animal research [43]. Therioepistemology formalizes two related shifts in thinking that are sufficiently tectonic to count as a paradigm shift. First, the shift from asking "what have we controlled?" to asking "what have we chosen to ignore, and at what cost?" And second, the shift from viewing research animals as little furry test tubes to viewing them as animal patients. Ultimately, by formalizing therioepistemology as a discipline, we can begin to discuss best practices that will improve the reproducibility and translatability of animal-based research, with concomitant benefits in terms of human health and animal well-being. We have tried to emphasize the need for a detailed analysis by the psychopharmacologist of environmental interactions or more exactly experimental conditions—behavior, for the understanding of the effects of a drug on behavior, and the relative inadequacy of the explanations of the type of interference of the drug with intermediate variables such as emotion, conditionality, etc. The major obstacle, however, remains the lack of knowledge about the relationships between behaviors and their biochemical support, and even more so between the action of drugs on behavior and their interference with neurochemical mechanisms. We believe that we can contribute by using specific agonists or antagonists of receptors of known drugs thus specifying the potential mechanism of action of psychotropic drugs.

It also seems to us that animal behaviorists do not use human results enough because of lack of psychiatric knowledge or inadequate collaboration with clinicians. For example, at a presentation in a psychiatry conference, we learned that men respond better than women to tricyclic antidepressants; while women responded better than men to the specific inhibitors of serotonin reuptake. From these results, we conducted in mice using the forced swimming test experiments that confirmed the results obtained in humans [46] mechanism of behavioral action [44, 45]. It would be necessary that the translational medicine takes this way in order to have various bridges from the animal to the man.

So in summary, multiple aspects of diseases and adapted animal models lead to difficulties for translational research in psychiatry [46]. The scientific community has insufficient access to research results; there is clinical research which is not red by behavioral science researchers. Also, current clinical trial standards do not prevent invalid results on non-clinically pertinent. Increasing the power of clinical trials may be accompanied by increased errors [47]. On the other end, apparently valid studies contain hidden errors; we have to try to look at carefully.

Is it a way to solve these problems leading to real translational medicine? In our opinion, it would be interesting to develop research-based guidance to support personalized treatments, modifying clinical trial designs to test treatments for clinical significance in individual patients [48]. As well, develop "true" biomarkers that challenge really clinical data. More difficult but absolutely fundamental, there is a necessity to publish negative data. More important in my mind is to design clinical trials for optimal information, not just regulatory approval. I hope that this entire proposal would permit to increase the efficacy of drugs development. It is therefore not a paradigm shift that is needed but a series of paradigm shifts and behavior shift of researchers.

References

1. Buffon G. Buffon's natural history, Vol. 4 of 10: containing a theory of the earth, a general history of man, of the brute creation, and of vegetables. Forgotten Books; 2018.
2. Bourin M. Animal models for screening anxiolytic-like drugs: a perspective. Dialogues Clin Neurosci. 2015;17:295–303.
3. Laviola G, Macrì S, Morley-Fletcher S, Adriani W. Risk-taking behavior in adolescent mice: psychobiological determinants and early epigenetic influence. Neurosci Biobehav Rev. 2003;27:19–31.
4. Palanza P, Parmigiani S, vom Saal FS. Effects of prenatal exposure to low doses of diethylstilbestrol, o, p'DDT, and methoxychlor on postnatal growth and neurobehavioral development in male and female mice. Horm Behav. 2001;40:252–65.
5. Laviola G, Terranova MTh. Developmental psychobiology of behavioural plasticity in mice: the role of social experiences in the family unit. Neurosci Biobehav Rev. 1998;23:197–213.
6. Rao TS, Cler JA, Mick SJ, Emmett MR, Farah JM Jr, Contreras PC, Iyengar S, Wood PL. Neurochemical interactions of competitive N-methyl-D-aspartate antagonists with dopaminergic neurotransmission and the cerebellar cyclic GMP system: functional evidence for a phasic glutamatergic control of the nigrostriatal dopaminergic pathway. J Neurochem. 1991;56:907–13.
7. Adriani W, Laviola G. Spontaneous novelty seeking and amphetamine-induced conditioning and sensitization in adult mice: evidence of dissociation as a function of age at weaning. Neuropsychopharmacology. 2002;27:225–36.
8. Honório de Melo Martimiano P, de Sa Braga Oliveira A, Ferchaud-Roucher V, Croyal M, Aguesse A, Grit I, Ouguerram K, Lopes de Souza S, Kaeffer B, Bolaños-Jiménez F. Maternal protein restriction during gestation and lactation in the rat results in increased brain levels of kynurenine and kynurenic acid in their adult offspring. J Neurochem. 2017;140:68–81.
9. Gross C, Hen R. Genetic and environmental factors interact to influence anxiety. Neurotox Res. 2004;6:493–501.
10. Kagan J, Snidman N. Early childhood predictors of adult anxiety disorders. Biol Psychiatry. 1999;46:1536–41.
11. Van Ameringen M, Mancini C, Oakman JM. The relationship of behavioral inhibition and shyness to anxiety disorder. J Nerv Ment Dis. 1998;186:425–31.
12. Harlow HF, Dodsworth RO, Harlow MK. Total social isolation in monkeys. Proc Natl Acad Sci USA. 1965;54:90–7.
13. Ruppenthal GC, Arling GL, Harlow HF, Sackett GP, Suomi SJ. A 10-year perspective of motherless-mother monkey behavior. J Abnorm Psychol. 1976;85:341–9.
14. Kalinichev M, Easterling KW, Plotsky PM, Holtzman SG. Long-lasting changes in stress-induced corticosterone response and anxiety-like behaviors as a consequence of neonatal maternal separation in Long-Evans rats. Pharmacol Biochem Behav. 2002;73:131–40.
15. Heisler LK, Chu HM, Brennan TJ, Danao JA, Bajwa P, Parsons LH, Tecott LH. Elevated anxiety and antidepressant-like responses in serotonin 5-HT1A receptor mutant mice. Proc Natl Acad Sci USA. 1998;95:15049–54.
16. Parks CL, Robinson PS, Sibille E, Shenk T, Toth M. Increased anxiety of mice lacking the serotonin1A receptor. Proc Natl Acad Sci USA. 1998;95:10734–9.
17. Pokorný J, Yamamoto T. Postnatal ontogenesis of hippocampal CA1 area in rats. I. Development of dendritic arborisation in pyramidal neurons. Brain Res Bull. 1981; 7:113–20.
18. Pokorný J, Yamamoto T. Postnatal ontogenesis of hippocampal CA1 area in rats. II. Development of ultrastructure in stratum lacunosum and moleculare. Brain Res Bull. 1981;7:121–30.
19. Chaaya N, Battle AR, Johnson LR. An update on contextual fear memory mechanisms: transition between Amygdala and Hippocampus. Neurosci Biobehav Rev. 2018;92:43–54.

20. Reagan LP. McEwen BS Controversies surrounding glucocorticoid-mediated cell death in the hippocampus. J Chem Neuroanat. 1997;13:149–67.
21. Sapolsky RM. Stress hormones: good and bad. Neurobiol Dis. 2000;7:540–2.
22. Bremner JD. Alterations in brain structure and function associated with post-traumatic stress disorder. Semin Clin Neuropsychiatry. 1999;4:249–55.
23. Sapolsky RM. Stress, glucocorticoids, and damage to the nervous system: the current state of confusion. Stress. 1996;1:1–19.
24. David DJ, Renard CE, Jolliet P, Hascoët M, Bourin M. Antidepressant-like effects in various mice strains in the forced swimming test. Psychopharmacology. 2003;166:373–82.
25. Augustsson H, Meyerson BJ. Exploration and risk assessment: a comparative study of male house mice (Mus musculus musculus) and two laboratory strains. Physiol Behav. 2004;81:685–98.
26. Holmes A, Parmigiani S, Ferrari PF, Palanza P, Rodgers RJ. Behavioral profile of wild mice in the elevated plus-maze test for anxiety. Physiol Behav. 2000;71:509–16.
27. Trullas R, Skolnick P. Differences in fear motivated behaviors among inbred mouse strains. Psychopharmacology. 1993;111:323–31.
28. Bourin M, Chenu F, Ripoll N, David DJ. A proposal of decision tree to screen putative antidepressants using forced swim and tail suspension tests. Behav Brain Res. 2005;164:266–9.
29. Ripoll N, David DJ, Dailly E, Hascoët M, Bourin M. Antidepressant-like effects in various mice strains in the tail suspension test. Behav Brain Res. 2003;143:193–200.
30. Febinger HY, George A, Priestley J, Toth LA, Opp MR. Effects of housing condition and cage change on characteristics of sleep in mice. J Am Assoc Lab Anim Sci. 2014;53:29–37.
31. Clarkson JM, Dwyer DM, Flecknell PA, Leach MC, Rowe C. Handling method alters the hedonic value of reward in laboratory mice. Sci Rep. 2018;8:2448. https://doi.org/10.1038/s41598-018-20716-3.
32. Cloutier S, LaFollette MR, Gaskill BN, Panksepp J, Newberry RC. Tickling, a technique for inducing positive affect when handling rats. J Vis Exp. 2018;135. https://doi.org/10.3791/57190.
33. Swierk L, Langkilde T. Does repeated human handling of study animals during the mating season affect their offspring? J Exp Zool A Ecol Integr Physiol. 2018;329:80–6.
34. Clénet F, Bouyon E, Hascoët M, Bourin M. Light/dark cycle manipulation influences mice behaviour in the elevated plus maze. Behav Brain Res. 2006;166:140–9.
35. Dalvi A, Rodgers RJ. GABAergic influences on plus-maze behaviour in mice. Psychopharmacology. 1996;128:380–97.
36. Bourin M. Animal models for screening anxiolytic-like drugs: a perspective. Dialogues Clin Neurosci. 2015;17:295–303.
37. Meeusen R, Roelands B. Central fatigue and neurotransmitters, can thermoregulation be manipulated? Scand J Med Sci Sports. 2010;20(Suppl 3):19–28.
38. Hascoët M, Colombel MC, Bourin M. Influence of age on behavioural response in the light/dark paradigm. Physiol Behav. 1999;66:567–70.
39. Bourin M, Colombel MC, Redrobe JP, Nizard J, Hascoët M, Baker GB. Evaluation of efficacies of different classes of antidepressants in the forced swimming test in mice at different ages. Prog Neuropsychopharmacol Biol Psychiatry. 1998;22:343–51.
40. David DJ, Bourin M, Hascoët M, Colombel MC, Baker GB, Jolliet P. Comparison of antidepressant activity in 4- and 40-week-old male mice in the forced swimming test: involvement of 5-HT1A and 5-HT1B receptors in old mice. Psychopharmacology (Berl). 2001;153:443–9.
41. Bourin M, Hascoët M. The mouse light/dark box test. Eur J Pharmacol. 2003;463:55–65.
42. Hascoët M, Bourin M. A new approach to the light/dark test procedure in mice. Pharmacol Biochem Behav. 1998;60:645–53.
43. Garner JP, Gaskill BN, Weber EM, Ahloy-Dallaire J, Pritchett-Corning KR. Introducing Therioepistemology: the study of how knowledge is gained from animal research. Lab Anim (NY). 2017;46:103–13.

44. Redrobe JP, MacSweeney CP, Bourin M. The role of 5-HT1A and 5-HT1B receptors in antidepressant drug actions in the mouse forced swimming test. Eur J Pharmacol. 1996;318(2–3):213–20.
45. Redrobe JP, Pinot P, Bourin M. The effect of the potassium channel activator, cromakalim, on antidepressant drugs in the forced swimming test in mice. Fundam Clin Pharmacol. 1996;10:524–8.
46. David DJ, Nic Dhonnchadha BA, Jolliet P, Hascoët M, Bourin M. Are there gender differences in the temperature profile of mice after acute antidepressant administration and exposure to two animal models of depression? Behav Brain Res. 2001;119:203–11.
47. Bourin M. New challenges for translational psychopharmacology. Front Psychiatry. 2010;18(1):3.
48. Lamberink HJ, Otte WM, Sinke MR, Lakens D, Glasziou PP, Tijdink JK, Vinkers CH. Statistical power of clinical trials increased while effect size remained stable: an empirical analysis of 136,212 clinical trials between 1975–2014. J Clin Epidemiol. 2018. pii: S0895-4356(18)30140-9.
49. Becker RE. Modifying clinical trial designs to test treatments for clinical significance in individual patients. Clin. Drug Invest. 2001;21:727–33.

Modeling Psychiatric Diseases with Induced Pluripotent Stem Cells

Eline van Hugte and Nael Nadif Kasri

Abstract

Neuropsychiatric disorders are a heterogeneous group of disorders that are challenging to model and treat, due to their underlying complex genetic architecture and clinical variability. Presently, increasingly more studies are making use of induced pluripotent stem cell (iPSC)-derived neurons, reprogrammed from patient somatic cells, to model neuropsychiatric disorders. iPSC-derived neurons offer the possibility to recapitulate relevant disease biology in the context of the individual patient genetic background. In addition to disease modeling, iPSC-derived neurons offer unprecedented opportunities in drug screening. In this chapter, the current status of iPSC disease modeling for neuropsychiatric disorders is presented. Both 2D and 3D disease modeling approaches are discussed as well as the generation of different neuronal cell types that are relevant for studying neuropsychiatric disorders. Moreover, the advantages and limitations are highlighted in addition to the future perspectives of using iPSC-derived neurons in the uncovering

E. van Hugte · N. Nadif Kasri (✉)
Department of Human Genetics, Radboudumc, Donders Institute for Brain, Cognition, and Behaviour, 6500 HB Nijmegen, The Netherlands
e-mail: n.nadif@donders.ru.nl

E. van Hugte
Academic Center for Epileptology Kempenhaeghe, Heeze, The Netherlands

N. Nadif Kasri
Department of Cognitive Neuroscience, Radboudumc, Donders Institute for Brain, Cognition and Behaviour, Nijmegen, The Netherlands

© Springer Nature Singapore Pte Ltd. 2019

Y.-K. Kim (ed.), *Frontiers in Psychiatry*, Advances in Experimental Medicine and Biology 1192, https://doi.org/10.1007/978-981-32-9721-0_15

of robust cellular phenotypes that consecutively have the potential to lead to clinical developments.

Keywords
Autism · Bipolar disorder · Disease modeling · iPSCs · Neuropsychiatric disorders · Neuronal differentiation · Schizophrenia · Human-induced pluripotent stem cells

Introduction

Neuropsychiatric disorders, including schizophrenia (SZ), ADHD, anxiety disorders, major depressive disorder (MDD), bipolar disorder (BPD), and autism spectrum disorders (ASD), are genetically and phenotypically highly heterogeneous and present a major challenge in clinical genetics and medicine. These disorders, which are typically caused by multiple mutations, have traditionally been modeled in animal models such as rodents, flies, and zebrafish. While modeling disease in animals has proven valuable to understand the function of individual disease-causing genes, it has been difficult to make the step from understanding gene functioning toward understanding of human disease mechanisms and progression [1–3]. The most important reason being that these models will never be able to recapitulate the exact genetic background of individual patients. Further, mechanistic insights into neuropsychiatric disorders have been inferred from postmortem studies. Whereas these studies are valuable to identify the molecular and cellular changes in individual brains, they however do not allow to investigate the developmental trajectory underlying the disease and hence do not allow to discriminate between cause or compensatory mechanisms.

In 2006, Yamanaka and his colleagues revolutionized the disease modeling field by successfully reprogramming first murine and later human fibroblasts to a pluripotent cell type, the so-called induced pluripotent stem cells (iPSCs) [4, 5]. IPSCs can be generated from patient's own somatic cells by the introduction of four reprogramming factors, OCT4, SOX2, KLF4, and c-MYC [4]. They are able to form the three embryonic germ layers, and therefore differentiate into each desirable terminal cell type. iPSCs therefore make it possible to study previously inaccessible cell types, including neurons and glial cells. Furthermore, since iPSCs are reprogrammed from the patient's own somatic cells, they retain the exact genetic makeup of the patient [4].

It is then not surprising that since this discovery, iPSCs have increasingly been used to model genetically and phenotypically complex neuropsychiatric disorders. Neuropsychiatric disorders greatly burden the well-being of patients and their families, economy and health care system [6]. All neuropsychiatric disorders have a genetic component. Twin studies estimate the heritability of SZ and BPD for example between 80 and 90% [7, 8]. The advances in genetic sequencing and genotyping techniques allow for the dissection of this genetic makeup [9]. Large-scale genome-wide association studies (GWAS) have revealed that the genetic architecture of complex disorders is defined by polygenic variants like single nucleotide polymorphisms (SNPs), deletions and duplications. These polygenic variants can include thousands of variants that all confer different risks [9–11]. In addition, patients can carry different combinations of disease-associated risk loci and not every genetic variant is highly penetrant, indicating that there are contributing influences from the environment or epigenetic factors [12, 13].

Despite the advances in GWAS, the understanding of the pathophysiology of many neuropsychiatric disorders remains restricted. Functional interpretation of GWAS data is still a hurdle. In most of the cases, disease-causing variants only have a statistical association with disease risk, no direct biological function can be inferred from it [2, 9, 10], and interpretation of the individual contribution of these disease loci is difficult due to their small effect sizes [2, 14]. Furthermore, most of the identified genetic variants inhabit the noncoding part of the genome, preventing any functional analysis in nonhuman animal models, since most noncoding parts of the genome are not highly conserved across species [15]. Thus, to understand neuropsychiatric disorders and complex diseases, the field should progress from disease-associated loci toward functional interpretation and understanding of disease development and pathology [16].

IPSC-derived neurons could be a valuable preclinical disease model for neuropsychiatric disorders to reach this goal. By differentiating patient-derived iPSCs into cell types affected by neuropsychiatric disorders, controlled in vitro experiments modeling the influence of multiple disease-associated risk loci are achievable [14]. In this chapter, we describe the promises and applications of iPSC-derived neurons for neuropsychiatric disorders. Different approaches to model neuropsychiatric patient cohorts using iPSC-derived neurons are highlighted, as well as the importance of choosing the appropriate cell type to study. Lastly, we discuss challenges and future perspectives of modeling neuropsychiatric disorders using human iPSC-derived neurons.

Approaches to Model Neuropsychiatric Cohorts

iPSC Models of Monogenetic Disorders

Historically, the first iPSC models for neurological disorders were investigating monogenetic causes of disease. These models have proven their usefulness for elucidating disease pathology, by confirming findings discovered in other disease models, and by uncovering new molecular and cellular phenotypes [2]. Consequently, highly penetrant, monogenetic mutations provide a well-defined first step to investigate the relationship between genetic risk and complex disease-relevant phenotypes [14].

Although rare, iPSC-derived models from families affected by a specific genetic locus, that confers susceptibility for distinct neuropsychiatric disorders, offer important insights into the relation of genetic risk and neuronal functioning

[14]. An example is a mutation in disrupted in schizophrenia 1 (DISC1), identified in a Scottish family, with affected family members suffering from SZ, BPD, and MDD [17]. In another family, a four base pair deletion in DISC1 was found to co-segregate with SZ and schizoaffective disorders [18]. In a study by Wen et al., human forebrain neurons were differentiated from iPSCs derived from patients affected by the DISC1 mutation. DISC1-deficient neurons showed aberrant presynaptic release and synaptic formation. Their findings suggested a model in which neuropsychiatric susceptibility genes, like DISC1, affect synaptic functioning by transcriptional dysregulation of numerous genes related to neuropsychiatric disorders in human neurons [19]. This study is a clear example of how a specific mutation affects the same cellular phenotypes in all family members, but the affected family members suffer from different clinically defined neuropsychiatric disorders [14].

Another example of monogenetic iPSC-derived models elucidating disease phenotypes comes from a syndromic, monogenetic form of ASD, i.e., Rett syndrome. Rett syndrome is caused by mutations in methyl CpG binding protein 2 (MECP2) [20]. iPSC-derived neurons from patients with Rett syndrome exhibited reduced soma sizes, altered dendritic spine morphology, and decreased number of excitatory synapses [21]. Moreover, MECP2 levels affect the expression of TRPC6, a cation channel that has been linked to non-syndromic autism, hinting toward common biological pathways [22]. A study by Tang et al. demonstrated that neurons derived from Rett syndrome patients show a reduced expression of K^+–Cl^- cotransporter 2 (KCC2), resulting in a delayed developmental switch of GABA from excitatory to inhibitory (reviewed by Ben-ari et al. [23, 24]). Consistent with this finding, their results were also reported in mouse models for ASD [25]. A significant decrease in KCC2 expression was also induced by a neuroligin 2 mutation, linked to SZ, and KCC2 expression levels were reduced in patients with SZ [26–28]. In addition, Na^+–K^+–$2Cl^-$ cotransporter1 (NKCC1), a chloride transporter with opposite function to KCC2, interacts with DISC1 [29]. Overall, these findings point toward a potential role for KCC1 in the pathogenesis for SZ, besides its role in Rett syndrome, suggesting that KCC1 can be a key factor in a variety of neuropsychiatric disorders [24]. Other examples of monogenetic neurodevelopmental disorders with neuropsychiatric phenotypes, like fragile X and Rett syndrome, are reviewed in Linda et al. [30]. These studies are clear examples of how iPSC-derived models uncover shared cellular phenotypes and common biological pathways, that result in different clinically defined neuropsychiatric disorders [14].

Cohorts with Multiple Disease-Associated Loci

Highly penetrant risk factors only account for a small subset of neuropsychiatric disease, which are in most cases sporadic with multiple affected disease loci of small effect. The proof of principle for modeling these sporadic neuropsychiatric disorders using iPSCs was provided without stratifying the patient cohort beforehand, i.e., comparing neurons from patients to age- and sex-matched controls [2]. In a study by Brennand et al., iPSC-derived neurons from SZ patients showed altered connectivity and gene expression [31]. In addition, Madison et al. showed that iPSC-derived neurons from BPD patients exhibited expression changes in genes responsible for neuronal plasticity and showed phenotypic differences in neurogenesis [32]. These results were initially unanticipated, given the diverse genetic background of the patient-derived neurons. Nevertheless, it could be that these different genetic variants converge on common underlying developmental pathways, responsible, for example, for synaptic functioning or cortical development [2, 33]. In the same way, comparisons between idiopathic neuropsychiatric patients to unaffected controls showed that distinct forms of neuropsychiatric disorders can also share common underlying molecular pathways [34]. The limitation of this approach is that it requires large sample sizes and that variability in underlying disease-associated loci can confound the effect [2].

Patient selection based on genotype or clinical features can provide a solution [35]. One approach is to stratify patients based on phenotypical variations, such as macrocephaly versus microcephaly in ASD patients [36]. Alternatively, patients can be selected based on their polygenic risk score, a score that summarizes the genetic risk alleles and their corresponding effect sizes per patient. Cohorts can then be stratified in four groups, namely, patients with high polygenic risk scores, patients with low polygenic risk scores, and unaffected controls with high and low polygenic risk scores [35]. This approach has the potential to facilitate the discovery of robust disease-associated phenotypes [31, 33, 35].

Generation of Isogenic IPSC Models

iPSC models can be generated that are genetically identical, except for the target loci of interest, i.e., isogenic. In an ideal situation, isogenic lines should be used whenever possible, to show the causal link between genotype and phenotype [33]. Even when focusing on single gene variants with large effect sizes, uncovering of disease phenotypes can be complicated by the genetic background of common variations in each case [33].

Isogenic models can be generated by genome editing tools like zinc finger nucleases, transcription activator-like effector nucleases and the relatively new technique clustered regularly interspaced palindromic repeats/CRISPR associated protein (CRISPR/Cas9) [37]. Isogenic lines have proven to be useful models to validate the effect of targeted mutations that are identified in patient cohorts [14, 38]. For example, by the generation of isogenic lines, either mimicking the mutation in control cell lines, or rescuing the mutation in patient lines, Wen et al. confirmed a direct causal link between DISC1 and a synaptic phenotype [19]. Furthermore, the CRISPR/Cas9 technique is able to edit multiple mutations at the same time, posing another upcoming advantage when studying a disease cohort that is influenced by many gene variants with small effect sizes [39].

Additionally, isogenic disease models can be obtained without gene editing, by employing cell-intrinsic properties. One example is making use of random X-inactivation in X-linked chromosomal disorders like Rett syndrome or autism, intellectual disability, and epilepsy caused by the PCDH19 mutation [40]. These disorders form a distinct group, with affected females expressing the heterozygous X-chromosomal disease mutation in a mosaic pattern. This expression pattern provides the opportunity to generate isogenic iPSC lines, in which either the mutated or the healthy X-chromosome is inactivated [41]. Likewise, isogenic iPSC lines can be generated for mutation in the mitochondrial DNA (mtDNA), which have been increasingly linked to psychopathology [42]. Each cell has thousands of mtDNA copies, and a mixture of wild-type and mutated mtDNA could be present, termed heteroplasmy. It has been demonstrated that during the reprogramming of fibroblasts, iPSC lines are generated with segregation of heteroplasmic mtDNA toward homoplasmy in a bimodal fashion, depending on the heteroplasmy level of the initial fibroblast that is reprogrammed [43, 44]. In this way, iPSCs can be produced with variable levels of heteroplasmy from individual patients, generating isogenic lines [43, 45].

The Importance of the Appropriate Cell Type for Disease Modeling

Equally to the selection of the appropriate patient cohort, it is important to select the right cell type for functional analysis [14]. Postmortem brain analysis, gene expression atlases, and brain imaging studies can give a primary indication which cell types to consider, and potentially reveal disease-related phenotypes [46, 47]. Brain imaging and postmortem studies of patients with neuropsychiatric diseases consistently show abnormalities in specific regions of the brain. For instance, patients with SZ and BPD show enlarged ventricles, pointing toward loss of cortical volume [48, 49]. Postmortem brain tissue from SZ patients shows decreased neuronal stem cell proliferation in the dentate gyrus, and cortical pyramidal neurons show reduced dendritic spine density [50, 51]. These observations have, for example, led to the investigation of neuroprogenitor cells, hippocampal dentate gyrus granule cells, and differentiated cortical cultures [46].

In addition to specific brain regions, specific neurotransmitter systems are affected in neuropsychiatric disorders. A well-known example is the implication of the serotonergic system in MDD, as selective serotonin reuptake inhibitors (SSRIs) can improve MDD symptoms [52]. Another example is the involvement of the glutamatergic system in SZ, based on the fact that ketamine and phencyclidine induce psychotic-like symptoms by blocking NMDA receptors [53]. Besides, GWAS for SZ identified genes like GRIN2A and GRM3, encoding for glutamate receptors subunits [11]. These studies show examples of the involvement of the excitatory system in neuropsychiatric disorders.

Alterations in either excitation or inhibition can result in a disruption of the excitatory/inhibitory (E/I) balance. Indeed, different neuropsychiatric disorders have the common comorbidity epilepsy, which has consistently been linked to altered E/I balance [54]. Currently, the focus has shifted toward the role of the inhibitory system and seems to be a point of convergence for multiple neuropsychiatric disorders [55]. In postmortem studies, consistent alterations in the inhibitory system have been revealed. For instance, ASD postmortem brains show reduced expression of GAD67 and GAD65, enzymes that synthesize GABA, the major inhibitory neurotransmitter [56]. Furthermore, in SZ there is an indication that reduced expression of interneuron markers are a result of a reduced activity of interneurons [55, 57]. Other studies implicating the role of the inhibitory system in neuropsychiatric disorders are reviewed in Selten et al. These studies support the evidence that the changes in inhibitory function lead to altered signal processing underlying neuropsychiatric disorders [55].

The latest developments in differentiation protocols allow for the generation of multiple neuronal subtypes of interest in the study of neuropsychiatric disorders, including cortical neurons, hippocampal dentate gyrus granule cells, dopaminergic neurons, GABAergic neurons and serotonergic neurons [58–61]. These protocols make use of developmental cues like morphogens or small molecules to modulate pathways that are important for development. In this way, these protocols follow the scheduling of regional commitment and neuronal specification throughout embryonic development. A significant downside of these iPSC differentiation protocols, however, is that they are less robust. Variability in culture conditions and clone to clone variability can affect the results of the differentiation process, adding a significant source of inconsistency [62]. Moreover, the differentiation process could also be disturbed by genetic variants, independent of the differentiation protocol used, complicating the understanding of phenotypic changes between control and disease cell cultures [2].

A solution to robustly control differentiation is the direct conversion of iPSCs toward the cell type of interest. This can be accomplished by forced expression of cell-fate specifying neuronal transcription factors such as NGN2 for cortical excitatory neurons or ASCL1/DLX for inhibitory neurons [63, 64]. Both protocols have proven to generate a mature, homogeneous, and robust population of excitatory and inhibitory neurons, respectively. A homogeneous population remains the benchmark when it comes to the investigation of cell autonomous molecular and cellular phenotypes, due to the lack of cellular heterogeneity as a confounding factor [2]. It can however be reasoned that protocols that differentiate neuronal subtypes in a more heterogeneous culture, recapitulating normal development, may better reflect the situation in vivo. This has been illustrated by Schafer et al., showing dysregulation of specific transcriptional networks important for neuronal maturation in ASD-derived neurons. These ASD-related changes traced back to altered chromatin accessibility in neural stem cells, reflecting a pathologically primed stage. This phenotype was conversely not picked up by generating neurons via forced expression of NGN2, skipping the neuronal stem cell stage [65]. Thus, the kind of experiment or phenotype to be investigated ultimately determines what the optimal trade-off is between the two types of protocols. Homogeneous populations of neurons could be better suitable for gene expression studies and high-throughput assays for drug development, while heterogeneous cultures could be used to investigate neuronal morphology or development [46].

Besides the protocols that use iPSCs as a starting point for neuronal differentiation, other protocols have emerged that make use of direct lineage conversion of somatic cells into neurons, using forced expression of neuronal transcription factors such as ASCL1, NGN2, Brn2, Myt1 l [66]. These protocols have led to the differentiation of fibroblasts into excitatory neurons, dopaminergic neurons, medium spiny neurons, and spinal motor neurons [66–69]. Direct differentiation protocols can possibly provide a more mature functional cell by recapitulating the transcriptional program of the initial somatic cell. The in vitro differentiated neurons could be compared to their in vivo counterparts, derived from primary tissue, and allow for the mapping of transcriptional and epigenetic differences in mature versus immature cell state. This could identify key regulators of neuronal maturity that could be used in the future to reprogram cells from an immature to a more mature state [2].

Nonautonomous interactions between neurons and other cell types such as astrocytes, oligodendrocytes, and microglia have been shown to play a role in neuropsychiatric disorders [2, 70]. For example, immunological mechanisms involving microglia have been linked to the pathophysiology of SZ and BPD [71, 72]. Synaptic pruning, taking place in the critical period of the adolescent brain, involving the removal of excess excitatory synapses by microglia, is believed to be abnormal in SZ [51, 73–75]. This excessive elimination of synapses is thought to be the cause of the loss of gray matter volume seen in SZ. Indeed, postmortem studies of SZ brain tissue show cortical infiltration and activation of microglia [76]. Moreover, brain imaging studies show strong evidence for the involvement of microglia in SZ [77]. Developments in differentiation protocols currently enable the generation of microglia from monocytes [78]. In this way, microglia can be cocultured with iPSC-derived or directly differentiated neurons from the same individual [46]. This example illustrates the importance of the incorporation of cell types other than neurons when investigating neuropsychiatric disorders.

From 2D to 3D Model Systems

Although previously discussed conventional two-dimensional (2D) models are relevant to discover cell autonomous phenotypes and non-cell autonomous interactions, these 2D models do not recapitulate the complex 3D structure of the real human brain [2]. A 3D in vivo setting to study human neuronal development and network integration can be achieved by generating interspecies chimeras by the transplantation of iPSC-derived neurons or neuroprogenitor cells into the mouse brain [79]. This approach is specifically useful when studying late-onset disorders, like most neuropsychiatric disorders, since it allows investigation of patient-derived cells in an in vivo environment over development and during extended time periods. An impressive example and proof of concept is the creation of a mouse with human glial cells by the transplantation of astrocyte/oligodendrocyte precursors into the neonatal mouse brain [79]. The human cells subsisted and integrated into host brain, and eventually substituted the complete glial cell population of the mice. This augmented learning and synaptic plasticity in the chimeric mice [79]. Equally, the transplantation of glial precursor cells from SZ patients into mouse brains induced SZ-like behaviors like anxiety, sleep disturbances, and antisocial behaviors. This implicates the involvement of glial cells in the pathophysiology of SZ and demonstrates the potential relevance of studying interspecies chimeras as a model system for neuropsychiatric disorders [2, 80].

Another approach to create 3D models to investigate neuropsychiatric disease is to generate iPSC-derived neuronal 3D cerebral organoids. These organoids have the potential to expand the range of phenotypes that can be investigated in vitro and more closely recapitulate the characteristics of the developing brain, such as organization of cortical layers, neuronal migration, axon guidance, and response to guidance cues, which all could not be studied in conventional 2D culture systems [81, 82]. Taking it one step further, the next generation of brain organoids is combining several cell lineages in 3D, called assembloids [83, 84]. These assembloids can be employed to investigate interactions between brain regions in vitro, the assembly of neuronal circuits, and complex cell–cell interactions. To examine these cell–cell interactions, organoids resembling the dorsal forebrain, containing glutamatergic neurons and organoids resembling ventral forebrain, containing GABAergic neurons, can be combined into a multi-region assembloid. GABAergic neurons migrate and form functional synapses onto glutamatergic neurons, creating a functional network. In this way, assembloids provide the opportunity to investigate GABAergic interneuron development and its implications in neuropsychiatric disorders such as ASD and SZ [83, 84]. In addition to assembling different types of neuronal organoids, other relevant cell types for neuropsychiatric disease can integrate into brain organoids. One approach is to make slight alterations in the differentiation protocol, allowing the development of microglia within cerebral organoids [85]. This permits the study of neuronal interactions with microglia, implicated in, for example, SZ, in a 3D context. Another approach is to triculture 3D assembloids, combining neurons, astrocytes and microglia to investigate the interactions of neurons and nonneuronal cells and imitate neuroinflammation [86].

Although significant advancements have been made in the generation of organoids, this field is still developing. While scarce, examples exist that show the potential of organoids to model neuropsychiatric disorders. For example, cerebral organoids derived from patients with syndromic ASD showed accelerated cell cycles and increased number of GABAergic inhibitory neurons [36], consistent with the hypothesis of the excitatory/inhibitory disbalance in ASD [87]. While advancements should be made in the understanding of 3D cerebral organoids in the next few years, it is highly likely that an increase in the use of 3D models for neuropsychiatric disorders will be witnessed [82].

Drug Discovery Using IPSC-Derived Model Systems

Patient affected by neuropsychiatric disorders often has a very different and individual response to medication. Equally, the heterogeneity of clinical phenotypes and drug response makes validation of cellular phenotypes difficult. Other than disease modeling, iPSCs have a significant potential for drug discovery and personalized medicine for neuropsychiatric disorders. Cellular phenotypes can be validated by coupling the treatment response of patients to response of the same drug in vitro [46]. An example came from a study by Mertens et al., where hippocampal DG granule cells were differentiated from iPSCs derived from patients with BPD [88]. These neurons exhibited a hyperexcitable phenotype, both by evoked and spontaneous action potentials. The authors suggest that this hyperexcitability is an early endophenotype of BPD. The cohort was further stratified into patients responding to lithium and patients who did not respond, revealing that the hyperexcitable phenotype in iPSC-derived neurons was only rescued by lithium in neurons derived from patients that responded to lithium treatment [88]. This study exemplifies the validity of iPSC-derived neurons as a model system for personalized medicine, but also confirms a cellular endophenotype by correlating drug responsiveness in vitro with treatment response by patients [46]. In a similar manner, a study by Vadodaria et al. stratified MDD patients in SSRI responders and nonresponders and showed that iPSC-derived forebrain neurons displayed serotonin-induced hyperactivity, only in nonresponders. This hyperactivity was a result of upregulated serotonin-receptor expression and could be rescued by blockade of these receptors by Lurasidone, a clinically defined compound [89]. Not only did this study uncover a phenotype and treatment target for SSRI resistance, it also showed the strength of iPSC-derived disease models as a stratification system between SSRI responders and nonresponders. Models like this can possibly be employed in the future to predict drug responsiveness of patients in the clinic.

Improving the IPSC Model System for the Study of Neuropsychiatric Disorders

iPSCs have proven their value for the uncovering of molecular and cellular phenotypes underlying disease, and offer a powerful instrument for drug screening and personalized medicine. While this model platform has the potential to overcome many problems associated with preclinical research for neuropsychiatric disorders, several limitations need to be addressed.

The present dogma that iPSC-derived models do not retain epigenetic modifications upon reprogramming and subsequent passages could be disadvantageous when studying neuropsychiatric disorders. Neuropsychiatric disorders like anxiety and MDD, are greatly influenced by environmental factors that can modify epigenetics. There is however evidence that low-passage iPSCs can retain some of the epigenetic memory of the original somatic donor cell [90]. Another option to circumvent the loss of epigenetic memory is to generate directly transdifferentiated neuronal cells of interest, since evidence suggests that transdifferentiated cells can retain limited amounts of the epigenetic memory of the original somatic cell [7, 91, 92].

In addition, iPSC-derived neurons are still representing a relatively immature state. While this is advantageous when modeling neurodevelopmental disorders, neuropsychiatric disease manifests frequently at a later time point. Therefore, iPSC-derived neurons are more useful to study developmental phenotypes that precede disease manifestation. Present research is nonetheless investigating methods to simulate aging and maturation [92, 93]. Another problem remains cell-to-cell variability across different batches. This can arise from accumulation of de novo mutations, mosaicism in initial donor cells, stochastic events during reprogramming or differences in patient genetics. While this heterogeneity in patient genetics can influence batch to batch differences, iPSC-derived models that are able to recapitulate this heterogeneity can, on the other hand, be employed as a system for patient-specific diagnostics or drug discovery [92].

Lastly, culturing iPSC-derived neurons remains time-consuming, and practical and technical reasons limit the sample size of most reported studies [7, 82]. However, especially in the study of neuropsychiatric disorders, an extensive amount of iPSC lines will be needed to discover the effect of disease-associated loci with small effect sizes [2]. An ideal strategy would be to generate large cohorts that include numerous patient-derived iPSC lines, replicating the approach and strengths of GWAS. In this way, collaborations between laboratories could pave the way toward more standardized protocols and high-throughput screening [92].

Conclusions

The previously described studies demonstrate the promises and applications of iPSC-derived models for neuropsychiatric disorders. iPSC-derived models provide a solution to the difficulties of modeling neuropsychiatric disorders using animal models and the inaccessibility of disease-relevant cell types. Moreover, iPSC-derived models have the possibility to move toward functional interpretation of large-scale GWAS studies, advancing from disease association toward uncovering disease-related phenotypes. Although promising progressions have been made, this progress represents only a minor part of the prospective that iPSC-derived models hold to discover underlying phenotypes for neuropsychiatric disorders, and several obstacles still need to be overcome. Nevertheless, it remains evident that iPSC-derived models will lead toward a better understanding of disease mechanisms and directed treatments.

References

1. Dawson TM, Golde TE, Lagier-Tourenne C. Animal models of neurodegenerative diseases. Nat Neurosci. 2018;21(10):1370–9.
2. Soldner F, Jaenisch R. Stem cells, genome editing, and the path to translational medicine. Cell. 2018;175(3):615–32.
3. Engle SJ, Blaha L, Kleiman RJ. Best practices for translational disease modeling using human iPSC-derived neurons. Neuron. 2018;100(4):783–97.
4. Takahashi K, Tanabe K, Ohnuki M, Narita M, Ichisaka T, Tomoda K, et al. Induction of pluripotent stem cells from adult human fibroblasts by defined factors. Cell. 2007;131(5):861–72.
5. Takahashi K, Yamanaka S. Induction of pluripotent stem cells from mouse embryonic and adult fibroblast cultures by defined factors. Cell. 2006;126(4):663–76.
6. Murray CJ, Barber RM, Foreman KJ, Ozgoren AA, Abd-Allah F, Abera SF, et al. Global, regional, and national disability-adjusted life years (DALYs) for 306 diseases and injuries and healthy life expectancy (HALE) for 188 countries, 1990–2013: quantifying the epidemiological transition. The Lancet. 2015;386(10009):2145–91.
7. Soliman M, Aboharb F, Zeltner N, Studer L. Pluripotent stem cells in neuropsychiatric disorders. Mol Psychiatry. 2017;22(9):1241.
8. Sullivan PF, Kendler KS, Neale MC. Schizophrenia as a complex trait: evidence from a meta-analysis of twin studies. Arch Gen Psychiatry. 2003;60(12):1187–92.
9. Lander ES. Initial impact of the sequencing of the human genome. Nature. 2011;470(7333):187.
10. McClellan J, King M-C. Genetic heterogeneity in human disease. Cell. 2010;141(2):210–7.
11. Ripke S, Neale BM, Corvin A, Walters JT, Farh K-H, Holmans PA, et al. Biological insights from 108 schizophrenia-associated genetic loci. Nature. 2014;511(7510):421.
12. Dolmetsch R, Geschwind DH. The human brain in a dish: the promise of iPSC-derived neurons. Cell. 2011;145(6):831–4.
13. Geschwind DH. Autism: many genes, common pathways? Cell. 2008;135(3):391–5.
14. Wen Z, Christian KM, Song H, Ming G-l. Modeling psychiatric disorders with patient-derived iPSCs. Curr Opin Neurobiol. 2016;36:118–27.
15. Cooper GM, Shendure J. Needles in stacks of needles: finding disease-causal variants in a wealth of genomic data. Nat Rev Genet. 2011;12(9):628.
16. Ward LD, Kellis M. Interpreting noncoding genetic variation in complex traits and human disease. Nat Biotechnol. 2012;30(11):1095.
17. Millar JK, Wilson-Annan JC, Anderson S, Christie S, Taylor MS, Semple CA, et al. Disruption of two novel genes by a translocation co-segregating with schizophrenia. Hum Mol Genet. 2000;9(9):1415–23.
18. Sachs N, Sawa A, Holmes S, Ross C, DeLisi L, Margolis R. A frameshift mutation in disrupted in schizophrenia 1 in an American family with schizophrenia and schizoaffective disorder. Mol Psychiatry. 2005;10(8):758.
19. Wen Z, Nguyen HN, Guo Z, Lalli MA, Wang X, Su Y, et al. Synaptic dysregulation in a human iPS cell model of mental disorders. Nature. 2014;515(7527):414.
20. Amir RE, Van den Veyver IB, Wan M, Tran CQ, Francke U, Zoghbi HY. Rett syndrome is caused by mutations in X-linked MECP2, encoding methyl-CpG-binding protein 2. Nat Genet. 1999;23(2):185.
21. Marchetto MC, Carromeu C, Acab A, Yu D, Yeo GW, Mu Y, et al. A model for neural development and treatment of Rett syndrome using human induced pluripotent stem cells. Cell. 2010;143(4):527–39.
22. Griesi-Oliveira K, Acab A, Gupta AR, Sunaga DY, Chailangkarn T, Nicol X, et al. Modeling non-syndromic autism and the impact of TRPC6 disruption in human neurons. Mol Psychiatry. 2015;20(11):1350.

23. Ben-Ari Y. Excitatory actions of gaba during development: the nature of the nurture. Nat Rev Neurosci. 2002;3(9):728.
24. Tang X, Kim J, Zhou L, Wengert E, Zhang L, Wu Z, et al. KCC2 rescues functional deficits in human neurons derived from patients with Rett syndrome. Proc Natl Acad Sci. 2016;113(3):751–6.
25. He Q, Nomura T, Xu J, Contractor A. The developmental switch in GABA polarity is delayed in fragile X mice. J Neurosci. 2014;34(2):446–50.
26. Hyde TM, Lipska BK, Ali T, Mathew SV, Law AJ, Metitiri OE, et al. Expression of GABA signaling molecules KCC2, NKCC1, and GAD1 in cortical development and schizophrenia. J Neurosci. 2011;31(30):11088–95.
27. Arion D, Lewis DA. Altered expression of regulators of the cortical chloride transporters NKCC1 and KCC2 in schizophrenia. Arch Gen Psychiatry. 2011;68(1):21–31.
28. Sun C, Zhang L, Chen G. An unexpected role of neuroligin-2 in regulating KCC2 and GABA functional switch. Mol Brain. 2013;6(1):23.
29. Kim JY, Liu CY, Zhang F, Duan X, Wen Z, Song J, et al. Interplay between DISC1 and GABA signaling regulates neurogenesis in mice and risk for schizophrenia. Cell. 2012;148(5):1051–64.
30. Linda K, Fiuza C, Kasri NN. The promise of induced pluripotent stem cells for neurodevelopmental disorders. Prog Neuropsychopharmacol Biol Psychiatry. 2018;84:382–91.
31. Brennand KJ, Simone A, Jou J, Gelboin-Burkhart C, Tran N, Sangar S, et al. Modelling schizophrenia using human induced pluripotent stem cells. Nature. 2011;473(7346):221.
32. Madison JM, Zhou F, Nigam A, Hussain A, Barker DD, Nehme R, et al. Characterization of bipolar disorder patient-specific induced pluripotent stem cells from a family reveals neurodevelopmental and mRNA expression abnormalities. Mol Psychiatry. 2015;20(6):703.
33. Hoffman GE, Schrode N, Flaherty E, Brennand KJ. New considerations for hiPSC-based models of neuropsychiatric disorders. Mol Psychiatry. 2018:1.
34. Doherty JL, Owen MJ. Genomic insights into the overlap between psychiatric disorders: implications for research and clinical practice. Genome Med. 2014;6(4):29.
35. Hoekstra SD, Stringer S, Heine VM, Posthuma D. Genetically-informed patient selection for iPSC studies of complex diseases may aid in reducing cellular heterogeneity. Front Cell Neurosci. 2017;11:164.
36. Mariani J, Coppola G, Zhang P, Abyzov A, Provini L, Tomasini L, et al. FOXG1-dependent dysregulation of GABA/glutamate neuron differentiation in autism spectrum disorders. Cell. 2015;162(2):375–90.
37. Jinek M, Chylinski K, Fonfara I, Hauer M, Doudna JA, Charpentier E. A programmable dual-RNA-guided DNA endonuclease in adaptive bacterial immunity. Science. 2012;337(6096):816–21.
38. Byrne SM, Mali P, Church GM. Genome editing in human stem cells. Methods Enzym. 2014;546. p. 119–38.
39. Cong L, Ran FA, Cox D, Lin S, Barretto R, Habib N, et al. Multiplex genome engineering using CRISPR/Cas systems. Science. 2013:1231143.
40. Redies C, Hertel N, Hübner CA. Cadherins and neuropsychiatric disorders. Brain Res. 2012;1470:130–44.
41. Hinz L, Hoekstra SD, Watanabe K, Posthuma D, Heine VM. Generation of isogenic controls for in vitro disease modelling of X-chromosomal disorders. Stem Cell Rev Rep. 2018:1–10.
42. Srivastava R, Faust T, Ramos A, Ishizuka K, Sawa A. Dynamic changes of the mitochondria in psychiatric illnesses: new mechanistic insights from human neuronal models. Biol Psychiatry. 2018.
43. Fujikura J, Nakao K, Sone M, Noguchi M, Mori E, Naito M, et al. Induced pluripotent stem cells generated from diabetic patients with mitochondrial DNA A3243G mutation. Diabetologia. 2012;55(6):1689–98.
44. Prigione A, Lichtner B, Kuhl H, Struys EA, Wamelink M, Lehrach H, et al. Human induced pluripotent stem cells harbor homoplasmic and heteroplasmic mitochondrial DNA mutations

while maintaining human embryonic stem cell-like metabolic reprogramming. Stem Cells. 2011;29(9):1338–48.
45. Hämäläinen RH, Manninen T, Koivumäki H, Kislin M, Otonkoski T, Suomalainen A. Tissue-and cell-type-specific manifestations of heteroplasmic mtDNA 3243A>G mutation in human induced pluripotent stem cell-derived disease model. Proc Natl Acad Sci. 2013;110(38):E3622–30.
46. Watmuff B, Berkovitch SS, Huang JH, Iaconelli J, Toffel S, Karmacharya R. Disease signatures for schizophrenia and bipolar disorder using patient-derived induced pluripotent stem cells. Mol Cell Neurosci. 2016;73:96–103.
47. Hawrylycz MJ, Lein ES, Guillozet-Bongaarts AL, Shen EH, Ng L, Miller JA, et al. An anatomically comprehensive atlas of the adult human brain transcriptome. Nature. 2012;489(7416):391.
48. Steen RG, Mull C, Mcclure R, Hamer RM, Lieberman JA. Brain volume in first-episode schizophrenia: systematic review and meta-analysis of magnetic resonance imaging studies. Br J Psychiatry. 2006;188(6):510–8.
49. Arnone D, Cavanagh J, Gerber D, Lawrie S, Ebmeier K, McIntosh A. Magnetic resonance imaging studies in bipolar disorder and schizophrenia: meta-analysis. Br J Psychiatry. 2009;195(3):194–201.
50. Reif A, Fritzen S, Finger M, Strobel A, Lauer M, Schmitt A, et al. Neural stem cell proliferation is decreased in schizophrenia, but not in depression. Mol Psychiatry. 2006;11(5):514.
51. Glantz LA, Lewis DA. Decreased dendritic spine density on prefrontal cortical pyramidal neurons in schizophrenia. Arch Gen Psychiatry. 2000;57(1):65–73.
52. Song J, Hou X, Hu X, Lu C, Liu C, Wang J, et al. Not only serotonergic system, but also dopaminergic system involved in albiflorin against chronic unpredictable mild stress-induced depression-like behavior in rats. Chem Biol Interact. 2015;242:211–7.
53. Javitt D. Negative schizophrenic symptomatology and the PCP (phencyclidine) model of schizophrenia. Hillside J Clin Psychiatry. 1987;9(1):12–35.
54. Engel J. Excitation and inhibition in epilepsy. Can J Neurol Sci. 1996;23(3):167–74.
55. Selten M, van Bokhoven H, Kasri NN. Inhibitory control of the excitatory/inhibitory balance in psychiatric disorders. F1000Research. 2018;7.
56. Fatemi SH, Halt AR, Stary JM, Kanodia R, Schulz SC, Realmuto GR. Glutamic acid decarboxylase 65 and 67 kDa proteins are reduced in autistic parietal and cerebellar cortices. Biol Psychiat. 2002;52(8):805–10.
57. Enwright JF, Sanapala S, Foglio A, Berry R, Fish KN, Lewis DA. Reduced labeling of parvalbumin neurons and perineuronal nets in the dorsolateral prefrontal cortex of subjects with schizophrenia. Neuropsychopharmacology. 2016;41(9):2206.
58. Cooper O, Hargus G, Deleidi M, Blak A, Osborn T, Marlow E, et al. Differentiation of human ES and Parkinson's disease iPS cells into ventral midbrain dopaminergic neurons requires a high activity form of SHH, FGF8a and specific regionalization by retinoic acid. Mol Cell Neurosci. 2010;45(3):258–66.
59. Yu DX, Di Giorgio FP, Yao J, Marchetto MC, Brennand K, Wright R, et al. Modeling hippocampal neurogenesis using human pluripotent stem cells. Stem Cell Rep. 2014;2(3): 295–310.
60. Shi Y, Kirwan P, Livesey FJ. Directed differentiation of human pluripotent stem cells to cerebral cortex neurons and neural networks. Nat Protoc. 2012;7(10):1836.
61. Lu J, Zhong X, Liu H, Hao L, Huang CT-L, Sherafat MA, et al. Generation of serotonin neurons from human pluripotent stem cells. Nat Biotechnol. 2016;34(1):89.
62. Soldner F, Jaenisch R. iPSC disease modeling. Science. 2012;338(6111):1155–6.
63. Yang N, Chanda S, Marro S, Ng Y-H, Janas JA, Haag D, et al. Generation of pure GABAergic neurons by transcription factor programming. Nat Methods. 2017;14(6):621.
64. Zhang Y, Pak C, Han Y, Ahlenius H, Zhang Z, Chanda S, et al. Rapid single-step induction of functional neurons from human pluripotent stem cells. Neuron. 2013;78(5):785–98.

65. Schafer ST, Paquola AC, Stern S, Gosselin D, Ku M, Pena M, et al. Pathological priming causes developmental gene network heterochronicity in autistic subject-derived neurons. Nat Neurosci. 2019:1.
66. Pang ZP, Yang N, Vierbuchen T, Ostermeier A, Fuentes DR, Yang TQ, et al. Induction of human neuronal cells by defined transcription factors. Nature. 2011;476(7359):220.
67. Pfisterer U, Kirkeby A, Torper O, Wood J, Nelander J, Dufour A, et al. Direct conversion of human fibroblasts to dopaminergic neurons. Proc Natl Acad Sci. 2011;108(25):10343–8.
68. Victor MB, Richner M, Hermanstyne TO, Ransdell JL, Sobieski C, Deng P-Y, et al. Generation of human striatal neurons by microRNA-dependent direct conversion of fibroblasts. Neuron. 2014;84(2):311–23.
69. Son EY, Ichida JK, Wainger BJ, Toma JS, Rafuse VF, Woolf CJ, et al. Conversion of mouse and human fibroblasts into functional spinal motor neurons. Cell Stem Cell. 2011;9(3):205–18.
70. Salter MW, Stevens B. Microglia emerge as central players in brain disease. Nat Med. 2017;23(9):1018.
71. Barbosa IG, Machado-Vieira R, Soares JC, Teixeira AL. The immunology of bipolar disorder. NeuroImmunomodulation. 2014;21(2–3):117–22.
72. Khandaker GM, Cousins L, Deakin J, Lennox BR, Yolken R, Jones PB. Inflammation and immunity in schizophrenia: implications for pathophysiology and treatment. Lancet Psychiatry. 2015;2(3):258–70.
73. Petanjek Z, Judaš M, Šimić G, Rašin MR, Uylings HB, Rakic P, et al. Extraordinary neoteny of synaptic spines in the human prefrontal cortex. Proc Natl Acad Sci. 2011;108(32):13281–6.
74. Feinberg I. Schizophrenia: caused by a fault in programmed synaptic elimination during adolescence? J Psychiatr Res. 1982;17(4):319–34.
75. Sellgren CM, Gracias J, Watmuff B, Biag JD, Thanos JM, Whittredge PB, et al. Increased synapse elimination by microglia in schizophrenia patient-derived models of synaptic pruning. Nat Neurosci. 2019:1.
76. Radewicz K, Garey LJ, Gentleman SM, Reynolds R. Increase in HLA-DR immunoreactive microglia in frontal and temporal cortex of chronic schizophrenics. J Neuropathol Exp Neurol. 2000;59(2):137–50.
77. Van Berckel BN, Bossong MG, Boellaard R, Kloet R, Schuitemaker A, Caspers E, et al. Microglia activation in recent-onset schizophrenia: a quantitative (R)-[11C] PK11195 positron emission tomography study. Biol Psychiat. 2008;64(9):820–2.
78. Ohgidani M, Kato TA, Kanba S. Introducing directly induced microglia-like (iMG) cells from fresh human monocytes: a novel translational research tool for psychiatric disorders. Front Cell Neurosci. 2015;9:184.
79. Espuny-Camacho I, Michelsen KA, Gall D, Linaro D, Hasche A, Bonnefont J, et al. Pyramidal neurons derived from human pluripotent stem cells integrate efficiently into mouse brain circuits in vivo. Neuron. 2013;77(3):440–56.
80. Windrem MS, Osipovitch M, Liu Z, Bates J, Chandler-Militello D, Zou L, et al. Human iPSC glial mouse chimeras reveal glial contributions to schizophrenia. Cell Stem Cell. 2017;21(2):195–208. e6.
81. Kadoshima T, Sakaguchi H, Nakano T, Soen M, Ando S, Eiraku M, et al. Self-organization of axial polarity, inside-out layer pattern, and species-specific progenitor dynamics in human ES cell-derived neocortex. Proc Natl Acad Sci. 2013;110(50):20284–9.
82. Quadrato G, Brown J, Arlotta P. The promises and challenges of human brain organoids as models of neuropsychiatric disease. Nat Med. 2016;22(11):1220.
83. Paşca SP. Assembling human brain organoids. Science. 2019;363(6423):126–7.
84. Birey F, Andersen J, Makinson CD, Islam S, Wei W, Huber N, et al. Assembly of functionally integrated human forebrain spheroids. Nature. 2017;545(7652):54.
85. Ormel PR, de Sá RV, van Bodegraven EJ, Karst H, Harschnitz O, Sneeboer MA, et al. Microglia innately develop within cerebral organoids. Nat Commun. 2018;9(1):4167.

86. Park J, Wetzel I, Marriott I, Dréau D, D'Avanzo C, Kim DY, et al. A 3D human triculture system modeling neurodegeneration and neuroinflammation in Alzheimer's disease. Nat Neurosci. 2018;21(7):941.
87. Rubenstein J, Merzenich MM. Model of autism: increased ratio of excitation/inhibition in key neural systems. Genes, Brain Behav. 2003;2(5):255–67.
88. Mertens J, Wang Q-W, Kim Y, Diana XY, Pham S, Yang B, et al. Differential responses to lithium in hyperexcitable neurons from patients with bipolar disorder. Nature. 2015;527(7576):95.
89. Vadodaria KC, Ji Y, Skime M, Paquola A, Nelson T, Hall-Flavin D, et al. Serotonin-induced hyperactivity in SSRI-resistant major depressive disorder patient-derived neurons. Mol Psychiatry. 2019:1.
90. Kim K, Doi A, Wen B, Ng K, Zhao R, Cahan P, et al. Epigenetic memory in induced pluripotent stem cells. Nature. 2010;467(7313):285.
91. Yang N, Ng YH, Pang ZP, Südhof TC, Wernig M. Induced neuronal cells: how to make and define a neuron. Cell Stem Cell. 2011;9(6):517–25.
92. Vadodaria KC, Amatya DN, Marchetto MC, Gage FH. Modeling psychiatric disorders using patient stem cell-derived neurons: a way forward. Genome Med. 2018;10(1):1.
93. Studer L, Vera E, Cornacchia D. Programming and reprogramming cellular age in the era of induced pluripotency. Cell Stem Cell. 2015;16(6):591–600.

Psychopharmacology and Psychotherapy Research

Kiwon Kim and Hong Jin Jeon

Abstract

With vigorous researches related to novel treatment in psychiatric field, paradigm shift is emerging with its design and assessment. Comprehension of the psychiatric phenomenon expanded beyond disease model, with dimensional approach. Assessment of patients' clinical state includes subjective reports related to problematic symptoms, functional change, and quality of life. They also include objective findings collected from mobile e-wearable monitoring system and advanced neuroimaging modalities. Novel treatment protocols are not just limited to pharmacology itself or psychotherapy itself, but the approach is integrated with stratification; pharmacological treatment enhanced by cognitive behavioral management and psychotherapeutic intervention has been emerged and studied for its impact. Numerous studies were conducted to understand placebo response and to differentiate this phenomenon with novel treatments. Trials to draw good adherence to research protocol with good compliance to treatment in real are strengthened by integrated approach, so-called psychopharmacology. With these paradigm shifts observed from recent

K. Kim
Department of Psychiatry, Veteran Health Service Medical Center,
Sungkyunkwan University School of Medicine, Seoul, Republic of Korea
e-mail: kkewni@gmail.com

H. J. Jeon (✉)
Department of Psychiatry, Depression Center, Samsung Medical Center,
Sungkyunkwan University School of Medicine, Seoul, Republic of Korea
e-mail: jeonhj@skku.edu

© Springer Nature Singapore Pte Ltd. 2019

Y.-K. Kim (ed.), *Frontiers in Psychiatry*, Advances in Experimental Medicine and Biology 1192, https://doi.org/10.1007/978-981-32-9721-0_16

researches, it is promising for great advance in quality of life and our mental health.

Keywords
Clinical design · Assessment · Digitalization · Integration · Placebo

Introduction

It took nearly 20 years to allow the use of medications on clinical setting, since basic discoveries on potential pharmacological targets on psychiatric disorders. Even though there was a huge progress on exceptions and improved safety issues, only few medications were introduced and applied to patients since the advent of first-generation psychotropics. This slow progress with lack of understanding on psychiatric disorders led researchers to frustration. Cold economic climate with regulatory restraints on pharmacological price and use, successive researches on other clinical domains, and limitation on biological tools on psychiatric disorders brought disengagement of big pharmaceutical companies on neuroscientific research.

However, it is too soon to give up. Research related to neuroscience is vigorous and vital, which is proved by dynamic national and supranational initiatives on research and design for enhanced treatment for psychiatric disorders. Considering numerous challenges and failures faced in all fields of researches on new drug, the myth of less effectiveness on psychotropics compared to other class drugs is relatively incorrect belief. From the trend emphasizing on quality of life and personalized medicine, psychotherapeutic research also highlighted evidence-based approval with various designs. For its change on perspective related to psychiatric conditions, therapeutic strategy has been changed into stratified approaches. And research to define treatment efficacy found breakthrough by complementary strategies combining psychotherapeutic approach with psychopharmacology. In this chapter, we summarized recent trends in psychopharmacological and psychotherapeutic researches which led to integrative paradigm shift, for future progress.

Innovative Changes in Clinical Design

In medical research, the most powerful clinical trial design, known as randomized controlled trial (RCT), is usually focused on the acute disease model and managed to evaluate treatment efficacy [1]. This clinical design has limitation with ignorance on previous treatment impact on modification of actual situation both possible on the course and on the individual clinical manifestations [2]. Researchers unintentionally cannot get objective evaluation for its design of current RCT with cross-sectional feature. Moreover, psychiatric disorders and situations which need clinical interest are usually chronic or non-disease specific dimensional problems [3]. From these backgrounds, innovative efforts on clinical design on psychopharmacology and psychotherapy research fields were assessed.

Strategies to Refine Assessment

To monitor the efficacy of psychopharmacological treatment and psychotherapeutic treatment, fine discriminating method should be applied to catch related changes in clinical condition [4–6]. Serial follow-up assessments after diagnosis, integrating multiple diagnostic categories, staging patient's clinical status, and collecting information through both subjective report and objective tests are included in new innovative trends of assessment. Previous assessment strategies did not fully contain the clinical status which led to clinician's decision of choice. Psychiatric diagnostic entities involve various phenotypes, which are much difficult with various biological mechanisms represented as same clinical manifestations. Assessments with various problematic changes on perception, cognition, and emotion were conducted on recent researches. As in depression, early evaluation of treatment on emotional processing can be a promising tool for clinical efficacy assessment and for adjustment of treatment target. These approaches fulfill the criteria for treatment efficacy biomarker in aspect of sensitivity, specificity, and relevance to clinical state, depression. Changes in emotional memory found to be relevant to both with depression itself and to treatment response, while changes in attention process found to be relevant to anxiolytics response [7, 8]. However, more validation is necessary for clinical use on treatment response biomarker, which can be applied to psychiatric researches and treatment development. Recent movements to expand assessment and differentiate each symptom were previously expressed as "negative symptoms" in schizophrenia also can be an innovative approach to evaluate novel treatment efficacy [9, 10].

However, neuropsychological changes detected by routine assessments are not enough to represent potential, actual biological changes, only showing same clinical symptoms. The dimensional characteristic of psychopathology is important as it refers to which refers to the limitation of categorical approach [11]. Trying to overcome this pitfall, efforts to subtype with more specific manifestations has been carried out, which still fail to bring more homogenous biological phenotypes in real clinical field. Trials to subtype have been evolved with various dimensional approaches, including subjective reports, neuroimaging findings, and genetic findings. Consortiums to discover characteristics that consistently affect biological changes has been built which integrate information by functional, structural neuroimage findings and genomic data [12–14].

Collecting information through various tests is coming from the epiphany as conventional measures and rating scales incapable of detect novel treatment actions. Evolution of readouts may deliver more sophisticated changes accompanied by treatment efficacy. Consideration of complicated evaluation on genetic manifestation brought new approach to develop biomarker through genetic stratification. Patients grouped by genetic stratification can be more biologically homogeneous, having strengths to suggest other biological phenotypes influencing clinical symptoms and functional decline [15, 16]. It could be applied as diagnostic stratification and treatment response assessment. Translational researches with these biological subgroups with strength of homogeneity would be promising for novel treatment efficacy exploration.

Information from patient report accompanied with clinician's evaluation has been stressed for its widened perspective to assess treatment efficacy, proposing novel therapeutic research agendas. Functional outcomes considering quality of life, which can represent real life, can compensate pitfalls of traditional assessment parallel to traditional treatment approaches [17]. There is a movement to find multimodal biomarker working as treatment response assessment, based on lack of surrogate marker for treatment efficacy [18]. Patients' subjective reports can work as a new viewpoint to clinicians' assessment in treatment researches. Open stance to hear patients' perspective on researches can strengthen patients' motivation and make a strong alliance between researcher (clinician) and participant (patient). Recently, more power is moving toward association of patient than before, making patients' opinion greater in the process of research design in psychiatric field. These associations are known as important partner in consortium-based researches, also having impact on clinical guideline developments. Patients' information can bring clear therapeutic needs, strengthened compliance on treatment protocol or research itself, and highly adjusted threshold on risk acceptance [19–21]. Also, this trend is accompanied with financial support as funding source coming from patient charities for novel psychiatric treatment researches [22, 23].

Beyond simple assessment on symptoms from patients, efforts to evaluate the wholesome function and quality of life of patients, representing real-world efficacy, are arising. Time of actual working or time spent on social relationship engagement can represent the genuine function of patient, which corresponds to real needs of patients [24, 25]. Clinically, recovery means regaining normal or usual act, while there are various definitions and appliances in researches [26]. It can be interpreted as dichotomous with recovery versus non-recovery, or as dimensional one with spectrum. Expression on recovery also can be different between clinician and patient with clinical discrepancies. Effort to narrow down this discrepancy between patient and clinician is growing, which is in line with previous researchers' report. They claimed nine themes found in trainees' concept of recovery, with "more strengthening on person than one's illness", "maintaining hope", "focusing on patient's goals", "collaborative care with patient and clinician", "trying to keep social function", "listening to the patient's own experience", "psychosocial interventions", "focusing on patient's empowerment", and "adherence to traditional stance" [27]. In these days, researchers trying to apply novel treatment approach on psychiatric illness use evaluation on quality of life from patients with anxiety disorder to psychotic disorder, even with cognitive disorder [28–31].

Strategies to Make Stratified Treatment Protocol

Multistage process and stratified treatment approaches are moving adaptive intervention in psychiatric research, based on patients' response after treatment. In a recent research, multiple intervention stages were applied with each stage corresponding to one of the decisions and leading to random reassign of the participants [32]. Adaptive design was also implemented in the STAR*D research, in case of the following nonresponse to initial treatment, change or augment medication with another type [33]. Before launching on large, multicenter interventional clinical trials, researchers would try initial proof of concept trial on novel treatment with smaller, homogenous, stratified participants through specific biomarkers [34–37].

Due to recent advances in neuropsychiatry, stratified treatment approach targeting etiological mechanism in psychiatric illness has been arisen. Stratified treatment has its base on the earlier terms individualized or personalized medicine, which better targets through genetic or endophenotypic assessments. We can find previous example from anticancer treatment such as breast cancer or leukemia treatment. Herceptin, known as treatment of choice for HER 2-positive breast cancer, was at first seem like a failure to broad criteria with all breast cancer, but finally found to be effective in HER 2-positive group. This phenomenon can suggest a hope for novel treatment discovery, suffering from the fiasco with broad spectrum/blockbuster medication with a chance to find a breakthrough from specific subgroup. Genotyping or subtyping through specific neuroimaging might retrieve failed treatment in psychiatric fields [38]. This approach is also in line with clinical design to narrow down participant group making homogenous group and with the concept of precision medicine. Even though advanced diagnostic and therapeutic technologies had integrated various parts of specialties, psychiatric field has not yet experienced their benefit. Special attention is focused on biomarker again, and it can be emphasized by the development of new technologies leading new biomarker discovery possible. Promising biomarkers can provide better and more accurate clarification including diagnosis and elaboration of patient's status (both functional and quality of life), expectation on prognosis, treatment suggestion, and prediction on response, which finally lead to new treatment development [39]. Endeavor to integrate the fields of biologic, psychological, and public health well-being with actual stakeholders is highly supported by international consortium, such like "Roadmap for Mental Health Research (ROAMER)" [40].

Multi-target and other network-driven strategies are also included in integrated and stratified treatment approach. In real clinical psychiatry, medications to treat illness, which are responsible for mood and cognition, impact widely on neuropsychiatric networks. For its unexpected impact on cerebral networks, interest of novel treatment discovery has been focused on high selectivity and potency for target with efficacy. Unfortunately, even with numerous targets that are clinically relevant, medications in clinical practice are not powerful enough, which needs medication combination [41, 42]. Sharing context with network thinking, focus on multi-target strategies is arising. As we can see the efficacy of clozapine in schizophrenia, combination of complementary mechanisms of action shows better appliance on multi-symptom psychiatric disorder [43]. It would be considered as ideal but developing both selective and multi-target agents might be the best innovative approach. Every novel treatment research now needs preclinical evaluation, involving screening to detect unforeseen events, having consideration of potential therapeutic activity [44]. Therefore, researches tried to find genuine innovative drug targets, leading to efficacy and therapeutic impact on psychiatric illness are the main trend in these days. Considering disappointing enigma remains, still after the human genome sequencing, various levels of targets are suggested as candidates. They include inverse agonist, allosteric ligands, appliance of biased transduction, G-protein independent signaling manipulation, GPCR-interacting proteins, and agents which can recognize heterodimeric assemblies by their own [45–47]. Other cellular processes are also promising candidates for novel treatment development, including axonal dynamics, dendritogenesis, mitochondrial energy modulation, glial transmission, myelination, cell adhesion and apoptosis, and neuroinflammation [48, 49]. However, finding the right target with cellular process does not promise breakthrough immediately, bringing necessities with right intracellular modulation [50].

Strategies to Keep High Adherence on Treatment Research

Maintaining high adherence to treatment protocol in research is important for its evaluation on efficacy. Compliance matters within liability of clinical trial these days. Speaking of adherence, therapeutic alliance between patient and clinician is a key element to keep this compliance. In consideration of their impact on patient's choice, alliance and appropriate psychoeducation to family and carer are also important element to keep adherence. Most simple issue related to adherence could be the complexity of the medication regimen. However, individual history of nonadherence, insight related to patient's condition and motivation to participate in treatment, fear related to adverse effects, cognitive decline to manage treatment regimen and patient's mood state can all influence on adherence to treatment itself and related research protocol. Hence, personalized approach with consideration of multiple risk factors would enhance treatment adherence, which led to novel treatment development in research [51].

Endeavor to make simple regimen and enhance convenience lead to long-acting injectable (LAI) type pharmacological treatment. LAI combined with personalized behavioral approach was reported to improve outcomes in patients with severe mental illness [52]. More fundamental effort to solve this problem is multifaceted interventions including motivational approach and psychoeducation. They were found out to improve medication adherence and treatment outcomes in bipolar patients [53]. As mentioned above, motivation can be emphasized with the self-regulation model, presenting illness perception as core impact on individual's coping and emotional response to their illness. An increasing number of researches are focusing on this model related to adherence, to determine individual's illness perception. People with high perception on treatment protocol, low perception on suffering events with mental illness, low perception of results, and thorough understanding of their illness showed a better adherence in bipolar disorder [54]. Illness perception and subjective experience on treatment are also important in schizophrenia, as stressed on another research with reducing trauma and increasing insight leading to treatment adherence [55]. The matter of therapeutic alliance and attitude toward illness can influence treatment adherence on depression, either. As emphasized before, clarification of multiple risk factors related to poor adherence, including stigma, misconceptions, and fears related to treatment and adverse events, is necessary before starting treatment. Another research group exercised the treatment initiation and participation program spending three sessions with 30-minute contacts for 6-weeks on people with depression. Participants in this group showed five times higher adherence for 6 weeks and three times higher adherence between 6 and 12 weeks compared to people without this approach. Because high adherence group showed better response on treatment, treatment initiation and participation program were suggested as promising intervention to improve adherence to treatment [56].

However, interpretation of these results should be carefully elaborated, as recent report suggested the presence of poor prognosis factors within the group of suboptimal adherent participants. Compared to participants with good adherence, suboptimal adherents showed significantly severe depressive symptoms, higher rate of hospitalizations, more prevalent suicidal ideation, comorbid physical pain, more frequent side effects, and experience of emotional maltreatment. They also showed higher association with unfavorable attitude toward treatment and poorer therapeutic relationship than optimal adherence group. This results importance of clarifying the probability of suboptimal adherence which are also related to difficult clinical characteristics, but both factors would lead to poor outcome and low adherence. Endeavor to overcome those characteristics related to poor adherence is necessary to get a good outcome and to develop novel effective treatment approach in research [57].

Findings from efficacy trials of intervention to build up therapeutic alliance and adherence in patients suggested psychoeducation before treatment initiation as a first-line mandatory approach. It is well known for its great impact on illness perception, not just on change related to knowledge or medication adherence alone [58].

Focusing on the psychoeducation and therapeutic alliance with patient alone is not enough to maintain treatment adherence, or to maintain good compliance to research protocol. Caregivers who influence their patient are also emphasized as a key factor related to compliance. Poor adherence to treatment can significantly affect the quality of life of their family. Carers also have a great impact on consumer attitudes toward therapy and adherence. A recent exploratory study to investigate the relationship between carer and consumer attitudes to therapy itself and consumer adherence behavior was conducted between participants with mental illness. The greater the difference between patient and carer attitudes, the lower the level of adherence was observed. Carer's attitudes can be emphasized in relation to patient perception and attitude to treatment, which should be considered thoroughly to improve compliance [58].

Advanced technologies in these days have empowered novel design in clinical researches. Remote monitoring or intervention with digital technologies is enhancing compliance and trial outcome [59, 60]. Even there has been a large advance in traffic, efforts to minimize actual visit are increasing [61, 62]. Web-based intervention is a complicated platform, for its issues related to lawsuit or limitation with physical examination but promising one to enhance compliance. Considering numerous informal patient self-help and self-diagnosis with Internet, web-based clinical intervention with limited boundaries would be helpful both for researches and for real clinical setting. Apart from the intervention, remote monitoring of symptoms or side effects is another effective and cost-efficient option for research follow-up. Recently, real-life compliance and health services impacts of monitoring on patients with mood disorder were reported to increase medication compliance, without increasing cost spent in mental health services [63]. Integrating mobile health technology into early psychosis care was attempted on clinical research field, brought positive result with feasibility and good validity to existing clinician-rated assessments [64]. Systematic results related to technology-based approaches on mood disorders also showed satisfaction and good feasibility accompanied with improvement on mood symptoms. Mobile technologies showed possibilities with impact on good adherence, available on symptom monitoring, side effect tracking, prescription refilling, and on raising motivation on treatment progress [65]. Digital technology is also useful with intervention, based on many studies related to computerized therapies. Social, cognitive, problem-solving games, and psychoeducation for patient with mood, anxiety disorder, and psychosis are included in computerized therapies. Emerging methods to enhance cognitive function with multiple virtualized techniques are proposed in recent researches [66–68]. Participants can be more active in these settings, but not all of them suit to this approach. Under the theme of smart healthcare, at-home electronic solutions and mobile devices minimize patient's inconvenience, therapy provided over the Internet is delivered quickly and cost-effectively [69]. However, as it was mentioned above, not all participants are satisfied with these technologies, which need researches to identify clinical situations appropriate for digital mediated strategies [70]. Larger size with controlled data on efficacy with specificity and long-term studies should be warranted for its use.

Careful Interpretation of Placebo and Nocebo

Placebo effect is common in experimental researches and in clinical situations influencing treatment outcomes. It can be differently viewed between three separated groups, including the clinical researcher's viewpoint, the placebo researcher's and the clinicians. Each group has separated viewpoints with different negotiation with placebo itself. The neurobiology of the nocebo and placebo phenomena is known to be associated with the reward pathways, involved with the μ-opioid and dopamine pathways. Previous neurobiological pathway studies related to placebo were usually based on experimental models of placebo analgesia. Molecular neuroimaging with positron emission tomography and the selective μ-opioid and dopamine radiotracers have been applied to researches to find placebo effect formation. Molecular imaging related to selective μ-opioid tracers' region of interests includes the rostral anterior cingulate, orbitofrontal and dorsolateral prefrontal cortex, anterior and posterior insula, nucleus accumbens, amygdala, thalamus, hypothalamus, and periaqueductal gray. They can also be interpreted as regions involved in pain, emotional regulation, and motivation. Findings from these studies confirm that specific neural circuits and neurotransmitter systems are in charge of the expectation of positive outcome during placebo trial, accompanied with physiological changes [71]. Based on these approaches, biological placebo effects are now understood with the concept of resilience, determined by genetic factors, and mediated by cognitive integration coupled with emotional expectations after treatment, finally maintained through learning mechanisms [72]. More and more researches are conducted to identify biological markers of placebo response, using neuroimaging and quantitative electroencephalography are in progress, developing more clinically efficient models [73].

These researches on placebo effect are now proposed in the participants with not just in pain but also in the field of mood disorders, anxiety disorders, schizophrenia, or even neurodegenerative movement disorders. But, especially in psychiatry, depressive disorder is known to be a placebo highly sensitive condition with placebo rates in clinical trials of 30–40%. It was also considered to be responsible for numerous failed antidepressant trials. Interestingly, recent systematic reviews overviewed this issue and found out completely opposite viewpoints. Systematic review in 2016 found that the placebo rates were stable for 25 years, and the apparent increase was confounded by difference in trial designs. The length of the study and the number of study centers were significant factors. Their review has suggested necessities of careful interpretation on the scientific literature and the future of psychopharmacology [74]. Another review oppositely reported that placebo response rates had grown steadily in the past 30 years [75]. But these two reviews had applied different datasets, different definition for placebo, which might be responsible for the opposite conclusions. Based on this perspective, another group reanalyzed the previous results with adjustment, suggesting no increase in placebo over the years, after 1990s [76]. How about placebo response with antipsychotics in schizophrenia patients? In a multivariable meta-regression, industry sponsorship was reported to be a significant moderator of effect sizes to increasing placebo response. Twice as many patients improved with antipsychotics as with placebo, but only a small group experienced a good response. Effect sizes were affected by industry sponsorship in a negative way, accompanied with increasing placebo response and no decreasing drug response. Even with the stable drug response over time, researchers should be aware of strength with smaller samples and better select patients in development of novel treatment [76].

Transdiagnostic approaches might point out common biological targets to regulate placebo and nocebo effect in the future. Through review of 31 meta-analyses and systematic reviews of more than 500 randomized placebo-controlled trials across psychiatry (depression, schizophrenia, mania, attention-deficit hyperactivity disorder, autism, psychosis, binge-eating disorder, and addiction), one research group tried to identify factors to be associated with increased placebo response. Only three were found to be associated with high placebo responses: low baseline severity of symptoms, more recent trials, and unbalanced randomization (more patients assigned to medication than placebo). It highlighted that predictors of the placebo response were still to be discovered, suggesting more complex mediators [77]. Recent trend in this research now involves interaction between psychological and physiological factors to find a core feature of nocebo and placebo response. Both theoretical and empirical perspectives on psychological mechanisms are traditionally suggested by two variables: conditioning related to learning theory and verbally induced expectations. Because psychological processes are in charge of link between the therapeutic context and placebo response, more sophisticated elaboration would be necessary to integrate experience-based and theoretically suggested psychological process. Clarification of psychological model on placebo response could enrich comprehension of placebo, suggesting the possibility of integration with evolved neurobiological model of placebo response [78].

Even some of significant mediators and moderators of nocebo and placebo response have been reported, elaboration with neurobiological pathways and the factors identified through experimental design are still necessary [79]. Recently, there is a new perspective approach to handle placebo with background of genetics. Using neuroimage findings on placebo, this research group presented a synthesis of recent genetic studies on the placebo effect with promising link between genetic variants in the dopamine, opioid, serotonin, and endocannabinoid pathways and placebo responsiveness. Such efforts may enable clinicians to personalize interventions to maximize treatment outcomes [80].

Trials to understand placebo and manipulate this reaction in research and reality are increasingly reported. However, conceptual clarity in placebo studies will only be settled when we first focus to how placebo terminology is in fact used within scientific researches. There are also an increasing number of trials to figure out the relationship between placebo response and psychotherapy questions starting with "What is the context of our analysis?" In a scientific perspective, researchers use placebo as "therapy theorized not to be effective for a clinical state by virtue of their intrinsic characteristics". In this way, common factors play charge on symptom improvement between psychotherapy and placebo response. Considering that this placebo effect cannot be equally replaced to nonphysical responses, more researches should be done with concentration on psychobiological mechanisms. This mechanism also can be elicited by psychotherapy, suggesting further researches with great potential on psychotherapy and placebo [81].

Sequential, Parallel Comparison Design (SPCD)

Within the perspective of placebo, patients can be divided into four groups with their intrinsic characteristics, to respond or not to respond to active treatment or placebo (Fig. 1a). Individuals responding to both drug and placebo have less characteristic necessary for the further researches. Individuals responding only to placebo and not to drug are usually negligibly small and are rarely observed. Participants who respond only to drug and not to placebo is the significant group,

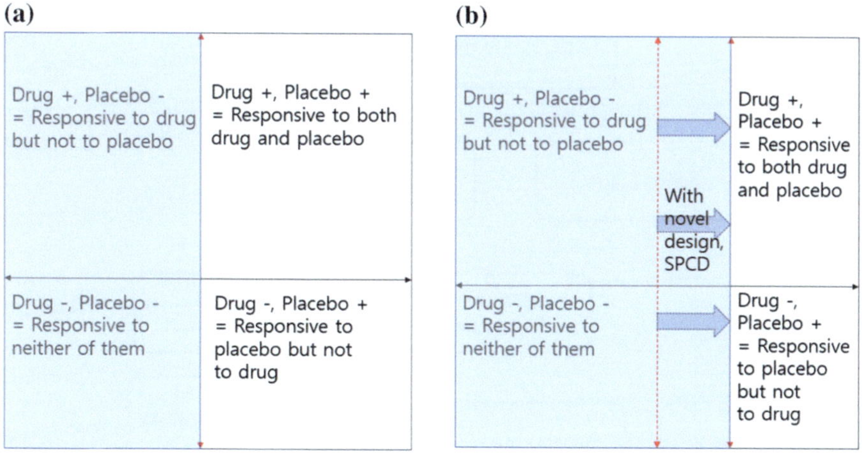

Fig. 1 Division of response to placebo and active drug and its change through SPCD. **a** Conventional division. **b** Changed division

which leads to the size of this group in research. Their numbers are affected by the numbers of the group of participants who have nonresponse both to drug and placebo and with the group of participants who have both responses to drug and placebo.

SPCD is a clinical trial design proposed by Dr. Fava and Schoenfeld in 2003 [82] to reduce both placebo response rate and the required sample size. This novel study design is cost-effective and efficient for researches in evaluating novel treatment efficacy.

In this design, researchers have common four assumptions on data analyzability and rate of response. First, every data-driven from participants will be analyzed from phase 1, while only data-driven by placebo nonresponder group will be analyzed from phase 2. Second, researchers accept low possibility in remission with placebo during the second phase in participants with nonrespond to placebo. Third, the difference between drug and placebo is expected to be greater in the second phase than the first phase. Fourth, participants are agreed to be dispositional in a half chance of active drug group through this research.

This novel design is promising to reduce the overall size of placebo response group, basically consisted of two phases of treatment (Fig. 2). In the first phase, it starts with unbalanced randomization, skewedly giving more participants randomized to placebo. Through this phase, a comparison between drug and placebo in a standard parallel comparison is possible, expecting smaller gap between drug and placebo and bringing cohort of placebo nonresponders. Then nonresponders with placebo are second randomized to either active drug or placebo. In this second phase, with greater gap between drug-placebo combined to smaller placebo response, comparison between drug and placebo is possible with a parallel comparison design. Due to participants who have already shown nonresponse to

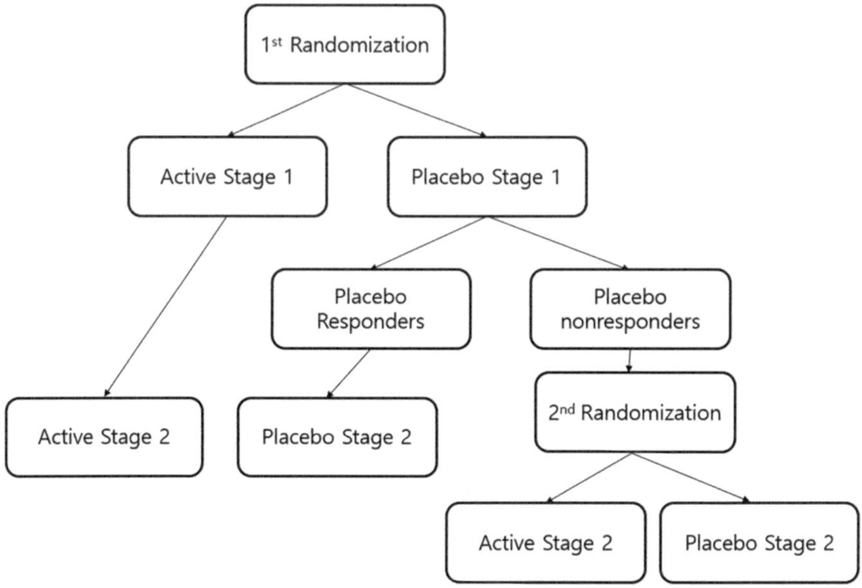

Fig. 2 Sequential, parallel comparison design (SPCD)

placebo, placebo response would be smaller than traditionally designed research (Fig. 1b). The data from both phases in a weighted fashion are used for analysis to maintain its effect size. SPCD is efficient with reducing the required sample size by 75–80% compared to the traditional two-arm clinical design [83, 84].

Integrated Treatment Research as Psychopharmacology

An area of clinical psychology which has a concern on psychological effects of medication, so-called psychopharmacology has been highlighted these days. The term "clinical psychopharmacology", was first derived from Krapelin, involving the clinical benefits of medication, the characteristics useful for prediction, negative aspects related to medication (adverse effects, toxicity, and iatrogenic conditions) and the interactions between medication and psychological variables [85]. Clinical psychopharmacology is expected to unify framework for the comprehension of clinical situation in medical and psychiatric settings. It clarifies the therapeutic use of medication and tries to differentiate from the study of the impact coming from substances with other purposes [86].

This point of view started from the major limitation of previous researches on pharmacological effect. The focused assessment usually involved the desired effects of a drug, which narrowed down the findings. This phenomenon brought a proposal of pharmacopsychometric triangle, consisting of the assessment of clinically expected effects, adverse effects, and subjectively reported quality of life [87]. This model was modified by another suggestion, consideration of three dimensions in the process of evidence-based decision with baseline risk of negative outcomes from state without treatment, responsiveness to treatment, and vulnerability to adverse effects [88]. Even their keen aspect and carefulness based on ethics, clinicians and researchers are likely to neglect the treatment-induced unwanted side effects and the patient's subjective experience of a change in terms of quality of life. The researchers should have a clear acknowledgment both to potential benefits of treatment with predictors of responsiveness and to potential adverse events by the treatment. A comprehensive clinimetric assessment of clinical variables with a view of psychopharmacology can mediate for thorough comprehension. This approach can be extended to several domains, encompassing the clinical benefits of medication, the characteristics useful for response prediction, the vulnerabilities induced by treatments, and the interactions between medication and psychological characteristics. It could be welcomed by all those who are disillusioned with the disappointing practical results of recent research in psychopharmacology, leading to strengthened reassessment.

References

1. Feinstein AR, Horwitz RI. Problems in the "evidence" of "evidence-based medicine". Am J Med. 1997;103(6):529–35.
2. Tomba E, Rafanelli C, Grandi S, Guidi J, Fava GA. Clinical configuration of cyclothymic disturbances. J Affect Disord. 2012;139(3):244–9.
3. Lavori PW, Dawson R. Adaptive treatment strategies in chronic disease. Annu Rev Med. 2008;59:443–53.
4. Tomba E, Bech P. Clinimetrics and clinical psychometrics: macro- and micro-analysis. Psychother Psychosom. 2012;81(6):333–43.
5. Moller HJ. Standardised rating scales in psychiatry: methodological basis, their possibilities and limitations and descriptions of important rating scales. World J Biol Psychiatry: Off J World Fed Soc Biol Psychiatry. 2009;10(1):6–26.
6. Castelnuovo G. New and old adventures of clinical health psychology in the twenty-first century: standing on the shoulders of giants. Front Psychol. 2017;8:1214.
7. Harmer CJ, Cowen PJ, Goodwin GM. Efficacy markers in depression. J Psychopharmacol. 2011;25(9):1148–58.
8. Dutcher JM, Creswell JD. Behavioral interventions in health neuroscience. Ann N Y Acad Sci. 2018;1428(1):51–70.
9. Malaspina D, Walsh-Messinger J, Gaebel W, Smith LM, Gorun A, Prudent V, et al. Negative symptoms, past and present: a historical perspective and moving to DSM-5. Eur Neuropsychopharmacol. 2014;24(5):710–24.
10. Bobes J, Arango C, Garcia-Garcia M, Rejas J, Group CSC. Prevalence of negative symptoms in outpatients with schizophrenia spectrum disorders treated with antipsychotics in routine clinical practice: findings from the CLAMORS study. J Clin Psychiatry. 2010;71(3):280–6.
11. Linden D. Biological psychiatry: time for new paradigms. Br J Psychiatry: J Ment Sci 2013;202(3):166–7.
12. 19th Workshop of the international stroke genetics consortium, April 28–29, 2016, Boston, Massachusetts, USA: 2016.001 MRI-defined cerebrovascular genomics-the CHARGE consortium. Neurol Genet. 2017;3(1 Suppl 1):S2–S11.
13. Schumann G, Loth E, Banaschewski T, Barbot A, Barker G, Buchel C, et al. The IMAGEN study: reinforcement-related behaviour in normal brain function and psychopathology. Mol Psychiatry. 2010;15(12):1128–39.

14. Thompson PM, Andreassen OA, Arias-Vasquez A, Bearden CE, Boedhoe PS, Brouwer RM, et al. ENIGMA and the individual: predicting factors that affect the brain in 35 countries worldwide. Neuroimage. 2017;145(Pt B):389–408.
15. Kirov G, Pocklington AJ, Holmans P, Ivanov D, Ikeda M, Ruderfer D, et al. De novo CNV analysis implicates specific abnormalities of postsynaptic signalling complexes in the pathogenesis of schizophrenia. Mol Psychiatry. 2012;17(2):142–53.
16. Uhlhaas PJ, Liddle P, Linden DEJ, Nobre AC, Singh KD, Gross J. Magnetoencephalography as a tool in psychiatric research: current status and perspective. Biol Psychiatry Cogn Neurosci Neuroimaging. 2017;2(3):235–44.
17. IsHak WW, Mirocha J, Christensen S, Wu F, Kwock R, Behjat J, et al. Patient-reported outcomes of quality of life, functioning, and depressive symptom severity in major depressive disorder comorbid with panic disorder before and after SSRI treatment in the star*d trial. Depress Anxiety. 2014;31(8):707–16.
18. Scarr E, Millan MJ, Bahn S, Bertolino A, Turck CW, Kapur S, et al. Biomarkers for psychiatry: the journey from fantasy to fact, a report of the 2013 CINP think tank. Int J Neuropsychopharmacol. 2015;18(10):pyv042.
19. Eichler HG, Bloechl-Daum B, Brasseur D, Breckenridge A, Leufkens H, Raine J, et al. The risks of risk aversion in drug regulation. Nat Rev Drug Discov. 2013;12(12):907–16.
20. Possemato K, Johnson EM, Beehler GP, Shepardson RL, King P, Vair CL, et al. Patient outcomes associated with primary care behavioral health services: a systematic review. Gen Hosp Psychiatry. 2018;53:1–11.
21. Csipke E, Williams P, Rose D, Koeser L, McCrone P, Wykes T, et al. Following the Francis report: investigating patient experience of mental health in-patient care. Br J Psychiatry: J Ment Sci. 2016;209(1):35–9.
22. Mathie E, Wythe H, Munday D, Rhodes G, Vicary P, Millac P, et al. Regional working in the east of England: using the UK national standards for public involvement. Res Involv Engagem. 2018;4:48.
23. Howe A. Principles of patient and public involvement in primary care research, applied to mental health research. A keynote paper from the EGPRN autumn conference 2017 in Dublin. Eur J Gen Pract. 2018;24(1):167–70.
24. Swildens WE, Visser E, Bahler M, Bruggeman R, Delespaul P, van der Gaag M, et al. Functional recovery of individuals with serious mental illnesses: development and testing of a new short instrument for routine outcome monitoring. Psychiatr Rehabil J. 2018;41(4):341–50.
25. Wojtalik JA, Smith MJ, Keshavan MS, Eack SM. A Systematic and meta-analytic review of neural correlates of functional outcome in schizophrenia. Schizophr Bull. 2017;43(6):1329–47.
26. de Zwart PL, Jeronimus BF, de Jonge P. Empirical evidence for definitions of episode, remission, recovery, relapse and recurrence in depression: a systematic review. Epidemiol Psychiatr Sci. 2018:1–19.
27. Gambino M, Pavlo A, Ross DA. Recovery in mind: perspectives from postgraduate psychiatric trainees. Acad Psychiatry. 2016;40(3):481–8.
28. Schawo S, Bouwmans C, van der Schee E, Hendriks V, Brouwer W, Hakkaart L. The search for relevant outcome measures for cost-utility analysis of systemic family interventions in adolescents with substance use disorder and delinquent behavior: a systematic literature review. Health Qual Life Outcomes. 2017;15(1):179.
29. Bateman DR, Srinivas B, Emmett TW, Schleyer TK, Holden RJ, Hendrie HC, et al. Categorizing health outcomes and efficacy of mHealth apps for persons with cognitive impairment: a systematic review. J Med Internet Res. 2017;19(8):e301.
30. Meerwijk EL, Parekh A, Oquendo MA, Allen IE, Franck LS, Lee KA. Direct versus indirect psychosocial and behavioural interventions to prevent suicide and suicide attempts: a systematic review and meta-analysis. Lancet Psychiatry. 2016;3(6):544–54.
31. Andrews G, Basu A, Cuijpers P, Craske MG, McEvoy P, English CL, et al. Computer therapy for the anxiety and depression disorders is effective, acceptable and practical health care: an updated meta-analysis. J Anxiety Disord. 2018;55:70–8.

32. Lei H, Nahum-Shani I, Lynch K, Oslin D, Murphy SA. A “SMART” design for building individualized treatment sequences. Annu Rev Clin Psychol. 2012;8:21–48.
33. Rush AJ, Trivedi MH, Wisniewski SR, Nierenberg AA, Stewart JW, Warden D, et al. Acute and longer-term outcomes in depressed outpatients requiring one or several treatment steps: a STAR*D report. Am J Psychiatry. 2006;163(11):1905–17.
34. Gelenberg AJ, Thase ME, Meyer RE, Goodwin FK, Katz MM, Kraemer HC, et al. The history and current state of antidepressant clinical trial design: a call to action for proof-of-concept studies. J Clin Psychiatry. 2008;69(10):1513–28.
35. Prisciandaro JJ, Tolliver BK. An item response theory evaluation of the young mania rating scale and the montgomery-asberg depression rating scale in the systematic treatment enhancement program for bipolar disorder (STEP-BD). J Affect Disord. 2016;205:73–80.
36. Pereira Junior Bde S, Tortella G, Lafer B, Nunes P, Bensenor IM, Lotufo PA, et al. The bipolar depression electrical treatment trial (BETTER): design, rationale, and objectives of a randomized, sham-controlled trial and data from the pilot study phase. Neural Plast. 2015;2015:684025.
37. Farmer C, Thurm A, Grant P. Pharmacotherapy for the core symptoms in autistic disorder: current status of the research. Drugs. 2013;73(4):303–14.
38. Owen DR, Rupprecht R, Nutt DJ. Stratified medicine in psychiatry: a worrying example or new opportunity in the treatment of anxiety? J Psychopharmacol. 2013;27(2):119–22.
39. Fernandes BS, Williams LM, Steiner J, Leboyer M, Carvalho AF, Berk M. The new field of ‘precision psychiatry’. BMC Med. 2017;15(1):80.
40. Schumann G, Binder EB, Holte A, de Kloet ER, Oedegaard KJ, Robbins TW, et al. Stratified medicine for mental disorders. Eur Neuropsychopharmacol. 2014;24(1):5–50.
41. Millan MJ. On ‘polypharmacy’ and multi-target agents, complementary strategies for improving the treatment of depression: a comparative appraisal. Int J Neuropsychopharmacol. 2014;17(7):1009–37.
42. Nikolic K, Mavridis L, Djikic T, Vucicevic J, Agbaba D, Yelekci K, et al. Drug design for CNS diseases: polypharmacological profiling of compounds using cheminformatic, 3D-QSAR and virtual screening methodologies. Front Neurosci. 2016;10:265.
43. Wong EH, Tarazi FI, Shahid M. The effectiveness of multi-target agents in schizophrenia and mood disorders: relevance of receptor signature to clinical action. Pharmacol Ther. 2010;126(2):173–85.
44. Ballon J, Stroup TS. Polypharmacy for schizophrenia. Curr Opin Psychiatry. 2013;26(2):208–13.
45. Duman RS, Voleti B. Signaling pathways underlying the pathophysiology and treatment of depression: novel mechanisms for rapid-acting agents. Trends Neurosci. 2012;35(1):47–56.
46. Duman RS. Neurobiology of stress, depression, and rapid acting antidepressants: remodeling synaptic connections. Depress Anxiety. 2014;31(4):291–6.
47. Zhou Y, Peng Y, Fang J, Sun W, Zhang G, Zhen L, et al. Effect of low dose ketamine on perioperative depressive symptoms in patients undergoing intracranial tumor resection (PASSION): study protocol for a randomized controlled trial. Trials. 2018;19(1):463.
48. Krystal JH, Abdallah CG, Averill LA, Kelmendi B, Harpaz-Rotem I, Sanacora G, et al. Synaptic loss and the pathophysiology of PTSD: implications for ketamine as a prototype novel therapeutic. Curr Psychiatry Rep. 2017;19(10):74.
49. Kim YK, Choi J, Park SC. A novel bio-psychosocial-behavioral treatment model in schizophrenia. Int J Mol Sci. 2017;18(4); pii: E734..
50. Miyashiro KY, Bell TJ, Sul JY, Eberwine J. Subcellular neuropharmacology: the importance of intracellular targeting. Trends Pharmacol Sci. 2009;30(4).203–11.
51. Arlt AD, Nestoriuc Y, Rief W. Why current drug adherence programs fail: addressing psychological risk factors of nonadherence. Curr Opin Psychiatry. 2017;30(5):326–33.
52. Sajatovic M, Ramirez LF, Fuentes-Casiano E, Cage J, Tatsuoka C, Aebi ME, et al. A 6-month prospective trial of a personalized behavioral intervention+ long-acting injectable antipsychotic in individuals with schizophrenia at risk of treatment nonadherence and homelessness. J Clin Psychopharmacol. 2017;37(6):702–7.

53. Pakpour AH, Modabbernia A, Lin CY, Saffari M, Ahmadzad Asl M, Webb TL. Promoting medication adherence among patients with bipolar disorder: a multicenter randomized controlled trial of a multifaceted intervention. Psychol Med. 2017;47(14):2528–39.
54. Averous P, Charbonnier E, Lagouanelle-Simeoni MC, Prosperi A, Dany L. Illness perceptions and adherence in bipolar disorder: an exploratory study. Compr Psychiatry. 2018;80:109–15.
55. Tessier A, Boyer L, Husky M, Bayle F, Llorca PM, Misdrahi D. Medication adherence in schizophrenia: the role of insight, therapeutic alliance and perceived trauma associated with psychiatric care. Psychiatry Res. 2017;257:315–21.
56. Sirey JA, Banerjee S, Marino P, Bruce ML, Halkett A, Turnwald M, et al. Adherence to depression treatment in primary care: a randomized clinical trial. JAMA Psychiatry. 2017;74(11):1129–35.
57. Baeza-Velasco C, Olie E, Beziat S, Guillaume S, Courtet P. Determinants of suboptimal medication adherence in patients with a major depressive episode. Depress Anxiety. 2018. https://doi.org/10.1002/da.22852. [Epub ahead of print].
58. Etain B, Scott J, Cochet B, Bellivier F, Boudebesse C, Drancourt N, et al. A study of the real-world effectiveness of group psychoeducation for bipolar disorders: is change in illness perception a key mediator of benefit? J Affect Disord. 2018;227:713–20.
59. Olfson M, Kroenke K, Wang S, Blanco C. Trends in office-based mental health care provided by psychiatrists and primary care physicians. J Clin Psychiatry. 2014;75(3):247–53.
60. Olfson M, Blanco C, Wang S, Laje G, Correll CU. National trends in the mental health care of children, adolescents, and adults by office-based physicians. JAMA Psychiatry. 2014;71(1):81–90.
61. Wolpert M, Dalzell K, Ullman R, Garland L, Cortina M, Hayes D, et al. Strategies not accompanied by a mental health professional to address anxiety and depression in children and young people: a scoping review of range and a systematic review of effectiveness. Lancet Psychiatry. 2019;6(1):46–60.
62. Davies EB, Morriss R, Glazebrook C. Computer-delivered and web-based interventions to improve depression, anxiety, and psychological well-being of university students: a systematic review and meta-analysis. J Med Internet Res. 2014;16(5):e130.
63. Simon J, Budge K, Price J, Goodwin GM, Geddes JR. Remote mood monitoring for adults with bipolar disorder: an explorative study of compliance and impact on mental health service use and costs. Eur Psychiatry. 2017;45:14–9.
64. Niendam TA, Tully LM, Iosif AM, Kumar D, Nye KE, Denton JC, et al. Enhancing early psychosis treatment using smartphone technology: a longitudinal feasibility and validity study. J Psychiatr Res. 2018;96:239–46.
65. Rootes-Murdy K, Glazer KL, Van Wert MJ, Mondimore FM, Zandi PP. Mobile technology for medication adherence in people with mood disorders: a systematic review. J Affect Disord. 2018;227:613–7.
66. Depp CA, Ceglowski J, Wang VC, Yaghouti F, Mausbach BT, Thompson WK, et al. Augmenting psychoeducation with a mobile intervention for bipolar disorder: a randomized controlled trial. J Affect Disord. 2015;174:23–30.
67. Hennemann S, Farnsteiner S, Sander L. Internet- and mobile-based aftercare and relapse prevention in mental disorders: a systematic review and recommendations for future research. Internet Interv. 2018;14:1–17.
68. Myin-Germeys I, Kasanova Z, Vaessen T, Vachon H, Kirtley O, Viechtbauer W, et al. Experience sampling methodology in mental health research: new insights and technical developments. World Psychiatry. 2018;17(2):123–32.
69. Rotondi AJ, Anderson CM, Haas GL, Eack SM, Spring MB, Ganguli R, et al. Web-based psychoeducational intervention for persons with schizophrenia and their supporters: one-year outcomes. Psychiatr Serv. 2010;61(11):1099–105.
70. Hatch A, Docherty JP, Carpenter D, Ross R, Weiden PJ. Expert consensus survey on medication adherence in psychiatric patients and use of a digital medicine system. J Clin Psychiatry. 2017;78(7):e803–12.

71. Zubieta JK, Stohler CS. Neurobiological mechanisms of placebo responses. Ann N Y Acad Sci. 2009;1156:198–210.
72. Pecina M, Zubieta JK. Molecular mechanisms of placebo responses in humans. Mol Psychiatry. 2015;20(4):416–23.
73. Gourion D, Mouchabac S. Placebo effect: clinical, biological and therapeutical involvements in depression. Encephale. 2016;42(1 Suppl 1):1S24–30.
74. Furukawa TA, Cipriani A, Atkinson LZ, Leucht S, Ogawa Y, Takeshima N, et al. Placebo response rates in antidepressant trials: a systematic review of published and unpublished double-blind randomised controlled studies. Lancet Psychiatry. 2016;3(11):1059–66.
75. Khan A, Fahl Mar K, Faucett J, Khan Schilling S, Brown WA. Has the rising placebo response impacted antidepressant clinical trial outcome? Data from the US food and drug administration 1987–2013. World Psychiatry. 2017;16(2):181–92.
76. Furukawa TA, Cipriani A, Leucht S, Atkinson LZ, Ogawa Y, Takeshima N, et al. Is placebo response in antidepressant trials rising or not? A reanalysis of datasets to conclude this long-lasting controversy. Evid Based Ment Health. 2018;21(1):1–3.
77. Weimer K, Colloca L, Enck P. Placebo eff ects in psychiatry: mediators and moderators. Lancet Psychiatry. 2015;2(3):246–57.
78. Geers AL, Miller FG. Understanding and translating the knowledge about placebo effects: the contribution of psychology. Curr Opin Psychiatry. 2014;27(5):326–31.
79. Dodd S, Dean OM, Vian J, Berk M. A review of the theoretical and biological understanding of the nocebo and placebo phenomena. Clin Ther. 2017;39(3):469–76.
80. Colagiuri B, Schenk LA, Kessler MD, Dorsey SG, Colloca L. The placebo effect: from concepts to genes. Neuroscience. 2015;307:171–90.
81. Blease CR. Psychotherapy and placebos: manifesto for conceptual clarity. Front Psychiatry. 2018;9:379.
82. Fava M, Evins AE, Dorer DJ, Schoenfeld DA. The problem of the placebo response in clinical trials for psychiatric disorders: culprits, possible remedies, and a novel study design approach. Psychother Psychosom. 2003;72(3):115–27.
83. Tamura RN, Huang X. An examination of the efficiency of the sequential parallel design in psychiatric clinical trials. Clin Trials. 2007;4(4):309–17.
84. Silverman RK, Ivanova A. Sample size re-estimation and other midcourse adjustments with sequential parallel comparison design. J Biopharm Stat. 2017;27(3):416–25.
85. Muller U, Fletcher PC, Steinberg H. The origin of pharmacopsychology: Emil Kraepelin's experiments in Leipzig, Dorpat and Heidelberg (1882–1892). Psychopharmacology. 2006;184(2):131–8.
86. Guidi J, Fava GA. Emerging trends in clinical psychology. Riv Psichiatr. 2014;49(6):227.
87. Bech P. Applied psychometrics in clinical psychiatry: the pharmacopsychometric triangle. Acta Psychiatr Scand. 2009;120(5):400–9.
88. Richardson WS, Doster LM. Comorbidity and multimorbidity need to be placed in the context of a framework of risk, responsiveness, and vulnerability. J Clin Epidemiol. 2014;67(3):244–6.

An Integrated Bio-psycho-social Approach to Psychiatric Disorders

Seon-Cheol Park and Yong-Ku Kim

Abstract
The biopsychosocial approach is a model of eclecticism, which consists of multidisciplinary academic fields, reacting against the "neuroscientification" of psychiatry. The biopsychosocial approach was proposed by George Engel following Adolf Meyer's psychobiological and Roy R. Ginker's eclectic approach to psychiatry. Although the use of the biopsychosocial approach is increasing, it has several limitations: First, specific practices cannot be guided by the biopsychosocial approach because it is considered to be "boundless psychiatry." Second, unlike an initial intention, the symptomatic use of psychotropic medications may be justified by the biopsychosocial approach. Third, the economic forces to enhance biological psychiatry cannot be hindered by the biopsychosocial approach. Hence, to overcome the limitations of the current biopsychosocial approach, potential new paradigms including evolutionary psychiatry, pragmatism, integrationism, and pluralism have been proposed. Above all, Eric Kandel presented the link between neuroscience and psychiatry from the perspective of integrationism. In accordance with integrationism and/or pluralism, based on the paradigm shift of the theoretical construct from chemical imbalance to dysfunctional circuit, next-generation treatments for mental disorders have been proposed by Thomas Insel. Thus, a more integrated biopsychosocial

S.-C. Park
Department of Psychiatry, Inje University Haeundae Paik Hospital, 875, Haeun-daero, Haeundae-gu, Busan 48108, Republic of Korea

Y.-K. Kim (✉)
Department of Psychiatry, College of Medicine, Korea University Ansan Hospital, 516, Gojan-dong, Ansan, Kyunggi Province 425-707, Republic of Korea
e-mail: yongku@korea.ac.kr; yongku@korea.edu

© Springer Nature Singapore Pte Ltd. 2019

Y.-K. Kim (ed.), *Frontiers in Psychiatry*, Advances in Experimental Medicine and Biology 1192, https://doi.org/10.1007/978-981-32-9721-0_17

approach to managing psychiatric disorders including schizophrenia and panic disorder may be proposed.

Keywords
Biopsychosocial approach · Eclecticism · Integrationism · Pluralism · Schizophrenia · Panic disorder

Introduction

The biopsychosocial approach has been the most representative model for conceptualizing the relationship between mind and body in the realm of psychiatry and was developed from the general systems theory. The biopsychosocial model is based on the following hypothesis: "Nature is arranged in a continuous hierarchy from small, less complex units to larger, more complex units [1]." In the general systems theory, each stage in the hierarchy appears as an organic dynamical whole, and each system has a distinctive relationship and properties according to its stages. On the other hand, nothing is isolated and all the systems are affected by the parts. Thus, an event should be viewed not as a specific dimension, but should be analyzed using a multidimensional approach instead, including interpersonal relationship, psychology, physiology, and biochemistry among others [1]. Based on the general systems theory, Schwartz [2] stated that the separation of mind and body is not categorically, but dimensionally different, and that the system is the whole created by the dynamic interaction of the elements. Bunge [3] explained that emergent properties are properties of the systems as a whole but not features of its individual components. Sperry [4] stated that mental phenomenon was closely related to material cerebral processing but differed from material cerebral processing in terms of the mind–body relationship. With the help of the emergent theory, the general systems theory influenced the biological psychosocial model, but the relationship between mind and body can be explained in terms of mind–body dualism [1].

According to two papers published in *Science* [5] and the *American Journal of Psychiatry* [6], the biopsychosocial model was formulated by George Engel (1913–1999), who was specialized in internal medicine, psychoanalysis, and psychosomatic medicine. However, George Engel was not a psychiatrist and never trained in psychiatry. Engel's biopsychosocial approach has been based on the criticisms of rigorous biomedicine. According to a paper in a family practice journal [7], Engel's criticisms on biomedicine were as follows: First, the correlation of a biochemical alteration into illness can be poor. Second, the relationship between the presence of a biological derangement and the meaning of the symptoms remains tenuous. Third, psychosocial factors can be regarded as important determinants of susceptibility, severity, and course of illness. Fourth, the presence of biological derangements cannot be inevitably linearly associated with the effects of illness. Fifth, psychosocial factors can affect the outcome of biological treatments. Sixth, the doctor–patient relationship can affect the clinical outcome. Finally, unlike in animal models, study methods can profoundly influence the patients and study subjects can influence the scientists. However, the failure to implement the biopsychosocial approach in psychiatry can contribute to the failure in establishing empirical concepts for reacting against the change of psychiatry into an aspect of biomedicine. Thus, to overcome the limitations of the biopsychosocial approach in psychiatry, the adoption of potential new paradigms as alternative methods has been currently proposed in psychiatry [8–10].

In this chapter, we review the theoretical backgrounds, clinical criticism, and potential new paradigms for the biopsychosocial approach and suggest a more integrated biopsychosocial approach to psychiatric disorders.

Theoretical Backgrounds for the Biopsychosocial Approach

The psychobiology of Adolf Meyer (1866–1950), the Swiss-born American psychiatrist, corresponded to the precursor of the biopsychosocial approach. From the viewpoint of Meyerian psychobiology, which opposes the concepts of categorical diseases and syndromes defined by Emil Kraepelin, psychiatric conditions were regarded as the reactions to life events. Moreover, in a Meyerian sense, in terms of psychiatric treatment, an understanding of the whole person from a longitudinal view of their entire life history was emphasized, not simply a cross-sectional assessment of their present conditions. The Meyerian approach considered that psychiatric treatment could be defined as a compromise between the most diverse theoretical views and methods in practical eclecticism [10, 11].

Roy R. Ginker (1900–1993), the American neuropsychiatrist, played a significant role in the intermediate link between the Meyerian psychobiological and the Engelian biopsychosocial approach [10]. According to Ginker, the term for the biopsychosocial model was coined and defined throughout the 1950s and the 1960s [12, 13]. Thus, the biopsychosocial approach originated from Ginker's concept for psychiatry. Ginker criticized the psychoanalytic dogmatists and interpreted the Freudian contributions in light of Thomas Kuhn's philosophy of science [10, 14]. From Ginker's viewpoint, although Freud's psychoanalysis had previously contributed to the psychiatry revolution, the analytic theory resulted in the stagnation in the realm of psychiatry. Thus, Ginker considered that psychiatry could be revolutionized by conceptualizing the biopsychosocial model. Ginker formulated the biopsychosocial model more definitely as follows: "The broad term *biopsychosocial* encompasses all aspects of the living organism. It indicates the inseparability of the environment from organic life and the relationship between human existence and its social and cultural products [13]." However, despite Ginker's priority of the term for the biopsychosocial approach, Engel never acknowledged this priority. Ginker's conceptualizing of psychiatry was based on the general systems theory, which denoted that nature was arranged in a consecutive hierarchy constituting of smaller and less complex units to bigger and more complex units. The general systems theory was introduced by the Austrian biologist Ludwig von Bertalanffy (1901–1972) and by the English-born American economist Kenneth Boulding (1910–1993) around the year 1955. In the general systems theory, each of the levels in the hierarchy was presented as an organic dynamical whole or system, and each system was distinctively related and characterized based on its levels. Moreover, for living organisms, there were no isolated units, all systems are affected by the parts, all levels were connected to each other, and one change inevitably resulted in another change [1, 12].

Rise and Fall of the Biopsychosocial Approach

Although the biopsychosocial approach was proposed as a reaction against the entire "scientification" of psychiatry, the rise of the psychosocial approach was definitely influenced by the rise of the empirical trends in the Diagnostic and Statistical Manual of Mental Disorders, the 3rd edition (DSM-III) [15] in the realm of psychiatry in the USA and around 1980. These trends can be attributed to the rise of psychoanalytic approach in the DSM-I [16] and DSM-II [17], and the placement of the biopsychosocial approach on the continuum between psychoanalytic and neurobiological dogmas. Moreover, there was a trend of the rise of the psychopharmacological approach compared to the decline of the psychoanalytic approach around the 1980s [8–10].

However, the biopsychosocial approach has been criticized in terms of several aspects as follows: First, despite the permission of all diverse treatment modalities, a specific practical guidance for clinical psychiatry cannot be provided by the biopsychosocial approach in terms of its nature of "boundless psychiatry." According to a metaphor from McHugh and Slaveney [18], the biopsychosocial approach is not like a recipe but rather a list of ingredients. Namely, in terms of understanding the psychiatric aspects in different conditions and in different circumstances, the biopsychosocial approach can list the relevant aspects of psychiatry but cannot guide a decision for providing a specific therapeutic approach [19]. Second, although the biopsychosocial approach was initially designed to protect psychotherapies and as a reaction against biological psychiatry, the approach cannot provide specific practical information for psychopharmacology through a valid and reliable method. Despite the initial purpose of conceptualizing the biopsychosocial approach, it has been known that the approach unexpectedly resulted in the justification of the symptomatic use of psychotropic medications because of its arbitrariness of eclecticism [10, 20]. Third, the economic forces to enhance the biological dogmatism have not been prevented from the biopsychosocial approach. It is speculated that the biomedical model may strongly influence the effect of the pharmaceutical industry on current psychiatric practices [10].

Potential New Paradigms for the Biopsychosocial Approach

Ghaemi [8] has proposed several potential new paradigms as alternative methods for the biopsychosocial approach, including evolutionary or Darwinian psychiatry, pragmatism, integrationism, and pluralism.

First, the evolutionary theory, known as Darwinian psychiatry, has been suggested as a new paradigm for psychiatry [21]. From the viewpoints of Darwinian psychiatry, the susceptibility gene for schizophrenia has been regarded as the inevitable trade-off for adaptations related to key species-related innovations in humans [22–24]. Crow [25] has speculated that schizophrenia can be conceptualized as the price that humans pay to acquire language, and shared common biological underpinnings of language and psychosis can be similar to a "big bang"-like genetic mechanism involving both the X and Y chromosomes. According to the hypothesis of Crow [23], because of many cases of non-recombination of Y chromosomes during meiosis, frequent genetic drift on the Y chromosome can develop. Thus, Crow [26] has hypothesized that one of these genes can be involved in both the speciation events of modern *Homo sapiens* and language ability. Additionally, Crow [24] has considered that Schneider's first-rank symptoms correspond to the extreme forms of language disorder and the extreme form of difficulty in distinguishing between speech production and speech perception in the sapiens-specific language circuit may be a central symptom of schizophrenia. Moreover, it has been suggested that the creative thinking associated with divergent thinking has been regarded as a compensatory reward of schizophrenia. It has been shown that creative thinking presents an inheritable pattern similar to that of schizophrenia, particularly in monozygotic twins [27]. Despite the advantages of the application of Darwinian psychiatry or evolutionary theory to psychiatry, skepticism for these scientific methods results from the dogmatic applications of the paradigm [9].

The second alternative method, pragmatism, emphasizes open-mindedness and a hesitance to accommodate any single system of thought, from the approach of James [28], or antipositivism, that states that the community of investigators get closer to the truth but never perceive the absolute truth, based on the approach by Pierce [29]. Thus, the utilitarianism of the method, seeking to combine different methods based on pragmatic ends and value judgments, can be one of the novel alternative paradigms for the biopsychosocial approach [30]. Although the pragmatic approach includes the important value of avoiding dogmatism, it has a limitation in that seeking to combine different methods is quite similar to extreme eclecticism [8].

The third novel approach stems from the failure of the dogmatic approaches of Carl Wernicke and Sigmund Freud, thereby needing an integration of different methods or paradigms and a flexible attitude in the realm of psychiatry [9]. Eric Kandel, the Nobel prize-winning neuroscientist and psychiatrist, is the most representative example of the modern integrationists. Kandel [31] showed that brain structure can be altered by the environment from his work on conditioned learning in the *Aplysia* where he demonstrated the two-way nature of interaction between brain and nature. Thus, it is postulated that neuroscience is linked to psychiatry, psychoanalysis, and psychotherapy. Namely, the post psychotherapy changes in brain function can be detected with neuroimaging. However, more systematic exploration is needed to better understand the differences between processes operating at conscious and unconscious levels. Furthermore, there is promise in pretherapeutic brain marker predictors for favorable treatment outcomes, including the pretreatment activity of the anterior cingulate and orbitofrontal cortices. Moreover, the improved outcome can be clearly achieved by combining novel pharmacological agents with psychotherapy [32]. It is presumed that the outcome of two different treatments including psychotherapy and pharmacotherapy can be predicted by specific and measurable biomarkers of the mental disorder, as well as neuroscientific connections between psychiatry, psychoanalysis, and psychotherapy [33]. According to Kandel's view [31–33], psychotherapy and psychoanalysis can be "neuroscientifically" justified and an approach for "neuro-psychoanalysis" can be developed [34].

Pluralism, the fourth approach, denotes neither an eclectic nor dogmatic perspective on psychiatry. The basic concept for pluralism means that all of psychiatry cannot be explained only by a single paradigm. From the perspective of pluralism, each of the methods or paradigms has strengths and weaknesses, and the method or paradigm that can be established as the best approach is the one with the greatest number of strengths and the lowest number of limitations. Thus, from the pluralistic perspective, having a point of perspective beyond the current confused eclecticism, as well as rejecting the dogmatisms of the biological, psychoanalytic, or postmodernistic perspectives, is important [8]. Karl Jaspers (1883–1969), the German psychiatrist and philosopher, took a stand in the pluralistic trend in psychiatry. Jaspers has criticized the theoretical incompleteness of both the opposite extreme poles of Carl Wernicke's dogma that "all mental illnesses are cerebral illnesses" and Sigmund Freud's dogma that "all mental illnesses are personality illnesses [35]." Moreover, Jaspers has initiated the phenomenological trend and the formulated rigorous distinction between explanation and understanding for evaluating the psychic phenomenon. Thus, Jaspers has proposed that, in terms of the phenomenological approach, explanation and understanding corresponds entirely to somatic and psychic approaches, respectively [36–40]. In a Jaspersian sense, psychiatry can be defined as a hybrid scientific discipline combining the methods of social science and natural science [41].

A More Integrated Biopsychosocial Approach to Psychiatric Disorders

In terms of precision medicine, consistent with the paradigm shift of the theoretical construct from chemical imbalance to dysfunctional circuit, Insel [42] has proposed next-generation treatments for mental disorders. According to Insel's proposal [42], the medication targets are changed from monoamines, cholinergic pathways, and γ-aminobutyric acid (GABA) receptors to glutamate receptors or transporters, neuropeptide receptors, neuroprotection or neurogenesis, synaptic plasticity, transcriptional or epigenetic modifiers, and neurophysiology. Additionally, the clinical targets are changed from psychosis, mood regulation, anxiety, and attention to a motivational state, attentional bias, executive function, anhedonia, hopelessness, social deficits, and working memory. Moreover, the clinical treatments are changed from medications, psychotherapy, and electroconvulsive therapy to targeted medications, structured psychotherapies, somatic therapies, and combination treatments. Thus, the next-generation treatments for mental disorders proposed by Insel [42], can be an alternative paradigm for overcoming the obstacles of the biopsychosocial approach from the viewpoints of integrationism and/or pluralism. Furthermore, according to Insel's view, the novel biopsychosocial approaches to schizophrenia and panic disorder can be discussed.

Novel Biopsychosocial Approach to Schizophrenia

Since the most common antipsychotic medications are focused on positive symptoms mainly associated with dopamine, the need for improving cognitive and negative symptoms is unsatisfied. However, cognitive and negative symptoms, rather than positive symptoms, are more significantly associated with psychosocial impairments in patients with schizophrenia. Glutamate, GABA, and other non-dopaminergic systems of the prefrontal cortex are proposed as the potential medication targets for schizophrenia. Additionally, epigenetic modifications with the balance of excitation and inhibition can be explained from the viewpoint of neurodevelopment and neural circuitry. Schizophrenia can be conceptualized as a neurodevelopmental disorder for the following reasons: Reduced elaboration of inhibitory pathways and excessive pruning of excitatory pathways in children contribute to alterations in the excitatory-inhibitory balance in the prefrontal cortex and reduced myelination in the development of schizophrenia. Thus, negative symptoms can be directly related with synaptic connectivity in schizophrenic patients. Moreover, preventive intervention for the prodromal neurodevelopmental changes can be established [43]. Hence, the non-dopaminergic systems involved in schizophrenia, rather than dopaminergic mechanisms, should be the focus of a novel biopsychosocial approach for the treatment of schizophrenia [44].

Using the clarification of clinical features and the advent of biomarkers and new cognitive methods, the stages of schizophrenia can be neurodevelopmentally established [43, 45]. Thus, based on the cognitive and negative symptoms, each of the stages for schizophrenia can be defined as follows: Stage I is defined by the characteristics of genetic vulnerability that correspond to the likelihood of being diagnosed with a genetic sequence and family history. Stage II is characterized by cognitive, behavioral, and social deterioration, and by diagnosis through cognitive assessment, neuroimaging, and the Structured Interview for Prodromal Syndrome. Stage II can be managed with cognitive training, treatment with polysaturated fatty acids, and social support. Stage III is defined by abnormal thought and behavior and a relapsing-remitting course, which are assessed through clinical interview. Stage III interventions include medications and psychosocial interventions. Stage IV is marked by continued deterioration, unemployment and homelessness, medical complications, and incarceration. Stage IV interventions include medication, psychosocial interventions, and rehabilitation services.

Novel Biopsychosocial Approach to Panic Disorder

A novel biopsychosocial approach to panic disorder can be proposed through a complete integration of behavioral, psychophysiological, neurobiological, and genetic characteristics. Serotonin, neuropeptides, glucocorticoids, neurotrophins, and other neurotransmitters have been the major targets for panic disorder-related molecular genetic studies. Despite the availability of current medications for panic disorder, there is still an unmet need for more effective, faster-acting, and more tolerable pharmacological interventions. Hence, novel pharmacological mechanism-based anti-panic therapeutics including glutamatergic receptor modulators, orexin receptor antagonists, and corticotropin-releasing factor 1 receptor antagonists have been suggested [46].

The dysfunctional "cross-talk" between emotional drive (limbic structure) and cognitive inhibition (prefrontal cortex) and the fear circuit (amygdala–hippocampus–prefrontal axis) have been continually regarded as the neural correlates of panic disorder. The pharmacotherapy for panic disorder may aim to improve learning during cognitive-behavioral therapy, targeted with plasticity in cortical structures. The effectiveness of pharmacotherapy has been demonstrated in both rodents and humans. The reductions of severe acrophobic symptoms are due to the use of pharmacotherapy together with D-cycloserine; a successful example of combination therapy. Thus, a novel biopsychosocial approach to panic disorder can be defined by targeting glutamate- and orexin-related molecular mechanisms associated with the fear circuit, which includes the amygdala–hippocampus–prefrontal cortex axis [47–49].

Conclusions

Under the influences of Meyer's psychobiology and Ginker's eclectic approach, the biopsychosocial approach has been developed to counteract the rise of biological psychiatry and/or neuroscience. However, because of its ambiguous eclecticism, the biopsychosocial approach has been devalued as "boundless psychiatry." Moreover, unintentionally, the biopsychosocial approach has justified the symptomatic use of psychotropic drugs in the realm of clinical psychiatry. Therefore, evolutionary psychiatry, pragmatism, integrationism, and pluralism have been proposed as novel alternative paradigms to compensate for the current eclecticism of the biopsychosocial approach in treating psychiatric disorders. Among the novel paradigms, Eric Kandel's link between neuroscience and psychotherapy enable the conceptualization of neuro-psychoanalysis. In terms of precision medicine, therapeutic targets and treatments can be more sophisticated and specialized in clinical psychiatry, and the integration of pharmacotherapy and psychotherapy is more emphasized. For example, since schizophrenia can be modeled as a neurodevelopment disorder, the integrations of psychotherapeutic and pharmacological treatments can be used at all the stages of schizophrenia. Moreover, in terms of the more integrated approach to psychiatric disorders, the fear circuit (amygdala–hippocampus–prefrontal axis) associated with glutamate- and orexin-related molecular mechanisms may be the treatment targets in panic disorder.

References

1. Yang BY, Hwang HS. The differentiation and integration of mind and body. Korean J Psychosom Med. 2000;8(1):110–21.
2. Schwartz GE. A systems analysis of psychology and behavior therapy. Implications for behavioral medicine. Psychother Psychosom. 1981;36(3–4):159–84.
3. Bunge M. Emergence and the mind. Neuroscience. 1977;2(4):501–9.
4. Sperry RW. A modified concept of consciousness. Psychol Rev. 1969;76(6):532–6.
5. Engel GL. The need for a new medical model: a challenge for biomedicine. Science. 1977;196(4286):126–36.
6. Engel GL. The clinical application of the biopsychosocial model. Am J Psychiatry. 1980;137(5):535–44.
7. Borrell-Carrio F, Suchman AL, Epstein RM. The biopsychosocial model 25 years later: principles, practice, and scientific inquiry. Ann Fam Med. 2004;2(6):576–82.
8. Ghaemi SN. Paradigms of psychiatry: eclecticism and its discontents. Curr Opin Psychiatry. 2006;19(6):619–24.
9. Ghaemi SN. The rise and fall of the biopsychosocial model. Br J Psychiatry. 2009;195(1):3–4.
10. Ghaemi SN. The rise and fall of the biopsychosocial model: reconciling art & science in psychiatry. Baltimore: The Johns Hopkins University Press; 2010.
11. Meyer A. The commonsense psychiatry of Adolf Meyer. New York: McGraw Hill; 1948.
12. Ginker RR Sr. "Open-system" psychiatry. Am J Psychoanal. 1966;26(2):115–28.
13. Ginker RR Sr. Training of a psychiatrist-psychoanalyst. J Am Acad Psychoanal. 1994;22(2):343–50.
14. Kuhn T. The structure of scientific revolutions. Chicago: University of Chicago Press; 1962.
15. American Psychiatric Association. Diagnostic and statistical manual of mental disorders. 3rd ed. Washington: American Psychiatric Association; 1980.
16. American Psychiatric Association. Diagnostic and statistical manual of mental disorders. 1st ed. Washington: American Psychiatric Association; 1952.
17. American Psychiatric Association. Diagnostic and statistical manual of mental disorders. 2nd ed. Washington: American Psychiatric Association; 1968.
18. McHugh RR, Slavney PR. The perspectives of psychiatry. Baltimore: Johns Hopkins University Press; 1983.
19. McLaren N. A critical review of the biopsychosocial model. Aust N Z Psychiatry. 1998;32(1):86–92.
20. Ghaemi SN. Toward a hippocratic psychopharmacology. Can J Psychiatry. 2008;53(3):189–96.
21. McGuire M, Troisi A. Darwinian psychiatry. New York: Oxford University Press; 1998.
22. Crow TJ. Schizophrenia as failure of hemispheric dominance for language. Trends Neurosci. 1997;20(8):339–43.
23. Pearlson GD, Folley BS. Schizophrenia, psychiatric genetics, and Darwinian psychiatry: an evolutionary framework. Schizophr Bull. 2008;34(4):722–33.
24. Park YC, Lee MS, Si TM, Chiu HFK, Kanba S, Chong MY, Tripathi A, Udomratn P, Chee KY, Tanra AJ, Rabbani G, Javel A, Kathiarachchi S, Myint WA, Cuoung TV, Sim K, Yang SY, Sartorius N, Tan CH, Shinfuku N, Park S-C. Psychotropic drug-prescribing correlates of disorganized speech in Asians with schizophrenia: the REAP-AP study. Saudi Pharm J. 2019;27(2):246–53.

25. Crow TJ. The 'big bang' theory of the origin of psychosis and the faculty of language. Schizophr Res. 2008;102(1–3):31–2.
26. Crow TJ. The origins of psychosis and the evolution of language: cerebral asymmetry. In: Hugdahl K, Westerhausen R, editors. The two halves of the brain: information processing in the cerebral hemispheres. Massachusetts: The MIT Press; 2010. p. 621–61.
27. Grigorenko EL, LaBuda MC, Carter AS. Similarity in general cognitive ability, creativity, and cognitive style in a sample of adolescent Russian twins. Acta Genet Med Gemellol. 1992;41(1):65–72.
28. James W. A pluralistic universe. Lincoln, NB: University of Nebraska Press; 1909.
29. Perice C. Selected writings. New York: Dover Publications; 1958.
30. Brendel D. Healing psychiatry. Cambridge, MA: MIT Press; 2006.
31. Kandel E. A new intellectual framework for psychiatry. Am J Psychiatry. 1998;155(4):457–69.
32. Etkin A, Pittenger C, Polan HJ, Kandel ER. Toward a neurobiology of psychotherapy: basic science and clinical applications. J Neuropsychiatry Clin Neurosci. 2005;17(2):145–58.
33. Kandel E. The new science of mind and the future of knowledge. Neuron. 2013;80(3):546–60.
34. Whitehead CC. Toward a 'new' paradigm of therapeutic action: neuro-psychoanalysis and downward causation. J Am Acad Psychoanal Dyn Psychiatry. 2005;33(4):637–56.
35. de Leon J. DSM-5 and the research domain criteria: 100 years after Jaspers' general psychopathology. Am J Psychiatry. 2014;171(5):492–4.
36. Yang BY, Moon HS. Karl Jaspers and general psychopathology. Korean J Psychopathol. 1992;1(1):12–21.
37. Yang BY, Hong SK. Psychopathology of delusion: the phenomenological approach of Karl Jaspers. Korean J Psychopathol. 1996;5(1):15–26.
38. Yang BY, Hur YS. Karl Jaspers and psychopathology. Korean J Psychopathol. 2002;11(2):3–11.
39. Ghaemi SN. Understanding mood disorders: Karl Jaspers' biological existentialism. In: Stanghellini G, Fuchs T, editors. One century of Karl Jaspers' general psychopathology. New York: Oxford University Press; 2013. p. 258–75.
40. Park S-C. Karl Jaspers' general psychopathology (Allgemeine Psychopathologie) and its implication for the current psychiatry. Psychiatry Investig. 2019;16(2):99–108.
41. de Leon J. Is psychiatry scientific? A letter to a 21st century psychiatry resident. Psychiatry Investig. 2013;10(3):205–17.
42. Insel TR. Next-generation treatments for mental disorders. Sci Transl Med. 2012;4(155):155psc19.
43. Insel TR. Rethinking schizophrenia. Nature. 2010;468(7321):187–93.
44. Kim YK, Choi J, Park S-C. A novel bio-psychosocial-behavioral treatment model in schizophrenia. Int J Mol Sci. 2017;18(4):E734.
45. Nestler EJ, Hyman SE. Animal models of neuropsychiatric disorders. Nat Neurosci. 2010;13(10):1161–9.
46. Park S-C, Kim Y-K. A novel bio-psychosocial-behavioral treatment model of panic disorder. Psychiatry Investig. 2019;16(1):4–15.
47. Dresler T, Guhn A, Tupak SV, Ehlis AC, Herrman MJ, Fallgatter AJ, Deckert J, Domschke K. Revise the revised? New dimensions of the neuroanatomical hypothesis of panic disorder. J Neural Transm. 2013;120(1):3–29.
48. Santos M, D'Amico D, Dierssen M. From neural to genetic substrates of panic disorder: insights from human and mouse studies. Eur J Pharmacol. 2015;759:127–41.
49. Choi KY, Kim YK. Plasticity-augemented psychotherapy for refractory depressive and anxiety disorders. Prog Neuropsychopharmacol Biol Psychiatry. 2016;70:134–47.

Early Identification of Psychiatric Disorders

Tae Young Lee, Minah Kim and Jun Soo Kwon

Abstract

Early detection and early intervention approach targets people at-risk stages who are very close to conversion to illness, or patients who have just transited the illness stage and have not, yet, become chronic. Rigorous efforts have been put in for early detection and early intervention in psychiatry. A high-risk population is identified by clinical manifestations that, as per their severity and suffering, do not yet meet the diagnostic criteria of psychiatric disorders. There have been attempts to break through the existing phenotype-based diagnostic system using biomarkers, but researchers have yet to overcome the heterogeneity of the disease. Nowadays, the clinical staging models in psychosis, bipolar, and depressive disorders have been proposed as a heuristic and a practical alternative to this. This model is evolving to integrate various demographic factors, clinical symptoms, cognitive functions, and biomarkers.

Keywords

Early identification · Heterogeneity · Clinical staging · Biomarkers · Psychiatric disorders

T. Y. Lee · M. Kim · J. S. Kwon (✉)
Department of Psychiatry, Seoul National University College of Medicine, 101 Daehak-ro, Chongno-gu, Seoul 03035, Republic of Korea
e-mail: kwonjs@snu.ac.kr

© Springer Nature Singapore Pte Ltd. 2019

Y.-K. Kim (ed.), *Frontiers in Psychiatry*, Advances in Experimental Medicine and Biology 1192, https://doi.org/10.1007/978-981-32-9721-0_18

Mental illness is one of the leading causes of disability worldwide [1]. They are not just an immense burden on those influenced and relatives, but also a social and economic burden on society. The World Health Organization (WHO) has given a high priority to prevention research on mental illness [2], the objective of which is to decrease the individual impairment, the stigmatization by the disease, and social expenses. Early detection and early intervention in psychiatric disorders have several simple principles; quick engagement, accurate assessment, evaluation of disease course, and stages of development, intervention considering bio-psycho-social aspects [3]. This approach targets people at-risk stages who are very close to conversion to illness, or patients who have just transited the illness stage and have not, yet, become chronic. Providing timely psychiatric help can assist the patients to recover more quickly, be more advantageous in their academical activities, regain their jobs, restore social ties, and eventually to live the fullest of lives. Hence, the consensus had been reached on the significance of early intervention in psychiatry, and studies began to focus on this topic. In the present article, we will look at the current situation for early detection on the major mental illnesses including psychosis, bipolar disorder, and depression, followed by the introduction of the clinical staging models for mental illness. This article first explores the current high-risk studies for various mental illnesses and aims to provide the future directions of early detection studies in psychiatry.

Early Identification of Psychiatric Disorders

Early detection and early intervention in mental illness have been well studied in psychosis. It is suggested that schizophrenia is "brewed" and the signs and the symptom preceded long before the actual diagnosis of the illness [4]. Previous studies have reported that the shorter the duration to the first treatment after the onset of schizophrenia, the better the response to the treatment and the less recurrence [5, 6]. Reducing the duration of untreated psychosis (DUP) in the first-episode psychosis has become a crucial goal for clinicians to alleviate the severity of symptoms and prevent chronification of the disease. Thus, it was hoped that intervention in the prodromal stage could further ameliorate, even delaying or preventing the onset of the disorder. Since the term "prodrome" is a retrospective concept, the terms "ultrahigh-risk" or "at-risk mental state" were introduced [7]. A large number of patients with schizophrenia already reported a variety of symptoms, including changes in perception, beliefs, cognition, mood, affect, and behavior that preceded psychosis, and only a small number of patients had developed the illness without any apparent symptoms [8]. Over the last 20 years, a number of psychosis studies have contributed to show the insight to detect and intervene individuals at high-risk for psychosis [7]. A high-risk population is identified by clinical manifestations that, as per their severity and suffering, do not yet meet the diagnostic criteria of schizophrenia. If necessary, the symptoms used to characterize the risk criteria are additionally combined with a decrease in general functioning and the genetic risk of the schizophrenia or comorbid schizotypal personality disorder [9].

The risk population are then administered with interventions to postpone or prevent a complete transition to major mental disorders. Meta-studies have reported that about one-third of high-risk individuals during the 3-year follow-up convert to psychosis [10]. The high false positive rate has raised ethical problems with the hope that the concept of psychosis high-risk group can effectively select candidates who can have future conversion [11]. It is known, however, that samples that have not yet converted still have a low level of functioning and a high rate of non-remission [12]. The presence of groups that do not transit to psychosis or remit from the high-risk state makes us to ask more questions about their characteristics. With the intensification of the high-risk group of psychosis studies, the incidence of psychotic outcomes gradually decreased [13]. However, a meta-analytic study has shown that intensive outreach campaigns predominantly targeting the general population and a higher proportion of self-referrals have been diluted in the pretest risk for psychosis studies [14]. This fact adds to the concern about the recruitment pathway of the current high-risk for psychosis diagnostic system. Therefore, the elaboration of diagnostic criteria for high-risk psychosis and objectification through biomarkers should be considered more seriously.

Bipolar disorder (BD) is a chronic and potentially debilitating illness. Notably, about 60% of patients with bipolar disorder mostly have their first episode in their teenage years, usually between ages 15 and 19 [15]. Previous studies suggest that the bipolar prodrome is characterized by symptoms associated with dysregulation in mood and energy, and the symptoms and the signs found are represented in the categories of "irritability and aggressiveness", "sleep disturbances", "mania-type symptoms and signs", "hyperactivity", "anxiety-related symptoms and signs", "mood swings", and "depression-type symptoms and signs" [16]. Therefore, there was an expectation that these clinical phenotypes could be utilized to study bipolar prodrome [17]. In the past, studies on subsyndromal bipolar spectrum disorders such as cyclothymia, bipolar disorder NOS, mood disorder NOS, or depressive disorder, as well as bipolar disorder offspring sample, have been extensively studied [18]. However, as it is in the high-risk for psychosis studies, new studies are being conducted on those with attenuated mania symptoms that have not yet been diagnosed as bipolar disorder. Based on the experience from the psychosis risk syndrome, bipolar at-risk (BAR) criteria were proposed [19]. It consists of the following three subcategories: group 1, subthreshold mania; group 2, depression with cyclothymic features; group 3, first degree relative with bipolar disorder. Preliminary results reported a 14.3% incidence of the BAR criteria positive group, which was significantly higher than the BAR criteria negative group [20]. Besides the BAR criteria, Semistructured Interview for the Bipolar At-Risk States (SIBARS) and the Bipolar Prodrome Symptom Scale (BPSS) were also proposed [21, 22]. Unlike the high-risk for psychosis, the mainstream was the bipolar offspring study [23]. Bipolar prodrome studies still have a mixture of clinical phenotype and genetic risk group studies, and the use of terms has not yet been clarified. Future studies may better integrate these research lines.

A similar concept also exists in depressive disorder studies. Although it is not the peak period of onset, the incidence of depression is still high in young people and about 28% experience depressive episodes before their 20 s [24]. When depression is chronic, recurrence is frequent, general functioning is deteriorated, and accompanying problems such as substance abuse are accompanied. Thus, it is imperative to detect depression early. The following three factors are the essential features of initial depressive disorder; [1] anxiety and irritability [2] subthreshold or minor depression [3] these symptoms are present in the first and subsequent episodes of major depression, which may inform efforts toward relapse prevention [25]. However, many studies that have been conducted by using the existing depression diagnostic scale simply refer to mild depression as a prodrome. In addition, most early detection studies are primarily for post-natal depression or differential diagnosis from dementia [26, 27]. Despite the presence of many screening instruments for depression, a few studies exist, which in a longitudinal study design prospectively observe the subthreshold depression than other psychosis studies.

More recently, a new concept covering all the possible outcomes of all major mental illnesses was proposed. For the first time, high-risk for psychosis individuals were examined for the possibility of developing into various mental illnesses [28]. The pluripotentiality of high-risk psychosis, however, is currently less supported. A recent study showed that high-risk psychosis is not a pluripotent group for various mental disorders but is indeed specific for psychotic disorders [29]. Besides, many of the illnesses, such as major depressive disorder, which are associated with the high-risk psychosis group, are not the consequences of being at the risk state, but in fact coexist even from the baseline [30]. In contrast, risk states for mood disorder or anxiety were relatively more pluripotential compared with high-risk for psychosis [31]. Recently, the concept of broad clinical high-risk mental state (CHARMS) was proposed [32]. It was made to reflect the transdiagnostic characteristics of the early stage of mental illness, and it reflects an "early shared pathway" or a form of pluripotency of the early clinical phenotypes of mental disorders [33]. Therefore, initial symptoms or signs of mental illness may not have a fixed trajectory to a specific psychiatric disorder and may evolve into a range of different mental illnesses [34]. Hence, the CHARMS criterion is expected to be a useful diagnostic scale that enables highly efficient detection and intervention from the early stage of mental illness that requires clinical help.

Heterogeneity Issues

Today, the early diagnosis of mental illness largely depends on phenotypes represented by symptoms, rather than the objective markers derived from the disease pathophysiology. Diagnostic systems based on these phenotypes, thus, inevitably lead to problems of heterogeneity. Several recent studies have raised serious questions for us in the differential diagnosis of mental illness. While there has been robust evidence of increased dopamine synthesis capacity of the patients with schizophrenia, it was not well known whether this synthesizing capacity of dopamine was related to the high-risk population before the onset of the psychosis.

Howes and his colleagues found evidence that the onset of psychosis is preceded by presynaptic dopaminergic dysfunction [35]. Moreover, in a subsequent study, they found that increased dopamine synthesis capacity was observed only in the high-risk group who later developed psychosis, and not in those who did not develop psychosis [36]. These findings have proven the value of dopamine synthesis capacity in the high-risk population as a useful biomarker for predicting the onset of psychosis and have been considered as highly innovative discoveries leading to early detection and early intervention in psychiatry. However, further studies threw the wet blanket on our expectations. In a study of people who experienced hallucinations but were not diagnosed with psychosis, no significant difference in dopamine synthesis capacity was found compared with healthy controls [37]. In addition, when the schizophrenia patients were divided into the groups with good response to antipsychotic drugs and those with poor, no significant group differences were present in their dopamine synthesis capacity [38]. In a test of the transdiagnostic dopamine hypothesis of psychosis, patients with bipolar disorder and schizophrenia showed the same pattern of dopamine synthesis ability in terms of symptom severity [39]. Taken together, these findings suggest that current diagnostic systems are not very efficient in making distinction between the psychiatric disorders in terms of dopamine synthesis capacity, there still have serious heterogeneity within the schizophrenia.

Clinical Staging

What is clinical staging? Clinical staging is, just, a heuristic and simplified type of diagnostic system [40]. Its utility has been well proven through the treatment of cancer, where the quality of life and expected survival rate depend on effective interventions. Staging of cancer is a system that assesses the development of the cancer by its growth and the spread. Contemporary practice allots a number from I to IV to a neoplasm growth, with I being a separated disease and IV being a disease that has spread to the furthest reaches of what the evaluation measures. The stage, for the most part, considers the extent of a tumor, regardless of whether it has intrusive neighboring organs, what number of territorial (adjacent) lymph nodes it has spread to (assuming any), and whether it has shown up in progressively far off areas (metastasized). The pathophysiology of cancer is different depending on the type of the cancer. However, the cancer staging system provides a very useful guide for making clinical decisions and this system has high applicability in various medical illnesses.

Clinical staging is distinguished from the existing diagnostic system for its consideration of the continuum of the illness course, as well as utilizing the features from multiple time points of the illness course. Clinical staging is based on the fact that the disease gradually progresses over time. All disorders are diagnosed making considerations of all possible situations, such as a condition with the mild symptoms that do not yet meet the diagnostic criteria, and a condition that has already converted to a specific disease and is now in a chronic phase [41].

There are two fundamental assumptions in clinical staging. First, patients in the early stages of the disease have a better response to treatment and a better prognosis than those in the later stages. Second, the treatment provided at the initial stage should be more benign and useful [42]. The cross-sectional and longitudinal diagnostic considerations of the disease will help the clinicians expand the scope of appropriate interventions and provide adequate treatment for the patient's status. This process is ultimately thought to help patients minimize risks and maximize benefits.

In schizophrenia, diagnostic systems prior to the DSM-IV identified the illness simply based on the symptoms, and, depending on their natures, the illness was divided into several subtypes, amongst which the residual subtype is an exception as it consists mostly of chronic patients. However, since the subtypes of schizophrenia were derived from the categorical classification, based on phenomenologies, rather than from the disease pathologic classification, many challenges were present when finding their usefulness in therapeutic response or actual biological researches. In the DSM-5, however, changes were made to include the attenuated psychosis syndrome in the diagnostic criteria, although it was included in the appendix with the limitation that further research is needed [43]. These changes can be considered to be more tailor-made because it allows clinicians to take into account various disease states by shifting away from the dichotomy of whether or not schizophrenia has developed. This strategy may still be useful in illnesses not only for schizophrenia, which is a progressive disease but also for a mood disorder, which recurs repeatedly. Primarily, Fava and Kellner proposed the first staging model of schizophrenia [44]. They referred to the first stage with depressed mood and negative symptoms as the prodromal symptom accompanied by deterioration of general functioning, the acute episode as the second stage, the residual symptom stage as the third stage, the chronic stage with the onset more than 6 months and less than 2 years as the fourth stage, and finally, the chronic phase of 2 years or more was classified into five stages. These disease stages were then refined by adding a premorbid stage and simplifying the residual and chronic stage [45]. Based on the experience with the high-risk population for psychosis, McGorry and his colleagues have presented an updated model that is not merely a demonstration of the natural course of schizophrenia, but rather an updated model that further subdivides the prodromal phase so that it can be implemented for early detection and early prevention of disease [46]. To do this, the authors proposed appropriate interventions for each stage. At stage 0, subjects show no apparent symptoms or signs. Then, subjects show mild or nonspecific symptoms or signs at stage 1a and present moderate subthreshold symptoms at stage 1b. At stage 2, subjects convert to first-episode psychosis, which can be diagnosed. Stage 3a refers to incomplete remission from the first-episode psychosis and stage 3b refers to recurrence or relapse of the disorder. Multiple relapses are objectively present at stage 3c. Finally, stage 4 refers to severe, persistent, or unremitting illness. The clinical staging model led to attempts to integrate and better explain the biological changes of schizophrenia. This is an attempt to link the clinical phenotypes to specific pathophysiological mechanism, and the cognitive markers, brain structural markers, mismatch negativity as a marker of brain function, sleep and chronobiological markers, neuroendocrine markers, and inflammatory and oxidative stress markers were proposed [47]. Validation of this model is still being conducted. So far, there was no difference between stage 0 and normal controls in the model, and, further, a significant gradual increase in clinical severity was observed with the progression of clinical stages from stage 0, stage 1a, and stage 1b [48]. This construct validity was also confirmed in schizophrenia spectrum disorders, as well as in stage 4 [49, 50].

Mood disorder was more elaborate than schizophrenia, reflecting the characteristics of the cyclic disease course. Stage 1a describes the subclinical level, with stage 1b describing an attenuated type, similar to those in schizophrenia, but from stage 3, the model is subdivided. Stage 3a refers to recurrence of subthreshold mood symptoms, stage 3b refers to first-threshold mood symptoms and stage 3c refers to multiple relapses [51]. However, although the main symptoms required for the diagnosis of high-risk for psychosis are limited to attenuated positive symptoms, stage 1a and stage 1b of hypothetical clinical staging model in bipolar disorder have a wide range of symptoms such as dysthymia, cyclothymia, mood fluctuation with comorbid symptoms of anxiety, substance abuse/misuse, conduct problems, suicidal thoughts/gestures, subthreshold psychotic symptoms, or a major depressive episode with bipolarity or admixtures of these symptoms and rapid changes between phase. It is also claimed that items such as a family history of bipolarity and suicide, a highly recurrent pattern of illness, early age of onset, atypical features (hypersomnia, hyperphagia, rejection sensitivity, leaden paralysis), abrupt onset or offset of episodes and brief episodes, postpartum episodes, severe premenstrual syndrome, psychotic features melancholia, a seasonal pattern, and prominent irritability should also be included in the category [52]. In stage 3, psychiatric conditions, mainly due to insufficient treatment, further complicate the discussion with the presence of bipolar II disorder, mixed state, and subsyndromal depressed state. However, the construct validity of the clinical staging model in bipolar disorder, like how it is in schizophrenia, has been confirmed throughout the stages of the high-risk to the refractory. Notably, the presence of a family history of bipolar disorder is the most significant risk factor for later transition to a bipolar disorder [53]. Biomarkers for predicting the onset of bipolar disorders include peripheral biomarkers such as antithyroid peroxidase antibodies, salivary cortisol levels, inflammation factors in blood, and cerebral metabolite concentrations ratio, but more research is needed to integrate with the clinical staging model. Staging model of depressive disorder is divided into stage 1a, which is characterized by mild functional impairment and fluctuation, but no depressive mood, and stage 1b, which is characterized by subsyndromal mood syndromes of clinical interest. In stage 3, there are three substages in which stage 3a refers to sleeping disorder, anxiety, irritability, libido but not mood symptoms; stage 3b refers to depressed mood, guilt, hopelessness; stage 3c refers to dysthymia [54].

However, clinical staging models based on specific mental disorder is being challenged today. High-risk for psychosis have specific criteria for psychosis, but other such prodromal stages for mental illness still have difficulties in maintaining their specificities [55]. Anxiety disorder is commonly observed at the prodromal stage of mood disorder, and mood disorders are also common in the prodromal stage of schizophrenia [56]. This pluripotentiality is widely observed in the high-risk state of anxiety disorders, depression, and bipolar disorder, and psychosis also occurs in these high-risk groups [31]. Therefore, considering the phenotyping diagnosis system of mental illnesses, it is an important issue that there exist overlapping areas between the symptoms and the illnesses. Ultimately, the staging model should be extended to the transdiagnostic model. Recently published meta-analytic investigation proposed the integrating model of clinical staging model of bipolar disorder [57]. This model shows stage 1 of non-mood disorders, stages 2, 3 of non-bipolar mood disorder, and stage 4 of mania/hypomania, together with possible substance use disorders. This model is becoming more convincing as the results of comorbid disorders and mood disorder clinical course are integrated. This type of integration eventually will evolve into a model encompassing both psychosis, bipolar disorder, unipolar disorder, and anxiety disorders [47]. When individuals at high-risk for psychosis, for example, convert to bipolar disorder, this model will help to resolve confusions whether the subjects had been mislabeled as a high-risk for bipolar disorder or had still been at a high-risk state of psychosis. However, the transdiagnostic approach in the psychiatric area still has several methodological inconsistencies [58]. Interestingly, a feasible model for predicting the onset of psychosis in high-risk groups has been published using empirical data, and recently a transdiagnostic model has also been proposed [59–61]. These models have excellent value in terms of making the theoretical model available in clinical practice. Therefore, it will be essential to verify the validity of the clinical staging model and elaborate it into a replicable model.

Ethical Considerations

Despite the good intentions, the problem of early detection of mental illness is caused by a high percentage of false positives. According to the current studies, only one-third of the subjects diagnosed with high-risk for psychosis are known to develop psychosis [10] and the results of prospective bipolar prodrome or depressive prodrome studies are not yet clear. However, preliminary results suggest it to be no more than one-third [19]. The small number of conversions limited the power of the study to identify associations with risk factors that have previously been reported to predict bipolar disorder [62]. This suggests that a large number of subjects will be diagnosed unnecessarily due to early detection. High ratio of false positives can lead to unnecessary social stigmatization that these people are dangerous, and can also cause anxiety to people that they may develop the illness at any time. Also, any medical effort to get out of a high-risk state can cause unnecessary medical costs or insurance problems [11]. Finally, those who are at high-risk for mental illness are mostly minors and need a guardian for medical decisions. Therefore, there is a risk that their autonomy will be restricted. Thus, the risks of stigmatization, side effects of medication, and restricted autonomy may need to be balanced against the therapeutic benefit of early intervention.

Conclusion

Rigorous efforts have been put in for early detection and early intervention in psychiatry. Early detection is tightly coupled with the development of the diagnostic system of mental illness. There have been attempts to break through the existing phenotype-based diagnostic system using biomarkers, but researchers have yet to overcome the heterogeneity of the disease. Nowadays, the clinical staging model has been proposed as a heuristic and a practical alternative to this. This model is evolving to integrate various demographic factors, clinical symptoms, cognitive functions, and biomarkers. It is expected that the clinical staging model that can be used in clinical practice will be further developed in the future.

Acknowledgements This research was supported by Basic Science Research Program through the National Research Foundation of Korea (NRF) funded by the Ministry of Science, ICT and Future Planning (Grant no. 2019R1A2B5B03100844).

References

1. Lopez AD, Murray CC. The global burden of disease, 1990–2020. Nat Med. 1998;4(11):1241–3.
2. Collins PY, Patel V, Joestl SS, March D, Insel TR, Daar AS, et al. Grand challenges in global mental health. Nature. 2011;475(7354):27–30.
3. McGorry PD, Killackey E, Yung A. Early intervention in psychosis: concepts, evidence and future directions. World Psychiatry Off J World Psychiatr Assoc WPA. 2008;7(3):148–56.
4. McGlashan TH, Zipursky RB, Perkins D, Addington J, Miller TJ, Woods SW, et al. The PRIME North America randomized double-blind clinical trial of olanzapine versus placebo in patients at risk of being prodromally symptomatic for psychosis. I. Study rationale and design. Schizophr Res. 2003;61(1):7–18.
5. Marshall M, Lewis S, Lockwood A, Drake R, Jones P, Croudace T. Association between duration of untreated psychosis and outcome in cohorts of first-episode patients: a systematic review. Arch Gen Psychiatry. 2005;62(9):975–83.
6. Perkins DO, Gu H, Boteva K, Lieberman JA. Relationship between duration of untreated psychosis and outcome in first-episode schizophrenia: a critical review and meta-analysis. Am J Psychiatry. 2005;162(10):1785–804.
7. Fusar-Poli P, Borgwardt S, Bechdolf A, Addington J, Riecher-Rossler A, Schultze-Lutter F, et al. The psychosis high-risk state: a comprehensive state-of-the-art review. JAMA Psychiatry. 2013;70(1):107–20.
8. Häfner H, Maurer K. Early detection of schizophrenia: current evidence and future perspectives. World Psychiatry. 2006;5(3):130–8.
9. Yung AR, Yuen HP, McGorry PD, Phillips LJ, Kelly D, Dell'Olio M, et al. Mapping the onset of psychosis: the comprehensive assessment of at-risk mental states. Aust N Z J Psychiatry. 2005;39(11–12):964–71.
10. Fusar-Poli P, Bonoldi I, Yung AR, Borgwardt S, Kempton MJ, Valmaggia L, et al. Predicting psychosis: meta-analysis of transition outcomes in individuals at high clinical risk. Arch Gen Psychiatry. 2012;69(3):220–9.
11. Mittal VA, Dean DJ, Mittal J, Saks ER. Ethical, legal, and clinical considerations when disclosing a high-risk syndrome for psychosis. Bioethics. 2015;29(8):543–56.

12. Addington J, Cornblatt BA, Cadenhead KS, Cannon TD, McGlashan TH, Perkins DO, et al. At clinical high risk for psychosis: outcome for nonconverters. Am J Psychiatry. 2011;168(8):800–5.
13. Yung AR, Yuen HP, Berger G, Francey S, Hung T-C, Nelson B, et al. Declining transition rate in ultra high risk (prodromal) services: dilution or reduction of risk? Schizophr Bull. 2007;33(3):673–81.
14. Fusar-Poli P, Schultze-Lutter F, Cappucciati M, Rutigliano G, Bonoldi I, Stahl D, et al. The dark side of the moon: meta-analytical impact of recruitment strategies on risk enrichment in the clinical high risk state for psychosis. Schizophr Bull. 2016;42(3):732–43.
15. Merikangas KR, Akiskal HS, Angst J, Greenberg PE, Hirschfeld RMA, Petukhova M, et al. Lifetime and 12-month prevalence of bipolar spectrum disorder in the National Comorbidity Survey replication. Arch Gen Psychiatry. 2007;64(5):543–52.
16. Skjelstad DV, Malt UF, Holte A. Symptoms and signs of the initial prodrome of bipolar disorder: a systematic review. J Affect Disord. 2010;126(1–2):1–13.
17. Correll CU, Penzner JB, Frederickson AM, Richter JJ, Auther AM, Smith CW, et al. Differentiation in the preonset phases of schizophrenia and mood disorders: evidence in support of a bipolar mania prodrome. Schizophr Bull. 2007;33(3):703–14.
18. Noto MN, de Souza Noto C, de Jesus DR, Zugman A, Mansur RB, Berberian AA, et al. Recognition of bipolar disorder type I before the first manic episode: challenges and developments. Expert Rev Neurother. 2013;13(7):795–806; quiz 807.
19. Bechdolf A, Nelson B, Cotton SM, Chanen A, Thompson A, Kettle J, et al. A preliminary evaluation of the validity of at-risk criteria for bipolar disorders in help-seeking adolescents and young adults. J Affect Disord. 2010;127(1–3):316–20.
20. Bechdolf A, Ratheesh A, Cotton SM, Nelson B, Chanen AM, Betts J, et al. The predictive validity of bipolar at-risk (prodromal) criteria in help-seeking adolescents and young adults: a prospective study. Bipolar Disord. 2014;16(5):493–504.
21. Fusar-Poli P, De Micheli A, Rocchetti M, Cappucciati M, Ramella-Cravaro V, Rutigliano G, et al. Semistructured interview for bipolar at risk states (SIBARS). Psychiatry Res. 2018;264:302–9.
22. Van Meter A, Guinart D, Bashir A, Sareen A, Cornblatt BA, Auther A, et al. Bipolar prodrome symptom scale—abbreviated screen for patients: description and validation. J Affect Disord. 2019;15(249):357–65.
23. Hafeman DM, Merranko J, Axelson D, Goldstein BI, Goldstein T, Monk K, et al. Toward the definition of a bipolar prodrome: dimensional predictors of bipolar spectrum disorders in at-risk youths. Am J Psychiatry. 2016;173(7):695–704.
24. Lewinsohn PM, Rohde P, Seeley JR. Major depressive disorder in older adolescents: prevalence, risk factors, and clinical implications. Clin Psychol Rev. 1998;18(7):765–94.
25. Hetrick SE, Parker AG, Hickie IB, Purcell R, Yung AR, McGorry PD. Early identification and intervention in depressive disorders: towards a clinical staging model. Psychother Psychosom. 2008;77(5):263–70.
26. Lee DTS, Chung TKH. Postnatal depression: an update. Best Pract Res Clin Obstet Gynaecol. 2007;21(2):183–91.
27. Swainson R, Hodges JR, Galton CJ, Semple J, Michael A, Dunn BD, et al. Early detection and differential diagnosis of Alzheimer's disease and depression with neuropsychological tasks. Dement Geriatr Cogn Disord. 2001;12(4):265–80.
28. Agius M, Zaman R, Hanafy D. An audit to assess the consequences of the use of a pluripotential risk syndrome: the case to move on from "psychosis risk syndrome (PRS)". Psychiatr Danub. 2013;25(Suppl 2):S282–5.
29. Woods SW, Powers AR, Taylor JH, Davidson CA, Johannesen JK, Addington J, et al. Lack of diagnostic pluripotentiality in patients at clinical high risk for psychosis: specificity of comorbidity persistence and search for pluripotential subgroups. Schizophr Bull. 2018;44(2):254–63.

30. Fusar-Poli P, Nelson B, Valmaggia L, Yung AR, McGuire PK. Comorbid depressive and anxiety disorders in 509 individuals with an at-risk mental state: impact on psychopathology and transition to psychosis. Schizophr Bull. 2014;40(1):120–31.
31. Lee TY, Lee J, Kim M, Choe E, Kwon JS. Can we predict psychosis outside the clinical high-risk state? A systematic review of non-psychotic risk syndromes for mental disorders. Schizophr Bull. 2018;44(2):276–85.
32. Hartmann JA, Nelson B, Spooner R, Paul Amminger G, Chanen A, Davey CG, et al. Broad clinical high-risk mental state (CHARMS): methodology of a cohort study validating criteria for pluripotent risk. Early Interv Psychiatry. 2017.
33. McGorry PD. Beyond psychosis risk: early clinical phenotypes in mental disorder and the subthreshold pathway to safe, timely and effective care. Psychopathology. 2014;47(5):285–6.
34. McGorry P. Early clinical phenotypes and risk for serious mental disorders in young people: need for care precedes traditional diagnoses in mood and psychotic disorders. Can J Psychiatry Rev Can Psychiatr. 2013;58(1):19–21.
35. Howes OD, Bose SK, Turkheimer F, Valli I, Egerton A, Valmaggia LR, et al. Dopamine synthesis capacity before onset of psychosis: a prospective [18F]-DOPA PET imaging study. Am J Psychiatry. 2011;168(12):1311–7.
36. Allen P, Luigjes J, Howes OD, Egerton A, Hirao K, Valli I, et al. Transition to psychosis associated with prefrontal and subcortical dysfunction in ultra high-risk individuals. Schizophr Bull. 2012;38(6):1268–76.
37. Howes OD, Shotbolt P, Bloomfield M, Daalman K, Demjaha A, Diederen KMJ, et al. Dopaminergic function in the psychosis spectrum: an [18F]-DOPA imaging study in healthy individuals with auditory hallucinations. Schizophr Bull. 2013;39(4):807–14.
38. Demjaha A, Murray RM, McGuire PK, Kapur S, Howes OD. Dopamine synthesis capacity in patients with treatment-resistant schizophrenia. Am J Psychiatry. 2012;169(11):1203–10.
39. Jauhar S, Nour MM, Veronese M, Rogdaki M, Bonoldi I, Azis M, et al. A Test of the transdiagnostic dopamine hypothesis of psychosis using positron emission tomographic imaging in bipolar affective disorder and schizophrenia. JAMA Psychiatry. 2017;74(12):1206–13.
40. McGorry PD, Purcell R, Hickie IB, Yung AR, Pantelis C, Jackson HJ. Clinical staging: a heuristic model for psychiatry and youth mental health. Med J Aust. 2007;187(7 Suppl):S40–2.
41. Scott J, Leboyer M, Hickie I, Berk M, Kapczinski F, Frank E, et al. Clinical staging in psychiatry: a cross-cutting model of diagnosis with heuristic and practical value. Br J Psychiatry. 2013;202(4):243–5.
42. McGorry PD, Hickie IB, Yung AR, Pantelis C, Jackson HJ. Clinical staging of psychiatric disorders: a heuristic framework for choosing earlier, safer and more effective interventions. Aust N Z J Psychiatry. 2006;40(8):616–22.
43. Tsuang MT, Van Os J, Tandon R, Barch DM, Bustillo J, Gaebel W, et al. Attenuated psychosis syndrome in DSM-5. Schizophr Res. 2013;150(1):31–5.
44. Fava GA, Kellner R. Staging: a neglected dimension in psychiatric classification. Acta Psychiatr Scand. 1993;87(4):225–30.
45. Lieberman JA, Perkins D, Belger A, Chakos M, Jarskog F, Boteva K, et al. The early stages of schizophrenia: speculations on pathogenesis, pathophysiology, and therapeutic approaches. Biol Psychiatry. 2001;50(11):884–97.
46. McGorry PD, Nelson B, Goldstone S, Yung AR. Clinical staging: a heuristic and practical strategy for new research and better health and social outcomes for psychotic and related mood disorders. Can J Psychiatry. 2010;55(8):486–97.
47. McGorry P, Keshavan M, Goldstone S, Amminger P, Allott K, Berk M, et al. Biomarkers and clinical staging in psychiatry. World Psychiatry. 2014;13(3):211–23.
48. Addington J, Liu L, Goldstein BI, Wang J, Kennedy SH, Bray S, et al. Clinical staging for youth at-risk for serious mental illness. Early Interv Psychiatry. 2019.

49. Berendsen S, van der Paardt J, van Bruggen M, Nusselder H, Jalink M, Peen J, et al. Exploring construct validity of clinical staging in schizophrenia spectrum disorders in an acute psychiatric ward. Clin Schizophr Relat Psychoses. 2018.
50. Godin O, Fond G, Bulzacka E, Schurhoff F, Boyer L, Myrtille A, et al. Validation and refinement of the clinical staging model in a French cohort of outpatient with schizophrenia (FACE-SZ). Prog Neuropsychopharmacol Biol Psychiatry. 2019;10(92):226–34.
51. Berk M, Hallam KT, McGorry PD. The potential utility of a staging model as a course specifier: a bipolar disorder perspective. J Affect Disord. 2007;100(1–3):279–81.
52. Faedda GL, Baldessarini RJ, Glovinsky IP, Austin NB. Pediatric bipolar disorder: phenomenology and course of illness. Bipolar Disord. 2004;6(4):305–13.
53. Vieta E, Salagre E, Grande I, Carvalho AF, Fernandes BS, Berk M, et al. Early intervention in bipolar disorder. Am J Psychiatry. 2018;175(5):411–26.
54. Cosci F, Fava GA. Staging of mental disorders: systematic review. Psychother Psychosom. 2013;82(1):20–34.
55. McGorry PD, Hartmann JA, Spooner R, Nelson B. Beyond the "at risk mental state" concept: transitioning to transdiagnostic psychiatry. World Psychiatry Off J World Psychiatr Assoc WPA. 2018;17(2):133–42.
56. Hartmann JA, Nelson B, Ratheesh A, Treen D, McGorry PD. At-risk studies and clinical antecedents of psychosis, bipolar disorder and depression: a scoping review in the context of clinical staging. Psychol Med. 2019;49(2):177–89.
57. Raouna A, Osam CS, MacBeth A. Clinical staging model in offspring of parents with bipolar disorder: a systematic review. Bipolar Disord. 2018;20(4):313–33.
58. Fusar-Poli P, Solmi M, Brondino N, Davies C, Chae C, Politi P, et al. Transdiagnostic psychiatry: a systematic review. World Psychiatry Off J World Psychiatr Assoc WPA. 2019;18(2):192–207.
59. Fusar-Poli P, Oliver D, Spada G, Patel R, Stewart R, Dobson R, et al. Real world implementation of a transdiagnostic risk calculator for the automatic detection of individuals at risk of psychosis in clinical routine: study protocol. Front Psychiatry. 2019;10:109.
60. Fusar-Poli P, Werbeloff N, Rutigliano G, Oliver D, Davies C, Stahl D, et al. Transdiagnostic risk calculator for the automatic detection of individuals at risk and the prediction of psychosis: second replication in an independent national health service trust. Schizophr Bull. 2018.
61. Fusar-Poli P, Rutigliano G, Stahl D, Davies C, Bonoldi I, Reilly T, et al. Development and validation of a clinically based risk calculator for the transdiagnostic prediction of psychosis. JAMA Psychiatry. 2017;74(5):493–500.
62. Ratheesh A, Cotton SM, Betts JK, Chanen A, Nelson B, Davey CG, et al. Prospective progression from high-prevalence disorders to bipolar disorder: exploring characteristics of pre-illness stages. J Affect Disord. 2015;1(183):45–8.

New Theoretical Frameworks for Research

Theoretical Psychiatry as a Link Between Academic and Clinical Psychiatry

Miro Jakovljevic and Ivan Jakovljevic

Nothing is as important as a good theory.
—Kurt Lewin

Abstract

From its beginning, psychiatry has been always characterized by different orientations in the forms of "schools", "fields", "branches", etc., some of which were dominating during different periods of time. Today psychiatry seems to be in cul de sac of a serious scientific crisis and in the midst of the paradigm clashes. Academic psychiatry has been more and more criticized to be more or less irrelevant to clinical practice. The new field called theoretical psychiatry is fundamental for further scientific and professional maturation of psychiatry at the twenty-first century. Theoretical psychiatry pursues knowledge and understanding of mental disorders, and it operates so through the formulation, testing, and evaluation of theories. Digital revolution is changing significantly all fields of science, medicine, and psychiatry changing regimes and methods of knowledge production. Big data approach promises to provide the scientific holy grail in psychiatry, a single overarching theory or multiple theories and models that unify all the scientific disciplines. Brain is place where biological, psychological, social, and spiritual mechanisms meet each other and interact. Theoretical psychiatry should give all psychiatrists a common language, build bridges over academic gaps, and creatively export insights across disciplinary borders.

Keywords

Theoretical psychiatry · Systems psychiatry · Academic psychiatry · Paradigms and perspectives in psychiatry

M. Jakovljevic (✉) · I. Jakovljevic
Department of Psychiatry, University Hospital Centre Zagreb, Zagreb, Croatia
e-mail: predstojnik_psi@kbc-zagreb.hr

© Springer Nature Singapore Pte Ltd. 2019

Y.-K. Kim (ed.), *Frontiers in Psychiatry*, Advances in Experimental Medicine and Biology 1192, https://doi.org/10.1007/978-981-32-9721-0_19

Introduction

The challenges of the twenty-first century, called the century of mind, the merger of infotech and biotech in medicine and cross-disciplinary interactions, are upon psychiatry. Psychiatry today faces with serious crisis and vigorous critiques of its theory and practice from many sides, although it has improved the lives of many people suffering from mental disorders [1]. The present crisis of psychiatry is predicated on "its ambiguous position as medical specialty, institution of social control, and secular arena for dealing with moral, spiritual, and existential problems" [2] as well as on its divided professional self and fragmented identities. The beginning of the twenty-first century has marked the strong need for a new field called "theoretical psychiatry" for the further scientific and professional maturation of the discipline [3]. In recent decades, huge progress in technology has resulted in a fascinating accumulation of new findings and concepts about all major mental disorders. However, the general sense in psychiatry and about psychiatry is that we are still far from an integrated and holistic understanding of mental health and mental disorders as well as from satisfactory treatment and prevention of mental disorders. From its very beginning, psychiatry has been characterized by different concepts, orientations, and approaches. These were formalized as "schools", like psychoanalysis and psychodynamic psychiatry, social psychiatry, medical psychiatry, biological psychiatry and pharmacopsychiatry, evolutionary psychiatry, etc., which bore influence during different eras. All models and explanations for mental disorders to date are more or less speculative and do not offer clear predictions in terms of prognosis, precise diagnosis, comorbidities, and treatment responses. Psychiatry has been frequently criticized because of weak science, of ethical lapses, and of pharmaceutical industry's corruption of the research enterprise with undisclosed conflicts of interest, inadequate research design, biased analyses, exaggerated claims, and misleading conclusions [4] Academic psychiatry is claimed to be without intellectual excitement and to be almost irrelevant to clinical practice and mental health field [5]. The motto of the X World Congress of psychiatry in Madrid 1996 "One World, One Language—Paving the Way of Better Perspective for Mental Health" has still remained only a nice dream. Be that as it may the new field called theoretical psychiatry based on philosophy of science and medicine, big data, and computational science is fundamental for further scientific and professional maturation of psychiatry and response to crisis. It has been argued that the gap between significant theoretical possibilities and poor situation in clinical practice as well as between academic and clinical psychiatry has been large so that the cross-disciplinary interactions and transdisciplinary systems approach are of paramount importance.

Fragmentary Identities and Divided Self of Psychiatry

> If a theory did not explain the fact, then there was something wrong with the theory.
>
> —Mehl (1989)

Theory and practice of psychiatry are fraught with debate, conflicts, confusion, and mutual annihilation and negation as well as with unnecessary and unproductive polarization, mistrust and disrespectfulness between proponents of different disciplines. Placed at the crossroads of biomedicine, the social sciences, and the humanities [6], psychiatry has several partial or fragmentary identities related to its biologic, psychodynamic, social, and culture subspecialties with many psychiatric schools. Many of these schools not only do not accept but criticize and negate the most basic tenets and treatment principles of the others. Disparate psychiatric branches are rooted in different modern and postmodern philosophical and scientific viewpoints about the nature of human beings and the nature of psychiatric problems [7–10]. Their proponents are talking different languages, making different assumptions and theories, use different criteria and methods and practice different systems of values. When looking at Table 1, one can say that psychiatry is a clash field where science, pop-science, pseudoscience, and antiscience confront one another.

At present we can identify several models of interactions between varied fields and schools in psychiatry: (1) the contact model: there is some overlap between the fields and schools in psychiatry which can interact into conflict or harmony; (2) the conflict model: differed fields and schools are in essential and principal conflict because they have used different information processing styles and methods; (3) the independence model: the fields and schools operate independently as separate disciplines which ARE quite removed from each other because they operate from different positions; (4) the dialogue model of pluralism: even the fields and schools are separated they communicate on the common ground of epistemology; and (5) the Integration model refers to transdisciplinary unification of different perspectives and paradigms. Psychiatry has yet to become a coherent field of scientific theory and one unified and standardized practice; rather it looks like a loosely assembled set of theoretical concepts and practices. Although many of the fragmentary psychiatric schools lack respect for, aggressively criticize, and negate the fundamental tenets and treatment principles of others, it is evident that there are common threads in many of these schools. The challenges of the present time push mainstream psychiatry to move beyond the narrow and fragmentary frameworks that characterized the discipline in the last century. It is wrong to pit biological, psychological, social, and spiritual concepts of mental disorders against one another as well as therapeutic procedures related to them. As different risk and ethiopathogenic factors may be complementary, rather than competitive excluding one another, so varied explanations and models may be interconnected and complementary because they are just different ways of talking about the same issue. The identification of etiopathogenesis is usually based on instrumental reasons to be able to obtain the desired outcome of treatment and depends on what are processes to be controlled. The systems

Table 1 Psychiatry in operation: syndrome of divided and fragmented professional identity

Medical psychiatry—Biological psychiatry—Biopsychiatry—Pharmacopsychiatry—Biophysical psychiatry—Biometaphorical psychiatry—Circuits-based psychiatry—Organic psychiatry—Clinical psychiatry—Descriptive psychiatry—Bihevioral medicine—Psychodinamic psychiatry—Existential psychiatry—Social psychiatry—Communal psychiatry—Holistic psychiatry—Humanistic psychiatry—Integrative psychiatry—Transcultural psychiatry (Intercultural psychiatry—Cultural psychiatry)—Etnopsychiatry—Interpersonal psychiatry—Narrative psychiatry—Neuropsychiatry—New brave psychiatry—Energetic psychiatry—Quantum psychiatry—Atomic psychiatry—Moral psychiatry—Spiritual psychiatry—General psychiatry—Special psychiatry
Modern psychiatry—Personalized psychiatry—Patients-friendly psychiatry—Patient-centered psychiatry—Positive psychiatry—Postmodern psychiatry—Postpsychiatry—Client friendly psychiatry
Law (forensic) psychiatry—Military psychiatry—War psychiatry—Industrial psychiatry—Community psychiatry—Pastoral psychiatry
Prenatal psychiatry—Child psychiatry—Adolescence psychiatry—Adult psychiatry—Developmental psychiatry—Gerontopsychiatry—Feminists psychiatry—Female psychiatry
Academic psychiatry—Experimental psychiatry—Scientific psychiatry—Evidence-based psychiatry—
Evolutionary psychiatry—Molecular psychiatry—Genetic psychiatry—Nutritional psychiatry—Metapsychiatry—Computational psychiatry—Theoretical psychiatry—Transdisciplinary integrative psychiatry—Multidimensional psychiatry—Systems psychiatry—Comparative psychiatry—Complementary psychiatry—Comprehensive psychiatry—Consultative (Liaison) psychiatry—Creative psychiatry—Dialectical psychiatry—Eclectic psychiatry—Mental health sciences
Ecological psychiatry—Public health psychiatry—Preventive psychiatry
American psychiatry—European psychiatry
Telepsychiatry (E-psychiatry)—Virtual psychiatry—Avatar psychiatry—Digital psychiatry
Antipsychiatry—Official psychiatry—Alternative psychiatry—Political psychiatry—Radical psychiatry—Critical psychiatry—Liberation psychiatry—Toxic psychiatry—Fragmentary psychiatry—Marketing-based psychiatry—Orthodox psychiatry—Private psychiatry—Folk psychiatry

approach suggests that the only way to understand the complex issues of mental disorders and human suffering is to approach them in a holistic multiperspective transdisciplinary way using different levels of explanation and both analytic-focused top-down processing and heuristic-global bottom-top processing [11]. The various perspectives and dimensions are interconnected and interdependent, and cannot be fully understood separately. According to the systems theory, the whole is more than the mere sum of its parts, and the interconnections between the parts add a specific and distinct quality and dimension to the whole. While the focus of the reductionistic, mechanistic, and formistic information processing is on the parts, the systems approach emphasizes the whole [12]. The application of the transdisciplinary systems approach helps us understanding mental disorders and their treatment and prevention in a better and more appropriate way and offers a broader range of more efficient treatment strategies and options.

The Problem of Truth and Validity in Psychiatry: Between Science and Pseudoscience

> Truth in science can be defined as the working hypothesis, best suited to open the way to the next better one.
>
> —Konrad Lorenz

> Truth is a fruit which should not be plucked until it is ripe.
>
> —Voltaire

Many clinicians are confronted with the fact that the evidence, concepts, and explanations that they learned in their school of medicine or during official specialized training programs have changed, often fundamentally. Further, in the last years, there has been an increasing number of warning reports "that in modern research, misrepresented, false and unuseful findings may be the majority or even the vast majority of published research claims" [13–18]. Some studies have also stressed a decreasing quality of published literature due to an increasing competition for grants and jobs, the current mania for publishing papers, and a disproportionate emphasis on quantity over quality in scientific outputs, huge administration, and overreliance on reductionism [17]. As psychiatry has been divided by different schools and practice of thoughts explaining mental disorders from some narrow perspectives, the problem of truth in psychiatric research and practice has become a serious question. Pseudoscience and evidence-biased medicine represent a serious threat to research and clinical practice with the problems and complications that are manifold and not easy to be resolved. Biomedical science as well as psychiatry in general is getting more and more complicated and should be understood as more as possible in an integrative, holistic, and transdisciplinary way. Each study should be evaluated in the context of what we really know and what we don't know about the object of study, existing data and what make sense from different paradigms and perspectives (see sections "Definiton and Some Key Terms" and "The Goals of Theoretical Psychiatry: Completing Full Circle, from Observation and Clinical Experience to Theory and Back"). To be able to get beyond pseudoscientific deceptions and spins, medical doctors need to be familiar with philosophy and psychology of science [11, 19, 20] different strategies of thinking and information processing and able to read between the lines.

Evidence-Based Versus Evidence-Biased Medicine in Psychiatry: Misunderstanding, Fraud and Spin

> After all, the ultimate goal of all research is not objectivity, but truth.
>
> —Helene Deutsch

> As so in life, it is not a case of true and false, but of true and more true.
>
> —Plato: Letters to my Son

Evidence-based medicine (EBM) as the highest standard of health care [21] became a very promising mainstream concept in psychiatry at the very beginning of twenty-first century. The antithesis of EBM is practice based on pseudoscience, tradition, vogue, marketing, and authority [22]. Isaacs and Fitzgerald [23] reported seven alternatives to EBM: eminence-based medicine, vehemence-based medicine, eloquence-based medicine, providence-based medicine, diffidence-based medicine, nervousness-based medicine, and confidence-based medicine. It is very important to differ personalized from de-personalized EBM. The philosophy of personalized or person-centered EBM encompasses five essential principles: (1) it is grounded on basic health care values (values-based medicine); (2) it requires that scientific evidence should be a base in making health care decisions; (3) it recognizes that the scientific evidence is complicated, hierarchical, often uncertain and ambiguous, and usually limited; (4) it assumes that other factors like patients' human rights, values, preferences, and choices are also important factors in medical decisions; (5) it argues that clinical expertise is an important component in medical decisions [24]. Here, a thorny questions arise: what is evidence and how do we know it is proper? According to the Encarta Concise Dictionary, Student Edition [25], evidence means "something that gives sign or proof of the existence or truth of something, or that helps somebody to come to a particular conclusion". The last meaning "something that helps somebody to come to a particular conclusion" means also facts, testimony, and proof in support of a statement, claim, or belief. But collected facts and data do not speak for themselves alone; they are subjects to varying thinking strategies and information processing depending on who is doing the information processing. It is important to note that the expectations from EBM have not been commonly fulfilled, the results have been often differed from practice-based evidence [26].

The problems with EBM in modern psychiatry and clinical psychopharmacology are manifold and are products of misunderstanding, fraud, and spin [27]. According to Marshal [27], fraud is defined as deliberate falsifications of study results, whereas spin is an attempt to mislead that falls short of actual falsification. Misunderstandings are usually results of mechanistic, formistic, reductionistic, and linear thinking. The existence of many different models, languages, and paradigms within competing, sometimes bitterly opposing schools of thought, is an important source of misunderstandings. The terms evidence-based practices (EBPs) and best practices are synonymous to many experts, yet they often have different meanings in real life [28]. EBPs refer to treatment modalities with scientifically proven effectiveness, while best practices refer to treatment modalities evaluated as most effective by a majority. EBPs and best practices sometimes overlap. Best practices may be biased by the actual beliefs, attitudes or theories of opinion makers in the field, by the prejudices of guild organizations or by the successful marketing of pharmaceutical industry. Proclaimed best practices are sometimes shown incorrect by reliable scientific research. On the other hand, the results of many randomized controlled trials (RCTs) and basic research are not confirmed in clinical practice and they may lead to the so-called science-biased practices. That's why the best evidence-based (RCTs) practices are only those confirmed by practice-based evidence (naturalistic studies, pragmatic trials). The majority of the large RCTs undertaken in clinical pharmacology are sponsored by the pharmaceutical industry with the aim to demonstrate to regulatory agencies the efficacy of investigated drug over placebo. However, there are significantly less natural, pragmatic or observational studies [29] demonstrating the effectiveness of drug in the real clinical context on the real patients' population which is more heterogenous and with severe disorders than those in registration RCTs. Furthermore, negative studies are very rarely published, so that what we call EBM in the form of guidelines and algorithms may be biased and become evidence-biased medicine [22].

Psychiatry is unique in being a creative network of natural sciences and humanistic sciences in which both are essential for gathering and integrating relevant, but very heterogenous data, information, and knowledge in order to construct a coherent model of mental disorders and their possible treatment. For a better understanding of EBM, it is very important to have in mind the concept of epistemological responsibility [30]. According to this concept in clinical practice, psychiatrists are expected to recognize the best possible diagnosis and create the best treatment plan for individual patients instead of strict following clinical guidelines blindly and so deferring their responsibility to them. EBM rule-based reasoning determined by clinical guidelines should be complemented by narrative reasoning and case-based reasoning.

At the end of the day, although all is not said and done, EBM is here to stay integrating evidence-based practice and practice-based evidence in the spirit of epistemological responsibility. Many health care opinion leaders have recognized EBM as "one of the most important medical milestones of the last 170 years placing the practice of medicine on a solid scientific basis" [31, 32]. It is important to have in mind that EBM should be the integration of (1) best research evidence (clinically relevant patient-centered research), with (2) clinical expertise (to identify unique health states and diagnosis, individual risks and benefits, as well as personal values and expectations, and with (3) patients values (unique preferences, concerns, and expectations) which should be integrated into effective clinical decision.

The Dominance of Nomothetic Versus Idiographic Knowledge

Epistemological problems in modern psychiatry are manifold. There is a re-emergence of the old clash in psychiatry between observation and empathy, general causal laws and diverse meaningful accounts, explanations, and understanding [33] that is, in fact, the clash between nomothetic and idiographic knowledge. The dominance of nomothetic knowledge impoverishes both scientific discourse and clinical practice. It is a well-known fact that what is statistically significant may not be clinically significant, and vice versa. Emphasizing nomothetic knowledge promotes a technical rationality in psychiatry which is not sufficient for addressing human health and disease problems. Clinicians in everyday practice treat individuals not groups, real patients not statistical ones. In everyday clinical practice, the question is not what a group of patients with the same diagnosis would benefit from, but rather what would be an optimal choice with the most benefit for an individual patient. Furthermore, the overall effect of any treatment depends on many complex internal and external factors. Each medication, for example, has its specific pharmacodynamic and non-specific psychological effects that may also significantly influence the final treatment outcome. That's why idiographic knowledge, characterized by tendency to specify and understand the meaning of contingent, accidental, and often subjective phenomena, is needed to estimate what the optimal treatment choice is for an individual patient [34]. An exclusive focus on idiographic knowledge is associated with a number of blind spots including difficulties in differentiating what is optimal for individual patients from what may be useful for a group of patients with the same diagnosis; harmful mistrust of professionals who operate on the basis of nomothetic evidence; and an oppositional discourse and blaming for problems [34]. The fact is that neither nomothetic nor idiographic knowledge alone is sufficient pillars for good clinical practice. For a rational and creative therapeutic approach, both nomothetic and idiographic knowledge are necessary evidence components.

Defining strictly and clearly specific mental disorders and specific treatments for them is a wishful thinking and exercise, and has not still become a successful practice in modern psychiatry. Within this framework, the priority is to diagnose and treat specific mental disorders with specific drugs and other methods that influence the psychopathophysiological processes associated with particular disorders. Clinical guidelines and research, focused on diagnostic groups which are very heterogeneous, are based on nomothetic knowledge. The key goal of research, for example, in biological psychiatry and psychopharmacology is to establish causal relationships between specific pathophysiological processes and specific mental diseases, and to choose rational treatment options on this basis; context, meaning, and reasons are neglected. RCTs have become a gold standard for evaluating drug efficacy and effectiveness and a cornerstone of evidence-based psychiatry [29]. They produce nomothetic knowledge and are characterized by tendency to generalize and derive laws that explain objective phenomena. Inferential statistics are used for evaluating the likelihood of intervention or drug X producing outcome Y in the form of the Number Needed to Treat (NNT) statistic. NNT refers to the number of patients who must be treated to achieve a positive outcome or prevent a negative outcome [34]. Therefore, the more effective treatments are those with a lower NNT. According to this concept, the optimal drug treatment is that with the lowest NNT, which "at least in theory, and sometimes in practice can be expressed as a deterministic flowchart" [34].

Science, Pseudoscience, and Antiscience in Psychiatry

> Science is a dominant force in our lives, and as such it presents many people a very ambivalent love-hate relationship.
>
> —Feist (2006)

Psychiatry has been commonly criticized as science in question or "quasi-scientific intellectual discipline that drew on the biological sciences" [35]. Science has usually been considered an objective, self-correcting, truthful and reliable human endeavor as well as the most important reliable source of knowledge and progress in medicine and psychiatry. However, we must always have in mind the gap between reality and our understanding of it and that there are no perfect and absolute scientific truths and knowledge. It may be useful to differ three guises of science: science as fact, science as hypothesis, and science as dogma [36]. In practice sometimes scientific hypothesis can be promulgated as fact, taught as dogma, and converted into philosophy, religion or mythology [36]. Generally speaking, science has progressed by forming hypotheses, which according to philosopher Karl Popper can never be proven, they can be falsified or rejected. In addition to this concept of falsifiability or falsificationism which promotes the deductivism as the standard of scientific thinking, there are still two major theories of scientific reasoning: inductivism and explanationism or holism [31]. Falsifiability is criticized as "both a naïve account of scientific practice and an unreliable guideline to demarcate between science and pseudoscience" [37]. Inductivism, dominant reasoning in medicine, represents information processing from a limited number of observations to wider, more probable generalizations, e.g., inferring diagnosis from a set of symptoms. Explanationism (holism) refers to the inference to best explanation so that "scientific evidence should be integrated and contrasted with the totality of our beliefs and knowledge" [31]. Antiscience is a position that rejects science as an objective method that can generate universal knowledge. Postmodernism which promotes the relativity, instability, and indeterminacy of meaning dismisses all trials to understand totalities or construct Grand Theory [11]. It is clear that scientific reductionism is not a good way to reach an understanding of a complex world of mental health and mental disorders, but the scientific approach is the best what we have. It is very important to distinguish scientific knowledge from its pseudoscientific look-alikes. Pseudoscience (a Greek word meaning false) and antiscience are positions of some branches within critical psychiatry and antipsychiatry. The seven key words of good science and research are: integrity, motivation, capacity, understanding, knowledge, experience, and creativity. Without integrity motivation is dangerous, without motivation capacity is impotent, without capacity understanding is limited, without understanding knowledge is meaningless, without knowledge experience is blind, without experience creativity is impossible, without creativity there is no progress. Demarcations of science from pseudoscience are very important from both theoretical and practical reasons. The first reason is theoretical and it goes to the epistemology and to the core of the nature of truth, evidence, and discovery ("How do we really know what we think that we know"). The second reason is political and economic because a huge amount of money has been spent on biomedical research. The false and misrepresented pseudoscientific data may all contribute to evidence-biased medicine and treatment inefficiencies as well as to the wasting of limited funding and investigators' work. The third reason is ethical because pseudoscience can be harmful to patients, sometimes fatally so, and undermine public confidence in scientific medicine.

Pseudoscience is non-science; it is invalid or fake science posing as real science involving varied fads and fallacies in the name of science. In medicine, pseudoscience represents any theory or method that claims falsely or mistakenly to be scientific or that is falsely or doubtfully regarded as scientific although they lack supporting evidence and plausibility. Term pseudoscience also refers to a field, practice, or body of knowledge claimed to be consistent with the norms of scientific information processing and research, but in reality, fails to meet these norms. In other words, pseudoscience is characterized by producing irreproducible, incorrect or falsified results, and non-useful research data. Pseudoscience can be product of misunderstanding and lack of education, fraud, and spin. Pseudoscience, fabrication, falsification, spin, and plagiarism are serious forms of scientific misbehavior that jeopardize the image of scientific journals and scientific community. While fabrication (making up data, results or cases) is evidently fraudulent scientific malpractice, pseudoscience lies somewhere between scientific fraud, bias, misunderstanding, and simple careless, and it is not easy to define it. With regard to scientific fraud and spin, the intention to deceive is a key element. Falsification is defined as willful or deliberate modifications of study results, while spinning is related to some kind of wishful thinking and subjective differences in research designing or interpreting. Researchers have great latitude in how they process data and report their results in the medical literature. Three common types of spin can be identified [27]: (1) spinning by selective reporting (e.g., not reporting disappointingly negative findings), (2) spinning using rating scales (e.g., evaluating outcome using multiple rating scales, or unpublished scales), (3) meta-spinning (reviewer's pessimistic or optimistic looking on inconsistent results of clinical trials). The distinction between real and artifact, true and false results and their interpretations is not an easy task. It is related to the applied mechanistic, formistic, contextual or systemic thinking or information processing strategies. Wishful mechanistic (single-cause or single-effect thinking) and formistic binary categorical (either-or thinking) strategies of information have produced a lot of oversimplifications, false beliefs or myths in some fields of medicine [22].

The boundaries and indicators separating science from pseudoscience and evidence-biased medicine are very fuzzy. Pseudoscience is like pornography: it is very hard to be defined, but one knows it when he sees it. According to Derry [38], defining characteristics of pseudoscience are: static or randomly changing ideas, vague mechanism to acquire understanding, loosely connected thought, lack of organized skepticism, and disregard of established results. Some scholars like Burke [39] are against using the term pseudoscience because it inherently creates framing issues as "us versus them" and "kto-kavo" ("who will eliminate whom"), pitting those who believe in "real" science against those who believe in "false" science. Psychiatry is a multilevel science with different kinds of logic and explanations that is not well integrated.

Validity, Reliability, Operationalism, and False Sense of Diagnostic Security

Human beings in general as well as psychiatrists in particular are feeling comfortable whenever they are able to recognize, explain, predict and control processes related to health and disease. DSM and ICD classifications have played a fundamental role in shaping modern psychiatric practice and research. Kendler [40] argued that our classification system has developed via two main paradigms, "the grand professorial principle" and "the consensus of experts". Current psychiatric classifications have been produced by the committees of expert convened by national and international bodies. There are two aspects of classification: (1) the classification of symptoms to form syndromes, and (2) categorization of patients in diagnostic groups. The traditional organization of diagnostic categories was: (1) according to clinical presentation into psychoses, neuroses, and personality disorders, and (2) with regard to aetiology into a. organic and symptomatic disorders that are secondary to known brain and somatic disease, and b. functional disorders whose biological basis is unknown. Providing diagnostic criteria for mental disorder they significantly increased diagnostic reliability, but diagnosis validity is still very low because they are too detached from the real nature of mental disorders. Both ICD-10/11 and DSM-5 still classify together very heterogenous mental disorders what stresses the importance of their clinical utility and diagnostic validity in psychiatric nosology (Table 2).

Theoretically, the purpose of diagnosis is to identify disorders having the same clinical characteristics, the same causes, the same underlying pathology, and the same likely response to treatment. Psychiatric diagnoses are usually disjunctive and syndrome diagnoses. Disjunctive diagnoses exist when underlying pathophysiology is less known and when there is no a gold pathognomic clinical sign or test that confirms the diagnosis. The problem is that a syndrome may have a different cause in different cases as well as that the multiple interacting factors may contribute to the aetiology of a single case. Concept of illness in psychiatry mainly relies on reported symptoms and observable behavior rather than objective signs and markers. Operational criteria and definitions of mental disorders have been applied in order to overcome diagnostic variabilities in psychiatry ("two psychiatrists, three varied diagnoses for the same case"). Although operational criteria were disparaged at the beginning as the "Chinese menu approach to diagnosis" [41] they have taken over official psychiatric classifications and become universally accepted as the method for defining psychiatric diagnoses. There are two reasons for that: they are highly reliable and easily applied using structured standardized interviews. However, operational criteria have many weaknesses: they are "top-down" with predetermined rules for their application, somewhat two dimensional, with rag bag categories such as atypical disorder and not otherwise specified disorder [41]. Last, but not least, diagnostic procedures predicated on symptoms and operationalized criteria produce temporally unstable or changeable diagnoses and high rates of comorbidity strongly related to severity of clinical picture [45].

Table 2 Terminology important for understanding psychiatric diagnosis and classification (adapted from Refs. [41–44])

Utility: the extent to which a classification or diagnosis is good for the purpose. It is associated with diagnostic interviews, rating scales (screening, observer, and self-rating questionnaires) and measuring psychopathology in specific subject groups
Reliability: is the agreement between users of a classification or diagnostic criteria. It reflects the consistency of the measure usually estimated with regard to three criteria: interrater, test–retest and internal or split-half reliability. *Interobserver or interrater reliability* is the extent to which two or more observers/raters agree about the diagnosis or classification of a given case on a given occasion. *Test–retest reliability* is the degree of agreement by a given observer in his/her diagnosis or classification of a given case over a period of time. *Internal consistency* is very important for the measures of a single dimension producing continuous scores when a single administration of a measure can be divided into two halves which can be scored separately and mutually compared
Validity: the extent to which the diagnosis or classification reflects the real world or that aspect of real world with which psychiatry is concerned. *Face validity*: the extent to which a diagnosis or classification appears to be of relevant features, which has consequences for the acceptability to its users. It is a common sense assessment of validity: whether the measures really test what is claimed to test. *Construct validity* is the most complex type of validity: the extent to which the construct that the measure seeks to address is real and coherent entity, and then to the salience of the measure to that construct, or the extent to which a classification relates to underlying theory or the extent to which a patient's symptoms reflect an underlying pathophysiological causal mechanism. *Criterion validity* refers to the conformity with an alternative, independent measure of the same construct. Criterion validity involves concurrent and predictive validity. *Concurrent validity* refers to the agreement with the criterion measure, when both measures are present at the same time. *Predictive validity*: the extent to which the diagnosis or classification allows us to predict prognosis (course and outcome) and response to treatment. *Content validity*: the demonstration that the defining characteristics of a given disorder are indeed enquired into and elicited before the diagnosis is made. This answers the question whether the specific content of the measures promotes a comprehensive and balanced evaluation
Consistency: a theory/classification which does not meet the requirement of consistency is worthless. *Internal consistency*: a theory/classification that is internally consistent is free of contradictions; *External consistency*: a theory/classification that is externally consistent is consistent with the data it's supposed to explain
Judging scientific concept/theory/construct on which classification is predicated: *testability*—whether there is some way to determine if a concept/theory/construct is valid/true; *fruitfulness*—the number of novel predictions or agreements (reliabilism) made; *scope*—the amount of diverse phenomena explained; *simplicity*—the number of assumptions made; *conservatism*—how well a theory or classification fits with existing knowledge
Evaluating theories: The TEST formula: (1) state the theory and check for consistency; (2) assess the evidence for the theory; (3) scrutinize alternative theories; and (4) test the theories with the criteria of adequacy

Theoretical Psychiatry or Metapsychiatry: Definitions and Aims

> Truth can be stated in a thousand different ways, yet each one can be true.
>
> —Swami Vivekananda

> On the mountains of truth you can never climb in vain: either you will reach a point higher up today, or you will be training your powers so that you will be able to climb higher tomorrow.
>
> —Friedrich Nietzsche

Psychiatry is a large and continually growing body of knowledge, and it is claimed to be scientific branch of medicine. However, defining science and what is scientific in psychiatry and medicine is more complex than defining psychiatry and medicine [26]. Science is self-correcting discipline that permanently changes its frontiers and whose aims are understanding, prediction, and control of processes and phenomena beyond levels achieved by common sense [46]. It helps to recognize and difference true facts from beliefs, fake data, and fashion in order to establish a foundation for sound clinical practice. Scientific psychiatry refers to the knowledge about mental disorders and their treatment based on scientific methods, not only on theories, opinions, random observation, and authorities.

Definiton and Some Key Terms

> Science is a way of thinking more than it is a body of knowledge.
>
> —Sagan (1934–1996)

Large progress and continually growing body of knowledge in neuroscience and humanity disciplines, huge number of different theories, models, and controversies make a strong need for highly systematized and comprehensive knowledge about mental disorders and their treatment. To avoid ambiguities and misunderstandings, clarification of terminology is useful (Table 3). Theoretical psychiatry can be defined as the field of hypotheses, models, and theories that describe and explain the mechanisms of mental disorders [47] and their treatment, of course. It can be described also as the epistemological system and study of major theoretical and philosophical aspects of psychiatry which employs analysis, abstractions, modeling, and generalizations to explain, clearly define, predict, and rationalize mental disorders and their treatment. Theorizing is a complex process that involves conceptualizing mental disorders and their treatment and prevention in terms of a set of

Table 3 Some useful terms and definitions (adapted from Refs. [48–56])

Science: an organized and systematic body of knowledge gained and verified through applications of scientific methods of empirical observations. *Normal science* works with paradigms whereas *revolutionary science* creates new paradigms
Meta-science: a theory or science of science; the systematic investigation of the scientific enterprise; the use of scientific methodology to study science itself
Philosophy of science: application of philosophical methods to philosophical problems as they arise in the context of science involving ethical issues, epistemological issues, and metaphysical issues
Scientific method: a method of acquiring knowledge by observing a phenomenon, formulating one or more hypotheses, further observing and experimenting, and refining and re-testing hypotheses and discovering laws and principles that govern phenomena,
Scientific explanation: an account of some fact, phenomenon, or event based on combination of facts, observations, and logic subjects to further proof
Scientism: "exclusive reliance on a narrow conception of science": "an excessive readiness to accept as authoritative any claim made by the sciences and to dismiss every kind of criticism of science or its practitioners as anti-scientific"
Sciosophy: "any system of thought not supported by scientific methods such as astrology, phrenology, numerology, etc."
Scientology: "a controversial human-development system with a spiritual and religious dimension developed by L. Ron Hubbard"
Epistemology: the study of knowledge, branch of philosophy dealing with issues about origin and legitimacy of knowledge and truth. "What do we really know and what we don't know"? How do we know what we know? What is true and what is reality?
Nomology: the science of law or laws; (2) the science of the laws of mind. "The branch of science and philosophy concerned with the laws or principles governing the operation of mind". *Nomological*: "relating to or expressing basic physical laws or rules of reasoning"
Mechanisms: "systems or sequences of causally interacting parts organized such that they produce the phenomenon to be explained"
Scientific explanation: "involves discovering underlying mechanisms that are responsible for the various phenomena that scientists aim to explain". *Etiological explanation*: a phenomenon is explained by the mechanism that consists of preceding causes of the phenomenon: *Constitutive explanation*: a phenomenon is explained by the mechanism that underlies or constitutes the phenomenon
Metaphysics: the branch of philosophy that studies the essence of things. This includes questions of being, becoming, existence, and reality
Metatheory: (1) the science of theories; (2) a set of general rules governing the construction of a theory; and (3) a theory about theory [49]
Meta-evaluation: the systemic evaluation of evaluation procedures including methodological rigor, utility, cost, relevance, scope, importance, credibility, and time-lines
Meta-analysis: "Integration of the findings of a number of research studies by means of statistical techniques focusing on the same research question leading to meaningful quantitative data" [49]
Metapsychology: in British: (1) the study of philosophical questions, such as the relation between mind and body, that go beyond the laws of experimental psychology, (2) any attempt to state the general laws of psychology; (3) another word for parapsychology; in American: speculation about origin, structure, function, etc. of the mind and about the relation between the mental and the physical, regarded as supplemental to psychology [51]

(continued)

Table 3 (continued)

Theoretical psychology: the study of major theoretical and philosophical aspects of psychology
The International Society for Theoretical Psychology (ISTP), founded in the early 1980s as an international forum for theoretical, meta-theoretical, and philosophical discussion in psychology in order to foster integration across areas and traditions of research, and to promote interdisciplinary and transdisciplinary approaches to psychological questions. It aims to serve as the stage for discussion of new theoretical ideas and conceptual frameworks, for the critical engagement of different theoretical approaches, and for discussions concerning the relation of theoretical psychology to other disciplines, to the history of psychology, and to the philosophy of knowledge
Theoretical neuroscience with its subfields computational and mathematical neuroscience is a branch of neuroscience which uses philosophy of science, mathematical models, theoretical analysis, and abstractions of the brain to understand the principles that rule the development, structure, physiology, and mental abilities of the brain
Metapsychiatry: (1) spiritual teaching and form of psychotherapy developed by psychiatrist Thomas Hora who described it as "scientific method of healing and education based on metaphysical concepts of man and the universe" [54]; (2) Psychiatric discipline that deals with conceptual basis of psychiatry, ontology, and epistemology of mental disorders at individual and collective level, and critical psychiatry, antipsychiatry, and postpsychiatry [55]
Philosophy of Psychiatry: an interdisciplinary field that explores philosophical questions relating to psychiatry and mental illness: (1) examination of psychiatry as a science applying the methods of the philosophy of science; (2) examination of the concepts and theories of mental disorders including the experience of mental disorder, and the normative questions related to it; (3) explanation of the links and discontinuities between the philosophy of mind and psychopathology
Philosophy of Mind: a branch of philosophy that studies the ontology, nature, and relationship of the mind to the body as well as mental events, mental functions, mental properties, nature of consciousness, and thought [56]
Computational psychiatry: psychiatry informed by computational neuroscience
Theoretical psychiatry could be defined as an interdisciplinary field that explores the theoretical and scientific basis of psychiatry as well as ontology, phenomenology, and epistemology of mental disorders

concepts and relationships among them in forms of theories, models, and hypothesis and their empirical testing. There is no good understanding and defining of mental disorders without appropriate conceptualizing mental health and its promotion. Unfortunately, in reality, there is no explicit field that can be recognized as "theoretical psychiatry", but there have been many attempts aiming to construct a theoretical perspective. At present theoretical psychiatry is a scientific infant that has much to learn from philosophy and psychology of science. The progress in psychiatry generally depends on the interplay between theory, research, and clinical practice.

Psychiatry as a field has a plentitude of data, hypotheses, diverse experiences, and stories, but there is a comparative lack of good scientific theories and models. Relatively few clinicians have time for theory because the clinical practice is so all-absorbing, and finding the time to consider epistemological issues is indeed a luxury [57]. Academic psychiatrists may have the time but are usually engaged with academic lectures and extending their own narrow research interests. Furthermore, academic psychiatry is dominated by rigid rules that define success and failure with short cycles for evaluation of performance and follow-up as well as by a fragmentary neurobiological paradigm, which both have stifled creativity [58]. The international ICD-10/11 and DSM-V classifications of mental disorders are descriptive and "atheoretical" tools, characterized by an empirical approach and demonstrating a limited understanding, rather than credible practices based on comprehensive theory.

The Goals of Theoretical Psychiatry: Completing Full Circle, from Observation and Clinical Experience to Theory and Back

> In theory, there is no difference between theory and practice. But, in practice, there is—A computer scientist.
>
> —Jaccard and Jacoby (2010)

The huge number of the most varied concepts, theories, and models may have two opposite meaning. For pessimists and close-minded psychiatrists that is a curse of conceptual cacophony and low scientific level, for optimists and open-minded psychiatrist that is a blessing of different fragments of a complex puzzle of mental disorders. Current psychiatry is like a story about six-man groping an elephant and coming up with the varying descriptions. This metaphor is pointing to the importance of theoretical psychiatry. Although the major schools in psychiatry have fundamental differences in approach and orientation, they all have legitimacy as parts of a science and philosophy in psychiatry, and they share a concern for discovering the aetiology, pathophysiology, and phenomenology of mental disorders and improving their treatment. Conceptual discord is a powerful disintegrating force within psychiatry, and the future of the discipline greatly depends on how conceptual heterogeneity will be resolved [59]. The truth is a common goal for both philosophy and science. Philosophy is the love of truth, science is the knowledge of truth [60], and epistemophilia is the love of knowledge. Philosophy particularly epistemology and axiology are not given enough respect in the mainstream psychiatry, particularly in medical psychiatry. As a result, many controversies persist regarding the appropriate methodological, epistemological, and ontological necessity for psychiatric explanations and therapies. Establishing a coherent transdisciplinary scientific narrative and a more theoretical foundation for psychiatry is a great challenge for the beginning of the twenty-first century. Really, "psychiatry needs its Higgs boson moment" [61] as well as a new neurophilosophy of the brain and mind [47]. Some authors propose constructs like dimensions, trajectories, thresholds, and context or predicaments. Dimensions are the parameters needed to describe the space within which processes, behaviors, or events occur, and within multidimensional space, process can follow various trajectories: moving towards attractors, loops, or limit circles, and exhibiting bifurcations or discontinuities when they cross particular thresholds [2, 62]. These dynamics may be related to psychobiological processes within the brain, between the person and the environment, as well as among people and in larger social systems [2]. Psychiatric nosology (ICD 10/11 and DSM-5) define and describe mental disorders abstracted out of context in terms of generalized processes and phenomena although local history contexts and contingencies constitute predicaments for individuals, families, and communities. The concept of transdisciplinary integrative psychiatry puts together person-centered psychiatry, precision and neuroscience-based psychiatry, value-based psychiatry, and narrative psychiatry in a complementary way. High infotech, modern network science, systems thinking, systems (complexity) theory, epistemology and philosophy of science are of paramount importance for psychiatry to develop explanatory concepts and models that successfully incorporate the incredible amount of data from neurosciences, mind sciences, social and spiritual sciences into a conceptual framework of mind–brain–body functions and dysfunctions. Creatively integrating the different theorctical perspectives brings us closer to a holistic understanding of the complex nature of mental health and mental disorders, and a more efficient treatment of mental health problems [3].

Theory Construction and Model-Building in Psychiatry

> Science is organized common sense where many a beautiful theory was killed by an ugly fact.
>
> —Thomas H. Huxley

> Science may be described as the art of systematic oversimplification.
>
> —Popper (1902–1994)

Theories are frameworks for observation and understanding of the world and its phenomena. They are usually generalized statements and explanations that depict connections between two or more phenomena, and commonly they should be translated into plausible models. In other words, theories promote understanding, but what we cannot model; usually we do not understand [63]. They guide research and organize its ideas at different levels; they are open to revision and progress into more accurate and comprehensive explanations. The relationship between theories and research is bidirectional. Science in psychiatry is predicated on the theory construction and it is about the nature of mental disorders and their treatment. As in other scientific disciplines [64] researchers in psychiatry "formulate theories, test theories, accept theories, reject theories, modify theories, and use theories as guides to understanding and predicting" mental disorders and to heal and cure them successfully and efficiently. Theories are bodies of interrelated principles and hypotheses that explain or predict a group of phenomena and have been largely verified by facts or data [49]. Mental disorders as well as mental health should be understood as complex and dynamic phenomena that appear unique and mostly obscure. "*The building blocks of our understanding are concepts and conceptual systems. Concepts are hypothetical, selective and generalized abstractions which encompass universes of possibilities. Most concepts are learned and acquired creations, reality oriented and socially shared. Constructs are higher order concepts which refer to instances that are constructed from the concepts at lower levels of abstraction. One type of construct is called a variable, and it has, or is composed of, different 'levels' or 'values'.*" [64]. For example, according to Cloninger's [60] psychobiological model personality is a construct which has two concepts temperament and character that are composed of seven variables: harm avoidance, novelty seeking, reward dependence and persistence (temperament) and self-directedness, cooperation, and self-transcendence. When concepts are put into relationship with one another they enable us to deeper understand a phenomena. The relationships can be various, for example, spatial, temporal, deterministic, kinship, legal ones, etc. Conceptual system (theory, model, or hypothesis) refers to the two or more concepts that are links to represent relationships. Scientific theories are in fact conceptual systems or mental maps created for identifying, organizing, explaining, predicting and ruling over some phenomena or processes related to mental health, brain dysfunctions, and mental disorders. Psychiatry is overcrowded with the most varied ideas, rules, principles, techniques, and methods, but short with comprehensive scientific theories. Symbols are external observable expressions that used to represent internal concepts that are internal mental representations. That's why it is important to make a distinction between our understanding of some phenomena, structures or processes as represented by our concepts we have in mind and our description of these phenomena and processes as represented by symbols, usually words, we use to express our thoughts. A number of different surface symbols could be used to communicate the same underlying meaning structure, and meaning should be always interpreted in context because a particular symbol may have different meanings in different contexts. The process of formulating conceptual systems and converting them into symbolic expressions is theorization or theory construction. Conceptualization may be predicated on what someone observes, imagines, considers what others said or done, or what he or she experienced or observed. It is also related to the applied mechanistic (single cause–single effect), formistic (binary, either-or, black or white), contextual, and systemic (seeing wholes) thinking or information processing strategies. Wishful mechanistic and binary categorical thinking and information processing have produced a lot of oversimplifications, false models and myths in aetiology, classification, and treatment of mental disorders [22]. Systems thinking is a framework for seeing the structure underlying complex situations and processes, interrelationships, and patterns of change rather than thing and statis snapshots [65]. The first step in theory or model construction is usually a generating idea about new explanatory constructs and the relationships between them regarding phenomenology or pathophysiology of some

Table 4 Some important terms in mathematical modeling and psychological theories (adapted from Refs. [25, 38, 49, 64])

Qualitative (categorical) and quantitative variables
A qualitative variable has different levels, values. The categories are labels that differentiate one group from another
A quantitative variable is one in which the numerical values imply an underlying dimension that is of theoretical interest and by means of statistical classification they are placed into different categories. Variable can be discrete or continuous. Discrete variable is one in which there are finite number of possible values. A continuous variable has infinite number of possible values because for every two points on a continuous scale we can find a point that lies in between them
Axioms and theorems
An axiom is a statement that represents a starting point from which other statements are logically derived
A theorem is: (1) a scientific proposition, axiom or premise that can be proved in a series of logical steps; (2) a statement generally accepted as true in psychology
Functions refer to numbers as inputs and outputs
Logarithmic, Exponential, Power, Polynomial, Trigonomic, and Multiple Variable
Models: imaginary simulations of the real natural systems we are trying to understand [38]
A deterministic model is one in which there is no random error operating and which performs the same way for any given set of conditions
A probabilistic model is one in which some degree of randomness is present
A just-identified model is one in which there is a unique solution (e.g., one and only one solution) for the values of each estimated parameter in the model
An under-identified model is one for which there is an infinite number of solution for one or more of the model parameters. This model is unsatisfactory
An overidentified model is one for which there is a unique solution for the model parameters, and there is more than one feature of the model that can be used to independently estimate the parameter values
Theory: a body of interrelated principles and hypotheses that explain or predict a group of phenomena and have been largely verified by facts or data [49]
Deductive theories are based on derivations of propositions to be tested by logical premises; they guide the design of the study and the interpretation of results
Inductive theories based on accumulation of evidence without deliberate attempt to achieve it; they are developing slowly, concept by concept, in a specific field
Functional theories based on small frequently modified hypotheses used as tools to permit more meaningful theoretical propositions
Holistic and general-system theories emphasize the importance of wholeness of the person, the unity and organization of behavior in shaping the parts of the whole, investigated by holistic approach based on many disciplines and specialties
Paradigm: (1) typical example of something; (2) model that form basis—an example that serves as pattern or model for something, especially one that forms the basis of methodology or theory; (3) relationships of ideas to one another—in the philosophy of science, a generally accepted model of how ideas relate to one another, forming a conceptual framework within which scientific research is carried out; (4) a set of assumptions limiting an area to be investigated scientifically and specifying the methods to be used for collecting and interpreting the data to be obtained

particular mental disorder or mental disorders in general. Brainstorming and De Bono's model of six hats [66] can be very useful. Creative ideas provide novel perspectives on mental disorders and mental health in ways that provide new insights. Creativity is a function of six resources: (1) intellectual abilities; (2) knowledge; (3) styles of thinking; (4) personality, (5) motivation, and (6) environment [67].

Causal thinking has been the most prominent approach in theoretical explanation and modeling in medicine and psychiatry. There are "six types of causal relationships and thinking about relationship between data scholars can think about: (1) direct causality, (2) indirect causality, (3) spurious causality, (4) moderated causality, (5) bidirectional causality, and (6) unanalyzed causality" [64]. In recent period traditional causal explanation has become the subject of many critics and mathematical (Table 4) and computational modeling has appeared offering a new hope. Systemic thinking and seeing circles of causality are crucial because reality, including also mental reality, is made up of circles although frequently we see straight lines. As divided field occupied by increasing complexity of mental disorders and their treatment, current psychiatry needs critical and systemic thinking as well as the art and practice of learning organization more than ever. Critical thinking refers to the systemic evaluation and formulation of beliefs, statements, or concepts by rational standards and reasons. It involves logic of truth and falsity of statements, the evaluation of arguments and evidence, the use of analysis and investigation, and the application of many other skills to decide what to believe or do [43, 44].

Scholars disagree about the best way to conceptualize mental health and mental disorders, as reflected in the psychiatric orientation, way of information processing, thinking and conceptualization, mentalization and mental constructionism, critical and hypothetical realism. There are many ways of obtaining and organizing knowledge about mental disorders and mental health, only one of which is science that consists of a conceptual realm on the one side, and an empirical realm, on the other. The conceptual realm involves the development of a conceptual system (concepts, constructs, and their relationship in the form of theories, models, and hypotheses) about a mental disorder which can be communicated to others. The empirical realm refers to the process whereby the worth of the conceptualization is assessed through conduct of scientific studies and empirical testing in clinical practice. According to Jaccard and Jacoby [64] in literature, one can identify different typologies of theories: constructive versus principle theories, reductive versus constructive theories, concatenated versus hierarchical theories at either molar or molecular levels, humanistic versus behavioristic, constructionist versus deconstructionist, functionalist versus structuralist, etc. Characteristics of a good theory are as follows: utility and explanatory power, consensual validation, agreement with known data and facts, understandability and communicability to other scientists, simplicity and parsimony with minimum concepts and principles (Occam's razor), novelty, testability, scope, amount of research generated [38, 64]. A good theory leads to new ideas, new applications, new connections to existing theories, prediction of unforeseen results and even new theories [38]. Psychiatry is abundant with many hypotheses and simple theories about mental disorders, but lacks comprehensive and predictive theories and models. According to Ghaemi [68], there are four basic conceptual approaches in psychiatry dogmatic, eclectic, pluralistic, and integrationist, whereas pluralistic and integrationist models are the most promising concepts.

The Four Paradigms in Psychiatry: From Paradigmatic to Holodigmatic Approach

According to Kuhn [69], paradigms are sets of beliefs that make up theoretical frameworks of a scientific school or discipline within which theories can be critically evaluated, modified, or revised and experiments performed in support of them are formulated. A paradigm can be described either with a set of ideas expressed as defined premises, by premises expressed as precepts, or by a set of precepts expressed as rules which represent the essence of those of premises or precepts [70]. The process of transformation in the thinking of scientists from one reality construct to another is defined as a paradigm shift. Contemporary clinical psychiatry rests on a simplified form of Cartesian dualism that posits two fundamentally irreducible ontological categories: a physical body/ brain and an embodied nonphysical soul/mind [71]. Four hierarchically related paradigms embodying different assumptions about phenomenological nature of mental health and mental disorders can be recognized: the body paradigm, the mind–body paradigm, the body–energy paradigm and the body–spirit paradigm [72].

Body paradigm is a framework in which an orthodox biological psychiatry and psychopharmacotherapy operate. The mind and mental functions develop as the genetically programmed maturation of the brain and neural circuits responding to ongoing experiences. The mind exists primarily as a by-product of brain activity and mental disorders are consequences of the disturbed brain functioning. Some scholars claim that the mental supervenes on the somatic/brain, others think it is realized by the somatic, and the thirds hold that the mental is constituted by the somatic. The widely spread view that mental disorders are caused by imbalance at the level of neurotransmitters is complemented by the model of neuroplasticity disruption and neuronal networks dysregulation. Effective treatment of mental disorders works by modifying structural and functional neuroplasticity and abnormal neural circuits functions.

Mind body paradigm is a framework in which the mind–body or psychosomatic medicine operates. The mental is real and irreducible to as well as it is non-identical with the physical/somatic, but it is physically influenced. The physical realm is causally closed, while the mental realm and the biological realm are not [53]. Acute and chronic stress is associated with dysregulation of hormones, immune dysfunctioning, and neurotransmitter imbalance that manifest as cognitive, affective, and behavioral symptoms of mental disorders. Increasing integration of mind–body practice with conventional treatments will probably result in significant improvements in patient's autonomy, improved outcomes, and reduced mental health costs.

Body–energy paradigm is based on scientifically validated forms of energy or information that are directed at the body/brain. Energy exists in various states and forms, with some more easily detectable than others (zero-point energy, life-fields). Representative energy-information treatment modalities that have been scientifically accepted by modern psychiatry are electroconvulsive therapy (ECT), transcranial magnetic stimulation (TMS), neurobiofeedback, vagal nerve stimulation, bright light therapy. Some recently introduced conceptual models assume that thoughts exist in fields, negative emotions are rooted in energy configurations, and psychological phenomena are fundamentally quantum physics events or processes [73]. Energy psychology and quantum psychiatry operate within body/brain–energy framework.

Body–spirit paradigm is a framework in which transpersonal psychology and spiritual psychiatry operate. According to this paradigm, human beings are "spiritual beings in a physical body". Body–spirit paradigm as well as body–energy is based on the beliefs that mind, body, and spirit can be described in terms of putative subtle energies [71].

According to the transdisciplinary integrative concept Body–Mind–Spirit–Energy are four mutually interconnected ontological dimensions/ domains of human beings in health and disease. This approach relies on the body–spirit, mind–body, and body–energy paradigms integrating them. A proper model of the nature of human being should understand each person as a unified entity comprised of body, mind, and spirit who is inseparable part of various social systems as well as an all-encompassing cosmic order [60].

The Seven Perspectives of Mental Disorders and Their Treatments

Different schools of thought assessing mental health and mental disorders have given rise to different perspectives of psychiatry: the medical or disease perspective, the dimensional or perspective of person, cognitive, behavioral, narrative, spiritual, and systems perspectives [74]. Each of these perspectives tends to analyze, understand, define, and treat mental disorders in different ways, each with their own merits, albeit commonly without enough success. Multidimensional profile from seven perspectives enables a holistic and deeper understanding of mental disorders and may identify clusters of mechanisms that are amenable to treatment increasing probability of more successful treatment outcome. In fact, it is a word about three perspectives, the disease perspective, the perspective of person, and the systemic perspective because the cognitive perspective, the behavioral perspective, the spiritual perspective, and the narrative perspective are components of the perspective of person. As these four sub-perspectives are associated with well-defined disciplines in psychiatry and offer specific treatment methods, they are described separately.

The Disease Perspective or Medical Perspective in Psychiatry

> The good physician will treat the disease, but the great physician will treat the patient.
> —William Osler

This perspective tries to work in psychiatry just as it does in somatic medicine. It implies that illness is something fundamentally different from normal state and function and is not just a variation in degree [75]. The disease/illness perspective rests on a logic that captures brain/mind abnormalities and includes this conceptual triad: clinical syndrome, pathological condition of brain/mind and aetiopathogenesis. The implication for practise is that disease/illness is to be prevented or cured [76]. However, there is no clear border between mental health and mental illness, mental health is not just absence of illness, nor mental illness is just the absence of mental health (Table 5).

Disease entities in somatic medicine are identified by their pathology and aetiology, not only by their syndrome clustering and operational definitions like in DSM-5 and ICD 10/11. There has been much controversy over the distinction between the concepts such as disease, illness, sickness, disorder, and maladaptation as well as between organic or symptomatic mental disorders, somatopsychological disorders, psychogenic disorders, and psychosomatic disorders in psychiatry. Psychiatric classifications use the term mental disorders to avoid the problems related to the use terms such as disease and illness. Disorder is defined

Table 5 Characteristics of mental health and mental disorder

Mental health
Mental health is defined as a state of complete physical, mental, social, and spiritual well-being in which every individual realizes his or her own potential, can cope with normal stresses of life, can work productively and fruitfully, and is able to make contribution to her or his community [77]. Mental health reflects a person's approach to life by communicating emotions, giving and receiving, working alone as well as with others, accepting authority, displaying a sense of humor, and coping successfully with emotional conflict [78]
Mental disorder/illness
Psychological disorders consist in malfunctioning of psychological mechanism [79]
"A mental disorder is a syndrome characterized by clinically significant disturbance in an individual's cognition, emotion regulation, or behavior that reflects a dysfunction in the psychological, biological, or developmental processes underlying mental functioning. Mental disorders are usually associated with significant distress in social, occupational, or other important activities. Socially deviant behavior (e.g. political, religious, or sexual) and conflicts that are primarily between the individual and society are not mental disorders unless the deviance or conflict results from a dysfunction in the individual, as described above" [80]
Mental disorders comprise a broad range of problems with different symptoms characterized by some combination of abnormal thoughts, emotions, behavior, and relationships with others [81]
"Mental illness reflects a person's inability to cope with stress, resulting in disruption, disorganization, inappropriate reactions, unacceptable behavior, and the inability to respond according to the person's expectations and the demands of society" [78]

as the existence of a clinically recognized set of symptoms and/or behaviors associated with subjective distress, functional impairment, dysfunction, and disability. However, it is not always easy to make a difference between normal inability and pathological disability, or what is pathological from what is part of normal subjective experience. Aetiology–pathophysiology–symptoms (AE–P–S) framework with clear knowledge of causes (aetiology) and mechanisms (pathophysiology) distinguishes disease entities from one another as well as from disorders and clinical syndromes. The medical model in terms of biological psychiatry defines mental disorders as harmful dysfunctions resulting from damages to a function and structure of the brain or as "diseases of dysfunctional brain–mind and mind–brain interaction" [82]. This perspective focuses on identifying symptoms of the disease linking symptoms to specific pathophysiological process involved and prescribing specific treatment. According to the biomedical model, mental disorders are brain diseases caused by chemical imbalance and neurotransmitter dysregulation, epigenetic dysregulation, brain circuits/systems disorganization, neuroglial failure, neurotoxicity, neurodegeneration, etc. Searching biological markers biological psychiatry attempts to define mental disorders as "real" medical conditions in contrast to "the problems of living" and "specific ways of being in the world". Treatment usually does not demand attention to the whole person and includes medications with neuropsychoactive actions and other biological therapies. Some structural and functional changes have been reported in the brain of patients with mental disorders, but without proper diagnostic validity and specificity. The assumption that disease captures the essence of illness is erroneous (disease without illness, and illness without disease).

According to Carpenter [83] a mental disorder or mental illness is a psychological or behavioral pattern of how a person feels, acts, thinks, and/or perceives which deviates from normal and is generally associated with distress, dysfunction, and/or disability. The medical model in terms of psychodynamic psychiatry defines mental disorder as an illness that is a subjective and interpersonal manifestation of dysfunctions which has to do with meaning, mentalization, interpretation of self and others, intrapersonal conflicts, and life scripts. Illness is a problem of the whole person, not of a single organ or organ system and is therefore subjectively defined. The medical model in terms of social psychiatry defines mental disorders as a sickness that represents community and health authority attitude. The attitudes of community or wider society "shape what individual victims feel has been done to them, and shape the vocabulary they use to describe this, whether or how they seek help, and their expectations of recovery" [84]. According to the structural strain theory, the origins of mental illness are in organization of society [85]. From this perspective, it is crucial to prevent secondary gain and adoption of a chronic sick role [86]. Some mental disorders may be normal reaction to pathological situation or traumatic event. Pathological normalcy and normal pathology are constructs referring to psychological aspects of alienation. At last, it seems that there is some truth in the Shepard's claim that "modern biological models of mental disorders perfectly reflect the atomized, de-socialized, individualistic, consumerist ethos of the twenty-first-century United States, the biochemical sense of self which now pervades popular culture, and the power of the pharmaceutical industry in modern medicine" [87].

The Perspective of Person in Psychiatry

> It is more important to know what kind of a patient has a disease than what kind of a disease a patient has. The good physician treats the disease; the great physician treats the patient who has the disease.
>
> —William Osler

The personal aspects of mental disorders have been frequently neglected in the practice of proclaimed evidence-based medicine in psychiatry. This perspective shifts from the strictly biological determinism to the appreciation of meaning and uniqueness in human behavior as well as to personhood and personality assessment in health and illness. Patients are much more than diagnosis and bundles of psychopathology. They are decision-making persons, similar to other people, but unique as well. As a person, each patient is a unity of unique temperament and character traits and physical constitution. Personality dynamics and types have an essential impact on understanding and treating mental disorders. The perspective of person (lat. persona—human being, a part in a drama, assumed character) is related to what or who someone is, to their personality with vulnerability and resilience characteristics, problems of living and specific ways of being in the world. Personality characteristics may have an important role not only in predisposing to mental disorders and response to the diagnosis, but also they may have a significant impact on the illness course, treatment response, and outcome. Psychiatry as the medical discipline should be set in the context of patients' physiological, psychological, social, cultural and spiritual needs, understandings and beliefs [88–90].

Personality and Its Subsystems in Health and Disease

> There can be fewer vocations more interesting than that of seeking to understand the human person.
>
> —Paul Tournier

A person is human being who is a psychobiological unit composed of many interdependent systems that are more than the sum of its parts because the parts do not function separately [60]. *Personality* (lat. personalitas) refers to the idiosyncrasies of the person, the characteristic set of cognitions, emotional responses, and behavioral patterns that evolves from biological and environmental factors. Many phenomenological and trait theories, psychodynamic and psychobiological theories of personality assume that there is an enduring pattern of characteristics, which is consistent across place and over time, and which is therefore capable of predisposing to certain behaviors in health and illness [91, 92]. *Temperament* or the biogenetic, subcortical–cortical backbone of personality represents the psychomotor characteristics and emotional reactivity of the individual which provide emotional significance to individual experience and behavior [92]. This personality subsystem is a percept-concept hybrid predicated in subcortical effects but associated with independent processes of conscious motivation. According to Cloninger's psychobiological model [60], the four temperament traits: harm avoidance, novelty seeking, reward dependence, and persistence refer to responses to prescriptive stimuli such as danger, stimulus/sensation seeking and epistemics, social approval, and frustrative non-reward, respectively. *Character* represents a subsystem of personality as a semi-stable configuration of mental (cognition, emotion, motivation) faculties and behavioral habits which are self-organized to mediate the evolutionary function of adaptation to the external environment, particularly social living [92]. In other words, character is what people make of themselves as the reflection of personal goals and values and there are three character traits: self-directedness, cooperativeness, and self-transcendence [60]. *Intelligence* (lat. intelligere—"to comprehend or perceive") refers to "the general mental ability to understand one's surroundings and solve novel problems figuratively to make sense of things, figure out what to do or catch on" (Gottfredson 1997 according to 92). *Self* represents a subsystem of personality predicated on process in which "I" (the mind) meets "me" and reflects back as self-representation [92]. The self-concept has been variously defined as perceptions, internal representations, mental schemas, roles, and reconstructive stories [93]. Self is construct which encapsulates a lot of biological, developmental, intra- and interpersonal, and contextual factors that are supposed to be associated with development, course, and outcome of mental disorders [94]. Self as the marriage of temperament and character can be measured by the Temperament and Character Inventory (TCI). *Theory of mind* (TOM) and *mentalization* refers to the person's capacity to understand own behavior and the behavior of others as a reflection of authentic and meaningful mental states, feelings, thoughts, memories, desires, and values. It also includes being able to make and detect deception, self-awareness, self-deception, and perspective-taking. Mental disorders are commonly associated with significant mentalization problems of implicit and explicit interpretation of self and others in terms of mental processes and subjective mental states. *Identity* (lat. identitas—sameness) is a subsystem of personality which refers to the autobiographical narrative of the self as "a sustained sense of oneself as a coherent, predictable, and authentic person" [92]. *Conscience* (lat. consciencia from suneidesis in Old Greek) refers to the subsystem of moral values based on emotion and rational associations. Metaphorically, it is the mind's inner censor of right and wrong. Within every person, there is a need for happiness, love, power, freedom, and purpose/meaning involved in coherent living.

The Vulnerability—Resilience Model

The vulnerability-resilience or stress-diathesis model rests on the fact that some individuals are more vulnerable while others are more resilient to mental distress. Mental illness and mental health are two ends of a one-dimensional continuum and they depend on a complex interaction of the three groups of factors: (1) "risk" or "vulnerability factors" (personality weakness) which enhance the likelihood of mental disorders, (2) "protective factors" that enhance the likelihood of recovery from trauma and stress, and (3) "generative or creativity factors" which increase revelatory learning, resource acquisition and development, accentuating personal growth. Treatment is focused on helping patients to use personality resources and strengths to increase their well-being and restore resilience so that they can cope with stress more successfully. The focus in contemporary psychiatry is still on the pathological dimension while a tendency to approach mental disorders in terms of challenges, opportunities, and posttraumatic growth is of recent date.

Resilience is a relatively new multidimensional psychobiological concept, essential for understanding of salutogenesis and pathogenesis as well as of therapeutic and healing mechanisms and responses. *Salutogenesis* (the Latin salus—health; the Greek genesis—origin) is related to healing that is natural process seen in all forms of life, and it is closely related to resilience. Resilience may be defined as a collection of protective and salutogenic factors that modulate the relationship between a stressful event, adversity or disease, and positive outcomes. Resilience is about the whole person, it includes biological, psychological, social, and spiritual dimension of human existence. It enables individuals and communities not only to survive and adapt to challenges and adversities but also to be better off and to grow and thrive (post-traumatic growth) in addition to overcoming a specific adversity. Resilience is a very complex process ranging from surviving to thriving. It includes positive transformation and personal growth, an indivisible part of mental health and health in general, well-being and quality of life as well as recovery and treatment outcome. It is very important to note that "some resilience factors contribute to the development of other resilience factors, and, in consistency with a cascade model, together they contribute to predict personal recovery" [95]. *Primary resilience* is related to maintaining equilibrium, balance, and mental health. The level of primary resilience has been regarded as a protective factor against developing illness what means that lack of resilience carries a risk for the appearance of mental disorders and somatic/neurologic diseases. It can be described as "bouncing back" and "rebounding after adversity" and as such it is related to disease prevention. The concept of primary resilience explains why many people do not become ill or do not develop a particular disorder although they are subject to the same kind of adversary events, even after a prolonged period of adversity, with psychological and physical burdens, that cause the disorder in other people [96]. *Secondary resilience* refers to the capability of individuals to cope with illness/disease and successfully recover. It is aimed to regain mental equilibrium and somatic balance after allostatic load and illness. The capability to achieve clinical, functional/social and personal recovery implies the presence of secondary resilience. Placebo response may be an expression of psychological and spiritual resilience [97]. In addition to clinical remission, secondary resilience may lead to personal growth and developing a meaningful life after mental illness. On the opposite side, lack of resilience determines onset, course, outcome, distress, and burden of mental illness [98]. *Tertiary resilience* enables patients to develop a healthy and productive way to live with their illness, helps them to adapt to limitations in life associated with illness and have positive and creative life attitudes. Proactive and more efficacious participation of patients with chronic illness and residual symptoms in their medical treatment is also an expression of tertiary resilience.

The model of primary, secondary, and tertiary resilience explains how appropriate resilience-enhancing interventions may help in obtaining favorable therapeutic response. The level of and pace by which personal recovery is established is a function of brain resilience, external resources like support, nature of illness, and chosen drug treatment. However, resiliency as a treatment target has been largely neglected in the field of therapeutics [99] so that the lack of favorable treatment outcome may be commonly related to the treatment focus only on symptoms and illness. The route of clinical, functional, and personal recovery lies not only in decreasing illness but also in enhancing resilience and increasing wellness [97]. Full personal recovery does not mean only the absence of symptoms of mental illness, but also the presence of resilience, quality of life and wellness. The concept of resilience enhancement promotes strengths and potentials for wellness which are present in patients instead of focusing only on their weakness and pathology. Each patient is unique, responsive, and responsible person and within every person, there is a force that drives them to strive to self-realization, self-understanding, self-transcendence, and a sense of coherence and control over their own life. Enhancing patients' resilience by emphasizing their strength and opportunities and covering up weakness is an ambitious goal that aims to promote positive mental health in spite of the presence of symptoms [100] and drug treatment failure. Good news is that resilience can be enhanced through learning and training. Resilience training can result in augmented neuroplasticity and balance of neural circuits that modulate reward and motivation, emotion regulation, cognitive reappraisal and executive function, novelty seeking, harm avoidance and fear response, self-directedness, cooperativeness and adaptive social behavior, and self-transcendence. Our five steps model of resilience-enhancing approach includes: (1) SWOT (strength, weakness, opportunities, threats) analysis; (2) Re-construct of disease and therapeutic narratives (DTN); (3) Construct of personal model of individual and family resilience (PMIFR); and (4) Put the PMIFR into operation and practice resilience; and (5) Practice personal recovery and creativity.

The Cognitive-Axiological Perspective in Psychiatry

> Men are not disturbed by things but by the views they take of them.
>
> —Epictetus

This perspective focuses on what someone is thinking about, perceives and learns about or assesses as valuable, i.e., which ideas and values in life should be followed. All that we are is a result of our thoughts and knowledge of ourselves and the world, and the defined values and goals we follow. What we think of us, that it is which significantly determines or indicates our fate so that all psychological stressors are cognitively mediated. The basic tenet of this perspective is that much of mental disorder is a result of errors, biases, and aberrations in cognitive functioning. From *the cognitive perspective*, aetiopathogenesis of mental disorders is associated with dysfunctional and conflicting cognitive strategies,

Table 6 Thinking errors in depression

– All-or-nothing (black or white) thinking—"I am no good"
– Negative automatic thoughts: I am not important, no one cares about me
– Overgeneralization: Never-ending pattern of defeat—"No one understands or cares about me": "No one loves me"
– Mental filter or selective darken abstraction: "I ruined everything"
– Disqualifying the positive: "I'm afraid to feel happy because I always feel bad afterward"
– Jumping to conclusions and arbitrary inferences: Mind reading and Fortune teller error
– Magnification and minimization (binocular trick): "I made a mistake, I am so stupid"
– Emotional reasoning: "I feel rejected, for sure they don't like me"
– Should and must statements
– Labeling and mislabeling: "I made a mistake, I am so dumb"
– Personalization: One sees oneself as a cause of bad event; "I was born bad", "It is waste of time someone talking to me"

misinterpretations and misrepresentations. Some mental disorders may be related to the explanatory style how an individual interprets the meaning of what happens and how explains both positive and negative (adversity) events. Pessimists tend to attribute the causes of negative events to permanent, uncontrollable, and pervasive factors, while conversely, optimists tend to attribute the causes of negative events to temporary, changeable, and specific factors [101]. Much of anxiety and depressive disorders may be created by errors or biases in thinking such as catastrophic thinking because our thoughts are important determinants of our actions. A negative cognitive style with deficit for retrieval of positive memories contributes to development and severity of some mental disorders, like depression, PTSD, etc. According to the cognitive model of depression [102], depressed individuals developed dysfunctional beliefs about being helpless or unlovable as a result of early learning experience (Table 6). The negative cognitive triad about self (negative self-precept), world (hostile and demanding), and future (the expectation of suffering and failure) with the vicious circle of low mood contributing to negative thinking and vice versa is an essential characteristic of depressive disorders. The negative automatic thoughts that prevail in depression are sustained through systematic distortions of information processing.

The cognitive theory of anxiety disorders postulates that underlying danger-oriented beliefs make the patients to overperceive possible threats in the environment and interpret ambiguous stimuli in a catastrophic way, underestimate their own coping abilities and resources, and engages in dysfunctional safety behaviors such as avoidance. Mental model includes three basic beliefs: (1) the world is a dangerous place, (2) I am not in control, and (3) things may become worse suddenly. When wrong, negative, catastrophic, self-limiting and self-defeating thoughts are corrected, mental health can be established again. Cognitive reframing may foster moving forward with life. Viewing traumatic stress and negative life events in the same time as challenges, turning points, or opportunities for growth creates a more positive frame of mind leading to positive mental health.

From *the axiological perspective,* mental disorders can be explained on the account of choosing a wrong life, social and spiritual values, but the question remains as to how much it entails a voluntary and conscious decision in comparison to subconscious autoprogramming. Exposure to traumatic experiences often leads to thinking and asking about meaning, values, and purpose within a personal and collective sense. Traumatic events commonly challenge one's core life values and beliefs about safety, self-worth, and the meaning of life. Individuals who are unable to resolve challenges to their moral and value beliefs might find themselves in a state of demoralization, disillusionment, nonsense, and social alienation. Demoralization associated with negative thinking and characterized by feelings of helplessness, hopelessness, subjective incompetence, and a loss of mastery and control, was found to be very common syndrome in depression, and PTSD. According to some opinions, PTSD results from "the shattering of basic assumptions" that people have about themselves and their world that is a consequence of "information shock" [103]. Beliefs, thoughts, attributions, cognitive schemas and general attitudes structure meaning of life events and influence emotional arousal. It seems that belief rigidity as well as specific negative beliefs constitute cognitive risks for PTSD. Albert Ellis's ABC (adversity–belief–consequence model) may help individuals with PTSD to distinguish activating event, their beliefs about activating event, and the emotional and behavioral consequences of those thoughts and take control over their thoughts and behavior [101].

The art of living, in hard and traumatic life situations, is to reveal their true meaning, values, and the purpose of life. Simply said, life places tasks before us, sometimes very painful ones, at other time incredibly difficult ones. The purpose of something dreadful happening to us is that something better in the future will happen. The good that comes from all bad things happening to us means that it helps us to achieve our best, our essence, our mission. According to M. Kirchenbaum [104], there are ten reasons why something happen to us: (1) to have a sense of belonging and to feel truly good in our world as though we were at home; (2) to completely accept ourselves; (3) to understand that we can liberate the fear that hinders us; (4) to gain the ability to forgive; (5) to recognize our true talents; (6) to discover true love; (7) to become stronger; (8) to learn to take pleasure in life and enjoy life; (9) to learn how to live with the feeling that we have a specific mission or vocation and purpose in life; and (10) to truly become good people. The purpose of our life does not consist solely in self-achievement, but also an autotranscendence or outdoing ourselves.

The Behavioral Perspective in Psychiatry

> We are what we repeatedly do.
>
> —Aristotle

While the perspective of disease/illness refers to what the patient has, the perspective of person to what the patient is, the cognitive perspective to what and how the patient thinks, value and respect, the behavioral perspective focuses on what and how the patient does. This is the perspective of behavioral medicine. From this perspective mental disorders are depicted as characteristic maladaptive behaviors developed during interactions with the interpersonal or physical environment. The conceptual triad involves physiological motivation (drive or need), different ways of learning and conscious or unconscious choice. Learning is usually defined as acquisition, modification, and elimination of behaviors and response pattern that is associated with environmental conditions in which occurs a connection between stimulus and response. Motivation may be related to innate or primary needs that urge the organism to act or to learned motives acquired by reward, reinforcement or punishment. Classical conditioning, operant conditioning, cognitive behavioral learning, and observational and imitational learning represent the basic types of learning. Mental or behavioral disorders emerge either because of unusual goals some people come to crave or because excess in their attempts to satisfy drives or needs, and develop through a process of learning, classical or operant conditioning as well as social learning. Two behavioral triads explain normal behaviors and provide a guide for the treatment of abnormal ones. Driven (internally or biologically motivated) behavior is composed of physiological drive, conditioning, and choice while socially learned behavior consists of antecedents (predisposing/precipitating factors), responses (choice/action factors) and consequences (reinforcing/sustaining factors) [76]. Social learning modifies the expression of driven behaviors based on physiological needs. Biologically driven behaviors are characterized by goal-oriented nature, intermittent occurring that is selectively provoked by stimuli that are later ignored, variations in intensity of drive which can be voluntarily suppressed. Some motivated behaviors follow a typical cycle: craving, consummation, decreased craving and the replacement by other activities, and reappearance of craving after spell of time. Cravings involve two components: a search for pleasure such as satiety or orgasm and a desire for release of inner tension such as hunger or sexual arousal [76]. At present, there are no valid or reliable ways of measuring the level of controllability or irresistibility of drives. Some behavior is usually considered as pathological when the object of the drive is socially unacceptable or when the strength of the drive is so intense that it causes suffering or injury to the patient or to the people around him [76]. Pathological behavior is self-sustaining with internal, biological and psychological and external, social reinforcers. Some mental disorders are abnormal or maladaptive behaviors (substance-abuse disorders, eating disorders, both anorexia and bulimia nervosa, impulse control disorders such as oniomania, kleptomania and pyromania, sexual disorders such as paraphilia) rather than diseases. The maladaptive preoccupation with concerns about personal and family safety, anxiety, irritability and pervasive and uncontrollable sense of threats and danger may be explained by classic fear conditioning

The learned helplessness theory explains depressive and anxiety disorders as reaction to the experience of uncontrollable stress. According to the evolutionary model depression serves the triple purpose: (1) to signal submission to dominant figures in a hierarchy conflict and internally communicate defeat, (2) to signal helplessness and communicate a need for help to potential caregivers, and (3) to disengage from commitments to futile or dangerous goals [105]. Mental disorders may reflect the pathology of learning mechanisms which normally uses past experience for better adaptation and improving harm avoidance abilities. Conscious and unconscious re-experiencing past traumatic situations and avoidance behavior represent the core processes which lost their adaptive value in mental disorders. Instead, they give way to phylogenetically primitive fear reactions as agitation/aggression, freezing, and dissociative states [105]. The traumatic (unconditioned) stimulus automatically evokes the post-traumatic (unconditioned) emotional response (fear, helplessness and/or horror) and/or dissociation. Conditioned stimuli which are reminders of the experienced traumatic event evoke similar conditioned emotional response, dissociation, and flashbacks as well as fear-induced avoidance and protective behaviors. The trauma re-experiencing symptoms can be a consequence of the primacy of traumatic over non-traumatic memory as a pathological exaggeration of an adaptive response to remember as much as possible about traumatic events in order to avoid similar threats and dangers in the future. In general, mental disorders may be the consequence of coincidental reinforcement of different behaviors, regardless of genetic predisposition, and related to combination of unconscious psychological needs, non-adaptive learning, and bad choices. Some aspects of mental disorders may result from what patients are doing wrong and from their wrong attitudes and beliefs about their symptoms or behavior.

The Spiritual/Transcendental/Transpersonal Perspective in Psychiatry

> What distinguishes man is his spiritual life.
>
> —Aristotle

This perspective represents a framework for the medicine of person and spiritual psychiatry. The key terms are spiritual self, self-transcendence, and healing. Throughout history, much of psychiatric care has been provided within a spiritual and faith context. Human beings exist not only in visible, physical world of senses but also in spiritual world of ideas or values which shape their life, behavior, and identity. In addition to scientific questions about the nature, mechanisms, and diagnosis of mental disorders, there are also spiritual questions about the meaning of mental disorders [106]. According to William Osler (1849–1919), the father of modern medicine, "faith has always been an essential factor in the practice of medicine". This is so whether we are speaking of faith in physician leading to compliance; faith in the efficacy of medical care, leading to positive expectations and, perhaps, to a salutary placebo effect; or faith in a divine being, leading to the psychosomatic benefits, or—as the religious themselves might claim—a divine blessing, or an expectation of such. For Paul Tournier (1898–1986), pioneer of medicine of the person, the twin pillars of medicine were science and faith, and an integration of body, mind, and spirit was necessary for health and wholeness [107]. Medicine of the person and person-centered psychiatry offer a well-grounded reason for incorporating spirituality and religion into psychiatrist assessment, diagnosis, case formulation, therapy, and as a component of psychiatric training and continuous professional development [108]. There are many ways to understand and define spirituality, as a human's characteristic and a personal experience. Spirituality can be viewed as "a distinctive, potentially creative and universal dimension of human experience arising both within the inner subjective awareness of individuals and within communities, social groups, and traditions [109] as well as" a facet of our individual humanity, linked to aspects uniqueness, meaning, identity, purpose, relationships, a sense of the holy, and the spirit and fire which drives us [110]. Ellison [111] argues that spiritual dimension does not exist in isolation from the mind and the body, and it is the spirit which synthesizes the total personality. Spirituality has been also argued as a belief system related to spiritual intelligence and transpersonal self-providing a person with meaning and purpose in life, a sense of the sacredness of life, a vision of the better world and connection to that which transcends the self. This transcendental connection might be to God, a higher power, a universal energy, cosmic law, the sacred, or to nature. Spirituality can be defined as a quality of human beings who are concerned or preoccupied with higher meaning or purpose in life rather than with affairs of the material world [34]. It is an integrating force for physical, biological, psychological, and social dimensions of human life and a potential source of strength and well-being. Spirituality may or may not be associated with a specific religion, but it is always related to the subjective experience of something sacred, transpersonal, transcendental and greater than self as well as to feelings of awe, reverence, and love. Transcendentality, vitality, meaningfulness, and connectedness are essential elements of a spiritual experience which can be understood in either secular or spiritual terms [34]. Vitality is an ability or powerful force of an organism to maintain its organic existence. It includes a creative attitude, being spirited, open to new experiences, and growing through inner exploration or meditation. Health, energy, and enthusiasm are secular terms, while soul, grace, and sanctity are spiritual terms related to vitality. Spirituality is associated with emerging of higher values and deeper meanings of life, frequently connected with a sense of mystery and awe. Art, science, and literature are secular terms, while faith, scriptures, and revelation are spiritual terms related to meaningfulness. Connectedness refers to a feeling of union or harmony with another being or thing which includes connection with a living, dead or imagined person, a cultural, ethnic or political group, humanity, nature or universe. Family, lovers and nature are secular terms, while God, fellowship, and church are spiritual terms related to connectedness.

Healthy and Pathological Spirituality

Spiritual beliefs are of great importance to many patients and may have a significant impact on both mental health and mental disorders. It is very important to make a distinction between path and pathology, between health promoting spirituality and pathological spirituality, between healthy minded and distorted or sick faith. Healthy spirituality in its many forms can bring patients (1) a mission-discovery process, sense of meaning, personal integrity, and purpose; (2) inspirations, values and the fuel to be good, do good and serve others; and (3) a deep enjoyment of life, clinical and personal recovery. Finding meaning in life is a fundamental challenge for everybody, including people with mental disorders. Personal recovery is a journey from alienation to a sense of meaning and purpose, from withdrawal and disengagement to engagement and active participation in life [34]. Healthy spirituality has been linked with higher self-esteem and optimism, more love, hope, positive thinking, and positive mood states as well as improved treatment outcome. The need to have uplifting experiences and to be part of something larger than oneself is essential to the personal recovery narratives of many people. Healthy spirituality is associated with placebo or white magic while pathological spirituality is associated with nocebo or black magic. According to Ferrer [37], spirituality is a basic transformative process that involves a shedding of narcissism, self-centeredness, self-separation, self-preoccupation, and so on. Spiritual narcissism represents a spiritual distortion and misuse of spiritual practice, energy, or experience, manifesting as ego-inflation with grandiosity fueled by spiritual energies, self-absorption with a preoccupation by own spiritual status and achievements, and spiritual materialism with the appropriation of spirituality to strengthen egoic way of life [37]. Healthy minded (salutogenic) and sick (distorted or pathogenic) expressions of faith are quite distinct—in the objects of faith, in the expectations of such faith, and in the observed effects and outcomes in the lives of the faithful [112]. Salutogenic faith motivated by intrinsic religion is associated with empathy, compassion, open-mindedness, self-esteem, altruism, and social responsibility. Healthy minded faith is the fuel that produces constructive social and cultural transformation—it inspires and directs acts of compassion, mercy, and justice [112]. Pathogenic or sick faith suggests itself as a font of psychopathology, which may have expressed somatic consequences. Certain expressions of distorted religious faith may serve as a source of or may reflect psychological conflict [112] and/or psychopathology. Distorted and pathogenic faith can indeed be an impediment to well-being and healing, no serious observer would deny this point [112]. According to Pargament [113], religious coping can be positive and negative and may serve five purposes: (1) spiritual (meaning, purpose, hope); (2) self-development (positive identity); (3) resolve (self-efficacy, comfort); (4) sharing (closeness, connectedness to community); and (5) restraint (helps keep emotions and behavior under control). Religious, spiritual, and therapy groups can be positive, health-promoting and life-affirming or manipulative, destructive, and harmful [114, 115].

Spirituality related to the sacred dimension, ultimate meaning, and purpose of life in health and disease is an important part of the transdisciplinary integrative psychiatry where Latin prefix "trans" means not only beyond but also through, across, and pervading. Trust in providence which is love and wisdom, belief in power which is greater than oneself, which is a source of significance and hope, ability to find meaning in suffering and illness, gratitude for life which is perceived as a gift, ability to forgive have stress coping values and protective effect on mental health. Spiritual psychiatry has been the medicine of the person, by the person, and for the person grounded on unconditioned love and empathy.

The Narrative Perspective in Psychiatry

> A man is always a story teller: he lives surrounded by his stories and the stories of others, he sees everything that happens to him through them; and he tries to live his life as if he were telling a story.
>
> —Jean-Paul Sartre

> Humans think in stories rather than in facts, numbers or equations.
>
> —Harrari (2018)

The narrative perspective is based on the narrative self and identity, logic of narratives (illness, therapeutic, recovery, quest, and restitution narratives), and distressed states of the soul, which are quite natural, understandable, and the result of adverse impressions and experiences [74, 76]. Narrative psychiatry is predicated on the conceptualization of human beings as narrators who live their lives in relationships and connect and cooperate with one another through the stories they create, tell, and live. Our identity is shaped by narratives or stories, both uniquely personal and culturally general. According to Gilbert [110], "great narratives of modernism, science and religion have been replaced by own personal consumeristic narratives so that we have become consumers, individualized to become pawns and pray to global capitalism". Human beings are immersed in narrative, first self creates personal stories, then over time personal stories modify and create self. The stories human beings tell about themselves not only describe themselves, their experiences and life philosophy but also shape their lives [116], telling and listening stories they recognize themselves in the stories of others, and others in their stories. One's ability to create, live and tell a coherent, hopeful and self-actualizing story of his or her life is a fundamental component of mental health and well-being. The story is a natural framework for a very different conclusion about how we live and what we do, and what is the meaning and purpose of everything. So we also give meaning to our lives and the world by the stories we tell to ourselves and each other; hence, we define our experiences, actions, and destiny.

A narrative perspective which focuses on the life story of a patient provides a better understanding, not only of the patient's actual mental state but also the significance, meaning, and processes contributing to the onset and maintenance of clinical symptoms. This perspective aims to understand and describe mental disorder by identifying the patient in life like a person with particular strengths and weakness (vulnerabilities) as well as with opportunities and threatens and helps him to create new life story. Mental disorders can be better understood through patients' personal history and narratives associated with networks of meaningful relationships that are embedded in societal and cultural systems. Patient stories are the background for diagnosis and the guide to managing treatment and measuring treatment response and effectiveness. Mental disorders often point to stories of lost love, faith, power, joy, identity or tragic story of "dread frozen in memory", existential despair and life irony. Person-centered narrative psychiatry is based on a deep and empathic understanding of the patients as persons with unique individual life stories and therapy involves their re-authoring and retelling the stories of their lives in creative and hopeful way. Through illness narratives, patients form their own explanations about the causes of their illnesses, in useful or harmful way. All therapies in psychiatry involve a therapeutic narrative and start with therapist listening to the patient's story and then helping him to recognize a new perspective on the problem and gain new coping and resilience skills [117]. Deconstructing hopeless and harmful into hopeful and useful narratives may help in achieving treatment success. Therapeutic narrative refers to explanations of how mental health medications or psychotherapy work. These explanations are different during acute, stabilizing, and maintenance phase of treatment. Treatment failure may be related to an inappropriate narrative in specific situations. Deconstructing narratives that fuel mental health problems and treatment failures is essential whenever desired treatment goals are not obtained. The restitution narrative presumes the illness to be cured or overcome so that the patient becomes the same or healthy again. While restitution story "yesterday I was healthy, today I am sick, but tomorrow I'll be healthy again" may work for some illness experience, it can be problematic in the context of some other mental disorders for which cure, or return to previous health as it was once, may not be forthcoming [118]. So, patients with severe major mental disorders need alternative narrative resources to preserve or reinstate sense of self, meaning, identity, hope, well-being, and mental health. In chaos narrative, the illness destroys the life of the patient. The quest narrative is characterized by the patient's search for meaning and the idea that something can be learned or gained from the illness experience [118]. The recovery narratives involve the four-component process: recognizing the problem, transforming the self through recovery narratives, reconciling with the system, and reaching out to others. Establishing a personal relationship with the patient should help the patient to find a new self as a person with a mental disorder who can recover from that disorder with a new perspective on life. The main focus is on the person, not on the symptoms and problems. This approach allows the patient to reconnect with his or her true healthy and positive self. Finding a new, true self is associated with a re-authoring life story, personal growth, self-actualization, and reaching one's full potential.

Narrative perspective emphasizes also the importance of personality organization and psychological life script for understanding the individual psychopathology. The psychological script contains the ongoing program for the person's life drama and tendencies to some mental disorders. From the narrative viewpoint, the patient's specific sad and defeating life story related to a mental disorder may reflect destructive self-attitudes and a particular unconscious loser life script [74]. Victimization is commonly associated with the shattering of at least three basic assumptions individuals hold about themselves and their world: (1) the belief in personal invulnerability; (2) the perception of the world as meaningful and comprehensible; and (3) the view of selfhood in positive terms. People normally operate on the basis of unchallenged and unquestioned positive assumptions about themselves and the world like, for example, "My world is predictable, safe, meaningful and just" and "bad things don't happen to good people like me". Recovery from mental disorders is related to the cognitive rebuilding of a viable assumptive world view which integrates the realms of vulnerability, meaning, and self-esteem [119] in order to create a new better life story. A new story gives the new meaning to life, but also fashions the relations with people. The story teaches how to behave and live in general as well as with experience of the traumatic events, discovers the true values that should be followed, and the meaning of life in adversary and happy times. The study of literature discovers universal themes with heroic figures whose journey of existential self-transformation and psychological and spiritual growth in the life cycle upon survival of traumatic experiences may be a form of transgenerationally transmitted experiences of collective narrative therapy. Metaphors have a healing power and are useful in narrative psychotherapy helping patients to change their loser's story into winner's story with a happy end.

The Systems Perspective in Psychiatry

> Pay attention to your thoughts, because they become words. Pay attention to your words, because they become actions. Pay attention to your actions, because they become habits. Pay attention to your habits, because they become your character. Pay attention to your character, because it is your fate.
>
> —Talmud

> The greatest mistake in the treatment of diseases is that there are physicians for the body and physicians for the soul, although the two cannot be separated.
>
> —Platon

This perspective represents a framework for systems neuroscience, integrative psychiatry, and person-centered psychiatry. The multidimensional features of mental disorders and their treatment require a systematic approach and a multidisciplinary paradigm to bridge substantial gap between different mental health disciplines and their paradigms. According to the general systems theory (GST), interconnections and interrelations exist between parts of entities that work together making a system which represents a coherent organization of that parts, the effect of which is greater than their simple sum. In this context, the genome operates within the context of the cell, the cell within the context of the body, the body within the context of the self, the self within the context of society, and the society within the context of the universe [60]. Any complex living system must have: (1) order defined as the global adaptive optimum; (2) must be adaptive that means capable of biomorphogenesis which leads to awareness, in humans it means self-understanding and nonlocal awareness which is basis for understanding of truth which lead to wisdom; and (3) must be flexible and creative in its self-organization [60]. From the systems perspective, mental disorders arise from various dysfunctions and interactions within and between complex systems that operate from the epigenetic molecular, cellular and neuronal networks level all the way up to the level of family, society, and culture. The system approach in psychiatry is an attempt to integrate the fragmented identities of psychiatry and different classes of neurobiological, psychodinamic, social, and cultural processes studied by different scientific disciplines into an organized whole.

Mental disorders in general and specifically may reflect the problems in many different, more or less related systems that can be considered on three levels. Macro-level addresses broad societal and cultural aspects of mental disorders. The meso-level is related to the family and social networks. The micro-level refers to the level of the individual as the person. Mental disorders and somatic diseases/illnesses can be conceptualized within different body, energy, mental, family, social, etc., systems. Experiential and behavioral manifestations of mental disorders are predicated on their underlying multilevel neurobiological systems as well as to cognitive-interpretative, narrative, behavioral, and interpersonal processes that are at play prior to the onset of mental disorder, during the pathogenesis, throughout the course of disorder and response to treatment [120]. There has been a substantial conflict between current knowledge and various concepts and models of mental disorders. The systems approach suggests that the only way to understand the complex issues of mental disorders and human suffering is to approach them in a holistic multiperspective transdisciplinary way using both analytic-focused top-down processing and heuristic-global bottom-top processing [11]. The various perspectives and dimensions are interconnected and interdependent, and cannot be fully understood separately. According to the systems theory, the whole is more than the mere sum of its parts, and the interconnections between the parts add a specific and distinct quality and dimension to the whole. While the focus of the reductionistic, mechanistic, and formistic information processing is on the parts, the systems approach emphasizes the whole [12]. The application of the transdisciplinary systems approach helps us understanding mental disorders and their treatment and prevention in a better and more appropriate way and offers a broader range of more efficient treatment strategies and options [121].

Systemic approach enables integration of different perspectives and paradigms into a holistic coherent model. Each of the seven perspectives presented here offers a different map which is complementary to others and all together outline the field of transdisciplinary integrative holistic mental health science and practice. All these perspectives can contribute to the formulation of clinical problem in a different way, so a simple one-to-one application of perspective to a case is not successful enough. How much each will contribute depends on the clinical characteristics of the case as well as on the treatment phase. At each particular case or phase of the treatment, the psychiatrist needs to select the primary perspective that best fits the patient and then integrate the other six perspectives into the formulation and treatment. When applied secondary to the disease perspective, the dimension perspective is mindful of the patient's personality strengths and weaknesses as the patient encounters the limitations imposed by the disease or posttreatment condition. The person perspective is focused on helping the patient to use personality resources to respond to the demands of the actual life situation and increase her or his well-being. The behavior perspective is secondary to the disease perspective for cases in which behavior pattern is associated with the disease process or condition. When the behavior perspective is the primary perspective, the focus is on stopping unwanted or starting wanted behaviors. The cybernetic perspective is related to a more or less self-inducing, self-regulating, and self-organizing aspect of behavior in illness and wellness. The cognitive perspective is of great importance for shared-vision of disease and its treatment. The spiritual perspective helps in finding sense of life as well as meaning in suffering and illness. The systems perspective gains momentum to transdisciplinary, integrative, context-sensitive psychiatry that responds more effectively and humanly to the challenges of mental disorders.

Conclusion

The challenges of the twenty-first century have marked the need for a new field called theoretical psychiatry, problem-oriented discipline which is about "the way we challenge in psychiatry, the way we integrate in psychiatry, the way we test a hypothesis in psychiatry, and the way we intervene in psychiatry". The content of this chapter has been shaped by the ideas, theories, researches, and clinical experiences of many scholars, experts and clinicians around the world, many of them found on Internet. It seems true what Stephen Hawking claimed that "we are all now connected by the Internet, like neurons in a giant brain". Systems thinking, systems (complexity) theory, brain network theory epistemology, and philosophy of science are pillars of the theoretical psychiatry. Creatively combining the different theoretical perspectives brings us closer to a holistic understanding of the complex nature of mental health and mental disorders, and a more efficient treatment of mental health problems. The traditional "one-size-fits-all" approach to diagnosis and treatment should be reconsidered and new diagnostic and therapeutic tools should be offered. Psychiatry seems to be moving toward an era of individualized and person-centered care and a type of treatment that should offer the right drug to the right patient. Although many concepts are hazy and confusing in psychiatry and many puzzles need to be resolved, the substantial groundwork for better understanding of mental disorders and their more successful treatment is being prepared. At the end of the day, one can say that due to computational science there is already a good theoretical and empirical background for a transdisciplinary, integrative, context-sensitive psychiatry that responds more effectively and humanly to the challenges of mental disorders.

References

1. Jakovljevic M. Psychiatry at crossroad between crisis and new identity. Psychiatria Danubina. 2012;24(suppl 3):267–71.
2. Kirmayer LJ, Lemelson R, Cummings C. Introduction: psychiatry at a crossroads. In: Kirmayer LJ, Lemelson R, Cummings C, editors. Re-visioning psychiatry: cultural phenomenology, critical neuroscience and global mental health. New York: Cambridge University Press; 2015. p. 1–38.
3. Jakovljevic M. Theoretical psychiatry: missing link between academic and clinical psychiatry for further scientific and professional maturation of psychiatry. Psychiatria Danubina 2013; 25:203–206.

4. Frances A, Carroll B. Barney Carroll: the conscience of psychiatry (obituaries). BMJ. 2018;362:k3016.
5. Kleinman A. Rebalancing academic psychiatry: why it needs to happen—and soon. Brirish J Psychiatry. 2012;201:421–2.
6. Kendler KS. The structure of psychiatric science. Am J Psychiatry. 2014;171:931–8.
7. Best S, Kellner D. Entropy, chaos and organism in postmodern science. In: Best S, Kellner S, editors. The postmodern turn. New York, London: The Guilford Press; 1997. p. 195–252.
8. Jakovljevic M. The brave new psychiatry: beyond modernism, antimodernism and postmodernism in psychiatry. Psychiatria Danubina. 2007;19:122–9.
9. Jakovljević M. The brave new psychiatry: a pluralistic integrating transdisciplinary approach in theory and practice. Psychiatria Danubina. 2007;19:262–69.
10. Jakovljevic M. Conceptual cacophony or different parts of a complex puzzle of mental disorders: transdisciplinary holistic integrative perspective. Psychiatria Danubina. 2011;23:232–6.
11. Feist GJ. The psychology of science and the origin of the scientific mind. New Haven, London: Yale University Press; 2006.
12. Jakovljevic M. The decade of the brain in biological psychiatry—biological psychiatry between conservation and change. Psychiatria Danubina. 1995;7:75–87.
13. Ioannidis JPA. Why most published research findings are false. PLOS Med. 2005;2:e124.A.
14. Ioannidis JPA. Why most published research findings are false. CHANCE. 2005;18:40–7.
15. Fanelli D. How many scientists fabricate and falsify research? A systematic review and meta-analysis of survey data. PLOS ONE. 2009;4(5):e5738.
16. Boutron I, Dutton S, Ravaud P, Altman DG. Reporting and interpretation of randomized controlled trials with statistically nonsignificant results. JAMA. 2010;303:2058–64.
17. Bowen A, Casadevall A. Increasing disparities between resource inputs and outcome, as measured by certain health deliverables, in biomedical research. PNAS. 2015;112:11335–40.
18. Ioannidis JPA. Why most clinical research is not useful. PLOS Med. 2016;13:e1002049.
19. Fullford KWMB, Thornton T, Graham G, editors. Oxford textbook of philosophy and psychiatry. New York: Oxford University Press; 2006.
20. Kendler KS, Parnas J, editors. Philosophical issues in psychiatry—explanation, phenomenology, and nosology. Baltimore: The John Hopkins University Press; 2008.
21. Sackett DL, Rosenberg WMC, Muir Gray JA, Richards WS. Evidence-based medicine: what it is and what it isn't? BMJ. 1996;312:71–2.
22. Jakovljevic M. Myths and facts in contemporary psychopharmacotherapy: evidence-based vs. evidence-biased treatment algorithms practice. Psychiatria Danubina 2007;19:342–9.
23. Isaacs D, Fitzgerald D. Seven alternatives to evidence-based medicine. BMJ. 1999;319:1618.
24. Drake RE. The principles of evidence-based mental health treatment. In: Drake RE, Merrens MR, Lynde DW, editors. Evidence-based mental health practice—a textbook. p. 45–65.
25. Dictionary Encharta Concise English. Student edition. London: Bloomsbury Publishing Inc; 2001.
26. de Leon J. Is psychiatry scientific? A letter to a 21st century psychiatry resident. Psychiatry Investig. 2013;10:205–17.
27. Marshall M: Randomized controlled trials—misunderstanding, fraud and spon. In: Priebe S, Slade M, editors. Evidence in mental health care. Hove, New York: Brunner, Routledge; 2002. p. 59–71.
28. Mueser KT, Drake RE. How does a practice become evidence-based? In: Drake RE, Merrens MR, Lynde DW, editors. Evidence-based mental health practice—a textbook. New York, London: W.W. Norton & Company; 2005. p. 217–41.
29. Jakovljevic M. New generation vs. first generation antipsychotics debate: pragmatic clinical trials and practice-based evidence. Psychiatria Danubina 2009;21:446–52.
30. van Baalen, Boon M. An epistemiological shift: from evidence-based medicine to epistemological responsibility. J Eval Clin Pract. 2015;21:433–9.

31. Djulbegovic B, Guyatt GH, Ashcroft RE. Epistemologic inquiries in evidence-based medicine. Cancer Control. 2009;16:158–68.
32. Djulbegovic B, Guyatt GH. Progress in evidence-based medicine: a quarter century on. Lancet. 20017;390:415–23.
33. Kecmanović D. DSM-5: the more it changes the more it is the same. Psychiatria Danubina. 2013;25:94–6.
34. Slade M. The primacy of personal recovery. In: Slade M, editor. Personal recovery and mental illness—a guide for mental health professionals. Cambridge: University Press; 2011. p. 45–56.
35. Wakefield JC. Darwin, functional explanation, and the philosophy of psychiatry. In: Adriaens PR, De Block A, editors. Maladapting minds—philosophy, psychiatry and evolutionary theory. New York: Oxford University Press; 2011.
36. Sims A. Neuroscience as belief. A Christian perspective. In: Cox J, Campbell AV, Fulford BKWM, editors. Medicine of the person—faith, science, and values in health care provision. London, Philadelphia: Jessica Kingsley Publishers; 2007. p. 207–24.
37. Ferrer JN. Revisioning transpersonal theory—a participatory vision of human spirituality. New York: State University of New York; 2002.
38. Derry GN. What science is and how it works. Princeton: Princeton University Press; 2002.
39. Burke KL. Stop using the word pseudoscience. American Science; 2016. www.americanscientist.org.
40. Kendler KS. Towards scientific psychiatric nosology: strengths and limitations. Arch Gen Psychiatry. 1990;47:969–73.
41. Farmer A, McGuffin P, Williams J. Measuring psychopathology. New York: Oxford University Press; 2002.
42. Fullford KWMB, Thornton T, Graham G. Natural classifications, realism, and psychiatric science. In: Fullford KWMB, Thornton T, Graham G, editors. Oxford textbook of philosophy and psychiatry. New York: Oxford University Press; 2006. p. 316–83.
43. Vaughn L. Inference for the best explanation. In: Vaughn L, editor. The power of critical thinking—effective reasoning about ordinary and extraordinary claims. New York, Oxford: Oxford University Press; 2008. p. 335–811.
44. Vaughn L. Judging scientific theories. In: Vaughn L, editor. New York, Oxford: Oxford University Press; 2008. p. 382–441.
45. Stephan KF, Bach DR, Fletcher PC, Flint J, Frank MJ, Friston KJ, Heinz A, Huys QJM, Owen MJ, Binder EB, Dayan P, Johnstone EC, Meyer-Lindenberg A, Montague PR, Schnyder U, Wang XJ. Charting the landscape priority problems in psychiatry, part 1: classification and diagnosis. Lancet Psychiatry. 2016;3:77–83.
46. Weller MPI, Eysenck MW. Changing frontiers. In: Weller MPI, Eysenck MW, editors. The scientific basis of psychiatry, 2nd ed. London: W.B. Sounders Company Ltd; 1992. p. 3–38.
47. Tretter F. Philosophical aspects of neuropsychiatry. In: Tretter F, Gebicke-Haerter PJ, Mendoza ER, Winterer G, editors. Systems biology in psychiatric research—from high-throughput data to mathematical modeling. Weinheim: Wiley-VCH Verlag GmbH & Co. KGaA; 2010. p. 3–25.
48. Kuhn T. The structure of scientific revolutions. Chicago: University of Chicago Press; 1962.
49. Corsini R. The dictionary of psychology. Brunner/Mazel: Taylor & Francis Group; 2002.
50. Huck S. Defending science within reason. Between scientism and cynicism. NY: Amherst Prometeus Books; 2007.
51. Webster's New World College Dictionary, 4th ed. Houghton Mifflin Harcourt; 1999
52. The Oxford English Dictionary, 3rd ed. Oxford University Press; 2010.
53. Krickel B. The mechanical world—the meta physical commitments of the new mechanistic approach. Springer Nature Switzerland; 2018.
54. Wikipedia: Metapsychiatry. https://en.wikipedia.org/wiki/Metapsychiatry.
55. Kecmanovic D. Metapsihijatrija (Metapsychiatry). Clio, Beograd; 2014.
56. Wikipedia: Philosophy of mind. https://en.wikipedia.org/wiki/Metapsychiatry

57. Bartlett SJ. Normality does not equal mental health—the need to look elsewhere for standards of good psychological health. Denver, Oxford: Praeger, Santa Barbara; 2011.
58. Priebe S, Burns T, Craig TKJ. The future of academic psychiatry may be social. Br J Psychiatry. 2013;202:319–20.
59. Kecmanović D. Conceptual discord in psychiatry: origin, implications and failed attempts to resolve it. Psychiatria Danubina. 2011;23:210–22.
60. Cloninger CR. Feeling good: the science of well-being. Oxford University Press; 2004.
61. Craddock N. Psychiatry needs its Higga boson moment. New Scientist. 2013;2914:29.
62. Gros C. Complex and adaptive dynamical systems: a primer. New York: Springer; 2013.
63. Anderson WS, Kreiman G. Neuroscience: what we cannot model, we do not understand. Curr Biol. 2011;8:R123–5.
64. Jaccard J, Jacoby J. Theory construction and model-building skills—a practical guide for social scientiss. New York: The Guilford Press; 2010.
65. Senge PM. The fifth discipline: the art and practice of the learning organisation. London: Random House; 2006.
66. De Bono E. Six thinking hats. London: Penguin Books Ltd.; 1999.
67. Sternberg R, Grigorenko E, Singer J. Creativity: from potential to realization. Washington, DC: American Psychological Association; 2004.
68. Ghaemi SN. The concepts of psychiatry: a pluralistic approach to the mind and mental illness. Baltimore, London: The John Hopkins University Press; 2003.
69. Kuhn T. The essential tension: selected studies in scientific tradition and change. Chicago: Univeresity Chicago Press; 1970.
70. Auerswald EH. Family healing and planetary healing—three paradigm in search of a culture. In: Prosky PS, Keith DV, editors. Family therapy as ab alternative to medication—an appraisal of pharmaland. New York, Hove: Brunner-Routledge; 2002.
71. Lake J. Textbook of integrative mental health care. New York, Stuttgart: Thieme; 2007.
72. Tataryn D. Paradigms of health and disease: a framework for classifying and understanding complementary and alternative medicine. J Altern Complement Med. 2002:877–92.
73. Gallo FP. Energy psychology: explorations at the interface of energy, cognition, behavior, and health. London, New York, Washington, DC: CRC Press, Boca Raton; 2005.
74. Jakovljevic M, Brajkovic L, Jaksic N, Loncar M, Aukst-Margetic B, Lasic D. Posttraumatic stress disorders from different perspectives: a transdisciplinary approach. Psychiatria Danubina. 2012;24:246–55.
75. Tyrer P, Steinberg D. Models for mental disorders: conceptual models in psychiatry. 3rd ed. Chicester: Willey; 1998.
76. McHugh PR, Slavne PR. The perspectives of psychiatry. 2nd ed. Baltimore, London: The John Hopkins University Press; 1998.
77. WHO: Promoting mental health. Concepts, emerging evidence, practice (Summsry Report) Geneva, 2004 (Google Scholar).
78. Shives LR. Basic concepts of psychiatric-mentalhealth nursing. 8th ed. Philadelphia: Walters Kluwer/Lippincott Williams, Wikins; 2012.
79. Wakefield JC. The concept of mental disorder: on the boundary between biological facts and social values. Am Psychol. 1992;47:373–88.
80. American Psychiatric Association (APA). Definition of a mental disorder. In: APA, editor. Diagnostic and statistical manual of mental disorders. Fifth edition DSM-5. Washington, DC, London: American Psychiatric Publishing; 2013.
81. WHO: mental disorders https://www.who.int.
82. Glannon W. Research domain criteria: a final paradigm for psychiatry? Front Hum Neurosci. 2015;9:488. https://doi.org/10.3389/fnhum.2015.00488.
83. Carpenter WT Jr. How diagnosis of schizophrenia impeded the advance of knowledge (and what to do about it). In: Silverstein SM, Moghaddam B, Wykes T, editors. Schizophrenia: evolution and synthesis, 49–62. Cambridge Massachusetts: Struengmann Forum Reports. The MIT Press; 2013.

84. Summerfield D. Cross-cultural perspective on the medicalization of human suffering. In: Rosen GM, editor. Post-traumatic stress disorder—issues and controversies. Willey; 2005. p. 233–45.
85. Thoits PA. Sociological approaches to mental illness. In: Horwitz AV, Shied TL, editors. A handbook for a study of mental health: social contexts, theories and systems. Cambridge: Cambridge University Press; 1999. p. 121–38.
86. Herbert JD, Sageman M. "First do no harm": emerging guidelines for the treatment of post-traumatic reactions. In: Rosen GM, editor. Post-traumatic stress disorder—issues and controversies. Willey; 2005. p. 213–32.
87. Shepard B. Risk factors and PTSD. In: Rosen GM, editor. Post-traumatic stress disorder—issues and controversies. Willey; 2005. p. 39–61.
88. Hilton C, Hilton M. Person in health care: Jewish approaches. In: Cox J, Campbell AV, Fulford BKWM, editors. Medicine of the person—faith, science, and values in health care provision. London, Philadelphia: Jessica Kingsley Publishers; 2007. p. 96–109.
89. Okasha: The individual versus the family: an Islamic and traditional societies perspective. In: Cox J, Campbell AV, Fulford BKWM, editors. Medicine of the person—faith, science, and values in health care provision. London, Philadelphia: Jessica Kingsley Publishers; 2007. p. 110–24.
90. Bhugra D. Hindu and Ayurvedic understanding s of the person. In: Cox J, Campbell AV, Fulford BKWM, editors. Medicine of the person—faith, science, and values in health care provision. London, Philadelphia: Jessica Kingsley Publishers; 2007. p. 125–38.
91. Weller MPI, Peck DF. Personality. In: Weller MPI, Eysenck MW, editors. The scientific basis of psychiatry, 2nd ed. London: W.B. Sounders Company Ltd; 1992. p. 407–14.
92. Svrakic D, Divac-Jovanovic M. The fragmented personality: an integrative, dynamic, and personalized approach to personality disorder. New York: Oxford University Press; 2019.
93. Bhar SS, Kyrios M. The self concept: theory and research. In: Kyrios M, Moulding R, Doron G, Bhar SS, Nedeljkovic M, Mikulincer M, editors. The self in understanding and treating psychological disorders. UK: Cambridge University Press; 2016. p. 1–7.
94. Kyrios M. The self in psychological disorders: an introduction. In: Kyrios M, Moulding R, Doron G, Bhar SS, Nedeljkovic M, Mikulincer M, editors. The self in understanding and treating psychological disorders. UK: Cambridge University Press; 2016. p. 1–7.
95. Echezarraga A, Calvete E, Gonzalez-Pinto AM, Las Hayas C. Resilience dimensions and mental health outcomes in bipolar disorder in a follow up study. Stress Health. 2018;34:115–26.
96. Kalisch R, Mueller MB, Tuescher O. A conceptual framework for the neurobiological study of resilience. Behav Brain Sci. 2015;62:1–79.
97. Jakovljevic M. Placebo and nocebo phenomena from the perspective of evidence based and person centered medicine. Hosp Pharmacol Int Multidiscip J. 2017;4:512–20.
98. Shrivastava A, De Sousa A, Sonavane S, Shah N. Resilience improves neurocognition and treatment outcomes in schizophrenia: a hypothesis. Open J Psychiatry. 2016;6:173–87.
99. Davidson JRT, Payne VM, Connor KM, Foa EB, Rothbaum BO, Hertzberg MA, Weisler RH. Trauma, resilience and saliostasis: effects of treatment in post traumatic stress disorder. Int Clin Psychopharmacol. 2005;20:43–8.
100. Bolos A. Considerations on assisted resilience and individualized therapy in bipolar affective disorder, with a clinical case exemplification. Clujul Medical. 2015;88:462–7.
101. Reivich KJ, Seligman MEP, McBride S. Master resilience training in the US army. Am Psychol. 2011;66:25–34.
102. Beck AT. Depression: clinical experimental and theoretical aspects. New York: Hoeber; 1967.
103. Orr SP, McNally RJ, Rosen G, Shalev AY. Psychophysiologic reactivity—implication for conceptualizing PTSD. In: Rosen GM, editor. Posttraumatic stress disorder—issues and controversies. Willey; 2005. p. 101–26.

104. Kirshenbaum M. Everything happens for a reason: finding the true meaning of the events in our lives; 2004.
105. Bruene M. Textbook of evolutionary psychiatry. New York: Oxford University Press; 2008.
106. Clark J. The Bible and medical practice. In: Cox J, Campbell AV, Fulford BKWM, editors. Medicine of the person—faith, science, and values in health care provision. London, Philadelphia: Jessica Kingsley Publishers; 2007. p. 56–69.
107. Pfeifer HR, Cox J. The man and his message. In: Cox JM Campbell AV, Fullford BKWM, editors. Medicine of the person—faith, science and values in health care provision. London, Philadelphia: Jessica Kingsley Publishers; 2007. p. 33–45.
108. Cox J, Verhagen PJ. Spirituality, religion and psychopathology: towards an integrative psychiatry. Int J Pers Cent Med. 2011;1:146–8.
109. Sims A, Cook CCH. Spirituality in psychiatry. In: Cook C, Powell A, Sims A, editors. Spirituality and psychiatry. London: The Royal College of Psychiatrists; 2009. p. 1–15.
110. Gilbert P. Spirituality and mental health: practical proposal for action. In: Cox J, Campbell AV, Fulford BKWM, editors. Medicine of the person—faith, science, and values in health care provision. London, Philadelphia: Jessica Kingsley Publishers; 2007. p. 141–55.
111. Ellison CW. Spiritual well-being: conceptualization and measurement. J Psychol Theol. 1983;11:4.
112. Levin J. How faith heals: a theoretical model. EXPLORE (NY). 2009;5:77–96.
113. Pargament KI. The psychology of religion and coping—theory, research and practice. New York, London: The Guilford Press; 1997.
114. Jakovljevic M. Psychiatry and religion: from divide to dialog and creative synergism in the search for humanistic self and global mental health promotion. In: Kurjak A, Chervenak A, editors. McCullough LB, Hasanovic A, co-editors. Science and religion: synergy and scepticism. New Delhi, London, Panama: Jaypee the Health Sciences Publisher; 2018.
115. Jakovljevic M. Empathy, sense of coherence and resilience: bridging personal, public and global mental health and conceptual synthesis. Psychiatria Danubina. 2018;30:380–4.
116. Lewis B. Narrative psychiatry—how stories can shape clinical practice. Baltimore: John Hopkins University Press; 2011.
117. Hamkins SE. The art of narrative psychiatry. New York: Oxford University Press; 2014.
118. Frank AW. The wounded story teller—body, illness and ethics. Chicago, London: The University of Chicago Press; 1995.
119. Amendolia R. A narrative constructivist perspective of treatment of PTSD with Ericksonian hypnosis and EMDR. The American Academy of Experts in Academic Stres, Inc.; 1998.
120. Kirmayer JK, Crafa D. What kind of science for psychiatry. Front Hum Neurosci. 2014;8:article 435, 1–12. https://doi.org/10.3389/fnhum.2014.00435.
121. Jakovljevic M, Jakovljevic I. Theoretical psychiatry and transdisciplinary integrative approach: stepping stones to precision and person-centered psychiatry. In: Kim YK, editor. Research methods and interventions in psychiatry—artificial intelligence, precision medicine, and other paradigm shifts.

A Transdisciplinary Integrative Approach for Precision Psychiatry

Miro Jakovljevic and Ivan Jakovljevic

Abstract

Theoretical psychiatry considers theoretical problems in psychiatry as well as the quality and effectiveness of mental health care. This chapter addresses the idea of predictive, preventive, precision, personalized, and participatory medicine in psychiatry from a theoretical transdisciplinary integrative perspective and systems networking. The aim of the chapter is to bring together some current ideas and concepts such as computational neuroscience, network theory, multi-omics profile, precision medicine, and person-centered psychiatry as a coherent system of theory and practice.

Keywords

Theoretical psychiatry · Precision psychiatry · Person centered psychiatry

Introduction

> It is sometimes important for science to know how to forget the things she is surest of.
>
> —Rostand (1958)

Psychiatry today is facing unprecedented challenges, old theories and concepts about mind and mental disorders are crumbling, and new ones are emerging. The concept of predictive, preventive, precision, person-centered, and participatory medicine (5PMed) is a new emerging paradigm in psychiatry. 5PMed promotes

M. Jakovljevic (✉) · I. Jakovljevic
Department of Psychiatry, University Hospital Centre Zagreb, Zagreb, Croatia
e-mail: predstojnik_psi@kbc-zagreb.hr

© Springer Nature Singapore Pte Ltd. 2019

Y.-K. Kim (ed.), *Frontiers in Psychiatry*, Advances in Experimental Medicine and Biology 1192, https://doi.org/10.1007/978-981-32-9721-0_20

the paradigmatic shift from delayed interventional to predictive and preemptive psychiatry, from blockbuster to person tailored psychiatry, from palliative to preventive psychiatry, and from illness to wellness. 5PMed is proclaimed as "the medicine of future that enables to predict individual predisposition before onset of the disease, to provide targeted preventive measures and create personalized treatment algorithms tailored to the person" [1]. As medical discipline psychiatry is an interdisciplinary field situated at the interface of natural sciences and humanistic sciences. It utilizes insights from the most varied disciplines such as biology, neuroscience, pharmacology, physics, anthropology, philosophy, ethics, axiology, psychology, sociology, informatics, etc. At the conceptual level, contemporary psychiatry can be divided into four approaches: dogmatic, eclectic, pluralist, and integrationist [2]. In practice, most psychiatrists are still, more or less, dogmatic claiming to be eclectic in theory. The pluralist and integrationist concepts are the most challenging, particularly from the perspective of personalized medicine and person-centered psychiatry. Transdisciplinary integrative approach within theoretical psychiatry tries to create an overarching theory that unifies all the scientific and humanistic disciplines dealing with human mind, mental health, and mental disorders. The idea is very attractive and challenging: to offer all mental health scientists and practitioners a common language, bridge over academic gaps and easily exchange insights across disciplinary borders. Personalized and precision medicine in psychiatry which in diagnostics, treatment, and prevention takes into account individual patient's variability in genes, environment, and lifestyle is still only a dream, but also an imminent reality at the information-gathering infancy stage [3, 4]. Dataism, the mindset or philosophy created by the emerging significance of Big Data and deeply entrenched in computer science (infotech) and biology (biotech)—[5, 6], is promising in achieving these goals.

Transdisciplinary holistic integrative psychiatry represents an attempt of creating a very complex system with compositional and dynamical integrative complexity and evolutionary contingency. To avoid misunderstanding it is important to have in mind that complex systems can be classified into aggregative and composite (component and integrative) systems [7]. Properties of aggregative systems depend linearly on component properties and actions of individual components simply add together. An example of aggregative system may be psychiatry as a field for the coexistence of multiple explanatory perspectives and treatment methods. In component systems, the properties and actions of the whole depend on the organization and activity of the parts, but the parts themselves are partially independent although there are interactions among them. In integrative systems, the components and their organization and actions influence the whole system but also depend on the features of the whole system. True multicomponent, multidimensional, multi and cross-level and transdisciplinary integration in psychiatry can be established only on the principles of complex integrative systems creation.

Paradigm Clash or Paradigm Shift for the Future: That's the Question Now

Many major theoretical and practical problems and quandaries in contemporary psychiatry are still waiting for their solutions. These problems can be roughly divided into three groups: the problems concerning nosology and diagnostics, problems related to aetiology and pathophysiology, and problems related to achieving the more stratified and personalized treatment. ICD and DSM classifications are criticized to be incoherent, heterogenous, and provincial and insufficiently informed by a biomedical understanding of mental illness [8]. Although treatment for major mental disorders is frequently effective, insufficient treatment outcome seems to be more commonly the rule than the exception [9, 10]. Choice of treatment and prescription of mental health medications are commonly haphazard, trials and errors, without valid and reliable parameters and biomarkers to guide it. Inspired by the famous list of Hilbert's problems in mathematics an international group of the most distinguished experts from different specialties [11] published their eclectic list of priority problems in psychiatry predicated on the discussion at a scientific symposium (Table 1). Finding answers to these questions/problems, and still to many others as well, and discovering the more appropriate directions of the future scientific and professional development of the individual, public, and global mental health care [12] is final cause and purpose of practicing theoretical psychiatry.

What is the crux in psychiatry it is its scientific background and coherence, transdisciplinarity and clinical utility. Theoretical psychiatry pursues knowledge and understanding of mental disorders, and it operates so through the formulation, testing, and evaluation of theories, clinical practice, and mental health care. Understanding the puzzle "how brain function enables the mind and mental functions and how brain dysfunctions lead to mental disorders" makes obvious the need for transdisciplinary integrative approach in neuroscience and psychiatry. The crucial component for a truly medical nosology in psychiatry is "a neurobiologically informed psychopathology" [14] or in other words functional psychopathology with a neuroscientific understanding of the mental functions, their regulation and dysregulation and underlying neuronal networks and brain systems. Of course, experimental psychopathology and psychodynamic psychopathology offer also very useful approaches to more valid assessment of mental disorders. Mental disorders are complex phenomena that demand complex, transdisciplinary multilevel and multiperspective explanations. One of the essential tasks of theoretical psychiatry is to bridge different theories and models originated from different levels of explanatory description, e.g., cellular, neural circuits and behavioral levels [15]. This bridging can be done in different ways, as vertical, bottom-up and top-down strategies and horizontal strategies of information processing. Bottom-up and

Table 1 Charting the landscape of priority problems in psychiatry (adapted from Refs. [11, 13])

Classification and diagnosis
Problem 1: Is mapping between mental states and brain states computable? (Classical computable mapping or quantum computable mapping)
Problem 2: What should be the status and role of symptoms in psychiatry? (Do symptoms define a series of diseases that can be disentangled with the help of neurobehavioral measures?)
Problem 3: Integration a dimensional perspective on general psychopathology with categorical definitions of disease entities (because our current categories provide only an approximation of much deeper patterns)
Problem 4: Show that the brain manifests disease in limited ways, possibly only three or four. (The complexity of the brain does not seem to be reflected in the number of diseases recognized by psychiatrists)
Problem 5: Bridging the comparative gap: can preclinical models help to establish diagnostic criteria based on observable signs?
Problem 6: What is the higher order structure of fundamental mechanisms relevant for diagnostics? (How to develop diagnostic strategies that identify clusters of mechanisms that are amenable to treatment, using multidimensional functional profiles that combine neural and behavioral indices to assess distinct identifiable mechanisms)
Problem 7: New approaches to patient stratification are needed for neuroscience research
Problem 8: Develop computational assays for symptom-guided reassembly of psychiatric nosology
Problem 9: Computational assessment of learning dysfunctions for a dimensional perspective on psychiatric disorders
Pathogenesis and aetiology
Problem 10: Derive a tractable account of the systems-level effect on the human brain of epidemiologically validated high-risk causal factors
Problem 11: What are the mechanisms of gene-environment interplay in psychiatry?
Problem 12: Understanding mechanisms of resilience
Problem 13: Can a mechanistic marker be found for diagnosis of schizophrenia and bipolar illness? What are the pitfalls?
Problem 14: What are the principles of cognitive-type microcircuits in a large-scale brain system, and how do their impairments explain mental disorders?
Problem 15: A Fokker–Planck equation for the brain
Problem 16: The problems of priors
Problem 17: Understanding psychiatric pathophysiology in terms of the computational processes that underlie inference
Treatment efficacy and efficiency
Problem 18. How to increase treatment effectiveness and efficiency?
Problem 19. Eliminating dangerous and ineffective treatments
Problem 20. More comprehensive understanding of therapeutic processes
Problem 21: Promotion of more precise, comprehensive (holistic) and personalized treatment guidelines
Problem 22. How to reconcile different views on the practice in psychiatry?
Psychiatry as a coherent field of scientific theory and one unified and standardized practice
Problem 23: How to integrate different but complementary branches, theories, and practices within psychiatry?
Problem 24. Does psychiatry need a new identity with new borders and new epistemics?
Problem 25. Transdisciplinary integrative 5P psychiatry: reality or Holy Grail?

top-down processing are also well-known therapeutic procedures in neuro-linguistic programming (NLP) psychotherapy. Bottom-up processing of information, also known as inductive reasoning and "small chunk" processing, is a method of gathering small chunks of information and building the conclusion. This approach is the method of piecing together of systems to create more complex systems. Bottom-up models integrate what is known on a lower level (e.g., activity of neurotransmitters and their receptors) to explain phenomena on a higher level, e.g., brain reward–punishment or dominance–submissive brain systems). Top-down processing, known as deductive reasoning or large chunk processing, is a method which starts with big picture or higher level and then analyzes the parts or goes down to lower level. Top-down models integrate what is known on a higher level to those what is known on lower level.

Treatment outcome for many mental disorders is considered as being unsatisfactory due to lack of preemptive mental health care, delayed interventions leading to palliative care, untargeted mental health medication, overdosed or hypodosed patients, low effectiveness of treatments, false beliefs and pessimistic treatment expectations, etc. Optimistic versus pessimistic expectations depends on paradigms which psychiatry and mental health services will put into practice. Transdisciplinary integrative approach with the multimodal diagnostics (i.e., deep, early and predictive diagnostics) and application of health- versus illness-specific biosignatures leading to valid diagnostic psychopathological entities, creation of extended patients' profiles and medical records with multidisciplinary questionnaires (integrative bioinformatics), and treatment tailored to the person gives hope for the future.

Research Domain Criteria (RDC) and Systems Neuroscience of Psychosis (SyNoPsis)

The nosologic insecurity and instability of diagnostic and classification systems in psychiatry (ICD, DSM) endanger its scientific image. Categorical ICD and DSM classifications pose a significant hindrance to discovering the real nature and origin of psychopathology as well as objective and specific diagnostic biomarkers for them. DSM nosology seems to be invalid because they cluster together disorders which are symptomatically similar, but ethiopathophysiologically different [16], and the use of artificial illness categories to map distinct biological substrates hindered the progress in the research of biological markers in psychiatry [17]. The purposes of classification and diagnosis in psychiatry are the same as in the other branches of medicine. Psychiatric diagnoses are still more syndromes than true disease entities because of the lack of demonstrable pathophysiology for majority of mental disorders. There is an increasing discussion about limitations of the current ICD-10/11 and DSM-5 classifications because of the atheoretical approach, law diagnostic validity and high rate of artificial comorbidities, lack of staging concepts and underlying pathophysiology to be a target for preventive or therapeutic action, etc. Reclassification of mental disorders predicated on computational neuroscience and mind/brain functional domains is expected to help in solving problems. Recent progress in basic neuroscience, technology, and informatics resulted in new approaches to understanding and forming classification in psychiatric research connecting biological, cognitive, emotional and behavioral components of normal and pathological functioning in order to recognize valid and reliable endophenotypes and biosignatures for mental disorders. The Research Domain Criteria (RDoC) is viewed as an innovative and transformative approach to understanding the nature of mental disorders and find their more successful treatment with implementation of neuroscience-based psychiatric classification and treatment. It represents a translational approach to organize multidisciplinary research and synergize basic neuroscience, clinical neuroscience, and psychiatry with strategic aims to reorient diagnostics and treatment of mental disorders [18, 19].

RDoC is focused on the investigation and characterization of relationships among the different research domains containing psychological constructs at the different levels of analysis, all that from biology to sociology (Table 2). Each construct is evaluated dimensionally at the continuum from normal to pathological to determine the full range of variation. It is claimed that the RDoC matrix has good potentials to be transformed into an ontological structure which could enable automatic data processing and facilitate intelligent diagnosis [22].

While RDoC relies on intermediate constructs to link brain and behavior, Systems Neuroscience of Psychosis (SyNoPsis) project aims at the direct link between a revised phenomenology of psychoses and brain circuits [14]. SyNoPsis project work is predicated on the following background: "(1) the core feature of schizophrenia is a fundamental interpersonal communication breakdown; (2) its symptoms are an expression of dysfunctional sensorimotor and corticostriatal brain systems specialized for distinct communication domains (candidate systems); (3) formal features allow the mapping of typical symptoms onto these systems; (4) the definitions of traditional psychopathology must be revised according to our knowledge about the functional neuroanatomy of the candidate systems;

Table 2 Units of analysis and functional domains of the NIMH Research Domain Criteria [20, 21]

Units of analysis: Genes, molecules, circuits, physiology, behavior, self-reports, paradigms
Functional domains
Negative valence systems: Acute threat (fear), potential threat (anxiety), sustained threat, loss, frustrating non-reward
Positive valence systems: Approach motivation, initial responsiveness to reward attainment, Sustained longer term responsiveness to reward attainment, reward learning, habit
Cognitive systems: Attention, perception, declarative memory, language, cognitive control, working memory
Social processing systems: Affiliation and attachment, Social communication, perception and understanding of self and others
Arousal and regulatory systems: Arousal, circadian rhythms, sleep–wakefulness

and (5) a neurobiologically informed psychopathology of psychoses will permit the generation of scientific hypotheses with falsifiable predictions about causal relationships between typical psychotic symptoms and dysfunctions of the candidate systems" [14].

In order to overcome the simplistic, symptom-centered approach in the DSM and ICD, several psychodynamic diagnostic manuals, the two most prominent ones are Psychodynamic Diagnostic Manual (PDM; PDM Task Force 2006) and Operationalized Psychodynamic Diagnosis (OPD; OPD Task Force 2008), have been offered aiming to increase the construct validity and provide clinical sophistication to categorical diagnosis [23]. Paraphrasing Sartorius [24] from the Preface to the ICD-10 psychiatric classification is a way of our understanding and defining mental disorders at a point of time. One can freely say that current psychiatric classifications based on phenomenological psychopathology and operational definitions are only "a best guess" and "working hypotheses" which represent our current knowledge about mental disorders and their treatment and as such they must be continually reviewed and reformulated in the light of new knowledge, concepts, and technology. Operational definitions of mental disorders with an aetiologically atheoretical perspective improved significantly interrater agreement and reliability for diagnosis in psychiatry and facilitated communication between clinicians.

Precision Psychiatry: Biomarkers, Endophenotypes, and Biosignatures—Is that All?

The era of precision medicine is imminent, but its use in psychiatry is still undeveloped [25], although the necessary technology to put it into operation is currently available. Precision medicine tends to use measurable health parameters or objective biomarkers to identify individuals at risk of a mental disorder, to improve diagnostic punctuality and to offer a personalized (patient-tailored) therapy. Theoretically, it is expected that is possible to combine clinical data with different neurobiological measures, single-nucleotide polymorphisms, and epigenetic mechanisms in the different populations of patients in order to identify profiles that refer to and predict individual clinical response to individual or personalized treatment. Three categories of biomarkers in psychiatry: single markers, endophenotypes, and biosignatures may come from molecular genetics, biochemistry, and neuroimaging. However, identifying endophenotypes, biomarkers, and biosignatures that would be applicable and useful for précising clinical diagnosis, monitoring disease regression, and predicting treatment response is still wishful, but a promising way of thinking. Endophenotypes are specific trait markers of an illness regardless of the phenotypic presence or absence of illness because they are heritable [3]. Biomarkers are measurable parameters that reflect biologic function or dysfunction, response to a treatment method, or predict the natural progression of illness. An attractive alternative to the single markers is the concept of biosignatures as the biological equivalent of a pathognomonic sign that could complement, augment, and make psychiatric diagnosis more valid and reliable.

For the time being, the reported biomarkers do not index pathophysiology or treatment responses and do not enable prediction and treatment selection in psychiatry. Stratified and personalized therapeutic interventions that are predicated on biomarkers are challenging concepts and treatment strategies.

Feature engineering for complexity of the mind. The first step in being able to classify which mental illness a patient has is determining the most informative set of markers either at behavioral, neurological, cellular or any other available measurement level. Quality of an illness marker can be evaluated based on two criteria: (a) sensitivity—does the marker always or at least often manifest in certain illness and (b) specificity—is it specific for that illness or does it manifest in other illnesses as well. Suggested psychiatric biomarkers unfortunately often fail at one or both of these criteria.

Biomedical data in its raw form is often unsuitable for statistical prediction. First expert has to interpret the data and pick certain features which can then be quantized. For example, EEG expert will analyze amplitudes of certain brain wave frequencies and then that can be correlated with certain diagnoses to see whether brain waves can be used as an illness marker. Problem with this approach is that there are infinite ways to combine raw data into features. Alpha, beta, gamma and delta waves for EEG are just one way to do it which is based on not on neuroscientific theory but rather on common sense. In genetics, we are trying to identify which genes might predispose some people to develop certain illnesses which is based on the mechanistic notion that there is a gene for everything. In reality, it is entirely plausible that different gene combinations are associated with specific illness in a way that is too complex for a human to understand.

A possible solution to this problem has been developed in area of machine learning. One of the most central endeavors in the field of artificial intelligence is the development of software that can process images and recognize objects in a way humans can. At first, scientists tried writing manual rules for object recognition. This has proven to be fruitless endeavor as there is no 1 on 1 relationship between 3D object and the image it projects on 2D camera. 3D object can project many different images depending on angle, distance, lighting, etc. Progress was made when scientists realized they are better off letting software develop its own rules for object recognition by implementing neural networks and other machine learning techniques. Artificial intelligence is trained on a large dataset of pictures which are labeled to inform the algorithm whether a specific picture contains the object. In other words, it is entirely possible to recognize 3D objects as a projection on 2D plane as evidenced by the ease at which humans do it. However, the rules are so complex and unintuitive that they have to be delegated to artificial intelligence. Parallel to that it is entirely possible that creation of biomedical data features relating to brain functioning should be relegated to artificial intelligence and expert should then analyze the most predictive feature AI has generated and try to explain it and integrate into existing theory.

Using a combination of predictors for nonlinear systems. In machine learning and Big Data, there is a very useful idea that unifying several weak classifiers which predict barely better than guessing can produce a strong classifier. The brain is a complex system and in complex systems, by definition, every variable is influenced my multitude of other variables. The more the causal factors are, there is less variance each individual factor can explain. Thus it is unlikely that behavior and psychological processes can be adequately explained by handful of variables. Instead of using individual markers, research paradigm needs to switch to using marker composites created by machine learning methods, which is related to the first idea that task of deriving biomedical markers should be relegated to artificial intelligence. Scientist should then strive to reverse engineer the markers created by AI and generate theoretical knowledge about brain functioning. This approach requires a large amount of data both in terms of patients and number of different measurements such as brain imaging or genetic testing which can be economically challenging. For that reason, researchers need to collaborate with practitioners and work on creating large, shared datasets so field as a whole can truly reap the benefits of Big Data revolution.

Transdisciplinary Integrative Precision, Personalized and Person-Centered Participatory Psychiatry: Hope for the Future

> There are two ways of acquiring knowledge, one through reason, the other by experiment.
>
> —Roger Bacon

Instead of relatively broad pathological diagnoses and nonspecific "one-size-fits-all" therapies, psychiatry is moving toward an era of individualized, precision, personalized and person-centered and participatory care that should offer the right treatment to the right patient in the right time. Current psychiatric therapies are actually effective and useful for many patients, but there are still high rates of partial response and treatment resistance [26]. Personalized, precision and person-centered psychiatry are commonly used as synonyms, in spite of the fact that these terms refer to the overlapping, but little distinctively different meanings and approaches. Personalized medicine considers clinical and personality characteristics of a patient in order to predict susceptibility to disease, aid in diagnosis, and tailor individualized treatment [25], precision medicine searches objective measures, biomarkers, endophenotypes or biosignatures, while person-centered medicine promotes a patient to be proactive as partner in the treatment choice. Using computation technology, *precision psychiatry* tends to detect and integrate data, information, and knowledge from biomedical research and clinical practice across many layers, from genes to behavior. *Personalized psychiatry* claims that mental disorders do not manifest only on a molecular but on personal-interpersonal level as well as that all patients have their own unique needs, life and illness histories, family and cultural backgrounds, and illness expressions, needs for specific therapies and treatment programs and unique way of recovery. *Person-centered or person-centric psychiatry* as the successor of the medicine of person [27] promotes the person who is responsible for his/her health as the key concept, the center and goal of health care in achieving mental, physical, social and spiritual well-being [28]. Mental health is not just mental capital to be consumed, it is a state of well-being created in each moment of life, and if it is not created then a person becomes sick [29]. In contrast to the traditional view of "patient" as a person who must wait patiently for the doctor to instruct them in what to do, person-centered psychiatry promotes personal mastery of patients to have a proactive and participatory role as partners in their treatment ("nothing about the patient without the patient)". The goals of 5-P person-centric psychiatry are to transdisciplinary address mental health and mental disorders in their totality and to render mental health care more precise, predictive, preventive, personalized and participatory in order to reduce the burden and negative impact of mental disorders.

From Paradigmatic to Holodigmatic Approach: Mental Symptoms are an Indication of Mental Disorders, Mental Disorders are an Indication of Brain Dysfunctions/Disorders, Brain Disorders are an Indication of Epigenetic Dysfunctions/Disorders

Mental health and mental disorders can and should be examined from different perspectives [30] and on different levels in system analysis (epi/genetic, biochemical, psychophysiological, cognitive, behavioral, family, social, and environmental). As health is a state of physical, mental, social, and spiritual well-being, transdisciplinary integrative psychiatry considers the whole person (body–mind–energy–spirit holodigm) in the understanding and treatment of mental disorders and the quest for positive mental health and well-being. From a complex systems perspective, mental disorders can be understood in multiple different ways such as in terms of dysfunctional neuronal circuits and brain systems, neuroglial and homeostatic failure, dysfunctional epigenetic mechanisms, symptom–symptom network dynamics, brain miscomputations, misrepresentations and aberrant mentalization, evolutionary mismatch, neurodevelopment disorders, toxic encephalopathies, pathopsychodynamics, dysfunctional self-dynamics, etc. Psychiatric theories of mental disorders seem to be mereological systems whose axioms are different, so that holodigmatic transdisciplinary integrative approach tends to create a sensible and functional network, vertical and horizontal, structures. According to mereology (Greek mere—part), discipline in philosophy and mathematical logic which studies the parts and the wholes they form, as well as system decomposition and parts, wholes and boundaries, everything is part of itself (reflexivity), that a part of a whole (transitivity), and that two distinct entities cannot each be a part of the other (antisymmetry), thus forming a partially ordered set or poset [31]. An example of a poset is a genealogical descendants of family. According to some opinions "psychological and biological networks may be related through mereological matryoshkas (a Russian doll) structure, where biological network structure are 'nested' in a psychological without one causing the other, just as one smaller matryoshka does not cause a larger matryoshka" [32]. This metaphor seems interesting; however, matryoshka is not living being, so that head of onion representing biological, psychological, social, and spiritual dimension of human beings or different levels of symptoms network could be a useful metaphor. Fortunately, computational models of mental disorders may operate without metaphor and they might identify transdiagnostic and diagnose specific computational mechanisms. There is hope that a computational approach could lead to better theory- and data-driven diagnoses as well as to more selective and successful treatment in psychiatry.

Computational psychiatry is predicated on multiple levels and types of computation with multiple types of data in order to contribute to the better understanding, prediction, and treatment of mental disorders [33]. Computational approach, in a narrow sense, refers to computer simulation in which components of a system are modeled and their simulated behavior is used to study the system at large. In a wider sense, the computational approach includes advanced statistical methods such as neural networks or numerical optimization techniques although it is debatable whether such techniques should be subsumed under that label. They are tools which can be used in computer simulations but they are also often used in other fields such as machine learning and statistics. In our context, it is important that simulations can be used to test scientific theories. For example, computational neuroscience aims to model a single neuron based on our knowledge of cellular biology and biochemical pathways. Model is then used to compare computational predictions with the actual behavior of the neuron. The convergence of simulated and empirical outcome supports the validity of theory being tested while their divergence can act as a focusing lens to guide future research and data collection toward specific areas of the system which are not yet sufficiently understood. There are four recognizable contexts in today's computational psychiatry: (1) dysfunctional brain connectivity; (2) dysfunctional network dynamics; (3) misrepresentation; and (4) aberrant information processing involving inference, information integration, and choice It tries to unite many dimensions and levels of description and explanation in mechanistic and rigorous way avoiding biologically reductionism and artificial nosological categories [34]. At a computational level, it describes the formal nature of mental disorders; at an algorithmic level, it depicts the method of solving the problem; and finally at an implementation level, it describes the physical realization of the method [34]. According to Huys et al. [33], computational psychiatry operates within three approaches: (1) data-driven, atheoretical data analysis based on machine learning (ML) widely construed, involving, but extending standard statistical methods, (2) theory-driven models which mathematically specify mechanistically interpretable relations between observable variables and postulated, theoretically meaningful hidden variables, and (3) combined theory- and data-driven approaches. Data-driven approaches have been applied to some clinically important issues such as automatic diagnosis and diagnostic classification (the symptom clusters mapping onto specific neuropsychophysiological substrates), understanding relations between symptoms (co-occurrence and sequential expression of symptoms), prediction of longitudinal illness course and treatment outcomes and treatment selection [33]. Theory-driven approach, based on a theoretical, commonly mechanistic understanding of brain functioning and mental disorders, can be classified synthetic (informed by relevant data from multiple sources, connectionist, and neural network models), algorithmic (exemplified by reinforcement learning—RL models), and optimal (linking observed behavior to the Bayes optimal solution) models (35, 33). Synthetic, biophysically realistic neural network models have been applied to link neuro/biological abnormalities to psychological and behavioral consequences. Application of algorithmic RL models to mental disorders (affect, motivation, and decision-making) is motivated by the next three facts: (1) decisions are central to mental disorders as the final common pathway preceding many abnormal behavior and bad decisions may have profound consequences for affected individuals; (2) psychoactive substance and medications influence neurotransmitters and neuromodulators in brain, and (3) unifying the previous two neurotransmitters/neuromodulators are central to computational models of decision-making and reward sensitivity [33, 35]. RL models can be model-free (MF) and model-based (MB) in which an internal/mental model of the world capture goal-directed behaviors and rely on cognitive and emotional (limbic) cortico-striato-thalamo-cortical (CSTC) loops [33]. Optimal behavior is related to comparing the future with the present, and RL models represent different ways in which past experience is used to estimate and predict future rewards and punishments [34].

A viable and useful paradigm must project a domain of reality which is congruent with a total reality projected by a holodigm, which reflects multiple evolving realities [36]. Disparate theories and models of explaining mental health and mental disorders imply the legitimacy of different assessment and treatment approaches. Concepts and models from seven perspectives are not mutually exclusive but rather different ways of looking at and contributing to a possibility of transdisciplinary understanding of the same phenomena. In general, due to focus on psychopathology, fragmentary, unidimensional, uniperspective, and unimodal therapeutic approach, treatment of mental disorders is unfortunately too often associated with partial remissions, frequent relapse or recurrence as well as with persistent residual symptoms, distress and low level of well-being, coherence of being, life satisfaction and quality of life. Many aspects of psychiatric treatment have been criticized as harmful, dehumanizing, exploitative, and only pathology oriented on decrease of symptoms and illness and not on the development of wellness, purpose of life, and creativity. Concurrent use of two or more treatments may be a sword with two sides regarding the compatibility of treatments so that the combined use of different treatments is a very important issue in every integrative approach in psychiatry. The combined use of incompatible treatments is contraindicated because of the unacceptable safety risks or a decrease in treatment efficacy and effectiveness. On the other side, the combined use of compatible treatments can result in neutral, additive, or synergistic effects.

Transdisciplinary holistic integrative treatment includes both disease and illness demotion and health and wellness promotion in the same time as well as combined parallel and sequential treatments with additive or synergistic effects.

It is aimed to stop multilevel pathogenesis of a mental disorder, but also to enhance salutogenesis and increase resilience and personal mastery of the patients including their somatic, psychological, social and spiritual well-being, better self-understanding, creativity and life satisfaction. In the acute phase of any mental disorder a disease/illness demotion treatment (e.g., psychopharmacotherapy combined with psychotherapy) is usually dominant, while in the stabilization phase, a wellness/health and resilience promotion treatment (well-being oriented therapy, life coaching, spiritual therapy, etc.) should be gaining a momentum and become dominant in prophylactic treatment phase. The treatment program for any mental disorder should be multimodal, pluralistic, multi-perspective, integrating, holistic and comprehensive, always including a set of strategies that address the specific needs of patients (person-centered psychiatry). Strictly individualized pharmacotherapy with adequate dosage of modern mental health medications, and if needed with rational combination of several drug classes (antipsychotics, antidepressants, mood stabilizers, anxiolytics, and hypnotics) should be only a cornerstone of holistic and integrating treatment. Treatment framework in practice is a human encounter focused on issues such as beliefs, images, hope, trust, dignity, encouragement, making sense, empowerment, empathy, and care. Well-being therapy and life coaching [37] should bolster salutogenic strengths, basic life skills, increase social efficacy, cooperativeness, self-esteem, and self-acceptance with separation of illness from identity, increase energy, coping efficacy, autonomy and independence, increase structure, increase activity, self-direction, and goal-orientation (proactivity). A healthy life-style package including nutritional advice and exercise may be also of the great importance for achieving full recovery, life satisfaction, happiness, and well-being by patients.

An interesting example is the person-centered integrative diagnosis (PID) model which is defined by (1) broad informational domains, covering both ill-health and positive health along three levels: (1) health status, contributors to health, and health experience, and values; (2) pluralistic descriptive procedures (categories, dimensions, and narratives); and (3) evaluation partnerships among clinicians, patients and families [38]. "The five P's approach to case formulation or case conceptualization" which includes the presenting problem, predisposing factors, precipitating factors, perpetuating factors, and protective/positive factors [39] may be very useful in clinical practice and research. Case formulation is the process of developing an explicit and clear understanding of patients and their problems that effectively guides treatment. This approach synthesizes the patient's experience with relevant clinical theory and research and builds the bridge between diagnostic assessment and treatment. When done well, case formulation provides an opportunity for a shared understanding of a patient's symptoms and difficulties answering the classic questions: "Why this problem? Why in this person? and Why just now?" [39]. It also offers a rationale and shared agenda for what to target and in what order [39].

A Network Theory, Multi-omics Illness Profile and Computational Neuroscience

If our brains were simple enough for us to understand them, we'd be so simple that we couldn't.
—Ian Stuart: The Collapse of Chaos: Discovering Simplicity in a Complex World

As mentioned earlier the traditional "one-size-fits-all" approach to diagnosis and treatment of mental disorders (so-called blockbuster medicine) has been criticized, and new diagnostic and therapeutic tools are offered. Psychiatry lags behind other medical disciplines in the application of concepts of precision medicine in research and clinical practice [25]. The purpose of diagnosis is to communicate information about symptoms, etiopathogenesis, prognosis, and optimal treatment. The DSM and ICD classifications have improved diagnosis reliability and communication between psychiatrists, but they do not have causality structure and treat mental disorders as syndromes in concepts of a minimal interpretation of medical model [40]. If psychiatry tends to be a real medical discipline, psychiatrists should use the strong interpretation of the medical model to define and describe psyhopathophysiological processes in mind–brain–body systems, just as diseases are explained by pathophysiological processes in somatic medicine. With a few exceptions, underlying aetiopathogenesis of mental disorders is still unknown or at best case controversial. The mind–brain–body trichotomy should not be present in research, understanding and explaining mental disorders. Huge body of research indicated that mental disorders are caused by, and cause in turn, many various physical, psychological, social, and spiritual factors. The core of transdisciplinary integrative systems psychiatry is to understand and describe mental and brain systems and their reciprocal communication involved in mental disorders at all levels, to dive into their complexities and to find effective methods of treatment. It is not to be expected that a single biomarker can impact the diagnosis and treatment of any mental illness. Only combination of multiple specific biomarkers obtained by multi-omics or panomics can identify aetiology, diagnostics, and prognostics of mental disorders (Table 3). Transdisciplinary systems approach that integrates many diverse inputs including neurobiological, phenomenological, environmental, and clinical information may produce plausible specific models for individual mental diseases. According to systems thinking, the genome operates within the context of the body, the body within the context of the self, the self within the context of society, and the society within the context of the universe [41]. The phenotype of an organism is the joint product of the genotype and the environment. It is fascinating that genes within a single body can compete as well as genes in different bodies can collaborate [42]. All systems are composed of various elements and their relations. Fortunately, we are able to understand reality by constructing maps and models, starting with qualitative concepts and ending up with corresponding mathematical models that can be tested by computer experiments within computational neuroscience [43].

Table 3 Domains related to precision psychiatry predicated on systems biology and computational psychiatry (adapted from Refs. [4, 23, 45, 47, 52])

Network theory: a study of graphs as a representation of symmetric and asymmetric relations in which nodes and edges have attributes. *Network models*: Originating from mathematical graph theory, network models represent complex systems as sets of discrete non-overlapping entities (nodes) and their pairwise relationships (edges) that can be summarized in the form of graph
Network: a physical system which can be represented by a graph consisting of nodes and edges
Nodes: anatomical areas at the macroscopic level, and single neurons or glial cells at the microscopic level in the brain, usually represent imaging voxels or regions of interest
Edges: connections between pairs of nodes; represent the functional or structural connectivity (often white matter fiber bundles between regions of interest or covariance in morphological measures)
Path: a connection route between two nodes
Path length: a property of a node in a network which is the average number of steps between a node and all other nodes in the network
Hub: a node identified as having a high number of edges connected to it, low clustering, a short path length with other nodes, and connecting different modes. Play a central role in the communication; e.g., the medial frontal cortex
Rich club structures: networks more densely connected among themselves, with more afferences than efferences
Clustering: density of connections between neighbors of nodes
Module: densely interconnected neighbors which form a cluster around the node
Large-scale brain networks: neural systems which are distributed across most of the brain.
Intrinsic connectivity network (ICN): a large-scale network of interdependent brain areas observed in subjects at rest
Small-world network (the optimal balance of both high local clustering and high global integrity): a graph in which most nodes are not neighbors of one another but most nodes can be reached from every other by a small number of steps. It is intermediate to that of total random model and a regular model and combines high clustering with short path length *Functional connectivity*: reflect the organization and interrelationship of spatially separated brain regions. It involves principal component analysis (PCS) and independent component analysis (ICA)
Coherence estimates the linear relationship in the frequency domain
Connectome: a large-scale brain network
Information segregation and integration in the brain networks: topological segregation of the brain refers to information processing among local communities or clusters of nodes that are highly interconnected; integration refers to the efficiency of global information communication or the ability to integrate distributed information in the network
Molecular biosignature
Panomics or multi-omics: proteomics, metabolomics, genomics, transcriptomics, epigenomics, microbiomics, other omics (GO–illness association, phenotype–illness associations, SNPs–illness association, CNVs–illness association, miRNA–illness association, illness–environment association, etc.)

(continued)

Table 3 (continued)

Neuropsychophysiological biosignature
Connectomics, neural circles and networks, brain systems: central executive system/network, memory formation and recollection system, default-mode system, central security (alarm) or harm avoidance system, approach-avoidance system, sleep–wakefulness (vigilance/alertness) system, motivational/reward–punishment system, decision-making system, appraisal-reappraisal/salience (trust–distrust) system, dominance-submission system, separation–attachment system, appetitive-aversive system, novelty-seeking/epistemic system, habituation–sensitisation (learning) system, empathy/mirror neuron system
Functional connectivity: the statistical interrelation of variable representing temporal changes in different network nodes as typically measured by fMRI
Structural connectivity: physical connectivity between brain areas measured with diffusion tensor imaging (DTI) in vivo or tracer studies on postmortem tissue
Neurodevelopment: *Periods of risk and opportunities for resilience*
Individual characteristics: *Clinical history, demographics, somatic comorbidities*
Personality characteristics or domains: *Psychosocial development, Temperament: harm avoidance, novelty-seeking, reward dependence, persistence; character: self-directedness, cooperativeness, self-transcendence*; *identity, resilience*
Environment: *Social status, stress, and trauma, lifestyle and culture*

GO—gene ontology, SNPs—single-nucleotide polymorphisms; CNVs—copy number variations

Mental Disorders as Disorders of Neuronal Network Coalition and Dynamics

Network neuroscience, dynamical systems theory (DST), a mathematical framework dealing with systems which evolve in time, be they at subcellular, network, behavioral or societal level [44] and network models in psychopathology are consistent and transparent frameworks for studying the core features of major mental disorders such as schizophrenia, bipolar disorder, depression, dementia, autism, etc. Predicated on these frameworks, a new concept termed pathoconnectomics [45] captures both complexity and individual variability of mental disorders and their treatment. According to Du et al. [46], more than 200 papers dealing with brain functional connectivity in the classification and prediction of brain disorders were published from 1990 and 2017. A network theory in psychiatry has been articulated in two forms: (1) in clinical psychopathology as concept that "the term mental disorder refers to a syndrome constellation of symptoms that hang together empirically, often for unknown reasons, and that psychiatric symptoms can cause each other" [47]; (2) in neuroscience as concept that mental disorders reflect aberrant or dysfunctional brain networks [48]. The systems approach network neuroscience and network theory of mental disorders integrate many insights from different paradigms and perspectives.

Criticizing explanatory reductionism in biological psychiatry Borsboom et al. [49] argue that "(1) mental disorders are massively multifactorial in their causal background; (2) many mechanisms that sustain disorders are transdiagnostic; and (3) mental disorders require pluralist explanatory accounts". According to the network approach to psychopathology, mental disorders can be depicted as complex networks states of causally interacting symptoms because mental disorders follow network structure in which some symptoms are more firmly connected than others [47, 49, 50]. Symptoms of mental disorders occur together not only because they reflect a common underlying mechanism, but also they influence each other. In other words, symptoms may have common causes, to be a result of individual developmental trajectories or of environmental adversity, but also they can cause one another forming a cascade of causal relationships. Symptoms that constitute the same mental disorder appear to be more strongly statistically and causally associated than the symptoms within different mental disorders. Interestingly enough, the interactions between symptoms can be depicted as a network in which symptoms are nodes while causal relations between them are connections between nodes [47]. The coupled sets of symptoms that are close in the network will tend to synchronize and form a self-sustaining cluster of symptoms, conditions and events which influence symptoms from outside the network, such as life adversities constitute the external field of symptoms and they may be symptom-specific or shared across symptoms [47]. This concept explains hysteresis effect, the self-sustaining symptom network that keeps itself activated even precipitating causes have disappeared [49]. Systems network theory offers a valid possibility for data integration from symptoms networks and brain circuits networks studies. As modern network science may provide an explicit study of billions of relationships in a single network model, it may enable a new nosology for clinical psychiatry that emphasizes the relationship between symptoms and syndromes and does not presume artificial separation between them [51].

Neurons make up intrinsically coherent neural networks that perform many brain functions, but neural networks also interconnect into more complex networks enabling the development of higher mental functions and complex learning and behaviors. It has been argued that the human brain has 100 billion neurons and that each neuron is connected with 10 thousand other neurons. The brain networks represent collections of brain regions (nodes) and connections (edges) that connect them [48]. Mental functions are represented by the joint activation of groups of neurons which form networks or assemblies by strengthening the connection between neurons that fire together commonly and persistently. From the recent time, the complex brain network characteristics in health and illness have become objects of mathematical deciphering. Major mental disorders reflect deficits in access, engagement, and disengagement of large-scale brain networks as well as disrupted information processing due to damage or dysfunction of individual nodes or edges. As according to DST the mental functions and processes are implemented in terms of the neural dynamics, mental illnesses may be viewed as "disorders of neural network dynamics which involve alterations of oscillations, synchronization among units of a system, attractor states, phase transitions, or deterministic chaos" [44]. So, schizophrenia may be a result of altered attractor dynamics when "weak and flat attractor landscapes, and the distractor susceptibility that come with it, lead to incoherent and disorganized thoughts as the network state meanders around mainly driven by the system noise, or lead to spontaneous pop-out of representations experienced as hallucinations" [44]. Overly strong attractor states in thought and action related to higher glutamate activity that amplifies recurrent excitation and thus strengthening attractors can explain recurring obsessions and compulsions in obsessive-compulsive disorder [44]. In schizophrenia found decreasing NMDA receptor density on GABA inhibitory interneurons which connect cortical pyramidal neurons led to weaker and broader attractor states suggesting that working memory in schizophrenia should be particularly sensitive to distractors [33]. Rumination and negative mood with lack of attention and impaired decision-making in major depressive disorder can be depicted as "strong attractor states in emotion and self-referential processing systems" related to " two network hubs with strong self-excitation but mutual inhibition" and when" either increasing the amount of self-excitation in one of two hubs or through an imbalance in the feedback between the two, one of the two attractor basins strongly expand at the expanse of the other" [44].

Alterations of the brain networks in the connectome have been reported in many major mental disorders indicating on biomarkers for illness diagnosis and prognosis as well as for the evaluation of treatment effectiveness [45]. A selective disruption of brain connectivity among central hubs and reduced regional centrality in hubs including the frontal association, parietal, limbic, and paralimbic brain areas based on structural studies, and including the frontal, temporal, parietal, limbic, and occipital areas based on functional studies have been reported in patients with schizophrenia [45]. In the patients with autism, the whole-brain network functional analysis indicate on hypoconnectivity in the so-called social network encompassing the default mode, attention and executive networks and hyperconnectivity in limbic region with the brain connectome dominated by reduced long-range connections and excessive short-range fibers [45]. Specific connectomic disturbances for individual mental disorders are very important for taxonomy and new classification of mental disorders. Patients with schizophrenia have abnormality primarily in the frontal-temporal regions and a randomized organization, ADHD patients' significant alterations in the prefrontal and orbitofrontal-striatal circuitry with an ordered tendency, and patients with autism spectrum disorders in the social network and subcortical regions [45].

The fundamental problem in mental disorders is a damage of the self-regulatory systems which are no able to distinguish relevant from irrelevant stimuli and to restore the balance. Starting from the concept of the mind-as-decision maker mental disorders can be described as a breakdown in the brain's ability to optimize decisions [53]. The old idea that neuronal connections and networks are crucial for brain function related to mental health as well as to mental disorders has been central to the emerging field of connectomics [47]. Major, for sure, but probably all mental disorders reflect "faulty wiring" or aberrant neuronal connectivity in the CNS. Schizophrenia as a dysconnectivity disorder has been an well-known idea. Extreme bottom-up processing is characteristic of psychosis so that psychotic patients are not able to understand the cause–effect relationships and outcomes of their actions. Some delusions may reflect a malfunctioning appraisal-reappraisal (suspicion) system. At the neuropsychophysiological level anxiety disorders reflects the hyperactivity or dysfunction of the security vigilance/alarm system with false alarms activations. Hyperactivity of the security alarm system is associated with chronic up-regulation of the hypothalamic–pituitary–adrenal (HPA) axis, which in turn, impairs integration of traumatic experiences and appropriate memory consolidation [54]. PTSD reflects the malfunction of a harm avoidance mechanism that normally uses past experiences to escape actual or future dangers, hazards, and damage [54]. Helplessness and hopelessness involve dysfunction of serotoninergic and noradrenergic systems. Imaging studies suggest that the amygdala is a key region in trustworthiness decisions [55]. Dysfunctions of motivational/reward–punishment system may be manifested like (1) motivational deficit such as anhedonia, apathy, akinetic mutism, abulia, avolition, psychomotor slowing, anergia, and social/affiliative amotivation, (2) motivational dysregulation such as addiction and impulsive-compulsive spectrum syndromes, and (3) motivational excess such as mania in bipolar disorder and some psychopathic personality traits [56]. The communication between right ("emotional") and left ("rational") brain was reported to be functionally impaired in some mental disorders such as schizophrenia and PTSD so that traumatic memories may be experienced as ego-alien [54].

Neuronal Network and Brain Systems Specific Interventions

Within the concept of functional psychopharmacology [57] and ECNP neuroscience-based nomenclature of psychotropic drugs, treatment interventions are expected to be psychoneurophysiologically specific targeting dysfunction of specific neuronal circuits and brain systems (Table 3) that are responsible for specific psychopathology. Some examples for illustration of the concept: pharmacological treatment of amotivational syndrome and anhedonia related to low activity of reward system with dopamine agonists, hallucinatory-delusional syndrome related to dysfunction of the appraisal/reappraisal information processing system with potent dopamine antagonists. According to DST alterations of neural network dynamics should be in the near future a central target of therapeutic intervention, although the channels of intervention would remain neurobiological and/or behavioral [44].

Mental Disorders Between Vicious and Virtuous Cycles—Circular Feedback Model of Mental Disorders

Circular feedback model (CFM) of mental disorders is a concept based on neuroscience, cognitive psychology, information processing which emphasize the circular and biopsychosocial nature of some mental disorders such as depression as well as the role of multiple factors that can trigger, exacerbate, or maintain depression and anxiety [58]. From the systems perspective, any mental disorder is part of the feedback process, not existing apart from it. There are two types of feedback processes: reinforcing and balancing [59]. Whenever things are growing, reinforcing or amplifying is at work. Two types of reinforcing feedback processes can be recognized which form "vicious" and "virtuous" cycles. Some reinforcing processes in mental disorders are "vicious cycles", in which processes start off badly and grow worse (the cuckoo's egg syndrome). Depression is a vicious circle of negative affectivity, cognition and behavior related to the negative cognitive triad about self (negative self-precept), world (hostile and demanding), and future (the expectation of suffering and failure). Personal, psychological and spiritual post-traumatic growth is strongly associated with processes that reinforce in desired directions of virtuous cycles. Balancing processes are associated with discovering the sources of stability and resistance [59]. Balancing processes underlies all goal-oriented behavior. The human mind–body system contains countless of balancing feedback processes that may heal our traumas and alert us to real threat. Organizations and societies have also myriad balancing feedback processes which may support healing, recovery, and resilience. The healing process and recovery are always balancing processes. The recovery occurs when a new balance between resilience from one side and damage and vulnerability from another side establishes enriching victim's mission, purpose, and quality of life. The road to full recovery lies not exclusively in alleviating the negative, but also in fostering the positive processes in a system. Bringing the person out of suffering and negative functioning is one form of success, but facilitating progression toward the restoration of positive functioning is also very important [41].

Concept of Staging: Preventive and Preemptive Versus Palliative Psychiatry

> The best way to predict the future is to create it.
>
> —Abraham Lincoln

Psychiatry nosology in DSM-5 and ICD-10/11 has been criticized by both contemporary neuroscience and the emerging field of global mental health with emphasis that psychiatric diagnoses could not be essentialist categories because biological mechanisms lead to a spectrum of symptoms, and health services need to respond to the various stages of mental disorder by providing stepped levels of care [60]. The nature of mental disorders is associated with three crucial problems for the definition of illness categories: the threshold problem, the boundary problem, and illness explanatory difficulties. *The threshold problem* is the point at which symptoms become distinguishable from normal experience and defined as psychopathology. *The boundary problem* refers to the valid boundary lines between mental disorders, e.g., between schizophrenia and bipolar disorder. Karl Jaspers (1913) suggested differentiation between (1) true diseases' with clear boundaries among themselves and with normality; (2) "circles" (schizophrenia, manic-depressive insanity) with clear boundaries with normality, but not among themselves; and (3) "types" (neuroses, abnormal personalities) without clear boundaries either among themselves or with normality. Charcot (1887–88) differentiated archetypes, as ideal types of disorder from formes frustes in which archetypes manifest atypically or partially in individual subject [40]. *Illness explanatory difficulties* are illustrated with the existence of a lot of theories and models of mental disorders.

The great problem with mental disorders is their complexity, frequently "silent" transition from health to mental illness and its unperceived course with a late recognition and delayed, commonly only palliative treatment. Staging is a proven strategy in somatic medicine which involves the links between biomarkers, clinical phenotypes and disease development as the base for personalized or stratified treatment. The pathophysiology and progression of any, particularly those lifelong major mental disorders is a developmental process of transition from one stage to the next in which: (1) the intensity of symptoms or dysfunctionality is greater in subsequent phases; (2) the progress into subsequent phase is followed by a typical clinical picture; and (3) the treatment in initial or early phases is associated with better initial response or prognosis, (4) in general the earlier treatment has a more favorable risk–benefit ratio in comparison with later treatments. Fava and Kellner [61] were the first who pointed to the staging as a neglected dimension in psychiatric classification. As the knowledge of pathophysiology of mental disorders has been developed, staging of major mental disorders such as schizophrenia, bipolar affective disorder, panic disorder, eating disorders, substance abuse disorders, etc., has been gaining momentum [62–68]. Almost all major mental disorders are progressive conditions which have (1) diathesis or an at-risk stage, (2) a prodrome phase, (3) a symptomatic phase, and (4) a residual state. There is ample research evidence that initial episodes of many mental disorders are frequently precipitated by stressful events and response to treatment very well with achieving complete remission, but the appearance of each next episode increases the probability of the future episode, neuroprogression, and illness progression. The term neuroprogression refers to the pathological reorganization of the brain systems which occurs as a consequence of different insults such as allostatic hyperload, neuroinflammation, oxidative stress, etc. Each new episode of mental illness induces a brain rewiring that causes more increasing vulnerability to stressful events. Concept of clinical and biosignature staging offers the potential to improve logic, timing, and type of therapeutic interventions and prediction of outcome in psychiatry. A comprehensive staging model is expected to promote early detection and effective personalized treatment. It is argued that brain network metrics could be used for defining biomarkers to differentiate illness stages. Small-world metrics have been reported to be useful in differentiation between brain functional changes in the early and late phases of Alzheimer's disease (AD) as well as between patients with AD and mild cognitive impairment [48].

According to the network theory, mental disorders are consequences of the presence of hysteresis, a phenomenon of self-sustaining feedback between symptoms in strongly connected symptom networks so that the symptoms continue to activate each other even in the absence of the triggering cause [47, 49]. Mental health is defined as the stable or equilibrium state of a weakly connected network, resilience is defined as the disposition of weakly connected networks to quickly return to their stable state of mental health, whereas vulnerability is depicted as the disposition of strongly connected networks to transition into a state of mental disorder upon perturbation in the external field. After an asymptomatic or premorbid phase in which the network is dormant (Phase 1), an external stress event activates some of the symptoms (Phase 2), which in turn activate connected symptoms (Phase 3), and finally there is no recovery because the network is self-sustaining strong connected and is stuck in its active state [47]. The network theory in psychopathology focuses on two aspects: (1) so-called warning signals that may predict the upcoming mental disorder for a specific patient, and (2) signs and features of group-level networks that may be useful in prognosis and course of mental disorders [50].

Staging Model Based on Neuroprogression and Allostatic Overload

Allostasis (from the Greek "allo" which means variable with connotation "remaining stable by being variable" and "stasis" which means "stand") is the active process of adaptation to everyday stress and challenges, predictable or unpredictable, good or bad, very important for health stability. Allostatic load and overload represent the cumulative pathophysiologic changes in brain and body leading to serious mental disorders and somatic diseases and their comorbidities. Major mental disorders seem to be neuroprogressive condition with declining neurotrophic support and augmented neuronal loss via apoptosis. They usually begin with an asymptomatic premorbid phase (stage 1) which is associated with vulnerability, specific risk factors, and low primary resilience. Low primary resilience is associated with stress vulnerability and development of allostatic hyperload. Morbid phase appears frequently after stressful life events associated with allostatic hyperload associated with dysregulation of the hypothalamic–pituitary–adrenal (HPA) axis, neuroinflammation of low intensity, increased oxidative stress, decreased neurotrophic support and increased neuronal apoptosis. The brain is vulnerable to oxidative stress because it uses a great amount of oxygen producing oxygen free radicals and reactive oxygen species, and has a limited antioxidant capacity [65] (Table 4).

Staging is particularly important in potentially severe disorders and disorders that may progress if they are untreated. Neurodegeneration is an example of chronic, long-term process which begins long before the appearance of clinically recognized symptoms, but unfortunately, there is an evident lack of targeted prevention in any age-category inchoate with early childhood, over adolescence until late adulthood [1]. Early treatment and phase-specific treatment of mental disorders could significantly improve treatment effectiveness and efficiency in psychiatry.

From the perspective of preemptive and preventive psychiatry studies of the fetal origins of mental health as well as mental disorders are formidable challenge [70–73]. According to "the developmental origins of health and disease" (DOHaD) hypothesis intrauterine signals which compromise fetal growth act to "program" tissue differentiation in a manner that shapes individual differences in the risk for chronic illness and mental disorders (ADHD, autism, anxiety, bipolar disorder, depression, schizophrenia, and substance abuse) over the lifespan [70]. The integrative models of fetal neurodevelopment suggest that antenatal maternal adversity operates through the biological pathways associated with fetal growth to program neurodevelopment. Compromised fetal development appears to establish a "meta-plastic" state that increases sensitivity to postnatal influences and risks for mental disorders. DOHaD studies provide an empirical basis for multidisciplinary

Table 4 A staging concept of mental disorders adapted from [62, 63, 69]

IDEAL PHASE: No vulnerability, no significant risk factors
1. PREMORBID PHASE (healthy but pathology-predisposed individuals)
Latent phase (*Stage 1*): Vulnerability, at-risk stage—Presymptomatic stage—Low primary resilience stage
Dysthymic, cyclothymic, hyperthymic, schizothymic, timid temperament; risky personality traits
2. PRODROMAL PHASE
Illness incipient phase (*Stage 2*): Prodrome/Prepsychotic: High-risk stage—Schizotypia: Neuroplasticity disorder and Neurodysplasticity
3. MORBID PHASE
FIRST EPISODE PHASE: Neuroinflammation of low intensity, Oxidative stress, Neurotoxicity
Stage 3a: Full syndrome—The first manic or depressive episode, the first psychotic episode, the first episode of panic attacks
Stage 3b: First relapse
Stage 3c: Residual symptoms
RECURRENT EPISODES PHASE: Neurotoxicity, neurodegenration
Stage 4: First Recurrence episode
ADVANCED ILLNESS PHASE (3 and more episodes)
Stage 5: Intermittent course
Stage 6: Intermittent or progressive course (Progression-Residual states-Neurotoxicity with increasing neuronal loss via apoptosis)
Stage 7: Late or End stage of illness (Treatment refractoriness—Marked neurodegeneration)

programs across obstetrics/gynecology, neonatology, pediatrics, neuroscience, psychiatry, and psychology and are essential for a comprehensive understanding of the relation between maternal health, fetal growth, and neurodevelopment and predisposition for mental disorders.

Staging Model and Specific Predisposing Temperament/ Personality Traits

Staging in most psychiatric disorders should involve evaluation of the patient's premorbid personality types and personality traits. Some clinical research has indicated the possibility of specific temperaments and personality traits associated with both the predisposition to and clinical manifestation of some mental disorders. In literature for a long time there exist a conceptualized link between major mental disorders and temperamental attributes and characteristics (Table 5). Kretschmer thought that endogenous psychoses are nothing but exaggerated forms of normal temperament. Many authors think about affective and schizothymic temperaments as proximate determinants in the heritability of bipolar disorder and schizophrenia. Hyperthymic and cyclothymic temperaments are considered as the

Table 5 Mental disorders and temperamental attributes [74–79]

Euthimic (sanguine) temperament: characterized by extroversion, activity, good humor, positive attitude, optimism, and enthusiasm
Schizotaxic temperaments
Schizotaxia (Meehl 1962)—a multifactorial polygenetic predisposition to developing schizotypia or schizophrenia, and for some a complete spectrum of schizophrenic disorders—an universal entry door for schizophrenia spectrum,
Schizothymia (Bleuler 1926)—split in emotions or the existence of a personality and emotional traits that include introversion, a closed nature, shyness—hypersensitive and not affectively disinterested, cold nor void
Schizoidia (E. Bleuler 1908)—the natural human tendency in directing the attention of some people to their own internal world distancing themselves from the external world—affectively disinterested, remote and cold
Schizoid temperament (Kretschmenr 1919): 1. affective timidity (fearfulness, bashfulness)—emotional hyperesthesia or hypersensitivity and 2. affective callousness and coldness –emotional anesthesia or affective coldness.
Schizotypia (Bleuler 1926—latent sch) personality traits and specific behavior that include a bizarre or eccentric and general unordinary communication, hypersensitivity toward criticism, excessive social anxiety, social isolation, distrustfulness, and ideas concerning relationships, repeated illusions, magical thoughts, and inappropriate expressions when communicating face to face—a specific doorway to schizophrenia
Schizophrenia diathesis: 1. schizothymic temperament, hypersensitivity personality traits, shyness and avoidant behaviors, schizoaffective disorder, schizophrenia 2. schizoid temperament
Affective temperaments
Thymopathic (Eugen Bleuler): an abnormal disruption of mood that shows a genetic susceptibility to manic-depressive psychosis (bipolar affective disorder)
Hyperthymic temperament: tendency to overactivity and an excessive emotional response
Dysthymic temperament: subdued emotional tone, inability to enjoy life, self-depreciation, tendency to sadness and low, depressive mood; prone to develop a severe depressive disorder under stress
Cyclothymic temperament: tendency to fluctuations of mood from periods of elation and excitement to periods of depression and underactivity
Anxious (timid) temperament: tendency to experience timidity and anxiety in new situations and meeting new people, emotional hypersensitivity
Choleric (irritable, angry) temperament: characterized by irritability, quick/short temper, dominance, decisiveness, goal-orientation, independence.
Phlegmatic temperament: easy-going individuals tend to relaxation, peacefulness and hiding emotions
Melancholic temperament: characterized by introversion, deep thinking, and feeling, striving for perfection and details
Anxiogenic diathesis: Timid temperament—Hypersensitivity personality traits—Avoidant personality traits/disorders—Acute stress reaction—PTSP—Social anxiety disorders—General anxiety disorder—Panic disorder—Depressive disorder—Schizoaffective disorder
Depression diathesis: Dysthymic temperament—Depressive personality characteristics/disorder—Dysthymia—Mild depressive disorder—Melancholia—Psychotic depressive disorder—Unipolar schizoaffective disorders

(continued)

Table 5 (continued)

Bipolar diathesis: 1. Cyclothymic temperament—Cyclothymic personality characteristic/disorder—Cyclothymia—Emotionally instability disorder—Bipolar II—Bipolar I—Psychotic bipolar—Schizoaffective bipolar disorder; 2. Hyperthymic temperament—Hyperthymic personality characteristics/disorder—Hypomania—Bipolar II—Bipolar I—Schizoaffective unipolar disorder

core in bipolar I and bipolar II subtypes, while irritable, angry profile may be a feature of the soft bipolar spectrum disorders [74]. Some studies suggest that irritability in major depression may be related to the bipolar spectrum and lithium responsivity.

Temperament refers to chemical, bioelectrical and neuropsychophysiological properties of central nervous system and it has been argued that temperament and mental disorders, particularly affective disorders, lie along continuum of neurobehavioral regulation [79]. Creating taxonomies arising from the continuum models may result in more valid alignment between illness categories and underlying psychopathophysiology. This might contribute to a better understanding of predisposition for some mental disorders and enable preemptive interventions. The spectrum concept of mental disorders was proposed with goals of better defining and recognizing nosological entities and conditions between categories. According to Ernst Von Zeller and Edward Griesinger (1871) whole psychopathology represent different manifestations of a single disorder called "Einheitspsychosis" (Unite Psychosis) and different mental disorders are just variations in the intensity of the "unite psychosis" [51].

Personalized Multi-omics Profiles, Comorbidity Mapping and Precision Medicine in Psychiatry

It has been argued that comorbidity is an intrinsic characteristic of mental disorders [47]. Due to shared genetic, molecular, neurophyschophysiological, environmental, and lifestyle risk factors, many mental disorders within themselves and with somatic diseases occur together in the same patient. Comorbidities are multi-interpretable phenomena which can be explained from different theoretical and conceptual perspectives [80, 81]. What is surprising at first glance many seemingly different disorders and diseases have common molecular mechanisms and strong comorbid associations [82]. From a traditional perspective, comorbid mental disorders are defined as a combination of different disorders. However, the network analysis indicates that mutual so-called bridged symptoms may spread activation from one disorder to another, e.g., from general anxiety disorder to major depressive disorder [50]. A complex integration of multi-omics, ontology, phenotypic and environmental information is fundamental for prediction of mental disorders and somatic diseases as well as for patients' stratification. The computational frameworks that integrate all available heterogenous and relevant data including miRNA–target interactions, miRNA–disease association, phenotype similarities of diseases, gene ontology (GO), single-nucleotide polymorphisms (SNPs), copy number variations (CNVs) and known disease–environmental associations to capture the complex relationships between phenotypes, genotypes and clinical comorbidity [82] offers a hope for the future of precision medicine. Precision medicine will probably be followed with fundamental changes in diagnostics and classifications of mental disorders and somatic diseases as well as in the health care enabling targeted preventive or therapeutic action at the earliest indications of risk, mental disorder or somatic disease.

Conclusion

The content of this chapter has been shaped by the ideas, theories, researches and clinical experiences of many scholars, experts and clinicians around the world, many of them found on Internet. The aim of the chapter is to offer an overarching model of a common language in psychiatry and bridging over theoretical gaps and disciplinary borders. Theoretical psychiatry predicated on transdisciplinary and trans-domain integrative approach and systems networking supports implementation of predictive, preventive, precision, person-centered and participatory medicine in psychiatry. It is a formidable challenge, not a hype but a hope for the future. Criticizing one explanatory reductionism from the perspective of another reductionistic perspective is usually counterproductive and has no sense. Creatively combining the different theoretical perspectives brings us closer to a holistic understanding of the complex nature of mental health and mental disorders, and to more efficient treatment of mental health problems. Psychiatry seems to be moving toward an era of a transdisciplinary, integrative, context-sensitive and person-centered mental health discipline that responds more effectively and humanly to the challenges of mental disorders. While we are waiting for establishing optimal 5P medicine in psychiatry and more personalized algorithms, it is possible to improve our clinical practice by using the knowledge and therapeutic methods from different disciplines in a creative and integrative way and practicing principles of the person-centered psychiatry.

References

1. Golubnitschaja O, Kinkorova J, Costigliola V. Predictive, preventive and personalized medicine as the hardcore of 'Horizon 2020': EPMA position paper. EPMA J. 2014;5:6.
2. Ghaemi SN. The concepts of psychiatry: a pluralistic approach to the mind and mental illness. Baltimore, London: The John Hopkins University Press; 2003.
3. Ozomaro U, Wahlestedt C, Nemeroff CB. Personalized medicine in psychiatry: problems and promises. BMC Med. 2013;11:132.

4. Fernandes BS, Williams LM, Steiner LM, Leboyer J, Carvalho AF, Berk M. The new field of 'precision psychiatry'. BMC Med. 2017;15:80.
5. Harari YN. Homo deus: a brief history of tomorrow. UK, London: Vintage, Penguin Random House; 2016.
6. Harari YN. 21 Lessons for the 21st century. UK, London: Vintage, Penguin Random House; 2018.
7. Bechtel W, Richardson RC. Discovering complexity decomposition and localization as strategies in scientific research. Princeton: Princeton University Press; 1993.
8. Thagard P. Mental illness from the perspective of theoretical neuroscience. Perspect Biol Med. 2008;51:335–52.
9. Jakovljevic M, Borovecki F. Epigenetics, resilience comorbidity and treatment outcome. Psychiatria Danubina. 2018;30:242–53.
10. Jakovljevic M. Creative, person-centered psychopharmacology for treatment resistance in psychiatry. Kim YK, editor. Treatment resistance in psychiatry—risk factors, biology, and management. Springer Nature Singapore Pte Ltd.; 2019. p. 273–93.
11. Stephan KE, Bach DR, Fletcher PC, Flint J, Frank MJ, Friston KJ, Heinz A, Huys QJM, Owen MJ, Binder EB, Dayan P, Johnstone EC, Meyer-Lindenberg A, Montague PR, Schnyder U, Wang XJ. Charting the landscape priority problems in psychiatry, part 1: classification and diagnosis. Lancet Psychiatry. 2016;3:77–83.
12. Jakovljevic M. Empathy, sense of coherence and resilience: bridging personal, public and global mental health and conceptual synthesis. Psychiatria Danubina. 2018;30:380–4.
13. Stephan KE, Binder EB, Breakspear M, Dayan P, Johnstone EC, Meyer-Lindenberg A, Schnyder U, Wang XJ, Bach DR, Fletcher PC, Flint J, Frank MJ, Heinz A, Huys QJM, Montague PR, Own MJ, Friston KJ. Charting the landscape priority problems in psychiatry, part 2: pathogenesis and aetiology. Lancet Psychiatry. 2016;3:84–90.
14. Strik W, Stegmayer K, Wačlther S, Dierks T. Systems neuroscience of psychosis: mapping schizophrenia symptoms onto brain systems. Neuropsychobiology. 2017;75:100–16.
15. Gerstner W, Sprekeler H, Deco G. Theory and stimulation in neuroscience. Science. 2012;338(60-6):5.
16. Pernu TK: Elimination, not reduction—lessons from the research domain criteria (RDoC) and multiple realization. Behav Brain Sci. 2017:e22. https://doi.org/10.1017/s0140525x18001139.
17. Elbau IG, Binder EB, Spoor maker VI: Symptoms are not the solution but the problem: why psychiatric research should focus on processes rather than symptoms. Behav Brain Sci. 2019:E7. https://doi.org/10.1017/s0140525x18001000.
18. Cuthert BN, Insel TR. Toward the future of psychiatric diagnosis: the seven pillars of RDoC. BMC Med. 2013;11:126.
19. Elmer GI, Brown FL, Shepard PD. Engaging research domain criteria (RDoC): neurocircuitry in search of meaning. Schizophrenia Bulletn. 2016;42:1090–5.
20. Sanislow CA, Pine DS, Quinn KJ, Kozak MJ, Garvey MA, Hainssen RK, Wang PS-E, Cuthbert BN. Developing constructs for psychopathology research: research domain criteria. J Abnorm Psychol. 2010;119:631–9.
21. Kelly JR, Clarke G, Cryan JF, Dinan TG. Dimensional thinking in psychiatry in the era of the research domain criteria (RDoC). Ir J Psychol Med. 2018;25:89–94.
22. Li F, Rao G, Du J, Xiang Y, Zhang Y, Selek S, Hamilton JE, Xu H, Tao C. Term standardization and normalization for research domain criteria (RDoC). In: 2018 International conference on healthcare informatics workshop. https://doi.org/10.1109/ichi-w.2018.00019.
23. Svrakic D, Divac-Jovanovic M. The fragmented personality: an integrative, dynamic, and personalized approach to personality disorder. New York: Oxford University Press; 2019.
24. Sartorius N. Preface. The ICD-10 classification of mental and behavioral disorders. In: Clinical descriptions and diagnostic guidelines. Geneva: World Health Organization; 1992.
25. Lydiard J, Nemeroff C. Precision psychiatry: biomarker-guided therapy for effective treatment and prevention. In: Kim YK, editor. Research methods and interventions in

psychiatry—artificial intelligence, precision medicine, and other paradigm shifts. Springer; 2020.
26. Kim YK, editor. Treatment resistance in psychiatry—risk factors, biology, and management. Springer Nature Singapore Pte Ltd; 2019.
27. Tournier P. Medicine de la Persone. Neuchatel: Delachauts et Niestle; 1940.
28. Mezzich JE, Botbol M, Christodoulou, Cloninger CR, Salloum IM, editors. Person centered psychiatry. Switzerland: Springer International Publishing; 2016.
29. Cloninger CR, Cloninger KM, Mezzich H. Holistic framework for ill health and positive health. In: Mezzich JE, Botbol M, Christodoulou, Cloninger CR, Salloum IM, editors. Person centered psychiatry. Switzerland: Springer International Publishing; 2016. p. 47–59.
30. Jakovljevic M, Jakovljevic I. Theoretical psychiatry: missing link between academic and clinic psychiatry for further scientific and professional maturation of psychiatry. In: Kim YK, editor. Research methods and interventions in psychiatry—artificial intelligence, precision medicine, and other paradigm shifts. Springer; 2020.
31. Wikipedia: Mereology. https://en.m.wikipedia.org.
32. Borsboom D, Cramer AOJ, Kalis A. Authors response: reductionism in retreat. Behav Brain Sci. 2019:e32. https://doi.org/10.1017/s0140525x18002091.
33. Huys QJN, Maia TV, Frank MJ. Computational psychiatry as a bridge from neuroscience to clinical applications. Nat Neurosci. 2016;19:404–13.
34. Adams RA, Huys QJM, Roiser JP. Computational psychiatry: towards a mathematically informed understanding of mental illness. J Neurol Neurosurg Psychiatry. 2015:1–11.
35. Huys Q. Computational psychiatry. In: Encyclopedia of computational neuroscience. New York: Springer Science + Business Media; 2014.
36. Mezzich JE, Botbol M, Christodoulou, Cloninger CR, Salloum IM. Introduction to the person-centered psychiatry. In: Mezzich JE, Botbol M, Christodoulou, Cloninger CR, Salloum IM, editors. Person centered psychiatry. Switzerland: Springer International Publishing; 2016. p. 1–15.
37. Williams P, Thomas LJ. Total life coaching—a compendium of resources. New York, London: W.W. Norton; 2005.
38. Auerswald EH. Family healing and planetary healing—three paradigm in search of a culture. In: Prosky PS, Keith DV, editors. Family therapy as ab alternative to medication—an appraisal of pharmaland. New York, Hove: Brunner-Routledge; 2002.
39. Macneil CA, Hasty MK, Conus P, Berk M. Is diagnosis enough to guide interventions in mental health? Using case formulation in clinical practice. BMC Med. 2012;10:111.
40. Stanford encyclopedia of philosophy: philosophy of psychiatry. https://plato.stanford.edu.
41. Cloninger CR. Feeling good: the science of well-being. Oxford University Press; 2004.
42. Dawkins R. The extended phenotype: the long reach of the gene. Oxford University Press; 1999.
43. Tretter F. Philosophical aspects of neuropsychiatry. In: Tretter F, Gebicke-Haerter PJ, Mendoza ER, Winterer G, editors. Systems biology in psychiatric research—from high-throughput data to mathematical modeling. Weinheim: Wiley-VCH Verlag GmbH & Co. KGaA; 2010. p. 3–25.
44. Durstewitz D, Huys QJM, Koppe G. Psychiatric illnesses as disorders of network dynamics. Transcontinental Computational Psychiatry Workgroup; 2018.
45. Cao M, Wang Z, He Y. Connectomics in psychiatric research: advances and apllications. Neuropsychiatr Dis Treat. 2015;11:2801–10.
46. Du Y, Fu Z, Calhoun VD. Classification and prediction of brain disorders using functional connectivity: promising but challenging. Front Neurosci. 2018;12:525.
47. Borsboom D. A network theory of mental disorders. World Psychiatry. 2017;16:5–13.
48. Menon V. Large-scale brain networks and psychopathology: a unifying triple network model. Trends Cogn Sci. 2011;15:483–506.
49. Borsboom D, Cramer AOJ, Kalis A. Brain disorders? Not really: why network structures block reductionism in psychopathology research. Behav Brain Sci. 2019;42(e2):1–63.

50. Fried EI, van Borkulo CD, Cramer AOJ, Boschloo L, Schoevers RA, Borsboom D. Mental disorders as networks of problems: a review of recent insights. Soc Psychiatry Psychiatr Epidemiol. 2017;52:1–10.
51. Goekoop R, Goekoop JG. A network view on psychiatric disorders: network clusters of symptoms as elementary syndromes of psychopathology. PLoS ONE. 2014;9(11):e112734.
52. Hulshoff Pol, Bullmore E. Neural networks in psychiatry. Eur Neuropsychopharmacol. 2013;23:1–6.
53. Moutossis M, Story GW, Dolan RJ. The computational psychiatry of reward: broken brains or misguided minds? Front Psychol. 2015;6:1445.
54. Bruene M. Textbook of evolutionary psychiatry. New York: Oxford University Press; 2008.
55. Benedetti F. The patient's brain—the neuroscience behind the doctor-patient relationship. New York: Oxford University Press; 2011.
56. Epstein J, Silbersweig D. The neuropsychiatric spectrum of motivational disorders. J Neuropsychiatry Clin Neurosci. 2015;27:7–18.
57. Van Praag HM. "Make-Believes" in psychiatry or the perils of progress. New York: Brunner/Mezel Publishers; 1993.
58. Alladin A. Handbook of cognitive hypnotherapy for depression—an evidence-based approach. Philadelphia: Wolters Kluwer/Lippincott Williams, Wilkins; 2007.
59. Senge PM. The fifth discipline: the art and practice of the learning organisation. London: Random House; 2006.
60. Stein DJ, Reed GM. Global mental health and psychiatric nosology: DSM-5, ICD-11 and RDoc. Bras J Psychiatry. 2019;41:3–4.
61. Fava GA, Kellner R. Staging: a neglected dimension in psychiatric classification. Acta Psychiatr Scan. 1993;87:225–30.
62. Cosci F, Fava GA. Staging of mental disorders: systematic review. Psychother Psychosom. 2013;82:20–34.
63. McGorry PD, Keshavan M, Goldstone S, Amminger P, Allott K, Berk M, Lavoie S, Pantelis C, Yung A, Wood S, Hickie I. Biomarkers and clinical staging in psychiatry. World Psychiatry. 2014;13:211–23.
64. Passos IC, Kapczinski NS, Quevedo J. Kauer-Sant'Anna M, Kapczinski F: Staging models and functional outcomes in bipolar disorder: clinical implications. Curr Treat Opinions Psych. 2015;2:205–17.
65. Muneer A. Staging models in bipolar disorder: a systematic review of literature. Clin Psychopharmacol Neurosci. 2016;14:117–30.
66. Wojciak P, Remlinger-Molenda A, Rybakowski J. Stages of the clinical course of schizophrenia—staging concept. Psychiatr Pol. 2016;50:717–30.
67. Berk M, Post R, Ratheesh A, Gliddon E, Singh A, et al. Staging in bipolar disorder: from theoretical framework to clinical utility. World Psychiatry. 2017;16.
68. Salagre E, Dodd S, Aedo A, Rpsa A, Amoretti S, Pinzon J, Reinares M, Berk M, Kapczinski FP, Vieta E, Grande I. Toward precision psychiatry in bipolar disorder: staging 2.0. Front Psychiatry. 2018;8:641.
69. McGorry PD, Purcell R, Hickie JB, Yung AR, Pantelis C, Jackson HJ. Clinical staging: a heuristic model for psychiatry and youth mental health. Med J Aust. 2007;187:s40.
70. O'Donell KJ, Meaney MJ. Fetal origins of mental health: the developmental origins of health and disease hypothesis. Am J Psychiatry. 2017;174:319–28.
71. Pettersson E, Larsson H, D'Onofrio B, Almqvist C, Lichtenstein P. Association of fetal growth with general and specific mental health conditions. JAMA Psychiatry. 2019;76:536–43.
72. Sonmez EO, Uguz F, Sahingoz M, Sonmez G, Kaya K, Camkurt MA, Gokmen Z, Basaran M, Gezginc K, Erdem SS, Dulger HK, Tasyurek. Effect of maternal depression on brain-derived neurotrophic factor levels in fetal cord blood. Clin Psychopharmacol Neurosci. 2019;17:308–13.

73. Kurjak A, Stanojevic M, Salihagic-Kadic A, Spalldi Barisic L, Jakovljevic M. Is four dimensional (4D) uktrasound entering a new field of fetal psychiatry. Psychiatria Danubina. 2019;31:131–40.
74. Gandotra S, Ram D, Kour J Praharaj SK. Association between affective temperaments and bipolar spectrum disorders: preliminary perspectives from a controlled family study. Psychopathology. 2011;44:216–24.
75. Clark LA. Temperament as a unifying basis for personality and psychopathology. J Abnorm Psychol 2005;114:505–21.
76. Horan WP, Blanchard JJ, Clark LA, Green MF. Affective traits in schizophrenia and schyzotypy. Schizophr Bull. 2008;34:856–74.
77. Mason OJ. The duality of schizotypy: it is both dimensional and categorical. Front Psychiatry. 2014;5:134.
78. Sakhvidi MN, Jafari L, Hosseini F. Classification of mental disorders based on temperament. Avicenna J Neuro Psych Pjysio. 2015;2(3):e35266.
79. Sulis W. Assessing the continuum between temperament and affective illness: psychiatric and mathematical perspectives. Phill. Trans. R. Soc. B. 2018;373:20170168.
80. Jakovljevic M. Psychopharmacotherapy and comorbidity: conceptual and epistemological issues, dilemmas, and controversies. Psychiatria Danubina. 2009;21:333–40.
81. Jakovljevic M, Reiner Z, Milicic D, Crncevic Z. Comorbidity, multimorbidity, and personalized psychosomatic medicine: epigenetics rolling on the horizon. Psychiatria Danubina. 2010;22:184–9.
82. Moni MA, Lio P. How to build personalized multi-omics comorbidity profiles. Front Cell Dev Biol. 2015;24(3):28.

Applications of Developmental Psychopathology

Gonzalo Salazar de Pablo, Julio David Vaquerizo Serrano, Sandra Gómez Vallejo, Javier Sánchez Cerezo and Carmen Moreno Ruiz

Abstract

Developmental psychopathology studies the basic mechanisms, including not only biological factors but also environmental and social factors that may interact with them, by means of which developmental pathways deviate toward pathological or typical outcomes. Family studies conducted during the last century show substantial evidence of heritability among psychiatric disorders. Besides, a large number of genes implicated in shaping the development of the central nervous system have been related to psychiatric conditions. In addition, there is a wide range of stressors and harmful agents that, when acting on sensitive developmental periods, might damage brain function and generate or precipitate psychopathology over time. All these factors have the potential to change the way disorders with a neurodevelopmental origin are expressed, including their age of appearance and clinical manifestations. Both symptoms and social impairment need to be considered in clinical evaluations, as treatment is unlikely to be effective if the problem has not been characterized

G. Salazar de Pablo · S. Gómez Vallejo
Institute of Psychiatry, Psychology, and Neuroscience, King's College London, London, UK

J. D. Vaquerizo Serrano · J. Sánchez Cerezo
Psychiatry Department, School of Medicine, Hospital General Universitario Gregorio Marañón, Universidad Complutense (UCM), Madrid, Spain

C. Moreno Ruiz (✉)
Child and Adolescent Psychiatry Department, School of Medicine, Hospital General Universitario Gregorio Marañón, Universidad Complutense (UCM), IiSGM, CIBERSAM, Madrid, Spaine-mail: cmoreno@hggm.es

© Springer Nature Singapore Pte Ltd. 2019

Y.-K. Kim (ed.), *Frontiers in Psychiatry*, Advances in Experimental Medicine and Biology 1192, https://doi.org/10.1007/978-981-32-9721-0_21

correctly or if the patients' particular characteristics, which change throughout development, are not taken into consideration.

Keywords
Developmental psychopathology · Neurodevelopmental disorders · Genetic factors · Environmental factors · Psychiatric disorders

Introduction

Developmental psychopathology is a relevant paradigm of psychopathological research encompassing the study of the development of psychiatric disorders by means of a life course perspective [1]. Different domains including perceptions, sleep, memory, language, and conduct are studied through this perspective throughout all stages of development.

When approaching the study of developmental psychopathology, biochemical, genetic, and neurophysiological factors, as well as other environmental and social factors, need to be taken into account. It is relevant to consider the interaction of biological factors with environmental factors that modify psychopathological development and expression as both of them have the potential to lead to an anomalous development.

The way clinical characteristics manifest in the child may vary according to child's stage of development. For instance, in depression, somatic symptoms are more frequent in childhood, while cognitive symptoms become prominent in adolescence. In other words, developmental stage can modulate and determine clinical expression of most psychiatric disorders.

In every developmental period, different physical, psychic, cognitive, and psychomotor particularities can be observed. The prenatal or intrauterine phase goes from the conception or fertilization to birth. This period is especially sensitive to infectious agents. Other risk factors that are detrimental to the development of the fetus in this period include toxicants exposure, or alcohol and drugs consumption, among other environmental stressors.

Most developmental disorders including attention deficit hyperactivity disorder (ADHD) and autism spectrum disorder (ASD) show their first manifestations during childhood. This is a period of dramatic growth and change in which most skills and cognitive functions are mastered. It is also a period of vulnerability in which the impact of some risk factors may lead to early-onset psychopathology. As an example, motor deficits and caretaker instability have been found to be associated with early-onset major depression, but not with depression that appears for the first time during adulthood [2].

Adolescence is a developmental stage in which important physical changes driven by overall physical growth spurt and sexual maturation are observed. This phase is also important for cognitive development and crucial for the acquisition of a stable sense of self. In this period, affective, psychotic or behavioral disorders may appear for the first time.

Genetic Factors

Family studies conducted during the last century show substantial evidence of heritability among psychiatric disorders, ranging between 30 and 90% depending on the disorder and the study design [3]. Studies based on large cohorts suggest a significant genetic overlap between different disorders, the most replicated findings including bipolar disorder (BD), schizophrenia and schizoaffective disorder [4]; BD and major depressive disorder (MDD); and ASD and ADHD [5].

Regarding genetic studies, in the last decade, the focus has turned to genome-wide association studies (GWAS), which study whether a genetic marker (usually a single nucleotide polymorphism–SNP-) appears more frequently in cases than in controls. Some of these studies have demonstrated quantified molecular evidence for genetic correlations between schizophrenia, BD, MDD, ADHD and ASD, being the strongest relationship the one between schizophrenia and BD [5]. Nevertheless, GWAS present limitations on identifying variants of small effect and to explain familial risk of common diseases [6, 7].

A recent study of 25 brain disorders concluded that psychiatric disorders shared a common genetic risks, in which genetic correlation with each other was limited [8]. Considering these results, some authors have suggested that common genetic influences break down the boundaries of psychiatric diagnoses [8–10]. One possible explanation for this might be that genes might confer risk for different psychiatric disorders. However, other factors are needed for the disease's clinical presentation, including environmental factors which result in epigenetic programming. This might also reflect the presence of phenotypes shared by more than one disorder, for instance, difficulties in executive function [8]. As a result, the interaction among epigenetics, pleiotropy, and symptom specific genes would lead the pathway to one disorder or another.

Regarding neurodevelopment, hundreds of genes have been involved in the development of the central nervous system (CNS). Both single rare genomic variations and common variants, frequently SNPs in coding and noncoding loci of the genome, have been associated with developmental disorders. However, few SNPs with a minor allele frequency of >1% have been associated with neurodevelopmental disorders (NDD), leading to identification of individually rare variants, either inherited or "de novo". These rare variants have been found to occur more frequently, especially for individual liability. For instance, copy-number variants (CNVs), which are duplications or deletions of more than 1000 base pairs that vary between the general population, are present in 10% of the cases of ASD, rate that rises to 22% in ASD associated with intellectual disability (ID) [11]. In addition, low-IQ ASD cases have been correlated with "de novo" mutations and less family history of psychiatric illness than high-functioning autism, where family history is more frequent [3, 11]. On the other hand, a recent study from a Swedish sample supports the idea that common variants have a substantial impact "*enmasse*" and that their contribution explains the biggest fraction of total heritability in ASD, rather than inherited rare variations [12].

The different genes identified to play a role in critical pathways for typical neurodevelopment have been related mostly to synapse formation, including NRXN1, NLGN3, NLGN4, CNTNAP2, and SHANK3, neuronal proliferation, growth, transcription and splicing, and chromatin remodeling (MET, TSC1/2) [3, 10, 13]. Moreover, mutations in the gene Gad1, essential for the functioning of the GABAergic circuit and therefore, for the appropriate functioning of brain circuits, have been associated also with some psychiatric disorders [14].

It seems that synapsis network may be altered from receptors and ion channels to proteins. The alteration of the synaptic function may alter the communication between neurons and lead to NDD or neurodegenerative disorders. Synaptic defects have been associated with defected genes in ASD, ID, and schizophrenia [15].

Even though different molecules and pathways have been identified, what it is yet to be discovered is where and when the events occur within the brain development (spatiotemporal gene expression). Expression of some of the genes described for some NDD increase during mid-gestation, when some early neuronal differentiation and synapse formation take place in prefrontal and temporal cortices [16].

As we mentioned above, to better understand the pathophysiology of NDD we need to consider the role of epigenetics. Epigenetics relates to long-lived and reversible modifications to nucleotides or chromosomes that do not change the genetic sequence but can alter gene expression and phenotype [17]. Epigenetics may help explain different phenotypic outcomes from similar gene mutations.

Several studies have demonstrated the influence of epigenetics on different processes, such as methylation of genes and acetylation of histones, regulation of the expression of genes, and its penetrance across individuals through the life span, as well as the contribution to cumulative risk. Epigenetic changes may persist across generations, and in animal studies, it has been demonstrated that environmental factors may interact with both maternal and offspring genetics [14].

To sum up, a broad number of genes, expressing differentially through development, have been associated with NDDs. Genes related to neuropsychiatric disorders play an essential role in the synapse formation as well as in the development of other neuronal pathways. Furthermore, the interaction between the genetic background and the environmental factors might explain the different outcomes or phenotypes observed.

Environmental Factors

Environmental non-genetic factors shape normal development and affect the expression of neurodevelopmental psychopathology. The individual needs to adapt to the environment and is especially dependent on it.

This influence is particularly important during sensitive developmental periods. A general environmental model suggests that the child's behavior is modulated by the environment in which it happens, as development of self-regulation ability is still undergoing [18]. It proposes that as long as the environment appears consistent, the child's behavior will be consistent as well, but, if the environment changes, so will the child's behavior [2]. Besides, each individual has certain particular characteristics that might match or not the environmental demands. In case demands are matched, the development will continue as expected. On the contrary, if the environmental demands cannot be fulfilled, there will be a deviation from normality. During periods of increased developmental plasticity, we may see this influence magnified [19].

The risk of developing psychopathological symptoms might be mediated by changes in patterns of emotional, behavioral, and cognitive functioning [20], in which environmental factors may have an impact. There is a wide range of different stressors and harmful agents which might damage brain function and may generate or precipitate psychopathology. These environmental factors are likely to have an influence on the psychopathology of children and adolescents through different moderators. Most of the brain development takes place intrauterus [21], and this period highly determines our life outside the uterus [22], so pregnancy and childbirth are especially critical times for neurodevelopment. Even before the child is born, he may be exposed to stress through infections by different agents such as bacteria, viruses, and parasites, which have been associated to fetal death, organ injury, and other effects on fetal development [23]. Exposure to hostile conditions during gestation may also result in a series of coordinated biological responses aimed at enhancing the probability of survival, but that could likewise increase the susceptibility to mental illnesses [24].

Exposure to psychosocial deprivation is also associated with elevations in numerous forms of impairment throughout the life course. Children who have been exposed to malnutrition show a higher prevalence of psychiatric disorders [25] and worse cognitive performance [26], what may consequently lead to worse family, school, and social performance.

The impact of other negative environmental factors is also relevant. It has been described that individuals reporting a history of childhood violence victimization suffer neurodevelopmental insults where brain function is impaired [27, 28], including deficits in clinically significant cognitive functions such as general intelligence and more specific measures of executive functioning, processing speed, memory, perceptual reasoning, and verbal comprehension [28]. Besides, child maltreatment has been linked with increased odds of a wide array of psychiatric disorders, including not only post-traumatic stress disorder but also internalizing and externalizing disorders.

Adolescence is a decisive phase of change and maturation. Adolescence is a peak age for both victimization and mental disorder onset [29], and for molecular changes in the developing brain that shape normal behavioral patterns [30]. In this stage of life, the relationship with others has a fundamental role and deviations from the expected patterns may have long-term consequences. As an example, school bullying has a role in a number of psychopathology outcomes in children and young people, especially anxiety disorders and depression, and it is suggested that its effects may be long-lasting [31].

Another important factor that needs to be taken into consideration is the family and cultural context. The way a baby or a child interacts with his relatives is based not only on the child's characteristics but also on parents/caregivers' previous experiences and affects, and the failure of caregivers to provide co-regulation of children's emotional experiences has enduring consequences in child development and predisposes to the emergence of psychopathology [32]. This may occur even in patients that suffer from disorders with an important genetic component. For instance, the way caregivers interact with children has been shown to have an impact on the development of communication skills in children with autism [33]. Insecure attachment and disorganized attachment, in particular, have prospectively been related to the development of a wide array of problems, including externalizing problems [34].

Finally, substance use may impact neurodevelopment, especially in the prenatal period, being prenatal alcohol exposure the cause of fetal alcohol spectrum disorders. Also, adolescence is a unique period of neurobiological vulnerability to alcohol and drugs [35], as adolescent alcohol exposure persistently impacts adult neurobiology and behavior [36], and cannabis use influences different domains as neurocognition, both acutely and non-acutely [37].

Neurodevelopmental Disorders—Some Examples

In the following section, we provide examples of psychiatric disorders including some with childhood onset and a clear neurodevelopmental origin, such as autism spectrum disorders, specific language impairment, or attention deficit hyperactivity disorders, and others, such as schizophrenia or bipolar disorder, in which developmental features are critical in early-onset cases.

Autism Spectrum Disorder and Social Communication Disorder

Autism Spectrum Disorder (ASD) is a group of pervasive developmental disorders with early childhood onset. Patients with ASD may be characterized by three main features: qualitative impairments in social interactions, qualitatively impaired verbal and nonverbal communication and restricted repetitive and stereotyped patterns of behavior or interests [1]. Recently, DSM-5 has modified the diagnostic criteria reducing the number of core domains from three to two. The new criteria include social communication impairment and restrictive repetitive behaviors (RRBs) [38]. This major change was decided based on repeated findings supporting that differentiation between social and communications criteria was arbitrary [39, 40].

Reported prevalence of ASD has changed over the last decades. While in the 1970s the prevalence was around 2/1000 population, nowadays the latest estimation from the Center for Disease Control and Prevention (CDC) in 2018 is 1.68%

of US children at age 8 [41]. Most authors relate this increase in prevalence with several factors like larger samples in the study, higher early detection, and broader criteria for the diagnosis [11, 41].

Although the etiology of ASD is complex, it has a strong genetic basis that seems to be explained more by multigene interactions, mainly in autistic individuals with unaffected family members, than by rare mutations with major effects. As in other neurodevelopmental disabilities, it is believed that gene–environment interaction is responsible for the final phenotype.

The estimated heritability of autism in a Swedish population-based cohort was found to be 0.80, suggesting that genetic factors explain most of the risk for autism [42, 43]. Several genes, such as NRXN, TSC1/2, CTNTAP2, and SHANK3, have been associated with the phenotype of autism, and their alteration suggests a crucial role of abnormalities in synapse formation as well as neuronal proliferation in this disorder [44]. It is important to note that, although heritability is high in a substantial fraction of ASD cases, most of them are not inherited, with mutations causing the phenotype not being present in the parental genome [45].

Besides the importance of synapse formation, several authors have found an alteration in different genes associated with the immune system. Studying brains postmortem, an increased transcription of these genes was observed. Additionally, several studies suggested the implication of chromosome 6 in the physiology of autism. It is indeed in this chromosome where the genes codifying HLA system are found. An altered HLA is a marker of some autoimmune diseases, and it is believed that, in the case of autism, it may show dysfunction in this system, which implies an important vulnerability to some infectious agents.

Environmental factors have been classically considered to be risk factors for ASD. These factors can contribute at different neurodevelopmental stages. For instance, maternal infection, maternal bleeding, and exposure to medication during pregnancy are relevant prenatal factors while birth injuries or a low APGAR [44], along with maternal hemorrhage, meconium aspiration, and anemia [46], have been associated with a higher risk of autism in the neonatal period. As for postnatal environmental factors, antibiotics use, acetaminophen use, ear infections, and early weaning were associated with increased risk of ASD as well [47].

Comorbidity is common in patients with ASD. According to previous studies, more than 70% of individuals with ASD have concurrent medical, developmental, or psychiatric conditions, which are frequently multiple [48]. Psychiatric comorbidity includes ADHD, which can reach 18% [49], and affective disorders, especially MDD, but also BD [50]. Although studies that examine the developmental trajectories of these comorbid conditions across the lifespan are sparse, a study observed that anxiety symptoms varied throughout the development, increasing from toddlerhood to childhood, then decreasing until young adulthood, where an increase along adulthood was found again [51].

ASD symptoms may present differently across development [52] and core symptoms may differ as well by age and gender [53]. The first symptoms of ASD usually appear early in childhood. Although there are still important delays between appearance of symptoms and diagnosis, children are nowadays more frequently diagnosed at an earlier age than before, likely due to factors such as a better knowledge of atypical development and greater awareness [54]. This allows early intervention, which has a great impact on children with autism [55]. However, less affected patients such as high-functioning children and females can remain undiagnosed longer. Core ASD symptoms often persist into adulthood. Although some improvement can be seen compared to adolescence [52, 56], social integration usually remains poor [56], and patients usually need the support of their families.

Regarding sex differences, males are affected more frequently than females, with a suggested 4:1 ratio. Some explanations for this differential prevalence include sex differences in clinical symptoms, with recent studies suggesting also that the biological male phenotype may itself confers a higher risk for ASD [57], while other results show that females may be underdiagnosed because they camouflage the symptoms [58]. Additionally, some studies have estimated a lower ratio (2:1) among those with comorbid ID and a higher ratio around 6:1 for high-functioning autism [59]. According to a previous metanalysis, boys may show less severe symptoms of repetitive and stereotyped behaviors, while males and females may not differ in the domain of social behavior and communication [53].

DSM-5 has created a new diagnosis to characterize children with social communication impairments with no presence of restricted rigid patterns of behavior, interests, and activities that are needed for ASD diagnosis, called social (pragmatic) communication disorder (SCD) [38].

SCD is a neurodevelopmental disorder, which impacts verbal and nonverbal communication for social purposes, ability of individuals to adapt their communication style and to follow conventional and cultural rules for conversation, as well as to understand implicit or ambiguous language [38]. Among observed difficulties, we may find in these patients alterations in theory of mind understanding and story content organization [60]. These difficulties may lead patients to impairments in their ability to communicate, social participation, and functioning at academic and professional level [38], as a result of a potential misinterpretation of what other people are willing to express or do. As a consequence, anxiety, social isolation, frustration, and behavioral problems may appear [61].

SCD diagnosis has been included in psychiatric nosology recently, so the characterization of its risk factors and physiopathology is still preliminary. The possibility of this social communication disorder being due to a failure in the brain's right hemisphere, which would impair simultaneous processing of verbal and visual information has been suggested [62].

As with other neurodevelopmental disorders, the first symptoms of SCD usually appear in the early developmental period [63], and the deficits are usually noticed when social communication demands surpass limited capacity. Around 5 years old, teachers and classmates may find out these difficulties, and this can confer susceptibility for being bullied or taken advantage of [61].

Diagnoses of ASD and SCD are clinical, ideally made by clinicians with experience in childhood disorders. Nevertheless, validated instruments such as "Autism Diagnostic Observation Schedule, second edition" (ADOS-2) [64] and

"Autism Diagnostic Interview-Revised" (ADI-R) [65] may be of help supporting the diagnostic process. Other comprehensive instruments such as the "Diagnostic Interview for Social and Communication Disorders" (DISCO) [66], which evaluates social communication and restricted and repetitive behavior separately, along with the psychiatric interview, may also be of help in the diagnosis of SCD, which will be established only after ASD is ruled out.

Specific Language Impairment

Language development can be altered by disorders such as hearing impairment, ID, congenital defects or acquired brain damage. When language delay is significant and another cause or associated clinical features are not found, patients receive the diagnosis of primary language disorder or specific language impairment (SLI) [67, 68]. This category is usually heterogeneous, and the severity and profile of the disorder may vary.

SLI is frequent in the general population [68, 69], and it has been stated that it may be the most prevalent developmental disorder [68]. It is known that deficits may appear beyond the linguistic domain, especially in the neuropsychological field, where there is evidence that limitation in phonologic working memory may be a core deficit in this disorder [70]. As for executive functions, children with SLI perform poorly on both inhibitory control tasks and cognitive flexibility tasks [71].

Complex neurogenetic and environmental factors contribute to the onset of this neurodevelopmental disease. Several specific genes have been found to be associated with language development. Among them, some genes including FOXP2 and CNTNAP2 influence language on its physiological acquisition, and others such as ATP2C2, CMIP, and NOP9 are related to specific language disorders [72]. This knowledge has led to the search for candidate genes that could potentially be causative of this disease. Unfortunately, for the moment a single gene that could be the cause of SLI has not been found. However, it has been suggested that regulatory mechanisms influence gene expression at particular developmental windows during child's development [69]. Other neural contributions that have been studied include abnormalities in fronto-corticocerebellar circuits [73]. Also, structural neuroimaging studies suggest that atypical patterns of asymmetry of language cortex, white-matter abnormalities, and cortical dysplasia may be associated with SLI as well [70].

While genes regulate developmental processes such as language acquisition, the environment may modulate genetic expression. Environmental variables related to poor early child language development include male gender, low level of education of the mother and family history of language or psychiatric problems, while bilingualism seems to be a protective factor [72]. As for SLI, it has been observed that children with SLI had younger mothers compared to children without SLI, were less often breastfed, and were more frequently not firstborns [68], while the significant association was not found between SLI and perinatal factors [68].

Most cases of SLI are diagnosed in childhood. However, some of these children do not outgrow their language difficulties. In fact, oral language deficits often persist [67, 74] and children with SLI are outperformed by their non-affected peers in the whole set of spoken language tests [67]. This suggests that neural networks related to their language may not normalize with time [75].

Emotional and behavioral needs have been reported in children with SLI [74], as well as social problems and behavioral difficulties [69, 74, 76]. In a longitudinal study, a decrease in behavioral and emotional problems was observed from childhood to adolescence [76]. While behavioral difficulties appeared to decrease to normative levels by adolescence, emotional problems were still evident in adolescence and social problems increased significantly [76]. It needs to be noticed that SLI is primarily a diagnosis made after the exclusion of other causes for delayed or deviant language, considering the evolutionary moment and main characteristics or difficulties.

Attention Deficit Hyperactivity Disorder

Attention deficit hyperactivity disorder (ADHD) is a neurodevelopmental heterogeneous disorder defined by a persistent pattern of inattention and hyperactivity that interferes with functioning or development [38, 77]. We may find patients with predominantly inattentive presentation, predominantly hyperactive-impulsive presentation or a combined presentation, which, from one another, show different patterns of cognitive and motivational deficits, some of which appear to be linked to delays in brain maturation [2, 78].

Multiple causal pathways and more than one developmental pathway are likely involved in ADHD [79]. Also, different genetic and environmental factors are found in this disorder, which may lead to one or more of the previously mentioned subtypes.

Early-onset ADHD has been associated with polygenic risk and neurocognitive deficits [80]. The high heritability (around 0.6–0.9) needs to be highlighted in this disease, especially in early-onset ADHD, stressing out the underlying importance of genetic factors and biological correlates. Among them, the involvement of brain dopamine systems in the pathogenesis of ADHD is suggested by the significant association found between ADHD and dopamine system genes, especially DRD4 and DRD5 [81]. The serotonergic neurotransmitter system, particularly serotonin receptor 1B, has been reported to be associated with the disorder too [82].

Individuals with ADHD present difficulties in cognitive functions, including alterations in several attentional domains [82]. Other disturbed cognitive functions include executive control of attention and response suppression, reward response, and state regulation implicating arousal or activation function [5].

An altered development of some neuroanatomical elements as the basal ganglia and the cerebellum appears in the disorder [79]. Besides, neuroimaging studies in children with ADHD disclose globally decreased brain volumes [83] and diminished volumes in the right frontal lobe, caudate nucleus, cerebellar hemisphere, and posterior-inferior lobules of the cerebellar vermis [82].

Although ADHD may be conceptualized as a final common manifestation of several neurobiological risks, these may be exacerbated or ameliorated by environmental factors [2, 84]. Prenatal stress has been linked to ADHD and pregnancy is a critical moment in which smoking, alcohol use, or drug use along with exposure to environmental toxins can lead to low birth weight or brain injuries. Additionally, the developmental progression is related to socialization contexts such as early parent–family interactions and later peer and school environments [67]. Among disease moderators, symptom patterns and severity of impairment are strongly associated with family adversity and parenting competence [2]. Childhood temperament, especially high levels of reactivity to the setting and low levels of regulation with great impulsivity and poor attention control have been associated with and may predispose to ADHD [79].

We have more and more data, not only about clinical expression, but also regarding comorbidity and residual adult life impairment states, that ADHD persists in adulthood [80, 85–87]. However, the clinical picture can change from childhood to adulthood and patients may shift from one subtype to another. Hyperactive-impulsive symptoms appear prominently during childhood, decreasing in frequency by adolescence, while inattentive symptoms that are usually noticed at school age are more stable throughout adolescence and adulthood [67]. ADHD in the general adult population is associated with substantial personal and individual burden [86], and patients who experience persistent symptoms that reach adulthood show greater psychiatric comorbidity and functional impairment [85]. Among peculiar characteristics and comorbidity that appears in late-onset ADHD, we find elevated rates of generalized anxiety disorder, conduct disorder, and substance use disorder [88]. ADHD is associated with elevated risk of several personality disorders as well, both cluster B and C, and these co-occurring disorders do not account for neither the impairments experienced by these adults or for their neuropsychological deficits [67].

Regarding sex differences, some studies suggest that young boys have higher rates of overt hyperactivity [77]. Studies of girls with ADHD suggest lower rates of conduct disorder compared with boys with ADHD and less externalizing behaviors, which may translate into less hyperactivity and impulsivity in adulthood [89]. Less disruptive behaviors lead to lower request for help and consequently underdiagnoses and absence of treatment. Thus, in this period, boys are more likely to be referred for childhood ADHD evaluation than girls [77]. On the other side, in late-onset ADHD the proportion of women referred is greater [88] compared to early-onset ADHD. The reason for this could be that they are more likely to show intellectual impairment and internalizing disorders as depressive disorders [90]. However, the fact that in adults women are more likely to seek psychiatric care and counseling in general [77] may influence as well.

Schizophrenia and Early-Onset Psychosis

Psychosis is a common and functionally disruptive clinical manifestation of many psychiatric, neurodevelopmental, and medical situations [91]. Psychotic symptoms have a high prevalence in children and adolescents, suggesting rates of 17% in children aged 9–12 years and 7.5% among adolescents aged 13–18 years [92]. These symptoms can be seen as either denoting an excess or distortion of normal function (positive symptoms) or a reduction or loss of normal function (negative symptoms) [93]. This high frequency and clinical heterogeneity, makes diagnosis in the early stages difficult, although most patients who are diagnosed with schizophrenia at early stages continue to meet criteria two years after [94].

Schizophrenia is a heterogeneous condition and a particularly devastating psychiatric disorder [95], that commonly involves severe cognitive deficits that compromise functional ability [96]. Although extremely rare before the age of 10, the incidence of schizophrenia rises gradually through adolescence to reach its peak in early adult life [93].

Early-onset schizophrenia (EOS) is a rare (1/5000–1/20,000) and severe chronic psychiatric disorder. Although all diagnostic criteria apply, positive symptoms of psychosis such as delusions, hallucinations, and disorganized speech or behavior before the age of 18 are necessary for the diagnosis. Patients have frequently deficits in cognition and communication, as well as neuromotor impairments. In addition, EOS is associated with predominant male gender, a positive family history of schizophrenia and other psychiatric conditions, particularly affective disorders [97].

Schizophrenia is a very highly heritable (80%) condition. Over 1,000 genes have been suggested. This makes schizophrenia one of the most studied disorders through a candidate gene approach [98]. Existing genetic data suggest that multiple genetic variants of varying frequencies contribute to the disorder [99]. A high heritability rate of schizophrenia has also been suggested in early adoption-twin studies and has been confirmed by family aggregation studies. Some of the candidate genes in the pathway of schizophrenia are NRG1, BDNF, DISC1, COMT, DTNBP1, and G72 (DAOA).

A polygenic basis of EOS involves hundreds of genes with small individual effects. In EOS pathogeny, recent studies have demonstrated the contribution of mutations in NRXN1 (2p16.3) and UPF3B (Xq24), which are crucial for development and brain function [97]. Some of the most consistent findings from schizophrenia Genome-Wide Association Studies (GWAS) implicate the role of major histocompatibility complex on chromosome 6, microRNA MIR137 and NGRN [100]. Recently, "de novo" heterozygous mutations (c.385G>A) in the ATP1A3 gene (19q13.2) have been described in this disorder [101].

The two-hit hypothesis of schizophrenia suggests that a combination of genetic susceptibility coupled with distinct developmental insults can lead to the onset of the full clinical syndrome [102]. According to this theory, environmental variables related to obstetric complications, abnormal presentation, and cesarean birth, as well as childhood viral infections or use of substances in the mother may be implicated [103, 104] in the early development of this disorder.

As for later events, migration and childhood abuse or neglect have been related to the diagnosis of schizophrenia. Migrants living in disadvantaged areas are exposed to a range of psychosocial difficulties including increased exposure to drugs, discriminatory experiences, violence, and crime [105]. Cannabis confers an overall increased risk for later schizophrenia, varying in a dose-dependent fashion and is associated with an average of 2.7 years earlier onset of psychotic symptoms [106, 107].

Although the same diagnostic criteria apply across the age range, there are developmental variations in phenomenology, with early-onset cases of schizophrenia characterized by a more insidious onset, negative symptoms, and fewer systematized or persecutory delusions. Schizophrenia in the pediatric population may vary from adult-onset form presentation because of developmental factors. In as much as 67% of children with childhood-onset schizophrenia, there are social, motor, and language premorbid disturbances as well as learning disabilities. Additionally, almost 30% met criteria for ASD prior to the onset of their psychotic symptoms [108]. EOS is characterized by greater disorganization and more negative symptoms, while in later-onset cases there is a higher frequency of systematized and paranoid delusions [109]. EOS usually runs a chronic course, with poor outcomes, progressive decline, and social functioning impairment. The short-term outcome for schizophrenia presenting in early life is worse than for first-episode adult patients. A number of long-term follow-up studies of EOS describe a typically chronic, continuous long-term course with severely impaired functioning in adult life. The predictors of poor outcome in adolescent-onset psychosis include premorbid social and cognitive impairments, a prolonged first psychotic episode, extended duration of untreated psychosis, and the presence of negative symptoms. Premorbid functioning and negative symptoms at onset provide better prediction for long-term outcomes than categorical diagnosis [110].

At-risk mental states (ARMS) describe the clinical characteristics related to highest risk of transition to psychosis or schizophrenia. ARMS are characterized by help-seeking behavior and the presence of attenuated positive symptoms of schizophrenia, brief-limited intermittent psychotic symptoms, or a combination of familial risk indicators with recent functional deterioration [93]. As treating ARMS leads to a significant reduction in the transition rate to a first-episode psychosis [111], identification and referral at this stage can be beneficial.

Bipolar Disorder

BD is characterized by the presence of manic or hypomanic episodes interspersed with periods of depression or euthymia. BD is considered rare during childhood, but the rates increase during adolescence and many bipolar adults recall that their first affective episode began before the age of eighteen [112]. According to the literature, there are no sex differences in rates of bipolar spectrum disorder or any of the bipolar subtypes [113]. However, sex differences are found regarding the clinical presentation, with males having greater rates of mania and females reporting greater rates of depression [114].

The number of children and adolescents diagnosed with BD has increased considerably over the last years [115]. A meta-analysis reported that the prevalence of bipolar spectrum disorders in youth is on average 1.8%, and for bipolar disorder I (BD-I) reached 1.2%, rates of BD-I being consistent among most countries [114, 116, 117].

Like many other psychiatric disorders, BD may be understood through a diathesis-stress model. BD seems to be a complex genetic disorder arising in response to a combination of environmental, genetic, and epigenetic factors [118].

Inherited factors are known to be important in BD. The risk of developing BD in a first degree relative is 5–10% and for monozygotic co-twins is around 40–70% [119]. In a Swedish population-based study they estimated a heritability of 59% [120]. Larger estimated heritability was found in twin studies [121], suggesting great importance of genetic factors in BD. In biological relatives of probands with BD, a significant increase in the risk of developing schizophrenia was found as well, which evidences the substantial genetic relationship between schizophrenia and BD [120].

Robust and replicable significant associations have been reported in genome-wide association studies at several common polymorphisms, including variants within the genes CACNA1C (related to ion channels in brain), ODZ4 (related to cell surface signaling and neuronal pathfinding), and NCAN (a brain-expressed extracellular matrix glycoprotein related behavioral abnormalities in mice) [119].

Many studies have demonstrated a higher risk of BD in the offspring of patients diagnosed with BD. In a recent meta-analysis of a high-risk cohort, 10.1% met criteria for a BD compared to 0.1% of control offspring [122] and the most robust risk factor predicting BD was a positive family history [123].

Some events associated with affective disorders appear at early developmental stages [2, 124], including psychomotor developmental deviations and general adjustment problems [124] and more and more studies underscore the importance of a developmental approach [124, 125], stating that pervasive developmental disorders increase the risk for both SZ and BD, especially in cases with normal intelligence [124]. A recent study about the differences between developmental pathology in BD and SZ underline that, although generally milder than in SZ, developmental pathology in BD is relevant and associated with early onset [125]. For instance, it is likely that sleep disorders confer an increased risk of bipolar disorder spectrum psychopathology [125].

The clinical presentation of pediatric BD has been controversial. Although diagnostic criteria for BD are the same for adults and for youths, the perception that pediatric BD is qualitatively different from adult BD still remains. According to a recent meta-analysis, the most prevalent manic symptoms in youth were increased energy, irritability, mood lability, distractibility, and goal-directed activity. Other frequent symptoms were euphoric/elated mood, pressured speech, hyperactive, racing thoughts, poor judgment, grandiosity, inappropriate laughter, decreased need for sleep, flight of ideas, increased productivity, increased creativity, uninhibited people-seeking, hypersexuality, hallucinations, and delusions [126]. BD is associated with a high risk of suicide and other adverse outcomes and is often initially misdiagnosed as depression [127]. Prospective studies of child and adolescent offspring of bipolar parents have found that BD begins with a depressive episode in 67–88%, often years before the first hypomanic or manic episode [128–130].

Bipolar youths tend to suffer different comorbid disorders, with the highest prevalence rate found for anxiety disorders (54%), ADHD (48%), disruptive behavior disorders (31%), and substance use disorders (31%) [131]. Besides being most prevalent comorbid disorders in BD, anxiety disorders and ADHD have been reported to precede and predict BD in family high-risk studies [132]. In a recent Danish study, it was found that the combination of ADHD and anxiety increased the risk of developing BD 30-fold compared with those with no prior history of ADHD or anxiety [133].

Diagnostic and Therapeutic Implications

Diagnosis, which in psychiatry is clinical, should be prompt. It is essential to attend to biological, psychological, and emotional factors associated with child development. A good diagnostic procedure should include an evaluation of personality, emotional intelligence, and affectivity as well as an exhaustive study of developmental milestones and of the current developmental status. It is important to consider any subtle alteration in development as a possible risk factor, especially in potentially vulnerable children such as those with a significant familiar psychiatric history or with an ID.

For an accurate diagnosis, we need to consider the peculiar characteristics of the different symptoms' appearance according to the evolutionary period. A predominance of somatic and behavioral response systems is the norm in childhood, while cognitive manifestations of those symptoms appear more typically in adolescence. As for anxiety, a change of trend may be found along development, with small children complaining of physical symptoms and adults suffering from repetitive thoughts and ruminations. Also, some disorders such as ADHD and ASD are more frequently diagnosed along childhood while psychotic and affective disorders are usually diagnosed later in life.

Although childhood is not risk-free, most symptoms associated with mental health disorders as BD and schizophrenia first emerge during adolescence, with a negative impact that can continue into adulthood. However, most of the children who will develop early-onset psychotic disorders already show alterations in development including abnormalities in language and motor development [134] and cognitive and adjustment to the environment impairments [124].

Especially in young children, who may not have enough cognitive level to understand and answer clinicians' questions, it is essential to collect information from different sources, including not only patients but also school and parents, and to integrate this information, which might not be totally concordant. In any case, the diagnostic process should include a mental state examination and, when needed, neuropsychological and supplementary tests, using age-appropriate instruments.

An accurate assessment and diagnostic formulation is essential. Internationally validated diagnostic criteria such as DSM-5 [38] and ICD-10 [135] are the diagnostic standard. Screening instruments and structured clinical diagnostic interviews are complementary to diagnostic interviews by trained clinicians with experience in child and adolescent psychiatry.

Treatment should be initiated as soon as possible, but it is unlikely to be effective if the clinical problem has not been characterized adequately. Although treatment targets are often symptomatic and may not point to the causal pathophysiological mechanisms, treatment plans should be comprehensive and consider psychiatric and somatic comorbidities as well as potential treatment complications. Early intervention and youth mental health research focusing on at-risk populations may provide insights on later severe mental illness development. Preliminary data from prevention programs for anxiety, psychosis, and depressive symptoms also point toward evidence of effectiveness of these interventions on promoting mental health outcomes [136].

According to established diagnoses, different interventions should be initiated. In ASD there are no pharmacological treatments for core symptoms, so stimulation and psychosocial intervention should be started [137]. In psychosis and bipolar disorder, atypical antipsychotic drugs are indicated as most of the evidence point to their usefulness [138]. In ADHD, evidence states that stimulants, the treatment of choice, and alpha-2 agonists, can reduce the symptoms of ADHD, but also environmental changes should be arranged [77].

Although other medications such as selective serotonin reuptake inhibitors (SSRIs) have proven effective for other pediatric disorders such as depression, anxiety or obsessive-compulsive disorders, research suggests that psychotherapy can have a significant and positive impact on children and adolescents with these mental health problems [139]. There are now several tested treatments that have proven to be useful, like cognitive behavioral therapy and family therapy, and others are under evaluation [93].

Development of pediatric psychopharmacology typically lags behind adult pharmacology. Comparatively, there are fewer studies in minors, and therefore children are frequently prescribed psychiatric drugs off-label, based on information derived from adult studies. This adds to the fact that pharmacokinetics, the way human body absorbs, distributes, and metabolizes drugs change along development, with rapid hepatic clearance often dominating at early ages. About pharmacodynamics, since a single drug may activate multiple receptor types with markedly different effects, developmental changes in the number and proportion of these receptors may account for differential drug effects on children and adults [114]. Differential effects are seen not only for therapeutic effects but also for adverse events [140], which makes frequent monitoring essential in minors undergoing psychopharmacological treatment. There are currently a number of international initiatives aimed at promoting pharmacological research in children and adolescents that will hopefully better inform the effective and safe use of psychiatric medications in children and adolescents.

References

1. Rutter M, Bishop DVM, Pine DS, Scott S, Stevenson J, Taylor E, et al. Rutter's child and adolescent psychiatry, 5th ed. 2009.
2. Lewis M, Rudolph KD editors. Handbook of developmental psychopathology, 3rd ed. 2014.
3. Kiser DP, Rivero O, Lesch K. Annual research review: the (epi) genetics of neurodevelopmental disorders in the era of whole-genome sequencing–unveiling the dark matter. J Child Psychol Psychiatry. 2015;3:278–95.
4. Cardno AG, Owen MJ. Genetic relationships between schizophrenia, bipolar disorder, and schizoaffective disorder. Schizophr Bull. 2014;40(3):504–15.
5. Group C, Consortium PG. Genetic relationship between five psychiatric disorders estimated from genome-wide SNPs. Nat Genet. 2013;45(9):984–94.
6. Korte A, Farlow A. The advantages and limitations of trait analysis with GWAS: a review. Plant Methods. 2013;9:29.
7. Boyle EA, Li YI, Pritchard JK. An expanded view of complex traits: from polygenic to omnigenic. Cell. 2017;169(7):1177–86.
8. Anttila V, Bulik-Sullivan B, Finucane HK, Walters RK, Bras J, Duncan L. Analysis of shared heritability in common disorders of the brain. Science. 2018;360(6395).
9. Smoller JW. Disorders and borders : psychiatric genetics and nosology. Am J Med Genet. 2013;162(B);559–78.
10. Hu WF, Chahrour MH, Walsh CA. The diverse genetic landscape of neurodevelopmental disorders. Annu Rev Genomics Hum Genet. 2014;15:195–213.
11. Parellada M, Penzol MJ, Pina L, Moreno C, González-Vioque E, Zalsman G, et al. The neurobiology of autism spectrum disorders. Eur Psychiatry. 2014;29(1):11–9.
12. Gaugler T, Klei L, Sanders SJ, Bodea CA, Goldberg AP, Lee AB, et al. Most genetic risk for autism resides with common variation. Nature. 2014;46(8):881–5.
13. Gilman SR, Iossifov I, Levy D, Ronemus M, Wigler MVD. Rare de novo variants associated with autism implicate a large functional network of genes involved in formation and function of synapses. Neuron. 2011;70:898–907.
14. Abbott PW, Gumusoglu SB, Bittle J, Beversdorf DQ. Psychoneuroendocrinology Prenatal stress and genetic risk: how prenatal stress interacts with genetics to alter risk for psychiatric illness. Psychoneuroendocrinology. 2018;90:9–21.
15. Buxbaum JDe Rubeis S HX. Synaptic, transcriptional, and chromatin genes disrupted in autism. Nature. 2014;515(7526):209–15.
16. State MW. The emerging biology of autism. Science. 2012;337:1301–4.
17. Lasalle JM, Powell WT, Yasui DH. Epigenetic layers and players underlying neurodevelopment. Trends Neurosci. 2013;36(8):460–70.
18. Winfree LT, Taylor TJ, He N, Esbensen FA. Self-control and variability over time: multivariate results using a 5-year, multisite panel of youths. Crime and Delinquency. 2006;52(2):253–86.
19. Danese A, McEwen BS. Adverse childhood experiences, allostasis, allostatic load, and age-related disease. Physiol Behav. 2012;106(1):29–39.
20. DeYoung CG, Cicchetti D, Rogosch FA. Moderation of the association between childhood maltreatment and neuroticism by the corticotropin-releasing hormone receptor 1 gene. J Child Psychol Psychiatry. 2011;52(8):898–906.

21. Andersen SL. Trajectories of brain development: point of vulnerability or window of opportunity? Neurosci Biobehav Rev. 2003;27:3–18.
22. Rees S, Harding R. Brain development during fetal life: influences of the intra-uterine environment. Neurosci Lett. 2004;361(6):111–4.
23. Adams Waldorf KM, McAdams RM. Influence of infection during pregnancy on fetal development. Reproduction. 2013;146(5):151–62.
24. Pallarés ME, Antonelli MC. Prenatal stress and neurodevelopmental plasticity: relevance to psychopathology. Adv Exp Med Biol. 2017;1015:117–129.
25. Yan X, Zhao X, Li J, He L, Xu M. Effects of early-life malnutrition on neurodevelopment and neuropsychiatric disorders and the potential mechanisms. Prog Neuropsychopharmacol Biol Psychiatry. 2018;20(83):64–75.
26. Lam LF, Lawlis TR. Feeding the brain–the effects of micronutrient interventions on cognitive performance among school-aged children: a systematic review of randomized controlled trials. Clin Nutr. 2017;36(4):1007–14.
27. Perry BD. The neurodevelopmental impact of violence in childhood. Child Adolesc Forensic Psychiatry. 2001.
28. Danese A, Moffitt TE, Arseneault L, Bleiberg BA, Dinardo PB, Gandelman SB, et al. The origins of cognitive deficits in victimized children: implications for neuroscientists and clinicians. Am J Psychiatry. 2017;174(4):349–61.
29. Schaefer JD, Moffitt TE, Arseneault L, Danese A, Fisher HL, Houts R, et al. Adolescent victimization and early-adult psychopathology: approaching causal inference using a longitudinal twin study to rule out noncausal explanations. Clin Psychol Sci. 2018;6(3):352–71.
30. Kyzar EJ, Floreani C, Teppen TL, Pandey SC. Adolescent alcohol exposure: burden of epigenetic reprogramming, synaptic remodeling, and adult psychopathology. Front Neurosci. 2016;10.
31. Fearon P. Reimagining the environment in developmental psychopathology: from molecules to effective interventions. J Child Psychol Psychiatry. 2018;59(4):299–302.
32. Humphreys KL, Zeanah CH. Deviations from the expectable environment in early childhood and emerging psychopathology. Neuropsychopharmacology. 2015;40(1):154–70.
33. Siller M, Sigman M. The behaviors of parents of children with autism predict the subsequent development of their children's communication. J Autism Dev Disord. 2002;32(2):77–89.
34. Fonagy P, Luyten P. Conduct problems in youth and the RDoC approach: a developmental, evolutionary-based view. Clin Psychol Rev. 2018;64:57–76.
35. Silveri MM, Dager AD, Cohen-Gilbert JE, Sneider JT. Neurobiological signatures associated with alcohol and drug use in the human adolescent brain. Neurosci Biobehav Rev. 2016;70:244–59.
36. Crews FT, Vetreno RP, Broadwater MA, Robinson DL. Adolescent alcohol exposure persistently impacts adult neurobiology and behavior. Pharmacol Rev. 2016;68(4):1074–109.
37. Crane NA, Schuster RM, Fusar-Poli P, Gonzalez R. Effects of cannabis on neurocognitive functioning: Recent advances, neurodevelopmental influences, and sex differences. Neuropsychology. 2013;23(2):117–37.
38. American Psychiatric Association. Diagnostic and statistical manual of mental disorders, 5th ed. (DSM-5); 2013 Diagnostic Stat Man Ment Disord. 4th ed. TR.
39. Lord C, Bishop S. Recent advances in autism research as reflected in Dsm-5 criteria for autism spectrum disorder. SSRN. 2015.
40. Doernberg E, Hollander E. Neurodevelopmental disorders (ASD and ADHD): DSM-5, ICD-10, and ICD-11. CNS Spectr. 2016;21(4):295–9.
41. Hill P. Editorial: the rising prevalence of autism. J Child Psychol Psychiatry. 2018;7:717–20.
42. Sandin S, Lichtenstein P, Kuja-Halkola R, Hultman C, Larsson HRA. The heritability of autism spectrum disorder analysis method B. J Am Med Assoc. 2017;318(12):1182–4.

43. Colvert E, Tick B, Mcewen F, Stewart C, Curran SR, Woodhouse E, et al. Heritability of autism spectrum disorder in a UK population-based twin sample. JAMA Psychiatry. 2015;72(5):415–23.
44. Chaste P, Leboyer M. Autism risk factors: genes, environment, and gene-environment interactions. Dialogues Clin Neurosci. 2012;14(3):281–92.
45. Beaudet AL. Autism: highly heritable but not inherited. Nat Med. 2007;13(5):534–6.
46. Gardener H, Spiegelman D, Buka SL. Perinatal and neonatal risk factors for autism: a comprehensive meta-analysis. Pediatrics. 2011;128(2):344–55.
47. Bittker SS, Bell KR. Acetaminophen, antibiotics, ear infection, breastfeeding, vitamin D drops and autism: an epidemiological study. Neuropsychiatr Dis Treat. 2018;14:1399–414.
48. Lai M-C, Lombardo MV, Baron-Cohen S. Autism. Lancet. 2014;383(9920):896–910.
49. Mannion A, Leader G, Healy O. An investigation of comorbid psychological disorders, sleep problems, gastrointestinal symptoms and epilepsy in children and adolescents with Autism Spectrum Disorder. Res Autism Spectr Disord. 2013;7(1):35–42.
50. Munesue T, Ono Y, Mutoh K, Shimoda K, Nakatani H, Kikuchi M. High prevalence of bipolar disorder comorbidity in adolescents and young adults with high-functioning autism spectrum disorder: a preliminary study of 44 outpatients. J Affect Disord. 2008;111(2–3):170–5.
51. Davis TE, Hess JA, Moree BN, Fodstad JC, Dempsey T, Jenkins WS, et al. Anxiety symptoms across the lifespan in people diagnosed with Autistic Disorder. Res Autism Spectr Disord. 2011;5(1):112–8.
52. Tillmann J, Ashwood K, Absoud M, Bölte S, Bonnet-Brilhault F, Buitelaar JK, et al. Evaluating sex and age differences in ADI-R and ADOS scores in a large European multi-site sample of individuals with autism spectrum disorder. J Autism Dev Disord. 2018;48(7):2490–505.
53. Van Wijngaarden-Cremers PJM, Van Eeten E, Groen WB, Van Deurzen PA, Oosterling IJ, Van Der Gaag RJ. Gender and age differences in the core triad of impairments in autism spectrum disorders: a systematic review and meta-analysis. J Autism Dev Disord. 2014;44(3):627–35.
54. Lai M, Lombardo M, Baron-Cohen S. Autism. Lancet. 2014;383(8):896–910.
55. Sacrey LAR, Bennett JA, Zwaigenbaum L. Early infant development and intervention for autism spectrum disorder. J Child Neurol. 2015;30(14):1921–9.
56. Howlin P, Moss P, Savage S, Rutter M. Social outcomes in mid- to later adulthood among individuals diagnosed with autism and average nonverbal IQ as children. J Am Acad Child Adolesc Psychiatry. 2013;52(6):572–81.
57. Ecker C, Andrews DS, Gudbrandsen CM, Marquand AF, Ginestet CE, Daly EM, et al. Association between the probability of autism spectrum disorder and normative sex-related phenotypic diversity in brain structure. JAMA Psychiatry. 2017;74(4):329–38.
58. Rynkiewicz A, Schuller B, Marchi E, Piana S, Camurri A, Lassalle A, et al. An investigation of the "female camouflage effect" in autism using a computerized ADOS-2 and a test of sex/gender differences. Mol Autism. 2016;21(7):10.
59. Kirkovski M, Enticott PG, Fitzgerald PB. A review of the role of female gender in autism spectrum disorders. J Autism Dev Disord. 2013;43(11):2584–603.
60. Ketelaars MP, Jansonius K, Cuperus J, Verhoeven L. Narrative competence in children with pragmatic language impairment: a longitudinal study. Int J Lang Commun Disord. 2016;51(2):162–73.
61. Simms MD, Jin XM. Autism, language disorder, and social (Pragmatic) communication disorder: DSM-V and differential diagnoses. Pediatr Rev. 2015;36(8):355–63.
62. McDonald S. Exploring the cognitive basis of right-hemisphere pragmatic language disorders. Brain Lang. 2000;75(1):82–107.
63. Baird G, Norbury CF. Social (pragmatic) communication disorders and autism spectrum disorder. Arch Dis Child. 2016;101(8):745–51.

64. Lord C, Risi S, Lambrecht L, Cook EH, Leventhal BL, DiLavore PC, et al. Autism diagnostic observation schedule (ADOS). J Autism Dev Disord. 2000;63(6):694–701.
65. Lord C, Rutter M, Le Couteur A. Autism diagnostic interview-revised: a revised version of a diagnostic interview for caregivers of individuals with possible pervasive developmental disorders. J Autism Dev Disord. 1994;24(5):659–85.
66. Brukner-Wertman Y, Laor N, Golan O. Social (Pragmatic) communication disorder and its relation to the autism spectrum: dilemmas arising from the DSM-5 classification. J Autism Dev Disord. 2016;46(8):2821–9.
67. Cicchetti D, Cohen DJ. Developmental psychopathology, Risk, disorder, and adaptation, vol. 3, 2nd ed. 2006.
68. Diepeveen FB, van Dommelen P, Oudesluys-Murphy AM, Verkerk PH. Specific language impairment is associated with maternal and family factors. Child Care Health Dev. 2017;43(3):401–5.
69. Rice ML. Children with specific language impairment and their families: a future view of nature plus nurture and new technologies for comprehensive language intervention strategies. Semin Speech Lang. 2016;37(4):310–8.
70. Webster RI, Shevell MI. Neurobiology of specific language impairment. J Child Neurol. 2004;19(7):471–81.
71. Pauls LJ, Archibald LMD. Executive functions in children with specific language impairment: a meta-analysis. Am J Speech Lang Pathol. 2016;59:1074–86.
72. Moriano-Gutierrez A, Colomer-Revuelta J, Sanjuan J, Carot-Sierra JM. Environmental and genetic variables related with alterations in language acquisition in early childhood. Rev Neurol. 2017;64(1):31–7.
73. Hodge SM, Makris N, Kennedy DN, Caviness VS, Howard J, McGrath L, et al. Cerebellum, language, and cognition in autism and specific language impairment. J Autism Dev Disord. 2010;40(3):300–16.
74. Helland WA, Helland T. Emotional and behavioural needs in children with specific language impairment and in children with autism spectrum disorder: The importance of pragmatic language impairment. Res Dev Disabil. 2017;70:33–9.
75. Plante E, Patterson D, Sandoval M, Vance CJ, Asbjørnsen AE. An fMRI study of implicit language learning in developmental language impairment. NeuroImage Clin. 2017;14:277–85.
76. St Clair MC, Pickles A, Durkin K, Conti-Ramsden G. A longitudinal study of behavioral, emotional and social difficulties in individuals with a history of specific language impairment (SLI). J Commun Disord. 2011;44(2):186–99.
77. Mattingly GW, Wilson J, Rostain AL. A clinician's guide to ADHD treatment options. Postgrad Med. 2017;129(7):657–66.
78. Gallo EF, Posner J. Moving towards causality in attention-deficit hyperactivity disorder: overview of neural and genetic mechanisms. Lancet Psychiatry. 2016;3(6):555–67.
79. Nigg JT, Goldsmith HH, Sachek J. Temperament and attention deficit hyperactivity disorder: the development of a multiple pathway model. J Clin Child. 2004;33(1):42–53.
80. Moffitt TE, Houts R, Asherson P, Belsky DW, Corcoran DL, Hammerle M, et al. Is adult ADHD a childhood-onset neurodevelopmental disorder? Evidence from a four-decade longitudinal cohort study. Am J Psychiatry. 2015;172(10):967–77.
81. Li D, Sham PC, Owen MJ, He L. Meta-analysis shows significant association between dopamine system genes and attention deficit hyperactivity disorder (ADHD). Hum Mol Genet. 2006;15(14):2276–84.
82. Curatolo P, Paloscia C, D'Agati E, Moavero R, Pasini A. The neurobiology of attention deficit/hyperactivity disorder. Eur J Paediatr Neurol. 2009;13(4):299–304.
83. Krain AL, Castellanos FX. Brain development and ADHD. Clin Psychol Rev. 2006;26(4):433–44.
84. Thapar A, Cooper M, Jefferies R, Stergiakouli E. What causes attention deficit hyperactivity disorder? Arch Dis Child. 2012;97(3):260–5.

85. Biederman J, Petty CR, Evans M, Small J, Faraone SV. How persistent is ADHD? a controlled 10-year follow-up study of boys with ADHD. Psychiatry Res. 2010;177(3):299–304.
86. Katzman MA, Bilkey TS, Chokka PR, Fallu A, Klassen LJ. Adult ADHD and comorbid disorders: clinical implications of a dimensional approach. BMC Psychiatry. 2017;17(1):302.
87. London AS, Landes SD. Attention deficit hyperactivity disorder and adult mortality. Prev Med (Baltim). 2016;90:8–10.
88. Agnew-Blais JC, Polanczyk GV, Danese A, Wertz J, Moffitt TE, Arseneault L. Evaluation of the persistence, remission, and emergence of attention-deficit/hyperactivity disorder in young adulthood. JAMA Psychiatry. 2016;73(7):713–20.
89. Wilens TE, Biederman J, Faraone SV, Martelon M, Westerberg D, Spencer TJ. Presenting ADHD symptoms, subtypes, and comorbid disorders in clinically referred adults with ADHD. J Clin Psychiatry. 2009;70(11):1557–62.
90. Gaub M, Carlson CL. Gender differences in ADHD: a meta-analysis and critical review: The scientific study of child and adolescent psychopathology: the attention deficit disorder syndrome (English). J Am Acad Child Adolesc Psychiatry. 1997;36(8):1036–45.
91. Arciniegas DB. Psychosis. Continuum (Minneap Minn). 2015;21(3):715–36.
92. Kelleher I, Connor D, Clarke MC, Devlin N, Harley M, Cannon M. Prevalence of psychotic symptoms in childhood and adolescence: a systematic review and meta-analysis of population-based studies. Psychol Med. 2012;42(9):1857–63.
93. Thapar A, Pine DS, Leckman JF, Scott S, Snowling MJ, Taylor E. Rutter's child and adolescent psychiatry, 6th ed. 2015.
94. Castro-Fornieles J, Baeza I, De La Serna E, Gonzalez-Pinto A, Parellada M, Graell M, et al. Two-year diagnostic stability in early-onset first-episode psychosis. J Child Psychol Psychiatry Allied Discip. 2011;52(10):1089–98.
95. Chen L, Selvendra A, Stewart A, Castle D. Risk factors in early and late onset schizophrenia. Compr Psychiatry. 2018;80:155–62.
96. Shafee R, Nanda P, Padmanabhan JL, Tandon N, Alliey-Rodriguez N, Kalapurakkel S, et al. Polygenic risk for schizophrenia and measured domains of cognition in individuals with psychosis and controls. Transl Psychiatry. 2018;8(1):78.
97. Fernandez A, Dor E, Maurin T, Laure G, Menard ML, Drozd M, Poinso F, Bardoni B, Askenazy F, Thümmler S. Exploration and characterisation of the phenotypic and genetic profiles of patients with early onset schizophrenia associated with autism spectrum disorder and their first degree relatives: a French multicentre case series study protocol (GenAuDiss). BMJ Open. 2018;8(7):e023330.
98. Gejman PV, Sanders AR, Kendler KS. Genetics of schizophrenia: new findings and challenges. Annu Rev Genomics Hum Genet. 2011;12:121–44.
99. Sullivan PF, Daly MJ, O'Donovan M. Genetic architectures of psychiatric disorders: the emerging picture and its implications. Nat Rev Genet. 2012;13:537.
100. Wright C, Turner JA, Calhoun VD, Perrone-Bizzozero N. Potential impact of miR-137 and its targets in schizophrenia. Front Genet. 2013;26(4):58.
101. Smedemark-Margulies N, Brownstein CA, Vargas S, Tembulkar SK, Towne MC, Shi J, et al. A novel de novo mutation in *ATP1A3* and childhood-onset schizophrenia. Mol Case Stud. 2016;2(5):a001008.
102. Feigenson KA, Kusnecov AW, Silverstein SM. Inflammation and the two hit hypothesis of schizophrenia. Neurosci Biobehav Rev. 2014;38:72–93.
103. Cannon M, Caspi A, Moffitt TE, Harrington H, Taylor A, Murray RM, et al. Evidence for early-childhood, pan-developmental impairment specific to schizophreniform disorder: results from a longitudinal birth cohort. Arch Gen Psychiatry. 2002;59(5):449–56.
104. Ordonez AE, Bobb A, Greenstein D, Baker N, Sporn A, Lenane M, et al. Lack of evidence for elevated obstetric complications in childhood onset schizophrenia. Biol Psychiatry. 2005;58(1):10–5.
105. Cantor-Graae E, Selten JP. Schizophrenia and migration: a meta-analysis and review. Am J Psychiatry. 2005;162(1):12–24.

106. Arseneault L, Cannon M, Witton J, Murray RM. Causal association between cannabis and psychosis: examination of the evidence. Br J Psychiatry. 2004;184:110–7.
107. Large M. Cannabis use and earlier onset of psychosi: a systematic meta-analysis cannabis use and earlier onset of psychosis. Arch Gen Psychiatry. 2011;68(6):555.
108. Driver DI, Gogtay N, Rapoport JL. Childhood onset schizophrenia and early onset schizophrenia spectrum disorders. Child Adolesc Psychiatr Clin N Am. 2013;22(4):539–55.
109. Kodish I, McClellan JM. Early onset schizophrenia. Dulcan's textbook of child and adolescent psychiatry, 2nd ed. 2016.
110. Fleischhaker C, Schulz E, Tepper K, Martin M, Hennighausen K, Remschmidt H. Long-term course of adolescent schizophrenia. Schizophr Bull. 2005;31(3):769–80.
111. Nieman DH, McGorry PD. Detection and treatment of at-risk mental state for developing a first psychosis: making up the balance. Lancet Psychiatry. 2015;2(9):825–34.
112. Perlis RH, Miyahara S, Marangell LB, Wisniewski SR, Ostacher M, DelBello MP, et al. Long-term implications of early onset in bipolar disorder: data from the first 1000 participants in the systematic treatment enhancement program for bipolar disorder (STEP-BD). Biol Psychiatry. 2004;55(9):875–81.
113. Duax JM, Youngstrom EA, Calabrese JR, Findling RL. Sex differences in pediatric bipolar disorder. J Clin Psychiatry. 2007;68(10):1565–73.
114. Merikangas KR, Cui L, Kattan G, Carlson GA, Youngstrom EA, Angst J. Mania with and without depression in a community sample of US adolescents. Arch Gen Psychiatry. 2012;69(9):943–51.
115. Moreno C, Laje G, Blanco C, Jiang H, Schmidt AB, Olfson M. National trends in the outpatient diagnosis and treatment of bipolar disorder in youth. Arch Gen Psychiatry. 2007;64(9):1032–9.
116. Stringaris A, Santosh P, Leibenluft E, Goodman R. Youth meeting symptom and impairment criteria for mania-like episodes lasting less than four days: an epidemiological enquiry. J Child Psychol Psychiatry. 2010;51(1):31–8.
117. Van Meter AR, Moreira ALR, Youngstrom EA. Meta-analysis of epidemiologic studies of pediatric bipolar disorder. J Clin Psychiatry. 2011;72(9):1250–6.
118. Petronis A. Epigenetics and bipolar disorder: new opportunities and challenges. Am J Med Genet C Semin Med Genet. 2003;123C(1):65–75.
119. Craddock N, Sklar P. Bipolar disorder 1-genetics of bipolar disorder. Lancet. 2013;381(9878):1654–62.
120. Lichtenstein P, Yip BH, Bjork C, Pawitan Y, Cannon TD, Sullivan PF, et al. Common genetic determinants of schizophrenia and bipolar disorder in Swedish families: a population-based study. Lancet. 2009;373(9659):234–9.
121. Kieseppä T, Partonen T, Haukka J, Kaprio J, Lönnqvist J. High concordance of bipolar I disorder in a nationwide sample of twins. Am J Psychiatry. 2004;161(10):1814–21.
122. Lau P, Hawes DJ, Hunt C, Frankland A, Roberts G, Mitchell PB. Prevalence of psychopathology in bipolar high-risk offspring and siblings: a meta-analysis. Eur Child Adolesc Psychiatry. 2017;27(7):823–37.
123. Gottesman II, Laursen TM, Bertelsen A, Mortensen PB. Severe mental disorders in offspring with 2 psychiatrically ill parents. Arch Gen Psychiatry. 2010;67(3):252–7.
124. Parellada M, Gomez-Vallejo S, Burdeus M, Arango C. Developmental differences between schizophrenia and bipolar disorder. Schizophr Bull. 2017;43(6):1176–1189.
125. Duffy A, Horrocks J, Doucette S, Keown-Stoneman C, McCloskey S, Grof P. The developmental trajectory of bipolar disorder. Br J Psychiatry. 2014;204(2):122–8.
126. Van Meter AR, Burke C, Kowatch RA, Findling RL, Youngstrom EA. Ten-year updated meta-analysis of the clinical characteristics of pediatric mania and hypomania. Bipolar Disord. 2016;18(1):19–32.
127. Grande I, Berk M, Birmaher B, Vieta E. Bipolar disorder. Lancet. 2016;387(10027):1561–72.

128. Axelson D, Goldstein B, Goldstein T, Monk K, Yu H, Hickey MB, et al. Diagnostic precursors to bipolar disorder in offspring of parents with bipolar disorder: a longitudinal study. Am J Psychiatry. 2015;172(7):638–46.
129. Mesman E, Nolen WA, Reichart CG, Wals M, Hillegers MHJ. The dutch bipolar offspring study: 12-year follow-up. Am J Psychiatry. 2013;170(5):542–9.
130. Hafeman DM, Merranko J, Axelson D, Goldstein BI, Goldstein T, Monk K, et al. Toward the definition of a bipolar prodrome: dimensional predictors of bipolar spectrum disorders in at-risk youths. Am J Psychiatry. 2016;173(7):695–704.
131. Frías Á, Palma C, Farriols N. Comorbidity in pediatric bipolar disorder: prevalence, clinical impact, etiology and treatment. J Affect Disord. 2015;15(174):378–89.
132. Duffy A, Alda M, Hajek T, Sherry SB, Grof P. Early stages in the development of bipolar disorder. J Affect Disord. 2010;121(1–2):127–35.
133. Meier SM, Pavlova B, Dalsgaard S, Nordentoft M, Mors O, Mortensen PB, et al. Attention-deficit hyperactivity disorder and anxiety disorders as precursors of bipolar disorder onset in adulthood. Br J Psychiatry. 2018;213(3):555–60.
134. Cannon TD, Van Erp TGM, Rosso IM, Huttunen M, Lönnqvist J, Pirkola T, et al. Fetal hypoxia and structural brain abnormalities in schizophrenic patients, their siblings, and controls. Arch Gen Psychiatry. 2002;59(1):35–41.
135. World Health Organization. The ICD-10 classification of mental and behavioural disorders. Int Classif. 1992.
136. Min JA, Lee CU, Lee C. Mental health promotion and illness prevention: a challenge for psychiatrists. Psychiatry Investig. 2013;10(4):307–16.
137. National Collaborating Centre for Mental Health (UK). The management and support of children and young people on the autism spectrum. NICE clinical guideline. 2013.
138. Datta SS, Kumar A, Wright SD, Furtado VA, Russell PS. Evidence base for using atypical antipsychotics for psychosis in adolescents. Schizophr Bull. 2014;40(2):252–4.
139. Kornør H, Bergman H, Hanssen-Bauer K, Soares-Weiser K, Tollefsen TK, Bjørndal A. Client feedback in psychological therapy for children and adolescents with mental health problems. Cochrane Database Syst Rev. 2018;8:CD011729.
140. Correll CU. Assessing and maximizing the safety and tolerability of antipsychotics used in the treatment of children and adolescents. J Clin Psychiatry. 2008;69(Suppl 4):26–36.

Concepts and Dysfunctions of Emotion in Neuropsychiatric Research

Zumrut Duygu Sen, Lejla Colic, Vanessa Kasties and Martin Walter

Abstract

This chapter aims to provide a perspective of the complex formation of emotion and its operational usage in neuroscience. In the first section, the essence and function of emotion will be introduced from different perspectives. After an overview of historical and ongoing debates in the second section, the neuroscientific findings regarding emotional instances in healthy subjects and psychiatric patients will be outlined throughout the third and fourth sections. In the last section, a comprehensive approach of the newly developing field of computational psychiatry to emotion will be introduced.

Keywords

Emotion · Affect · Affective reactivity · Affective bias · Theory of basic emotions · Theory of constructed emotions

Z. D. Sen · V. Kasties · M. Walter (✉)
Department of Psychiatry and Psychotherapy, University of Tübingen, 72076 Tübingen, Germany
e-mail: martin.walter@uni-tuebingen.de

L. Colic · M. Walter
Leibniz Institute for Neurobiology, 39118 Magdeburg, Germany

M. Walter
Max Planck Institute for Biological Cybernetics Tübingen, 72076 Tübingen, Germany

M. Walter
Center for Behavioral Brain Sciences, 39106 Magdeburg, Germany

M. Walter
Department of Psychiatry and Psychotherapy, University of Jena, D-07743 Jena, Germany

© Springer Nature Singapore Pte Ltd. 2019

Y.-K. Kim (ed.), *Frontiers in Psychiatry*, Advances in Experimental Medicine and Biology 1192, https://doi.org/10.1007/978-981-32-9721-0_22

Introduction

Emotion is one of the phenomena that everyone has an intuitive understanding of. A feeling of certainty accompanies our own subjective experience such as for other complex subjective phenomena, for example, being alive or conscious. Think for a moment how much emotions shape our daily subjective and interpersonal experiences, our behaviour and decisions, even in situations when we are not aware of them. However, defining emotion is not as easy as it seems at first glance. Is emotion only how we subjectively experience? What about the things that we do not experience consciously, yet elicit peripheral physiological reactions? It is widely accepted that emotion has a complex formation, namely, that it has multiple components, which cannot be reduced to one another. For example, facial and bodily expressions are somatic components of emotion; thereby they present our affective state to the world and play a dominant role in social functions. But are a smile and happiness the same thing? Obviously, they are not the "same" thing, but they could possibly be components or results of a "same" thing.

It may appear trivial at a first glance, but, as the previous example showed, it is important to delineate the elements that constitute emotion and the elements that result from emotion. For example, if the subjective experience was merely a result of emotion (an epiphenomenon), an emotional instance could happen without an accompanying observational change. Therefore, even though the subjective experience could guide to tracing emotion, it cannot serve as a reliable indicator of an emotional instance, because it would not be an essential element of emotion. Imagine a situation, where a person feels happy, but measures of the autonomic nervous system (ANS), in contrast, indicate a threatening condition, while the person's face shows no emotion. Is she fooling herself, thinking that she is happy, although she is afraid? Many neuroscientists would agree that this individual is much more fearful than happy. Nevertheless, can we claim that feeling fearful is not a necessary element of the emotion of fear? For some researchers, the answer would simply be "yes". However, other theorists would prioritize subjective experience of the individual in their considerations. Indeed, the main historical and current debate on the nature of emotion is principally about the demonstration of the necessary and sufficient components. Interestingly, answers to these questions are becoming a part of the competition between technology companies in the recognition and categorization of emotional instances that will open the barrier between artificial intelligence and social cognition.

Although there is no consensus on the nature of emotion, many philosophers and scientists agree that emotion has a vital role in the functioning of an organism [1, 2]. If we agree that emotion is elicited by either an internal or external incentive, we also need to accept its intentional character. In other words, emotion presents a complex form of information about significant stimuli, which furthermore shapes our attention, memory and behavioural actions. Since antiquity, there has been a notion that the interference of emotion with cognition and the influence on behavioural actions should be overcome. Yet, these interactions are not constant throughout a lifetime. Humans learn to utilize emotion according to their immediate and future needs, following the underlying biological development from birth to adulthood. In that sense, psychiatric disorders impede patients' ability to produce appropriate affective responses and/or to utilize emotional information.

Regardless of the diagnostic category, psychiatric patients manifest symptoms related to the experience, the expression and the regulation of emotion [3–6]. Along with the subjective burden that patients suffer from, problems in emotional processing interfere with bodily, cognitive and social functionality and contribute to the progression of the disease [7], to the treatment response [8, 9] or relapses [10]. The complexity of psychiatric disorders calls for better diagnostic and prognostic approaches that would enable clinicians to detect risk groups and plan individual treatment programmes. Thus, components of emotional processing are good candidates as state or trait markers, which might even indicate the psychiatric disorder before the full-scale disease manifestation is presented or predict the treatment response before the clinical response is achieved. A striking example is the prediction of antidepressant response to selective serotonin reuptake transporters (SSRI) by measures of emotional processing that are acquired before or in the early phase of the treatment [11, 12].

What is Emotion?

The nature of emotion has been a matter of debate since antiquity, and in principle, it all boils down to two reasons why human beings generate emotions: being alive and possessing a complex nervous system. Some researchers argue that emotion, as humans experience it, is specific to humans [13], while others propose that animals as well experience emotion [14]. Darwin argued that emotions are mental states that cause stereotypic bodily expressions [15]. His aim was to point out the homologous behaviours between humans and other animals and to underscore the common natural selection processes behind animal (including human) evolution [15]. The prominence of an instinctual aspect of emotion was strengthened by Mc Dougall [16]. He interpreted the experience of emotion as a combination of the affective quality of instinctive processes and the sum of bodily changes resulting from it [16]. His position continues to be defended by basic emotion theorists today. They argue that every emotional instance consists of a specific combination of fixed components that involve facial expressions, subjective feelings and autonomic changes or brain activities, all of which share common structural and functional mechanisms installed through evolution [14, 17].

It is evident that emotional experiences are accompanied by physiological changes. When we are angry or afraid, our heart rate and blood pressure are increased, our pupils are dilated and we start to sweat. In the nineteenth century, James wrote a critique called "What is an Emotion?" [18]. Contrary to his successors who assumed that mental events have temporal priority over emotional expressions and bodily changes, he postulated that physiological activity in response to an event produces emotion [18]. Independent of that, Lange wrote that emotions result from a bodily vasomotor response [19]. Moreover, Watson drew attention to the reflex-like physiological patterns in emotion and defined emotion as a hereditary pattern of reactions of visceral and glandular systems [20]. The most famous criticism of the priority of physiological reactions in emotion was provided by Cannon in his article

"The James-Lange Theory of Emotions: A Critical Examination and an Alternative Theory" [21]. According to Cannon, the body cannot be the cause of emotion because visceral changes are not fast enough and are too ambiguous to be interpreted as specific emotions [21]. Likewise, some research points to the incoherence between the self-reports of discrete emotions and physiological measures [22, 23]. Furthermore, many contemporary researchers challenge the idea of common labels across participants for discrete instances of emotion [24, 25].

Lang and colleagues support a framework built upon on motivational circuits that are shared between man and animal [26]. Instead of labelling discrete emotions such as anger or happiness, Lang put forward a simple affective language based on motivational tendencies—approach and avoidance [24, 26]. When a threatening object is encountered there is a tendency to avoid, whereas an appealing stimulus evokes a tendency to approach it. In this context, activation of the appetitive system is associated with positive affect and activation of the aversive system induces a negative affect [24]. The intensity of the motivational mobilization depends on the need and the probability of an expected outcome of a stimulus [24]. Factor analysis studies of verbal reports of phenomena of emotion disclosed two main factors explaining most of the experimental variance [27–29]. The first factor is named hedonic valence, which refers to a sign of internal state that is directional [24]. The second one is named emotional arousal, which refers to the intensity of activation. This view is commonly called a dimensional view of emotion, which proposes that there is rather a dimensional emotional space compared to the discrete emotions, which the classical view proposes [30].

At the beginning of the twentieth century, Allport argued that peripheral physiological reactions vary only according to pleasant and unpleasant affect [31]. However, he claimed that afferent sensory information from body postures and facial expressions facilitates the discrimination of affective responses according to the discrete kind of emotions [31]. Interestingly, behavioural measures like facial expressions and ANS measures show high correlation, unlike self-reports and ANS measures [32]. Allport put forward a set of possible configurations of facial muscle movements that involve six basic elements: pain, grief, surprise, fear, anger, disgust, pleasure and various "attitudes" (a group of neutral expressions). He put forward the idea of a one to one correspondence between facial movements and discrete emotion categories, setting a hallmark of modern basic emotion theory.

The vital role of common facial expressions across individuals on social interaction and survival has been stressed since Darwin. When we see a fearful expression even on the face of a stranger, we effortlessly recognize fear. This non-verbal communication can incite further thoughts or behavioural actions, such as helping the other. If emotional expressions are products of evolution, it is expected that human beings present similar facial expression in similar conditions that can be recognized by other human beings. In other words, facial expressions should be universal. Following these assumptions, Ekman conducted numerous studies on the topic of the universality of facial emotions. Contrary to social constructivists, who propose that emotion and facial expressions are culture-dependent, he showed that basic emotional expressions are universal [33]. However, a recent cross-cultural study has reported conflicting results with the universality of expressions specific to basic discrete emotions [34].

Furthermore, we could ask: are instances of emotion only instinctual reflexes or are there additional mental mechanisms modulating or even causing emotion? As early as in 1897, Wundt put forward that "inceptive feelings", which are caused either by an external or internal source, are distinguished by "ideational processes" and consequently different emotions occur [35]. Later, Newman, Perkins [36] underscored the indispensability of mental representations of the objects or related goals. They argued that only if a goal is present and meaning is attributed to the stimulus, cues from the body contribute to the production of an emotional state [36]. By highlighting the role of assigning meaning to a bodily cue, they influenced the modern appraisal theory. Appraisal theory was also inspired by Young [37], who described emotions as an acute disturbance in behaviour, experience and visceral functioning originating from mental events, and later on by Arnold [38], who defined emotions as intentional states referring to an object or a situation, in which different emotions are produced through meaning analysis of a particular stimulus.

In the last decade, the existence of mere emotional instances and emotion-related brain circuits has been challenged. Barrett [39] suggests that hedonic valence and arousal are basic features of consciousness, that are not specific to emotion, and underlines the degeneracy in the emotion-related brain circuits not only between discrete emotional types but also for different functions of the brain, including emotion, interoception and visceromotor actions. Indeed, regions including the amygdala, ventral striatum, insula, OFC, ACC and mPFC are collectively called "visceromotor areas", and are also proposed to be involved in the circuits related to emotion and the formation of the self [39]. Subsequently, followers of constructive emotion theory object to the existence of emotion-specific mechanisms in the brain, and in turn argue that emotion is produced dynamically [39]. This outlines the existence of two separate frameworks within the field of emotion research—one highlights "domain-general mechanisms", as proposed by Barrett and colleagues, and the other is centred on "emotion-specific mechanisms", as pursued by representatives of both the classical and dimensional viewpoints.

Affective Reactivity

Affective reactivity refers to short-term affective responses following an emotion-inducing stimulus [40]. The severity and timing of affective reactions to stimuli vary across individuals and are widely accepted to be indicative of durability, psychological functioning and well-being [41, 42]. Affective reactivity is also commonly referred to as emotional reactivity. Here, we preferred to use "affective" instead of "emotional", because of the relatively clear reference of affect to rapid and short-term changes in emotional expression of patients in the field of psychiatry [43], whereas the definitions of emotion and emotional instances/episodes are more arguable.

Affective reactivity manifests in different components of emotion. Affective response domains can be operationally defined as systems that are reactive to affective stimuli and enable us to measure the resulting reactivity. In this framework, three domains of affective responses are usually mentioned [41]:

(a) subjective/experiential (self-report scales) (b) behavioural/expressive/motor (e.g. facial expression coding, electromyographic (EMG) recordings of the facial muscles, affect-modulated startle reflex), (c) somatic, which involves peripheral and central nervous system responses, and endocrine responses. In the following paragraphs, affective reactivity in subjective, behavioural and ANS domains will be outlined. There are also plenty of findings from EEG and imaging studies about the activity in brain regions and networks, which will be discussed in a separate section about affective biases in perception and attention.

Affective Reactivity in the Subjective Domain

Self-reported questionnaires like positive and negative affect scale and self-assessment manikin (SAM; [29]) have been used as operational measures of subjective affective states. Because the experiential component of emotion is not directly accessible by objective methods, participants are asked to choose one of the discrete emotions or to rate the valence and arousal according to how they feel during the induction of affect. Proponents of each emotion theory have a particular view about the experiential component of emotion. For that reason, they are using different kinds of stimuli and types of scales. For example, researchers akin to basic emotions, mostly induce affect by presenting facial expressions or words and probe the subjective experience of the participant using scales, in which participants choose one of the discrete emotion categories or families of emotion kinds [45]. In 1976, Ekman and colleagues developed a frequently used validated set of facial expressions for the purpose of affect induction in the laboratory [46]. The set consists of 60 black and white pictures of facial expressions. Later, the Karolinska Emotional Directed Faces database was developed [47]. This database contains 490 pictures of 70 men and women, each displaying six basic emotions and a neutral facial expression. The development of validated sets of affective stimuli provided standardized affective stimuli that can reliably evoke reactions in different affective domains and contributed to improving emotion research.

Dimensional theorists generally prefer to use scales, by which they can probe the dimensions of affectivity. For example, participants are asked to rate their feelings during affect induction in terms of valence and arousal. Lang, Bradley [48] developed a broad and validated database of affective pictures called International Affective Picture System (IAPS), and a database of affective audio recordings and words (IADS; [49]), single words (ANEW; [50]) and descriptive sentences (ANET; [51]). Normative valence, arousal and dominance ratings for each item in these datasets was calculated by using SAM in validation studies separately for each set, even for different cultures [29].

Plenty of studies examined affective reactivity in Major Depressive Disorder (MDD) patients. The cognitive theory of depression suggests that dysfunctional processing of affective stimuli, especially negative-valence stimuli play role in the development and continuation of depressive state [52, 53]. Others emphasize the attenuation of the affective response to positive stimuli and associate this with symptoms like anhedonia, fatigue and sleep problems. Furthermore, a distinctive role of the reward system in attenuation of affect was proposed [54, 55]. Therefore, it would be expected that, because of the ongoing depressive mood, processing of positive stimuli should be attenuated, while processing of negative stimuli should be facilitated in MDD. Contrary to these expectations, some studies reported that the affective reactions of depression patients probed over different affective response domains are comparable to healthy controls or suppressed for both positive and negative stimuli [56–59]. Rottenberg, Gross [56] proposed an emotion-context insensitivity in depression and argued that the affective reactivity in depression is reduced in response to both the rewarding and threatening stimuli. Accordingly, a meta-analysis showed that depression patients reported less positive mood, more depressed feelings and anhedonia, but short-term instantaneous responses to positive and negative stimuli were both attenuated [41]. Further, the variability of affect, collected with experience sampling method (ESM), was the same for depressed and control subjects for 90 min intervals but was lower in MDD patients for shorter intervals [60]. Therefore, some authors suggest that MDD patients are less likely to respond to stimuli known to induce affect in healthy people in short time intervals, and mood disturbances might be driven by long-term cognitive processes such as rumination, as well as other contributors of depressive state such as circadian rhythm dysfunctions instead of immediate affective reactions to the world [61].

Affective Reactivity in the Behavioural Response Domain

Analysis of Facial Expressions

Facial expressions, the necessary constituents of social behaviour, exhibit the individual's affective state and intentions to observers. They are considered as one of the behavioural or motor components of emotion [41]. Facial expressions are operationally measured via coding of facial expressions by researchers Lang, Greenwald [62] (e.g. FACS; [63]), or via electromyographic (EMG) measures of facial muscles [62, 64, 65]. By using FACS analysis, it was found that schizophrenia and depression patients exhibited less spontaneous facial expressions than control subjects [66].

Affect Modulated Startle Reflex

Startle reflex is a common motor response, found in mammals, to an unexpected and intense stimulus, characterized by series of quick muscle contractions, presumably to protect the body from a sudden attack and/or to facilitate the escape response [67]. Whilst the startle reflex is stereotypical, the magnitude of the reflex is variable and reflects the state of the organism [68–72]. Years of research have shown that the startle response also varies as a function of affective state [73–77]. Rodent studies have shown that after conditioning with an aversive stimulus, rats show higher amplitude in startle response in the presence of a non-aversive stimulus such as light or a tone [78]. The increase in amplitude of the startle reflexes in the presence of the conditional stimulus is called the "fear-potentiated startle reflex" and is considered to be an operational measure of fear [77]. Studies have confirmed the presence of fear-potentiated startle reflex also in humans [79]. The startle reflex is generally measured over the EMG activity of the orbicularis oculi [80]. The startle reflex is potentiated not only in the presence of a threat-relevant stimulus but also with other unpleasant or highly arousable stimuli [81]. Moreover, the presence of pleasant stimuli can lead to an attenuation in the amplitude of startle reflex [81]. This phenomenon is known as affective modulation of startle reflex [82].

Affective modulation of the startle reflex can be atypical in psychiatric disorders. For instance, when phobia-relevant stimuli are presented to phobia patients the affect modulate startle is potentiated [75]. On the contrary, depressive patients have a decreased amplitude in the startle reflex during a pleasant or unpleasant stimulus, which is comparable to the reaction to a neutral stimulus [57, 58, 83]. Furthermore, this atypical pattern in affect-modulated startle reflex in depression can change after SSRI treatment [84]. Affect-modulated startle paradigms are also commonly used in substance abuse research. The presence of alcohol-related cues in pictures presented to alcohol use disorder patients attenuates their startle response [85, 86]. Curiously, in the early phase of withdrawal, alcohol-relevant cues attenuate the startle reflex, suggesting they are still interpreted as tempting; however, later on, alcohol-relevant cues potentiate the startle reflex [85]. These findings indicate that the affective modulation of the startle reflex is associated with disease state and can thus serve as a treatment marker.

Affective Reactivity in the Autonomic Nervous System

Another research measure that has been commonly applied is the effect of affect induction on the autonomic nervous system. The ANS can be investigated in terms of cardiorespiratory parameters including heart rate, respiration rate or heart rate variability (HRV). These are controlled by both the sympathetic and the parasympathetic branches of the ANS. Moreover, pupil reactivity and skin conductance response (SCR) can be also utilized, which are under control of the sympathetic branch of the ANS.

Currently, there is a debate if the ANS reactivity differs across discrete emotions. Some evidence for specific patterns of ANS reactivity was reported [87–89]. For example, fear and anger could be differentiated when cardiovascular and respiratory measures were used concomitantly [90]. However, a meta-analysis has reported inconsistencies across studies [91], and instead, the authors argued that the variability in ANS reactivity between studies was produced by the different affect induction methods instead of discrete emotions. On the other hand, there is plenty of evidence that ANS measures vary, especially in response to arousal and hedonic valence of the stimuli [73, 92]. For example, heart rate deceleration is observed initially when a novel affective picture is presented and the decrease in heart rate is higher for unpleasant pictures than pleasant ones [82].

Skin conductance response is a widely used and reliable measure of affect-related autonomic arousal [93]. It shows high correlation with self-rating of arousability [82]. The electrodes connected to hands monitor the activity in the eccrine sweat glands, which are only innervated by the sympathetic branch of the ANS [93].In addition to the threat-relevant stimuli, any other stimulus that has a motivational significance can evoke SCR even when the stimulus is masked and not consciously processed by the participant [94].

Altered skin conductance response has also been discussed as a potential diagnostic marker of depression. Depression patients have decreased SCR amplitudes in response to arousing stimuli [95, 96]. Compared to this, the amplitude of the SCR was reported to be higher in general anxiety disorder than in major depression after stress induction [97]. Additionally, SRC may also serve as a treatment marker; lower amplitudes of SCR during a sad movie clip have been associated with a better response to antidepressants [98].

A number of studies indicate abnormalities of the HRV at rest or during affect induction in major depression [99]. It is widely assumed that HRV reflects the integrated function of both sympathetic and parasympathetic branches of ANS on the heart [100–102]. Evidence suggests that HRV can be used as a probe to mirror the physiological reactivity of the body to affective stimuli [101, 103]. Moreover, HRV is associated with the ability to regulate emotion [11, 104]. Various parameters of the HRV change after antidepressant treatment [105] and are also associated with other types of intervention [106, 107].

The Interaction Between Affective Processing and Cognition

Traditionally, cognitive functions and emotion are considered as two discrete functions of the brain, but it is now commonly acknowledged that the presence of an affective stimulus alters the processing of any kind of information. Affective bias on attention, memory and decision making were repeatedly reported in healthy subjects [108–110] and was shown to differ between different psychopathologies [111–115]. In this part, the interaction between affective processing and cognition will be outlined over behavioural, neurophysiological and especially neuroimaging findings. Similar to behavioural studies, neuroimaging studies employ several different approaches to emotion theories. Despite the inconsistencies in methodology, certain canonical regions and circuits have been found repeatedly [116–118]. The principle of emotion circuits was first formulated in animal research, where fear regulation and response are commonly tested in rodents [119]. In these animal studies, core components of the fear circuit were delineated, encompassing the amygdala and the mPFC [120]. This cortico-limbic circuit has been extensively investigated in humans, too, particularly in the context of fear formation, regulation and deviations in conditions such as anxiety and PTSD [121, 122]. Moreover, regions of this circuit, such as rostral anterior cingulate cortex, have been implicated in emotion regulation [123–126]. Consequentially, these areas have been frequently investigated in the context of mood disorders such as unipolar and bipolar depression [3, 127–129].

Affective Bias in Perception and Attention

The interaction between attention and affect is usually examined with visual search paradigms. Participants are asked to find an affective target among an array of non-affective distractors. Affective targets, especially threat-related ones, can be detected faster than neutral targets [108, 130]. Another common paradigm used to delineate the interaction of affect and attention is the dot-probe task. A neutral probe (e.g. a dot or letter) is preceded by an affective predictive cue such as a picture. Participants are asked to detect the location of the probe. If the location of the preceding affective predictive cue matches with the probe, the reaction time is faster [131, 132]. In addition, the meaning of the affective cue plays a substantial role in attention. For example, people who have snake phobia detect snake pictures faster, while the ones with spider phobia detect spider pictures faster [108].

Remarkably, the facilitative effect on attention can be observed even if the affective cue is presented for such a brief period that it cannot be perceived in full awareness [133]. It was therefore suggested that such stimuli modulate further processing even if they were not consciously perceived [134]. The processing of affective stimuli without awareness indicates its pre-attentive feature [135]. For that reason, some authors suggest a modulatory role of emotion in early perceptual processing [135, 136]. Accordingly, EEG studies have shown that affective stimuli evoke higher event-related potentials (ERPs) in the early stages of stimulus processing [131, 137]. In support of these findings, the enhancing effect of affective stimuli on neural activity could also be shown in early sensory cortices, as well as in associative cortices in fMRI studies [138–141].

Prior research generally confirms the role of the amygdala on the processing of both the overt and the masked affective stimuli. First, the enhancement in activity of early sensory cortices is correlated with the increase in amygdala activity [142–145]. Second, the change in the amygdala activity in response to masked stimuli differs between the affective and the non-affective stimuli [146, 147]. More direct evidence comes from patients with amygdala lesions. Contrary to healthy subjects, patients who suffer from selective amygdala and hippocampus lesions do not exhibit higher activity in fusiform gyrus when viewing fearful faces [148]. Strikingly, patients who only suffer from hippocampal lesions but have an intact amygdala show a similar pattern as controls, which indicates that the amygdala might mediate the changed fusiform gyrus behaviour [148]. The reciprocal connections of the amygdala with sensory and associative cortices shed light on the joint functions of the amygdala, other limbic areas and prefrontal regions during the processing of affective salient stimuli. On the other hand, some studies reported that patients with unilateral or bilateral amygdala damage still demonstrate some facilitatory effect of affect on attention [149, 150]. These findings indicate that other brain regions might be involved in the processing of affective stimuli as well. Indeed, the activity in the anterior cingulate cortex (ACC) and orbitofrontal cortex (OFC) correlate with the effect of affect on attention [151–153].

From the evolutionary perspective, threat-related stimuli are expected to be more significant for the organism, therefore more salient. However, evidence from behavioural, EEG and imaging studies indicate that the influence of an affective stimulus on attention depends on how arousing the stimulus is, rather than on its hedonic valence [154]. In contrast, there are also results suggesting that negative stimuli might induce higher detection rates and higher activity in the early sensory areas than positive ones, even though both types of stimuli were equally arousing [144, 155, 156]. Generally, it is thought that healthy subjects attend more to pleasant stimuli as compared to MDD patients [157]. Instead of focusing on indistinct negative versus positive stimuli, other authors drew attention to the differential effect of discrete emotions like fear or disgust [158].

Although some brain areas are preferentially responsive to specific kinds of discrete emotions or dimensions of emotion, affective stimuli induce distributed and overlapping activations in many areas depending on the type and modality of stimulus [118, 159, 160]. Contrary to its commonly acknowledged role in processing threat, it was recently indicated that the amygdala is rather involved in dealing with unexpected information [30, 161, 162]. Likewise, although anterior insula activation is associated with the discrete emotion disgust [163–166] increased anterior insula activity was also reported for other conditions such as anger, or happiness [167]. Furthermore, even though studies have demonstrated involvement of the rostral ACC and ventral striatum in processing laughter [116, 160, 168], the ventral striatum also contributes to processing of negative stimuli [169, 170]. Similarly, the rostral ACC is responsive to sadness [171], but also to other discrete emotions [172, 173]. It has been suggested that overlapping activations in brain regions in response to different affective stimuli may indicate general functions like detection of salience, appraisal of relevance or novelty, which are accompanied by an alteration in sensorimotor, autonomic and cognitive domains [174].

Depression and anxiety disorder patients show disorder-specific affective attentional biases [175]. While anxiety disorder patients detect threat-relevant stimuli like angry or fearful faces, a bias toward sad faces is seen in depression patients [111–114]. Moreover, it is hypothesized that anxiety is related to biases in attention allocation, whereas depression patients have difficulties associated with biases in sustained attention [111, 113]. Accordingly, the differential features of affective attentional bias are apparent in eye-tracking studies, where anxiety disorder patients initially orient to threat-relevant stimuli, while depression patients show increased sustained attention to negative stimuli and decreased sustained attention to positive stimuli [176].

Contrary to behavioural findings, ERP indices of affective attentional biases indicate that threat-relevant stimuli exert bias both in attention allocation and sustained attention in anxiety disorders [177–180]. Depression patients show reduced amplitudes in late ERP indices, a sustained attentional associate, with either positive or negative stimuli instead of only depression-relevant stimuli [181, 182].

One of the prominent hypotheses about affective processing in MDD states that limbic areas are hyperactive at rest or in response to affectively salient stimuli

[183–186]. Amygdala's influence on the pre-attentive and the early phases of attention cannot be effectively regulated by the lateral PFC, right inferior frontal gyrus and dorsal ACC via shifting attention away from the affective information in MDD patients [187–189]. In line with this hypothesis, neuroimaging findings mostly point to an increase in amygdala activity in response to negative stimuli relative to controls [190, 191]. Studies have shown that after antidepressant treatment, the activity of amygdala becomes comparable to controls [190, 191]. Interestingly, the effects of antidepressants can be detected even in the early phases of treatment [12]. Moreover, basal amygdala reaction to negative stimuli can predict the treatment response 8 weeks later [12].

Affective Bias in Memory

Besides prompting immediate effects on perception and attention, affective stimuli can modulate memory formation and recall [110]. As with attention, healthy subjects display a bias for positive stimuli [192]. Patient populations that have been mostly investigated in this framework are depression patients. The affective bias is most evident in encoding tasks with a self-referential component [193, 194]. In such tasks, depressive patients show negative encoding more frequently than controls [115], although the recall of positive cues seems to be modulated as well [195]. The affective state might thus contribute to the formation of memory in mood disorder patients [196] indicating a mood-congruent bias. In tasks, where the implicit memory is tested, depressed subjects have shown similar results to controls [197]. Some authors have thus argued that mood-congruent biases are observable with conceptual and not perceptual memory tasks [198]. However, a recent meta-analysis showed a negative bias in depressive patients during implicit memory as well [199]. Moreover, other memory tasks, such as prospective memory tasks, show only mild interaction with mood state [200].

In healthy controls, amygdala–hippocampus interaction is involved in general recall, while the orbitofrontal cortex is activated during mood-congruent recall [201]. During encoding and retrieval of happy faces, depression patients display enhanced recruitment of amygdala/hippocampus area, as well as medial frontal gyrus [202], and posterior cingulate cortex [203]. Intriguingly, certain features of affective bias in memory are also observed in remitted patients [203] or after pharmacological treatment [204]. Strikingly, the bias toward negative stimuli could be induced via tryptophan depletion in controls [205, 206].

Looking at a dimensional approach, arousal and valence seem to activate distinct circuits in emotion memory; arousal is dependent on the amygdala–hippocampus activation [207, 208]. Valence, on the other hand, activates frontal regions for positive, and occipital, and temporal regions for negative stimuli [209].

Emotion Regulation

Emotion regulation can be broadly defined as a set of consciously executed strategies that are used during different stages of emotions, or on different components of emotions [210]. The process of emotion regulation can be set during monitoring (awareness), evaluation or modulation of behavioural response [210, 211]. As with other aspects of emotion, there is no unique classification of emotion regulations strategies. Koole [212] proposed a dichotomy of targets (attention, emotion-relevant knowledge and bodily manifestations) and psychological functions (hedonic needs, goal-orientation, person-orientation), where different combinations can be paired. Others have focused on fear-circuitry modulations, like extinction and reconsolidation [213]. The most investigated strategy is reappraisal, a method which reinterprets emotion-inducing situations [214, 215]. For example, in experimental tasks subjects might be instructed to re-evaluate negative pictures as "not real" or "better than it seems" [216]. Another common strategy is suppression, which aims to inhibit or reduce emotion experience [214, 215].

In general, it is thought that prefrontal regions mediate regulation of the immediate response to affect inducing stimuli of the sub-cortical regions, but emotion regulation strategies activate distinct regions [217]. For instance, when subjects were asked to use reappraisal to downregulate their emotions, researchers found activation of cognitive prefrontal areas and decreased amygdala/hippocampus, and insula activity. In contrast, when applying emotion suppression, subjects showed increased activity in prefrontal cortex, but also insula [123–125, 216]. Enhancing current emotions (upregulating) via reappraisal, on the other hand, resulted in both increased amygdala and prefrontal cortex activity [124, 216]. The distinction in amygdala activity between downregulation and upregulation mirrors the goal of the regulation strategy. Contrarily, activation of prefrontal regions like dorsomedial prefrontal cortex reflects the domain-general cognitive aspect of regulation strategies [123, 218].

Importantly, the emotion regulation is also observable in the ANS response [219, 220], and subjects with mood vulnerabilities show decreased HRV during emotion regulation [221].

Dysfunctions in emotion regulation are a hallmark of psychiatric disorders [222]. Certain regulation strategies, like suppression, are also more present in psychopathologies than in healthy controls [223]. On the brain level, bipolar patients show hyperactivation, while schizophrenia patients show hypoactivation of prefrontal regions during downregulation/reappraisal, indicating disorder-specific deficits [224]. Moreover, a hypofunction of the frontoparietal network during reappraisal was found in a recent meta-analysis in mood and anxiety patients [225]. In contrast, other regions like insula showed increased activity indicating an increase in the experience of emotions [225].

Lastly, emotion regulation strategies can be utilized to change mood state [226], and so may serve as additional treatment in clinical practice [227], e.g. for anxiety disorders [228]. Recently, other approaches such as mindfulness have also been introduced in clinical practice [229, 230].

Future Perspectives

A fundamental debate was started in recent years, following the development of Bayesian understanding of the brain functions. The conventional view that understands emotion as a/or a part of the response of the organism after the perception of emotion-inducing stimuli has been challenged [231, 232]. It is put forward that an instance of emotion is a result of prediction error processing as any other psychological phenomena. Prediction signals based on past experiences serve as Bayesian filters for incoming sensory input to drive action, construct perception and emotion. Considering the vitality of the maintenance of the homeostatic state of the organism, the brain is proposed to have an internal model of "the body in the world"; a model that is dynamically formatted with internal and external sensations [231]. The representation and utilization of the internal sensations are called interoception, which is suggested to arise from allostatic processes [39, 233–236]. Moreover, it is proposed that affective states mirror the precision of the interoceptive predictions [236–238]. Positive emotions are associated with events resolving interoceptive uncertainty, whereas negative emotions are related to events that are expected to enhance it [239, 240].

In the light of the incongruent scientific findings on the affective reactivity and affective bias in MDD patients, an explanation of the relation between affect and mood is crucial. Some findings indicate that MDD patients show a blunted affective reactivity both to positive and negative stimuli [56, 83, 84, 181, 182], whereas most of the studies show processing bias to negative stimuli [114, 176, 206]. Bayesian decision theory proponents Huys, Daw [241] argue that in MDD patients, biases are seen in the interpretation of ambiguous information rather than in the immediate evaluation. In psychiatry practice, mood is defined as common, slow-changing sensation states that do not have an immediate relationship with stimuli, whereas affect is the rapid and short-term response to stimuli [43]. This definition in the frame of clinical practice might correspond to the Bayesian inference model of the mood and the affect that has been put forward by Clark, Watson [232]. Accepting that the affective state is associated with the precision of interoceptive predictions, Clark and colleagues [232] proposed that mood refers to the predictions about the affective state. Based on this predictive coding framework of interoception, affect and mood, the discrepancy in the findings of a study in different response domains of affective reactivity and affective bias in cognition is to be re-conceptualized.

Conclusion

This chapter sought to introduce basic concepts of emotion theories, accompanying methods and exemplary findings in psychiatry patients. As it was seen, often, methodological approaches are not clear cut within an outline of a single theory, and findings from different research modalities complement each other by tapping into distinctive psycho-biological measures of emotions, and related components. In sum, the field of emotion research is wide and highly dynamic, yet it provides a crucial element in investigating psychiatric disorders and can be used both in basic science and clinical practice.

References

1. Damasio A. Fundamental feelings. Nature. 2001;413(6858):781.
2. Fischer AH, Manstead AS. Social functions of emotion. In: Handbook of emotions, vol. 3; 2008. p. 456–68.
3. Phillips ML. Understanding the neurobiology of emotion perception: implications for psychiatry. Br J Psychiatry. 2003;182:190–2.
4. Harrison A, Sullivan S, Tchanturia K, Treasure J. Emotional functioning in eating disorders: attentional bias, emotion recognition and emotion regulation. Psychol Med. 2010;40(11):1887–97.
5. Gur RE, Kohler CG, Ragland JD, Siegel SJ, Lesko K, Bilker WB, et al. Flat affect in schizophrenia: relation to emotion processing and neurocognitive measures. Schizophr Bull. 2006;32(2):279–87.
6. Leppanen JM. Emotional information processing in mood disorders: a review of behavioral and neuroimaging findings. Curr Opin Psychiatry. 2006;19(1):34–9.
7. Leboyer M, Kupfer DJ. Bipolar disorder: new perspectives in health care and prevention. J Clin Psychiatry. 2010;71(12):1689–95.
8. Steketee G, Chambless DL. Does expressed emotion predict behaviour therapy outcome at follow-up for obsessive-compulsive disorder and agoraphobia? Clin Psychol Psychother. 2001;8(6):389–99.
9. Kim EY, Miklowitz DJ. Expressed emotion as a predictor of outcome among bipolar patients undergoing family therapy. J Affect Disord. 2004;82(3):343–52.
10. Butzlaff RL, Hooley JM. Expressed emotion and psychiatric relapse: a meta-analysis. Arch Gen Psychiatry. 1998;55(6):547–52.
11. Williams DP, Cash C, Rankin C, Bernardi A, Koenig J, Thayer JF. Resting heart rate variability predicts self-reported difficulties in emotion regulation: a focus on different facets of emotion regulation. Front Psychol. 2015;6:261
12. Godlewska BR, Browning M, Norbury R, Cowen PJ, Harmer CJ. Early changes in emotional processing as a marker of clinical response to SSRI treatment in depression. Translational Psychiatry. 2016;6:e957.
13. Barrett LF. Categories and their role in the science of emotion. Psychol Inq. 2017;28(1):20–6.
14. Panksepp J. Affective consciousness: core emotional feelings in animals and humans. Conscious Cogn. 2005;14(1):30–80.

15. Darwin C. The expression of the emotions in man and animals. London: J. Murray; 1872.
16. Mc Dougall W. An introduction to social psychology. Boston: John W. Luce; 1908/1921.
17. Ekman P. An argument for basic emotions. Cogn Emot. 1992;6(3–4):169–200.
18. James W. What is an Emotion? Mind. 1884;9(34):188–205.
19. Lange CG. The emotions. In: Haupt IA, editor. The emotions, vol. 1. A series of reprints and translations. Baltimore, MD, US: Williams & Wilkins Co; 1885/1922. p. 33–90.
20. Watson JB. A schematic outline of the emotions. Psychol Rev. 1919;26(3):165.
21. Cannon WB. The James-Lange theory of emotions: a critical examination and an alternative theory. Am J Psychol. 1927;39(1/4):106–24.
22. Barrett LF, Satpute AB. Historical pitfalls and new directions in the neuroscience of emotion. Neurosci Lett. 2017.
23. Siegel EH, Sands MK, Van den Noortgate W, Condon P, Chang Y, Dy J, et al. Emotion fingerprints or emotion populations? A meta-analytic investigation of autonomic features of emotion categories. Psychol Bull. 2018;144(4):343.
24. Lang PJ, Bradley MM. Emotion and the motivational brain. Biol Psychol. 2010;84(3):437–50.
25. Barrett LF, Gendron M, Huang Y-M. Do discrete emotions exist? Philos Psychol. 2009;22(4):427–37.
26. Lang PJ. The emotion probe: studies of motivation and attention. Am Psychol. 1995;50(5):372.
27. Osgood CE, Suci GJ, Tannenbaum PH. The measurement of meaning. University of Illinois press; 1957.
28. Russell JA, Mehrabian A. Evidence for a three-factor theory of emotions. J Res Pers. 1977;11(3):273–94.
29. Bradley MM, Lang PJ. Measuring emotion: the self-assessment manikin and the semantic differential. J Behav Ther Exp Psychiatry. 1994;25(1):49–59.
30. Barrett LF. Are emotions natural kinds? Perspect Psychol Sci. 2006;1(1):28–58.
31. Allport FH. Social psychology. New York: Houghton Mifflin; 1924.
32. Mandler G, Mandler JM, Kremen I, Sholiton RD. The response to threat: Relations among verbal and physiological indices. Psychol Monogr: Gen Appl. 1961;75(9):1.
33. Ekman P, Friesen WV. Constants across cultures in the face and emotion. J Pers Soc Psychol. 1971;17(2):124.
34. Gendron M, Roberson D, van der Vyver JM, Barrett LF. Perceptions of emotion from facial expressions are not culturally universal: evidence from a remote culture. Emotion. 2014;14(2):251.
35. Wundt W. Outlines of psychology (CH Judd, trans.). Leipzig: Wilhelm Engelmann. 1897. p. 1.
36. Newman EB, Perkins FT, Wheeler RH. Cannon's theory of emotion: a critique. Psychol Rev. 1930;37(4):305–326.http://dx.doi.org/10.1037/h0074972.
37. Young PT. Emotion in man and animal; its nature and relation to attitude and motive. 1943, Oxford, England: Wiley.
38. Arnold MB. Emotion and personality, vol. I. Psychol Asp. 1960.
39. Barrett LF. The theory of constructed emotion: an active inference account of interoception and categorization. Soc Cogn Affect Neurosci. 2017;12(1):1–23.
40. Nelson BD, Shankman SA, Olino TM, Klein DN. Defining reactivity: how several methodological decisions can affect conclusions about emotional reactivity in psychopathology. Cogn Emot. 2011;25(8):1439–59.
41. Bylsma LM, Morris BH, Rottenberg J. A meta-analysis of emotional reactivity in major depressive disorder. Clin Psychol Rev. 2008;28(4):676–91.
42. Cole PM, Zahn-Waxler C, Fox NA, Usher BA, Welsh JD. Individual differences in emotion regulation and behavior problems in preschool children. J Abnorm Psychol. 1996;105(4):518.
43. Walker HK, Hall WD, Hurst JW. Clinical methods: the history, physical, and laboratory examinations, 3rd ed. Boston: Butterworths; 1990.

44. Watson D, Clark LA, Tellegen A. Development and validation of brief measures of positive and negative affect: the PANAS scales. J Pers Soc Psychol. 1988;54(6):1063.
45. Harmon-Jones C, Bastian B, Harmon-Jones E. The discrete emotions questionnaire: a new tool for measuring state self-reported emotions. PLoS ONE. 2016;11(8):e0159915.
46. Ekman P, Friesen WV. Pictures of facial affect. Consulting Psychologists Press; 1976.
47. Lundqvist D, Flykt A, Öhman A. The Karolinska directed emotional faces (KDEF). CD ROM from department of clinical neuroscience, psychology section, Karolinska Institutet. 1998;91:630.
48. Lang P, Bradley M, Cuthbert B. International affective pictures system (IAPS): digitized photographs, instruction manual and affective ratings. Gainesville: University of Florida, NIMH Center for the Study of Emotion and Attention; 1998.
49. Bradley M, Lang PJ. The International affective digitized sounds (IADS): stimuli, instruction manual and affective ratings. NIMH Center for the Study of Emotion and Attention; 1999.
50. Bradley MM, Lang PJ. Affective norms for English words (ANEW). Gainesville, FL: The NIMH Center for the Study of Emotion and Attention, University of Florida; 1999.
51. Bradley MM, Lang PJ. Affective norms for English text (ANET): affective ratings of text and instruction manual. Techical Report D-1, University of Florida, Gainesville, FL; 2007.
52. Santesso DL, Steele KT, Bogdan R, Holmes AJ, Deveney CM, Meites TM, et al. Enhanced negative feedback responses in remitted depression. NeuroReport. 2008;19(10):1045.
53. Joormann J, Gotlib IH. Selective attention to emotional faces following recovery from depression. J Abnorm Psychol. 2007;116(1):80.
54. Russo SJ, Nestler EJ. The brain reward circuitry in mood disorders. Nat Rev Neurosci. 2013;14:609.
55. Heshmati M, Russo SJ. Anhedonia and the brain reward circuitry in depression. Curr Behav Neurosci Rep. 2015;2(3):146–53.
56. Rottenberg J, Gross JJ, Gotlib IH. Emotion context insensitivity in major depressive disorder. J Abnorm Psychol. 2005;114(4):627–39.
57. Allen NB, Trinder J, Brennan C. Affective startle modulation in clinical depression: preliminary findings. Biol Psychiat. 1999;46(4):542–50.
58. Dichter GS, Tomarken AJ, Shelton RC, Sutton SK. Early-and late-onset startle modulation in unipolar depression. Psychophysiology. 2004;41(3):433–40.
59. Deldin PJ, Keller J, Gergen JA, Miller GA. Cognitive bias and emotion in neuropsychological models of depression. Cogn Emot. 2001;15(6):787–802.
60. Koval P, Pe ML, Meers K, Kuppens P. Affect dynamics in relation to depressive symptoms: variable, unstable or inert? Emotion. 2013;13(6):1132.
61. Rottenberg J, Hindash AC. Emerging evidence for emotion context insensitivity in depression. Curr Opin Psychol. 2015;4:1–5.
62. Lang PJ, Greenwald MK, Bradley MM, Hamm AO. Looking at pictures: affective, facial, visceral, and behavioral reactions. Psychophysiology. 1993;30(3):261–73.
63. Ekman P, Matsumoto D, Friesen WV. Facial expression in affective disorders. 2005.
64. Cacioppo JT, Petty RE, Losch ME, Kim HS. Electromyographic activity over facial muscle regions can differentiate the valence and intensity of affective reactions. J Pers Soc Psychol. 1986;50(2):260.
65. Cacioppo JT, Martzke JS, Petty RE, Tassinary LG. Specific forms of facial EMG response index emotions during an interview: from darwin to the continuous flow hypothesis of affect laden information processing. J Pers Soc Psychol. 1988;54(4):592.
66. Trémeau F, Malaspina D, Duval F, Corrêa H, Hager Budny M, Coin-Bariou L, et al. Facial expressiveness in patients with schizophrenia compared to depressed patients and nonpatient comparison subjects. Am J Psychiatry. 2005;162(1):92–101.
67. Koch M. The neurobiology of startle. Prog Neurobiol. 1999;59(2):107–28.
68. Hoffman HS, Fleshler M. Startle reaction: modification by background acoustic stimulation. Science. 1963;141(3584):928–30.

69. Walker DL, Davis M. Anxiogenic effects of high illumination levels assessed with the acoustic startle response in rats. Biol Psychiat. 1997;42(6):461–71.
70. Reijmers LG, Peeters BW. Effects of acoustic prepulses on the startle reflex in rats: a parametric analysis. Brain Res. 1994;661(1–2):274–82.
71. Hoffman HS, Ison JR. Reflex modification in the domain of startle: I. Some empirical findings and their implications for how the nervous system processes sensory input. Psychol Rev. 1980;87(2):175.
72. Davis M. Neurochemical modulation of sensory-motor reactivity: acoustic and tactile startle reflexes. Neurosci Biobehav Rev. 1980;4(2):241–63.
73. Lang PJ, Bradley MM, Cuthbert BN. Emotion, attention, and the startle reflex. Psychol Rev. 1990;97(3):377.
74. Bradley MM, Cuthbert BN, Lang PJ. Startle reflex modification: Emotion or attention? Psychophysiology. 1990;27(5):513–22.
75. Vrana SR, Constantine JA, Westman JS. Startle reflex modification as an outcome measure in the treatment of phobia: two case studies. Behav Assess. 1992.
76. Vaidyanathan U, Patrick CJ, Cuthbert BN. Linking dimensional models of internalizing psychopathology to neurobiological systems: affect-modulated startle as an indicator of fear and distress disorders and affiliated traits. Psychol Bull. 2009;135(6):909.
77. Grillon C, Baas J. A review of the modulation of the startle reflex by affective states and its application in psychiatry. Clin Neurophysiol. 2003;114(9):1557–79.
78. Brown JS, Kalish HI, Farber I. Conditioned fear as revealed by magnitude of startle response to an auditory stimulus. J Exp Psychol. 1951;41(5):317.
79. Grillon C, Davis M. Fear-potentiated startle conditioning in humans: explicit and contextual cue conditioning following paired versus unpaired training. Psychophysiology. 1997;34(4):451–8.
80. Blumenthal TD, Cuthbert BN, Filion DL, Hackley S, Lipp OV, Van Boxtel A. Committee report: guidelines for human startle eyeblink electromyographic studies. Psychophysiology. 2005;42(1):1–15.
81. Vrana SR, Spence EL, Lang PJ. The startle probe response: a new measure of emotion? J Abnorm Psychol. 1988;97(4):487.
82. Bradley MM, Lang PJ, Cuthbert BN. Emotion, novelty, and the startle reflex: habituation in humans. Behav Neurosci. 1993;107(6):970.
83. Kaviani H, Gray J, Checkley S, Raven P, Wilson G, Kumari V. Affective modulation of the startle response in depression: influence of the severity of depression, anhedonia, and anxiety. J Affect Disord. 2004;83(1):21–31.
84. Sen ZD, Temucin CM, Başar K, Ulug BD, Gökcen O, Özer S. Alteration of the affective modulation of the startle reflex during antidepressant treatment. Psychophysiology. 2019:e13286.
85. Leménager T, Hill H, Reinhard I, Hoffmann S, Zimmermann US, Hermann D, et al. Association between alcohol-cue modulated startle reactions and drinking behaviour in alcohol dependent patients—results of the PREDICT study. Int J Psychophysiol. 2014;94(3):263–71.
86. Rubio G, Borrell J, Jiménez M, Jurado R, Grüsser SM, Heinz A. Variables involved in the cue modulation of the startle reflex in alcohol-dependent patients. Addict Biol. 2013;18(1):170–80.
87. Christie IC, Friedman BH. Autonomic specificity of discrete emotion and dimensions of affective space: a multivariate approach. Int J Psychophysiol. 2004;51(2):143–53.
88. Ekman P, Levenson RW, Friesen WV. Autonomic nervous system activity distinguishes among emotions. Science. 1983;221(4616):1208–10.
89. Stemmler G, Heldmann M, Pauls CA, Scherer T. Constraints for emotion specificity in fear and anger: the context counts. Psychophysiology. 2001;38(2):275–91.
90. Stemmler G. Physiological processes during emotion. The regulation of emotion. Psychology Press; 2004. p. 48–85.

91. Cacioppo J, Klein D, Berntson G, Hartfield E. The psychophysiology of emotion. The handbook of emotions. New York: The Guilford Press; 1993. p. 119–42.
92. Lang PJ, Bradley MM, Cuthbert BN. Emotion and motivation: measuring affective perception. J Clin Neurophysiol. 1998;15(5):397–408.
93. Sequeira H, Hot P, Silvert L, Delplanque S. Electrical autonomic correlates of emotion. Int J Psychophysiol. 2009;71(1):50–6.
94. Balderston NL, Helmstetter FJ. Conditioning with masked stimuli affects the timecourse of skin conductance responses. Behav Neurosci. 2010;124(4):478.
95. Iacono WG, Lykken DT, Peloquin LJ, Lumry AE, Valentine RH, Tuason VB. Electrodermal activity in euthymic unipolar and bipolar affective disorders: a possible marker for depression. Arch Gen Psychiatry. 1983;40(5):557–65.
96. Ward NG, Doerr HO, Storrie MC. Skin conductance: a potentially sensitive test for depression. Psychiatry Res. 1983;10(4):295–302.
97. Pruneti C, Lento R, Fante C, Carrozzo E, Fontana F. Autonomic arousal and differential diagnosis in clinical psychology and psychopathology. J Psychopathol. 2010;16:43–52.
98. Fraguas R Jr, Marci C, Fava M, Iosifescu DV, Bankier B, Loh R, et al. Autonomic reactivity to induced emotion as potential predictor of response to antidepressant treatment. Psychiatry Res. 2007;151(1–2):169–72.
99. Kemp AH, Quintana DS, Gray MA, Felmingham KL, Brown K, Gatt JM. Impact of depression and antidepressant treatment on heart rate variability: a review and meta-analysis. Biol Psychiat. 2010;67(11):1067–74.
100. Malik M, Camm A. Components of heart rate variability-what they really mean and what we really measure. Am J Cardiol. 1993;72:821–2.
101. Mccraty R, Shaffer F. Heart rate variability: new perspectives on physiological mechanisms, assessment of self-regulatory capacity, and health risk. Glob Adv Health Med. 2015;4(1):46–61.
102. Shaffer F, McCraty R, Zerr CL. A healthy heart is not a metronome: an integrative review of the heart's anatomy and heart rate variability. Front Psychol. 2014;5:1040.
103. Thayer JF, Hansen AL, Saus-Rose E, Johnsen BH. Heart rate variability, prefrontal neural function, and cognitive performance: the neurovisceral integration perspective on self-regulation, adaptation, and health. Ann Behav Med. 2009;37(2):141–53.
104. Appelhans BM, Luecken LJ. Heart rate variability as an index of regulated emotional responding. Rev Gen Psychol. 2006;10(3):229–40.
105. Licht CM, de Geus EJ, van Dyck R, Penninx BW. Longitudinal evidence for unfavorable effects of antidepressants on heart rate variability. Biol Psychiat. 2010;68(9):861–8.
106. Chambers AS, Allen JJ. Vagal tone as an indicator of treatment response in major depression. Psychophysiology. 2002;39(6):861–4.
107. Karavidas MK, Lehrer PM, Vaschillo E, Vaschillo B, Marin H, Buyske S, et al. Preliminary results of an open label study of heart rate variability biofeedback for the treatment of major depression. Appl Psychophysiol Biofeedback. 2007;32(1):19–30.
108. Flykt A, Caldara R. Tracking fear in snake and spider fearful participants during visual search: a multi-response domain study. Cogn Emot. 2006;20(8):1075–91.
109. Leith KP, Baumeister RF. Why do bad moods increase self-defeating behavior? Emotion, risk tasking, and self-regulation. J Pers Soc Psychol. 1996;71(6):1250.
110. Elliott R, Zahn R, Deakin JW, Anderson IM. Affective cognition and its disruption in mood disorders. Neuropsychopharmacology. 2011;36(1):153.
111. Joormann J, Arditte KA. The relational theory of attention: implications for the processing of emotional stimuli in psychological disorders. Aust Psychol. 2013;48(6):399–401.
112. Bar-Haim Y, Lamy D, Pergamin L, Bakermans-Kranenburg MJ, Van Ijzendoorn MH. Threat-related attentional bias in anxious and nonanxious individuals: a meta-analytic study. Psychol Bull. 2007;133(1):1.
113. Gotlib IH, Joormann J. Cognition and depression: current status and future directions. Annu Rev Clin Psychol. 2010;6:285–312.

114. Peckham AD, McHugh RK, Otto MW. A meta-analysis of the magnitude of biased attention in depression. Depress Anxiety. 2010;27(12):1135–42.
115. Dobson KS, Shaw BF. Specificity and stability of self-referent encoding in clinical depression. J Abnorm Psychol. 1987;96(1):34.
116. Vytal K, Hamann S. Neuroimaging support for discrete neural correlates of basic emotions: a voxel-based meta-analysis. J Cogn Neurosci. 2010;22(12):2864–85.
117. Lindquist KA, Wager TD, Kober H, Bliss-Moreau E, Barrett LF. The brain basis of emotion: a meta-analytic review. Behav Brain Sci. 2012;35(3):121–43.
118. Kober H, Barrett LF, Joseph J, Bliss-Moreau E, Lindquist K, Wager TD. Functional grouping and cortical–subcortical interactions in emotion: a meta-analysis of neuroimaging studies. Neuroimage. 2008;42(2):998–1031.
119. LeDoux J. The emotional brain, fear, and the amygdala. Cell Mol Neurobiol. 2003;23(4–5):727–38.
120. Giustino TF, Maren S. The role of the medial prefrontal cortex in the conditioning and extinction of fear. Front Behav Neurosci. 2015;9:298.
121. Etkin A, Egner T, Kalisch R. Emotional processing in anterior cingulate and medial prefrontal cortex. Trends Cogn Sci. 2011;15(2):85–93.
122. Etkin A, Wager TD. Functional neuroimaging of anxiety: a meta-analysis of emotional processing in PTSD, social anxiety disorder, and specific phobia. Am J Psychiatry. 2007;164(10):1476–88.
123. Buhle JT, Silvers JA, Wager TD, Lopez R, Onyemekwu C, Kober H, et al. Cognitive reappraisal of emotion: a meta-analysis of human neuroimaging studies. Cereb Cortex. 2014;24(11):2981–90.
124. Frank D, Dewitt M, Hudgens-Haney M, Schaeffer D, Ball B, Schwarz N, et al. Emotion regulation: quantitative meta-analysis of functional activation and deactivation. Neurosci Biobehav Rev. 2014;45:202–11.
125. Goldin PR, McRae K, Ramel W, Gross JJ. The neural bases of emotion regulation: reappraisal and suppression of negative emotion. Biol Psychiat. 2008;63(6):577–86.
126. Ochsner KN, Knierim K, Ludlow DH, Hanelin J, Ramachandran T, Glover G, et al. Reflecting upon feelings: An fMRI study of neural systems supporting the attribution of emotion to self and other. J Cogn Neurosci. 2004;16(10):1746–72.
127. Grimm S, Boesiger P, Beck J, Schuepbach D, Bermpohl F, Walter M, et al. Altered negative BOLD responses in the default-mode network during emotion processing in depressed subjects. Neuropsychopharmacology. 2008;34:932.
128. Price JL, Drevets WC. Neural circuits underlying the pathophysiology of mood disorders. Trends Cogn Sci. 2012;16(1):61–71.
129. Kupfer DJ, Frank E, Phillips ML. Major depressive disorder: new clinical, neurobiological, and treatment perspectives. Lancet. 2012;379(9820):1045–55.
130. Gerritsen C, Frischen A, Blake A, Smilek D, Eastwood JD. Visual search is not blind to emotion. Percept Psychophys. 2008;70(6):1047–59.
131. Pourtois G, Grandjean D, Sander D, Vuilleumier P. Electrophysiological correlates of rapid spatial orienting towards fearful faces. Cereb Cortex. 2004;14(6):619–33.
132. Armony JL, Dolan RJ. Modulation of spatial attention by fear-conditioned stimuli: an event-related fMRI study. Neuropsychologia. 2002;40(7):817–26.
133. Mogg K, Bradley BP. Orienting of attention to threatening facial expressions presented under conditions of restricted awareness. Cogn Emot. 1999;13(6):713–40.
134. Esteves F, Dimberg U, Öhman A. Automatically elicited fear: Conditioned skin conductance responses to masked facial expressions. Cogn Emot. 1994;8(5):393–413.
135. Dolan RJ. Emotion, cognition, and behavior. Science. 2002;298(5596):1191–4.
136. Phelps EA, Ling S, Carrasco M. Emotion facilitates perception and potentiates the perceptual benefits of attention. Psychol Sci. 2006;17(4):292–9.

137. Dolan RJ, Heinze H, Hurlemann R, Hinrichs H. Magnetoencephalography (MEG) determined temporal modulation of visual and auditory sensory processing in the context of classical conditioning to faces. Neuroimage. 2006;32(2):778–89.
138. Pessoa L, McKenna M, Gutierrez E, Ungerleider L. Neural processing of emotional faces requires attention. Proc Natl Acad Sci. 2002;99(17):11458–63.
139. Grandjean D, Sander D, Pourtois G, Schwartz S, Seghier ML, Scherer KR, et al. The voices of wrath: brain responses to angry prosody in meaningless speech. Nat Neurosci. 2005;8(2):145.
140. Lang PJ, Bradley MM, Fitzsimmons JR, Cuthbert BN, Scott JD, Moulder B, et al. Emotional arousal and activation of the visual cortex: an fMRI analysis. Psychophysiology. 1998;35(2):199–210.
141. Ethofer T, Bretscher J, Gschwind M, Kreifelts B, Wildgruber D, Vuilleumier P. Emotional voice areas: anatomic location, functional properties, and structural connections revealed by combined fMRI/DTI. Cereb Cortex. 2011;22(1):191–200.
142. Peelen MV, Atkinson AP, Andersson F, Vuilleumier P. Emotional modulation of body-selective visual areas. Soc Cogn Affect Neurosci. 2007;2(4):274–83.
143. Pessoa L, Kastner S, Ungerleider LG. Attentional control of the processing of neutral and emotional stimuli. Cogn Brain Res. 2002;15(1):31–45.
144. Morris JS, Friston KJ, Büchel C, Frith CD, Young AW, Calder AJ, et al. A neuromodulatory role for the human amygdala in processing emotional facial expressions. Brain: J Neurol. 1998;121(1):47–57.
145. Sabatinelli D, Bradley MM, Fitzsimmons JR, Lang PJ. Parallel amygdala and inferotemporal activation reflect emotional intensity and fear relevance. Neuroimage. 2005;24(4):1265–70.
146. Vuilleumier P, Armony JL, Driver J, Dolan RJ. Effects of attention and emotion on face processing in the human brain: an event-related fMRI study. Neuron. 2001;30(3):829–41.
147. Whalen PJ, Rauch SL, Etcoff NL, McInerney SC, Lee MB, Jenike MA. Masked presentations of emotional facial expressions modulate amygdala activity without explicit knowledge. J Neurosci. 1998;18(1):411–8.
148. Vuilleumier P, Richardson MP, Armony JL, Driver J, Dolan RJ. Distant influences of amygdala lesion on visual cortical activation during emotional face processing. Nat Neurosci. 2004;7(11):1271.
149. Bach DR, Talmi D, Hurlemann R, Patin A, Dolan RJ. Automatic relevance detection in the absence of a functional amygdala. Neuropsychologia. 2011;49(5):1302–5.
150. Piech RM, McHugo M, Smith SD, Dukic MS, Van Der Meer J, Abou-Khalil B, et al. Attentional capture by emotional stimuli is preserved in patients with amygdala lesions. Neuropsychologia. 2011;49(12):3314–9.
151. Lucas N, Vuilleumier P. Effects of emotional and non-emotional cues on visual search in neglect patients: evidence for distinct sources of attentional guidance. Neuropsychologia. 2008;46(5):1401–14.
152. De Martino B, Kalisch R, Rees G, Dolan RJ. Enhanced processing of threat stimuli under limited attentional resources. Cereb Cortex. 2008;19(1):127–33.
153. Schwabe L, Merz CJ, Walter B, Vaitl D, Wolf OT, Stark R. Emotional modulation of the attentional blink: the neural structures involved in capturing and holding attention. Neuropsychologia. 2011;49(3):416–25.
154. Fusar-Poli P, Placentino A, Carletti F, Landi P, Allen P, Surguladze S, et al. Functional atlas of emotional faces processing: a voxel-based meta-analysis of 105 functional magnetic resonance imaging studies. J Psychiatry Neurosci. 2009.
155. Dima DC, Perry G, Messaritaki E, Zhang J, Singh KD. Spatiotemporal dynamics in human visual cortex rapidly encode the emotional content of faces. Hum Brain Mapp. 2018;39(10):3993–4006.

156. Fox E, Lester V, Russo R, Bowles R, Pichler A, Dutton K. Facial expressions of emotion: Are angry faces detected more efficiently? Cogn Emot. 2000;14(1):61–92.
157. Beck AT. Depression: Clinical, experimental, and theoretical aspects. University of Pennsylvania Press; 1967.
158. Vermeulen N, Godefroid J, Mermillod M. Emotional modulation of attention: fear increases but disgust reduces the attentional blink. PLoS ONE. 2009;4(11):e7924.
159. Phan KL, Wager T, Taylor SF, Liberzon I. Functional neuroanatomy of emotion: a meta-analysis of emotion activation studies in PET and fMRI. NeuroImage. 2002;16(2):331–48.
160. Murphy FC, Nimmo-Smith I, Lawrence AD. Functional neuroanatomy of emotions: a meta-analysis. Cogn, Affect, Behav Neurosci. 2003;3(3):207–33.
161. Berridge K, Kahneman D, Diener E, Schwarz N. Well-being: the foundations of hedonic psychology. New York: Russell Sage Foundation; 1999. p. 527–59.
162. Holland PC, Gallagher M. Amygdala circuitry in attentional and representational processes. Trends Cogn Sci. 1999;3(2):65–73.
163. Jabbi M, Bastiaansen J, Keysers C. A common anterior insula representation of disgust observation, experience and imagination shows divergent functional connectivity pathways. PLoS ONE. 2008;3(8):e2939.
164. Sprengelmeyer R, Rausch M, Eysel UT, Przuntek H. Neural structures associated with recognition of facial expressions of basic emotions. Proc R Soc Lond B. 1998;265(1409):1927–31.
165. Gorno-Tempini ML, Pradelli S, Serafini M, Pagnoni G, Baraldi P, Porro C, et al. Explicit and incidental facial expression processing: an fMRI study. Neuroimage. 2001;14(2):465–73.
166. Jehna M, Neuper C, Ischebeck A, Loitfelder M, Ropele S, Langkammer C, et al. The functional correlates of face perception and recognition of emotional facial expressions as evidenced by fMRI. Brain Res. 2011;1393:73–83.
167. Schienle A, Stark R, Walter B, Blecker C, Ott U, Kirsch P, et al. The insula is not specifically involved in disgust processing: an fMRI study. NeuroReport. 2002;13(16):2023–6.
168. Delgado MR. Reward-related responses in the human striatum. Ann N Y Acad Sci. 2007;1104(1):70–88.
169. Calder AJ, Keane J, Lawrence AD, Manes F. Impaired recognition of anger following damage to the ventral striatum. Brain. 2004;127(9):1958–69.
170. Klucken T, Tabbert K, Schweckendiek J, Merz CJ, Kagerer S, Vaitl D, et al. Contingency learning in human fear conditioning involves the ventral striatum. Hum Brain Mapp. 2009;30(11):3636–44.
171. Mayberg HS, Liotti M, Brannan SK, McGinnis S, Mahurin RK, Jerabek PA, et al. Reciprocal limbic-cortical function and negative mood: converging PET findings in depression and normal sadness. Am J Psychiatry. 1999;156(5):675–82.
172. Blair R, Morris JS, Frith CD, Perrett DI, Dolan RJ. Dissociable neural responses to facial expressions of sadness and anger. Brain. 1999;122(5):883–93.
173. Killgore WD, Yurgelun-Todd DA. Activation of the amygdala and anterior cingulate during nonconscious processing of sad versus happy faces. Neuroimage. 2004;21(4):1215–23.
174. Pourtois G, Schettino A, Vuilleumier P. Brain mechanisms for emotional influences on perception and attention: what is magic and what is not. Biol Psychol. 2013;92(3):492–512.
175. Gibb BE, McGeary JE, Beevers CG. Attentional biases to emotional stimuli: key components of the RDoC constructs of sustained threat and loss. Am J Med Genet Part B: Neuropsychiatr Genet. 2016;171(1):65–80.
176. Armstrong T, Olatunji BO. Eye tracking of attention in the affective disorders: a meta-analytic review and synthesis. Clin Psychol Rev. 2012;32(8):704–23.
177. O'Toole LJ, DeCicco JM, Berthod S, Dennis TA. The N170 to angry faces predicts anxiety in typically developing children over a two-year period. Dev Neuropsychol. 2013;38(5):352–63.

178. Rossignol M, Campanella S, Bissot C, Philippot P. Fear of negative evaluation and attentional bias for facial expressions: an event-related study. Brain Cogn. 2013;82(3):344–52.
179. Solomon B, DeCicco JM, Dennis TA. Emotional picture processing in children: an ERP study. Dev Cogn Neurosci. 2012;2(1):110–9.
180. Staugaard SR. Threatening faces and social anxiety: a literature review. Clin Psychol Rev. 2010;30(6):669–90.
181. Bruder GE, Kayser J, Tenke CE. Event-related brain potentials in depression: clinical, cognitive and neurophysiologic implications. In: The Oxford handbook of event-related potential components. New York: Oxford University Press; 2012:563–92.
182. Proudfit GH, Bress JN, Foti D, Kujawa A, Klein DN. Depression and event-related potentials: Emotional disengagement and reward insensitivity. Curr Opin Psychol. 2015;4:110–3.
183. Bishop SJ. Neural mechanisms underlying selective attention to threat. Ann N Y Acad Sci. 2008;1129(1):141–52.
184. De Raedt R, Koster EH. Understanding vulnerability for depression from a cognitive neuroscience perspective: a reappraisal of attentional factors and a new conceptual framework. Cogn, Affect, Behav Neurosci. 2010;10(1):50–70.
185. Disner SG, Beevers CG, Haigh EAP, Beck AT. Neural mechanisms of the cognitive model of depression. Nat Rev Neurosci. 2011;12:467.
186. Frewen PA, Dozois DJ, Joanisse MF, Neufeld RW. Selective attention to threat versus reward: meta-analysis and neural-network modeling of the dot-probe task. Clin Psychol Rev. 2008;28(2):308–38.
187. Ochsner KN, Gross JJ. The cognitive control of emotion. Trends Cogn Sci. 2005;9(5):242–9.
188. Ochsner KN, Bunge SA, Gross JJ, Gabrieli JD. Rethinking feelings: an FMRI study of the cognitive regulation of emotion. J Cogn Neurosci. 2002;14(8):1215–29.
189. Pourtois G, Spinelli L, Seeck M, Vuilleumier P. Temporal precedence of emotion over attention modulations in the lateral amygdala: Intracranial ERP evidence from a patient with temporal lobe epilepsy. Cogn, Affect, & Behav Neurosci. 2010;10(1):83–93.
190. Victor TA, Furey ML, Fromm SJ, Öhman A, Drevets WC. Relationship between amygdala responses to masked faces and mood state and treatment in major depressive disorder. Arch Gen Psychiatry. 2010;67(11):1128–38.
191. Sheline YI, Barch DM, Donnelly JM, Ollinger JM, Snyder AZ, Mintun MA. Increased amygdala response to masked emotional faces in depressed subjects resolves with antidepressant treatment: an fMRI study. Biol Psychiat. 2001;50(9):651–8.
192. Bradley BP, Mogg K, Williams R. Implicit and explicit memory for emotion-congruent information in clinical depression and anxiety. Behav Res Ther. 1995;33(7):755–70.
193. Williams JM, Broadbent K. Autobiographical memory in suicide attempters. J Abnorm Psychol. 1986;95(2):144.
194. Hammen C, Zupan BA. Self-schemas, depression, and the processing of personal information in children. J Exp Child Psychol. 1984;37(3):598–608.
195. van Vreeswijk MF, de Wilde EJ. Autobiographical memory specificity, psychopathology, depressed mood and the use of the autobiographical memory test: a meta-analysis. Behav Res Ther. 2004;42(6):731–43.
196. Teasdale JD, Dent J. Cognitive vulnerability to depression: an investigation of two hypotheses. Br J Clin Psychol. 1987;26(2):113–26.
197. Denny EB, Hunt RR. Affective valence and memory in depression: dissociation of recall and fragment completion. J Abnorm Psychol. 1992;101(3):575.
198. Barry ES, Naus MJ, Rehm LP. Depression and implicit memory: Understanding mood congruent memory bias. Cogn Ther Res. 2004;28(3):387–414.
199. Gaddy MA, Ingram RE. A meta-analytic review of mood-congruent implicit memory in depressed mood. Clin Psychol Rev. 2014;34(5):402–16.
200. Arnold NR, Bayen UJ, Böhm MF. Is prospective memory related to depression and anxiety? A hierarchical MPT modelling approach. Memory. 2015;23(8):1215–28.

201. Fitzgerald DA, Arnold JF, Becker ES, Speckens AEM, Rinck M, Rijpkema M, et al. How mood challenges emotional memory formation: an fMRI investigation. NeuroImage. 2011;56(3):1783–90.
202. Arnold JF, Fitzgerald DA, Fernández G, Rijpkema M, Rinck M, Eling PA, et al. Rose or black-coloured glasses? Altered neural processing of positive events during memory formation is a trait marker of depression. J Affect Disord. 2011;131(1–3):214–23.
203. van Wingen GA, van Eijndhoven P, Cremers HR, Tendolkar I, Verkes RJ, Buitelaar JK, et al. Neural state and trait bases of mood-incongruent memory formation and retrieval in first-episode major depression. J Psychiatr Res. 2010;44(8):527–34.
204. Harmer D, Phil CJ, O'Sullivan U, Favaron E, Massey-Chase R, Ayres R, Reinecke A, et al. Effect of acute antidepressant administration on negative affective bias in depressed patients. Am J Psychiatry. 2009;166(10):1178–84.
205. van der Veen FM, Evers EA, Deutz NE, Schmitt JA. Effects of acute tryptophan depletion on mood and facial emotion perception related brain activation and performance in healthy women with and without a family history of depression. Neuropsychopharmacology. 2007;32(1):216.
206. Klaassen T, Riedel W, Deutz N, Van Praag H. Mood congruent memory bias induced by tryptophan depletion. Psychol Med. 2002;32(1):167–72.
207. Kensinger EA, Corkin S. Two routes to emotional memory: Distinct neural processes for valence and arousal. Proc Natl Acad Sci. 2004;101(9):3310–5.
208. Sharot T, Delgado MR, Phelps EA. How emotion enhances the feeling of remembering. Nat Neurosci. 2004;7(12):1376.
209. Mickley Steinmetz KR, Kensinger EA. The effects of valence and arousal on the neural activity leading to subsequent memory. Psychophysiology. 2009;46(6):1190–9.
210. Gross JJ, Thompson RA. Emotion regulation. Conceptual Foundations; 2007.
211. Gross JJ. The emerging field of emotion regulation: an integrative review. Rev Gen Psychol. 1998;2(3):271–99.
212. Koole SL. The psychology of emotion regulation: an integrative review. Cogn Emot. 2009;23(1):4–41.
213. Hartley CA, Phelps EA. Changing fear: the neurocircuitry of emotion regulation. Neuropsychopharmacology. 2010;35:136.
214. Gross JJ, John OP. Individual differences in two emotion regulation processes: implications for affect, relationships, and well-being. J Pers Soc Psychol. 2003;85(2):348.
215. Cutuli D. Cognitive reappraisal and expressive suppression strategies role in the emotion regulation: an overview on their modulatory effects and neural correlates. Front Syst Neurosci. 2014;8:175.
216. Ochsner KN, Ray RD, Cooper JC, Robertson ER, Chopra S, Gabrieli JD, et al. For better or for worse: neural systems supporting the cognitive down-and up-regulation of negative emotion. Neuroimage. 2004;23(2):483–99.
217. Dörfel D, Lamke J-P, Hummel F, Wagner U, Erk S, Walter H. Common and differential neural networks of emotion regulation by detachment, reinterpretation, distraction, and expressive suppression: a comparative fMRI investigation. Neuroimage. 2014;101:298–309.
218. McRae K, Hughes B, Chopra S, Gabrieli JDE, Gross JJ, Ochsner KN. The neural bases of distraction and reappraisal. J Cogn Neurosci. 2009;22(2):248–62.
219. Pavlov SV, Reva NV, Loktev KV, Tumyalis AV, Korenyok VV, Aftanas LI. The temporal dynamics of cognitive reappraisal: cardiovascular consequences of downregulation of negative emotion and upregulation of positive emotion. Psychophysiology. 2014;51(2):178–86.
220. Jonckheere JD, Rommel D, Nandrino J, Jeanne M, Logier R, editors. Heart rate variability analysis as an index of emotion regulation processes: Interest of the Analgesia Nociception Index (ANI). 2012 Annual international conference of the IEEE engineering in medicine and biology society; 2012 28 Aug.–1 Sept. 2012.
221. Di Simplicio M, Costoloni G, Western D, Hanson B, Taggart P, Harmer C. Decreased heart rate variability during emotion regulation in subjects at risk for psychopathology. Psychol Med. 2012;42(8):1775–83.

222. Phillips ML, Ladouceur CD, Drevets WC. A neural model of voluntary and automatic emotion regulation: implications for understanding the pathophysiology and neurodevelopment of bipolar disorder. Mol Psychiatry. 2008;13(9):833.
223. Aldao A, Nolen-Hoeksema S, Schweizer S. Emotion-regulation strategies across psychopathology: a meta-analytic review. Clin Psychol Rev. 2010;30(2):217–37.
224. Morris RW, Sparks A, Mitchell PB, Weickert CS, Green MJ. Lack of cortico-limbic coupling in bipolar disorder and schizophrenia during emotion regulation. Transl Psychiatry. 2012;2:e90.
225. Picó-Pérez M, Radua J, Steward T, Menchón JM, Soriano-Mas C. Emotion regulation in mood and anxiety disorders: a meta-analysis of fMRI cognitive reappraisal studies. Prog Neuropsychopharmacol Biol Psychiatry. 2017;79:96–104.
226. Kalisch R, Wiech K, Critchley HD, Seymour B, O'doherty JP, Oakley DA, et al. Anxiety reduction through detachment: subjective, physiological, and neural effects. J Cogn Neurosci. 2005;17(6):874–83.
227. Berking M, Wupperman P, Reichardt A, Pejic T, Dippel A, Znoj H. Emotion-regulation skills as a treatment target in psychotherapy. Behav Res Ther. 2008;46(11):1230–7.
228. Mennin DS. Emotion regulation therapy: an integrative approach to treatment-resistant anxiety disorders. J Contemp Psychother. 2006;36(2):95–105.
229. Geschwind N, Peeters F, Drukker M, van Os J, Wichers M. Mindfulness training increases momentary positive emotions and reward experience in adults vulnerable to depression: a randomized controlled trial. J Consult Clin Psychol. 2011;79(5):618.
230. Goldin PR, Gross JJ. Effects of mindfulness-based stress reduction (MBSR) on emotion regulation in social anxiety disorder. Emotion. 2010;10(1):83.
231. Barrett LF. Functionalism cannot save the classical view of emotion. Soc Cogn Affect Neurosci. 2016;12(1):34–6.
232. Clark JE, Watson S, Friston KJ. What is mood? A computational perspective. Psychol Med. 2018;48(14):2277–84.
233. Craig A. How do you feel? An interoceptive moment with your neurobiological self. Princeton: Princeton University Press; 2015.
234. Barrett LF, Simmons WK. Interoceptive predictions in the brain. Nat Rev Neurosci. 2015;16(7):419.
235. Chanes L, Barrett LF. Redefining the role of limbic areas in cortical processing. Trends Cogn Sci. 2016;20(2):96–106.
236. Seth AK, Friston KJ. Active interoceptive inference and the emotional brain. Philos Trans R Soc B: Biol Sci. 2016;371(1708):20160007.
237. Wager TD, Kang J, Johnson TD, Nichols TE, Satpute AB, Barrett LF. A Bayesian model of category-specific emotional brain responses. PLoS Comput Biol. 2015;11(4):e1004066.
238. Joffily M, Coricelli G. Emotional valence and the free-energy principle. PLoS Comput Biol. 2013;9(6):e1003094.
239. Barrett LF, Satpute AB. Large-scale brain networks in affective and social neuroscience: towards an integrative functional architecture of the brain. Curr Opin Neurobiol. 2013;23(3):361–72.
240. Gu X, Hof PR, Friston KJ, Fan J. Anterior insular cortex and emotional awareness. J Comp Neurol. 2013;521(15):3371–88.
241. Huys QJ, Daw ND, Dayan P. Depression: a decision-theoretic analysis. Annu Rev Neurosci. 2015;38:1–23.

Resilience as a Psychopathological Construct for Psychiatric Disorders

Amresh Shrivastava, Avinash De Sousa and Pragya Lodha

Abstract

Understanding of the psychopathology of various psychiatric disorders is evolving, with newer avenues of research enlightening us from genetics, epigenetics, functional neurobiology, neural circuits, hormones and social/environmental determinants. We are now aware that neurobiological factors are contributing to the development of psychiatric illnesses coupled with their interaction with psychosocial factors. Resilience is defined as the ability to bounce back after an adversity or life event that was traumatic and life-changing. It is a factor that is a unique psychopathological construct as it is a biopsychosocial factor which determines an individual's response to an illness and recovery from the same. Resilience is a human capacity to adapt swiftly and successfully to stress and to revert to a positive state. There has been now a paradigm shift in the understanding of resilience with respect to stress risk vulnerability and such dimensions of psychopathology. Resilience is a factor that must be evaluated in every patient and that shall help us determine the outcome

A. Shrivastava
Parkwood Institute, The Western University, Mental Health Care Building, 550 Wellington Road, London, ON N6C 0A7, Canada
e-mail: dr.amresh@gmail.com

A. De Sousa (✉)
Department of Psychiatry, Lokmanya Tilak Municipal Medical College, Mumbai, India
e-mail: avinashdes888@gmail.com

Carmel, 18, St. Francis Road, off SV Road, Santacruz West, 400054 Mumbai, India

P. Lodha
Desousa Foundation, Mumbai, India
e-mail: pragya6lodha@gmail.com

© Springer Nature Singapore Pte Ltd. 2019

Y.-K. Kim (ed.), *Frontiers in Psychiatry*, Advances in Experimental Medicine and Biology 1192, https://doi.org/10.1007/978-981-32-9721-0_23

of psychiatric disorders and will also be a determinant in the occurrence of relapses. Early identification of vulnerable patients shall lead to the implementation of resilience-based interventions in these populations and shall prevent against future occurrence of these disorders. In this chapter, we posit the construct of resilience as a psychopathological construct for mental disorders.

Keywords
Resilience · Psychopathology · Recovery · Construct · Psychiatric disorder · Neurobiology · Biopsychosocial

Introduction

The present chapter evaluates the concept of resilience as psychopathological construct in the development of psychiatric disorders. The role of resilience is multipronged from being an antecedent factor, to a factor in the genesis and pathophysiology of psychiatric disorder and emerging as both a preventive and promoting factor in recovery and outcome from psychiatric disorder. It is a single factor that is intertwined in all phases and during the entire course of a psychiatric disorder and one which is determined genetically, by neurobiology and psychosocial determinants. The chapter evaluates the role of resilience as a unique construct of psychopathology from a clinical perspective.

The Concept of Resilience

Resilience is defined as the ability of an individual to bounce back after trauma or stressful event to previous levels of functioning and even better levels of functioning [1]. Resilience is a construct that plays a role in mediating stress and protecting the individual from adverse events or stimuli that may challenge one's psychobiological homeostasis [2]. While resilience may mediate the stress response of trauma, a high level of resilience works as a protective factor, and a lower level of resilience increases the vulnerability for developing a psychiatric disorder [3]. Resilience is a construct with inherent biological, social and psychological factors that affect it and consequently lead to substantial ramifications [4]. Resilience has also been implicated as a marker of psychopathology in various disorders and plays a role in the recovery from trauma and stressful events [5].

The stress-diathesis model explains the biopsychosocial vulnerability and predisposition of risk that an individual develops towards a psychiatric disorder. Some individuals may withstand the worst stressors and still not undergo a psychological collapse. Resilience thus reflects a positive spectrum of adaptation and maladaptation in response and exposure to risk factors [6].

Resilience is a protective factor against the development of mental disorder and a risk factor for a number of clinical conditions. Main features describing resilience include good outcomes regardless of high-risk statuses, constant competence under stress and recovery from a trauma considered as challenging for growth, which helps making future hardships more tolerable [7]. Resilience is a multidimensional concept variously defined as a personal trait protective against mental disorders and a dynamic process of adaptation to challenging life conditions. It expresses psychological flexibility (bounceback) face to highly stressful events operating relevant changes: it is devoted to obtain new and positive homeostasis and help the individual emerge from adversities stronger and more resourceful [8].

Resilience has now received a lot of attention in studies on major psychiatric disorders like schizophrenia, substance abuse and bipolar disorder [9]. Resilience is now emerging as a vital concept in preventive psychiatry and mental health promotion with resilience promoting interventions being discussed in various preventive mental health community programmes [10]. Resilience as one of the predictors of sound mental health is now receiving the attention as a construct that could be developed in an individual for better combat from illnesses [11]. Resilience is understood as a concept that has emerged from positive psychology and is looked at as a construct for describing and explaining unexpected positive outcomes in an individual, despite high risks of maladjustment and exposures to psychosocial adversities [12].

Psychosocial Factors and Resilience

There are several psychosocial factors that determine the 'level' of resilience in an individual; however, many of these factors have not been identified. Out of the many, there is a reasonable possibility that there may be underlying cognitive features and neurobiological factors that couple onto psychosocial factors and play a significant role in determining how appropriate and sufficient a response to stress or trauma is and/or how well the symptoms are minimized or maximized in a specific case [13]. Resilience is most widely understood as a human capacity to adapt swiftly and successfully to stressful or traumatic events and to be able to revert to a positive state of being and function [14].

The extent to which resilience may act as a preventive measure also depends on various psychosocial factors that an individual is exposed to. The primary psychosocial factors include active coping, cognitive flexibility and social support available to an individual [15]. Active coping is one's lifelong acquired ability to utilize psychological and behavioural resources to cope with trauma or stressors [16]. Active coping is an effective means of resilience which develops better given an individual is placed in an environment that encourages healthy development of intrinsic psychological and behavioural traits. Cognitive flexibility adds to resilience strength by allowing the person to adapt to rising challenges that come from changing circumstances. The ability to bend one's cognition to better deal with trauma arises from acquired knowledge which developed throughout one's life, and it forms the bases for flexibility in cognitive resilience [17]. The social environment of an individual is equally important as it allows the individual to respond to stressful events and adversities in a positive manner. The quality of social support available, the number of individuals the person is networked with and the significance of social relationships that the individual has are all crucial determinants of the strength of resilience [18]. These are a combination of intrinsic (coping style and cognitive flexibility) and extrinsic (social support) factors that have a crucial interplay in determining the overall resilience in an individual [15].

There are many other psychosocial factors that have been implicated across various studies in the development and maintenance of resilience. These include loneliness [19], early life adversity [20], unemployment [21], financial status and poverty [22] and the presence of psychopathology in the family [23].

The better developed these factors in an individual, better and stronger is the resilience in that individual. Resilient individuals are also characterized by dispositional optimism and high positive emotionality [24]. Apart from the above mentioned, other psychosocial characteristics associated with stress resilience include a sense of purpose in life, a moral compass, spirituality and the ability to find meaning in the midst of trauma [25]. There is also a role of family resilience that contributes to individual resilience which is the resilience of various family members when summed up. Thus, even if a particular individual within a family may have low resilience, the overall resilience of the family being strong protects him from vulnerability [26]. An extension of this concept is the role of community resilience that results in the protection of the weaker sections of society and nurturing them in times of crisis. This has been particularly noted in times of how a community reacts in the wake of disasters and natural calamities [27].

Neurobiological Factors in Resilience

In neurobiological paradigms, resilience is seen as an independent domain and there is increasing neurobiological evidence regarding the existing molecular structures and neural circuits that are implicated in protection of stress-related neuropsychiatric disorders. There are various areas of the brain involved in responding to stress and thus are equally implicated in the capacity of resilience in an individual. There is an interplay of hormones, neurotransmitters and neuropeptides which are found implicated in the acute psychobiological responses to stress. Differences in the function, balance and interaction of these factors underlie inter-individual variability in stress resilience [25].

The corticotropin-releasing hormone is released by the hypothalamus, in response to stress, which activates the hypothalamus–pituitary–adrenal axis (HPA axis) and leads to the release of cortisol. Although the short-term actions of cortisol are protective and promote adaptation, sustained exposure to abnormally high levels of cortisol can be harmful, leading to hypertension, immunosuppression, cardiovascular disease and other health problems [28]. Glucocorticoids (GCs) are situated to either promote or prevent adaption to stress. They are considered to be important regulators of basal and stress-related homeostasis that influence a wide array of genes in almost every organ and tissue [29] (Fig. 1).

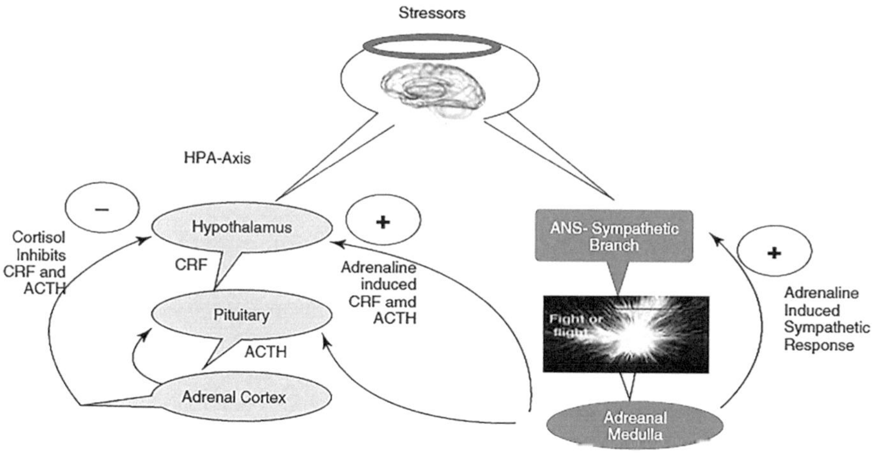

Fig. 1 CRH and the stress response

Noradrenergic systems as well as serotonergic and dopaminergic systems are also related to acute stress response [30]. Neuropeptide Y (NPY) is widely distributed in the brain, has anxiolytic-like effects in rodents and is thought to enhance cognition under stressful conditions. NPY also counteracts the anxiogenic effects of CRH in the amygdala, the hippocampus, the hypothalamus and the locus coeruleus, and resilience might involve maintaining a balance between NPY and CRH levels during stress [31]. There is evolving literature in the psychoneuroendocrinology of resilience where multiple hormones like galanin [32], testosterone [33] and estrogens [34] being implicated in the genesis and maintenance of resilience.

Brain-derived neurotrophic factor (BDNF) is one of the crucial elements that affect different brain regions. Though its role has not been well researched in resilience, it has been seen that raised levels of BDNF have been found in nucleus accumbens in states of chronic stress and depression. There is now research to indicate that BDNF is a protective factor that plays a role in development or resilience and low BDNF levels increases the vulnerability to develop a psychiatric disorder [35].

Neuronal systems underlying positive traits of an individual are also being explored. Current investigations have shown the possible exploratory role of neuronal systems in treatment and prevention of mental disorders [36]. Additionally, it has also been identified that changes in neuroplasticity [37], HPA axis response to stress and neurotransmissions of dopamine, serotonin and norepinephrine play an important role in maintaining homeostasis of resilience plasticity [38].

Complex interactions between an individual's genetic make-up and their specific history of exposure to environmental stressors have also been found to determine the degree of adaptability of neurochemical stress response systems to new adverse exposures, as well as the function of the neural circuitry involved in stress responses [39].

A growing number of brain imaging studies have attempted to define the brain circuits that mediate distinct aspects of mood and emotion under normal circumstances and in various pathological conditions which are indicative of low resilience. Several limbic regions have been identified in the forebrain, which are highly interconnected and function as a series of integrated parallel circuits that regulate emotional states. There are three main circuits, that of fear, rewards and emotional regulation [40].

Conditions characterized by diminished resilience on exposure to a traumatic stressor, involve abnormal fear learning and an underlying dysfunction in the neural circuitry of fear, comprising the amygdala, the hippocampus and the ventromedial prefrontal cortex (vmPFC) [41]. The neural circuitry of fear is found to be important in resilience, but it has not been studied well in resilient individuals and thus, greater evidence is awaited to draw concrete parallels. Trait optimism is linked to resilience, and it is hypothesized that it may also be related to reward circuit function [42]. In a study, researchers scanned participants who were imagining positive and negative future events. The phenomenon of optimism bias was studied. It is the tendency to expect future events to be positive—which was found associated with higher activation in the amygdala and the rostral anterior cingulate cortex (ACC) when imagining positive events than when imagining negative events. Thus, the level of activation in the rostral ACC was positively correlated with dispositional optimism [43]. The circuit of emotional regulation includes regions of amygdala, hippocampus, subgenual ACC and PFC function. In addition, there are circuits for social behaviour as well those comprise mirror neurons (cortical neurons), limbic system and the vmPFC. The capacity for empathy enables individuals to generate appropriate emotional responses in social contexts and might be related to social competence, which is a characteristic of resilient individuals [44].

There is a need for future research to look at genetic and epigenetic factors that may be responsible for the development of resilience. There is also a need to study and identify neural circuits that play a role and may thus elicit a neural circuitry and pathway that can be implicated in the development of resilience.

Resilience as a Psychopathological Construct

The understanding of resilience is still in its developing phases; recent investigations reveal that there are various mechanisms including factors such as genetic, epigenetic, developmental, psychological and neurochemical that underlie the development of and building of resilience. These are also the factors that predict (vulnerability to stress and) susceptibility to psychiatric disorders in the face of stress and trauma [45].

Resilience is an individualized and contextualized construct that builds depending on the experience of an individual. The internal and external sources of support and protection available to the individual also help determine the extent of the level of resilience that may prevail in an individual. Among other crucial factors, early experiences in the course of life promote the development of healthy brain architecture supporting cognitive flexibility that allows the brain to continue to change with ongoing experiences. There is no one agreed-upon definition of resilience but the construct is understood under various lenses depending on the context of study [46].

It is essential to remember that resilience can be a protective factor against the development of mental disorders and also be a risk factor for a number of clinical conditions. It has been well established that resilience is lower in individuals suffering from a psychiatric condition or disorder and that higher levels of resilience may prevent the development of an illness or minimize the severity of illness [47].

Resilience is being studied as a marker for the response of an individual to treatments; it defines how fast an individual may recover from psychiatric illness. It also defines the response and participation of an individual in psychotherapy and the response of a patient to therapeutic interventions. Thus, resilience is a predictor of psychopathology; it is a psychopathological construct itself and can also serve as a marker for risk and recovery towards psychiatric disorders.

Resilience-Based Interventions and Psychopathology

Studies to determine the effectiveness of resilience-enhancement in youth have been done in Russia. The study offered combined recreational sport and psychosocial rehabilitation for 94 participants who were taken hostage in the 2004 school tragedy. The results indicated that significant intraparticipant mean increase in resilience at follow-up assessment, and greater self-reported improvement in resilience processes for participants who experienced more traumatic events. Therefore, resilience can be modified and enhanced by metallization and cognitive training [48].

There is another study where researchers have examined the effect of mindfulness training on resilience mechanisms in active-duty marines preparing for deployment [49]. The mindfulness training condition was in the form of 8 weeks of Mindfulness-based Mind Fitness Training (MMFT), with 20 hours of the classroom instruction and daily homework exercises. MMFT emphasizes interceptive awareness, attention control and tolerance of present-moment experiences. The main outcome measures were heart rate, breathing rate, plasma neuropeptide Y concentration, score on the response to stressful experiences scale, and brain activation as measured by functional magnetic resonance imaging. The results showed that mechanisms related to stress recovery could be modified in healthy individuals prior to stress exposure [49].

Resilience programmes can also be developed in order to increase the effectiveness of a treatment and improve the outcome of the treatment process. In this, respect and intervention can increase resilience and as such, high resilience can enhance the effectiveness of a treatment. This has a strong clinical implication for physical disorder as well as mental disorders, especially in how the treatment is developed and delivered [50]. Similarly, resilience plays an important role in the overall well-being of individuals [51]. Some resilience researchers believe that personal resilience is significant for responding to workplace and adult adversity [52], and when someone experiences childhood adversity, they seem to demonstrate a certain level of posttraumatic growth.

Resilience is now being thought of as a paradigm to be introduced in nursing curricula, medical curricula, corporate wellness training and as part of psychiatric rehabilitation programmes in substance abuse and other psychiatric disorders. It is also an integral part of emotional intelligence interventions that may increase coping resources and enhance social skills for nurses and doctors, which may benefit their long-term occupational health [52]. Resilience has a pivotal role in the wake of trauma and natural disasters. In one study, with Palestinian school children, it has been reported that resilience-based interventions did not increase the level of resilience. The effect of the intervention was moderated by maternal attachment responses and/or family atmosphere [53]. Therefore, it must be noted that the level of adversity which creates a beneficial level of posttraumatic resilience training is finite and is modified by various familial and psychosocial factors.

There is also a newer emerging role of resilience enhancement training with at-risk children and adolescents and school-based resilience enhancement interventions are being aimed at targeting high-risk groups and in the prevention of development of psychopathology in childhood and adolescence. This in turn may also help in the prevention of adult psychopathology [54].

Conclusion

Resilience, an individual's capacity for successful adaptation to acute stress, trauma or more chronic forms of adversity, is determined by the functional capacity of the brain structures that are involved in the integrated circuits that mediate mood and emotion. This is reflected in an individual's psychological make-up. An individual's level of resilience is better if they have a greater adaptive functioning of fear, reward, emotion regulation or social-behaviour circuits. This helps them to better face fears, experience positive emotions, increase their ability to search for positive ways to reframe stressful events and derive benefit from supportive relationships. In conclusion, resilience is an active process, and not just the absence of (psycho)pathology, and it can be promoted by enhancing protective factors. Greater research is required to understand stress resistance and resilience in individuals. There are several other factors that may better help to determine resilience levels in an individual.

It is not yet known whether resilience is a modifiable or non-modifiable factor. The role of psychopharmacological treatments in the enhancement of resilience is still unknown. Thus, we need future research to elucidate the neurobiology of resilience and mechanisms that promote it in order to enhance resilience as a modifiable psychopathological construct in the recovery and outcome from psychiatric disorders.

References

1. Norris FH, Tracy M, Galea S. Looking for resilience: understanding the longitudinal trajectories of responses to stress. Soc Sci Med. 2009;68:2190–8.
2. Bonanno GA. Loss, trauma, and human resilience: have we underestimated the human capacity to thrive after extremely aversive events? Am Psychol. 2004;59:20–8.
3. Tusaie K, Dyer J. Resilience: a historical review of the construct. Holistic Nurs Pract. 2004;18(1):3–10.
4. Southwick SM, Vythilingam M, Charney DS. The psychobiology of depression and resilience to stress: implications for prevention and treatment. Annu Rev Clin Psychol. 2005;1:255–91.
5. Rutter M. Annual research review: resilience–clinical implications. J Child Psychol Psychiatry. 2013;54(4):474–87.
6. Stratta P, Capanna C, Dell'Osso I, Carmassi C, Patriarca S, Di Emidio G, Riccardi CA, Rossi A. Resilience and coping in trauma spectrum symptoms prediction: a structural equation modeling approach. Pers Individ Diff. 2015;77:55–61.
7. Agaibi CE, Wilson JP. Trauma, PTSD, and resilience: a review of the literature. Traum Viol Abuse. 2005;6(3):195–216.
8. Fletcher D, Sarkar M. Psychological resilience: a review and critique of definitions, concepts, and theory. Eur Psychol. 2013;18(1):12–8.
9. Windle G. What is resilience? A review and concept analysis. Rev Clin Gerontol. 2011;21(2):152–69.
10. Leve LD, Harold GT, Chamberlain P, Landsverk JA, Fisher PA, Vostanis P. Practitioner review: children in foster care–vulnerabilities and evidence-based interventions that promote resilience processes. J Child Psychol Psychiatry. 2012;53(12):1197–211.
11. Almedom AM. Resilience, hardiness, sense of coherence, and posttraumatic growth: all paths leading to "light at the end of the tunnel"? J Loss Trauma. 2005;10(3):253–65.
12. Yates TM, Masten AS. Fostering the future: resilience theory and the practice of positive psychology. Posit Psychol Pract. 2004;30:521–39.
13. Ong AD, Bergeman CS, Bisconti TL, Wallace KA. Psychological resilience, positive emotions, and successful adaptation to stress in later life. J Pers Soc Psychol. 2006;91(4):730–8.
14. Youssef CM, Luthans F. Positive organizational behavior in the workplace: the impact of hope, optimism, and resilience. J Manag. 2007;33(5):774–800.
15. Shrivastava A, Desousa A. Resilience: a psychobiological construct for psychiatric disorders. Indian J Psychiatry. 2016;58:38–43.
16. Steinhardt M, Dolbier C. Evaluation of a resilience intervention to enhance coping strategies and protective factors and decrease symptomatology. J Am Coll Health. 2008;56(4):445–53.
17. Dumon M, Provost MA. Resilience in adolescents: protective role of social support, coping strategies, self-esteem, and social activiies on experiences of stress and depression. J Youth Adolesc. 1999;28(3):343–63.
18. Ozbay F, Johnson DC, Dimoulas E, Morgan CA III, Charney D, Southwick S. Social support and resilience to stress: from neurobiology to clinical practice. Psychiatry (Edgmont). 2007;4(5):35–40.
19. Adams KB, Sanders S, Auth EA. Loneliness and depression in independent living retirement communities: risk and resilience factors. Aging Ment Health. 2004;8(6):475–85.
20. Daskalakis NP, Bagot RC, Parker KJ, Vinkers CH, de Kloet ER. The three-hit concept of vulnerability and resilience: toward understanding adaptation to early-life adversity outcome. Psychoneuroendocrinology. 2013;38(9):1858–73.
21. Sojo V, Guarino L. Mediated moderation or moderated mediation: relationship between length of unemployment, resilience, coping and health. Span J Psychol. 2011;14(1):272–81.
22. Bradley RH, Whiteside L, Mundfrom DJ, Casey PH, Kelleher KJ, Pope SK. Early indications of resilience and their relation to experiences in the home environments of low birthweight, premature children living in poverty. Child Dev. 1994;65(2):346–60.

23. Flouri E, Tzavidis N, Kallis C. Adverse life events, area socioeconomic disadvantage, and psychopathology and resilience in young children: the importance of risk factors' accumulation and protective factors' specificity. Eur Child Adolesc Psychiatry. 2010;19(6):535–46.
24. Segovia F, Moore JL, Linnville SE, Hoyt RE, Hain RE. Optimism predicts resilience in repatriated prisoners of war: a 37-year longitudinal study. J Traum Stress. 2012;25(3):330–6.
25. Feder A, Nestler EJ, Charney DS. Psychobiology and molecular genetics of resilience. Nat Rev Neurosci. 2009;10(6):446–57.
26. Walsh F. Family resilience: a framework for clinical practice. Fam Process. 2003;42(1):1–8.
27. Walsh F. Traumatic loss and major disasters: strengthening family and community resilience. Fam Process. 2007;46(2):207–27.
28. Franklin TB, Saab BJ, Mansuy IM. Neural mechanisms of stress resilience and vulnerability. Neuron. 2012;75(5):747–61.
29. Reul JM, Collins A, Saliba RS, Mifsud KR, Carter SD, Gutierrez-Mecinas M, Qian X, Linthorst AC. Glucocorticoids, epigenetic control and stress resilience. Neurobiol Stress. 2015;1:44–59.
30. Berger M, Sarnyai Z. "More than skin deep": stress neurobiology and mental health consequences of racial discrimination. Stress. 2015;18(1):1–10.
31. Cohen H, Liu T, Kozlovsky N, Kaplan Z, Zohar J, Mathé AA. The neuropeptide Y (NPY)-ergic system is associated with behavioral resilience to stress exposure in an animal model of post-traumatic stress disorder. Neuropsychopharmacology. 2012;37(2):350–63.
32. Sciolino NR, Smith JM, Stranahan AM, Freeman KG, Edwards GL, Weinshenker D, Holmes PV. Galanin mediates features of neural and behavioral stress resilience afforded by exercise. Neuropharmacology. 2015;89:255–64.
33. Charney DS. Psychobiological mechanisms of resilience and vulnerability: implications for successful adaptation to extreme stress. Focus. 2004;161(3):195–391.
34. Wei J, Yuen EY, Liu W, Li X, Zhong P, Karatsoreos IN, McEwen BS, Yan Z. Estrogen protects against the detrimental effects of repeated stress on glutamatergic transmission and cognition. Mol Psychiatry. 2014;19(5):588–98.
35. Rothman SM, Mattson MP. Activity-dependent, stress-responsive BDNF signaling and the quest for optimal brain health and resilience throughout the lifespan. Neurosci. 2013;239:228–40.
36. Burgdorf J, Colechio EM, Stanton P, Panksepp J. Positive emotional learning induces resilience to depression: a role for NMDA receptor-mediated synaptic plasticity. Curr Neuropharmacol 2017;15(1):3–10.
37. McEwen BS. In pursuit of resilience: stress, epigenetics, and brain plasticity. Ann NY Acad Sci. 2016;1373(1):56–64.
38. Espallergues J, Teegarden SL, Veerakumar A, Boulden J, Challis C, Jochems J, Chan M, Petersen T, Deneris E, Matthias P, Hahn CG. HDAC6 regulates glucocorticoid receptor signaling in serotonin pathways with critical impact on stress resilience. J Neurosci. 2012;32(13):4400–16.
39. Southwick SM, Charney DS. The science of resilience: implications for the prevention and treatment of depression. Science. 2012;338(6103):79–82.
40. LeDoux JE. Emotion circuits in the brain. Ann Rev Neurosci. 2000;23(1):155–84.
41. Kim JJ, Jung MW. Neural circuits and mechanisms involved in Pavlovian fear conditioning: a critical review. Neurosci Biobehav Rev. 2006;30(2):188–202.
42. Ganzel BL, Kim P, Glover GH, Temple E. Resilience after 9/11: multimodal neuroimaging evidence for stress-related change in the healthy adult brain. Neuroimage. 2008;40(2):788–95.
43. Maynard AS, Monk JD, Booker KW. Building empathy through identification and expression of emotions: a review of interactive tools for children with social deficits. J Creat Ment Health. 2011;6(2):166–75.
44. Iarocci G, Yager J, Elfers T. What gene–environment interactions can tell us about social competence in typical and atypical populations. Brain Cogn. 2007;65(1):112–27.

45. Wu G, Feder A, Cohen H, Kim JJ, Calderon S, Charney DS, Mathé AA. Understanding resilience. Front Behav Neurosci. 2013;7:10.
46. Juster RP, McEwen BS, Lupien SJ. Allostatic load biomarkers of chronic stress and impact on health and cognition. Neurosci Biobehav Rev. 2010;35(1):2–16.
47. Masten AS. Ordinary magic: resilience processes in development. Am Psychol. 2001;56(3):227–54.
48. Makhnach AV. Resilience in Russian youth. Int J Adolesc Youth. 2016;21(2):195–214.
49. Johnson DC, Thom NJ, Stanley EA, Haase L, Simmons AN, Shih PA, Thompson WK, Potterat EG, Minor TR, Paulus MP. Modifying resilience mechanisms in at-risk individuals: a controlled study of mindfulness training in Marines preparing for deployment. Am J Psychiatry. 2014;171(8):844–53.
50. Jha AP, Morrison AB, Parker SC, Stanley EA. Practice is protective: mindfulness training promotes cognitive resilience in high-stress cohorts. Mindfulness. 2017;8(1):46–58.
51. Mguni N, Bacon N, Brown JF. The wellbeing and resilience paradox. London: The Young Foundation; 2012.
52. Brennan EJ. Towards resilience and wellbeing in nurses. Br J Nurs. 2017;26(1):43–7.
53. Shah R. Protecting children in a situation of ongoing conflict: is resilience sufficient as the end product? Int J Disaster Risk Reduct. 2015;14:179–85.
54. Prince-Embury S, Saklofske DH, editors. Resilience in children, adolescents, and adults: translating research into practice. Springer Science & Business Media; 2012.

Role of Inflammation in Psychiatric Disorders

Sang Won Jeon, Ho-kyoung Yoon and Yong-Ku Kim

Abstract

Psychiatric disorders are too multifactorial to be defined as a primarily inflammatory disorder, and increased inflammatory response is not specific to mental disorder only. Numerous factors are involved in neuroinflammation, and there also are many confounding variables, making it difficult to obtain consistent outcomes. Therefore, it is necessary to specify genetic, physiological, and epidemiological attributes of particular population groups vulnerable to inflammatory response as well as the disease subtypes. As of now, definitive inflammatory markers for psychiatric disorders have not been identified, but they could be very useful in patients with minimal vulnerability. In addition, it is possible to use inflammatory markers as depression biomarkers in subtypes of depression, which can serve as bases to develop medications to treat the disorder.

Keywords

Inflammation · Neuroinflammation · Cytokine · Psychiatric disorder · Depression · Medication

S. W. Jeon
Department of Psychiatry, Kangbuk Samsung Hospital, Sungkyunkwan University School of Medicine, 29, Saemunan-ro, Jongno-gu, Seoul 03181, Republic of Korea
e-mail: sangwonyda@hanmail.net

H. Yoon · Y.-K. Kim (✉)
Department of Psychiatry, College of Medicine, Korea University, Ansan Hospital, 123, Jeokgeum ro, Danwon-gu, Ansan-si, Gyeonggi-do 15355, Republic of Korea
e-mail: yongku@korea.ac.kr; yongku@korea.edu

H. Yoon
e-mail: hkhkgogo@korea.ac.kr

© Springer Nature Singapore Pte Ltd. 2019

Y.-K. Kim (ed.), *Frontiers in Psychiatry*, Advances in Experimental Medicine and Biology 1192, https://doi.org/10.1007/978-981-32-9721-0_24

Introduction

It was a fresh idea that inflammation may be contributory to the pathophysiology of psychiatric disorders, and much research has so far been conducted based on the idea. However, whether the scientific bases to link psychiatric disorders to inflammation are a fact or hypothesis continues to be debatable. Though neuroinflammation seems to be associated with various psychiatric disorders such as schizophrenia [1], dementia [2], and anxiety disorder [3], depression (a highly prevalent condition that causes social issues including suicide) is of the greatest interest among them [4]. The first question we may ask is: "Is depression an inflammatory disorder?" The question concerns whether inflammation indeed induces depression or some pathologies present in depression are inflammation. The second question is: "Is there a subtype of depression related to inflammation?" Namely, the question concerns the possibility that a subtype of depression specific to inflammation may exist, even if inflammatory process cannot explain the entirety of developing depression [5]. The last question is: "Is it possible to develop potential therapeutic agents related to inflammation that can be applied in the treatment of depression?" Below, we aim to organize available evidence to answer the question.

Evidence for the Linkage Between Inflammation and Depression

The immune system modulates the nervous system through diverse cytokines produced in immune cells. Cytokines are proteins produced in inflammatory response, primarily found in monocytes, lymphocytes, T cells, B cells, natural killer cells, and fibroblasts, and they deliver information from a cell to another by reacting to cell membrane receptors like a neurotransmitter, or intracellular receptors like a hormone. Representative cytokines are interleukin (IL), interferon (IFN), chemokine, tumor necrosis factor (TNF), and transforming growth factor (TGF). Of those, IL-1, IL-2, IL-6, IFN-γ, and TNF-α are pro-inflammatory cytokines, and IL-4, IL-10, IL-11, IL-13, and TGF-β are anti-inflammatory cytokines [6]. Pro-inflammatory cytokines activate cyclooxygenase-2 (COX-2) to increase prostaglandin E2 (PGE2) and activate inflammatory cells, triggering an inflammatory process. But an increase in cytokines in the cerebrospinal fluid is not considered to be a marker for such condition as encephalitis, because they can be increased in the central nervous system not only by inflammation but also by other diverse factors like stress and ischemia [7].

Cytokines keep in balance through a mutual interaction. As examples, IL-10 suppresses the production of TNF, and a cytokine exists which is antagonistic to IL-1 receptors (i.e., IL-1 receptor antagonist, IL-1ra). Chronic inflammation seems to induce a variety of disorders including depression, because it increases pro-inflammatory cytokines and decreases anti-inflammatory cytokines, disturbing balance among cytokines [8]. Such change is commonly observed in various inflammatory and infectious diseases as well as in metabolic diseases such as atherosclerosis and obesity.

Cytokines generated by the peripheral immune system are directly transported to the central nervous system through the blood–brain barrier in a relatively loose area like choroid plexus or through the vagus nerve. They also have an indirect impact on the brain by decomposing nitric oxide and PGE2 in blood vessels [6]. Cytokines are produced in cerebral neurons and glial cells. Hence, a change in cytokines both induces and reflects an abnormality in the central nervous system. In addition, cytokine receptors are densely distributed in the areas important in mental functioning such as hypothalamus, hippocampus, locus coeruleus, and prefrontal cortex, by which the linkage to psychiatric disorder like depression is explained [9].

IL-1 and IL-2, i.e., representative pro-inflammatory cytokines, on the one hand, increase the synthesis and circulation of serotonin, norepinephrine, and dopamine. On the other hand, they increase the activation of indoleamine-2,3-dioxygenase (IDO), a protein that decomposes tryptophan (a precursor of serotonin), thus inhibiting the production of serotonin [10]. Additionally, in combination with pro-inflammatory cytokines (such as IL-6), norepinephrine and dopamine reduce immune function by facilitating the secretion of CRF and activating the sympathetic nervous system. In this process, the temperature of the central nervous system rises, causing what is called "sickness behavior" [7]. Sickness behavior refers to behavioral change occurring in an individual with inflammation and is accompanied by weakness, depressive symptoms, anxiety, somnolence, loss of appetite, decreased concentration, etc. The hypothesis that depression is a type of sickness behavior has been derived on the basis of the findings that in patients with depression, pro-inflammatory cytokines decreased and anti-inflammatory cytokines increased in blood [11], and PGE2 increased in cerebrospinal fluid [12].

Cytokines are mutually involved in the secretion of neurotransmitters in the central nervous system. Particularly, they modulate the activation of the hypothalamic–pituitary–adrenal (HPA) axis and affect the occurrence of various psychiatric symptoms including depression and anxiety, by working on the hypothalamus, an area playing a central role in the endocrine system. Recently, cytokines were reported to facilitate neuronal differentiation and remodeling in the brain. Accordingly, interest in their role in neurodegenerative diseases is increasing. In the brain of patients with chronic depression, in particular, the ventricles expand, and the volumes of the hippopotamus, prefrontal lobe, and amygdala decrease due to increased apoptosis, and the likelihood for dementia to develop increases [13]. Chronic inflammatory response is speculated to be involved in the process. It has been reported that pro-inflammatory cytokines reduce neuroplasticity by increasing quinolinic acid, a powerful agonist of N-methyl-D-aspartate (NMDA) receptor.

Is Depression an Inflammatory Disorder?

The question of whether depression is an inflammatory disorder began with the study findings that the incidence of depression was high in patients with inflammatory disease and that inflammatory markers increased in patients with major depressive disorder without physical symptoms [14]. The symptoms occurring in as high as 90% of patients administered with an IFN therapy to treat hepatitis C or cancer, i.e., fatigue, depression, restricted affect are an especially strong clue that depression may be an inflammatory disorder. It remains controversial among researchers whether these symptoms should be considered as the symptoms of sickness behavior or chronic fatigue syndrome or the symptoms of depression. However, a study found that more than 50% of patients treated with high-dose IFN-α met the diagnostic criteria for major depressive disorder within 3 months [15] and that IL-6 and TNF-α increased following IFN-α administration, which showed a strong association with increase in depression symptoms [16]. It has been reported that the occurrence of fatigue and depression symptoms induced by IFN-α administration depends on genetic polymorphism of serotonin transporter genes and IL-6 genes [17] and that IFN-α decreases the utilization rate of serotonin by increasing IDO activation [18]. More direct evidence will be obtained in autopsy research on the brain tissues of patients with depression. But previously, studies have demonstrated an increase in the expression of TNF, an important protein in apoptosis, in the frontal lobe [19].

To summarize the study findings so far, it seems that inflammatory change affects the pattern of neurotransmission in the brain which plays a crucial role in depression and the mechanism of action of antidepressants, and that further, it is related to neurogenesis and neuroplasticity. Accordingly, inflammation plays a role of causing or prolonging depression, and in the future, it could be used as a marker for depression diagnosis, treatment response, and prognosis prediction. Based on epidemiologic studies, elevated IL-6, IL-1ra, and CRP levels are likely to be predictors of the occurrence of depression [20].

According to a recently conducted meta-analysis, IL-6, TNF-α, TNF-β1, IFN, and C-reactive protein (CRP) are the inflammatory markers in depression showing an increase relatively consistently [21]. These pro-inflammatory cytokines have the effect of increasing glucocorticoid resistance in the periphery, resulting in HPA axis dysfunction, and decrease the utilization rate of serotonin by converting tryptophan to kynurenine rather than to serotonin [10].

However, depression is too multifactorial to be defined as a primarily inflammatory disorder, and increased inflammatory response is not specific to depression only. Clearly, stress increases pro-inflammatory cytokines. But it is difficult to say that inflammatory markers extracted from blood samples completely represent the state of the central nervous system. In addition, it cannot be determined whether depression occurs due to increased inflammation or inflammation increases as a consequence of depression. It is speculated that as a part of the far-flung "super-network" involving allostatic load, inflammatory response probably has an effect of immune response element amplification, interlinked with desensitization of glucocorticoid negative feedback, decreased conduction of the parasympathetic nervous system, reduction of a brain-derived neurotrophic factor (BDNF), increased activity in the anterior cingulate cortex, hippocampal atrophy, increased adipose tissues, atherosclerosis, etc. [22]. Thus, the increase in inflammation and the incidence of depression may not form a direct causal relationship with each other. In many studies, increase and decrease in the level of inflammatory markers have been observed only in a small scale and inconsistently. It is possible that these problems have arisen, because depression is currently diagnosed based only on phenomenological symptoms, and hence, intrinsic heterogeneities and diverse environmental factors across patients are not controlled in diagnosing the disorder.

Does the Linkage Between Inflammation and Depression Remain Merely Hypothetical?

Clinical studies conducted so far support the idea that the activation of the immune system may be involved in the etiology and pathophysiology of depression via the production of pro-inflammatory cytokines, but the cytokine hypothesis in depression is still controversial. According to the cytokine hypothesis, depletion of such neurotransmitters as serotonin, activation of the HPA axis (both observed in depression), and manifestation of depressive symptoms are secondary to an increase of pro-inflammatory cytokines. However, the hypothesis does not specify whether in depression, an increase of pro-inflammatory cytokines due to immune system activation is a causative factor or an epiphenomenon that plays the role of sustaining and prolonging depressive symptoms regardless of the etiology of depression. However, based on the findings that in cancer patients, IFN-α used in immunotherapy induces depression and the termination of the therapy or the administration of an antidepressant ameliorates depression symptoms, it is highly likely that pro-inflammatory cytokines are a causative factor in cases in which depression occurs due to a medical illness or immunotherapy. However, it is unclear whether or not pro-inflammatory cytokines are a causative factor for depression in cases other than those mentioned above [23].

Second, with respect to the mechanism of action of antidepressants, not all antidepressants are associated with decrease in pro-inflammatory cytokines [24]. It may be that regardless of the level of cytokines decomposed in the periphery, antidepressants interfere with the process in which peripheral cytokines work on the central nervous system and not that the drugs' mechanism of action is to suppress the synthesis and decomposition of cytokines. Accordingly, the efficacy of antidepressants may not be in directly inhibiting immune system activation but in indirectly modulating immune function.

The increased levels of cytokines in depression shown in previous studies are in a lower range in comparison to immune disease. Thus, a question arises as to whether or not a low volume of cytokines can work on the central nervous system to affect the pathophysiology. One previous study compared the effects of low and high levels of peripheral cytokines in sleep–wake behavior and reported that the effects were similar [25]. However, it has not yet been resolved whether a similar effect would be observed in depression.

Currently, a large number of preclinical and clinical trials exist which support inflammation as an etiology of depression, but the precise mechanism in which inflammation mediates depression symptoms has not been clearly identified. Thus, it is difficult to conclude whether inflammation is a cause of depression or an immunological by-product following the occurrence of depression. Nevertheless, the cytokine hypothesis in depression is of great significance as a hypothesis formulated from a new viewpoint to explain psychological and neurophysiological changes associated with depression.

Does Inflammatory Process Contribute to the Development of a Specific Subtype of Depression?

Animal experimental studies demonstrating that cytokines induce symptoms similar to depression and are associated with HPA axis activation, CRF production, and neurotransmitter metabolism present a possibility for the link between the pathophysiology of depression and cytokines. However, in human studies, different researchers have reported different findings. Some human studies found no significant difference in in vivo or in vitro levels of IL-1β, IL-2, IL-6, and TNF-α [26], or rather reduced IL-1, IL-2, and IL-6 levels in depression patient [24]. On the one hand, these findings could be attributed to different research methods. On the other hand, it is speculated that depression may be immunologically heterogeneous and varying changes in immune function may occur depending on the patient's specific state. For instance, a recent report that serum IL-1β level increased in typical depression whereas it did not change in atypical depression suggests that the extent of the involvement of inflammation may differ according to depression subtype [27]. It implies that the pathophysiology of depression and the mechanism of action of antidepressants may vary according to depression subtype and also that the biological mechanism of depression may differ depending on the type of depression symptoms.

Change in cytokines is observed not only in depression but also in other mental disorders (schizophrenia, bipolar disorder, eating disorder, and post-traumatic stress disorder). As an example, IL-6 shows abnormality both in schizophrenia and depression [28]. These findings may be due to a problem intrinsic to the current diagnostic system, which is based only on clinical symptoms. But it is also possible for behavioral and biological changes associated with cytokines to be linked to biological etiology common across various psychiatric disorders.

Depression is a syndrome having a collection of diverse symptoms, recognized as a complex psychiatric disorder with heterogeneous etiologies. Therefore, a single mechanism to explain all aspects of depression may not exist. That is, some patients may present depression in association with the dysfunction of cellular proteins produced in the process of complex intracellular signal conduction of neurotransmitters such as 5-HT and NE, while in some other patients depression may develop as a consequence of the overproduction of cerebral CRF induced by continued activation of central CRF systems due to stress in childhood and increased stress response in adulthood [29]. In still other patients, depression may occur as the HPA axis is activated and the circulation of neurotransmitters is increased by an increase of cytokines due to immune system activation. In a depression patient, these factors may function simultaneously or one of them plays a dominant role.

Potential Antidepressants Concerning the Modulation of Inflammation

A drug preventing or alleviating hypercortisolism could be a potent antidepressant. From this perspective, researchers paid attention to metyrapone, a cortisol synthesis inhibitor, as a treatment for depression [30]. Metyrapone has been reported to show antidepressant effect in open-label clinical trials conducted with patients with treatment-resistant depression [30]. Additionally, a double-blind randomized controlled trial conducted with patients with severe major depressive disorder observed that a combination therapy using metyrapone (1 g/day) and nefazodone or fluvoxamine had a significant antidepressant effect over 3 weeks, compared with a placebo [31]. Since that study, an additional double-blind randomized controlled study was initiated but the outcome has not yet been reported (NCT01375920). Oxytocin, a drug highlighted as a potential antidepressant, is also known as an inhibitor of the HPA axis [32]. Currently, a randomized controlled trial is underway to investigate the antidepressant effect of oxytocin nasal administration in patients with treatment-resistant depression (NCT01239888).

An increase of inflammatory cytokines in the brain increases IDO such that tryptophan (a precursor of serotonin) is metabolized into kynurenine, which can consequently induce depression by reducing serotonin level [29]. In addition, inflammatory cytokines such as TNF-α and IL-1β overactivate microglia and reduce neuroplasticity, consequently inducing depression by interfering with the function of the neural circuits involved in mood regulation [33]. The overactivation of microglia also works to activate glutamate receptors, thus inducing depression [34]. Furthermore, inflammatory cytokines may induce depression by activating the HPA axis and suppressing the inhibitory feedback loop to increase cortisol and decrease neuroplasticity [35]. As discussed so far, inflammatory response in the brain is related to depression, and accordingly, anti-inflammatory agents like non-steroidal anti-inflammatory drugs (NSAID), TNF-α antagonist, omega 3, and curcumin could work as antidepressants.

Of NSAIDs, aspirin (acetyl-salicylic acid), known as an irreversible inhibitor of COX-1 and COX-2, has been confirmed in many clinical trials to have antidepressant effect when used in combination with an existing antidepressant [36]. Additionally, several clinical trials with depressed patients showed that combination therapy consisting of celecoxib (a selective COX-2 inhibitor) and an existing antidepressant drug had antidepressant effect [37].

Of TNF-α antagonists, infliximab is currently used to treat autoimmune disease, but it is also studied as a treatment for depression [38]. In a randomized controlled study conducted to test for the antidepressant effect of infliximab IV administration in patients with treatment-resistant depression, the study drug (infliximab 5 mg/kg) did not show a significant difference from a placebo, but its antidepressant effect was significant in a patient group with high level of inflammatory marker (a hsCRP value of 5 mg/dL or higher) [39].

Omega 3 is an unsaturated fat and has anti-inflammatory effects, suppressing the production of inflammatory cytokines [40]. Hence, it may have an antidepressant effect, and the effect has indeed been confirmed in several studies in which antidepressant effect of combination therapies with SSRIs and omega 3 was investigated in patients with major depressive disorder [41, 42]. Curcumin is a natural substance with powerful anti-inflammatory and anti-oxidative effects, and so it may show antidepressant effect, as well [43]. The antidepressant effect of curcumin was reported in animal research, but it has not been clearly demonstrated in clinical studies [44, 45].

Doxycycline and minocycline, both tetracycline antibiotics, have powerful anti-inflammatory effect, and thus, it is speculated that they may also show antidepressant effect [46]. The antidepressant effect of doxycycline has been shown in animal research, but clinical research has not yet been conducted on the drug [47]. Regarding minocycline, clinical research was conducted in bipolar disorder, and many additional clinical trials are underway in major depressive disorder and bipolar depression [48].

Conclusions

In the past 50 or so years, monoamine hypothesis has been the main framework in which to explain psychiatric disorders. But the mental disorders are much more multifactorial than postulated by monoamine hypothesis, and therapies to correct the pathology pathways have not been developed. In the situation, an understanding of immunity-inflammatory response can provide a more global approach to psychiatric disorders. Inflammation in the central nervous system is involved in the pathophysiology of psychiatric disorders, by inducing changes in cytokines and neurotransmitters as allostatic load mediated by microglia and affecting neuronal apoptosis and regeneration and neuroplasticity. Accordingly, research has continued to explore inflammatory markers which could be used as markers for psychiatric disorders and the effects of anti-inflammatory drugs in the disorders. From the perspective of psychoneuroimmunology, the immune, endocrine, and nervous systems in human body closely interact with one another, and inflammation works as allostatic load to induce changes in the systems. Depression can occur as a consequence. Sickness behavior manifest in inflammatory state are very similar to depression symptoms, and the findings that cytokine therapy to treat various illnesses increases the occurrence of depression have shed light on the linkage between inflammation and depression. In light of this, studies have been mainly conducted on changes in cytokines and the HPA axis, and cytokines such as IL-1, IL-2, IL-6, IFN-γ, and TNF-α, hormones such as CRF and glucocorticoid, and CRP were proposed as inflammatory markers.

The inflammatory response seems to contribute to the incidence of depression and interfere with the recovery from the disorder by affecting the synthesis and transportation of neurotransmitters, glucocorticoid resistance, neural regeneration, etc. Typically, however, neuroinflammation is chronic and progresses very slowly.

Moreover, numerous factors are involved in neuroinflammation, and there also are many confounding variables, making it difficult to obtain consistent outcomes. Therefore, it is necessary to specify genetic, physiological, and epidemiological attributes of particular population groups vulnerable to inflammatory response as well as the disease subtypes. From a therapeutic perspective, research should be conducted to develop interventions to reduce the loss and unnecessary overactivation of microglia, and decrease oxidative stress. As of now, definitive inflammatory markers for psychiatric disorders have not been identified, but they could be very useful in patients with minimal vulnerability. In addition, it is possible to use inflammatory markers as depression biomarkers in subtypes of depression, which can serve as bases to develop medications to treat the disorder.

References

1. Na KS, Jung HY, Kim YK. The role of pro-inflammatory cytokines in the neuroinflammation and neurogenesis of schizophrenia. Prog Neuropsychopharmacol Biol Psychiatry. 2014;48:277–86.
2. Heneka MT, Carson MJ, El Khoury J, Landreth GE, Brosseron F, Feinstein DL, et al. Neuroinflammation in Alzheimer's disease. Lancet Neurol. 2015;14(4):388–405.
3. Furtado M, Katzman MA. Neuroinflammatory pathways in anxiety, posttraumatic stress, and obsessive compulsive disorders. Psychiatry Res. 2015;229(1–2):37–48.
4. Kim YK, Na KS, Myint AM, Leonard BE. The role of pro-inflammatory cytokines in neuroinflammation, neurogenesis and the neuroendocrine system in major depression. Prog Neuropsychopharmacol Biol Psychiatry. 2016;64:277–84.
5. Raison CL, Miller AH. Is depression an inflammatory disorder? Curr Psychiatry Rep. 2011;13(6):467–75.
6. Kronfol Z, Remick DG. Cytokines and the brain: implications for clinical psychiatry. Am J Psychiatry. 2000;157(5):683–94.
7. Silverman MN, Pearce BD, Biron CA, Miller AH. Immune modulation of the hypothalamic-pituitary-adrenal (HPA) axis during viral infection. Viral Immunol. 2005;18(1):41–78.
8. Marques AH, Cizza G, Sternberg E. Brain-immune interactions and implications in psychiatric disorders. Braz J Psychiatry. 2007;29(Suppl 1):S27–32.
9. Anisman H, Merali Z. Cytokines, stress and depressive illness: brain-immune interactions. Ann Med. 2003;35(1):2–11.
10. Christmas DM, Potokar J, Davies SJ. A biological pathway linking inflammation and depression: activation of indoleamine 2,3-dioxygenase. Neuropsychiatr Dis Treat. 2011;7:431–9.
11. Maes M, Stevens W, DeClerck L, Bridts C, Peeters D, Schotte C, et al. Immune disorders in depression: higher T helper/T suppressor-cytotoxic cell ratio. Acta Psychiatr Scand. 1992;86(6):423–31.
12. Song C, Lin A, Bonaccorso S, Heide C, Verkerk R, Kenis G, et al. The inflammatory response system and the availability of plasma tryptophan in patients with primary sleep disorders and major depression. J Affect Disord. 1998;49(3):211–9.
13. Leonard BE, Myint A. Inflammation and depression: is there a causal connection with dementia? Neurotox Res. 2006;10(2):149–60.
14. Zunszain PA, Hepgul N, Pariante CM. Inflammation and depression. Curr Top Behav Neurosci. 2013;14:135–51.
15. Musselman DL, Lawson DH, Gumnick JF, Manatunga AK, Penna S, Goodkin RS, et al. Paroxetine for the prevention of depression induced by high-dose interferon alfa. N Engl J Med. 2001;344(13):961–6.

16. Capuron L, Gumnick JF, Musselman DL, Lawson DH, Reemsnyder A, Nemeroff CB, et al. Neurobehavioral effects of interferon-alpha in cancer patients: phenomenology and paroxetine responsiveness of symptom dimensions. Neuropsychopharmacology. 2002;26(5):643–52.
17. Bull SJ, Huezo-Diaz P, Binder EB, Cubells JF, Ranjith G, Maddock C, et al. Functional polymorphisms in the interleukin-6 and serotonin transporter genes, and depression and fatigue induced by interferon-alpha and ribavirin treatment. Mol Psychiatry. 2009;14(12):1095–104.
18. Raison CL, Borisov AS, Majer M, Drake DF, Pagnoni G, Woolwine BJ, et al. Activation of central nervous system inflammatory pathways by interferon-alpha: relationship to monoamines and depression. Biol Psychiatry. 2009;65(4):296–303.
19. Dean B, Tawadros N, Scarr E, Gibbons AS. Regionally-specific changes in levels of tumour necrosis factor in the dorsolateral prefrontal cortex obtained postmortem from subjects with major depressive disorder. J Affect Disord. 2010;120(1–3):245–8.
20. Glover AT, Shaw SM, Williams SG, Fildes JE. Can inflammation be an independent predictor of depression? Brain Behav Immun. 2010;24(2):173 author reply 4–5.
21. Dowlati Y, Herrmann N, Swardfager W, Liu H, Sham L, Reim EK, et al. A meta-analysis of cytokines in major depression. Biol Psychiatry. 2010;67(5):446–57.
22. Krishnadas R, Cavanagh J. Depression: an inflammatory illness? J Neurol Neurosurg Psychiatry. 2012;83(5):495–502.
23. de Beaurepaire R. Questions raised by the cytokine hypothesis of depression. Brain Behav Immun. 2002;16(5):610–7.
24. Weizman R, Laor N, Podliszewski E, Notti I, Djaldetti M, Bessler H. Cytokine production in major depressed patients before and after clomipramine treatment. Biol Psychiatry. 1994;35(1):42–7.
25. Pollmacher T, Haack M, Schuld A, Reichenberg A, Yirmiya R. Low levels of circulating inflammatory cytokines–do they affect human brain functions? Brain Behav Immun. 2002;16(5):525–32.
26. Brambilla F, Maggioni M. Blood levels of cytokines in elderly patients with major depressive disorder. Acta Psychiatr Scand. 1998;97(4):309–13.
27. Ravindran AV, Griffiths J, Merali Z, Anisman H. Circulating lymphocyte subsets in major depression and dysthymia with typical or atypical features. Psychosom Med. 1998;60(3):283–9.
28. Jeon SW, Kim YK. Neuroinflammation and cytokine abnormality in major depression: cause or consequence in that illness? World J Psychiatry. 2016;6(3):283–93.
29. Jeon SW, Kim YK. Inflammation-induced depression: its pathophysiology and therapeutic implications. J Neuroimmunol. 2017;313:92–8.
30. Sigalas PD, Garg H, Watson S, McAllister-Williams RH, Ferrier IN. Metyrapone in treatment-resistant depression. Ther Adv Psychopharmacol. 2012;2(4):139–49.
31. Jahn H, Schick M, Kiefer F, Kellner M, Yassouridis A, Wiedemann K. Metyrapone as additive treatment in major depression: a double-blind and placebo-controlled trial. Arch Gen Psychiatry. 2004;61(12):1235–44.
32. Cardoso C, Kingdon D, Ellenbogen MA. A meta-analytic review of the impact of intranasal oxytocin administration on cortisol concentrations during laboratory tasks: moderation by method and mental health. Psychoneuroendocrinology. 2014;49:161–70.
33. Miller AH, Haroon E, Raison CL, Felger JC. Cytokine targets in the brain: impact on neurotransmitters and neurocircuits. Depress Anxiety. 2013;30(4):297–306.
34. Hashimoto K, Malchow B, Falkai P, Schmitt A. Glutamate modulators as potential therapeutic drugs in schizophrenia and affective disorders. Eur Arch Psychiatry Clin Neurosci. 2013;263(5):367–77.
35. Jeon SW, Kim YK. The role of neuroinflammation and neurovascular dysfunction in major depressive disorder. J Inflamm Res. 2018;11:179–92.
36. Berk M, Dean O, Drexhage H, McNeil JJ, Moylan S, O'Neil A, et al. Aspirin: a review of its neurobiological properties and therapeutic potential for mental illness. BMC Med. 2013;11:74.

37. Faridhosseini F, Sadeghi R, Farid L, Pourgholami M. Celecoxib: a new augmentation strategy for depressive mood episodes. A systematic review and meta-analysis of randomized placebo-controlled trials. Hum Psychopharmacol. 2014;29(3):216–23.
38. Ekdahl CT. Microglial activation-tuning and pruning adult neurogenesis. Front Pharmacol. 2012;3:41.
39. Raison CL, Rutherford RE, Woolwine BJ, Shuo C, Schettler P, Drake DF, et al. A randomized controlled trial of the tumor necrosis factor antagonist infliximab for treatment-resistant depression: the role of baseline inflammatory biomarkers. JAMA Psychiatry. 2013;70(1):31–41.
40. Calder PC. The relationship between the fatty acid composition of immune cells and their function. Prostaglandins Leukot Essent Fatty Acids. 2008;79(3–5):101–8.
41. Gertsik L, Poland RE, Bresee C, Rapaport MH. Omega-3 fatty acid augmentation of citalopram treatment for patients with major depressive disorder. J Clin Psychopharmacol. 2012;32(1):61–4.
42. Jazayeri S, Tehrani-Doost M, Keshavarz SA, Hosseini M, Djazayery A, Amini H, et al. Comparison of therapeutic effects of omega-3 fatty acid eicosapentaenoic acid and fluoxetine, separately and in combination, in major depressive disorder. Aust N Z J Psychiatry. 2008;42(3):192–8.
43. Seo HJ, Wang SM, Han C, Lee SJ, Patkar AA, Masand PS, et al. Curcumin as a putative antidepressant. Expert Rev Neurother. 2015;15(3):269–80.
44. Lopresti AL, Maes M, Maker GL, Hood SD, Drummond PD. Curcumin for the treatment of major depression: a randomised, double-blind, placebo controlled study. J Affect Disord. 2014;167:368–75.
45. Sanmukhani J, Satodia V, Trivedi J, Patel T, Tiwari D, Panchal B, et al. Efficacy and safety of curcumin in major depressive disorder: a randomized controlled trial. Phytother Res. 2014;28(4):579–85.
46. Levine J, Cholestoy A, Zimmerman J. Possible antidepressant effect of minocycline. Am J Psychiatry. 1996;153(4):582.
47. Mello BS, Monte AS, McIntyre RS, Soczynska JK, Custodio CS, Cordeiro RC, et al. Effects of doxycycline on depressive-like behavior in mice after lipopolysaccharide (LPS) administration. J Psychiatr Res. 2013;47(10):1521–9.
48. Savitz J, Preskorn S, Teague TK, Drevets D, Yates W, Drevets W. Minocycline and aspirin in the treatment of bipolar depression: a protocol for a proof-of-concept, randomised, double-blind, placebo-controlled, 2x2 clinical trial. BMJ Open. 2012;2(1):e000643.

The Frontiers of Suicide

Sheikh Shoib and Yong-Ku Kim

Abstract

Suicide is a serious public health problem in many countries and has always been a source of apprehension and quest to human mankind, which can be resolved with due diligence. Suicide is a hidden and silent epidemic, with many causative factors. Studying and researching on various causative factors have always been the subject of significance for the researchers. Psychiatric illnesses happen to be the primary reasons for the majority of suicide mortality cases. Not only this, there has been a consistent increase in the no. of cases of mental disorders as well as attempted and completed suicide cases. If one looks at global scenario, an approximate 70,000 people commit suicide, and further alarming is the fact that the rate of suicide attempt cases has gone up to 250% during last 18 years in conflict zones. Poisoning, hanging and self-immolation are some of the common methods to commit suicide. Physical and mental illness, disturbed emotional relationships and economic difficulties were the major reasons for suicide with the predisposed population being women, student and farmers. As per WHO's approximation, there is one suicide every minute and an attempted suicide every third second. It implies that the number of killed due to suicide is greater than that of the ones killed due to the armed conflict.

Keywords

Suicide · Suicidal ideation · Mental illness · Risk factors

S. Shoib (✉)
Department of Health, SR Gung Tenga Pora Nawab Bazar, Srinagar 190002, Jammu and Kashmir, India
e-mail: sheikhshoib22@gmail.com; Sheikhshoib22@yahoo.com
Y.-K. Kim Department of Psychiatry, College of Medicine, Korea University, Seoul, South Korea

© Springer Nature Singapore Pte Ltd. 2019

Y.-K. Kim (ed.), *Frontiers in Psychiatry*, Advances in Experimental Medicine and Biology 1192, https://doi.org/10.1007/978-981-32-9721-0_25

Introduction and Epidemiology of Suicide

Suicide is one of the major mental health and public health problems in the world. It is estimated that about one million people die each year. The incidence of suicide worldwide is 16 per 100,000 or one death every 40s [1]. It is estimated that the number of people who attempt suicide is 20 times more than the people who commit suicide. Further, according to the World Health Organisation (WHO), the number of suicide deaths is more as compared to deaths due to war and homicide [2]. In India, over 100,000 people die by committing suicide. The suicide rate has increased from 6.4 in 1982 to 10.5 in 2002 [3]. The numbers of recorded suicide cases have been approximately 14.5 in 100000 populations. In 1 year, the suicide rate is less in women compared to men being 18.9 cases per 100000 in men and 10.6 in 100000 in women [3]. Suicide ranks as the 11th leading cause of death in the United States. The WHO estimates that one suicide occurs approximately every minute and an attempt at the suicide is made every third second; thus, a number of people killed by suicide are more than the armed conflict [1]. The rate of completed suicides cases increases with age [4]. Suicide is a multidimensional phenomenon which has different meanings in different cultures and places [5]. A review of the literature shows that the attempted suicide rates vary from 100 to 300 per 100,000, with a preponderance of those attempting suicides being females [6]. Suicide is among the top 10 causes of death in India and among the top 3 causes of death in those who are between 16 and 25 years of age [7]. The national suicide rate for 2001 was 10.6 per 100,000 populations, a 14.5% increase from the statistics of 1991 [8].

Definition of Suicide and Process

Suicide does not have one universally accepted definition. It can be defined as intentional self-inflicted death [9]. Suicide can be defined as a self-inflicted death with evidence that the person intended to die. It can be preceded by suicidal ideation or suicidal attempts or without them [10, 11]. Scheidman (1985) defines it as 'the conscious act of self-induced annihilation, best understood as multidimensional malaise in a needful individual who defines an issue for which the suicide act is perceived as the best solution [9, 12]. As stated by Shneidman (1985), suicide is a multidimensional malaise. In fact, a more complete understanding of the etiology of suicide requires a multidimensional model comprising psychosocial, biologic, genetic, psychiatric and temperament/personality domains [12]. Suicide is not a random or pointless act; on the contrary, it is a way out of a problem [10]. In 1975, Clinard and Meier defined suicide in a legal sense as the destruction of oneself, self-killing or self-murder [13].

Retterstol (1993) defined suicide and gave a more detailed definition as an act with a fatal outcome, that is deliberately initiated and performed by the deceased him- or herself, in the knowledge or expectation of its fatal outcome, the outcome being considered by the actor as instrumental in bringing about desired changes in consciousness and/or social conditions. The spectrum of suicidal behaviour ranges from low-level suicide ideation, which is occasionally thinking about suicide through, to a deliberate action that actually results in death [14].

Attempted Suicide

It can be defined as self-harm where there is an intention to kill oneself but it will not lead to death [15]. Attempted suicide can be defined as action that is intended at self-harm/the possibility of self-harm in a way that it does not ensure his/her survival. However, many a time, in the case of attempted/completed suicide, it is vague as to if the person really wanted to die. The four different types of behaviours pertaining to suicide can be termed as suicide, attempted suicide, suicidal gesture and accidents [16]. Suicide may be defined as a situation in which the person intends to die and actually dies. Attempted suicide, as the name suggests, is an attempt/intention to die by a person but he actually does not die. Suicidal gestures, on the other hand, are the gestures which merely indicate a sign or some inkling to but with no real intention to do so, and therefore, the person does not die. Accidents refer to the situations in which the person does not want to die, but does [16].

Stages of Suicide

Suicide is not merely an event but it's a process which can be presented in various and different intensities in the individuals who attempt suicide [17]

I. **Wish to be dead**: Person endorses thoughts about a wish to be dead or not alive anymore, or wish to fall asleep and not wake up.
II. **Suicidal thoughts**: General non-specific thoughts of wanting to end one's life/commit suicide, 'I've thought about killing myself' without general thoughts of ways to kill oneself/associated methods, intent or plan.
III. **Suicidal thoughts with method (without specific plan or intent to act)**: Person endorses thoughts of suicide and has thought of a least one method during the assessment period. This is different than a specific plan with time, place or method details worked out.
IV. **Suicidal intent (without specific plan)**: Active suicidal thoughts of killing oneself and patient reports, having some intent to act on such thoughts.
V. **Suicide intent with specific plan**: Thoughts of killing oneself with details of plan fully or partially worked out and the person has some intent to carry it out
VI. **Suicidal behaviour can be described as a situation in which the person has started taking concrete steps/actions to end his/her life**: It can be in the form of acquiring pills, getting a gun, leaving all valuables in name of

others, or swallowing pills/actually pulling the trigger. These are definitive actions taken by a person to ensure that he/she does not survive from that point onwards.

Factors Affecting Suicide and Recent Changing Trends

Suicide is a phenomenon influenced by individual, family, psychosocial and cultural factors. Many factors contribute to person attempting suicide. Suicide can be seen as a psychological phenomenon, social phenomenon and as a phenomenon associated with psychiatric disorders, genetic and biologic problems [17–21].

Factors Affecting Suicide

Risk factor can be defined as the relation between some features of a person and a heightened probability of occurrence of some disease/allied problems. The etiological factors responsible for the development of suicide phenomenon can be identified with the study of risk factors. The identification of risk factors is an important step in the prevention of suicide. The knowledge about risk factors can provide a basis for suicide prevention programmes and interventions [22]. A combination of individual, relationship, community and societal factors contributes to the risk of suicide. Risk factors are characteristics associated with suicide, which can directly lead to suicide or indirectly influence the process of suicide. There are various factors which increase the risk of suicide [22]. These factors can be kept in three categories: biopsychosocial, environmental and sociocultural.

Biological

Biological causes account for most suicides and attempted suicides [9]. These causes include mental health disorders such as:

I. Depression.
II. Bipolar disorder.
III. Schizophrenia.
IV. Anxiety disorders.
V. Personality disorders.
VI. Substance abuse.
VII. Childhood abuse or trauma.
VIII. Family history of suicide.
IX. Previous suicide attempts.
X. Having a chronic disease.
XI. Feelings of hopelessness.
XII. Impulsive and aggressive tendencies.

Environmental and Sociocultural Causes

There is also an increased risk of suicide due to environmental and social causes. Environmental factors that increase the risk for suicide often occur due to a stressful life event or a series of such incidents. This may include the loss of a person, pet or job. This may also include social factors like being isolated or feeling unaccepted by others or being unable to adjust to others [19, 20].

Other causes include:

I. Social loss, such as the loss of a significant relationship.
II. Recent loss of a family member or close friend.
III. Access to lethal means, including firearms and drugs.
IV. Being exposed to suicide.
V. Being a victim of harassment, bullying or physical abuse.
VI. Difficulty seeking help or support.
VII. Lack of access to mental health or substance abuse treatment.
VIII. Following belief systems that accept suicide as a solution to personal problems.
IX. Exposure to suicidal behaviour.
X. Problems in sexual orientation, religious beliefs and gender identity.
XI. Legal or disciplinary problems.
XII. Local epidemics of suicide.
XIII. Being a victim of cyberbullying.

Protective Factors

The factors that restrain/stop a person from committing suicide are very significant. Though such factors have not been well researched/studied in detail, a comprehensive understanding of these could facilitate a decrease in suicide rates. Some of the preventive factors could include:

(a) Clinical care (physical, mental and abuse disorders).
(b) Timely clinical interventions and support groups.
(c) Community and significant others' support.
(d) Healthcare providers' care.
(e) Teaching conflict management.
(f) Faith and positive belief systems [23, 24].

Suicidal Behaviour and Process of Suicide

Suicidal behaviour and process of suicide can be better explained by three-step theory of suicide. The three-step theory is based on the ideation of action framework. The theory is based on the key constructs that are hopelessness, connectedness and suicide capacity. This theory provides a good insight into suicide, suicidal behaviour and prediction of suicide [25].

Step 1. Development of Suicidal Ideation: The first step is the development of suicidal ideation. Suicidal ideation begins with psychological or physical pain. The pain can be of physical suffering, social isolation, burdensomeness and low sense of belongingness. The mere presence of pain will not lead to suicidal ideation. However, if the pain is associated with the hope that it can be reduced, then the individual may not think about suicide. But if the pain is associated with hopelessness and there is no way to diminish it, the person may think of suicide. Suicide is motivated by the presence of many other factors including burdensomeness, lack of belongingness, a desire for help or communication and impulsivity, social isolation and negative self-perception. Pain and hopelessness in combination are what lead to suicidal ideation [26].

Step 2. Strong versus Moderate Ideation: The second step towards suicide occurs when the person experiences a deepening sense of pain and loose connectedness. The connectedness can be explained as the feeling of being connected to other people or having a goal or interest or any sense of purpose in life. If the person experiences pain and hopelessness but has connectedness with people, then the person will have moderate ideation. If the person is in pain, hopelessness and the connectedness is absent, the person will have a desire to end his life. The second step towards potentially lethal suicidal behaviour occurs when pain exceeds connectedness. In contrast to other theories, the three-step theory explains that connectedness protects the person from developing suicidal ideation due to pain and hopelessness. In other theories, like joiners' interpersonal theory, it is explained that disrupted connectedness can contribute directly to suicidal ideation [27].

Step 3. Progression from Ideation to Attempts: In the third step, the individuals have strong ideation which can often lead to suicide attempt. Most individuals with ideation do not make a suicide attempt; the final step of the suicidal ideation addresses the conditions under which strong ideation leads to a suicide attempt. Joiner explained that fear of death is a strong instinct that stops the person from attempting suicide. The individuals can attempt suicide only if they cross the barrier. The wish to die is developed by painful events that decrease tolerance to pain, individuals differ in different pain sensitivity. Practical variables like easy accessibility play a major role in suicide attempts [27].

Suicidal Behaviour and Models of Suicidal Process

Suicidal behaviour can be described as the series of behaviours which reflect death ideation initially, and suicide attempts in between and completed suicide in some cases. It is pertinent to note here that the behaviours are different for suicide ideators, suicide attempters and suicide completers. However, in almost all cases, suicide ideation precedes suicidal actions and states. It was found that suicide ideators and attempters manifest similar psychological phenomena. Suicidal ideation is important in the prediction of suicide. Different concepts or models have been developed to understand the relationship among these phenomena [28]. These are the suicidal pyramid, the suicidal career and the suicidal process.

- **Suicidal pyramid**: Suicidal pyramid is based on the assumption that there is a systematic progression from suicide ideation, suicide attempts to completed suicide. A cross-sectional and longitudinal study by researchers placed the individuals as per the prevalence of exhibited suicidal behaviour [29]. A distinction was made bet

The suicidal career: It can be defined as a sequence of events that happen in individual's life that kind of shape his/her suicidal career. This concept implies that suicide is not always reactive to a particular time/incident at one point but is rather, a result of series of incidents in one's life experiences of a person:

(a) when the profiles for suicides, nonfatal suicide attempters and natural deaths were observed a given set of variables will differ, (b) it is necessary to predict that life-threatening behaviours can have histories, i.e. as being processes in evolution (c) death cannot be instantaneous and reactive to a situation. There should be a relevant biography or a set of biographies which mediates reaction to stress and helps to specify which individuals can be kept in high-risk groups and who are most likely to react to a stressful situation with self-destructive behaviour. (d) The lethality of the act can be acute (the short-term probability of self-destruction) or chronic (the long-term probability of self-destruction). (e) The self-destructive behaviours should be spanned into causal models that span suicidal subjects' lives from birth to death; the behaviours should be analysed using analysis of variance and regression techniques. These informative studies are better than merely labelling individuals as members of high risk groups, (f) the concept of suicidal careers implies that death styles can be mapped on relevant indices like the numbers of previous acts of self-destruction [30].

- **Suicidal process**: Suicidal process approach posits that suicidal behaviour of an individual is a result of transition from suicidal thoughts to completed action, which is prompted by underlying/hidden suicidal tendencies. Further, this can begin at any point in life and is generally heightened by negative/stressful life experiences. These can either be chronic or temporary or may even never show up actively.
- The suicidal process is believed to be entrenched in early childhood. The process may begin with momentary thoughts about suicide which may disappear and return after a while and eventually progress into a completed action. The important fact observed in the studies conducted is that most of the unplanned and planned suicides occur within 1 year of onset of suicide ideation. Kessler et al. state that suicide ideation in psychiatric patients can predict the occurrence of eventual suicide. Moreover, suicide ideation significantly predicts eventual suicide in psychiatric patients [31, 32]. In a 20-year prospective study, Brown et al. found that there are higher levels of suicidal ideation in completed

suicide committed by patients with psychiatric illness and the suicide risk can be estimated [31]. Rudd and joiner in 1998 studied about the continuum of suicidality and the continuous process of suicidal behaviour presence and intensity of suicidal ideation with other associated factors [33]. Rudd and Joiner provide the description of the suicide continuum thus: (a) nonexistent: no identifiable suicidal ideation; (b) mild: the suicidal ideation of limited frequency, intensity and duration, and it is without any identifiable plans. There will be no intent to die subjectively or objectively. There may be mild dysphoria with good self-control that can be subjective or objective with few risk factors and identifiable protective factors. (c) Moderate: there will be frequent suicidal ideation, with limited intensity and duration and there will be some specific plans but without any intent subjectively or objectively. There is good self-control that can be subjective or objective with limited Dysphoria and presence of some risk factors but with identifiable protective factors. (d) Severe: there will be frequent, intense and enduring suicidal ideation with specific plans and no subjective intent but some objective markers of intent like easy accessibility to lethal methods. There will be evidence of impaired self-control which is subjective and/or objective with severe dysphoria and presence of multiple risk factors and few protective factors. (e) Extreme: there will be frequent, intense and enduring suicidal ideation with specific plans and clear subjective and objective intent with impaired self-control that can be subjective and objective with, severe dysphoria and presence of many risk factors with no protective factors [34].

Association Between Suicide and Psychiatric Disorders

Suicide is one of the major causes of death all over the world [23]. The multifactorial etiology of suicide has not been well understood and research is going on to understand the underlying causal factors. Most of the studies show the association of suicide with a mental health problem [35, 36]. People who commit suicide have a diagnosable mental disorder; studies suggest 90% of people who die by suicide have a diagnosable mental disorder [37, 38]. Mental disorders in general are associated with suicidal behaviour, and research is being done on which disorders exhibit more association with suicidal behaviour [39, 40]. When examined in this way, virtually all mental disorders are associated with suicidal behaviour [35].

- **Depression**: As per WHO's approximation, presently there are 121 million people who suffer from depression globally. Further, it is believed that it would become the second most common reason for disability by 2020. If one looks at the high prevalence of depression and its correlation with suicide, it becomes imperative that this issue needs to be addressed immediately to reduce the mortality rate. For instance, the lifetime risk for unipolar depression is 15%. It can be thus inferred that the crucial and causal role of depression in suicide has limited validity in Asia. Even those who were depressed had only mild to moderate symptomatology for a brief duration and majority committed suicide during their very first episode [43, 44].

- **Bipolar disorder**: The patients with bipolar disorder have a significant association with suicide. 10–15 percent of patients have a significant risk of committing suicide. Nearly 20–50 percent has a history of suicide attempt [45]. Suicide is most commonly seen in patients with bipolar 2 disorder than in bipolar 1 disorder [46].
- **Schizophrenia**: The patients suffering from schizophrenia have 10–15 percent lifetime risk of suicide. The recent studies show that the risk of suicide is more common in the first years of illness. Studies conducted by Palmar BA& Panbratzy VS showed a lifetime risk of 5.6 percent [45]. There are various factors like previous suicide attempt, comorbid depression, substance abuse, poor compliance and personal loss. Suicide is more commonly observed in patients with positive symptoms. The most common cause of death in those with schizophrenia is suicide [47].
- **Alcohol dependence**: The risk of suicide in alcohol abusers is estimated to be 7 percent [19]. The associations between substance use and suicidal behaviour are due to several underlying mechanisms. Depression is the most common comorbid disorder in substance abuse. It increases the individual's vulnerability for suicide. The rate of suicide increases due to the presence of social, marital and family personality disorders. The other factors including unemployment and physical illness also play a role in suicidal behaviour. There is an increased risk of suicide with patients who are acutely intoxicated. The increased rate is due to underlying risk factors like impulsiveness, childhood abuse and personality disorders. Acute intoxication may increase impulsiveness and trigger suicidal acts. Alcohol or other substances are used by the patient to get courage to commit the act or reduce fear pain and less conscious about the surroundings. The alcohol abuse of patient can have an impact on the family and increase vulnerability for suicide. The alcohol abuse of patient can have an impact on the family and increase vulnerability for suicide [47, 48].

Understanding Changes in Variables Influencing Suicide

Nearly a million people worldwide die annually by committing suicide, and suicide prevention is among the primary global and public health objectives. Central risk factors for suicide are a previous suicide attempt and mood disorders [46, 49]. Although mood episodes, suicidal ideation and suicide attempts are major indicators of risk, numerous other factors are also likely to have an influence. Psychological factors including hopelessness, impulsivity and other personality traits and adult and childhood negative life events presumably affect the diathesis of suicidal behaviour [50].

- **Suicide intent**: Suicide intent is commonly defined as the seriousness or intensity of the wish of a patient to terminate his or her life [51]. Several studies were conducted on the effect of suicide intent on suicide. The findings show that suicide attempters with high intent scores may have a higher risk of completed suicide than attempters with lower intent [52]. Several studies have investigated the relationship between depression and suicide intent [52]. Hopelessness is the key mediating variable between depression and suicide intent [53].

- **Hopelessness**: Hopelessness has emerged as a powerful cognitive variable connecting depression and suicidal behaviour. Early empirical studies revealed a significant association between hopelessness and degree of suicide intent in suicide attempters. Looked deeply, hopelessness has been found to be more related to suicide intent as compared to depression [54]. In fact, studies have proved that hopelessness is more likely to be the governing element in suicide attempters, ideators as well as completed suicides. A 10-year study of 165 hospitalized patients also led to the conclusion that hopeless was a quite significant indicator of eventual suicide [55]. Studies have reported that hopelessness is predictive of actual suicide both in psychiatric out-patients and in hospitalized suicidal ideators [2, 51, 56]. However, the role of hopelessness in the relationship between alcoholism and suicide attempts is not clear [57–59]. In fact, Beck found that only a diagnosis of alcoholism predicted eventual suicide in a sample of hospitalized suicide attempters [60]. On the other hand, in patients with affective disorders the degree of hopelessness appeared to be an important factor predicting eventual suicide, although its significance may depend on the history of drug and alcohol abuse. Thus, the role of hopelessness may vary between mental disorders [61].
- **Impulsiveness**: Impulsiveness can be defined as a tendency to respond quickly to a given stimulus, without deliberation and evaluation of the consequences [62]. The role of serotonin and its major metabolite 5-hydroxyindole acetic acid (5-HIAA) in depression, impulsiveness and suicidal behaviour has been studied for the past few years and has prompted great interest in suicide research. In a number of post-mortem studies, low levels of serotonin or 5-HIAA have been found in the brainstem of suicide victims. Low cerebrospinal fluid (CSF) 5-HIAA levels have been associated with acts characterized by violence, aggressiveness and impulsivity. Particularly several studies show that low CSF 5-HIAA indicates a particularly high risk of suicide following a previous attempt [63]. Studies show subjects who attempt suicide with impulsivity are less depressive, less hopeless and have a less strong intent to die than non-impulsive attempters. Biological markers of impulsiveness have been identified as one of the risk factors for completed suicide [64].

Current Challenges

Publicity and Internet, social and cluster suicides and cyberbullying are the major current challenges. Research shows that extensive media coverage of suicide is associated with increase in the rate of suicide [65].

- **Suicide cluster (adapted from Larkin and Beautrais 2012)**: A series of three or more closely grouped death which is linked by space or social relationships constitutes suicide cluster. In the absence of transparent social connectedness, evidence of space and time linkages is required to define a cluster. In the presence of a strong demonstrated social connection, only temporal significance is required [66, 67].
- **Suicide contagion**: It is more likely to occur where a suicide involves a person with similar characteristics (e.g. gender, age, social circumstances) to other people who have died (so-called horizontal transmission). Such deaths may have occurred within an individual's social network or among people they became aware of, through media or other influences and new methods of suicide are publicized there is a death of a celebrity by suicide (or other cause). This is known as 'vertical transmission' [67].
- **Suicide contagion**: Suicide can be contagious too. Social learning may be an important factor in both familial and non-familial transmission of suicidal behaviours. The concept of suicide contagion is based on the infective disease model and assumes that a suicidal behaviour by one person may facilitate the occurrence of subsequent, similar behaviours by other researchers [22]. The process is implemented via imitation. Theories of imitation have been postulated to explain clustering of suicides and DSH behaviours. Studies conducted primarily in adolescents revealed that up to 5% of all suicides may be the result of suicide clustering and that exposure to DSH behaviours in family and friends was predictive of DSH and suicide ideation [22].
- **Role of media**: The most important factor influencing cluster suicides is media. The news of suicides is spread through print television, newspapers and Internet including social media. The suicide clusters or contagion can occur if the news is highlighted more frequently and if it is shown on the first page. The way of presentation of news like description of methods and detailed descriptions of the individuals who committed suicide [68, 69].
- **The Papageno effect**: A vast majority of the studies about the possible effects of media reporting on suicidal contagion have been conducted. The findings show that there is a definite association between the role of media and mass suicide. There is also a risk of reporting bias showing only positive reports. Studies conducted in 1990 showed that there is no significant association between media reports and suicide [70–72]. Media can provoke suicides by Werther effect or even make it less by the Papageno effect. The modelling or copycat mechanisms are observed in people who are more influenced by acts done by highly influential people or public figures. It implies that the individuals with same age, gender and demographic background also play a role in this modelling. People get influenced by seeing the suicide acts in a dramatic way with repetition. Many a time, the facts tend to be exaggerated by media and are often not the official reports which can be reliable. Many studies go on to show the negative impact of media's dramatic representation and advertisement of lethal and dangerous suicide methods, which don't happen to occur in real life frequently. It must be understood that media reports in no way are an authentic

representation of official suicide data [70]. The description of suicide in movies also plays a major role in suicide clusters. Research has been done on the influence of movies on suicide. Movies portray suicide in giving vivid and graphic details of methods of suicide. These can create misunderstanding of nature and context and can trigger mass suicides through identification and copy cat mechanisms. Suicides in movies generally don't have any guidelines or age restrictions [71].

- **The Werther effect**: The rate of suicides is directly proportional to the duration of the content and prominence of the media coverage. In 1774, after the release of a novel called 'The sorrows of Young Werther', there was a substantial increase in the number of suicides. It was believed that this book had somehow prompted the negative/suicidal behaviours among people and therefore was banned in many European countries. In the contemporary world, similar concerns have been raised by the unjudicious use of Internet as there are some sites that promote suicide, dangerous behaviours and even online games prompting self-harm to its users. In a similar fashion, media can also act as an instrument of preventing suicides contagion by its effective and correct usage. The lack of an apparent copycat effect following Cobain's death is hypothesized to be due to various aspects of the media coverage and the intense activity of the crisis centre and community outreach interventions in Seattle that occurred following Cobain's suicide [72].
- **Bullying**: Bullying is the use of strength or influence on someone and forcing them to do something. Bullying is commonly seen in children with conduct disorder and childhood depression. It can be continued in adulthood. Studies showed that there is increased number of suicides in bullies and also the victims [73, 74]. The studies of Brent et al. show that boys who were bully showed more number of suicide attempts than girls and in contrast girls who were victims did more suicide attempts [24]. The young people now are exposed to cyberbullying through Internet social sites, gaming sites, emails and even text messages. There is an increase in the rate of suicides due to social contagion, in spite of strenuous efforts children and adolescents are drawn towards the dangerous suicide games and resulting in mass suicides [75].
- **Sexual orientation**: Sexual orientation is significant as it plays a major role in adolescents and young people. Young people with same-sex orientation are at greater risk to attempt suicide [41]. Studies show that people with same-sex orientation experience a negative family reaction and family rejection or rejection from the society. There is eight times more risk of committing suicide in these individuals [42]. Provision of gay/lesbian/bisexual support groups can resolve gender identity and help in prevention of suicide [76].

References

1. World Health Organization. Figures and facts about suicide mental and behavioural disorders. World Health Organisation: Department of Mental Health; 1999.
2. ReddyMS. Suicide incidence and epidemiology. Indian J Psychol Med. 2010;32:77–82.

3. Accidental deaths and suicides in India. National Crime Research Bureau. Ministry of Home Affairs. Government of India; 2002.
4. Krug EG, Mercy JA, Dahlberg LL, Zwi AB. The world report on violence and health. Lancet. 2002 5;360(9339):1083–8.
5. Goldston DB, Molock SD, Whitbeck LB, Murakami JL, Zayas LH, Hall GC. Cultural considerations in adolescent suicide prevention and psychosocial treatment. Am Psychol. 2008;63(1):14.
6. Shoib S, Dar MM, Bashir H, Qayoom G, Arif T. Psychiatric morbidity and the socio-demographic determinants of patients attempting suicide in Kashmir valley: a cross-sectional study. Int J Health Sci Res. 2012;2(7):45–53.
7. Latha KS, Bhat SM. D'Souza Suicide attempters in a general hospital unit in India. Their socio-demographic and clinical profile, emphasis on cross-cultural aspects. Acta Psychiatr Scand. 1996;94:26–30.
8. Comer RJ. Suicide. In: Abnormal psychology. 6th ed. New York: Worth Publishers; 2007. p. 278–307.
9. Kaplan HI, Sadock BJ. Synopsis of psychiatry, 8th ed. New York: Lippincott Williams and Willikins; 1998. Chapter 3, p. 864–72.
10. Jacobs DG, Baldessarini RJ, Conwell Y, Fawcett JA, Horton L, Meltzer H, et al. Suicide behavior practice guidelines for assessment and treatment of patients with suicidal behavior. Am J Psychiatry 2003;160:3–60.
11. Andreason NC, Black DW. Introductory textbook of psychiatry, 8th ed. Washington DC: American Psychiatry Press; 1995. Chapter 20, p. 511–24.
12. Shneidman E. Definition of suicide. New York: Wiley; 1985. p. 202–13.
13. De Leo D, Burgis S, Bertolote JM, Kerkhof AJ, Bille-Brahe U. Definitions of suicidal behavior: lessons learned from the WHO/EURO Multicentre Study. Crisis. 2006;27(1):4–15.
14. Retterstol N. Suicide: a Europ perspective. Cambridge: Cambridge University Press; 1993.
15. Nock MK, Borges G, Bromet EJ, Cha CB, Kessler RC, et al. Suicide and suicidal behavior. Epidemiol Rev. 2008;30:133–54.
16. Taylor S. The sociology of suicide. London: Longman; 1988.
17. WHO Media Centre–Suicide risk high for young people: World Suicide Prevention Day: 2008–2012. 2009:2.
18. Bazrafshan MR, Sharif F, Molazem Z, Mani A. Cultural concepts and themes of suicidal attempt among Iranian adolescents. Int J High Risk Behav Addict. 2015;4(1).
19. Keyvanara M, Haghshenas A. Sociocultural contexts of attempting suicide among Iranian youth: a qualitative study. Eastern Mediterr Health J. 2011;17(6).
20. Tovilla-Zárate CA, González-Castro TB, Juárez-Rojop I, Pool García S, Velázquez-Sánchez MP, Villar-Soto M, et al. Study on genes of the serotonergic system and suicidal behavior: protocol for a case-control study in Mexican population. BMC Psychiatry. 2014;14:29.
21. Petersen L, Sørensen TI, Kragh Andersen P, Mortensen PB, Hawton K. Genetic and familial environmental effects on suicide attempts: a study of Danish adoptees and their biological and adoptive siblings. J Affect Disord. 2014;155:273–7.
22. Klerman GL. Clinical epidemiology of suicide. J Clin Psychiatry. 1987;48:33–8.
23. Goldsmith SK, Pellmar TC, Kleinman AM, Bunney WE. Reducing suicide: a national imperative. National Academies Press; 2002.
24. McLean, J, Maxwell, M, Platt, S, Harris, F, Jepson, R. Scottish Government Social Research. Edinborough. 2008.
25. Klonsky ED, May AM. The three-step theory (3ST): a new theory of suicide rooted in the "ideation-to-action" framework. Int J Cogn Therapy. 2015;8(2):114–29.
26. Klonsky ED, May AM. Differentiating suicide attempters from suicide ideators: a critical frontier for suicidology research. Suicide Life Threat Behav. 2014,44(1):1–5.
27. Klonsky ED, May AM, Saffer BY. Suicide, suicide attempts, and suicidal ideation. Annu Rev Clin Psychol. 2016;28(12):307–30.

28. Bedrosian RC, Beck AT. Cognitive aspects of suicidal behavior. Suicide Life Threat Behav. 1979;9(2):87–96.
29. De Leo D, Heller T. Social modeling in the transmission of suicidality. Crisis. 2008;29(1):11–9.
30. Maris RW. Pathways to suicide-a survey of self destructive behaviours. Baltimore M.D: John Hopkins University Press; 1989.
31. Brown GK, Beck AT, Steer RA, Grisham JR. Risk factors for suicide in psychiatric outpatients: a 20-year prospective study. J Consult Clin Psychol. 2000;68(3):371.
32. Fawcett J, Scheftner W, Clark D, Hedeker D, Gibbons R, Coryell W. Clinical predictors of suicide in patients with major affective disorders: a controlled prospective study. Am J Psychiatry. 1987;144:35–40.
33. Rudd MD, Joiner TE. The assessment, management, and treatment of suicidality: toward clinically informed and balanced standards of care. Clin Psychol Sci Pract. 1998;5:135–50.
34. U.S. Department of Health and Human Services (HHS) Office of the Surgeon General and National Action Alliance for Suicide Prevention. Washington DC: HHS; 2012.
35. Harris EC, Barraclough B. Suicide as an outcome for mental disorders. A meta-analysis. Br J Psychiatry. 1997;170(3):205–28.
36. Pokorny AD. Prediction of suicide in psychiatric patients. Report of a prospective study. Arch Gen Psychiatry. 1983;40, 249–57.
37. Beautrais AL, Joyce PR, Mulder RT, Fergusson DM, Deavoll BJ, Nightingale SK. Prevalence and comorbidity of mental disorders in person making serious attempts: a case control study. Am J Psychiatry. 1996;153(8):1009–4.
38. Wunderlich U, Bronisch T, Wittchen HU. Comorbidity patterns in adolescents and young adults with suicide attempts. Eur Arch Psychiatry Clinic Neurosci. 1998;248:87–95.
39. Bostwick JM, Pankratz VS. Affective disorders and suicide risk: a reexamination. Am J Psychiatry. 2000;157(12):1925–32.
40. Weissman MM, Klerman GL, Markowitz JS, Ouellette R. Suicidal ideation and suicide attempts in panic disorder and attacks. N Engl J Med. 1989;321(18):1209–14.
41. Murray CJL, Lopez AD. The global burden of disease: a comprehensive assessment of mortality and disability from vijaykumar: suicide & mental disorders 373 diseases, injuries and risk factors in 1990 and projected to 2020. Cambridge Harvard University Press.
42. Lester D. Seasonality in suicide and economic growth. Percept Mot Skills. 1993;77:10.
43. Phillips MR, Yang G, Zhang Y, Wang L, Ji H, Zhou M. Risk factors for suicide in China: a national case-control psychological autopsy study. Lancet. 2002;360:1728–36.
44. Vijayakumar L. Suicide and mental disorders in Asia. Int Rev Psychiatr. 2005;17:109–14.
45. Palmar BA, Panbratzy VS, Bostwich JM. The lifetime risk of suicide in schizophrenia-a reexamination. Arch Gen Psychiatr. 2005;62:247–53.
46. Hawton K, Sutton L, Haw C, Sinclair J, Harriss L. Suicide and attempted suicide in bipolar disorder. A systematic review of risk factors. J Clin Psychiatr 2005; 66:693–704.
47. Inskip H, Harris C, Barraclough B. Lifetime risk of suicide for affective disorder, alcoholism and schizophrenia. Br J Psychiatry. 1998;172(1):35–7.
48. Rossow I. Substance use and suicidal behaviour-Prevention and treatment of suicidal behaviour. In: Hawton K, editor. UK: Oxford University Press; 2005. p. 261–77.
49. Cavanagh JT, Carson AJ, Sharpe M, Lawrie SM. Psychological autopsy studies of suicide: a systematic review. Psychol Med. 2003;33(3):395–405.
50. Mann J.J. Neurobiology of suicidal behaviour. Nat Rev Neurosci. 2003;4:819–28.
51. Beck AT, Schuyler D, Herman I. Development of suicidal intent scales. Charles Press Publishers;1974.
52. Linehan MM. Suicidal people: one population or two? In: Mann JJ, Stanley M, editors. Psychobiology of suicidal behavior, vol. 487;1986. P. 16–33 Ann NY Acad Sci.
53. Weishaar ME, Beck AT. Hopelessness and suicide. Int Rev Psychiatry.1992;4(2):177–84.
54. Beck AT, Weissman A, Kovacs M. Alcoholism, hopelessness and suicidal behaviour. J Stud Alcohol. 1976:37:66–77.

55. Keller F, Wolfersdorf M. Hopelessness and the tendency to commit suicide in the course of depressive disorders. Crisis J Crisis Interv Suicide Prev. 1993;14:173–77.
56. Beck AT, Schuyler D, Herman I. Development of suicidal intent scales. In: Beck AT, Resnick H, Lettieri D, editors. The prediction of suicide. Bowie, MD: Charles Press Publishers Inc.;1974. p. 45–56.
57. Nielsen AS, Stenager E, Bille-Brahe U. Attempted suicide, suicidal intent and alcohol. Crisi s. 1993;14:32–8.
58. Beck AT, Steer RA, Kovacs M, Garrison B. Hopelessness and eventual suicide: a 10-year prospective study of patients hospitalized with suicidal ideation. Am J Psychiatry. 1985;142(5):559–63.
59. Beck AT, Steer RA. Clinical predictors of eventual suicide: a 5-to 10-year prospective study of suicide attempters. J Affect Disord. 1989;17(3):203–9.
60. Young MA, Fogg LF, Scheftner WA, Fawcett JA. Interactions of risk factors in predicting suicide. Am J Psychiatry. 1994;151:434–5.
61. Evans J, Platts H, Liebenau A. Impulsiveness and deliberate self-harm: a comparison of 'first-timers' and'repeaters '. Acta Psychiatr Scand. 1996;93(5):378–80.
62. Nordstrom P, Samuelsson M, Asberg M, et al. CSF 5-HIAA predicts suicide risk after attempted suicide. Suicide Life Threat Behav. 1994;24:1–9.
63. Cornelius JR, Salloum IM, Mezzich J, Cornelius MD. Disproportionate suicidality in patients with comorbid major depression and alcoholism. Am J Psychiatry. 1995;152(3):358.
64. Gould MS. Suicide and the media. Ann N Y Acad Sci. 2001;932(1):200–24.
65. Larkin GL, Beautrais AL. Geospatial mapping of suicide clusters. Auckland: The National Centre of Mental Health Research, Information and Workforce Development; 2012.
66. Ladwig KH, Kunrath S, Lukaschek K, Baumert J. The railway suicide death of a famous German football player: impact on the subsequent frequency of railway suicide acts in Germany. J Affect Disord. 2012;136(1–2):194–8.
67. Westerlund M, Schaller S, Schmidtke A. The role of mass-media in suicide prevention. In: Wasserman D, Wasserman C, editors. Oxford textbook of suicidology and suicide prevention: a global perspective. Oxford, UK: Oxford University Press; 2010. p. 515–23.
68. Ashton JR, Donnan S. Suicide by burning as an epidemic phenomenon: an analysis of 82 deaths and inquests in England and Wales in 1978–9. Psychol Med. 1981;11(4):735–9.
69. Stack S. The effect of the Jonestown suicides on American suicide rates. J Soc Psychol. 1983;119:145–6.
70. Wasserman IM. Imitation and suicide: a reexamination of the Werther effect. Am. Sociol. Rev. 1984;49(3), 427–36.
71. Kessler RC, Downey G, Milavsky JR, Stipp H. Clustering of teenage suicides after television news stories about suicides: a reconsideration. Am J Psychiatry. 1988;145(11):1379.
72. Sisask M, Värnik A. Media roles in suicide prevention: a systematic review. Int J Environ Res Public Health. 2012;9:123–38.
73. Stack S. Celebrities and suicide: a taxonomy & analysis, 1948–1983. Am Sociol Rev. 1987;52:401–12.
74. Lecrubier Y. The influence of comorbidity on the prevalence of suicidal behaviour Eur Psychiatry. 2001;16(7):395–9.
75. Klomek AB, Sourander A, Niemelä S, Kumpulainen K, Piha J, Tamminen T, Almqvist F, Gould MS. Childhood bullying behaviors as a risk for suicide attempts and completed suicides: a population-based birth cohort study. J Am Acad Child Adolesc Psychiatry. 2009;48(3):254–61.
76. Ryan C, Huebner D, Diaz RM, Sanchez J. Family rejection as a predictor of negative health outcomes in white and Latino lesbian, gay, and bisexual young adults. Pediatrics. 2009;123:346–52.

Novel Interventions

Interventions Addressing the Telomere-Telomerase System

Ather Muneer

Abstract

Major psychiatric disorders are linked to early mortality and patients afflicted with these ailments demonstrate an increased risk of developing physical diseases that are characteristically seen in the elderly. Psychiatric conditions like major depressive disorder, bipolar disorder, and schizophrenia may be associated with accelerated cellular aging, indicated by shortened leukocyte telomere length (LTL), which could underlie this connection. Telomere shortening occurs with repeated cell division and is reflective of a cell's mitotic history. It is also influenced by cumulative exposure to inflammation and oxidative stress as well as the availability of telomerase, the telomere-lengthening enzyme. Precariously short telomeres can cause cells to undergo senescence, apoptosis, or genomic instability; shorter LTL correlates with compromised general health and foretells mortality. Important data specify that LTL may be reduced in principal psychiatric illnesses, possibly in proportion to exposure to the ailment. Telomerase, as measured in peripheral blood monocytes, has been less well characterized in psychiatric illnesses, but a role in mood disorder has been suggested by preclinical and clinical studies. In this manuscript, the most recent studies on LTL and telomerase activity in mood disorders are comprehensively reviewed, potential mediators are discussed, and future directions are suggested. An enhanced comprehension of cellular aging in psychiatric illnesses could lead to their re-conceptualizing as systemic ailments with manifestations

A. Muneer (✉)
Islamic International Medical College, Riphah International University, 274 Peshawar Road, Rawalpindi, Pakistan
e-mail: muneerather2@gmail.com

© Springer Nature Singapore Pte Ltd. 2019

Y.-K. Kim (ed.), *Frontiers in Psychiatry*, Advances in Experimental Medicine and Biology 1192, https://doi.org/10.1007/978-981-32-9721-0_26

both inside and outside the brain. At the same time, this paradigm shift could identify new treatment targets, helpful in bringing about lasting cures to innumerable sufferers across the globe.

Keywords
Mood disorders · Leukocyte telomere length · Telomerase activity · Aging · Mortality

Introduction

There is emerging evidence that patients suffering from major psychiatric disorders have accelerated aging, are prone to systemic ailments seen in the elderly, and have increased morbidity and mortality [1]. Chronological age is determined by calendar units, while biological age is characterized in physiological terms and is underlined by disease processes. Principal psychiatric disorders which are epitomized by major depressive disorder (MDD), bipolar disorder (BD) and schizophrenia are often associated with comorbid medical conditions such as atherosclerotic diseases, cancer, dementia, metabolic irregularities, type II diabetes mellitus, and osteoporosis [2]. While, no doubt that lifestyle factors and socioeconomic adversities play a role, psychiatric ailments themselves may be contributing toward this phenomenon. This outpacing of biological versus chronological aging raises the possibility that major psychiatric disorders are associated with increased senescence at the level of the organism, or more specifically the cellular level [3].

Biological aging is signified by cellular aging, and an important marker of the latter is telomere length or more distinctively leukocyte telomere length (LTL) as measured in peripheral blood mononuclear cells (PBMC) [4]. Composed of DNA and proteins, telomeres cap the chromosomal ends of the double-stranded DNA and serve a protective purpose. In primates, telomeres are composed of multiple, noncoding repeats of the nucleotide sequence, TTAGGG. These shorten with each cell division because of incomplete replication of the telomere end due to the so-called end-replication problem [5]. Approximately, 50–100 nucleotides are lost per DNA replication cycle and this attrition is increased with pathological conditions like oxidative stress, inflammatory signaling, exposure to certain cytotoxins, and persistently raised stress hormones [6]. The enzyme telomerase which is comprised of telomerase reverse transcriptase (TERT) and telomerase RNA component (TERC) can rebuild telomeres by adding small DNA fragments called "Okazaki fragments" on the lagging strand. However, when this delicate balance is perturbed, telomeres become precariously short and cells undergo replicative arrest or become genomically unstable. Cells damaged in this manner can malfunction in specific ways, for example, p53 or tumor suppressor protein may become active, impairing oxidative defense mechanisms, inducing mitochondrial dysfunction and apoptosis [7]. Additionally, precancerous cells with critically short telomeres are inherently unstable genomically and exhibit end to end fusion of DNA, thus facilitating tumor progression. To extend this argument further, rare Mendelian disorders involving genes implicated in telomere maintenance and preservation cause a variety of human diseases involving different organ systems and lead to premature death [8].

Telomeres are not dormant structures and telomere length is determined by both genetic heritability and epigenetic influences which act throughout the lifespan. In this regard, longitudinal studies have shown that the weighted average telomere attrition rate is 40.7 base pairs/year [9]. Leukocyte telomere length is determined by innate genetic factors, with heritability estimates ranging from 0.36 to 0.84 [10]. With aging, telomeres undergo shortening influenced by environmental factors, but this is not so invariably. Certain individuals sustain and even lengthen average LTL over time. While the reasons for this are obscure at present, relatively long telomeres tend to shorten and comparatively short telomeres increase in length over time. A supposed mechanism in this respect may be the preferential activation of such repair pathways as the catalyst telomerase, the telomere-lengthening enzyme [11].

In numerous studies shortening of telomeres has been consistently associated with serious medical ailments chief among which are coronary artery disease, diabetes mellitus, and various types of cancers [12]. Moreover, according to current research evidence critically short telomeres are indicative of increased suffering from chronic diseases, as well as premature mortality. Even as there are a few negative reports, the bulk of the epidemiological evidence suggests that baseline LTL links with medical illnesses cross-sectionally, and prospectively foretells the development of grave medical conditions leading to early mortality. In this regard, one study in men showed that relatively hastened shortening of LTL over two and a half years was longitudinally associated with increased mortality from cardiovascular disease (CVD) on a 12-year follow-up period [13].

So, the next obvious question should be what mechanism underlies this association? With the present state of knowledge, a definitive answer cannot be given in this respect. The shortened telomeres may be causally involved in the development of medical diseases, or general pathogenic features such as inflammation and oxidative stress may be acting as core etiologic factors. In all likelihood, it is the combination of both phenomena which act as contributory factors [14]. A well-cited meta-analysis determined that the risk of CVD was connected to shared variations in a designated cluster of genes with known association to telomere preservation and maintenance [15]. While this population-based genetic investigation holds up an etiologic role for deregulated telomere maintenance in CVD, factors such as increased inflammation and oxidative stress are perhaps responsible for both effects. LTL is an indicator of the cell's accumulating mitotic history, as well as its exposure to damaging signals such as inflammatory cytokines and free radicals [16]. In this vein, it can be hypothesized that increased cell death due to shortened telomeres can diminish regenerating stem and progenitor cells, such as hematopoietic cells, endothelial progenitor cells, and neural stem/progenitors cells. This ostensibly impairs cellular replacement and repair processes, directly contributing to disease progression [17]. Furthermore, aged immune cells secrete pro-inflammatory cytokines (IL-6, TNF-α) which could be responsible for a vicious cycle of inflammation, oxidative stress, and telomere shortening [18].

Potential Confounders in the Interpretation of Leukocyte Telomere Length

Shortened LTL is a sensitive index of cellular aging and an increasing number of studies demonstrate that principal psychiatric disorders are connected to this phenomenon [19]. Nonetheless, as data are heterogeneous, several limitations have to be considered when interpreting the finding from studies on this subject.

(i) When examining the peripheral blood film, it must be understood that there are young versus aging leukocytes (monocytes, lymphocytes), with varying telomere lengths.
(ii) The relationship between LTL and TL in other tissues is not fully ascertained, and while they generally correlate positively, this fact is not fully established.
(iii) Different laboratories utilize diverse methodologies with respect to DNA extraction and assaying techniques.
(iv) Perhaps, even trivial DNA degradation can produce false results.
(v) Variables that are subject-specific can introduce bias in interpreting the association of LTL to the disease processes being investigated. Some examples include sex, age, early life adversity, psychological resiliency, lifestyle factors, latent, or active viral infections (CMV, herpes, etc.).
(vi) The presence of psychiatric comorbidities in the study subjects. For instance, patients with MDD have high rates of anxiety spectrum disorders and it may be difficult to tease apart the effect of either condition on LTL.
(vii) Substance use disorders are frequently associated with major psychiatric disorders. In this regard, it has been shown that heavy alcohol use is by itself linked to telomere shortening in PBMC.
(viii) Many studies investigating LTL in psychiatric disorders did not control for medical comorbid conditions which are by themselves associated with shortened LTL.
(ix) Finally, several psychotropic and non-psychopharmacological agents can influence LTL, and not statistically controlling for these medications can introduce bias in the results.

In spite of these caveats, recent meta-analyses have found robust effect size of LTL shortening for psychiatric disorders as a whole compared with controls [20]. With respect to mood disorders, such differences were found in patients with all mood states and in studies using different methods for measuring telomere length.

Major Depressive Disorder

Patients with MDD suffer from unipolar depressive episodes; onset is usually in early adulthood and the disorder tends to follow a chronic course with remissions and relapses throughout the lifespan. Subjects affected by MDD likely show a variable trajectory with some patients having relatively mild and self-limited episodes, while others suffering from severe exacerbations have physical and neuropsychiatric comorbidities and varying degrees of impairment in the psychosocial realms of functioning [21]. Several authors have expressed the view that MDD is a syndrome of premature aging, so that the patients not only have excess of physical conditions, but brain also demonstrates early aging, which in a subset of cases is manifested as the dementia syndrome [22].

It was in 2006 that Simon and coauthors first showed that leukocyte telomere length was decreased in mood disorder subjects. In their seminal study, the total number of subjects was 88 (44 patients and 44 controls) and out of them, 15 had MDD without any associated anxiety spectrum disorder. Cases with mood disorders had mean LTL which was 660 base pairs (bp) shorter as compared to age-matched nonpsychiatric healthy controls (HC), while in the MDD subgroup mean LTL was 770 bp shorter than in HC. The statistical analysis utilized by the authors demonstrated that this difference carried a large effect size which corresponded to approximately 10 years of accelerated aging in the mood disorder subjects and as such was highly significant. The limitations of this study were that structured clinical interviews were not given to HC and the confounding effect of current and past psychotropic medications was not controlled for [23].

In pursuance of this argument, the Dutch study by Verhoeven et al. (2014) is highly important. The data was from the Netherlands Study of Depression and Anxiety (NESDA) and in this longitudinal cohort study 1095 current MDD patients, 802 remitted MDD patients, and 510 HC were included. It was shown that both currently depressed and remitted MDD groups had significantly shorter LTL as compared to HC. Further, the NESDA study demonstrated that the mean LTL in the currently depressed and remitted MDD subjects did not significantly differ from each other. The difference in LTL between the depressed groups and HC remained significant after controlling for such confounding factors as age, sex, education level, alcohol use disorders, body mass index, physical diseases, and exercise. In the currently depressed group, LTL was inversely correlated to the severity and duration of depression over the last 4 years, pointing to a dose–response relationship between mood episodes and shortening of the telomere length in PBMC. Furthermore, the authors hypothesized that the lack of difference in LTL between currently depressed and remitted patients was because of the fact that MDD episodes left a persisting signature on the LTL [24]. However, there is the likelihood that LTL is already short in patients who are susceptible to developing depressive episodes because of genetic and epigenetic influences and thus represents a risk factor for the disorder. Support for this assumption is provided by the study of Gotlib et al. (2015) who showed that LTL was short in girls whose mothers suffered from major depressive episodes and this signified hereditary and environmental factors in LTL maintenance [25].

The subject of dose response relationship between LTL and MDD is of great interest and thus requires further examination. The literature is mixed on this aspect as some studies support the hypothesis while others don't. The prospective study of Shalev et al. (2014) found that in men the duration of internalizing disorders (including major depression) at ages 11–38 predicted LTL at age 38 in a dose–response mode [26]. In contrast, the study by Jodczyk et al. (2014) showed that the diagnosis of major depression between ages 16 and 25 did not foretell shortened LTL between ages 28 and 30. However, in that study exact assessment of the total exposure to the duration of depressive episodes was not cumulatively measured [27]. Hoen et al. (2011) studied leukocyte telomere length in patients with coronary artery disease (CAD). Their study had a large cohort of 952 patients at baseline and 608 subjects at 5 years of follow-up. Two hundred and six participants had major depression at baseline. Compared to nondepressed CAD patients, those with MDD had shorter telomere length in PBMC. However, at 5 years of follow-up, this difference was no longer statistically significant when controlling for confounders such as BMI, smoking, diabetes mellitus, ejection fraction, statin use, antidepressant use, level of physical activity, and comorbid anxiety. Although this was a negative study, it is possible that CAD by itself influenced LTL, covering-up the effect of MDD [28]. In yet another study, Hoen et al. (2013) investigated LTL in a population-based sample and reported that baseline anxiety disorders but not MDD predicted telomere length in PBMC at two years of follow-up [29].

It can be reasonably concluded that the bulk of studies on the relationship between LTL and MDD show an association, supported by the following facts:

(i) Studies having a greater number of subjects ($\geq$40) generally demonstrated a statistically significant shortening of LTL in MDD subjects.
(ii) Negative studies either had fewer subjects or investigated late-life depression.
(iii) Negative studies had several limitations, for example, a study in elderly depressed subjects did not control for such pathologies as vascular conditions, or the combined effect of other lifetime diseases which cumulatively may be responsible for telomere shortening [30].
(iv) In some investigations, the premature loss to the study of elderly depressed individuals with advanced cell aging who may already have died was not taken into account.
(v) On the other hand, a study conducted in subjects with anxiety spectrum disorder seemed to contradict the above argument. It revealed that as compared to younger patients, LTL was shortened only in older subjects aged between 48 and 87 years. The authors assumed that the cumulative effect of suffering from a lifetime of anxiety disorders corresponded to persistent stress with neurobiological sequelae, accelerated aging, and greater telomere attrition [31].
(vi) A negative study showed that MDD subjects receiving antidepressant medications had significant shortening of LTL, while those on no medications did not. The authors conjectured that MDD cases on psychotropic medications were more severely depressed and hence also demonstrated shortened telomeres in PBMC [32].
(vii) Another small study (17 MDD and 16 HC) found no significant difference in mean LTL. Nevertheless, it did reveal increased expression of $p16^{INK4a}$

and stathmin genes in the group with major depression, which are markers of cellular aging, telomere maintenance, microtubule functioning, biological aging, and cell cycle regulation [33].

(viii) Finally, another small-scale study (18 MDD, 17 HC) was negative in an overall approach. However, on further analysis, significantly shorter LTL was found in the subgroup of MDD cases with long duration of illness ($\geq$9.2 years). Moreover, a dose–response relationship was found in MDD cases who had long periods of untreated major depression, but the number of individuals was too small to draw a meaningful conclusion [34].

A well-cited review examined the studies on LTL and MDD and calculated the effect sizes (ES) for studies which had robust methodology as manifested by the use of structured clinical interviews. Of the studies included in the review, the ES varied from 0.04 to 0.98 (mean Cohen's $d = 0.41$; weighted mean Cohen's $d = 0.23$) and this represented a small ES. It is worth mentioning that the smallest ES was noted in studies which included elderly participants and this could have introduced potential bias in the overall calculations. The majority of the studies utilizing dimensional diagnostic scales for depression failed to find a correlation between LTL and MDD. Although the reasons for this discrepancy between studies using categorical criteria versus dimensional scales are not known, it can be speculated that the latter took into account only short-term depression ratings (1–2 weeks), lacked criteria for illness duration and severity, had absence of assessment of psychosocial functioning, and in general their participants suffered from milder forms of depression [35]. Overall, examining the relevant literature in entirety, it can be justified that MDD is independently associated with shortened telomeres in the PBMC and this linkage is robust when the duration and severity of the illness are accounted for [36]. However, as alluded to above, the effect size in this regard is modest.

Bipolar Disorder

It must be recognized that there are fewer papers on the subject of LTL and BD, as merely a handful of studies had examined the issue of leukocyte telomere length in bipolar subjects. The following is a summary of the extant literature on this topic.

(a) Elvsashagen et al. (2011) were among the first to study LTL in bipolar subjects and their sample consisted of BD type II cases. They defined short telomeres as <3000 bp and found that these were significantly increased in cases versus controls using one-tailed tests, with a trend toward shortened absolute LTL in BD type II patients. The total lifetime number of depressive episodes, as opposed to hypomanic episodes was significantly related to shortening of telomeres in PBMC (statistically significant with two-tailed analysis) and in the authors' view, this indicated the presence of dose–response relationship in the sufferers [37]. The definition of short telomeres as $\leq$3000 bp was somewhat arbitrary, but consistent with the fact that telomeres shorter than 3800 bp were inherently instable as measured by array-comparative genome hybridization analysis.

(b) Rizzo et al. (2013) only studied euthymic females with BD type I, while reasons for excluding men were not clear. The bipolar patients had notably raised IgG antibodies to the cytomegalovirus (CMV) and also exhibited statistically significant shortened LTL. Since CMV antibody titers were inversely correlated to telomere length, it was assumed that the association between LTL and BD was because of infection with the cytomegalovirus. Furthermore, the total duration of BD corrected for age was not statistically correlated to LTL [38].

(c) Martinsson et al. (2013) studied the effect of lithium therapy on LTL in BD subjects. Their study was intriguing, as it found increased LTL in lithium-treated cases compared to controls. BD patients who had received lithium over most part of the last 2½ years and those who showed clinical response to this agent had significantly longer LTL than nonresponders. The authors hypothesized that lithium was responsible for telomerase activation and this effect was obvious in patients who had therapeutic response to this medication [39]. Hence, the protective effect of lithium ion on telomere maintenance and preservation was demonstrated.

(d) The study by Lima et al. (2015) is important for a number of reasons. First, BD patients regardless of subtype were recruited. Second, it was a rather large study with 85 BD subjects and 95 carefully matched HC. Finally, it employed real-time quantitative PCR, a time-tested technique to quantify LTL. As a whole, BD subjects showed statistically significant shortening of LTL as compared to HC, and while the duration of illness and medication use were not controlled for, the resulted implied decreased telomere length in BD [40].

(e) Finally, the study by Barbe-Tuana et al. (2016) helps us in understanding the relationship between early and late stages of BD and telomere attrition. Twenty-six euthymic BD cases and 34 HC were included, and it was demonstrated that telomeres were significantly reduced in length using real-time PCR in both early and late stage cases. Shortened LTL, an indicator of accelerated aging, was shown to be associated with BD and this could partially explain the increased prevalence of age-related medical conditions in this disorder [41].

It can be concluded from the above discussion that like MDD, bipolar disorder is also a disease of accelerated aging with significantly reduced life expectancy in both conditions and undoubtedly accurate quantification of LTL represents a reliable measure of this effect.

Ltl—The Dose–Response Relationship

In order to address this issue, it is important to keep in consideration that serious mental disorders are commonly associated with inflammation and oxidative stress and prolonged exposure to psychiatric illnesses results in enhanced contact with the latter factors, thus causing greater telomere attrition. Conversely, if LTL shortening precedes major psychiatric disorders posing as a risk factor, there might be a fixed level of LTL shortening despite the degree of actual exposure ("premature" rather than "accelerated" telomere shortening). These premises are not in opposition to each other, since it is likely that vulnerable individuals have shortened telomeres preceding the onset of psychiatric illness and have further reductions of telomere length with greater cumulative exposure to the illness but such notions await further clarification [42].

A meticulous review of the existing literature is suggestive of a dose–response relationship in mood disorders, so that mounting exposure to affective episodes leads to increased telomere attrition. In the Dutch study by Verhoeven et al. (2014), the severity of depressive episodes and cumulative exposure over the past 4 years were inversely correlated with LTL [24]. The prospective study by Shalev et al. (2014) showed that only male subjects had decreased telomere length in a dose–response manner [26]. The small-scale MDD study by Wolkowitz et al. (2011) found that LTL shortening was correlated with lifetime duration of depression, in particular poorly treated or untreated depression [34]. Martinsson et al. (2013) in their study in BD subjects demonstrated that LTL shortening was associated with prior depressive rather than manic episodes [39]. Finally, a negative study is worth mentioning as it did report significantly shorter telomeres in depressed subjects compared to controls but was unable to find a relationship between MDD severity and chronicity with LTL shortening [43].

Factors Mediating Telomere Shortening

Major psychiatric disorders like mood disorders and schizophrenia show biological abnormalities that cut across diagnostic categories, and importantly, include increased inflammation, oxidative stress, serum cortisol aberrations, and autonomic system anomalies [44]. These biochemical irregularities likely cause telomere attrition, so that the latter may be related to specific biological processes or endophenotypes rather than psychiatric diagnoses per se [45]. While this supposition remains to be fully established, it can help explain the heterogeneity of findings in specific diagnostic groups, as well as the apparent inconsistencies in LTL studies among patients belonging to different categorical diagnoses. Furthermore, it is worth remembering that major psychiatric illnesses are often associated with such lifestyle factors as irregular sleep schedules, unhealthy dietary habits, inadequate physical activity, cigarette smoking, and alcohol abuse. These aggravating factors may, by themselves, lead to increased DNA loss from the telomeres during cell division as well as in nondividing cells [46]. Additionally, there may be shortened telomeres in some psychiatric patients prior to disease onset, where decreased LTL represents an existing anomaly and acts as a risk factor. In this scenario, epigenetic reprogramming of telomere maintenance is conceivably acting as a mediating factor in stress-related psychiatric conditions [47].

Inflammation and Oxidative Stress

Deficiency of telomerase, the enzyme responsible for telomere lengthening, causes shortened telomeres in mitotic cells such as leukocytes, stem/progenitor cells as well as dividing neurons in the dentate gyrus and subventricular zone. Chronic viral infection, exemplified by cytomegalovirus, is being increasingly associated with shortened LTL and this effect is possibly due to selective expansion of leukocytes and a predominance of senescent T cells (e.g., CD8+CD28−) [48]. Furthermore, inflammation and oxidative stress are two key determinants of LTL attrition which act independently from telomere shortening that ensues from curtailed DNA end replication in frequently dividing cells. These factors are often increased in severe psychiatric illnesses and mature resting cells including neurons can acquire a senescent phenotype if exposed to them [49]. In this regard, it must be kept in mind that inflammation together with increased oxidative stress can become mutually reinforcing with increasing damage that promotes accelerated cell aging. The effect of inflammation on LTL is likely caused by its association with increased immune cell replication during inflammation, as well as by pathways leading from inflammation to oxidation. Pro-inflammatory cytokine levels are inversely correlated with LTL in MDD, in individuals with histories of early life stress and in healthy individuals with high C-reactive protein levels or elevated overall inflammatory load [50]. Oxidative stress probably has an even more fundamental role in LTL shortening, since telomeric DNA is very sensitive to free radicals and this coupled to relatively inefficient repair of oxidative damage results in severe telomere attrition. Accordingly, there is evidence that oxidative stress markers are inversely correlated with LTL in MDD [51].

Role of Stress Hormones—Cortisol and Catecholamines

The relationship between serum cortisol levels and LTL is not very clear cut. With regards to basal cortisol concentration, a study showed no significant association while another reported an inverse relationship to LTL [52]. In studies conducted in Cushing's syndrome patients, LTL and cortisol levels were not related cross-sectionally but LTL significantly lengthened after remission from the active disease [53]. Studies are more, but not always, consistent in showing inverse relationships between LTL and dynamic aspects of cortisol secretion (e.g., waking-associated increases in cortisol or cortisol responses provoked by psychological stress) as opposed to basal, resting or even circadian cortisol levels [54]. In contrast, shortened LTL has been associated with increased urinary catecholamine concentrations or increased sympathetic nervous system activity more unequivocally. Individuals with increased inflammation, higher cortisol awakening responses, and increased heart rates displayed progressively shorter telomeres as the number of such irregularities increased [52]. Lastly, certain anabolic hormones may be related to LTL. Stress-stimulated salivary testosterone levels were positively correlated with buccal cell TL, but resting, basal, and circadian testosterone levels were not [55]. Also, higher anabolic/catabolic ratios (higher dehydroepiandrosterone sulfate and insulin-like growth factor-I levels, along with lower cortisol, catecholamine, and IL-6 levels) in elderly subjects were associated with relatively longer LTL [56].

Psychotropic Medications and LTL

Two studies in MDD [24] and one in BD type II [37] found no significant difference in LTL between those who were currently receiving psychoactive medication compared to those who were not, and one study found no difference between those on high dose versus low dose antidepressants [43]. These findings must be interpreted cautiously, however, since only current or recent medication use was assessed, not the cumulative duration of prior medication use. A study reported that a mixed group of patients who had severe psychiatric illnesses requiring hospitalization had longer LTL than controls, but only if they had been prescribed psychotropic medications. However, this finding is difficult to interpret, since cases did not necessarily have current mental illnesses (patients were included if they had been psychiatrically hospitalized over approximately the preceding four decades). Further, patients were not randomized to medication and there may have been a survivor selection bias [57]. As mentioned earlier, Martinsson et al. (2013) reported significantly increased LTL in individuals with BD treated with lithium compared to controls; they hypothesized that lithium may increase telomerase activity, but this supposition was untested in humans [39]. The literature search found one study which directly assessed the impact of antipsychotic medication on telomere length in animals. Mice administered atypical antipsychotics for 2 weeks (from the age of 8 weeks on) had lengthened hippocampal TL compared to untreated mice; typical antipsychotic medications did not share this effect [58].

Telomerase Activity (TA) in Mood Disorders—Overview

The enzyme telomerase represents the principal means by which telomeres are maintained and their lengths replenished. Inadequate TA in replicating or injured cells decreases the capacity to repair shortened telomeres and increases vulnerability to premature cellular senescence, apoptosis, or genomic instability [59]. Most normal human somatic cells have very little, if any, detectable TA and therefore, possess limited capacity for cellular division. By distinction, germ-line/stem/progenitor cells have characteristically high TA, and so is the case with various rapidly dividing and cancerous cells [60].

Clinical and animal studies have established the significance of balanced telomerase activity for cellular health and successful aging. In humans inherited telomerase deficiency causing a twofold decline in gene dosage is associated with malignancies and several other diseases. On the other hand, too much TA can also be harmful such that mutations that increase expression of TERT, the catalytic subunit of telomerase, by twofold result in enhanced risk for certain cancers. This supports the notion of "just right" telomerase activity for appropriate physiological functioning throughout human life [61]. In this vein, the latest studies have started to examine the ratio of TA to LTL, since higher ratios, particularly in the presence of lower telomere length, may point to severe cell stress or a failed attempt by telomerase at telomere maintenance [62]. Separately from their role in telomere preservation, telomerase and TERT may have a major function in cellular health through other mechanisms such as angiogenesis, mitochondrial working, neurogenesis, decreased excitotoxicity, and apoptosis, even though most of this data is from animal experiments, and its human importance is as yet undecided [63]. Mature mice totally lacking in TA exhibited obtunded, dysfunctional telomeres and a senescent phenotype. Interestingly, in these animals re-activating telomerase experimentally for a short period of 4 weeks lengthened telomeres, diminished DNA damage signaling, and reversed degenerative phenotypes through multiple organs, including the brain [64]. Such findings indicate that telomerase not only repairs certain types of age-associated cellular damage but also leads to its reversal.

Preclinical investigations raise the likelihood that brain TA may be relevant to the depression phenotype in mice because of the following observations [3, 65]:

(1) Chronic mild stress (CMS) diminished hippocampal TA.
(2) Treatment with desipramine, a tricyclic antidepressant, reversed the CMS-induced decreases in hippocampal TA.
(3) Inhibition of TA in the hippocampus leads to blighted neurogenesis and "depression-like" behaviors.
(4) By contrast, overexpression of intra-hippocampal TA increased neurogenesis, produced "antidepressant-like" behaviors and precluded CMS-induced behavioral alterations.
(5) Irradiation ablation of the dentate gyrus prevented the "antidepressant-like" effects of telomerase overexpression.

This is robust evidence that hippocampal TA in mice is linked to the modulation of "depression-like" behaviors and possibly "antidepressant-like" effects mainly by promoting adult neurogenesis in the dentate gyrus. Nevertheless, a noteworthy limitation is that telomeres and telomerase are regulated differently in rodents and humans so that caution is warranted when extending these findings to man [66].

TA—A Review of Studies in Psychiatric Disorders

There is a paucity of studies on TA in psychiatric conditions but this area is better investigated in the framework of psychological stress. In this regard, a study in premenopausal women reported that compared to low-stress mothers, severely stressed caregiving mothers in good overall health and without clinical depression had lower resting PBMC TA (basal TA) [67]. Another study investigated basal or resting TA in peripheral monocytes and lymphocytes of elderly women, half of whom were stressed dementia caregivers while the other half were low-stress controls. Basal TA was lower in dementia caregivers as compared to controls, but exposure to acute laboratory stress transiently increased resting telomerase activity in all of the participants, both in proportion to the cortisol response to the stressor and (in the low-stress women only) to the degree of anticipatory threat [68]. Further studies indicated that basal PBMC TA could be upregulated in stressful conditions or clinical depression. For instance, in another study in caregivers of dementia sufferers, several subjects had signs of clinical depression and demonstrated short LTL but with increased basal PBMC TA [69]. Likely reasons for these conflicting findings could be that the caregiver mothers in the study by Epel et al. (2004) were premenopausal whereas those in the Damjanovic et al. (2007) study were postmenopausal and estrogen, a recognized regulator of TERT, was involved in these effects. Moreover, few of the caregiver mother subjects (Epel et al. 2004), but many dementia caregivers in the study by Damjanovic et al. (2007) had clinical depression. Nonetheless, the etiology for the opposing effects on basal PBMC TA was unclear and it was considered that the increased TA in the latter study was an unsuccessful attempt to compensate for the excessive loss of telomeres. The same rationalization was given in a small-scale study in MDD, in which unmedicated individuals with major depression had significantly increased basal PBMC TA [70]. In addition, Tessyier et al. demonstrated that expression of TERT mRNA, while not significantly different in MDD and control groups, was positively correlated with depression and anxiety severity ratings in the combined sample of MDD subjects and controls [33]. A study in patients with schizophrenia reported a nominally significant decrease in basal PBMC TA in cases compared to controls [71]. Two genetic studies deserve mentioning; in a study in Han Chinese subjects, researchers investigated NVL gene variants in cases with MDD and schizophrenia and compared these to nonpsychiatric controls. NVL (nuclear valosin containing protein/p97-Like), a member of the AAA-ATPase (ATPases associated with various cellular activities) family, encodes a novel hTERT (human telomerase reverse transcriptase) interacting protein NVL2 which is a telomerase component essential for holoenzyme assembly. The investigators were able to show that NVL gene may contain overlapping common genetic polymorphisms acting as risk factors for both MDD and schizophrenia, highlighting the role of telomerase in the pathogenesis of major psychiatric disorders [72]. In an interesting study, a genetic polymorphism in hTERT gene associated with shortened telomeres was investigated in patients with major depression, BD type I subjects, current episode depressed, and healthy controls. It was shown that telomere length, as measured in saliva, was shorter in depressed subjects compared to controls and that rs2736100 minor allele in hTERT gene was associated with depression among those without experience of childhood adversity, and with number of depressive episodes in BD1 patients responding well to lithium. The results suggested that genetic variation in hTERT gene, the catalytic subunit of telomerase, may influence the vulnerability to depression [73].

It can be surmised that changes in telomerase activity may include genetic alterations in the enzyme, while common pathogenic factors like oxidative stress and inflammation also influence TA. Likewise, stress associated cortisol changes may have major effects on TA and, in this regard, the underlying mechanisms are just beginning to be elucidated. In a recently published study, cases with MDD on routine drug treatment were randomized to 12 weeks of yoga- and meditation-based lifestyle intervention (YMLI) or no such treatment and several neuroplasticity and cellular health biomarkers were measured. It was found that depression scores significantly decreased in the intervention group compared to controls and that this was associated with increased serum BDNF levels in the former. Increased sirtuin 1 and telomerase activity and decreased cortisol significantly predicted this association (all $p < 0.05$) [74].

Psychotropic Medications and TA

In this section, first preclinical data is described in relation to psychotropic medications and TA and the results are interpreted with respect to depression-like behaviors in murine models. In a recently published study, it was found that depressive phenotype induced by CMS in rats was reversed by desipramine (a TCA) and this was associated with restored TA as measured by increased TERT expression along with a reduction in oxidative damage to animal liver [65]. The study concluded that antidepressant administration was able to rescue age-related phenotypes in depressed individuals induced by chronic stress. Wei et al. (2015) reported short telomeres and reduced TERT expression and TA in the hippocampus of Flinders sensitive line rats, which are a genetic model of depression, compared to Flinders-resistant line rats. They also found that lithium administration for 6 weeks significantly increased TERT expression and TA in the hippocampus of the Flinders sensitive line rats, thereby normalizing their baseline abnormalities [75].

Although still few in number, clinical studies are now providing a clearer picture with respect to TA and psychotropic agents. In a small-scale study in BD type I patients, it was shown that compared to controls, medication-free subjects with manic episodes had shortened LTL at baseline, but this increased after treatment with lithium plus antipsychotics. Whole blood TERT gene expression levels were upregulated in mania and remission compared to controls and this effect was speculated to be a compensatory attempt by the body to restore LTL [76]. An interesting study in bipolar disorder found that in cases, LTL was positively correlated with lithium therapy when treatment lasted for a duration of more than 2 years. Moreover, there was increased expression of telomerase gene in neural progenitor cells derived from lithium-treated patients [77]. A large-scale study sheds light on recently recognized hTERT SNP rs2736100, shortened LTL and depression. It was found that telomere length was decreased in cases versus controls and the rs2736100 minor allele was associated with MDD among those without experience of childhood adversity, and with number of depressive episodes in BD types I patients responding well to lithium. While it was an original report on hTERT gene variation in mood disorders, it demonstrated that the newly documented SNP was associated with depressive recurrences in bipolar disorder even in patients who had response to lithium [73]. In the study by Martinsson et al. (2013), it was shown that both BD type I and II patients had lengthening of LTL compared to controls with long-term lithium treatment ($\geq$30 months) and this effect was positively correlated with lithium response [39]. In a small-scale study, Wolkowitz et al. (2012) reported that unmedicated MDD subjects who had relatively low basal PBMC TA at baseline (prior to treatment and compared to the entire MDD group), and who had the greatest increases in basal PBMC TA over the course of treatment, showed superior antidepressant response to 8 weeks of sertraline treatment. Across the entire sample (responders and nonresponders to treatment), however, antidepressant treatment was not associated with significant changes in basal PBMC TA. These findings raise the possibility that depressed individuals with relatively low basal PBMC TA while unmedicated (compared to other depressed individuals) stand to gain the most from exogenous telomerase activation, and that telomerase activation may be a novel mechanism of action of some antidepressants [70]. Further studies are urgently needed to assess the role of telomerase in psychiatric disorders, identify new mechanisms of action of psychopharmacological/psychological treatments, and most importantly, answer the key question whether cellular aging can be reversed or slowed down?

Peripheral Aging Biomarkers and Brain

In this section, the relationship of LTL and PBMC TA with brain tissue is examined as it is crucial to establish whether key peripheral biomarkers of aging bear any association with brain functioning. Since telomere length is frequently interrelated across certain tissues including skeletal muscle, skin, subcutaneous fat, and cerebral cortex it is conceivable that LTL is linked to TL in some brain tissues. Additionally, the rates of telomere shortening over time are also comparable across tissues, at least for leukocytes, skeletal muscle, skin, and subcutaneous fat [78]. Further, to the extent that LTL is shortened by inflammation and oxidative stress, these systemic factors may affect telomeres in brain cells also, since the latter are highly susceptible to such conditions. Regardless of these speculations, it is unknown whether LTL and TL are correlated in brain cells [79].

In this vein, two postmortem studies of TL in cerebellar gray matter and occipital cortex found no significant differences between MDD subjects and controls, although correlations with LTL were not assessed [80, 81]. However, compared to areas like dentate gyrus in hippocampus where actively dividing neuronal precursor cells are found, cerebellar and occipital gray matter are presumably not much influenced by mitosis-related telomere shortening. Among the glial cells, oligodendrocytes are exquisitely sensitive to oxidative stress. Szebeni and colleagues, studying the autopsied brains of individuals who had MDD reported shortened TL, decreased TERT expression, and lowered antioxidant enzymes in oligodendrocytes in two white matter regions implicated in MDD [82].

Neuroimaging research is beginning to shed light on the correlation between LTL and hippocampal volume in MDD. Wikgren et al. found that, whereas shorter LTL was associated with greater subcortical atrophy and more white matter hyperintensities, shorter LTL was related to *larger* hippocampal (HC) volume. This was discovered in non-demented apolipoprotein E ε3/ε3 carriers, but not in non-demented apolipoprotein E ε4 carriers. The authors interpreted their finding in the former group as being consistent with greater overall cellular proliferation in leukocytes as well as the hippocampus. While this would lead to relatively shorter LTL due to more frequent mitoses in the leukocytes, HC volume would increase due to enhanced neurogenesis in the dentate gyrus [83]. Nonetheless, a more recent MRI study found the opposite relationship; it discovered that in apolipoprotein E ε3/ε3 carriers LTL was directly correlated with HC volume, which was interpreted as evidence of "coordinated chromosomal and neural aging" [84]. A relatively recent population-based study found that LTL was positively linked to total cerebral volume, including white and cortical matter gray volume, as well as with HC size and volumes of several other subregions. These correlations were generally more robust in relatively older individuals, but remained significant after adjusting for multiple covariates, including age, gender, and cardiovascular risk factors [85].

Very few studies have examined relationships between basal PBMC TA and brain TA or brain structural volumes. To date, only one study has assessed the relationship of basal PBMC TA to HC volume in MDD. In an interesting study, Wolkowitz et al. (2015) reported a significant positive correlation between basal PBMC TA and hippocampal volume in a small group of unmedicated individuals with MDD but not in healthy controls. The authors interpreted these findings as being consistent with the fact that increased neurogenesis in the dentate gyrus resulted from greater TA in the hippocampus in MDD subjects [86].

In conclusion, it is implicit that only peripheral markers of cell aging (LTL, PBMC TA) are obtainable in living humans. Nonetheless, it is essential to establish their relationship to neural processes involved in psychiatric illnesses; otherwise, their ultimate value will be limited. As alluded to above, in the case of clinical populations there are enough promising leads to justify further trials comparing peripheral and central markers using various forms of neuroimaging. For investigation purposes, autopsied brain specimens can provide corroborating evidence.

Averting or Undoing Cellular Aging

There is sufficient evidence that shortened LTL accompany major psychiatric disorders and this raises the exciting prospect that appropriate treatment or prevention of psychiatric illnesses might lead to telomere lengthening and delay cellular aging. Only few pharmacologic studies have investigated this issue; nonetheless, several behavioral and psychological intervention studies in nonpsychiatric populations have examined the matter. In the later subjects, researchers have determined the effects of interventions on basal PBMC TA or LTL. The behavioral techniques range from intensive lifestyle modification, mindful eating, mindfulness-based stress reduction, and various types of meditation, even though it is not known whether the findings could be extended to psychiatric populations. An additional limitation is that these studies have often been non-randomized and not adequately controlled; however, in general, these have found intervention-associated increases in basal PBMC TA [87]. Furthermore, some investigators reported "dose–response" relationships, such that enhancement in mental well-being, greater sense of purpose in life and superior adherence to behavioral interventions correlated with larger increases in basal PBMC TA [88]. One study found that retreat participants meditating for 6 h daily for 3 months had greater PBMC basal TA at the end of the 3 months than did a waitlist control group [89]. In a sample of breast cancer patients, Lengacher et al. (2014) showed that mindfulness-based stress reduction for 6 weeks significantly increased PBMC basal TA compared to a waitlist control group. Unlike some other studies, the investigators controlled for basal PBMC TA at the start of the study, but the active group was heterogeneous in terms of treatment received and time since treatment completion. Additionally, the treatment program was rather short and thus only short-term assessment of basal PBMC TA was possible [90]. In a first of its kind non-randomized study in low-risk prostate cancer patients, Ornish et al. (2008) showed that 3 months of wide-ranging lifestyle alterations resulted in significant increases in basal PBMC TA along with reduction in psychological distress. Potential limitations of this study were the lack of a control group, the fact that only 30 of 126 eligible patients agreed to participate in the study after learning the details and an all-male cohort [91]. A 5-year follow-up of 10 of the original participants from that study showed that their LTL significantly increased from baseline relative to controls who were only provided active surveillance. This study was limited by small sample size (10 subjects in the intervention group and 25 controls) and non-randomized design [92].

The literature on exercise and telomere length is rather unclear but by and large indicates that exercise is coupled to a telomere-protective phenotype in leukocytes and skeletal muscles [93]. One study found that aspects of "multisystem resiliency" defined by positive lifestyle (e.g., social support, good emotional regulation, sleep, and exercise), collectively but not individually, statistically diminished the negative relationship between MDD and LTL. Analyses in this study were cross sectional, thus causality could not be deduced and data on diet, another lifestyle factor that could influence LTL, were not offered [94]. The same group of investigators previously found, in a group of caregiving and non-caregiving postmenopausal women, that highly stressed women had shorter telomeres, but only if they were inactive, signifying a protective effect of exercise. As that study was cross-sectional, it was hard to infer causal relationships, particularly since the more highly stressed women were less likely to be physically active [95].

Similarly, a prospective study of healthy postmenopausal women followed over the course of 1 year found that major life stresses during the study year were associated with significant telomere shortening over the period, but that this effect was significantly attenuated in women with positive health behaviors (leisure-time physical activity, healthy dietary practices, and good sleep quality) [96]. This study was important since it was one of the few prospective longitudinal studies to examine stress-related changes in TL and possible moderators of this correlation. However, since the measures of stress and of health behaviors were self-reported this could have possibly influenced the results. Further, health behaviors may have been falsely associated with LTL if physical diseases (that could, themselves, reduce health behaviors) had the primary relationship with LTL. One additional small-scale study involving telephone-based psychological stress-reduction counseling in 22 women with cervical cancer found no significant overall change in LTL following four months of counseling, but did observe that changes in distress ratings over that time period were inversely correlated with changes in LTL. This study, however, had no control group, and four months of study time may be too short to determine significant changes in LTL [97].

The behavioral/psychological/lifestyle intervention literature linking these measures to basal PBMC TA and LTL is attention-grabbing and points in the presumed direction. However, it is limited by the small-scale, non-randomized, non-blinded design of the studies, as well as, the short duration of the interventions. Despite the fact that the biochemical intermediaries in this regard remain to be determined, the presumed effects of certain of these interventions on cellular aging appear to be mediated by such psychological factors as improvements in stress arousal and lessening of threat cognitions and ruminative thought.

Conclusion

The study of telomere biology in psychiatric illnesses is in its initial stages and, while firm conclusions cannot be drawn, the evidence is suggestive of accelerated cellular aging in major mental disorders. Admittedly, there are discrepancies between studies and the possible reasons include varied subject demographics (e.g., age, gender, race, socioeconomic status, history of childhood adversity), dissimilar study designs, differences in duration or severity of the investigated illness, diverse specimen processing, and disparate assaying protocols. Moreover, variations in moderators of LTL and basal PBMC TA are often not assessed and, among others, these comprise of genetic risk-alleles, cognitive threat appraisal, pessimistic outlook, arousal and regulatory system activation, and stress resiliency factors. In this respect, recent data suggest that "high risk" genetic polymorphisms in the serotonin and dopamine systems may interact with early life adversity to affect adult LTL. Short LTL is unlikely to be specific to any one categorical psychiatric illness and is more likely related to underlying transdiagnostic biological abnormalities or behavioral dimensions/phenotypes. Combining LTL measures with psychiatric disorder evolution may inform clinical practice as the current evidence is suggestive of progressive telomere attrition with repeated mood episodes.

Further, a study of cell aging in psychiatric illnesses and of its moderators and mediators may have preventive and protective value. However, the overarching question is whether peripheral LTL and PBMC basal TA are reflective of brain processes relevant to mental illness.

Evidence of significant LTL shortening in psychiatric illnesses is cause for concern, since shorter LTL has been linked to current and future medical illnesses and to premature mortality, although causality has not been demonstrated. More research is needed to define possible roles of telomerase and TERT in psychiatric illnesses, but initial preclinical and clinical findings are intriguing and indicative of a potential role in hippocampal neurogenesis and the action of psychotropic medications. Risk factors and mechanisms for accelerated cell aging in humans are just beginning to be understood, and longitudinal studies will be needed to infer causality as well as to address the important questions of timing, prevention, and reversibility of cell aging. If LTL attrition is related to psychiatric disease in a "dose–response" relationship, it will be important to determine whether lessening the "dose" of the disease by adequate treatment will help preserve LTL or reverse telomere attrition. In a rather dismal situation, findings that certain positive lifestyle changes are correlated with lowered degrees of cell aging provide reasons for optimism, although many of the studies are small-scale, open-label, or inadequately controlled. Thus, more prospective longitudinal, large-scale, well-controlled studies are required. Since subjective stress ratings and the anticipation of threat bear closer relationships to cell aging than do objective stressors, it is conceivable that psychotherapy and stress coping mechanisms might also attenuate stress associated cell aging, but this has not been well studied.

Measuring LTL and basal PBMC TA may someday prove to be useful biomarkers in personalized medicine for staging disease progression and disease risk and selecting treatments, but there is insufficient research yet to support this assumption. Further, inadequate calibration of assay methods across labs and the lack of accepted "normal ranges" for LTL and basal PBMC TA make it premature for cell aging markers to enter clinical use at this time. The relatively small effect sizes reported in positive studies, as well as the lack of diagnostic specificity of LTL and PBMC basal TA changes also argue against the use of such markers as diagnostic tools in isolation from other measures. As the mechanistic relationships between psychiatric illnesses, biological aging, and comorbid physical illnesses become clearer, psychiatric illnesses may come to be understood as *systemic* illnesses with specific mental manifestations rather than as purely brain diseases, thus expanding the range of therapeutic targets and diminishing the stigma associated with these illnesses.

References

1. Liu MY, Nemes A, Zhou QG. The emerging roles for telomerase in the central nervous system. Front Mol Neurosci. 2018;11:160.
2. Manoliu A, Bosch OG, Brakowski J, Brühl AB, Seifritz E. The potential impact of biochemical mediators on telomere attrition in major depressive disorder and implications for future study designs: a narrative review. J Affect Disord. 2018;225:630–46.

3. Xie X, Chen Y, Ma L, Shen Q, Huang L, Zhao B, et al. Major depressive disorder mediates accelerated aging in rats subjected to chronic mild stress. Behav Brain Res. 2017;329:96–103.
4. Maurya PK, Rizzo LB, Xavier G, Tempaku PF, Zeni-Graiff M, Santoro ML, et al. Shorter leukocyte telomere length in patients at ultra high risk for psychosis. Eur Neuropsychopharmacol. 2017;27:538–42.
5. Zhu Y, Liu X, Ding X, Wang F, Geng X. Telomere and its role in the aging pathways: telomere shortening, cell senescence and mitochondria dysfunction. Biogerontology. 2018 Epub ahead of print.
6. Barnes RP, Fouquerel E, Opresko PL. The impact of oxidative DNA damage and stress on telomere homeostasis. Mech Ageing Dev. 2018 Epub ahead of print.
7. Sui B, Hu C, Jin Y. Mitochondrial metabolic failure in telomere attrition-provoked aging of bone marrow mesenchymal stem cells. Biogerontology. 2016;17:267–79.
8. Bertuch AA. The molecular genetics of the telomere biology disorders. RNA Biol. 2016;13:696–706.
9. Révész D, Milaneschi Y, Terpstra EM, Penninx BW. Baseline biopsychosocial determinants of telomere length and 6-year attrition rate. Psychoneuroendocrinology. 2016;67:153–62.
10. Müezzinler A, Zaineddin AK, Brenner H. A systematic review of leukocyte telomere length and age in adults. Ageing Res Rev. 2013;12:509–19.
11. Schmidt JC, Cech TR. Human telomerase: biogenesis, trafficking, recruitment, and activation. Genes Dev. 2015;29:1095–105.
12. Bhattacharyya J, Mihara K, Bhattacharjee D, Mukherjee M. Telomere length as a potential biomarker of coronary artery disease. Indian J Med Res. 2017;145:730–7.
13. Epel ES, Merkin SS, Cawthon R, Blackburn EH, Adler NE, Pletcher MJ, et al. The rate of leukocyte telomere shortening predicts mortality from cardiovascular disease in elderly men. Aging. 2008;1:81–8 Albany NY.
14. Nilsson PM, Tufvesson H, Leosdottir M, Melander O. Telomeres and cardiovascular disease risk: an update 2013. Transl Res. 2013;162:371–80.
15. Codd V, Nelson CP, Albrecht E, Mangino M, Deelen J, Buxton JL, et al. Identification of seven loci affecting mean telomere length and their association with disease. Nat Genet. 2013;45:422–7.
16. Zhang J, Rane G, Dai X, Shanmugam MK, Arfuso F, Samy RP, et al. Ageing and the telomere connection: an intimate relationship with inflammation. Ageing Res Rev. 2016;25:55–69.
17. Liu L. Linking telomere regulation to stem cell pluripotency. Trends Genet. 2017;33:16–33.
18. Cai Z, Yan LJ, Ratka A. Telomere shortening and Alzheimer's disease. Neuromolecular Med. 2013;15:25–48.
19. Huang YC, Wang LJ, Tseng PT, Hung CF, Lin PY. Leukocyte telomere length in patients with bipolar disorder: an updated meta-analysis and subgroup analysis by mood status. Psychiatry Res. 2018;270:41–9.
20. Darrow SM, Verhoeven JE, Révész D, Lindqvist D, Penninx BW, Delucchi KL, et al. The association between psychiatric disorders and telomere length: a meta-analysis involving 14,827 persons. Psychosom Med. 2016;78:776–87.
21. Ten Have M, de Graaf R, van Dorsselaer S, Tuithof M, Kleinjan M, Penninx BWJH. Recurrence and chronicity of major depressive disorder and their risk indicators in a population cohort. Acta Psychiatr Scand. 2018;137:503–15.
22. Nunes PV, Nascimento CF, Kim HK, Andreazza AC, Brentani HP, Suemoto CK, et al. Low brain-derived neurotrophic factor levels in post-mortem brains of older adults with depression and dementia in a large clinicopathological sample. J Affect Disord. 2018;241:176–81.
23. Simon NM, Smoller JW, McNamara KL, Maser RS, Zalta AK, Pollack MH, et al. Telomere shortening and mood disorders: preliminary support for a chronic stress model of accelerated aging. Biol Psychiatry. 2006;60:432–5.
24. Verhoeven JE, Revesz D, Epel ES, Lin J, Wolkowitz OM, Penninx BW. Major depressive disorder and accelerated cellular aging: results from a large psychiatric cohort study. Mol Psychiatry. 2014;19:895–901.

25. Gotlib IH, LeMoult J, Colich NL, Foland-Ross LC, Hallmayer J, Joormann J, et al. Telomere length and cortisol reactivity in children of depressed mothers. Mol Psychiatry. 2015;20:615–20.
26. Shalev I, Moffitt TE, Braithwaite AW, Danese A, Fleming NI, Goldman-Mellor S, et al. Internalizing disorders and leukocyte telomere erosion: a prospective study of depression, generalized anxiety disorder and post-traumatic stress disorder. Mol Psychiatry. 2014;19:1163–70.
27. Jodczyk S, Fergusson DM, Horwood LJ, Pearson JF, Kennedy MA. No association between mean telomere length and life stress observed in a 30 year birth cohort. PLoS One. 2014;9:e97102.
28. Hoen PW, de Jonge P, Na BY, Farzaneh-Far R, Epel E, Lin J, et al. Depression and leukocyte telomere length in patients with coronary heart disease: data from the Heart and Soul Study. Psychosom Med. 2011;73:541–7.
29. Hoen PW, Rosmalen JG, Schoevers RA, Huzen J, van der Harst P, de Jonge P. Association between anxiety but not depressive disorders and leukocyte telomere length after 2 years of follow-up in a population-based sample. Psychol Med. 2013;43:689–97.
30. Schaakxs R, Verhoeven JE, Oude Voshaar RC, Comijs HC, Penninx BW. Leukocyte telomere length and late-life depression. Am J Geriatr Psychiatry. 2015;23:423–32.
31. Kananen L, Surakka I, Pirkola S, Suvisaari J, Lonnqvist J, Peltonen L, et al. Childhood adversities are associated with shorter telomere length at adult age both in individuals with an anxiety disorder and controls. PLoS One. 2010;5:e10826.
32. Needham BL, Mezuk B, Bareis N, Lin J, Blackburn EH, Epel ES. Depression, anxiety and telomere length in young adults: evidence from the National Health and Nutrition Examination Survey. Mol Psychiatry. 2015;20:520–8.
33. Teyssier JR, Chauvet-Gelinier JC, Ragot S, Bonin B. Up-regulation of leucocytes genes implicated in telomere dysfunction and cellular senescence correlates with depression and anxiety severity scores. PLoS One. 2012;7:e49677.
34. Wolkowitz OM, Mellon SH, Epel ES, Lin J, Dhabhar FS, Su Y, et al. Leukocyte telomere length in major depression: correlations with chronicity, inflammation and oxidative stress–preliminary findings. PLoS One. 2011;6:e17837.
35. Lindqvist D, Epel ES, Mellon SH, Penninx BW, Révész D, Verhoeven JE, et al. Psychiatric disorders and leukocyte telomere length: underlying mechanisms linking mental illness with cellular aging. Neurosci Biobehav Rev. 2015;55:333–64.
36. Russo P, Prinzi G, Proietti S, Lamonaca P, Frustaci A, Boccia S, et al. Shorter telomere length in schizophrenia: evidence from a real-world population and meta-analysis of most recent literature. Schizophr Res. 2018 Epub ahead of print.
37. Elvsashagen T, Vera E, Boen E, Bratlie J, Andreassen OA, Josefsen D, et al. The load of short telomeres is increased and associated with lifetime number of depressive episodes in bipolar II disorder. J Affect Disord. 2011;135:43–50.
38. Rizzo LB, Do Prado CH, Grassi-Oliveira R, Wieck A, Correa BL, Teixeira AL, et al. Immunosenescence is associated with human cytomegalovirus and shortened telomeres in type I bipolar disorder. Bipolar Disord. 2013;15:832–8.
39. Martinsson L, Wei Y, Xu D, Melas PA, Mathe AA, Schalling M, et al. Long-term lithium treatment in bipolar disorder is associated with longer leukocyte telomeres. Transl Psychiatry. 2013;3:e261.
40. Lima IM, Barros A, Rosa DV, Albuquerque M, Malloy-Diniz L, Neves FS, et al. Analysis of telomere attrition in bipolar disorder. J Affect Disord. 2015;172:43–7.
41. Barbé-Tuana FM, Parisi MM, Panizzutti BS, Fries GR, Grun LK, Guma FT, et al. Shortened telomere length in bipolar disorder: a comparison of the early and late stages of disease. Rev Bras Psiquiatr. 2016;38:281–6.
42. Verhoeven JE, van Oppen P, Révész D, Wolkowitz OM, Penninx BW. Depressive and anxiety disorders showing robust, but non-dynamic, 6-year longitudinal association with short leukocyte telomere length. Am J Psychiatry. 2016;173:617–24.

43. Hartmann N, Boehner M, Groenen F, Kalb R. Telomere length of patients with major depression is shortened but independent from therapy and severity of the disease. Depress Anxiety. 2010;27:1111–6.
44. Solana C, Pereira D, Tarazona R. Early senescence and leukocyte telomere shortening in schizophrenia: a role for cytomegalovirus infection? Brain Sci. 2018;8:E188.
45. Nguyen TT, Eyler LT, Jeste DV. Systemic biomarkers of accelerated aging in schizophrenia: a critical review and future directions. Schizophr Bull. 2018;44:398–408.
46. Osler M, Bendix L, Rask L, Rod NH. Stressful life events and leucocyte telomere length: do lifestyle factors, somatic and mental health, or low grade inflammation mediate this relationship? Results from a cohort of Danish men born in 1953. Brain Behav Immun. 2016;58:248–53.
47. Provenzi L, Scotto di Minico G, Giorda R, Montirosso R. Telomere length in preterm infants: a promising biomarker of early adversity and care in the neonatal intensive care unit? Front Endocrinol (Lausanne). 2017;8:295.
48. Wilson SJ, Woody A, Padin AC, Lin J, Malarkey WB, Kiecolt-Glaser JK. Loneliness and telomere length: immune and parasympathetic function in associations with accelerated aging. Ann Behav Med. 2018 Epub ahead of print.
49. Wolkowitz OM, Reus VI, Mellon SH. Of sound mind and body: depression, disease, and accelerated aging. Dialogues Clin Neurosci. 2011;13:25–39.
50. Wolkowitz OM, Epel ES, Reus VI, Mellon SH. Depression gets old fast: do stress and depression accelerate cell aging? Depress Anxiety. 2010;27:327–38.
51. Tolahunase M, Sagar R, Dada R. Impact of yoga and meditation on cellular aging in apparently healthy individuals: a prospective, open-label single-arm exploratory study. Oxid Med Cell Longev. 2017;2017:7928981.
52. Révész D, Verhoeven JE, Milaneschi Y, de Geus EJ, Wolkowitz OM, Penninx BW. Dysregulated physiological stress systems and accelerated cellular aging. Neurobiol Aging. 2014;35:1422–30.
53. Aulinas A, Ramírez MJ, Barahona MJ, Valassi E, Resmini E, Mato E, et al. Telomere length analysis in Cushing's syndrome. Eur J Endocrinol. 2014;171:21–9.
54. Tomiyama AJ, O'Donovan A, Lin J, Puterman E, Lazaro A, Chan J, et al. Does cellular aging relate to patterns of allostasis? an examination of basal and stress reactive HPA axis activity and telomere length. Physiol Behav. 2012;106:40–5.
55. Drury SS, Shirtcliff EA, Shachet A, Phan J, Mabile E, Brett ZH, et al. Growing up or growing old? Cellular aging linked with testosterone reactivity to stress in youth. Am J Med Sci. 2014;348:92–100.
56. Vasunilashorn S, Cohen AA. Stress responsive biochemical anabolic/catabolic ratio and telomere length in older adults. Biodemography Soc Biol. 2014;60(2):174–84.
57. Savolainen K, Räikkönen K, Kananen L, Kajantie E, Hovatta I, Lahti M, et al. History of mental disorders and leukocyte telomere length in late adulthood: the Helsinki Birth Cohort Study (HBCS). J Psychiatr Res. 2012;46:1346–53.
58. Toriumi K, Miyashita M, Ichikawa T, Kobori A, Nohara I, Arai M, et al. JSNP excellent presentation award for CINP2014. Nihon Shinkei Seishin Yakurigaku Zasshi. 2015; 35: 61–62 Article in Japanese.
59. Boccardi V, Paolisso G. Telomerase activation: a potential key modulator for human healthspan and longevity. Ageing Res Rev. 2014;15:1–5.
60. Choudhary B, Karande AA, Raghavan SC. Telomere and telomerase in stem cells: relevance in ageing and disease. Front Biosci (Schol Ed). 2012;4:16–30.
61. Ozturk MB, Li Y, Tergaonkar V. Current insights to regulation and role of telomerase in human diseases. Antioxidants (Basel). 2017;6:E17.
62. Zalli A, Carvalho LA, Lin J, Hamer M, Erusalimsky JD, Blackburn EH, et al. Shorter telomeres with high telomerase activity are associated with raised allostatic load and impoverished psychosocial resources. Proc Natl Acad Sci USA. 2014;111(12):4519–24.
63. Vaváková M, Ďuračková Z, Trebatická J. Markers of oxidative stress and neuroprogression in depression disorder. Oxid Med Cell Longev. 2015;2015:898393.

64. Jaskelioff M, Muller FL, Paik JH, Thomas E, Jiang S, Adams AC, et al. Telomerase reactivation reverses tissue degeneration in aged telomerase-deficient mice. Nature. 2011;469:102–6.
65. Xie X, Chen Y, Wang Q, Shen Q, Ma L, Huang L, et al. Desipramine rescues age-related phenotypes in depression-like rats induced by chronic mild stress. Life Sci. 2017;188:96–100.
66. Zhou QG, Liu MY, Lee HW, Ishikawa F, Devkota S, Shen XR, et al. Hippocampal TERT regulates spatial memory formation through modulation of neural development. Stem Cell Rep. 2017;9:543–56.
67. Epel ES, Blackburn EH, Lin J, Dhabhar FS, Adler NE, Morrow JD, et al. Accelerated telomere shortening in response to life stress. Proc Natl Acad Sci USA. 2004;101:17312–5.
68. Epel ES, Lin J, Dhabhar FS, Wolkowitz OM, Puterman E, Karan L, et al. Dynamics of telomerase activity in response to acute psychological stress. Brain Behav Immun. 2010;244:531–9.
69. Damjanovic AK, Yang Y, Glaser R, Kiecolt-Glaser JK, Nguyen H, Laskowski B, et al. Accelerated telomere erosion is associated with a declining immune function of caregivers of Alzheimer's disease patients. J Immunol. 2007;179:4249–54.
70. Wolkowitz OM, Mellon SH, Epel ES, Lin J, Reus VI, Rosser R, et al. Resting leukocyte telomerase activity is elevated in major depression and predicts treatment response. Mol Psychiatry. 2012;17:164–72.
71. Porton B, Delisi LE, Bertisch HC, Ji F, Gordon D, Li P, et al. Telomerase levels in schizophrenia: a preliminary study. Schizophr Res. 2008;106:242–7.
72. Wang M, Chen J, He K, Wang Q, Li Z, Shen J, et al. The NVL gene confers risk for both major depressive disorder and schizophrenia in the Han Chinese population. Prog Neuropsychopharmacol Biol Psychiatry. 2015;62:7–13.
73. Wei YB, Martinsson L, Liu JJ, Forsell Y, Schalling M, Backlund L, et al. hTERT genetic variation in depression. J Affect Disord. 2016;189:62–9.
74. Tolahunase MR, Sagar R, Faiq M, Dada R. Yoga- and meditation-based lifestyle intervention increases neuroplasticity and reduces severity of major depressive disorder: a randomized controlled trial. Restor Neurol Neurosci. 2018;36:423–42.
75. Wei YB, Backlund L, Wegener G, Mathé AA, Lavebratt C. Telomerase dysregulation in the hippocampus of a rat model of depression: normalization by lithium. Int J Neuropsychopharmacol 2015;18:pyv002.
76. Köse Çinar R. Telomere length and hTERT in mania and subsequent remission. Rev Bras Psiquiatr. 2018;40:19–25.
77. Squassina A, Pisanu C, Congiu D, Caria P, Frau D, Niola P, et al. Leukocyte telomere length positively correlates with duration of lithium treatment in bipolar disorder patients. Eur Neuropsychopharmacol. 2016;26:1241–7.
78. Daniali L, Benetos A, Susser E, Kark JD, Labat C, Kimura M, et al. Telomeres shorten at equivalent rates in somatic tissues of adults. Nat Commun. 2013;4:1597.
79. Dlouha D, Maluskova J, Kralova Lesna I, Lanska V, Hubacek JA. Comparison of the relative telomere length measured in leukocytes and eleven different human tissues. Physiol Res. 2014;63(Suppl 3):S343–50.
80. Teyssier JR, Ragot S, Donzel A, Chauvet-Gelinier JC. Telomeres in the brain cortex of depressive patients. Encephale. 2010; 36: 491–494 Article in French.
81. Zhang D, Cheng L, Craig DW, Redman M, Liu C. Cerebellar telomere length and psychiatric disorders. Behav Genet. 2010;40:250–4.
82. Szebeni A, Szebeni K, DiPeri T, Chandley MJ, Crawford JD, Stockmeier CA, et al. Shortened telomere length in white matter oligodendrocytes in major depression: potential role of oxidative stress. Int J Neuropsychopharmacol. 2014;17:1579–89.
83. Wikgren M, Karlsson T, Soderlund H, Nordin A, Roos G, et al. Shorter telomere length is linked to brain atrophy and white matter hyperintensities. Age Ageing. 2013;43:212–7.
84. Jacobs EG, Epel ES, Lin J, Blackburn EH, Rasgon NL. Relationship between leukocyte telomere length, TA and hippocampal volume in early aging. JAMA Neurol. 2014;71:921–3.

85. King KS, Kozlitina J, Rosenberg RN, Peshock RM, McColl RW, Garcia CK. Effect of leukocyte telomere length on total and regional brain volumes in a large population-based cohort. JAMA Neurol. 2014;71:1247–54.
86. Wolkowitz OM, Mellon SH, Lindqvist D, Epel ES, Blackburn EH, Lin J, et al. PBMC telomerase activity, but not leukocyte telomere length, correlates with hippocampal volume in major depression. Psychiatry Res. 2015;232:58–64.
87. Deng W, Cheung ST, Tsao SW, Wang XM, Tiwari AF. Telomerase activity and its association with psychological stress, mental disorders, lifestyle factors and interventions: a systematic review. Psychoneuroendocrinology. 2016;64:150–63.
88. Schutte NS, Malouff JM. A meta-analytic review of the effects of mindfulness meditation on telomerase activity. Psychoneuroendocrinology. 2014;42:45–8.
89. Jacobs TL, Epel ES, Lin J, Blackburn EH, Wolkowitz OM, Bridwell DA, et al. Intensive meditation training, immune cell TA, and psychological mediators. Psychoneuroendocrinology. 2011;36:664–81.
90. Lengacher CA, Reich RR, Kip KE, Barta M, Ramesar S, Paterson CL, et al. Influence of mindfulness-based stress reduction (MBSR) on TA in women with breast cancer (BC). Biol Res Nurs. 2014; 16: 438–447.
91. Ornish D, Lin J, Daubenmier J, Weidner G, Epel E, Kemp C, et al. Increased TA and comprehensive lifestyle changes: a pilot study. Lancet Oncol. 2008;9:1048–57.
92. Ornish D, Lin J, Chan JM, Epel E, Kemp C, Weidner G, et al. Effect of comprehensive lifestyle changes on telomerase activity and telomere length in men with biopsy-proven low-risk prostate cancer: 5-year follow-up of a descriptive pilot study. Lancet Oncol. 2013;14:1112–20.
93. Arsenis NC, You T, Ogawa EF, Tinsley GM, Zuo L. Physical activity and telomere length: impact of aging and potential mechanisms of action. Oncotarget. 2017;8:45008–19.
94. Puterman E, Epel ES, Lin J, Blackburn EH, Gross JJ, Whooley MA, Cohen BE. Multisystem resiliency moderates the major depression-telomere length association: findings from the Heart and Soul Study. Brain Behav Immun. 2013;33:65–73.
95. Puterman E, Lin J, Blackburn E, O'Donovan A, Adler N, Epel E. The power of exercise: buffering the effect of chronic stress on telomere length. PLoS One. 2010;5:e10837.
96. Puterman E, Lin J, Krauss J, Blackburn EH, Epel ES. Determinants of telomere attrition over 1 year in healthy older women: stress and health behaviors matter. Mol Psychiatry. 2015;20:529–35.
97. Biegler KA, Anderson AK, Wenzel LB, Osann K, Nelson EL. Longitudinal change in telomere length and the chronic stress response in a randomized pilot biobehavioral clinical study: implications for cancer prevention. Cancer Prev Res (Phila). 2012;5:1173–82.

Neuromodulation and Cognitive Control of Emotion

Meysam Amidfar, Young-Hoon Ko and Yong-Ku Kim

Abstract

Recently, noninvasive brain stimulation (NIBS) methodologies, including TMS and tDCS, have been considered as efficacious, safe, and innovative treatments and alternatives to conventional therapies for some psychiatric disorders. Developing evidence suggests that applying rTMS and tDCS over the cognitive control network (CCN), particularly the dorsolateral prefrontal cortex (DLPFC), may improve core symptoms in various psychiatric disorders via direct impact on the cognitive control processes involved in emotion regulation. Therefore, neuromodulation of brain regions involved in the cognitive control of emotion by NIBS approaches could contribute to a paradigm shift in psychiatry. The available evidence suggests that development of effective treatment alternatives to enhance cognition is critical for patients with psychiatric disorders. The purpose of this chapter is to review the cognition enhancing properties of tDCS and TMS and the impact of these treatments on cognitive control processes, especially those related to emotion regulation in psychiatric disorders.

Keywords

TMS · tDCS · Neuromodulation · Cognition · Psychiatric disorders · Paradigm shift

M. Amidfar
Shiraz Neuroscience Research Center,
Shiraz University of Medical Sciences, Shiraz, Iran
e-mail: amidfar.m@gmail.com

Y.-H. Ko · Y.-K. Kim (✉)
Department of Psychiatry, College of Medicine, Korea University, Seoul, South Korea
e-mail: yongku@korea.edu; yongku@korea.ac.kr

© Springer Nature Singapore Pte Ltd. 2019

Y.-K. Kim (ed.), *Frontiers in Psychiatry*, Advances in Experimental Medicine and Biology 1192, https://doi.org/10.1007/978-981-32-9721-0_27

Introduction

Over the past two decades, the therapeutic potential of noninvasive brain stimulation (NIBS) techniques has received increased attention for treatment of psychiatric disorders [1]. NIBS techniques, including transcranial direct current stimulation (tDCS) and transcranial magnetic stimulation (TMS), have shown therapeutic potential to alleviate symptoms and improve cognitive performance in a range of psychiatric disorders [2, 3]. For example, brain stimulation techniques that selectively target brain circuitry related to cognitive and emotional functions, such as the dorsolateral prefrontal cortex (DLPFC), appear to improve cognitive function in depression and schizophrenia [4, 5]. Cognitive impairment is a core feature of many neuropsychiatric disorders and a major index of clinical outcomes, and it has a substantial impact on quality of life and disability related to disease [2]. Although current drugs improve specific symptoms of psychiatric disorders, such as delusions, anxiety, and depression, cognitive deficits are not generally alleviated and may even be exacerbated [6, 7]. Therefore, there is a need for effective therapeutic approaches that restore and enhance cognitive performance in psychiatric disorders [2]. The use of NIBS techniques to stimulate the DLPFC in various psychiatric conditions has been shown to have beneficial clinical effects, such as improvement of core symptoms and cognitive deficits [2, 3]. The DLPFC is one of the brain regions of the cognitive control network (CCN) that contributes to regulation of emotions [8]. The CCN is a regulatory system that enables the individual to accomplish goals by modulation of other cognitive and emotional functions. It encompasses multiple brain areas, including the dorsal anterior cingulate cortex (dACC), posterior parietal regions and DLPFC [9]. TMS and tDCS have been established as neuroscientific tools for the investigation of brain function and as major advancements for the treatment of many psychiatric disorders. Therefore, in this literature review, we aimed to determine whether the use of NIBS approaches for the neuromodulation of brain regions involved in the cognitive control of emotion, especially the DLPFC, enhances cognition and improves core symptoms in psychiatric disorders.

Emotion Regulation in Psychiatry Disorders

Emotion regulation plays a critical role in both physical and mental health, and emotion dysregulation results in various types of psychopathology [10]. Through emotion regulation strategies, people can completely or partially modify the nature, magnitude, and duration of various emotional responses, such as happiness, joy, sadness, and anger [11, 12]. Two important strategies of emotion regulation that have received extensive experimental attention are cognitive reappraisal and expressive suppression. Both are associated with specific mood-related problems, including trait anxiety and depressed mood in healthy adults [13]. Cognitive reappraisal involves a cognitive change to neutralize negative emotional impacts or amplify positive emotional aspects of a future event via a shift in thinking about that event. It is considered an antecedent-focused strategy because it happens before the full experience of distress [13]. Expressive suppression is a response-focused strategy that inhibits an emotional response that is already occurring, but it does not generally decrease subjective emotional distress [13]. Individuals with better interpersonal functioning and greater subjective and objective well-being show more positive expression and fewer negative emotions, probably as a result of their tendency to use cognitive reappraisal instead of expressive suppression [14]. Deficits of emotion regulation play an important role in a range of psychiatric disorders, such as depression, schizophrenia, anxiety disorders, posttraumatic stress disorder (PTSD), substance abuse, attention-deficit/hyperactivity disorder (ADHD) and autism spectrum disorder (ASD) [10, 15–20]. The CCN plays an important role in the processes of emotion regulation by assigning resources to goal-directed behaviors that are probably crucial for reappraisal [8, 9]. Moreover, mental manipulation of emotional information is modulated by the DLPFC [8, 21].

The Cognitive Control of Emotion

Cognitive control is required for the regulation of emotional responses to aversive events and is thus a crucial capability for mental and physical health [22]. In contrast to other strategies for emotion regulation, cognitive transformation of emotional experience or reappraisal of an aversive event, such as reappraisal of negative photos, results in activation of brain regions related to working memory, cognitive control, and self-monitoring, including the lateral prefrontal cortex (LPFC) and the medial prefrontal cortex (MPFC), as well as decreased activation of brain areas involved in emotion processing, including the medial orbitofrontal cortex (MOFC) and the amygdala [22–28]. Similarities of particular activated regions of the LPFC and MPFC by reappraisal, working memory, and response-selection tasks suggest that an overlapping set of prefrontal regions that are responsible for the consciousness maintaining information and preventing interference from competing inputs also control cognitive regulation of feelings and thoughts [22, 24, 26, 29–35]. With regard to cognitive-emotional interaction in psychiatric disorders and the multiple hypothetical pathways by which cognitive behavior regulates pathological emotions, improved understanding of the functional roles of specific brain regions can help define targets for neuromodulatory interventions [36–38]. The notion of a relationship between cognition and emotion originates from early reports that indicated damage to certain parts of the brain can change both cognitive and emotional behaviors [39]. Therefore, separating the brain into cognitive and affective regions is intrinsically challenging and eventually indefensible [39]. Cognitive dysfunctions, including deficits in various mental processes such as attention, decision-making, problem-solving, language, memory, and executive functions, are prevalent in many psychiatric disorders that have been attributed to deficits in brain networks underlying cognitive regulatory functions [40]. Executive dysfunctions comprising abnormalities in three main components, including inhibitory functions, working memory, and cognitive flexibility or their combination as higher order executive functions such as problem-solving and planning, are the most widely recognized cognitive deficits in psychiatric disorders [40]. Executive functions may contribute to psychiatric disorders through neural circuitry overlapping with emotional regulation processes that impact psychiatric functioning and circuit abnormalities relevant to executive functions and emotional regulation [40]. Multiple neural circuits within the PFC are responsible for coordinating different aspects of cognition and cognitive regulation of behavior and emotions [40]. Neuroimaging studies have identified two main cognitive control networks in the brain, including a cingulo-opercular network containing the dACC, anterior insula, and anterior PFC; and a frontoparietal network containing the DLPFC and posterior parietal cortices [41, 42]. Integration of emotion and cognitive control processes probably occurs in the prefrontal cortex, especially the anterior cingulate, and neurotransmission by dopamine and serotonin is also involved in these functions [43]. Cognitive functions, such as attention, are controlled by the prefrontal and parietal cortices, especially via the lateral aspect of the PFC that is essential for the maintenance and manipulation of information and performance of "cognitive control" operations [26, 44, 45]. The LPFC, especially the DLPFC, is often considered a purely cognitive region, but functional studies using functional magnetic resonance imaging (fMRI) have provided evidence that the LPFC could be a site for interaction and integration of cognition and emotion [39, 46–48]. Indeed, the DLPFC is involved in an extensive range of cognitive functions, including executive functions, attention, social cognition, psychomotor speed, and memory, suggesting it a reasonable therapeutic target with strong potential to influence cognition [2]. In addition, among a complex cortical and subcortical network, the DLPFC has been frequently connected with decision-making processes [49]. Other PFC subregions that are robustly involved in affective function, such as the ACC, ventromedial PFC (VMPFC), and OFC, might also be areas for functional integration of cognition and emotion [50]. A relationship between reduced activity of cognitive control brain regions and psychiatric psychopathology is supported by several findings, such as decreased DLPFC activity in depression and decreased recruitment of dACC and DMPFC in PTSD [38, 51, 52].

Noninvasive Brain Stimulation as an Intervention for Psychiatric Disorders

Transcranial Magnetic Stimulation (TMS)

Repetitive transcranial magnetic stimulation (rTMS) is a noninvasive neuromodulation technique that uses an electric coil placed near the scalp to produce a magnetic field pulse that penetrates the brain and induces an electric field in the underlying region of the cerebral cortex [36]. The rapidly changing electromagnetic fields induced by rTMS can depolarize local neurons up to a depth of

2 cm in the underlying region of the cerebral cortex and consequently modulate cortical and subcortical function by generation of action potentials and alteration of regional activity within the cortex [36, 53, 54]. Low-frequency ($\leq$1 Hz) rTMS usually has inhibitory effects and decreases cortical excitability and metabolism in focal areas, while high-frequency ($\geq$5 Hz) rTMS or HF-rTMS has the opposite effects [36, 54]. Mechanisms underlying the therapeutic effects of rTMS within directly stimulated brain regions have been attributed to structural changes, including increased gray matter volumes; functional changes, including increased regional cerebral blood flow and neurotransmitter levels; and changes in interconnected brain regions [55, 56]. A systematic review and meta-analysis of randomized, double-blind, and sham-controlled trials has demonstrated that HF-rTMS applied to the left prefrontal cortex has antidepressant properties [57]. Although most randomized-controlled trials (RCTs) have demonstrated the effectiveness of HF-rTMS applied to the left DLPFC or low-frequency (LF) rTMS applied to the right DLPFC in the treatment of MDD and treatment-resistant depression (TRD), a systematic review and meta-analysis found that LF right-sided rTMS shows greater promise because it has fewer side effects and offers greater protection against seizures [1, 58]. rTMS has been approved for the treatment of MDD and for auditory verbal hallucinations in schizophrenia, but it might also be beneficial in the treatment of PTSD, obsessive-compulsive disorder (OCD), and schizophrenia [1]. Two systematic reviews and meta-analyses revealed that active rTMS significantly improves negative symptoms in schizophrenia compared to sham rTMS [59] and that rTMS is an effective treatment for negative symptoms in schizophrenia [60]. A systematic review of several meta-analytical studies found that rTMS is an effective therapeutic intervention for both positive and negative symptoms of schizophrenia, although the authors recommended further studies to more precisely clarify the role of rTMS in the treatment of this illness [61]. An exploratory meta-analysis of randomized and sham-controlled trials on rTMS for treatment of OCD demonstrated that HF-rTMS protocols and rTMS targeted at the DLPFC do not have greater efficacy than sham rTMS, but LF-rTMS protocols and rTMS employed over non-DLPFC regions, particularly the supplementary motor area (SMA) or OFC, might be promising approaches with the greatest potential efficacy [62]. Another exploratory meta-analysis of randomized, double-blind, and sham controlled trials suggested that active rTMS applied to the DLPFC might be effective for PTSD, but larger scale RCTs will be needed to investigate the differential efficacy of targeting the left or right DLPFC and performing HF-rTMS or LF-rTMS protocols [63]. In addition, a recent systematic review and meta-analysis suggested that both HF- and LF-rTMS may alleviate the symptoms of PTSD [64]. Moreover, HF-rTMS over the right or left DLPFC could reduce cravings for various substances, such as cocaine [65, 66], alcohol [67] and nicotine [68–70]. Modulation of the right PFC with rTMS could be an alternative therapeutic approach for ADHD because chronic conventional treatments are intolerable for many patients with ADHD, and reduced excitability of the right PFC is involved in the pathophysiology of ADHD [71]. Indeed, improvement in ADHD symptoms has been reported after HF-rTMS delivered to

Table 1 Characteristics of the included double-blind, randomized, sham-controlled studies of the cognitive effects of tDCS in patients with psychiatric diseases

Psychiatry disease	Active(n)/sham(n)	Characteristics of patients	Anode/Cathode placement	Location of stimulation	Duration of stimulation	Cognitive tasks	Results	References
MDD	11/11	14 MDD patients were medicated with anti-depressants, primarily with SSRIs	F3/right deltoid muscle	Left DLPFC	One session of anodal tDCS (1 mA, 35 cm^2) for 20 min	TMT-A, TMT-B, MWT, VLMT	Anodal tDCS improved working memory performance in both patients and control subjects	Wolkenstein and Plewnia (2013)
SCZ	19/18	37 outpatients with schizophrenia or schizoaffective psychosis who were regular cigarette smokers	F3/FP2	Left DLPFC	Five sessions of tDCS (1 mA, 35 cm^2) for 20 min	MCCB	Active tDCS significantly improved working memory and attention-vigilance	Smith et al. (2015)
SUD	6/6	12 subjects with addiction to nicotine smoking	F4/F3	Right DLPFC	Five sessions of tDCS (2 mA, 35 cm^2) for 30 min	Ultimatum game	Active tDCS modulated some processes of decision-making behaviors	Fecteau et al. (2014)
ADHD	18/19	37 adult ADHD patients	F3/FP2	Left DLPFC	Three sessions of tDCS (2 mA, 35 cm^2) for 20 min	CPT, SST	Active tDCS could improve impulsivity symptoms	Allenby et al. (2018)
ASD	6/6	12 adults with high-functioning ASD	F3/F4	Bifrontal DLPFC	Three sessions of tDCS (1.5 mA, 35 cm^2) for 40 min	Backward spatial span, BDS, spatial n-back, and letter n-back	Active bifrontal tDCS improved WM task performance	Van Steenburgh et al. (2017)

VLMT, verbal learning memory test; DSF, digit span forward; DSB, digit span backward; COWAT, controlled oral word association test; MWT, multiple-choice word fluency test; TMT-A, trail making test, part A; TMT-B, trail making test, part B; SSRIs, selective serotonin reuptake inhibitors; F3, left dorsolateral prefrontal cortex (10/20 EEG system); F4, right dorsolateral prefrontal cortex (10/20 EEG system); FP2, right supraorbital region (10/20 EEG system); MCCB, MATRICS Consensus Cognitive Battery; CANTAB, Cambridge Neuropsychological Test Automated Battery; SUD, substance use disorder; ADHD, attention-deficit/hyperactivity disorder; SCZ, schizophrenia; OCD, obsessive-compulsive disorder; CPT, Conners Continuous Performance Task; SST, stop signal task; WM, working memory; MDD, major depression disorder; ASD, autism spectrum disorder the right PFC [71]. It has also been suggested that rTMS may be both an innovative research tool and a potential treatment for amelioration of some core symptoms associated with ASD [72] (Table 1).

Transcranial Direct Current Stimulation (tDCS)

tDCS is a neuromodulation technique that may produce therapeutic effects by increasing or decreasing cortical excitability, via anode or cathode scalp electrodes [3]. tDCS involves the application of a weak, constant, low-intensity electrical current (usually 1 or 2 mA) over the scalp that flows through the brain from the anode electrode to the cathode electrode and generates bidirectional polarity-dependent alterations in cortical excitability [73, 74]. Neurophysiological evidence suggests that the anode electrode increases cortical excitability, and the cathode electrode reduces cortical excitability [75]. tDCS appears to be a safe and promising tool in the treatment of conditions with abnormal cortical activity because of its ability to control cortical excitability and its potential to induce long-lasting effects, especially in patients experiencing too many side effects of medication or those with treatment-resistant symptoms [3]. The results of a systematic review of RCTs and meta-analyses showed that active tDCS was superior to sham tDCS for treatment of acute depression [76]. A systematic review of preliminary data suggests that tDCS is a promising technique for the treatment of bipolar disorders, especially during depressive episodes. It may also improve neurocognition and sleep quality in euthymia and help to prevention of relapse [77]. Another recent systematic review and meta-analysis found that bifrontal tDCS improves symptoms of bipolar depression via stimulation of the left DLPFC and inhibition of the right DLPFC, suggesting tDCS is a promising therapeutic modality for bipolar depression [78]. However, future RCTs are required to determine the effectiveness of active versus sham tDCS in patients with bipolar depression [78]. Moreover, reversing this bifrontal montage as an add-on in bipolar depression may also reduce manic symptoms during a manic state [79]. Several case reports and RCTs have investigated the clinical effects of cathodal-tDCS applied over the left temporoparietal junction (TPJ) or right DLPFC coupled with anodal tDCS applied over the left PFC or right supraorbital region on auditory verbal hallucinations (AVH) and other schizophrenic symptoms in adult patients with schizophrenia, and all reported sustained improvement of global symptoms and the safety and tolerability of the technique [3]. It has been hypothesized that modulation of the abnormal brain network underlying obsessive-compulsive (OC) symptoms, including the orbitofronto-striato-pallido-thalamic network, by the use of tDCS over abnormal brain regions could alleviate OC symptoms [80]. Several case reports, open-label studies, and a randomized-controlled study investigating the clinical effects of tDCS in patients with OCD proposed tDCS as a tool to alleviate OC symptoms, as well as comorbid depression and anxiety in patients with treatment-resistant OCD. However, additional sham-controlled studies are required to verify these results [80]. Preliminary results summarized in a systematic review proposed that tDCS is a promising therapeutic tool for obsessive-compulsive symptoms in patients with treatment-resistant OCD, but further sham-controlled studies are needed to confirm these findings [80]. Several RCTs have reported mixed results related to the effect of tDCS applied over the DLPFC on cravings for tobacco, alcohol, methamphetamine, cocaine, and marijuana [81]. Three double-blind RCTs showed a significant reduction in cue-induced nicotine craving and a greater reduction in cigarette consumption in both protocols of active tDCS, including anode-right/cathode-left and anode-left/cathode-right, applied over the DLPFC compared with sham treatment [82–84]. In addition, reduced substance craving has been reported with both active tDCS conditions (anode-left/cathode-right and anode-right/cathode-left) over the DLPFC compared with sham tDCS for alcohol abusers [85], marijuana users [86], and cocaine-dependent users [87]. Only active anodal left/cathodal right tDCS over the DLPFC could diminish cue-induced food craving [88]. Although tDCS has been proposed as a promising therapeutic option for reduction of substance abuse craving, a systematic review suggested that more data is needed before considering tDCS as an effective technique for SUDs [89]. A double-blind, sham-controlled randomized clinical trial of tDCS in adult patients with ADHD showed that application of anode-left/cathode-right over the DLPFC improved ADHD symptoms, and this improvement persisted after the stimulation ended [90]. Based on the hypothesis that anodal tDCS over the left hemisphere has beneficial effects by increasing hypoactivation in the brain of individuals with autism, two RCTs found that anodal tDCS over the DLPFC improves language acquisition and other autistic symptoms compared with sham stimulation [91–93] (Tables 1 and 2).

The Effects of tDCS and rTMS on the Cognitive Control of Emotion in Psychiatric Disorders

Depression

Major depression (MD) is characterized by emotional impairments and cognitive deficits, including deficits in executive functions, psychomotor skills, attention, perception, and different types of memory [94–97]. Anodal tDCS applied to the left DLPFC ameliorates deficient cognitive control in patients with major depression disorder (MDD), leading to improvements in working memory performance and attentional bias. Thus, NIBS might be an efficient approach for treatment of MDD by recovery of deficient cognitive control because deficient cognitive control has been suggested as a contributing factor to the onset and maintenance of depression [98]. A recent study assessed cognitive control by presenting a delayed response working memory task with emotional pictures versus neutral pictures during the delay period in 22 MDD patients and 22 healthy control subjects [98]. Anodal tDCS to the DLPFC improved working memory performance in both patients and control subjects. Attentional bias, as measured by the distractive

Table 2 Characteristics of included double-blind, randomized, sham-controlled studies of the cognitive effects of rTMS in patients with psychiatric diseases

Psychiatry disease	Active(n)/sham(n)	Characteristics of patients	Characteristics of stimulation	Location of stimulation	Duration of stimulation	Cognitive tasks	Results	References
MDD	29/30	59 patients with major depression	10 Hz, 20 trains, 55-s intertrain, 110% resting MT	Left DLPFC	10 rTMS sessions within 2 weeks	CAMCOG, MMSE, Digit span F/B, Digit symbol modalities, Grooved pegboard	No significant between-group differences	Mogg et al. (2008)
SCZ	13/12	27 medicated schizophrenia patients	20 Hz, 25 trains, 30-s intertrain, 90% resting MT	Bilateral DLPFC	20 treatments during 4 weeks	N-back task	Significant improvement of working memory	Barr et al. (2013)
PTSD	10/10/10	30 patients medicated by antidepressants or benzodiazepines	20 Hz, 40 trains, intertrain 28-s, 80% resting MT	Bilateral DLPFC	10 rTMS sessions within 2 weeks	Stroop Test, DSF, Raven Colored Progressive Matrices, and Wisconsin Card Sorting Test	No significant association with cognitive improvement	Boggio et al. (2010a)
ADHD	6/7	13 patients with adult ADHD	20-Hz, 42 trains, intertrain 30-s, 100% resting MT	Right DLPFC	1 session	PANAS, VASs, CANTAB	Real rTMS improved attention based on the PANAS score	Bloch et al. (2010)
ASD	20/20	40 older youth and young adults with ASD	20 Hz, 25 trains, 30-s intertrain, 90% resting MT	Bilateral DLPFC	20 treatments during 4 weeks	CANTAB, spatial working memory task	Real rTMS improved executive function deficits	Ameis et al. (2017)

MMSE, mini-mental-state examination; VLT, verbal learning task; TRD, treatment-resistant depression; COWAT, Controlled Oral Word Association Test; MT, motor threshold; SCZ, schizophrenia; PTSD, posttraumatic stress disorder; DSF, digit span forward; DSB, digit span backward; PANAS, Positive and Negative Affect Schedule; VASs, visual analogue scales; CANTAB, Cambridge Neuropsychological Test Automated Battery; ADHD, attention-deficit/hyperactivity disorder; SCZ, schizophrenia; PTSD, posttraumatic stress disorder; MDD, major depression disorder; ASD, autism spectrum disorder effects of emotional stimuli on working memory performance, was entirely eliminated by anodal tDCS in MDD patients [98]. In addition, it has been hypothesized that the antidepressant efficacy of TMS might be a result of enhancement of the CCN [8]. With regard to involvement of the DLPFC in cognitive control, the mechanisms contributing to the antidepressant effects of rTMS may be directly related to the enhancement of cognitive control, specifically the cognitive control processes involved in emotion regulation [8]. A placebo-controlled trial of adjunctive rTMS of the left DLPFC showed no significant group × time interactions for any of the cognitive test scores in both real and sham rTMS after a 2-week course or during a 4-month follow-up [99]. Several clinical trials of rTMS applied to the left or right DLPFC for the treatment of depression reported mixed results related to improvement of working memory, verbal memory, executive functioning, processing speed, attention, psychomotor speed, and concentration [2, 100].

Schizophrenia

Cognitive deficits strongly predict functional abilities of patients with schizophrenia, but despite the beneficial therapeutic effects of numerous antipsychotic medications on positive symptoms, these medications are not effective for cognition [101]. Despite advances in pharmacological treatment, psychotic symptoms, negative symptoms, and cognitive deficits will remain in almost 30% of patients with schizophrenia, demonstrating a need to develop alternative therapeutic approaches to improve treatment-resistant symptoms [3]. tDCS has shown positive effects on cognitive performance in patients with schizophrenia, including significant improvements in working memory and attention-vigilance after active tDCS compared to sham tDCS [102]. A 4-week randomized double-blind sham-controlled pilot study reported that HF-rTMS targeted bilaterally to the left and right DLPFC significantly improved working memory performance in patients with schizophrenia compared with sham treatment, suggesting bilateral rTMS is a novel, efficacious, and safe treatment for working memory deficits in schizophrenia [4]. A systematic review and meta-analysis of 30 randomized, sham-controlled studies that administered rTMS to the DLPFC in patients with neuropsychiatric conditions found no robust effect of active rTMS on the majority of cognitive domains, including executive functioning, attention, and verbal memory, compared to sham treatment. However, working memory performance improved in a small number of trials conducted in patients with schizophrenia, demonstrating that this advantage may be specific to the clinical population [103].

Obsessive-Compulsive Disorder (OCD)

It has been hypothesized that the reduced ability of OCD patients to inhibit intrusive thoughts, impulses, or images and repetitive motor responses may be associated with hyperactivity in the medial and lateral frontal areas, such as the supplementary motor area (SMA), anterior cingulate, DLPFC, and orbito-fronto-striatal and parietal regions [104]. tDCS may have pro-cognitive effects in patients with OCD [3]. Similar to second- or third-line pharmacological strategies, rTMS might be an effective treatment for OC symptoms but without the long-standing metabolic adverse effects [62]. Although available data reported by numerous open and randomized, sham-controlled trials about the efficacy of rTMS in the treatment of OCD have shown inconsistent and contradictory findings related to the stimulation areas targeted, stimulus parameters, study design and sample size, rTMS targeted at the supplementary motor area (SMA) or possibly the OFC appears to be the most promising treatment for OCD in terms of potential efficacy [105]; targeting the DLPFC does not appear to be effective. Future open-label and randomized-controlled trials are needed to clarify the therapeutic role of rTMS and tDCS in improvement of cognitive functions in patients with OCD.

Posttraumatic Stress Disorder (PTSD)

tDCS and rTMS have been considered as potential therapeutic options for the treatment of PTSD, particularly for those patients that fail to respond to conventional treatments [106, 107]. The results of one pilot study in four PTSD patients with poor working memory demonstrated that a combination of computerized working memory training and tDCS led to clinically significant improvements in cognitive and emotional performance measures accompanied by the normalization of posttreatment electroencephalographic (EEG) recordings [108]. Another pilot study evaluated the effect of tDCS during extinction learning versus consolidation of extinction learning on the improvement of extinction recall in veterans with war zone-related PTSD that were randomized to sham or anodal tDCS over the VMPFC combined with a fear-conditioning paradigm [109]. The results showed that tDCS during extinction consolidation led to moderately better extinction memory during extinction recall than tDCS during extinction learning, suggesting tDCS during consolidation of fear extinction may enhance extinction recall [109]. Despite a high level of variability in findings reported to date, TMS applied over the right DLPFC and MPFC has been suggested as a promising therapeutic approach for treatment of PTSD because psychotherapy, pharmacotherapy, and complementary medicine therapies have shown inadequate efficacy in patients with PTSD [106]. One double-blind, placebo-controlled phase II trial investigated the efficacy of 20 Hz rTMS of either the right or left DLPFC as compared to sham rTMS for the alleviation of PTSD–associated symptoms and cognitive functions and reported that the right and left rTMS were both associated with nonsignificant improvements in neuropsychological performance but significant alleviation of core PTSD symptoms; HF-rTMS of the right DLPFC appeared to be the optimal treatment strategy [110]. The efficacy of rTMS and tDCS in treating cognitive impairments in patients with PTSD requires further RCTs.

Substance Use Disorder

Substance use disorders are characterized by intense desire for the drug, regardless of adverse effects, and reduced ability to control that desire [111]. Cue-induced craving is a significant predictor of relapse, and intensity of craving is a way to evaluate substance use disorder [112]. In addition, addiction is characterized by risky decision-making as a prominent behavioral phenotype that plays an important role in the maintenance and relapse of substance use and abuse [113, 114]. It has been suggested that the DLPFC, by integrating cognitive and emotionally relevant information, is involved in decision-making, and modulation of DLPFC activity can improve the level of inhibitory control of emotional impulses, which is essential for suitable decision-making behaviors [115]. Neuromodulation of the DLPFC with tDCS and rTMS may modify decision-making processes and suppress drug craving in substance user patients with decision-making impairments [115]. The DLPFC is involved in decision-making processes, and individuals with substance use disorders show decision-making impairments, such as demonstrating riskier behaviors during decision-making tasks [3]. In addition, the prefrontal- cortical network, especially the DLPFC, is involved in inhibitory control mechanisms during confrontation of patients with addictive substances [116]. Increased activity of the PFC and, consequently, enhancement of cognitive control, can diminish automatic impulses and thus decrease substance abuse behavior [117]. Investigation of the effect of tDCS applied over the right DLPFC on nicotine intake and decision-making behaviors in tobacco smokers showed that active stimulation significantly reduced the number of cigarettes smoked compared to sham stimulation and induced beneficial decision-making processes and reward sensitive effects [83]. Thus, tDCS appears to be a promising therapy to decrease cravings and modulate cognitive functions associated with consumption and relapse [3]. Further RCTs are required to explore the role of rTMS and tDCS in the treatment of cognitive functions in SUDs.

Attention-Deficit Hyperactivity Disorder (ADHD)

Neuropsychological tasks have revealed several types of cognitive impairments in ADHD, particularly deficits in response inhibition (inhibition of a prepotent response in support of a suitable response), which may be attributed to abnormality in the fronto-striatal circuitry in the brains of ADHD patients [118]. It has been hypothesized that administration of NIBS to the fronto-striato-cerebellar network combined with cognitive training could improve cognition in ADHD [2]. A within-subject crossover study showed that three sessions of active anodal tDCS applied over the left DLPFC, with concurrent training in working memory, improves impulsivity symptoms in adults with ADHD compared to working memory training with sham stimulation, suggesting tDCS may contribute to treatment of impulsivity in ADHD [119]. Neuroimaging studies in ADHD subjects revealed a significant association between cognitive deficits, impulsive decision-making, and response inhibition and reduced activity in brain regions of the CCN, particularly the DLPFC, compared to healthy controls [120–123]. Therefore, modulation of cognitive control circuits and enhancement of DLPFC activity by tDCS may improve impulse control in ADHD [119]. Interestingly, a crossover double-blind randomized, sham-controlled pilot study in adult patients with ADHD showed that a single session of HF-rTMS directed to the right DLPFC significantly improves attention compared to sham rTMS [124]. However, further RCTs are needed to clarify the effect of neuromodulation approaches on cognitive deficits and their effect on alleviation of core symptoms of ADHD.

Autism

There is a pressing need for effective treatments of ASD because no specific treatment has been approved for this population, and behavioral and pharmacologic therapies cause adverse effects and unsatisfactory outcomes [125, 126]. NIBS techniques, including tDCS and rTMS, have been proposed as treatment options for autism [2]. Children and adolescents with ASD usually show impairment of working memory (WM) that likely underlies core deficits in cognitive functions, including language, general intelligence, and reasoning [127–129]. WM depends on prefrontal activity, and poor WM performance in individuals with ASD has been attributed to prefrontal hypoactivation during working memory tasks in adults with ASD [130–133]. A sham-controlled crossover study in adults with high-functioning autism revealed that active bifrontal tDCS applied over the DLPFC could improve working memory (WM) performance and some core deficits [134]. Preliminary data reported by one randomized-controlled trial suggested that bilateral, 20-Hz rTMS applied to the DLPFC may be an efficacious intervention for treatment of executive functions deficits in ASD [135]. However, more controlled clinical trials are warranted to evaluate the true potential of rTMS and tDCS in the treatment of ASD.

Conclusion

Despite advances in psychopharmacology and psychotherapeutic interventions for treatment of psychiatric disorders, many patients with these diseases do not respond to conventional treatments. Therefore, development of new therapeutic approaches is warranted. Application of NIBS techniques over the DLPFC node of the CCN may through direct effects on cognitive control processes involved in emotion regulation, improve core symptoms of psychiatric disorders. Beneficial effects of targeting the DLPFC by tDCS and rTMS in the treatment of psychiatric disorders is probably due to the direct effects of these interventions on the cognitive processes regulated by the CCN, as well as behavioral manifestations of CCN function. Among the cognitive functions, working memory has received the most attention in studies of the cognitive-enhancing effects of neuromodulation of psychiatric disorders, although other cognitive domains have also been assessed. Exploring the effects of rTMS and tDCS on working memory and other processes associated with DLPFC/CCN function, such as cognitive inhibition, self-regulation, and problem-solving, could be useful in the development of effective treatment alternatives as a paradigm shift in psychiatry. In addition, future studies that analyze the synergistic effects of NIBS protocols and other conventional interventions, such as computerized tasks, behavioral paradigms, drug treatment, psychotherapy, and cognitive training, is warranted in patients with various psychiatric conditions.

References

1. Andrade J, Mota D, Ferreira S, Araújo A. Neuromodulation in Psychiatric disorders: recent findings and clinical implications;2017.
2. Demirtas-Tatlidede A, Vahabzadeh-Hagh AM, Pascual-Leone A. Can noninvasive brain stimulation enhance cognition in neuropsychiatric disorders? Neuropharmacology. 2013;64:566–78.
3. Mondino M, Bennabi D, Poulet E, Galvao F, Brunelin J, Haffen E. Can transcranial direct current stimulation (tDCS) alleviate symptoms and improve cognition in psychiatric disorders? World J Biol Psychiatry. 2014;15(4):261–75.
4. Barr MS, Farzan F, Rajji TK, Voineskos AN, Blumberger DM, Arenovich T, et al. Can repetitive magnetic stimulation improve cognition in schizophrenia? Pilot data from a randomized controlled trial. Biol Psychiat. 2013;73(6):510–7.
5. Guse B, Falkai P, Wobrock T. Cognitive effects of high-frequency repetitive transcranial magnetic stimulation: a systematic review. J Neural Transm. 2010;117(1):105–22.
6. Hill SK, Bishop JR, Palumbo D, Sweeney JA. Effect of second-generation antipsychotics on cognition: current issues and future challenges. Expert Rev Neurother. 2010;10(1):43–57.
7. Millan MJ. Multi-target strategies for the improved treatment of depressive states: conceptual foundations and neuronal substrates, drug discovery and therapeutic application. Pharmacol Ther. 2006;110(2):135–370.
8. Lantrip C, Gunning FM, Flashman L, Roth RM, Holtzheimer PE. Effects of transcranial magnetic stimulation on the cognitive control of emotion: potential antidepressant mechanisms. J ECT. 2017;33(2):73–80.
9. Niendam TA, Laird AR, Ray KL, Dean YM, Glahn DC, Carter CS. Meta-analytic evidence for a superordinate cognitive control network subserving diverse executive functions. Cogn Affect Behav Neurosci. 2012;12(2):241–68.
10. Koole SL. The psychology of emotion regulation: an integrative review. Cogn Emot. 2009;23(1):4–41.
11. Gross JJ. Handbook of emotion regulation. Guilford Publications;2013.
12. Ochsner KN, Silvers JA, Buhle JT. Functional imaging studies of emotion regulation: a synthetic review and evolving model of the cognitive control of emotion. Ann N Y Acad Sci. 2012;1251(1):E1–24.
13. Dennis TA. Interactions between emotion regulation strategies and affective style: implications for trait anxiety versus depressed mood. Motiv Emotion. 2007;31(3):200–7.
14. Gross JJ, John OP. Individual differences in two emotion regulation processes: implications for affect, relationships, and well-being. J Pers Soc Psychol. 2003;85(2):348.
15. Bradley SJ. Affect regulation and the development of psychopathology. Guilford Press;2003.

16. Flack WF, Laird JD. Emotions in psychopathology: theory and research. Oxford University Press on Demand;1998.
17. Gratz KL, Roemer L. Multidimensional assessment of emotion regulation and dysregulation: Development, factor structure, and initial validation of the difficulties in emotion regulation scale. J Psychopathol Behav Assess. 2004;26(1):41–54.
18. Martel MM. Research review: a new perspective on attention-deficit/hyperactivity disorder: emotion dysregulation and trait models. J Child Psychol Psychiatry. 2009;50(9):1042–51.
19. Rottenberg JE, Johnson SL. Emotion and psychopathology: bridging affective and clinical science. Am Psycholog Assoc. 2007.
20. Samson AC, Phillips JM, Parker KJ, Shah S, Gross JJ, Hardan AY. Emotion dysregulation and the core features of autism spectrum disorder. J Autism Dev Disord. 2014;44(7):1766–72.
21. Miller EK. The prefontral cortex and cognitive control. Nat Rev Neurosci. 2000;1(1):59.
22. Ochsner KN, Bunge SA, Gross JJ, Gabrieli JD. Rethinking feelings: an FMRI study of the cognitive regulation of emotion. J Cogn Neurosci. 2002;14(8):1215–29.
23. Adolphs R. The neurobiology of social cognition. Curr Opin Neurobiol. 2001;11(2):231–9.
24. Davidson RJ, Irwin W. The functional neuroanatomy of emotion and affective style. Trends Cogn Sci. 1999;3(1):11–21.
25. Gross JJ. Emotion regulation: affective, cognitive, and social consequences. Psychophysiology. 2002;39(3):281–91.
26. Miller EK, Cohen JD. An integrative theory of prefrontal cortex function. Annu Rev Neurosci. 2001;24(1):167–202.
27. Richards JM, Gross JJ. Emotion regulation and memory: the cognitive costs of keeping one's cool. J Pers Soc Psychol. 2000;79(3):410.
28. Rolls ET. Emotion explained. USA: Oxford University Press;2005.
29. Alexander GE, DeLong MR, Strick PL. Parallel organization of functionally segregated circuits linking basal ganglia and cortex. Annu Rev Neurosci. 1986;9(1):357–81.
30. Cabeza R, Nyberg L. Imaging cognition II: an empirical review of 275 PET and fMRI studies. J Cogn Neurosci. 2000;12(1):1–47.
31. Courtney SM, Petit L, Maisog JM, Ungerleider LG, Haxby JV. An area specialized for spatial working memory in human frontal cortex. Science. 1998;279(5355):1347–51.
32. Knight RT, Staines WR, Swick D, Chao LL. Prefrontal cortex regulates inhibition and excitation in distributed neural networks. Acta Physiol (Oxf). 1999;101(2–3):159–78.
33. Ochsner KN, Feldman Barrett L. A multiprocess perspective on the neuroscience of emotion. Emotion: current issues and future directions. 2001. p. 38–81.
34. Petit L, Courtney SM, Ungerleider LG, Haxby JV. Sustained activity in the medial wall during working memory delays. J Neurosci. 1998;18(22):9429–37.
35. Smith EE, Jonides J. Storage and executive processes in the frontal lobes. Science. 1999;283(5408):1657–61.
36. Gershon AA, Dannon PN, Grunhaus L. Transcranial magnetic stimulation in the treatment of depression. Am J Psychiatry. 2003;160(5):835–45.
37. Mayberg HS, Lozano AM, Voon V, McNeely HE, Seminowicz D, Hamani C, et al. Deep brain stimulation for treatment-resistant depression. Neuron. 2005;45(5):651–60.
38. Taylor SF, Liberzon I. Neural correlates of emotion regulation in psychopathology. Trends Cogn Sci. 2007;11(10):413–8.
39. Pessoa L. On the relationship between emotion and cognition. Nat Rev Neurosci. 2008;9(2):148.
40. Etkin A, Gyurak A, O'Hara R. A neurobiological approach to the cognitive deficits of psychiatric disorders. Dialogues Clin Neurosci. 2013;15(4):419.
41. Dosenbach NU, Fair DA, Miezin FM, Cohen AL, Wenger KK, Dosenbach RA, et al. Distinct brain networks for adaptive and stable task control in humans. Proc Natl Acad Sci. 2007;104(26):11073–8.
42. Dosenbach NU, Visscher KM, Palmer ED, Miezin FM, Wenger KK, Kang HC, et al. A core system for the implementation of task sets. Neuron. 2006;50(5):799–812.

43. Holroyd CB, Coles MG. The neural basis of human error processing: reinforcement learning, dopamine, and the error-related negativity. Psychol Rev. 2002;109(4):679.
44. Botvinick MM, Braver TS, Barch DM, Carter CS, Cohen JD. Conflict monitoring and cognitive control. Psychol Rev. 2001;108(3):624.
45. Corbetta M, Shulman GL. Control of goal-directed and stimulus-driven attention in the brain. Nat Rev Neurosci. 2002;3(3):201.
46. Davidson RJ, Pizzagalli D, Nitschke JB, Kalin NH. Parsing the subcomponents of emotion and disorders of emotion: perspectives from affective neuroscience. Handb Affect Sci. 2003;2:8–24.
47. Heller W, Nitscke JB. Regional brain activity in emotion: a framework for understanding cognition in depresion. Cogn Emot. 1997;11(5–6):637–61.
48. Mitchell RL, Phillips LH. The psychological, neurochemical and functional neuroanatomical mediators of the effects of positive and negative mood on executive functions. Neuropsychologia. 2007;45(4):617–29.
49. Krain AL, Wilson AM, Arbuckle R, Castellanos FX, Milham MP. Distinct neural mechanisms of risk and ambiguity: a meta-analysis of decision-making. Neuroimage. 2006;32(1):477–84.
50. Damasio AR. The feeling of what happens: body and emotion in the making of consciousness. New York Times Book Review. 1999;104:8-.
51. Brody AL, Barsom MW, Bota RG, Saxena S, editors. Prefrontal-subcortical and limbic circuit mediation of major depressive disorder. Semin Clin Neuropsychiatry. 2001.
52. Liberzon I, Martis B. Neuroimaging studies of emotional responses in PTSD. Ann N Y Acad Sci. 2006;1071(1):87–109.
53. George MS, Post RM. Daily left prefrontal repetitive transcranial magnetic stimulation for acute treatment of medication-resistant depression. Am J Psychiatry. 2011;168(4):356–64.
54. Rosa MA, Lisanby SH. Somatic treatments for mood disorders. Neuropsychopharmacology. 2012;37(1):102.
55. Lan MJ, Chhetry BT, Liston C, Mann JJ, Dubin M. Transcranial magnetic stimulation of left dorsolateral prefrontal cortex induces brain morphological changes in regions associated with a treatment resistant major depressive episode: an exploratory analysis. Brain Stimul. 2016;9(4):577–83.
56. Noda Y, Silverstein W, Barr M, Vila-Rodriguez F, Downar J, Rajji T, et al. Neurobiological mechanisms of repetitive transcranial magnetic stimulation of the dorsolateral prefrontal cortex in depression: a systematic review. Psychol Med. 2015;45(16):3411–32.
57. Berlim M, Van den Eynde F, Tovar-Perdomo S, Daskalakis Z. Response, remission and drop-out rates following high-frequency repetitive transcranial magnetic stimulation (rTMS) for treating major depression: a systematic review and meta-analysis of randomized, double-blind and sham-controlled trials. Psychol Med. 2014;44(2):225–39.
58. Chen J, Zhou C, Wu B, Wang Y, Li Q, Wei Y, et al. Left versus right repetitive transcranial magnetic stimulation in treating major depression: a meta-analysis of randomised controlled trials. Psychiatry Res. 2013;210(3):1260–4.
59. Osoegawa C, Gomes JS, Grigolon RB, Brietzke E, Gadelha A, Lacerda AL, et al. Non-invasive brain stimulation for negative symptoms in schizophrenia: an updated systematic review and meta-analysis. Schizophr Res. 2018.
60. Shi C, Yu X, Cheung EF, Shum DH, Chan RC. Revisiting the therapeutic effect of rTMS on negative symptoms in schizophrenia: a meta-analysis. Psychiatry Res. 2014;215(3):505–13.
61. Hovington CL, McGirr A, Lepage M, Berlim MT. Repetitive transcranial magnetic stimulation (rTMS) for treating major depression and schizophrenia: a systematic review of recent meta-analyses. Ann Med. 2013;45(4):308–21.
62. Berlim MT, Neufeld NH, Van den Eynde F. Repetitive transcranial magnetic stimulation (rTMS) for obsessive–compulsive disorder (OCD): an exploratory meta-analysis of randomized and sham-controlled trials. J Psychiatr Res. 2013;47(8):999–1006.

63. Berlim MT, Van den Eynde F. Repetitive transcranial magnetic stimulation over the dorsolateral prefrontal cortex for treating posttraumatic stress disorder: an exploratory meta-analysis of randomized, double-blind and sham-controlled trials. Can J Psychiatry. 2014;59(9):487–96.
64. Yan T, Xie Q, Zheng Z, Zou K, Wang L. Different frequency repetitive transcranial magnetic stimulation (rTMS) for posttraumatic stress disorder (PTSD): a systematic review and meta-analysis. J Psychiatr Res. 2017;89:125–35.
65. Camprodon JA, Martínez-Raga J, Alonso-Alonso M, Shih M-C, Pascual-Leone A. One session of high frequency repetitive transcranial magnetic stimulation (rTMS) to the right prefrontal cortex transiently reduces cocaine craving. Drug Alcohol Depend. 2007;86(1):91–4.
66. Politi E, Fauci E, Santoro A, Smeraldi E. Daily sessions of transcranial magnetic stimulation to the left prefrontal cortex gradually reduce cocaine craving. Am J Addict. 2008;17(4):345–6.
67. Mishra BR, Nizamie SH, Das B, Praharaj SK. Efficacy of repetitive transcranial magnetic stimulation in alcohol dependence: a sham-controlled study. Addiction. 2010;105(1):49–55.
68. Amiaz R, Levy D, Vainiger D, Grunhaus L, Zangen A. Repeated high-frequency transcranial magnetic stimulation over the dorsolateral prefrontal cortex reduces cigarette craving and consumption. Addiction. 2009;104(4):653–60.
69. Johann M, Wiegand R, Kharraz A, Bobbe G, Sommer G, Hajak G, et al. Repetitiv transcranial magnetic stimulation in nicotine dependence. Psychiatr Prax. 2003;30(Suppl 2):129–31.
70. Li X, Hartwell KJ, Owens M, LeMatty T, Borckardt JJ, Hanlon CA, et al. Repetitive transcranial magnetic stimulation of the dorsolateral prefrontal cortex reduces nicotine cue craving. Biol Psychiat. 2013;73(8):714–20.
71. Zangen A. T033 Right prefrontal rTMS for the treatment of ADHD: electrophysiological correlates and prognostic biomarkers. Clin Neurophysiol. 2017;128(3):e11.
72. Oberman LM, Enticott PG, Casanova MF, Rotenberg A, Pascual-Leone A, McCracken JT, et al. Transcranial magnetic stimulation in autism spectrum disorder: challenges, promise, and roadmap for future research. Autism Res. 2016;9(2):184–203.
73. Nitsche MA, Paulus W. Excitability changes induced in the human motor cortex by weak transcranial direct current stimulation. J Physiol. 2000;527(3):633–9.
74. Priori A, Berardelli A, Rona S, Accornero N, Manfredi M. Polarization of the human motor cortex through the scalp. NeuroReport. 1998;9(10):2257–60.
75. Nitsche MA, Doemkes S, Karakose T, Antal A, Liebetanz D, Lang N, et al. Shaping the effects of transcranial direct current stimulation of the human motor cortex. J Neurophysiol. 2007;97(4):3109–17.
76. Shiozawa P, Fregni F, Benseñor IM, Lotufo PA, Berlim MT, Daskalakis JZ, et al. Transcranial direct current stimulation for major depression: an updated systematic review and meta-analysis. Int J Neuropsychopharmacol. 2014;17(9):1443–52.
77. Dondé C, Neufeld NH, Geoffroy PA. The impact of transcranial direct current stimulation (tDCS) on bipolar depression, mania, and euthymia: a systematic review of preliminary data. Psychiatric Q. 2018:1–13.
78. Dondé C, Amad A, Nieto I, Brunoni AR, Neufeld NH, Bellivier F, et al. Transcranial direct-current stimulation (tDCS) for bipolar depression: a systematic review and meta-analysis. Prog Neuropsychopharmacol Biol Psychiatry. 2017;78:123–31.
79. Schestatsky P, Janovik N, Lobato MI, Belmonte-de-Abreu P, Schestatsky S, Shiozawa P, et al. Rapid therapeutic response to anodal tDCS of right dorsolateral prefrontal cortex in acute mania. Brain Stimul Basic Transl Clin Res Neuromodulation. 2013;6(4):701–3.
80. Brunelin J, Mondino M, Bation R, Palm U, Saoud M, Poulet E. Transcranial direct current stimulation for obsessive-compulsive disorder: a systematic review. Brain Sci. 2018;8(2):37.
81. Kekic M, Boysen E, Campbell IC, Schmidt U. A systematic review of the clinical efficacy of transcranial direct current stimulation (tDCS) in psychiatric disorders. J Psychiatr Res. 2016;74:70–86.

82. Boggio PS, Liguori P, Sultani N, Rezende L, Fecteau S, Fregni F. Cumulative priming effects of cortical stimulation on smoking cue-induced craving. Neurosci Lett. 2009;463(1):82–6.
83. Fecteau S, Agosta S, Hone-Blanchet A, Fregni F, Boggio P, Ciraulo D, et al. Modulation of smoking and decision-making behaviors with transcranial direct current stimulation in tobacco smokers: a preliminary study. Drug Alcohol Depend. 2014;140:78–84.
84. Fregni F, Liguori P, Fecteau S, Nitsche MA, Pascual-Leone A, Boggio PS. Cortical stimulation of the prefrontal cortex with transcranial direct current stimulation reduces cue-provoked smoking craving: a randomized, sham-controlled study. J Clin Psychiatry. 2008;69(1):32–40.
85. Boggio PS, Sultani N, Fecteau S, Merabet L, Mecca T, Pascual-Leone A, et al. Prefrontal cortex modulation using transcranial DC stimulation reduces alcohol craving: a double-blind, sham-controlled study. Drug Alcohol Depend. 2008;92(1–3):55–60.
86. Boggio PS, Zaghi S, Villani AB, Fecteau S, Pascual-Leone A, Fregni F. Modulation of risk-taking in marijuana users by transcranial direct current stimulation (tDCS) of the dorsolateral prefrontal cortex (DLPFC). Drug Alcohol Depend. 2010;112(3):220–5.
87. Gorini A, Lucchiari C, Russell-Edu W, Pravettoni G. Modulation of risky choices in recently abstinent dependent cocaine users: a transcranial direct-current stimulation study. Front Hum Neurosci. 2014;8:661.
88. Fregni F, Orsati F, Pedrosa W, Fecteau S, Tome FA, Nitsche MA, et al. Transcranial direct current stimulation of the prefrontal cortex modulates the desire for specific foods. Appetite. 2008;51(1):34–41.
89. Lupi M, Martinotti G, Santacroce R, Cinosi E, Carlucci M, Marini S, et al. Transcranial direct current stimulation in substance use disorders: a systematic review of scientific literature. J ECT. 2017;33(3):203–9.
90. Cachoeira CT, Leffa DT, Mittelstadt SD, Mendes LST, Brunoni AR, Pinto JV, et al. Positive effects of transcranial direct current stimulation in adult patients with attention-deficit/hyperactivity disorder a pilot randomized controlled study. Psychiatry Res. 2017;247:28–32.
91. Amatachaya A, Auvichayapat N, Patjanasoontorn N, Suphakunpinyo C, Ngernyam N, Aree-uea B, et al. Effect of anodal transcranial direct current stimulation on autism: a randomized double-blind crossover trial. Behav Neurol. 2014;2014.
92. Gómez L, Vidal B, Maragoto C, Morales LM, Berrillo S, Vera Cuesta H, et al. Non-invasive brain stimulation for children with autism spectrum disorders: a short-term outcome study. Behav Sci. 2017;7(3):63.
93. Schneider HD, Hopp JP. The use of the Bilingual Aphasia Test for assessment and transcranial direct current stimulation to modulate language acquisition in minimally verbal children with autism. Clin Linguist Phon. 2011;25(6–7):640–54.
94. Davidson RJ, Pizzagalli D, Nitschke JB, Putnam K. Depression: perspectives from affective neuroscience. Annu Rev Psychol. 2002;53(1):545–74.
95. Marazziti D, Consoli G, Picchetti M, Carlini M, Faravelli L. Cognitive impairment in major depression. Eur J Pharmacol. 2010;626(1):83–6.
96. Rock P, Roiser J, Riedel W, Blackwell A. Cognitive impairment in depression: a systematic review and meta-analysis. Psychol Med. 2014;44(10):2029–40.
97. Wagner S, Doering B, Helmreich I, Lieb K, Tadić A. A meta-analysis of executive dysfunctions in unipolar major depressive disorder without psychotic symptoms and their changes during antidepressant treatment. Acta Psychiatr Scand. 2012;125(4):281–92.
98. Wolkenstein L, Plewnia C. Amelioration of cognitive control in depression by transcranial direct current stimulation. Biol Psychiat. 2013;73(7):646–51.
99. Mogg A, Pluck G, Eranti S, Landau S, Purvis R, Brown R, et al. A randomized controlled trial with 4-month follow-up of adjunctive repetitive transcranial magnetic stimulation of the left prefrontal cortex for depression. Psychol Med. 2008;38(3):323–33.
100. Moreines JL, McClintock SM, Holtzheimer PE. Neuropsychologic effects of neuromodulation techniques for treatment-resistant depression: a review. Brain Stimul. 2011;4(1):17–27.

101. Green MF. Cognitive impairment and functional outcome in schizophrenia and bipolar disorder. J Clin Psychiatry. 2006;67:3–8 Discussion 36–42.
102. Smith RC, Boules S, Mattiuz S, Youssef M, Tobe RH, Sershen H, et al. Effects of transcranial direct current stimulation (tDCS) on cognition, symptoms, and smoking in schizophrenia: a randomized controlled study. Schizophr Res. 2015;168(1):260–6.
103. Martin DM, McClintock SM, Forster J, Loo CK. Does therapeutic repetitive transcranial magnetic stimulation cause cognitive enhancing effects in patients with neuropsychiatric conditions? A systematic review and meta-analysis of randomised controlled trials. Neuropsychol Rev. 2016;26(3):295–309.
104. Menzies L, Chamberlain SR, Laird AR, Thelen SM, Sahakian BJ, Bullmore ET. Integrating evidence from neuroimaging and neuropsychological studies of obsessive-compulsive disorder: the orbitofronto-striatal model revisited. Neurosci Biobehav Rev. 2008;32(3):525–49.
105. Jaafari N, Rachid F, Rotge J-Y, Polosan M, El-Hage W, Belin D, et al. Safety and efficacy of repetitive transcranial magnetic stimulation in the treatment of obsessive-compulsive disorder: a review. World J Biol Psychiatry. 2012;13(3):164–77.
106. Clark C, Cole J, Winter C, Williams K, Grammer G. A review of transcranial magnetic stimulation as a treatment for post-traumatic stress disorder. Curr Psychiatry Rep. 2015;17(10):83.
107. D'Urso G, Mantovani A, Patti S, Toscano E, de Bartolomeis A. Transcranial direct current stimulation in obsessive-compulsive disorder, posttraumatic stress disorder, and anxiety disorders. J ECT. 2018;34(3):172–81.
108. Saunders N, Downham R, Turman B, Kropotov J, Clark R, Yumash R, et al. Working memory training with tDCS improves behavioral and neurophysiological symptoms in pilot group with post-traumatic stress disorder (PTSD) and with poor working memory. Neurocase. 2015;21(3):271–8.
109. van 't Wout M, Longo SM, Reddy MK, Philip NS, Bowker MT, Greenberg BD. Transcranial direct current stimulation may modulate extinction memory in posttraumatic stress disorder. Brain Behav. 2017;7(5):e00681.
110. Boggio PS, Rocha M, Oliveira MO, Fecteau S, Cohen RB, Campanhã C, et al. Noninvasive brain stimulation with high-frequency and low-intensity repetitive transcranial magnetic stimulation treatment for posttraumatic stress disorder. J Clin Psychiatry. 2010;71(8):992.
111. Kalivas PW, Volkow ND. The neural basis of addiction: a pathology of motivation and choice. Am J Psychiatry. 2005;162(8):1403–13.
112. Hanlon CA, Jones EM, Li X, Hartwell KJ, Brady KT, George MS. Individual variability in the locus of prefrontal craving for nicotine: implications for brain stimulation studies and treatments. Drug Alcohol Depend. 2012;125(3):239–43.
113. Fishbein DH, Eldreth DL, Hyde C, Matochik JA, London ED, Contoreggi C, et al. Risky decision making and the anterior cingulate cortex in abstinent drug abusers and nonusers. Cogn Brain Res. 2005;23(1):119–36.
114. Garavan H, Stout JC. Neurocognitive insights into substance abuse. Trends Cogn Sci. 2005;9(4):195–201.
115. Fecteau S, Fregni F, Boggio PS, Camprodon JA, Pascual-Leone A. Neuromodulation of decision-making in the addictive brain. Subst Use Misuse. 2010;45(11):1766–86.
116. Bechara A. Decision making, impulse control and loss of willpower to resist drugs: a neurocognitive perspective. Nat Neurosci. 2005;8(11):1458.
117. Houben K, Nederkoorn C, Wiers RW, Jansen A. Resisting temptation: decreasing alcohol-related affect and drinking behavior by training response inhibition. Drug Alcohol Depend. 2011;116(1–3):132–6.
118. Epstein JN, Casey B, Tonev ST, Davidson MC, Reiss AL, Garrett A, et al. ADHD-and medication-related brain activation effects in concordantly affected parent–child dyads with ADHD. J Child Psychol Psychiatry. 2007;48(9):899–913.
119. Allenby C, Falcone M, Bernardo L, Wileyto EP, Rostain A, Ramsay JR, et al. Transcranial direct current brain stimulation decreases impulsivity in ADHD. Brain Stimul. 2018.

120. Cortese S, Kelly C, Chabernaud C, Proal E, Di Martino A, Milham MP, et al. Toward systems neuroscience of ADHD: a meta-analysis of 55 fMRI studies. Am J Psychiatry. 2012;169(10):1038–55.
121. Francx W, Oldehinkel M, Oosterlaan J, Heslenfeld D, Hartman CA, Hoekstra PJ, et al. The executive control network and symptomatic improvement in attention-deficit/hyperactivity disorder. Cortex. 2015;73:62–72.
122. Ortiz N, Parsons A, Whelan R, Brennan K, Agan ML, O'Connell R, et al. Decreased frontal, striatal and cerebellar activation in adults with ADHD during an adaptive delay discounting task. Acta Neurobiol Exp(Wars). 2015;75(3):326–38.
123. Salavert J, Ramos-Quiroga JA, Moreno-Alcázar A, Caseras X, Palomar G, Radua J, et al. Functional imaging changes in the medial prefrontal cortex in adult ADHD. J Atten Disord. 2018;22(7):679–93.
124. Bloch Y, Harel E, Aviram S, Govezensky J, Ratzoni G, Levkovitz Y. Positive effects of repetitive transcranial magnetic stimulation on attention in ADHD Subjects: a randomized controlled pilot study. World J Biolog Psychiatry. 2010;11(5):755–8.
125. Myers SM, Johnson CP. Management of children with autism spectrum disorders. Pediatrics. 2007;120(5):1162–82.
126. Oswald DP, Sonenklar NA. Medication use among children with autism spectrum disorders. J Child Adolesc Psychopharmacology. 2007;17(3):348–55.
127. Joseph RM, McGrath LM, Tager-Flusberg H. Executive dysfunction and its relation to language ability in verbal school-age children with autism. Dev Neuropsychol. 2005;27(3):361–78.
128. Kane MJ, Hambrick DZ, Tuholski SW, Wilhelm O, Payne TW, Engle RW. The generality of working memory capacity: a latent-variable approach to verbal and visuospatial memory span and reasoning. J Exp Psychol Gen. 2004;133(2):189.
129. Wang Y, Zhang Y-b, Liu L-l, Cui J-f, Wang J, Shum DH, et al. A meta-analysis of working memory impairments in autism spectrum disorders. Neuropsychol Rev. 2017;27(1):46–61.
130. D'Esposito M, Postle BR, Rypma B. Prefrontal cortical contributions to working memory: evidence from event-related fMRI studies. Executive control and the frontal lobe: Current issues: Springer; 2000. p. 3–11.
131. Koshino H, Kana RK, Keller TA, Cherkassky VL, Minshew NJ, Just MA. fMRI investigation of working memory for faces in autism: visual coding and underconnectivity with frontal areas. Cereb Cortex. 2007;18(2):289–300.
132. Luna B, Minshew N, Garver K, Lazar N, Thulborn K, Eddy W, et al. Neocortical system abnormalities in autism an fMRI study of spatial working memory. Neurology. 2002;59(6):834–40.
133. Vogan VM, Morgan BR, Lee W, Powell TL, Smith ML, Taylor MJ. The neural correlates of visuo-spatial working memory in children with autism spectrum disorder: effects of cognitive load. J Neurodev Disord. 2014;6(1):19.
134. Van Steenburgh JJ, Varvaris M, Schretlen DJ, Vannorsdall TD, Gordon B. Balanced bifrontal transcranial direct current stimulation enhances working memory in adults with high-functioning autism: a sham-controlled crossover study. Mol Autism. 2017;8(1):40.
135. Ameis SH, Daskalakis ZJ, Blumberger DM, Desarkar P, Drmic I, Mabbott DJ, et al. Repetitive transcranial magnetic stimulation for the treatment of executive function deficits in autism spectrum disorder: clinical trial approach. J Child Adolesc Psychopharmacol. 2017;27(5):413–21.

Psychobiotics

Alper Evrensel, Barış Önen Ünsalver
and Mehmet Emin Ceylan

Abstract

Psychobiotics are live bacteria that directly and indirectly produce positive effects on neuronal functions by colonizing into the intestinal flora. Preliminary studies, although in limited numbers, have found that these bacteria have anxiolytic and antidepressant activities. No research has yet been published on the antipsychotic efficacy of psychobiotics. However, these preliminary studies have opened up new horizons and raised the idea that a new class is emerging in psychopharmacology. About 70 years have passed since the discovery of chlorpromazine, and while the synaptic transmission is understood in almost all details, there seems to be a paradigm shift in psychopharmacology. In recent years, the perspective has shifted from synapse to intestinal microbiota. In this respect, germ-free and conventional animal experiments and few human studies were examined in a comprehensive manner. In this article, after a brief look at the history of contemporary psychopharmacology, the mechanisms of the gut–brain relationship and the evidence of metabolic, systemic, and neuropsychiatric activities of psychobiotics were discussed in detail. In conclusion, psychobiotics seem to have the potential for treatment of neuropsychiatric

A. Evrensel (✉)
Department of Psychiatry, Uskudar University, NP Brain Hospital, Saray Mah. Ahmet Tevfik İleri Cad. No: 18 PK, 34768 Ümraniye, Istanbul, Turkey
e-mail: alperevrensel@gmail.com

B. Ö. Ünsalver
Department of Medical Documentation and Secretariat, Vocational School of Health Services, Uskudar University, Istanbul, Turkey
e-mail: onenunsalver@gmail.com

M. E. Ceylan
Departments of Psychology and Philosophy, Uskudar University, Istanbul, Turkey
e-mail: m.eminceylan@yahoo.com

© Springer Nature Singapore Pte Ltd. 2019
Y.-K. Kim (ed.), *Frontiers in Psychiatry*, Advances in Experimental Medicine and Biology 1192, https://doi.org/10.1007/978-981-32-9721-0_28

disorders in the future. However, there are many questions and we do not know the answers yet. We anticipate that the answer to these questions will be given in the near future.

Keywords
Probiotics · Psychobiotics · Microbiota · Psychopharmacology · Gut–brain axis · Dysbiosis

Introduction

In modern medicine, there have been four important developments in the field of treatment. These are: Louis Pasteur and Robert Koch's bringing in vaccine to the medicine, the discovery of penicillin, use of oral contraceptives, and psychopharmacology [1]. The term "psychopharmacology" has been first visible in the literature under the title of pharmacologist David Macht's article [2]. Although morphine, codeine, cocaine, bromine, barbiturates, and amphetamines have been discovered until 1920 and have been used in various psychiatric indications, the discovery of chlorpromazine in 1951 is the milestone of psychopharmacology as a modern scientific discipline [3]. The discovery of chlorpromazine not only made a revolutionary difference in the treatment of psychotic patients but also paved the way for an understanding of synaptic transmission and subsequently identified neurotransmitters in the central nervous system [4].

Understanding the synaptic transmission, determining the neurobiological basis of psychiatric disorders and controlling the disease symptoms with drugs is the first paradigm shift in psychopharmacology. With the discovery of chlorpromazine, psychiatry has become one of the other medical disciplines, and psychiatrists have become true physicians rather than being therapists listening only to the problems of their patients [3].

A paradigm shift is being experienced in recent years. In the etiopathogenesis of neuropsychiatric disorders, in addition to synaptic neurotransmission, the role of intestine–microbiota–immune system–brain interactions have begun to be better understood [5]. In recent years, much evidence has been obtained on bidirectional interaction between intestinal microbiota and brain, impaired microbiota composition (dysbiosis), leaky gut, effects of germ-free conditions on neurodevelopment, and neuroinflammation in preclinical and clinical studies [5]. In the light of this evidence, microbiota-based treatments have emerged. The main heading of this treatment is probiotics or, in other words, psychobiotics.

Current Paradigm in Psychopharmacology

Although the discovery of chlorpromazine has been a short while ago, about 70 years, the roots of modern psychopharmacology go back further. Roughly two periods can be defined before the discovery of chlorpromazine. In the first period (1800s), pharmacological agents were used for the first time in the control of behavioral disorders through the discovery of drugs such as morphine, potassium bromide, chloral hydrate, hyoscine, and paraldehyde. With the introduction of thiamine (vitamin B1), nicotinic acid (niacin, vitamin B3), and penicillin in the first half of the twentieth century, neuropsychiatric disorders related to beriberi disease, psychosis related to pellagra and dementia caused by syphilis have been taken under control, respectively, and the incidence in psychiatric hospitals has decreased significantly. In addition, barbiturates were discovered in this period, and the antimanic property of lithium (1949) was noticed [4]. In order to understand the paradigm shift created by chlorpromazine, it is useful to take a brief look at the period before chlorpromazine.

Before Chlorpromazine

The first fruit of modern psychopharmacology is for sure morphine. Friedrich Wilhelm Adam Sertürner succeeded in decomposing morphine from opium in 1805 and observed that this substance caused sleep in animals, therefore, by being inspired by the ancient Greek god of sleep *Morpheus*, he named it morphine [1]. Morphine has been used for rapid control of aggression and agitation in 1860s [4]. Again, during these years, potassium bromide has been used in the treatment of epilepsy and anxiety [6]. Chloral hydrate is also used for sedation and behavior control, like morphine and potassium bromide [4].

Barbiturates produced in 2500 different derivatives since being synthetized in 1903 have been the leading drug in the treatment of insomnia for nearly 50 years [7]. In the treatment of schizophrenia; electrically induced convulsions (electroconvulsive therapy, ECT) have been used since the mid-1930 s, and camphor or pentetrazol (cardiazol) induced convulsions and insulin coma therapy have been applied throughout the 1940s [4]. From a different angle, these treatment methods are also a reflection of the desperation in psychopharmacology.

Chlorpromazine

Phenothiazine, a precursor of chlorpromazine, was produced in 1883 for use in the blue dye industry [4]. Paul Charpentier et al., working in the French pharmaceutical company Rhone-Poulenc laboratories, succeeded in synthesizing chlor promazine from the phenothiazine on December 11, 1951, and announced their general anesthetic properties [8]. Surgeon Henri Laborit and his colleagues used it for anesthetic purposes, and published its sedative activity without loss of con sciousness in an article named “A new vegetative stabilizer” [9]. The first case treated with chlorpromazine at the psychiatry ward (Jacques Lh, a male, agitated, psychotic manic patient) has been reported at Laborit’s request [10]. One month after this article, on March 22, 1952, Pierre Deniker and Jean Delay started clinical research and published six articles in 6 months [3]. Chlorpromazine has been recognized and used in psychiatry because of these articles. In November 1952, chlorpromazine was registered in France (inspired by "large in action") as *Largactil* [3].

After Chlorpromazine

Simultaneously with the introduction of chlorpromazine, the mechanisms of synaptic transmission have begun to be understood. At this point, the discovery of spectrophotofluorometer (SPF) is a revolutionary progress [11]. Electrical and chemical activity in the synaptic range has been observed through SPF, and six neurotransmitters (acetylcholine, dopamine, serotonin, gamma-aminobutyric acid, norepinephrine, and substance P) have been identified [3]. In 1963, it has been understood that potent antipsychotics block catecholamine receptors [12]. The information obtained by observing the effect of chlorpromazine on neuronal transmission through SPF pave the way for both the understanding of schizophrenia physiopathology and the discovery of other drugs [13]. Today, around 112 psychotropic drugs (32 antipsychotics, 25 antidepressants, 20 anxiolytics/sedatives/hypnotics, 4 chemical dependency adjuncts, 4 monoamine oxidase inhibitors, 6 mood stabilizers, 8 stimulants, and 17 miscellaneous drugs) are in clinical use (*Last updated: July 10, 2018)* [14].

Paradigm Shift in Psychopharmacology

Over the last half century, enormous knowledge to illuminate the etiopathogenesis of neuropsychiatric disorders and to offer treatment options has been accumulated. The vast majority of this information is associated with neuronal transmission. Dopamine hypothesis in psychotic disorders, serotonin hypothesis in anxiety/depression still remains valid [15]. However, dopamine is not the only neurotransmitter associated with psychosis. Increased evidence in recent years show that serotonin and glutamate are associated with dopamine, the problems in these two neurotransmitter systems (serotonin hyperactivity in 5-HT_2A receptors on glutamatergic neurons in the cerebral cortex and NMDA receptor hypoactivity in GABAergic interneurons in the prefrontal cortex) cause dopamine hyperactivity in D_2 receptors at mesolimbic pathway [16]. Similarly, serotonin plays an important role in peripheral tissues (gastrointestinal system, hematopoiesis, bone metabolism, metabolic homeostasis) and especially in immune system functions [17, 18]. Gardner and Boles, using the term "mitochondrial psychiatry", proposed a new model of serotonergic insufficiency, mitochondrial dysfunction, and inflammation in the pathogenesis of depression and affective spectrum disorders [19, 20]. Han et al. argued that commensal microorganisms living in the intestines communicate with the host mitochondria and thus, it is possible to live long and healthy [21]. Anderson also emphasized the importance of the relationship between mitochondria and melatonin, inflammation, sirtuin, tryptophan metabolites, DNA repair, and oxidative/nitrosative stress [22].

The focus of psychopharmacology is shifting from synaptic neurotransmission to the peripheral system, intestinal microbiota, mitochondria, immune system, and neuroinflammation. In this period, psychobiotics stand out as the new class of psychotropic drugs [23, 24].

Gut–Brain Communication

Human body is a complex ecosystem where our eukaryotic own cells and prokaryotic commensal microorganisms live together [5]. The term "psychobiotics" has been proposed to describe viable microorganisms colonized in the intestines that have positive effects on neuropsychiatric functions in various ways [24]. Psychobiotics interact with the body and especially the brain through the endocrine, metabolic, and immune system by virtue of the hormones, metabolites, and immune factors they secrete [25]. This bidirectional interaction between the gut and the brain, defined as the "gut–brain axis", is not yet fully elucidated [26]. However, according to the information obtained from a large number of animal and human studies, psychobiotics are effective on brain functions by secreting neuroactive metabolites (serotonin, catecholamine, gamma-aminobutyric acid [GABA], acetylcholine) as well as short-chain fatty acids (SCFAs) [27]. In the presence of dysbiosis, pathogenic live bacteria or bacterial components (endotoxins with lipopolysaccharide and peptidoglycan structure) that are involved in the systemic circulation due to leaky gut can cause a low-grade immune reaction. These components that pass through the blood–brain barrier can induce neuroinflammation by activating microglia [28]. This interaction is discussed below in detail.

Microbiota–Immune System Interaction

In recent years, with the emergence of evidence that the brain has its own lymphatic drainage system, the neuroinflammation hypothesis has become more important [29]. The main function of the immune system is to find and destroy germs. Molecular elements of microorganisms (nucleic acids, cell wall components in lipopolysaccharide, flagella, etc.) activate immune system cells [25]. Microbiota bacteria contact the immune system through pattern recognition receptors (PRRs). The most important member of the PRRs is toll like receptors (TLRs) [26].

Anti-inflammatory cytokines such as interleukin-10 (IL-10) are produced when PRRs are activated by commensal bacteria [30]. For example, *Bifidobacterium infantis* and *Lactobacillus GG* increase the level of IL-10 in human and reduce the level of pro-inflammatory cytokine and repair impaired blood–brain barrier permeability due to inflammation [31]. In addition, beneficial bacteria block the pathogenic pro-inflammatory process by activation of TLR-2 and TLR-4 [32]. Pro-inflammatory cytokines induced by pathogens can affect neuronal functioning in several ways. For example, they can change the level of neurotransmitters in the brain [33]. They may also induce prostaglandin synthesis that provokes pro-inflammatory process through another way [34]. Psychobiotics play an important role in reducing low-level neuroinflammation by decreasing the levels of pro-inflammatory cytokines in the systemic circulation.

However, in a trial by Bercik et al. on mice infected with *Trichuris muris* parasites, *Bifidobacterium longum NCC3001* reduced anxiety levels and normalized low hippocampal BDNF levels without any change in cytokine levels [35]. This finding suggests that non-cytokine mechanisms are also present in the probiotic effect.

Microbiota–Enteric Nervous System Interaction

Myenteric neurons are located just below the enterocytes and can be in direct contact with microbiota bacteria as it is close to the lumen [36]. Microbiota may be effective on the electrical activity of the enteric nervous system. For example, Bifidobacterium longum exhibits anxiolytic activity by reducing the action potential of myenteric neurons [37]. In the study of Ma et al., psychobiotic treatment (*Lactobacillus reuteri*) prevented hyperexcitation of dorsal root ganglion neurons in the colon due to noxious stimulation [38]. The excitability levels of myenteric afterhyperpolarization neurons of germ-free mice were found to be lower [39]. In another study, various abnormalities of the enteric nervous system of germ-free mice (increased number of myenteric nitrergic neurons, less neuronal density in ganglions) has been found [40]. By altering ion transport in the colon mucosa and submucosa, microbiota also affects the operation of the myenteric nervous system and peristalsis [41]. Microbiota bacteria play an important and decisive role in the enteric nervous system not only on neurons but also on homeostasis and control of glial cells in lamina propria [42].

In addition, there are neurotransmitters in the intestinal bacteria metabolites. *Bacillus* produces dopamine and noradrenaline, *Bifidobacteria and Lactobacillus* produce GABA, *Escherichia* produces noradrenaline and serotonin, and *Enterococcus* and *Streptococcus* produce serotonin [43]. Although there is no evidence, these neuroactive amines have the potential to affect synaptic activity in the enteric nervous system.

The Role of Nervus Vagus

Nervus vagus is the most important linkage regulating parasympathetic activity between brain and gastrointestinal system and plays a direct role in immune system functions [44]. Stress [45], nutrition [46], and exercise [47] affect vagal activity. Nervus vagus stimulation produces anti-inflammatory [44], analgesic [48], antiepileptic [49], antidepressant, and anxiolytic [50] effects. On the contrary, there are indications that antidepressant and anxiolytic drugs also affect vagal activity [51].

The effectiveness of psychobiotics has not been observed in many vagotomy applied animal experiments [37, 52]. However, in an experiment on mice, Bercik et al. also found that neurobehavioral effect of *Lactobacillus* continued after vagotomy [53]. Apparently, nervus vagus is one of the pathways responsible for psychobiotic activity in the gut–brain axis.

The Effect of Stress on Leaky Gut

The intestinal epithelial cell (enterocytes) is the largest mucosa in the human body due to its ciliary structure, and its total surface area can reach the size of a tennis court. Tight junction proteins that bind enterocytes make the mucosa intact. The mucous layer secreted by the mucosa is a strong physical barrier between bacteria and the toxic material and the host. Leaky gut may occur for various reasons (stress, glucocorticoids, dysbiosis, and endotoxins). The pathogenic microorganisms and their toxic metabolites in the lipopolysaccharide structure enter into the bloodstream and cause a pro-inflammatory effect. The role of this process in the etiopathogenesis of depression is quite clear [54]. Psychobiotics caused leaky gut-reducing and anti-inflammatory effect in animal experiments [55–57].

The leaky gut effect of glucocorticoids is reversed by psychobiotics and ultimately it results in an anti-inflammatory effect, which is significant in neuropsychiatric disorders accompanied by low-level neuroinflammation. However, it was found that psychobiotics had beneficial effects in cases where leaky gut and associated neuroinflammation are not observed [58, 59].

Bacteria-Induced Active Metabolites and Short-Chain Fatty Acids (SCFAs)

The human genome does not encode enzymes that digest plant-derived polysaccharides. The digestion of these foods in the diet is carried out by enzymes synthesized by microbiota [60]. Digestion of plant-derived polysaccharide fibers results in production of SCFAs (acetate, butyrate, propionate, lactate) [61]. SCFAs are absorbed from the colon, enter into the systemic circulation, and reach the liver and muscles, and then they are involved in the metabolic functions and are perhaps the most important microbiota metabolites [61]. A small amount of SCFAs reaches the central nervous system by crossing the blood–brain barrier [62]. Although the effect of this small amount on synaptic transmission is not yet clear, experimentally it has been shown that high-dose fatty acids have an agonist effect on free fatty acid receptors [63] and change neuromodulation via epigenetic mechanisms [64, 65]. This may have important consequences for neuropsychiatric functions.

Microbiota bacteria metabolites are not limited to SCFAs. Metabolites of microbiota, which are involved in blood circulation, have a very important role in neuroimmune disorders and neuroinflammation [66]. For example, the role of metabolites synthesized from tryptophan (such as serotonin and antioxidant-featured indoxyl sulfate and indole-3-propionic acid) has been shown in germ-free animal experiments [67]. The plasma tryptophan amounts of germ-free mice are higher than that of normal (microbiota-containing) mice, whereas the plasma serotonin levels of normal mice are approximately three times higher [68]. This result may be interpreted as the effect of microbiota bacteria have an effect on the metabolism of enterochromaffin cells secreting serotonin in the intestine.

However, the information obtained so far in this area is unfortunately insufficient and is limited to animal experiments. After it is uncovered in detail from which substrate which metabolites are produced by human microbiota bacteria and which effects they have on all organ systems, its role in maintenance of health and etiopathogenesis of diseases will be clarified.

Psychobiotics

In the scientific literature, Ilya Ilyich Mechnikov (Elie Metchnikoff) is the first person who mentioned the beneficial effects of commensal microorganisms on the host, without using the name psychobiotics [69]. The Nobel Prize in Physiology or Medicine 1908 was awarded to Mechnikov in recognition of his work on immunity [70]. Probably the first research has been published in 1910 on the efficacy of probiotic bacteria in the treatment of depression [71]. After a long period of silence, as summarized in Fig. 1, extensive clinical and preclinical studies have been started to be conducted in recent years on the effectiveness of probiotics [23, 72].

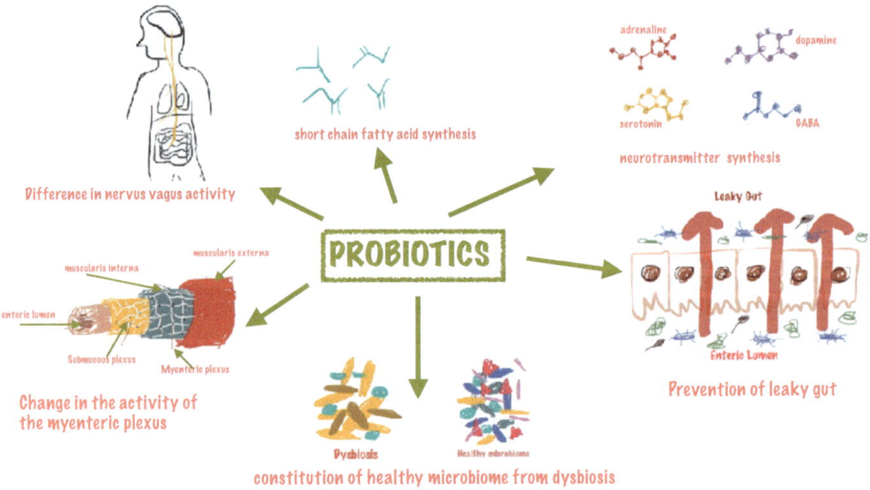

Fig. 1 Overview of the effectiveness of probiotics

Animal Studies

Two of the effective studies conducted in this field have been performed by Desbonnet et al. In the first of these, adult male Sprague-Dawley rats were divided into two groups; 12 for experiment group and 8 for control group. *Bifidobacterium infantis* was administered orally to the experiment group for 14 days. Then, forced swim test and various blood analyses (cytokine, plasma tryptophan, brain monoamine, vasopressin, corticotrophin-releasing factor) were applied to the rats. *Bifidobacterium infantis* showed antidepressant-like activity in serological analyzes, even though there was no change in the test performance of two groups [73]. In their second study, the researchers tested the antidepressant activity of this psychobiotic bacterium by comparing it with citalopram (a selective serotonin reuptake inhibitor antidepressant). At the end of the experiment, no difference has been found between citalopram and *Bifidobacterium infantis* antidepressant activity [74].

Psychobiotics can cause a decrease in anxiety scores as well as in depression scores. In an experiment by Bravo et al., BALB/c mice applied *Lactobacillus rhamnosus JB-1* showed lower scores in forced swim test and elevated plus maze test forced swim test [75]. Additionally, $GABA_{B1b}$ receptor expression levels decreased in amygdala and hippocampus and increased in cingulate and prelimbic areas in the experiment group. These findings can be interpreted as psychobiotics can exhibit anxiolytic activity by modulating inhibitory neurotransmitter (GABA) functions. Janik et al. measured the efficacy of *Lactobacillus rhamnosus JB-1* on brain neurotransmitter levels of BALB/c mice with magnetic resonance spectroscopy (MRS) [76]. GABA, glutamate, and aspartate levels were found to be high in mice fed with psychobiotics. This is the first study showing that probiotics increased the level of central glutamate. Another interesting finding is the difference between the duration of neurotransmitters' elevation and staying high after probiotic administration.

In another study with similar design, the efficacy of *Mycobacterium vaccae* as a psychobiotic bacterium was tested using the Hebb–Williams style complex maze [77]. Test performances of the probiotic applied group were higher than the control group. However, the life span of this effect was limited to 1 week. This finding is significant even if it contradicts some previous studies on the persistent colonization of exogenous probiotics in the intestine [78, 79]. In the study of Liang et al., the efficacy of *Lactobacillus helveticus* and citalopram was compared in various biochemical analyzes and anxiety tests (elevated plus maze and open field test) [80]. Psychobiotic-fed rats had lower levels of anxiety and higher memory performance, lower hypothalamo pituitary activity, and higher anti-inflammatory markers than control group. These findings are similar to citalopram activity.

In a study by Moya-Perez et al., the C57Bl/6 J male pups under the model of chronic stress induced by maternal separation were divided into two groups [81]. The effects of *Bifidobacterium pseudocatenulatum CECT 7765* were evaluated on the 21st day and on the 41st day of postpartum through various analyzes and tests (corticosterone, neurotransmitters, cytokines, fecal microbiota analysis, elevated plus maze, open field, acute immobilization). As a result of the experiment, it was found that the bacteria weakened the acute stress response, showed anti-inflammatory effect, decreased the level of anxiety and repaired intestinal dysbiosis.

In a very recent study, the antidepressant-like efficacy of *Lactobacillus rhamnosus JB-1* and fluoxetine was tested on two different mouse strains (BALB/c and Swiss Webster) [82]. The tail suspension test and the corticosterone response to an acute restraint stressor were applied to the laboratory animals. *Lactobacillus rhamnosus JB-1* and fluoxetine showed antidepressant-like behavior in both tests in BALB/c mice (n = 46). However, Swiss Webster mice (n = 36) did not respond to both treatments. In the light of these results, it is seen that the selection of laboratory animal strain in the studies on psychobiotics is of importance on the results.

Human Studies

Although the findings from animal experiments give hope for human studies, they may not always meet expectations. Therefore, it is necessary to increase studies on human and discuss their results. Let's take a look at the small number of human studies in this context.

In an early study on patients with irritable bowel syndrome, 75 participants (48 females, 27 males) received *Lactobacillus salivarius UCC4331, Bifidobacterium infantis 35624* or placebo for 8 weeks. The rates of interleukin 10/interleukin 12 before and after treatment were compared. It was found that this ratio, which was low before treatment and indicated a pro-inflammatory condition, was normalized in the group receiving *Bifidobacterium infantis 35624* [31]. In the following period, many studies showing the probiotic efficacy of the *Lactobacillus* family has been published. One of these, published in 2007, investigated the effect of *Lactobacillus casei Shirota* or a placebo-containing milky drink on mood and cognitive functions after drinking for 3 weeks [83]. No statistically significant difference was found between the two groups as the result of self-report tests (POMS, Wechsler Memory Scale, NART). However, the group with the lowest mood scores was found to be happier after probiotic supplementation. Surprisingly, the probiotic group had lower scores in the tests that measure cognitive functions.

Messaoudi et al. tested the effects of *Lactobacillus helveticus R0052* and *Bifidobacterium longum* on mood and cognition in two randomized controlled trials (RCT) [58, 84]. In experiment group, self-report tests (Hopkins Symptom Checklist-90, Hospital Anxiety and Depression Scale, Coping Checklist) showed a decrease in depression and anxiety scores, but the cognitive function scores did not decrease.

Different *Bifidobacterium* and *Lactobacillus* strains were given to healthy volunteers in another RCT [85]. The participants applied Leiden Index of Depression Sensitivity, The Beck Depression Inventory, The Beck Anxiety Inventory (n = 40) showed a decrease in sad mood levels. Similarly, healthy medical faculty students (n = 47) who had a *Lactobacillus* bacterium (*L. casei strain Shirota*) for 8 weeks before the exam were found to have lower plasma cortisol levels compared to placebo [86]. Another study that examined the effects of another *Lactobacillus* bacterium (*L. gasseri OLL2809 LG2809*) on student-athletes (n=44) found decreased symptoms of fatigue and improved mood [87]. These findings can be interpreted as psychobiotics may be useful in reducing performance anxiety.

In the study of Tillish et al., healthy volunteers who were given fermented milk containing psychobiotics (*Bifidobacterium animalis subsp Lactis, Streptococcus thermophiles, Lactobacillus bulgaricus, and Lactococcus lactis subsp Lactis*) were examined under functional magnetic resonance imaging (fMRI). In the experiment group, reduced activity in the emotional and somatosensory centers (insula, somatosensory cortex, and periaqueductal cortex) was found [88]. The findings of this study were duplicated by an RCT on patients with irritable bowel syndrome (n=44) [89].

However, in a recent study on healthy male volunteers, *Lactobacillus rhamnosus* were found to be similar to placebo in terms of stress and cognitive tests (Cambridge Neuropsychological Test Automated Battery, Socially Evaluated Cold Pressor Test, electroencephalography, cortisol, and cytokine analysis). This finding may be caused by the small sample size (n=29) [90].

Future Directions and Knowledge Gaps

Although promising evidence has been obtained about psychobiotics, there are still many uncertainties. For example, it cannot be said that psychobiotics change the composition of microbiology permanently and comprehensively. However, there are only short-term studies in this field. There is a need for long-term (months or even years) studies in which fecal samples are examined. Moreover, the composition of the microbiota is dynamic and changes with aging [91]. There are no studies on the effectiveness of psychobiotics in different age groups. Another issue is which dose to be applied in which indication. Psychobiotics may also have a therapeutic range, as in medicines. There may be unresponsiveness due to low dose applications, or there may be side effects caused by high doses. In addition, it is not clear when physiological, metabolic, immunological, and neuropsychiatric effects occur after oral administration of psychobiotics. The number of preliminary studies is quite small. Another problem is; after ending the application of psychobiotics, it is not clear how long the activity lasts. Studies on this subject are limited. This information should be clarified by monitoring clinical and biological parameters for a long time.

The simultaneous increase of the neurotransmitters that interact with each other in the brain (for example, GABA and glutamate) suggests that the total effect can vanish [76]. Local and specific effects shall be emphasized and functional responses of neurotransmitter changes shall be monitored. Evidence of improvement in cognitive functions through psychobiotics has been obtained from preclinical studies [77]. Even if it is not repeated, there is even a study reporting that it causes cognitive impairment [83]. The lack of information in this area needs to be addressed through additional human studies.

Another important problem is the uncertainty about why some bacterial strains have a psychobiotic effect while others do not. Germ-free animal experiments are ideal for separating the neurophysiological effects of a single bacterial strain. However, it should be clarified whether the psychobiotic effect is caused by the synergy of singular or bacterial strains (*quorum sensing*) [92]. Factors that can affect psychobiotic outcome and their impact strength are not known clearly. Some of them may be age, gender, diet, and genotype.

However, there is no study on the interactions of psychobiotics and psychotropic drugs. Most psychotropic drugs have antibiotic activity [93]. For this reason, adding psychobiotics to psychotropics may vanish positive effects. There is a need to clarify which psychotropic has an antibiotic effect on which psychobiotic.

Conclusion

Jacques Lh, the first patient treated with chlorpromazine, received total of 855 mg of medication during his 20-day treatment [10]. Today we know that this dose (40–50 mg/day) is too low for antipsychotic activity [15]. In addition to its role in the synaptic transmission, other activities of chlorpromazine have been discovered. New evidence suggests that it (and many other psychotropics) has antibiotic/antifungal activity [94]. For example, SSRIs kill bacteria through efflux pump inhibition, MAO inhibitors through cell wall synthesis inhibition and tricyclic antidepressants through DNA gyrase inhibition by antiplasmic action (antimicrobial effect) [93]. Here, it is important to remind that the first antidepressant molecule is an antituberculosis drug, iproniazid (an MAO inhibitor) [95]. Perhaps today, distinguished and active psychotropics change the microbiota composition in addition to their monoaminergic and synaptic effects (dysbiosis or restoration) [54].

Thus, about 70 years of the reign of psychotropics has begun to be questioned [96]. The paradigm in psychopharmacology may be changing. The viewpoint seems to shift from synapse to intestinal microbiota and immune system [97]. Although more than 100 years have passed since the discovery of psychotropic activities, the power and role of psychobiotics have just begun to be understood. However, there are many questions to be answered. By answering these questions in the coming years, psychobiotics may be used as a new class of psychotropics.

Acknowledgements We would like to thank Dr. Barış Önen Ünsalver for preparing the image used in this review.

References

1. Köknel O. At the fiftieth anniversary the history of modern psychopharmacology and psychopharmacological researches in Turkey and in the world. Bull Clin Psychopharmacol. 2002;12(4):201–10.
2. Macht DL. Contributions to psychopharmacology. Johns Hopkins Hosp Bull. 1920;31:167–73.

3. Ban TA. Fifty years chlorpromazine: a historical perspective. Neuropsychiatr Dis Treat. 2007;3(4):495–500.
4. Ban TA. Pharmacotherapy of mental illness. A historical analysis. Progress Neuro Psychopharmacol Biol Psychiatry. 2001;25(4):709–727.
5. Evrensel A, Önen Ünsalver B, Ceylan ME. Gut-brain axis and psychiatric disorders. Curr Psychiatry Rev. 2018;14(3):178–186.
6. Ewing JA, Grant WJ. The bromide hazard. South Med J. 1965;58:148–52.
7. Lehmann HE. Before they called it psychopharmacology. Neuropsychopharmacology. 1993;8(4):291–303.
8. Charpentier P, Gailliot P, Jacob R, Gaudechon J, Buisson P. Recherches sur les diméthylaminopropyl-N phénothiazines substituées. Comptes rendus de l'Académie des sciences (Paris). 1952;235:59–60.
9. Laborit H, Huguenard P, Alluaume R. Un noveau stabilisateur vegetatif (le 4560 RP). La Presse Médicale. 1952;60:206–8.
10. Hamon J, Paraire J, Velluz J. Remarques sur l'action du 4560 RP sur l'agitation maniaque. Annales Médico-psychologiques (Paris). 1952;110:331–5.
11. Bowman RL, Caulfield PA, Udenfriend S. Spectrophotometric assay in the visible and ultraviolet. Science. 1955;122(3157):32–3.
12. Carlsson A, Lindqvist M. Effect of chlorpromazine and haloperidol on formation of 3-methoxytyramine and normetanephrine on mouse brain. Acta Pharmacologica et Toxicologica (Copenh). 1963;20:140–4.
13. Bennett MR. Monoaminergic synapses and schizophrenia: 45 years of neuroleptics. J Psychopharmacol. 1998;12(3):289–304.
14. Psychotropic Medications-Consent Drug List Classes of Medications Frequently Used for Psychiatric Indication. Texas Department of State Health Services. https://www.dshs.texas.gov/mhprograms/consentdruglist/.
15. Stahl SM. Stahl's essential psychopharmacology: neuroscientific basis and practical applications. 4th ed. New York: Cambridge University Press; 2013.
16. Stahl SM. Beyond the dopamine hypothesis of schizophrenia to three neural networks of psychosis: dopamine, serotonin, and glutamate. CNS Spectr. 2018;23(3):187–91.
17. Spohn SN, Mawe GM. Non-conventional features of peripheral serotonin signalling - the gut and beyond. Nat Rev Gastroenterol Hepatol. 2017;14(7):412–20
18. Wu H, Denna TH, Storkersen JN, Gerriets VA. Beyond a neurotransmitter: the role of serotonin in inflammation and immunity. Pharmacol Res. 2018. (in press).
19. Gardner A, Boles RG. Is a "mitochondrial psychiatry" in the future? A Rev Curr Psychiatry Rev. 2005;1(3):255–71.
20. Gardner A, Boles RG. Beyond the serotonin hypothesis: mitochondria, inflammation and neurodegeneration in major depression and affective spectrum disorders. Prog Neuropsychopharmacol Biol Psychiatry. 2011;35(3):730–43.
21. Han B, Sivaramakrishnan P, Lin CJ, Neve IAA, He J, Tay LWR, et al. Microbial genetic composition tunes host longevity. Cell. 2017;169(7):1249–62.
22. Anderson G. Linking the biological underpinnings of depression: Role of mitochondria interactions with melatonin, inflammation, sirtuins, tryptophan catabolites, DNA repair and oxidative and nitrosative stress, with consequences for classification and cognition. Prog Neuropsychopharmacol Biol Psychiatry. 2018;80(Pt C):255–66.
23. Sarkar A, Lehto SM, Harty S, Dinan TG, Cryan JF, Burnet PWJ. Psychobiotics and the manipulation of bacteria-gut-brain signals. Trends Neurosci. 2016;39(11):763–81.
24. Dinan TG, Stanton C, Cryan JF. Psychobiotics: a novel class of psychotropic. Biol Psychiat. 2013;74(10):720–6.
25. Evrensel A, Ceylan ME. Microbiome: the missing link in neuropsychiatric disorders. EMJ Innov. 2017;1(1):83–8.
26. Evrensel A, Ceylan ME. The gut-brain axis: the missing link in depression. Clin Psychopharmacol Neurosci. 2015;13(3):239–44.

27. Evrensel A, Ceylan ME. Fecal microbiota transplantation and its usage in neuropsychiatric disorders. Clin Psychopharmacol Neurosci. 2016;14(3):231–7.
28. Rea K, Dinan TG, Cryan JF. The microbiome: a key regulator of stress and neuroinflammation. Neurobiol Stress. 2016;4:23–33.
29. Louveau A, Smirnov I, Keyes TJ, Eccles JD, Rouhani SJ, Peske JD. Structural and functional features of central nervous system lymphatic vessels. Nature. 2015;523(7560):337–41.
30. Chu H, Mazmanian SK. Innate immune recognition of the microbiota promotes host-microbial symbiosis. Nat Immunol. 2013;14(7):668–75.
31. O'Mahony L, McCarthy J, Kelly P, Hurley G, Luo F, Chen K, et al. Lactobacillus and bifidobacterium in irritable bowel syndrome: symptom responses and relationship to cytokine profiles. Gastroenterology. 2005;128(3):541–51.
32. Zhou W, Lv H, Li MX, Su H, Huang LG, Li J, et al. Protective effects of bifidobacteria on intestines in newborn rats with necrotizing enterocolitis and its regulation on TLR2 and TLR4. Genetics Mol Res. 2015;14(3):11505–14.
33. Miller AH, Haroon E, Raison CL, Felger JC. Cytokine targets in the brain: impact on neurotransmitters and neurocircuits. Depress Anxiety. 2013;30(4):297–306.
34. Felger JC, Lotrich FE. Inflammatory cytokines in depression: neurobiological mechanisms and therapeutic implications. Neuroscience. 2013;246:199–229.
35. Bercik P, Verdu EF, Foster JA, Macri J, Potter M, Huang X, et al. Chronic gastrointestinal inflammation induces anxiety-like behavior and alters central nervous system biochemistry in mice. Gastroenterology. 2010;139(6):2102–12.
36. Foster JA, McVey Neufeld KA. Gut–brain axis: how the microbiome influences anxiety and depression. Trends Neurosci. 2013;36(5):305–12.
37. Bercik P, Park AJ, Sinclair D, Khoshdel A, Lu J, Huang X, et al. The anxiolytic effect of Bifidobacterium longum NCC3001 involves vagal pathways for gut–brain communication. Neurogastroenterol Motil. 2011;23(12):1132–9.
38. Ma X, Mao YK, Wang B, Huizinga JD, Bienenstock J, Kunze W. Lactobacillus reuteri ingestion prevents hyperexcitability of colonic DRG neurons induced by noxious stimuli. Am J Physiol Gastrointest Liver Physiol. 2009; 296(4): G868–G875.
39. McVey Neufeld KA, Mao YK, Bienenstock J, Foster JA, Kunze WA. The microbiome is essential for normal gut intrinsic primary afferent neuron excitability in the mouse. Neurogastroenterol Motil. 2013;25(2):183–8.
40. Collins J, Borojevic R, Verdu EF, Huizinga JD, Ratcliffe EM. Intestinal microbiota influence the early postnatal development of the enteric nervous system. J Neurogastroenterol Motil. 2014;26(1): 98–107.
41. Lomasney KW, Houston A, Shanahan F, Dinan TG, Cryan JF, Hyland NP. Selective influence of host micro-biota on cAMP-mediated ion transport in mouse colon. J Neurogastroenterol Motil. 2014;26(6):887–8.
42. Kabouridis PS, Lasrado R, McCallum S, Chng SH, Snippert HJ, Clevers H, et al. Microbiota controls the homeosta-sis of glial cells in the gut lamina propria. Neuron. 2015;85(2):289–95.
43. Lyte M. Probiotics function mechanistically as delivery vehicles for neuroactive compounds: microbial endocrinology in the design and use of probiotics. BioEssays. 2011;33(8):574–81.
44. Bonaz B, Sinniger V, Pellissier S. The vagus nerve in the neuro-immune axis: Implications in the pathology of the gastrointestinal tract. Front Immunol. 2017;8:1452.
45. Mezzacappa ES, Kelsey RM, Katkin ES, Sloan RP. Vagal rebound and recovery from psychological stress. Psychosom Med. 2001;63:650–7.
46. de Haan JJ, Hadfoune M, Lubbers T, Hodin C, Lenaerts K, Ito A, et al. Lipid-rich enteral nutrition regulates mucosal mast cell activation via the vagal anti-inflammatory reflex. Am J Physiol Gastrointest Liver Physiol. 2013;305(5): G383–G391.
47. Weberruss H, Maucher J, Oberhoffer R, Müller J. Recovery of the cardiac autonomic nervous and vascular system after maximal cardiopulmonary exercise testing in recreational athletes. Eur J Appl Physiol. 2018;118(1):205–11.

48. Kirchner A, Birklein F, Stefan H, Handwerker HO. Left vagus nerve stimulation suppresses experimentally induced pain. Neurology. 2000;55(8):1167–71.
49. Galbarriatu L, Pomposo I, Aurrecoechea J, Marinas A, Agúndez M, Gómez JC, et al. Vagus nerve stimulation therapy for treatment-resistant epilepsy: a 15-year experience at a single institution. Clin Neurol Neurosurg. 2015;137:89–93.
50. Shah AP, Carreno FR, Wu H, Chung YA, Frazer A. Role of TrkB in the anxiolytic-like and antidepressant-like effects of vagal nerve stimulation: comparison with desipramine. Neuroscience. 2016;322:273–86.
51. Abdel Salam OM. Fluoxetine and sertraline stimulate gastric acid secretion via a vagal pathway in anaesthetised rats. Pharmacol Res. 2004;50(3):309–16.
52. Malick M, Gilbert K, Daniel J, Arseneault-Breard J, Tompkins TA, Godbout R, et al. Vagotomy prevents the effect of probiotics on caspase activity in a model of postmyocardial infarction depression. J Neurogastroenterol Motil. 2015;27(5):663–71.
53. Bercik P, Denou E, Collins J, Jackson W, Lu J, Jury J, et al. The intestinal microbiota affect central levels of brain-derived neurotropic factor and behavior in mice. Gastroenterology. 2011;141(2):599–609.
54. Evrensel A, Ceylan ME. Gut-microbiota-brain axis and depression. In: Kim YK, editor. Understanding depression. Singapore: Springer; 2018. p. 197–207.
55. Zareie M, Johnson-Henry K, Jury J, Yang PC, Ngan BY, McKay DM, et al. Probiotics prevent bacterial translocation and improve intestinal barrier function in rats following chronic psychological stress. Gut. 2006;55(11):1553–60.
56. Ait-Belgnaoui A, Durand H, Cartier C, Chaumaz G, Eutamene H, Ferrier L, et al. Prevention of gut leakiness by a probiotic treatment leads to attenuated HPA response to an acute psychological stress in rats. Psychoneuroendocrinology. 2012;37(11):1885–95.
57. Reandon S. Gut-brain link grabs neuroscientists. Nature. 2014;515(7526):175–7.
58. Messaoudi M, Lalonde R, Violle N, Javelot H, Desor D, Nejdi A, et al. Assessment of psychotropic-like properties of a probiotic formulation (Lactobacillus helveticus R0052 and Bifidobacterium longum R0175) in rats and human subjects. Br J Nutr. 2011;105(5):755–64.
59. Schmidt K, Cowen PJ, Harmer CJ, Tzortzis G, Errington S, Burnet PW, et al. Prebiotic intake reduces the waking cortisol response and alters emotional bias in healthy volunteers. Psychopharmacology. 2015;232(10):1793–801.
60. Qin J, Li R, Raes J, Arumugam M, Burgdorf KS, Manichanh C, et al. A human gut microbial gene catalogue established by metagenomic sequencing. Nature. 2010;464(7285):59–65.
61. Tan J, McKenzie C, Potamitis M, Thorburn AN, Mackay CR, Macia L. The role of short-chain fatty acids in health and disease. Adv Immunol. 2014;121:91–119.
62. Lei E, Vacy K, Boon WC. Fatty acids and their therapeutic potential in neurological disorders. Neurochem Int. 2016;95:75–84.
63. Sun J, Wang F, Hong G, Pang M, Xu H, Li H, et al. Antidepressant-like effects of sodium butyrate and its possible mechanisms of action in mice exposed to chronic unpredictable mild stress. Neurosci Lett. 2016;618:159–66.
64. Stilling RM, Dinan TG, Cryan JF. Microbial genes, brain & behaviour–epigenetic regulation of the gut–brain axis. Genes Brain Behav. 2014;13(1):69–86.
65. Erny D, Hrabě de Angelis AL, Jaitin D, Wieghofer P, Staszewski O, David E, et al. Host microbiota constantly control maturation and function of microglia in the CNS. Nat Neurosci. 2015;18(7):965–77.
66. Morris G, Berk M, Carvalho A, Caso JR, Sanz Y, Walder K, et al. The role of the microbial metabolites including tryptophan catabolites and short chain fatty acids in the pathophysiology of immune-inflammatory and neuroimmune disease. Mol Neurobiol. 2017;54(6):4432–51.
67. Yano JM, Yu K, Donaldson GP, Shastri GG, Ann P, Ma L, et al. Indigenous bacteria from the gut micro-biota regulate host serotonin biosynthesis. Cell. 2015;161(2):264–76.

68. Wikoff WR, Anfora AT, Liu J, Schultz PG, Lesley SA, Peters EC, et al. Metabolomics analysis reveals large effects of gut microflora on mammalian blood metabolites. Proc Natl Acad Sci. 2009;106(10):3698–703.
69. Mackowiak PA. Recycling Metchnikoff: probiotics, the intestinal microbiome and the quest for long life. Front Public Health. 2013;1:52.
70. Scott KA, Ida M, Peterson VL, Prenderville JA, Moloney GM, Izumo T, et al. Revisiting Metchnikoff: Age-related alterations in microbiota-gut-brain axis in the mouse. Brain Behav Immun. 2017;65:20–32.
71. Philips JGP. The treatment of melancholia by the lactic acid bacillus. J Ment Sci. 1910;56(234):422–31.
72. Bambury A, Sandhu K, Cryan JF, Dinan TG. Finding the needle in the haystack: systematic identification of psychobiotics. Br J Pharmacol. 2017. (in press).
73. Desbonnet L, Garrett L, Clarke G, Bienenstock J, Dinan TG. The probiotic Bifidobacteria infantis: an assessment of potential antidepressant properties in the rat. J Psychiatr Res. 2008;43(2):164–74.
74. Desbonnet L, Garrett L, Clarke G, Kiely B, Cryan JF, Dinan TG. Effects of the probiotic Bifidobacterium infantis in the maternal separation model of depression. Neuroscience. 2010;170(4):1179–88.
75. Bravo JA, Forsythe P, Chew MV, Escaravage E, Savignac HM, Dinan TG, et al. Ingestion of Lactobacillus strain regulates emotional behavior and central GABA receptor expression in a mouse via the vagus nerve. Proc Natl Acad Sci. 2011;108(38):16050–5.
76. Janik R, Thomason LAM, Stanisz AM, Forsythe P, Bienenstock J, Stanisz GJ. Magnetic resonance spectroscopy reveals oral Lactobacillus promotion of increases in brain GABA, N-acetyl aspartate and glutamate. NeuroImage. 2016;125:988–95.
77. Matthews DM, Jenks SM. Ingestion of Mycobacterium vaccae decreases anxiety-related behavior and improves learning in mice. Behav Process. 2013;96:27–35.
78. Alander M, Satokari R, Korpela R, Saxelin M, Vilpponen-Salmela T, Mattila-Sandholm T, et al. Persistence of colonization of human colonic mucosa by a probiotic strain, Lactobacillus rhamnosus GG, after oral consumption. Appl Environ Microbiol. 1999;65(1):351–4.
79. Frese SA, Hutton AA, Contreras LN, Shaw CA, Palumbo MC, Casaburi G, et al. Persistence of Supplemented Bifidobacterium longum subsp. infantis EVC001 in Breastfed Infants. mSphere. 2017;2(6): pii: e00501–17.
80. Liang S, Wang T, Hu X, Luo J, Li W, Wu X, et al. Administration of Lactobacillus helveticus NS8 improves behavioral, cognitive, and biochemical aberrations caused by chronic restraint stress. Neuroscience. 2015;310:561–77.
81. Moya-Perez A, Perez-Villalba A, Benitez-Paez A, Campillo I, Sanz Y. Bifidobacterium CECT 7765 modulates early stress-induced immune, neuroendocrine and behavioral alterations in mice. Brain Behav Immun. 2017;65:43–56.
82. McVey Neufeld KA, Kay S, Bienenstock J. Mouse strain affects behavioral and neuroendocrine stress responses following administration of probiotic Lactobacillus rhamnosus JB-1 or traditional antidepressant fluoxetine. Front Neurosci. 2018;12:294.
83. Benton D, Williams C, Brown A. Impact of consuming a milk drink containing a probiotic on mood and cognition. Eur J Clin Nutr. 2007;61(3):355–61.
84. Messaoudi M, Violle N, Bisson JF, Desor D, Javelot H, Rougeot C. Beneficial psychological effects of a probiotic formulation (Lactobacillus helveticus R0052 and Bifi-dobacterium longum R0175) in healthy human volunteers. Gut Microbes. 2011;2(4):256–61.
85. Steenbergen L, Sellaro R, van Hemert S, Bosch JA, Colzato LS. A randomized controlled trial to test the effect of multispecies probiotics on cognitive reactivity to sad mood. Brain Behav Immun. 2015;48:258–64.
86. Kato-Kataoka A, Nishida K, Takada M, Kawai M, Kikuchi-Hayakawa H, Suda K, et al. Fermented milk containing Lacto-bacillus casei strain Shirota preserves the diversity of the gut microbiota and relieves abdominal dysfunction in healthy medical students exposed to academic stress. Appl Environ Microbiol. 2016;82(12):3649–58.

87. Sashihara T, Nagata M, Mori T, Ikegami S, Gotoh M, Okubo K, et al. Effects of Lactobacillus gasseri OLL2809 and α-lactalbumin on university-student athletes: a randomized, double-blind, placebo-controlled clinical trial. Appl Physiol Nutr Metab. 2013;38(12):1228–35.
88. Tillisch K, Labus J, Kilpatrick L, Jiang Z, Stains J, Ebrat B, et al. Consumption of fermented milk product with probiotic modulates brain activity. Gastroenterology. 2013;144(7):1394–401.
89. Pinto-Sanchez MI, Hall GB, Ghajar K, Nardelli A, Bolino C, Lau JT, et al. Probiotic Bifidobacterium longum NCC3001 reduces depression scores and alters brain activity: A pilot study in patients with irritable bowel syndrome. Gastroenterology. 2017;153(2):448–459.e8.
90. Kelly JR, Allen AP, Temko A, Hutch W, Kennedy PJ, Farid N, et al. Lost in translation? The potential psychobiotic Lactobacillus rhamnosus (JB-1) fails to modulate stress or cognitive performance in healthy male subjects. Brain Behav Immun. 2017;61:50–9.
91. Claesson MJ, Cusack S, O'Sullivan O, Greene-Diniz R, de Weerd H, Flannery E, et al. Composition, variability, and tempo-ral stability of the intestinal microbiota of the elderly. Proc Natl Acad Sci. 2011;108(Suppl 1):4586–91.
92. Abisado RG, Benomar S, Klaus JR, Dandekar AA, Chandler JR. Bacterial quorum sensing and microbial community interactions. MBio. 2018;9(3): pii: e02331–17.
93. Macedo D, Filho AJMC, Soares de Sousa CN, Quevedo J, Barichello T, Junior HVN, et al. Antidepressants, antimicrobials or both? Gut microbiota dysbiosis in depression and possible implications of the antimicrobial effects of antidepressant drugs for antidepressant effectiveness. J Affect Disord. 2017;208:22–32.
94. Ishijima SA, Abe S. A novel murine candidiasis model with severe colonization in the stomach induced by n-acetylglucosamine-treatment and its scoring system based on local characteristic stomach symptoms. Med Mycol J. 2015;56(4):E31–9.
95. López-Muñoz F, Alamo C. Monoaminergic neurotransmission: the history of the discovery of antidepressants from 1950s until today. Curr Pharm Des. 2009;15(14):1563–86.
96. Patra S. Psychobiotics: A paradigm shift in psychopharmacology. Indian J Pharmacol. 2016;48(4):469–70.
97. Sato-Kasai M, Kato TA, Ohgidani M, Mizoguchi Y, Sagata N, Inamine S, et al. Aripiprazole inhibits polyI:C-induced microglial activation possibly via TRPM7. Schizophr Res. 2016;178(1–3):35–43.

Digital Interventions for Mental Disorders: Key Features, Efficacy, and Potential for Artificial Intelligence Applications

David Daniel Ebert, Mathias Harrer, Jennifer Apolinário-Hagen and Harald Baumeister

Abstract

Mental disorders are highly prevalent and often remain untreated. Many limitations of conventional face-to-face psychological interventions could potentially be overcome through Internet-based and mobile-based interventions (IMIs). This chapter introduces core features of IMIs, describes areas of application, presents evidence on the efficacy of IMIs as well as potential effect mechanisms, and delineates how Artificial Intelligence combined with IMIs may improve current practices in the prevention and treatment of mental disorders in adults. Meta-analyses of randomized controlled trials clearly show that therapist-guided IMIs can be highly effective for a broad range of mental health problems. Whether the effects of unguided IMIs are also clinically relevant, particularly under routine care conditions, is less clear. First studies on IMIs for the prevention of mental disorders have shown promising results.

D. D. Ebert (✉)
Department of Clinical Psychology, Vrije Universiteit Amsterdam,
Van der Boechorststraat 1, 1881 BT Amsterdam, The Netherlands
e-mail: d.d.ebert@vu.nl

M. Harrer
Clinical Psychology and Psychotherapy, Friedrich-Alexander-University
Erlangen-Nuremberg, Erlangen, Germany
e-mail: mathias.harrer@fau.de

J. Apolinário-Hagen
Faculty of Psychology, Department of Health Psychology,
FernUniversität in Hagen, Hagen, Germany
e-mail: jennifer.apolinario-hagen@fernuni-hagen.de

H. Baumeister
Clinical Psychology and Psychotherapy,
University of Ulm, Ulm, Germany
e-mail: harald.baumeister@uni-ulm.de

© Springer Nature Singapore Pte Ltd. 2019

Y.-K. Kim (ed.), *Frontiers in Psychiatry*, Advances in Experimental Medicine and Biology 1192, https://doi.org/10.1007/978-981-32-9721-0_29

Despite limitations and challenges, IMIs are increasingly implemented into routine care worldwide. IMIs are also well suited for applications of Artificial Intelligence and Machine Learning, which provides ample opportunities to improve the identification and treatment of mental disorders. Together with methodological innovations, these approaches may also deepen our understanding of how psychological interventions work, and why. Ethical and professional restraints as well as potential contraindications of IMIs, however, should also be considered. In sum, IMIs have a high potential for improving the prevention and treatment of mental health disorders across various indications, settings, and populations. Therefore, implementing IMIs into routine care as both adjunct and alternative to face-to-face treatment is highly desirable. Technological advancements may further enhance the variability and flexibility of IMIs, and thus even further increase their impact in people's lives in the future.

Keywords
Internet interventions · eHealth · Mental disorders · Psychotherapy · Prevention · Artificial intelligence · Machine learning

Introduction

Mental disorders are highly prevalent, with lifetime and 12-month prevalence rates estimated to range between 18–36% and 10–19%, respectively [1, 2]. Worldwide, mental disorders are one of the leading causes of disability [3]. They are associated with an enormous disease burden, including poorer quality of life [4], worse educational attainment and social role functioning [4–8], increased risk of developing chronic somatic conditions and related mortality [10], as well as suicidality [11, 12]. The World Health Organization (WHO) estimates that by 2030, depression alone will be the largest contributor to the total burden of disease worldwide [13].

It is well established that psychological interventions are an effective procedure for the treatment of various mental disorders, including depression [14–16], anxiety disorders [17–19], posttraumatic stress disorder [20, 21], personality disorders [22, 23], psychosomatic disorders [24–26], or sexual dysfunction disorders [27], to name only a few indications. However, research also documents a large treatment gap, as the majority of individuals suffering from a mental disorder remain untreated. In developed countries alone, treatment coverage of mental disorders only ranges from 23 to 46% [28–30]. Structural shortfalls, such as a lack of therapists within proximity, or long waiting times, do not appear to be the sole reason for such low treatment rates. Research suggests that psychological barriers to treatment may also play an important role. Despite the availability of effective treatment, many individuals prefer to deal with mental health issues on their own or are afraid of stigmatization [31, 32]. Furthermore, even in successfully treated patients, relapse remains frequent [33, 34], and cost-effective measures for aftercare are needed.

Technological advancements in the last decades have enabled the development and provision of Internet and Mobile-based psychological Interventions (IMIs [35]). Many limitations of traditional psychotherapeutic interventions could potentially be overcome with such interventions. IMIs are free from the restraints of location and time, and allow for reaching participants who would not make use of mental health treatments otherwise. They may therefore be a viable means to optimize current practices in mental health treatment. Nevertheless, the implementation of IMIs also raises new questions concerning effectiveness, safety, as well as patient and professional preferences.

In this chapter, we will provide an overview on the status and future development of IMIs. We aim to define the core elements of IMIs and review available evidence for their efficacy and mechanisms of change in treating and preventing mental disorders. We will also describe the important ethical and professional considerations associated with IMIs. Lastly, we illustrate current advances and future directions of the field, such as the application of data-driven methods and Machine Learning, and discuss their implications.

Key Features of Internet- and Mobile-Based Interventions

A basic feature of all IMIs is the transference of therapeutic processes to a digital environment. Like face-to-face psychotherapy, IMIs aim to modify individuals' emotions, cognitions, and behaviors, and promote their generalization to the daily life of users through established psychotherapeutic techniques. There is, however, a wide range of possibilities for using IMIs for the prevention and treatment of mental disorders, including mobile applications, stand-alone self-help interventions, or IMIs integrated into conventional on-site psychotherapy (i.e., blended concepts). IMIs can be categorized according to the technology they use, the extent of human support, the theoretical basis, and with respect to their areas of application. In Fig. 1, key aspects of IMIs are presented.

Theory Base

It is of utmost importance to resort to evidence-based psychotherapeutic models and techniques when developing and implementing IMIs. Among all psychotherapeutic approaches, most empirical evidence currently supports the efficacy of cognitive behavioral (CBT) techniques for common mental disorders [36, 37]. Furthermore, most CBT manuals are highly structured, standardized, and focus on specific strategies and concrete behavior. This is probably why most evaluated IMIs have used CBT principles as their treatment rationale [38]. Such approaches are often referred to as *iCBT* (Internet-based cognitive behavioral therapy [39]) or *cCBT* (computerized cognitive behavioral therapy). Nevertheless, IMIs are not

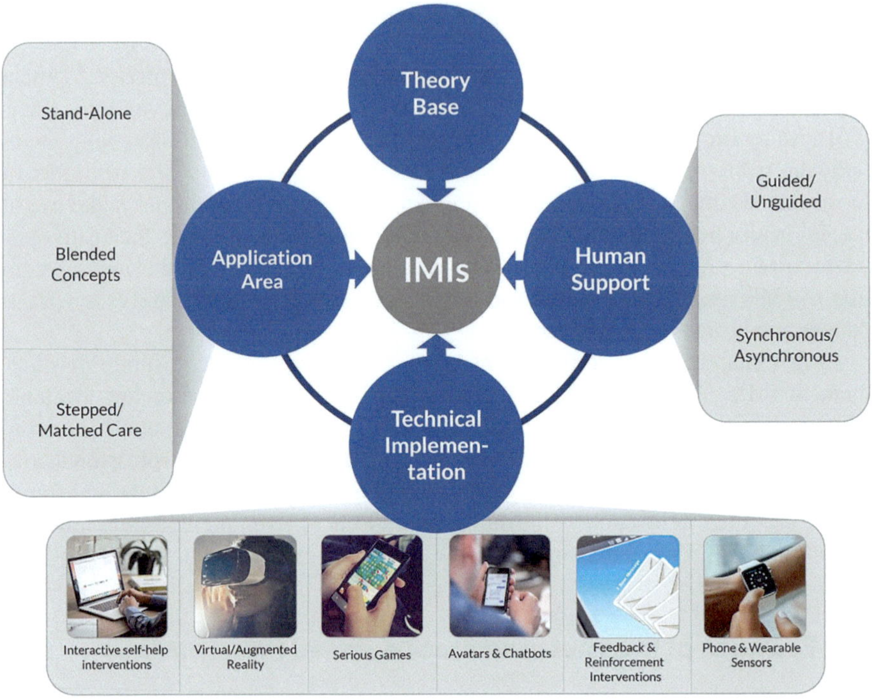

Fig. 1 Key aspects of Internet- and mobile-based interventions

restricted to any type of specific therapy approach. In recent years, other psychotherapeutic theories have also been used as the basis of IMIs, such as psychodynamic or mindfulness-based approaches [40–45].

Technical Implementation

For the implementation of IMIs, developers rely on a diverse and rapidly growing technical repertoire. Such modes of delivery include:

1. Interactive self-help interventions providing evidence-based psychological strategies through web-based platforms and/or mobile apps.
2. Virtual or augmented reality interventions for exposure to feared stimuli [46–48].
3. Serious games, training psychological strategies in a video game format [49].
4. Avatar-led therapy sessions [50] or chatbot-mediated interventions [51].
5. Automated memory, feedback, and reinforcement interventions delivered through apps, e-mails, text messages, or short prompts, which support the participant in incorporating intervention content into everyday life [52].

6. Phone and wearable sensors, as well as apps to monitor symptoms or motivate health behavior, such as homework completion or healthy behaviors to support the therapeutic process [53].

Videoconferencing [54] and other telehealth services, which primarily assist communication between therapist and patient may also be classified as IMIs. However, such services share greater similarities with conventional on-site psychotherapy and will therefore not be discussed here.

Human Support

IMIs can integrate varying degrees of human support. A commonly used method is guided self-help approaches. Such interventions provide multimedia-based self-help material, so that most tasks and techniques can be performed independently. A clinical psychologist, health professional or trained lay health worker then provides feedback or guidance in a regular interval. A major aim of human support in stand-alone IMIs is to foster adherence to intervention contents [55–57]. To reach this, communication can happen either synchronously (per chat or video) or asynchronously (e.g., via e-mail). Asynchronous communication formats are more commonly used and usually take from a few minutes to a few hours per participant and intervention. For users, the processing of self-help material, execution of exercises, and correspondence with a therapist can be very intense. IMI users may thus invest much greater time than the supporting therapist. Nevertheless, asynchronous contact and time-independent communication allow for increased flexibility and autonomy for both participants and therapists.

Area of Application

Possible applications of IMIs in the mental health field are manifold. IMIs can be used, among other things, for mental health promotion, prevention and treatment of mental disorders, prevention of relapse or recurrence of mental disorders, or chronic illness management. From a public health perspective, IMIs can be considered a promising approach to increase the accessibility of evidence-based psychotherapeutic techniques in the general population due to their low threshold for accessibility, location and time independence, and anonymous usability [58]. IMIs can be applied as stand-alones, within a stepped-care approach, or as blended treatment approaches, in which IMI and conventional face-to-face psychotherapy components are combined.

Stand-alone IMIs. A large asset of stand-alone IMIs is their independence from space and time. Stand-alone IMIs can be accessed through the Internet or a smartphone app at any time, and everywhere, thus facilitating the access to evidence-based interventions for individuals with limited mobility or people living in underserved areas. Such approaches could also overcome common psychological barriers by helping people with difficulties expressing themselves or individuals not appreciating social or human contact [59]. People who are not inclined to use conventional face-to-face psychotherapy due to reasons such as fear of stigma could access IMIs as an alternative. Despite the increasing social acceptance of psychotherapy, having mental health problems still causes a feeling of shame in many individuals, resulting in help-seeking barriers [60]. Using distant technologies such as IMIs may be a suitable way to address such issues. Stand-alone IMIs can also be disseminated as massive open online interventions (MOOIs) to offer free behavioral health services worldwide at minimal costs [61]. They could therefore be an innovative instrument to increase access to evidence-based interventions in countries with defunct or insufficient health care services around the globe [62].

Blended Care. In blended concepts, face-to-face psychotherapies are combined with IMIs [63]. The extent to which treatment is provided through online-based components can vary in such concepts. In some blended concepts, face-to-face treatment remains the primary delivery mode, and only some treatment elements are supported through digital components. For example, traditional psychotherapy can be supported with mobile-based exercises to facilitate adherence to homework assignments. Conversely, IMIs can also serve as the major part of the treatment, and face-to-face meetings are scheduled as a support element. In blended care concepts, IMIs may either replace parts of psychological treatment, especially parts of the treatment which do not require mediation by a psychotherapist. This enables that time during therapy sessions can be used more efficiently, providing more time for face-to-face process work. However, conventional treatment could also be augmented through application of IMIs to improve its effectiveness. IMIs can be used in such contexts to provide exercises for the participant to work on between sessions, or to support the integration of behavior changes into daily life. Instruments to achieve this are mobile-based diaries, ultrashort prompts aimed at training previously learned strategies, or mobile coaches leading patients through difficult situations. This allows expanding the "therapeutic arm" into patients' everyday experiences and behaviors. Another promising application of blended concepts might be the delivery of psychological IMIs in chronic somatic care [64–68].

Stepped and Matched Care. Stepped-care approaches adapt the degree of therapeutic support based on previous intervention effects, while matched care matches patients based on baseline characteristics such symptom severity or comorbidity indicators to specific treatment formats such as self-help-guided self-help or blended care. For depression, for example, self-help IMIs can be offered in a stepped-care approach as a first step to individuals showing prodromal symptoms of the disease to prevent the onset of a major depressive episode [69, 70]. Intensive therapeutic support is provided in case patients do not respond to IMIs. As step-down interventions, IMIs can be used for relapse prevention and chronic care in remitted patients to stabilize treatment effects [71–76].

Efficacy

Internet Interventions Compared to Control Groups

The substantial potential of IMIs to prevent and treat mental and behavioral disorders is documented by countless randomized controlled trials (RCTs) which have been conducted within the last 20 years. The efficacy of guided self-help interventions is particularly well established. Common mental disorders, such as depression and anxiety disorders, have been studied frequently in IMI research. In such studies, IMIs have been found to be highly efficacious when compared to untreated controls (e.g., [77–82]). A recent meta-analysis, comprising ten studies, found a statistically significant pooled effect of Hedges' $g=0.90$ (95% CI 0.73-1.07, $p<0.001$) favoring IMIs compared to waitlist controls in the treatment of major depression [83]. Compared to waitlist controls, a recent Cochrane review on Internet-based therapist-assisted self-help for anxiety disorders found a large effect of $g=1.06$ (95%CI: 0.92–1.29, $p<0.001$; 28 studies) favoring IMIs [82].

Table 1 Efficacy of IMIs based on selected meta-analytic reviews

Target population	Study authors	SMD	[95% CI]	*k*	*N*	I^2
Adults						
Major depression	Königbauer et al. (2017)	0.90	[0.73; 1.07]	10	727	0
Panic disorder	Olthuis et al. (2015)	1.52	[0.48; 2.56]	6	323	93
Social phobia	Olthuis et al. (2015)	0.92	[0.74; 1.09]	8	661	48
Generalized anxiety disorder	Olthuis et al. (2015)	0.80	[0.42; 1.19]	6	394	69
Posttraumatic stress disorder	Kuester et al. (2016)	0.95	[0.56; 1.43]	8	936	91
Insomnia	Zachariae et al. (2015)	1.09	[0.74; 1.45]	8	1071	83
Eating disorders	Melioli et al. (2016)	0.31[a]	[0.21; 0.42]	16	1643	0
Hazardous alcohol use	Riper et al. (2014)	0.20	[0.13; 0.27]	16	5612	27
Obsessive-compulsive disorder	Own calculations[b]	0.90	[0.61; 1.19]	3	122	0
Chronic pain	Buhrmann et al. (2016)	0.42	[0.28; 0.55]	15	2213	54
Irritable bowel syndrome	Own calculations[c]	0.74	[0.37; 1.11]	4	353	58
Children and Adolescents						
Depression	Ebert et al. 2015	0.76	[0.41,1.12]	4	796	61
Anxiety	Ebert et al. 2015	0.68	[0.45,0.92]	7	796	0

[a]Purging
[b]Own calculations (Hedges' g using Comprehensive Meta-Analyses 2.0) based on primary study results of Andersson et al. (2012), Herbst et al. (2014), and Lenhard et al. (2017)
[c]Own calculation of between-group effect sizes, based on studies reported in Hedman et al., 2012. SMD = standardized mean difference (Cohen's d/Hedges' g); CI = confidence interval; k = number of primary randomized trials; N = number of participants in primary studies included in the meta-analysis

Meta-analytic evidence also supports the effectiveness of IMIs for a plethora of other indications, including posttraumatic stress disorder (PTSD), sleep disorders, eating disorders, chronic pain, or substance abuse [84–88]. Table 1 presents current meta-analytical evidence on the effects of IMIs for different disorders compared to untreated controls.

Individual RCTs also provide promising evidence for the efficacy of IMIs for a range of other disorders, such as obsessive-compulsive [89–92], dissociative [93, 94], body dysmorphic [95], and bipolar disorders [96], as well as for sexual dysfunction [97–101], tinnitus [102–111], complicated grief [112–114], pathological gambling [115, 116], fatigue [117, 118], as well as suicidal thoughts and behaviors [119].

However, for some indications, the substantial heterogeneity of effects between studies should be considered. Not all IMIs result in similar effects, and more research is warranted to disentangle which intervention works for whom, and in which context. Many of the studies mentioned above used waitlist control designs to determine the effectiveness of IMIs, which is a rather weak control group design. Treatment expectancies have been discussed as an artifact in such trials, because participants with delayed access to treatment might be less motivated to initiate behavior changes [120]. This may cause such studies to overestimate the effects of IMIs compared to what can be expected in routine care.

Internet Interventions Compared to Face-to-Face Treatments

The potential of IMIs is also illustrated by their efficacy compared to face-to-face therapy. In 2014, Andersson and colleagues synthesized evidence from 13 RCTs focusing on various disorders (including depression, phobias, tinnitus, anxiety disorders, and sexual dysfunction disorders), and found no differences in effect size for direct comparisons between guided IMIs and face-to-face psychotherapy ($g=0.01$; 95% CI 0.13–0.12; [111]). These results were further corroborated by a Cochrane review for IMIs targeting anxiety disorders in adults ($g=0.06$; 95% CI 0.37–0.25; in the direction of favoring face-to-face; [82]), a meta-analysis by Andersson and colleagues for the treatment of depression ($g=0.12$, 95% CI 0.06–0.30; in the direction of favoring guided IMIs [121]), and an update of the initial meta-analysis in 2018 by Carlbring, Andersson, and colleagues [122]. Even though the number of RCTs evaluating IMIs in direct comparison to face-to-face psychotherapy is still limited, all current meta-analyses indicate that IMIs and conventional psychotherapy are equivalent in their effects. However, these results do only count for individuals who are willing to participate in Internet-based or on-site treatment. IMIs, which usually have a strong self-help focus, are not an option for some patients [123–126], while face-to-face treatment is not an attractive intervention format for all either [31]. A recent survey among 641 patients with depressive symptoms in primary care, for example, revealed that younger individuals with higher levels of education could be more likely to prefer IMI-based treatments than other patient groups [127].

There is a small, but steadily increasing number of studies suggesting that blended concepts can also serve as an effective approach to increase the effects of current state-of-the-art treatments [63]. Sethi and colleagues [128] found, for example, that blended face-to-face and iCBT was superior to both face-to-face therapy and iCBT alone in the treatment of depression and anxiety in youth. A meta-analysis by Lindhiem and colleagues [129] supports that increasing the effectiveness of psychological interventions through blended concepts might be possible. Synthesizing 10 RCTs, they found that psychological interventions were considerably more effective for various indications when on-site sessions were supplemented by mobile-based treatment components (SMD = 0.27, $p < 0.05$). Preliminary evidence also suggests that blended concepts can reduce the time clinicians have to invest for each client without compromising treatment efficacy [130, 131]. However, Kenter and colleagues [132], comparing blended CBT in routine care to routine face-to-face data in a naturalistic study, in which therapist was not asked to follow a specific standardized concept, found that, while outcomes were equivalent, blended concepts were associated with a higher number of sessions and more time invested by the therapist. This was the case because Internet based contents were only provided in addition to face-to-face therapy. Such findings underline the importance of developing a stringent implementation model aimed at allowing therapists to delegate tasks to digitalized tools. More valuable insight on the potential of blended concepts can be expected from a number of large international studies investigating the topic [133–136] which have currently been finalized, or are still ongoing.

Prevention of Mental Disorders

Ebert, Cuijpers, Muñoz, and Baumeister [137] identified 10 RCTs focusing on the effect of IMIs on the incidence of mental disorders in asymptomatic or subclinical populations. Six of these studies reported positive results, with the number needed to treat (NNT) to avoid one additional disorder onset ranging from 9.3 to 41.2. Since then, Buntrock, Ebert, and colleagues found a guided IMI to be effective in reducing the risk of developing depression by 41% within 1 year [70, 138, 139]. They also found the intervention to show an acceptable cost-benefit ratio [140]. Findings on the prevention of relapse in patients with fully or partially remitted depression using IMIs are mixed. While Holländare and colleagues [141] found that guided iCBT led to significantly reduced relapse rates during a 2-year follow-up compared to controls, Klein and colleagues [142] did not find such effects when evaluating an IMI providing less human support to patients, despite favorable short-term results.

In sum, current evidence indicates that, potentially, prevention of mental disorders using IMIs is possible. Furthermore, there is a large body of evidence supporting the effectiveness of IMIs in promoting health behavior, such as the reduction of problematic alcohol consumption [143, 144], improving sleep [85, 145, 146], and reducing work-related or academic stress [147–152], all of which might contribute to preventing mental disorders from a broader perspective as well.

Nonetheless, the research on preventive IMIs is still in its infancy, making it impossible to draw definite conclusions. For both the prevention of depression and anxiety, observed effects of psychological preventive interventions in general tend to be lower in studies with follow-up periods longer than 1 year, indicating that current available psychological interventions might potentially only delay, and not prevent the onset of the disorder [153, 154]. As by now, whether IMIs actually have the potential to prevent mental disorders can thus not be ultimately decided.

Cost-Effectiveness

Mental disorders result in enormous societal costs [155–157], both directly (e.g., through general practitioner visits, medication, social services, stationary treatment) and indirectly (e.g., through sickness leave, worse job functioning, or mortality). In European countries alone, the total costs of depression are estimated to sum up to as much as 118 billion Euro (app. 137 billion US-Dollars [157]). As health care resources are limited, and thus have to be used economically, health economic analyses of IMIs allow to analyze if intervention costs stand in proportion to their effectiveness [158]. There are two recent systematic reviews supporting the cost-effectiveness of guided IMIs [159, 160]. Paganini and colleagues found a net benefit of 3,088–22,609 Euro (3,581–26,288 US-Dollars) per participant for guided depression prevention IMIs. These findings support the implementation of IMIs from a health economic perspective. However, current evidence does not allow to come to ultimate conclusions if IMIs are the most economical of all available treatment approaches for mental disorders [159].

Routine Care

Accumulating evidence suggests that IMI-based treatment can also result in clinically relevant improvements for many disorders when implemented into routine care [161–168]. In a recent paper, Titov and colleagues [169] describe that in Sweden, Denmark, Norway, Canada, and Australia, online clinics providing guided IMIs have already been successfully implemented, and have shown to be effective in treating patients under routine conditions. Whether unguided IMIs also have this potential is much less clear. Meta-analyses have shown that unguided IMIs can be effective, for example, in addressing anxiety disorders [82], or depression [170]. It should be considered, however, that such evidence is based on RCTs, in which researchers indirectly provide a high level of structure and positive regard to subjects as part of the study procedures. These factors are unlikely to be found in routine clinical care. Human support functions as an adherence-promoting element in IMIs. It is therefore likely that the effect sizes of IMIs which do not provide any human contact are overestimated in RCTs when compared to their potential in routine care [171]. Lin and colleagues [172], for example, found that adherence to an unguided IMI for chronic pain was remarkably low when the intervention was not provided in a clinical trial context, with virtually none of the patients completing the intervention. In a pragmatic study, Littlewood and colleagues [173] found no additional benefit of an unguided IMI compared to standard treatment. However, when the same intervention was delivered with additional telephone support in addition to general practitioner care, it was found to be superior to the pure unguided intervention [174]. Based on current evidence, preference in routine care should thus be given to guided interventions. Nevertheless, unguided interventions do also offer some advantages, including their low costs. Research by Muñoz and colleagues [61] suggests that unguided IMIs can be used as MOOIs to facilitate the provision of evidence-based mental health interventions worldwide. Such interventions could be combined within digital apothecaries [175] to promote behavioral health and address mental disorders through prevention and treatment, especially in countries with limited access to free health care. Overall, more research is warranted into how digital health care services can be optimally integrated into routine care [176]. *ImpleMentAll*, a current large project investigating the implementation of IMIs into routine care in 11 European countries, may provide more insight into this matter in the future.

Acceptance-Facilitating Interventions

Despite the increasing implementation of IMIs into routine care, uptake is still often low, with rates ranging from 3 to 25% [177–181]. Psychological factors are a common barrier toward seeking treatment [31], and thus may also apply to IMIs. Research also documents that many individuals in the general population have little knowledge about IMIs [125, 126]. Many misperceive IMIs as being less effective than conventional therapies [126], including therapists [182, 183], although previously presented evidence clearly shows that this is not the case. In a survey of the German general population ($N=646$), Apolinário-Hagen and colleagues [126] found that awareness of Internet-based treatment is associated with higher preference for IMIs. This finding points to the importance of developing measures to increase awareness of and knowledge about the efficacy of IMIs in the public to raise their acceptance. Recent developments, such as acceptance-facilitating interventions (AFIs) using brief, highly scalable educational videos have been shown to be a valid strategy to enhance the acceptability of Internet interventions in clinical practice [124, 184–187]. AFIs can be easily disseminated through official health care information channels. Therefore, they might be a promising approach to increase the public acceptance of IMIs and their utilization.

Limitations and Possible Negative Effects

Although IMIs may have immense benefits for individual patients and the health care system, risks and potential negative effects must also be properly considered. As by now, however, no reliable information is available on potential contraindications for IMIs, warranting further research. An individual patient data meta-analysis by Ebert and colleagues [188] found that guided IMIs significantly reduce the risk for symptom deterioration compared to controls. In this analysis, no patient subgroup with increased risk for deterioration was detected. However, it was found that education level moderated effects on deterioration, meaning that participants with a low level of education should be more closely monitored, as they might show a greater risk for deterioration. In the context of stand-alone methods with no human support, it has often been argued that emergencies, such as acute suicidality, cannot be addressed appropriately in IMIs. Suicidality is therefore often considered as an exclusion criterion for stand-alone IMIs. Studies show, however, that IMIs can be used effectively in treating suicidal patients, and can decrease suicidal thoughts and behaviors [189, 190]. Further research should therefore explore in which contexts IMIs can be used safely in patients showing suicidal ideation. Overall, although initial studies address the subject of negative effects of IMIs [188, 191–193], little more can be said about this topic right now. In addition, there is currently no consensus about the reporting of adverse events in clinical trials of IMIs. Other negative effects of IMIs, including imprecise diagnoses, lowered health-related self-efficacy or negative attitudes toward psychological interventions in patients who do not profit from IMIs, excessive demands on IMI users, increased dependence on technology, or the delivery of potentially harmful techniques may be possible, but further research is needed to shed more light on this topic.

Mechanisms of Change

In the past, IMI research has primarily focused on the question if such formats can be effective at all. Current research is increasingly concerned with establishing mechanisms of change in IMIs. However, such research questions are still in their beginnings, and it cannot be fully answered what makes IMIs work. As most IMIs integrate evidence-based psychotherapeutic techniques into their rationale, one can assume that the mechanisms of change underlying those techniques are an active component in IMIs. Yet, confirming specific effects of an active component is notoriously difficult in psychotherapy research (cf. [194] p. 241ff.). A recent systemic review by Domhardt et al. [195] on IMIs for anxiety suggests, seemingly, that the different psychotherapeutic techniques are not active ingredients of IMIs (neither CBT vs. other techniques, nor disorder-specific vs. transdiagnostic techniques). Furthermore, cognitive factors (e.g., negative automatic thoughts, rumination, or dysfunctional attitudes), emotion regulation, expectancy, attributional style, coping strategies, perfectionism, therapeutic alliance, and treatment credibility have been examined as mechanisms of change [108, 196–205]. It seems plausible to assume that nonspecific factors equal to those of face-to-face psychotherapy also contribute to the efficacy of IMIs. Nevertheless, when focusing on mechanisms of change in IMIs, features which are special to such interventions should also be properly considered. Here, we will describe three mechanisms through which IMIs may contribute to individuals' mental health and well-being: self-empowerment, reinforcement mechanisms, and human guidance.

Self-empowerment

Previously, we described that IMIs do not have to be inferior to on-site psychological interventions in section "Internet Interventions Compared to Face-to-Face Treatments". One explanation for this finding could be that IMIs, while providing less human contact, put a stronger emphasis on self-empowerment. Empowering patients through IMIs may therefore foster users' self-management skills, with then translate to treatment benefits. Alpay and colleagues [206] delineate specific components of empowerment which IMIs may use to foster self-management, such as increasing health or disorder literacy, promoting coping self-efficacy, and specific problem-solving skills, as well as providing heuristics for decision-making. There is evidence showing that IMIs can contribute to the empowerment of patients with mental disorders [207] or chronic diseases [208, 209]. However, whether amelioration of self-management is indeed a mechanism of change contributing to the efficacy of IMIs is still largely unclear. Further process studies are needed to gain a better understanding of the mechanisms underlying patient empowerment in IMIs.

Reinforcement Mechanisms

In contrast to traditional psychotherapy, IMIs allow for sending automatized reminders, feedback, and reinforcements (so-called *prompts*) to increase the application of previously learned strategies and techniques into daily life. One way to implement this is through ultra-brief exercises sent to users' mobile phones via SMS. Another important way is reminders to support the usage of IMIs, which can increase treatment adherence. At the same time, such applications are very economical and easily programmed. A meta-analysis by Cowpertwait and Clarke [210] supports the relevance of prompts in driving IMI effectiveness. In subgroup analyses, they found that IMIs for depression in which reminders were integrated showed larger effect sizes ($g=0.49$) than interventions without reminders ($g=0.24$).

Human Support

Technology-mediated human support (guidance) can be considered a more well-studied mechanism of effective IMIs. Meta-analytic research suggests that guided IMIs show greater effects than IMIs with no human support element [80, 147, 152, 210, 211]. In these analyses, guided IMIs also show lower attrition rates and higher numbers of completed modules ($g=0.52$). For the treatment of depression, guidance seems to be particularly influential. Richards and Richardson (2012) found a small effect of unguided IMIs for depression ($g = 0.36$), but high effects ($g = 0.78$) for guided interventions. A more recent meta-analysis [83] could not corroborate this finding, but this might have been due to limited power. No significant differences in effects between unguided and guided IMIs could be found in the treatment of anxiety in one meta-analysis [82], while a more recent analysis suggests that guided IMIs are superior over unguided IMIs [195]. This suggests that the effect of guidance might be more complex than the simple generic superiority hypothesis suggests.

Evidence from systematic reviews indicated that the efficacy of IMIs increases with the degree of human guidance provided. Yet, Andersson and colleagues [212, 213] estimated that exceeding 100 min of support per participant in a 10-week IMI does not translate to higher treatment effects anymore. It is still unclear, however, which dosage of guidance is optimal at which stage of treatment. Further research is needed to discern which characteristics of human support are most important: is it the quantity, or the quality (viz., coach qualification) of human support that matters? Which communication medium (e-mail, face-to-face, video chat, phone) and communication mode (asynchronous or synchronous) should be used, and does this vary for different disorders and patient subgroups?

Despite this lack of in-depth knowledge concerning the intricate details of human support in IMIs, there is evidence showing that therapeutic alliances of high quality can be achieved with such interventions. Although IMIs provide less therapeutic contact, and social and nonverbal cues are often completely absent, the quality of therapeutic relationships in IMIs is often comparable to face-to-face treatment [214–219].

Some models have been proposed to conceptualize and optimize human support in IMIs, such as the supportive accountability [220] and efficiency model [56]. According to the supportive accountability model, human support can be used to increase adherence to an intervention by providing accountability through a coach who is perceived as caring, trustworthy, and competent. The model states that patient motivation plays a crucial role in determining the importance of guidance. While patients who are intrinsically motivated to use an IMI do not require as much feedback and guidance to adhere to the intervention, patients for which this is not the case need more human support to uphold extrinsic motivation to stay engaged. The efficiency model of support, on the other hand, might be a valuable framework to guide researchers in developing optimally tailored, efficient support systems for IMIs.

Ethical and Professional Considerations

As a novel approach to the prevention and treatment of mental disorders, new risks and opportunities arise from the implementation of IMIs. Since mental and physical health plays an important role in many individuals' lives, there is a large demand for freely available technological instruments to promote one's own well-being. Thus, many start-ups and corporations have started to capitalize on this interest by providing an ever-growing number of commercial Internet portals, mobile health apps, or online life coaching services. For costumers, however, it is often hard to see which services are based on sound empirical evidence, and which ones are not. Recent studies have shown that less than 4% of all commercial apps for symptoms of depression [221] and anxiety [222] have been subject to rigorous clinical examination like the studies we presented in 3.1–3.5 [223]. To protect costumers from dubious products and to provide assistance in finding evidence-based, effective services, consolidated quality standards for commercially available IMIs are needed [224]. Such standards have been recently established in countries such as the Netherlands, the United Kingdom, and Germany [224].

On the other hand, the enormous potential of strictly evidence-based IMIs also suggests that withholding such interventions as a complementary treatment option for patients is also ethically questionable. As we have shown before, IMIs can lead to comparable outcomes as state-of-the-art services and have the potential to reach individuals who may not want to use conventional treatment approaches. Therefore, one should differentiate if IMIs are used as a supplement to, or a replacement of current treatments. Using IMIs to improve existing health care infrastructure might be beneficial to both patients and the health care system. Aiming to replace conventional treatments with IMIs because they are more economic, of course, is much more critical from an ethical perspective.

Future Advances: Machine Learning and Data-Driven Applications

Most IMIs developed today still resemble conventional face-to-face psychological interventions in many ways. Typical IMIs contain bundles of various psychological techniques and modules (e.g., psychoeducation, behavioral activation, cognitive restructuring, progressive muscle relaxation) adapted from evidence-based face-to-face interventions. Such modules are then assembled into a "treatment package" and delivered on a digital platform. Although the evidence we presented before clearly documents the effectiveness of such approaches, recent advances in the field of statistical learning may allow to make IMIs even more pervasive and impactful in the lives of their users in the future.

The last decades have seen the advent and proliferation of "Artificial Intelligence", especially in terms of the development of Machine Learning methods [225]. Machine Learning algorithms automatically determine ("learn") parameters from data to optimally fulfill a certain task. They also have a tight focus on predicting outcomes in unseen data and are thus often referred to as *predictive modeling* methods [226, 227]. Applications of predictive modeling have been facilitated by the increased volume, variety, and velocity (the *Three V's* [228]) of data collected, stored, and processed. In this respect, IMIs have an enormous advantage compared to traditional face-to-face psychotherapy, as they easily allow for the collection of unprecedented amounts of fine-grained patient and process data [229]. IMIs are therefore a key field in which Machine Learning and

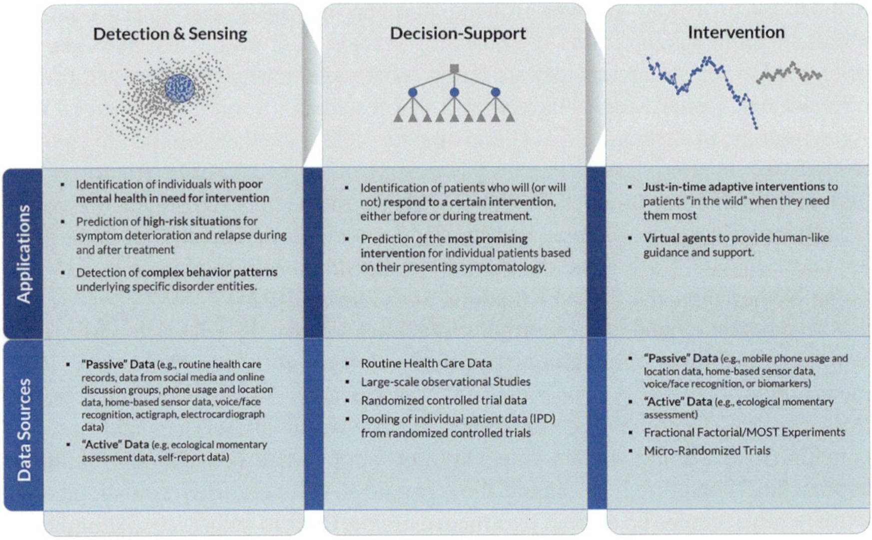

Fig. 2 Overview of artificial intelligence applications in psychological intervention research

data-driven approaches can be applied to advance the prevention and therapy of mental disorders as a whole.

In the following, we will provide a brief overview of recent advances in data mining and Machine Learning to (i) detect mental disorders, (ii) generate data-driven decision-support systems, and (iii) optimize intervention approaches, focusing on Internet- and mobile-based measures. Presented findings were gathered by searching two large bibliographical databases (*PubMed* and *PsycInfo*) on August 26th, 2018, and screening previous review articles on overlapping topics. Nevertheless, our overview is not meant to be exhaustive (Fig. 2).

Detection and Sensing

Although there is no dearth of evidence-based, effective measures to treat mental disorders, research still documents a large treatment gap in the general population, with 54–77% of all mental disorders remaining undiagnosed and untreated [28–30]. As mental disorders have a long-standing negative impact on individuals' physical health, mortality, and social functioning [4–10], early detection of mental disorder symptoms is of paramount importance. Due to their capability to accrue meaningful information from large and complex data sets, Machine Learning approaches could therefore be successfully applied to overcome current limitations in the:

- Identification of individuals with poor mental health in need for intervention.
- Prediction of high-risk situations for symptom deterioration and relapse during and after treatment.

- Detection of complex behavior patterns underlying specific disorder entities.

Predictive models for the identification and prediction of poor mental health can be derived from two types of data sources: *passive data* (such as routine health care records, data from social media and online discussion groups, phone usage and location data, home-based sensor data, voice/face recognition, actigraph, and electrocardiograph data), and *active data* (such as ecological momentary assessment (EMA) data, self-report data).

Identification of individuals with poor mental health. Although the field is still in its infancy, a rising number of studies suggests that Machine Learning applications may play a pivotal role in detecting mental health issues in the general population in the future [230]. Social media in particular provides an unprecedented opportunity to crawl real-life mental health-related data, extract and select relevant features, and train valid models to detect untreated individuals facing mental health issues on the Internet using Machine Learning algorithms. Already, social media data have been used to determine symptoms of depression [231–240], stress and well-being [241–244], PTSD [245, 246], eating disorders [247, 248], or suicidality [249–254], to name a few indications. Burnap and colleagues [253], for example, developed a machine classifier to distinguish factual reporting of suicide on Twitter from content reflective of suicidal ideation in users. Employing face detection, color and metadata analysis, Reece and Danforth [240] extracted statistical features from more than 40.000 photos posted by 166 individuals on Instagram. With this data, they could train a model predicting the prevalence of depression with higher accuracy than general practitioners. The model also outperformed clinicians when only photos posted by individuals before they were diagnosed with depression were used. Kornfield and colleagues [255], using a bag-of-words-based Machine Learning approach, were able to identify struggling alcohol abuse patients in an online discussion forum with an excellent accuracy of AUC = 0.92.

Routine health care data sets, on the other hand, can also be used to better detect mental health issues [256–261], and improve their diagnosis [262, 263]. Chekroud and colleagues [259], for example, obtained large data sets from US National Health Surveys. Using various Machine Learning techniques, they developed a prediction model to identify individuals with depression who do not receive needed treatment. The learned model reached a prediction accuracy of 72% in detecting untreated patients. This approach may be particularly promising, as patients identified with such models could be offered IMI-based services to overcome structural or psychological barriers to treatment on a population level. Jiménez-Serrano and colleagues [264], for example, developed a classification model for mothers at high risk for postpartum depression the first week after childbirth, along with a tailored mobile-based app to more easily provide interventions to identified cases.

Prediction of high-risk situations for symptom deterioration and relapse. As technological advances have expanded the ability to collect and process real-time behavioral and biological data, predictive modeling is also a promising approach to forecast high-risk situations for mood decrements, symptom deterioration, or relapses. Such approaches may be used either during treatment, or for relapse prevention in remitted patients. Portable sensors play an important role in this development. *Wearables*, such as Smart Watches or Fitness Trackers, can serve as accelerometers, pedometers, actigraphs, GPS trackers, or electrocardiographs, thus producing enormous amounts of diverse personal data to predict mood states and behavior [265, 266]. Smartphones, a ubiquitous companion for most people, can also be turned into mobile sensors using applications developed for this purpose in recent years (e.g., *Purple Robot* [267] or *Beiwe* [268]). By aggregating such information, general or personalized predictive models [229, 269] can be learned to extrapolate mood states [269], refine mental disorder diagnoses [270, 271], or predict relapse [272, 273], to name only a few applications. Predictive modeling has already been applied to detect symptoms of depression [274–277], stress [278–280], and autism [281], to predict sleep quality [282] or relapse risk in patients with substance addictions [270] and psychosis [283]. Pratap and colleagues [275], for example, collected passive phone usage data from 241 patients with depression. Results of this study show that personalized Machine Learning models can reach promising results in predicting mood changes in depressive patients. Depp and colleagues [284], using daily self-report data from 86 individuals with bipolar disorder, were able to reach an excellent AUC (area under the curve; defining the quality of a screening assessment to predict the presence of the respective condition) of 0.91 in predicting heightened suicidality in patients, thus doubling the predictive accuracy compared to only using baseline symptomatology data.

Detection of complex behavior patterns underlying specific disorder entities. Predictive modeling approaches may not only provide ample opportunity to improve the detection of mental health issues as well as the timely prediction of high-risk states, but may also have a paradigm-shifting impact on the way mental disorders are conceptualized. Hofmann and colleagues [285] propose that in the future, mental disorders may be seen less as latent disease entities, but as complex networks of symptoms and behaviors, which can change abruptly once a critical threshold is reached (i.e., through the onset of a disorder or relapses). Fine-grained time series data is needed to predict such complex network transitions. Hofmann and colleagues therefore stimulate that personalized predictive models based on experience sampling may be used to achieve this, and may thus help to better understand and predict the complex underpinnings of mental disorders (for applications, see [286, 287]).

Decision Support

Despite an abundance of evidence showing that psychological interventions can be effective in treating mental disorders, there is still little evidence of differential effects between various treatment types [288]. On a patient level, the benefits of psychological interventions are highly heterogeneous, with 18–62% of all patients not responding to the evidence-based treatment they receive [289, 290].

Among depressive patients, only 30% of all patients achieve remission through the first treatment they receive, but roughly 70% after several courses of treatment [291]. Thus, many patients either have to try out multiple treatments until they find the one approach that works for them, or become demoralized by many failed treatment attempts and drop out without any substantial improvements [292]. Knowledge of which intervention strategy works best for which individual thus has an enormous potential to improve the real-life impact of current psychological treatments. Machine Learning approaches have the potential to support clinical decisions on which intervention might be best suited for which patient. Such approaches may be used for the:

- Identification of patients who will (or will not) respond to a certain intervention, either before or during treatment.
- Prediction of the most promising intervention for individual patients based on their presenting symptomatology.

Data to derive such decision-support algorithms can either stem from *clinical trials* (such as RCTs and individual patient data sets containing data from many similar clinical trials [293]) or large-scale observational studies.

Identification of intervention (non)responders. Currently, Machine Learning has already been used to derive risk stratification tools through which patients at risk for nonresponse [294–299], chronic mental illness trajectories [300], or grave emergencies, such as suicide [301] can be identified. Lenhard and colleagues [296], for example, used RCT baseline data of 61 adolescents to establish a Machine Learning algorithm predicting response to an IMI for obsessive-compulsive disorder (OCD). While conventional logistic regression approaches could not detect any predictors, Machine Learning algorithms identified responders with 75–83% accuracy. Månsson and colleagues [297] could predict long-term responders of an IMI for social anxiety disorder using Support Vector Machines and fMRI data, reaching an accuracy of 92%. However, this finding could not be replicated in a similar study by Sundermann and colleagues [299] when aiming to predict response to CBT for panic disorder. In another application, Perlis [302], drawing data from the STAR*D cohort, developed a risk stratification tool predicting treatment-resistant depression cases.

However, risk stratification based on Machine Learning may also be used in patients who already receive treatment. IMIs are well suited for this purpose, as virtually all communication with patients is stored digitally on the Internet and thus usable for prediction. Hoogendoorn and colleagues [295], for example, analyzed the writings of patients undergoing guided IMI treatment for social anxiety disorder and were able to predict the treatment outcome with 78% precision halfway through therapy. Mikus and colleagues [303], using EMA data of depressive patients from the *E-COMPARED* project, were able to predict mood fallbacks in treated patients with similar accuracy.

Prediction of the most promising intervention for individual patients. Predictive modeling may also be used to derive concrete decision-support systems to determine which treatment may best fit which patient based on baseline data. For example, Chekroud and colleagues [304], using a large database of depression patients ($N=4041$), used elastic net regression and gradient boosting machines to derive a parsimonious model predicting response to antidepressive medication, thereby requiring only 25 questionnaire items as input. The model was found to accurately generalize to new samples and showed high specificity. Along with further research [305], the model has since been implemented into a web-based decision-support service for primary care [306]. In another recent study, Bremer and colleagues [307] used data of 350 patients in an RCT comparing treatment-as-usual to blended care for depression (see, in section "Area of Application"). They derived a personalized treatment recommendation model through which patients can be allocated to the most suitable treatment type based on baseline data. Results show that such recommendations can be used to decrease treatment costs (5.42%) without having to sacrifice treatment effects to a substantial degree (1.98%).

Kessler [292] recently delineated future pathways to establish reliable decision-support systems for the treatment of depression using Machine Learning. While current approaches have mostly been based on RCT data comprising no more than 200–800 participants, much larger data sets are needed to establish valid allocation systems based on predictive modeling. He therefore stimulates that large pragmatic trials and observational studies should be conducted. On this basis, prediction models could be established to determine which patient subsets might be suited for inexpensive first-line monotherapies, such as exercise or minimally guided IMIs, and which patients will likely need more intensive care. Another possibility may be to compound large databases using individual patient data (IPD) from previous RCTs [293]. However, this requires studies to provide a shared set of baseline and outcome variables to be used for modeling. Recently, Ebert and Cuijpers [308] thus proposed that researchers should include broad assessments of variables which may predict differential treatment outcomes when conducting clinical trials in the future.

Intervention

Irrespective of being digital or face-to-face, current psychological treatments are still largely incapable of providing real-time support for patients between sessions, especially in situations in which individuals may be most at risk. Previously, we described the enormous potential of Machine Learning methods in the prediction of mood and behavior, and in deciding which interventions might be most promising throughout varying contexts. These capabilities may therefore be combined in data-driven psychological interventions to:

- Deliver just-in-time adaptive interventions to patients "in the wild" when they need them most.
- Develop virtual agents to provide human-like guidance and support.

Data needed for such applications may again be categorized as either passive (e.g., mobile phone usage and location data, home-based sensor data, voice/face recognition, or biomarkers) or active (e.g., self-report data). In addition, data can be obtained from innovative research designs such as micro-randomized trials [309] or multiphase optimization strategy (MOST) designs [310], which allow for a data-driven understanding of which intervention components are effective, and when.

Just-In-Time Adaptive Interventions. To prevent symptom fallbacks from occurring, and to provide help in real time to patients "in the wild", *Just-In-Time Adaptive Interventions* (JITAIs [311–317]) have gained increasing attention in recent years. JITAIs aim to predict changes in an individual's status to deliver personalized support when a person needs it most, or is most likely to be receptive [312]. Previously, we described the potential of Artificial Intelligence in predicting mood and behavior (see, in section "Detection and Sensing"), as well as for guiding treatment decisions (see, in section "Decision Support"). In JITAIs, these components are combined to create a personalized digital intervention framework. Nahum-Shani and colleagues [312] named four core components of JITAIs: *decision points*, *interventions*, *tailoring variables*, and *decision rules* (for a model, see Fig. 3). Juarascio and colleagues [317] recently delineated ways in which JITAIs could be integrated to optimize existing psychological treatments in the future. As JITAIs are not aimed at providing in-depth therapeutic content, they could primarily be implemented to bridge the time between longer intervention sessions, thereby picking up previously learned techniques. JITAIs could thus be used in conjunction with standard IMI or face-to-face treatment, either during the first weeks of treatment to promote skill-building and uphold adherence, or after termination of treatment to prevent relapse. On the other hand, JITAIs may also be used as low-intensity entry points to treatment and could be enhanced with full-blown

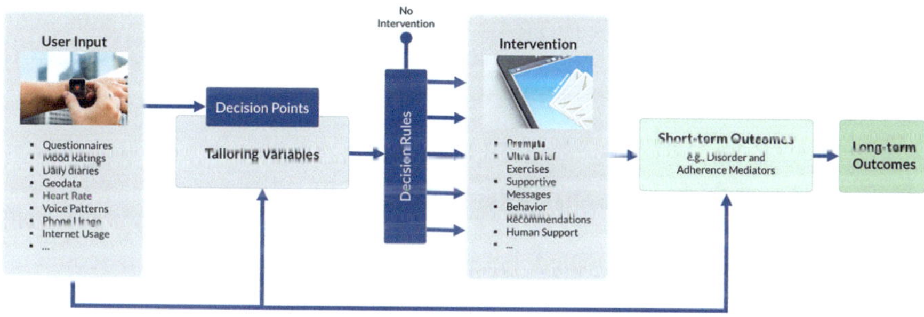

Fig. 3 Generic model of a just-in-time adaptive-intervention (adapted from Nahum-Shani et al. [339]). Description. *Decision points* are either prespecified time intervals and dates or triggered by user input, such as sensor data (e.g., reaching a critical skin conductance threshold). *Interventions* may range from quick alerts and nudges to tailored advice, coping strategies, or eCoach referral, and can be delivered through the Internet. *Tailoring variables* can be personalized progressively by various types of user input. They build the basis for *decision rules*, demarcating which intervention should be provided for which patient, and when

IMIs or psychotherapy in case more help is needed. There is first evidence showing that JITAIs can be successfully applied to mental and behavioral health promotion [318–323]. For example, Wahle and colleagues [318] evaluated a just-in-time intervention app for depression in $N = 126$ individuals. They used GPS, activity, and passive phone usage data to establish predictive models, and delivered CBT-based intervention components (e.g., behavioral activation, breathing exercises) at real time to individuals predicted to be at risk for depression. Participants in this feasibility study who kept using the intervention showed significant decreases in depressive symptoms at posttest. Gustafson and colleagues [321] conducted an RCT evaluating *A-CHESS*, a just-in-time intervention for patients after alcohol use disorder treatment. The intervention tracked geolocation data to trigger alerts when patients neared a high-risk location for relapse. The intervention was shown to have significant benefits on alcohol use compared to treatment-as-usual at posttest and 4-month follow-up.

Virtual Agents. Previously, we described that human support plays an important role in IMIs in section "Human Support" and may be one reason why unguided IMIs are often less effective than guided interventions (see section "Routine Care"). However, even guided IMIs aim to provide human support in a very efficient manner, and mostly resort to highly standardized and adherence-focused communication. Potentially, such communication patterns could also be emulated by Artificial Intelligence to provide patients with a feeling of accountability and assistance, and thus promote adherence. If feasible, this would greatly enhance the value of unguided treatment and free the dissemination of effective IMIs even more from restraints of time and money. First tentative steps have been undertaken in recent years to study the potential of such automated *virtual agents* in mental health research [51, 324–329]. Fitzpatrick and colleagues [51], for example, evaluated *Woebot*, a digital chatbot providing CBT principles through brief daily conversations. The agent uses decision trees and natural language processing to tailor response to patients, and to provide accountability through empathic responses to user input. In an RCT among 70 young adults, *Woebot* was found to significantly reduce symptoms of depression compared to an active control group receiving psychoeducation.

Limitations and Ethical Considerations

Although the evidence presented before points at the enormous potential of Machine Learning, methodological and ethical issues should also be properly considered. As described, Machine Learning approaches require large observational data sets to enable valid personalized predictions. However, even large data are fraught with biases [330, 331], especially when taken from routine care [293], and can lead to a multiplication of spurious correlations and ecological fallacies. This is further aggravated by the fact that currently, many predictive modeling studies in medicine are of suboptimal quality, either in their methodological approach [332], reporting [333], or by using weak comparators to assess prediction fidelity [334]. It should also be noted, as Pearl and MacKenzie [335] recently emphasized, that Machine Learning techniques are purely correlational. Thus, they cannot establish causal relationships [229], which is the core aim of most psychological research. Machine Learning should thus not be seen as a panacea for all limitations in current mental health research. Chen and Asch [336] recently suggested that Machine Learning may have reached its "peak of inflated expectations" in medicine. This, however, should be seen in a positive light, as it allows a soberer look at where Machine Learning can be applied in a meaningful way, and where limitations and risks prevail.

Ethical aspects should play an important role in such considerations. Data security, for example, is an important issue in social media, as has only recently been illustrated by the data breaches involving Facebook [337]. Another risk is that Machine Learning approaches may further fortify social inequality in mental health care. Even today, some minorities and people with a low socioeconomic background are missing or underrepresented in health care data, which may lead to imprecise or false predictions for such individuals, for example within decision-support systems [338].

Overall, Khoury and Ioannidis [330] emphasize that the implementation of Machine Learning and Personalized Medicine approaches, despite their enormous capacities, cannot substitute rigorously conducted clinical research. From an ethical viewpoint, the recent advances in mental health and invention research should also be accompanied by a societal discourse to ensure that Artificial Intelligence can promote mental health for all.

Conclusions

IMIs are flexible, technically diverse methods which lend themselves to a variety of application areas as well as indications of varying degrees of severity. As empirical findings on the impact of human support show, IMIs are seen less as a substitute for conventional psychotherapeutic interventions, and should rather be understood more as a useful addition to the current treatment spectrum. IMIs have an ability to reach target groups in a way not yet achieved by classical on-site activities, and on the other hand, can excellently accompany conventional psychotherapy and thereby reduce cost or increase effectiveness. Findings suggest that stand-alone IMIs can be effective in routine conditions, although further research is needed. As IMIs make it easier to collect and process fine-grained patient data in real time, they are well suited as instruments in innovative Machine Learning applications. Despite important limitations and ethical considerations, Artificial Intelligence-enhanced IMIs thus have an enormous potential to advance mental health care in the next decades.

References

1. Kessler RC, Aguilar-Gaxiola S, Alonso J, Chatterji S, Lee S, Ormel J, Ustün TB, Wang PS. The global burden of mental disorders: an update from the WHO World Mental Health (WMH) surveys. Epidemiol Psichiatr Soc NIH Public Access; 2009;18(1):23–33 PMID:19378696.
2. Steel Z, Marnane C, Iranpour C, Chey T, Jackson JW, Patel V, Silove D. The global prevalence of common mental disorders: a systematic review and meta-analysis 1980–2013. Int J Epidemiol. 2014;43(2):476–93 PMID:24648481.
3. Whiteford HA, Degenhardt L, Rehm J, Baxter AJ, Ferrari AJ, Erskine HE, Charlson FJ, Norman RE, Flaxman AD, Johns N, Burstein R, Murray CJ, Vos T. Global burden of disease attributable to mental and substance use disorders: findings from the Global Burden of Disease Study 2010. Lancet. 2013;382(9904):1575–86. https://doi.org/10.1016/S0140-6736(13)61611-6.
4. Saarni SI, Suvisaari J, Sintonen H, Pirkola S, Koskinen S, Aromaa A, Lönnqvist J. Impact of psychiatric disorders on health-related quality of life: general population survey. Br J Psychiatry. 2007;190:326–32.
5. Hysenbegasi A, Hass SL, Rowland CR. The impact of depression on the academic productivity of university students. J Ment Health Policy Econ. 2005;8(3):145.
6. Eisenberg D, Golberstein E, Hunt JB. Mental health and academic success in college. B E J Econom Anal Policy. 2009;9(1). https://doi.org/10.2202/1935-1682.2191.
7. Ishii T, Tachikawa H, Shiratori Y, Hori T, Aiba M, Kuga K, Arai T. What kinds of factors affect the academic outcomes of university students with mental disorders? A retrospective study based on medical records. Asian J Psychiatr. 2018;32:67–72. https://doi.org/10.1016/j.ajp.2017.11.017.
8. Kessler RC, Foster CL, Saunders WB, Stang PE. Social consequences of psychiatric disorders, I: educational attainment. Am J Psychiatry Am Psychiatr Assoc. 1995;152(7):1026.
9. Fergusson DM, Woodward LJ. Mental health, educational, and social role outcomes of adolescents with depression. Arch Gen Psychiatry Am Med Assoc. 2002;59(3):225–31.
10. Cuijpers P, Smit F. Excess mortality in depression: a meta-analysis of community studies. J Affect Disord. 2002;72(3):227–36 PMID:12450639.
11. Nock MK, Hwang I, Sampson N, Kessler RC, Angermeyer M, Beautrais A, Borges G, Bromet E, Bruffaerts R, De Girolamo G. Cross-national analysis of the associations among mental disorders and suicidal behavior: findings from the WHO World Mental Health Surveys. PLoS Med Public Libr Sci. 2009;6(8):e1000123.
12. Nock MK, Hwang I, Sampson NA, Kessler RC. Mental disorders, comorbidity and suicidal behavior: results from the national comorbidity survey replication. Mol Psychiatry. 2009;15:868. http://dx.doi.org/10.1038/mp.2009.29 Macmillan Publishers Limited.
13. World Health Organization. The Global Burden of Disease: 2004 update. 2008 PMID:15572474.
14. Barth J, Munder T, Gerger H, Nüesch E, Trelle S, Znoj H, Jüni P, Cuijpers P. Comparative efficacy of seven psychotherapeutic interventions for patients with depression: a network meta-analysis. Focus (Madison) Am Psychiatric Assoc. 2016;14(2):229–43.
15. Cuijpers P, van Straten A, Andersson G, van Oppen P. Psychotherapy for depression in adults: a meta-analysis of comparative outcome studies. J Consult Clin Psychol. 2008;76(6):909–22 PMID:19045960.
16. Cuijpers P, Smit F, Bohlmeijer E, Hollon SD, Andersson G. Efficacy of cognitive–behavioural therapy and other psychological treatments for adult depression: meta-analytic study of publication bias. Br J Psychiatry RCP. 2010;196(3):173–78.
17. Hofmann SG, Smits JAJ. Cognitive-behavioral therapy for adult anxiety disorders: a meta-analysis of randomized placebo-controlled trials. J Clin Psychiatry. 2008;69(4):621–32 PMID:18363421.

18. Cuijpers P, Sijbrandij M, Koole S, Huibers M, Berking M, Andersson G. Psychological treatment of generalized anxiety disorder: a meta-analysis. Clin Psychol Rev. 2014;34(2):130–40 Elsevier.
19. Mayo-Wilson E, Dias S, Mavranezouli I, Kew K, Clark DM, Ades AE, Pilling S. Psychological and pharmacological interventions for social anxiety disorder in adults: a systematic review and network meta-analysis. Lancet Psychiatry. 2014;1(5):368–76 PMID:26361000.
20. Bisson JI, Ehlers A, Matthews R, Pilling S, Richards D, Turner S. Psychological treatments for chronic post-traumatic stress disorder: systematic review and meta-analysis. Br J Psychiatry Camb Univ Press. 2007;190(2):97–104.
21. Powers MB, Halpern JM, Ferenschak MP, Gillihan SJ, Foa EB. A meta-analytic review of prolonged exposure for posttraumatic stress disorder. Clin Psychol Rev. 2010;30(6):635–41 Elsevier.
22. Leichsenring F, Leibing E. The effectiveness of psychodynamic therapy and cognitive behavior therapy in the treatment of personality disorders: a meta-analysis. Focus (Madison) Am Psychiatric Assoc. 2005;160(3):1223–1428.
23. Cristea IA, Gentili C, Cotet CD, Palomba D, Barbui C, Cuijpers P. Efficacy of psychotherapies for borderline personality disorder: a systematic review and meta-analysis. Jama Psychiatry Am Med Assoc. 2017;74(4):319–28.
24. Kleinstäuber M, Witthöft M, Hiller W. Efficacy of short-term psychotherapy for multiple medically unexplained physical symptoms: a meta-analysis. Clin Psychol Rev. 2011;31(1):146–60 Elsevier.
25. Olatunji BO, Kauffman BY, Meltzer S, Davis ML, Smits JAJ, Powers MB. Cognitive-behavioral therapy for hypochondriasis/health anxiety: a meta-analysis of treatment outcome and moderators. Behav Res Ther. 2014;58:65–74 Elsevier.
26. Glombiewski JA, Sawyer AT, Gutermann J, Koenig K, Rief W, Hofmann SG. Psychological treatments for fibromyalgia: a meta-analysis. PAIN®. 2010;151(2):280–95 Elsevier.
27. Frühauf S, Gerger H, Schmidt HM, Munder T, Barth J. Efficacy of psychological interventions for sexual dysfunction: a systematic review and meta-analysis. Arch Sex Behav Springer; 2013;42(6):915–33.
28. Auerbach RP, Alonso J, Axinn WG, Cuijpers P, Ebert DD, Green JG, Hwang I, Kessler RC, Liu H, Mortier P, Nock MK, Pinder-Amaker S, Sampson NA, Aguilar-Gaxiola S, Al-Hamzawi A, Andrade LH, Benjet C, Caldas-de-Almeida JM, Demyttenaere K, Florescu S, de Girolamo G, Gureje O, Haro JM, Karam EG, Kiejna A, Kovess-Masfety V, Lee S, McGrath JJ, O'Neill S, Pennell B-E, Scott K, Ten Have M, Torres Y, Zaslavsky AM, Zarkov Z, Bruffaerts R. Mental disorders among college students in the World Health Organization world mental health surveys. Psychol Med. 2016 Oct 3 [cited 2017 Mar 20];46(14):1–16 PMID:27484622.
29. Kessler RC, Berglund PA, Bruce ML, Koch RJ, Laska EM, Leaf PJ, Manderscheid RW, Rosenheck RA, Walters EE, Wang PS. The prevalence and correlates of untreated serious mental illness. Health Serv Res. 2001;36(6):987–1007.
30. Bijl RV, de Graaf R, Hiripi E, Kessler RC, Kohn R, Offord DR, Ustun TB, Vicente B, Vollebergh WAM, Walters EE. The prevalence of treated and untreated mental disorders in five countries. Health Aff Project HOPE-The People-to-People Health Foundation, Inc.; 2003;22(3):122–33.
31. Andrade LH, Alonso J, Mneimneh Z, Wells JE, Al-Hamzawi A, Borges G, Bromet E, Bruffaerts R, de Girolamo G, de Graaf R, Florescu S, Gureje O, Hinkov HR, Hu C, Huang Y, Hwang I, Jin R, Karam EG, Kovess-Masfety V, Levinson D, Matschinger H, O'Neill S, Posada-Villa J, Sagar R, Sampson NA, Sasu C, Stein DJ, Takeshima T, Viana MC, Xavier M, Kessler RC. Barriers to mental health treatment: results from the WHO World Mental Health surveys. Psychol Med. 2014;44(6):1303–17 PMID:23931656, Cambridge University Press.

32. Clement S, Schauman O, Graham T, Maggioni F, Evans-Lacko S, Bezborodovs N, Morgan C, Rüsch N, Brown JSL, Thornicroft G. What is the impact of mental health-related stigma on help-seeking? A systematic review of quantitative and qualitative studies. Psychol Med. 2015 [cited 2017 May 10];45(1):11–27 PMID:24569086.
33. Kessler RC, Berglund P, Demler O, Jin R, Merikangas KR, Walters EE. Lifetime prevalence and age-of-onset distributions of DSM-IV disorders in the national comorbidity survey replication. Arch Gen Psychiatry. 2005;62(6):593–602 PMID:18188442.
34. Kupfer DJ, Frank E, Phillips ML. Major depressive disorder: new clinical, neurobiological, and treatment perspectives. Lancet. 2012;379(9820):1045–55 Elsevier.
35. Ebert DD, Van Daele T, Nordgreen T, Karekla M, Compare AT, Zarbo C, Brugnera A, Oeverland S, Trebbi G, Jensen K, Kaehlke F, Baumeister H. Internet- and mobile-based psychological interventions: applications, efficacy, and potential for improving mental health. Eur Psychol 2018. https://doi.org/10.1027/1016-9040/a000318.
36. Craske MG, Stein MB. No psychotherapy monoculture for anxiety disorders; Authors' reply. Lancet. 2017;389(10082):1883. https://doi.org/10.1016/s0140-6736(17)31208-4 Elsevier.
37. Cuijpers P, Cristea IA, Karyotaki E, Reijnders M, Huibers MJH. How effective are cognitive behavior therapies for major depression and anxiety disorders? A meta-analytic update of the evidence. World Psychiatry. 2016;15(3):245–58 Wiley Online Library.
38. Berking M, Rief W, editors. Klinische Psychologie und Psychotherapie. Epidemiology. Berlin Heidelberg: Springer;2012. ISBN:9783642130175 PMID:15003161.
39. Andersson G, Carlbring P, Lindefors N. History and current status of ICBT. In: Lindefors N, Andersson G, editors. Guid internet-based treat psychiatry. Cham: Springer International Publishing; 2016. p. 1–16. https://doi.org/10.1007/978-3-319-06083-5_1.
40. Andersson G, Paxling B, Roch-Norlund P, Ostman G, Norgren A, Almlöv J, Georén L, Breitholtz E, Dahlin M, Cuijpers P, Carlbring P, Silverberg F. Internet-based psychodynamic versus cognitive behavioral guided self-help for generalized anxiety disorder: a randomized controlled trial. Psychother Psychosom. 2012;81(6):344–55 PMID:22964540.
41. Johansson R, Hesser H, Ljótsson B, Frederick RJ, Andersson G. Transdiagnostic, affect-focused, psychodynamic, guided self-help for depression and anxiety through the internet: study protocol for a randomised controlled trial. BMJ Open. 2012;2(6):1–6 PMID:23257775.
42. Donker T, Batterham PJ, Warmerdam L, Bennett K, Bennett A, Cuijpers P, Griffiths KM, Christensen H. Predictors and moderators of response to internet-delivered Interpersonal Psychotherapy and Cognitive Behavior Therapy for depression. J Affect Disord. 2013;151(1):343–51.
43. Mak WWS, Chan ATY, Cheung EYL, Lin CLY, Ngai KCS. Enhancing Web-based mindfulness training for mental health promotion with the health action process approach: randomized controlled trial. J Med Internet Res. 2015;17(1):e8 PMID:25599904.
44. Lin J, Lüking M, Ebert DD, Buhrman M, Andersson G, Baumeister H. Effectiveness and cost-effectiveness of a guided and unguided internet-based acceptance and commitment therapy for chronic pain: study protocol for a three-armed randomised controlled trial. Internet Interv 2015;2(1). https://doi.org/10.1016/j.invent.2014.11.005.
45. Johansson R, Björklund M, Hornborg C, Karlsson S, Hesser H, Ljótsson B, Rousseau A, Frederick RJ, Andersson G. Affect-focused psychodynamic psychotherapy for depression and anxiety through the Internet: a randomized controlled trial. PeerJ. 2013;1(Suppl 2):e102- PMID:23862104.
46. Garcia-Palacios A, Hoffman H, Carlin A, Furness TA, Botella C. Virtual reality in the treatment of spider phobia: a controlled study. Behav Res Ther. 2002;40(9):983–93 PMID:12296495.
47. Freeman D, Haselton P, Freeman J, Spanlang B, Kishore S, Albery E, Denne M, Brown P, Slater M, Nickless A. Automated psychological therapy using immersive virtual reality

for treatment of fear of heights: a single-blind, parallel-group, randomised controlled trial. Lancet Psychiatry. 2018;5(8):625–32 Elsevier.

48. Pot-Kolder RMCA, Geraets CNW, Veling W, van Beilen M, Staring ABP, Gijsman HJ, Delespaul PAEG, van der Gaag M. Virtual-reality-based cognitive behavioural therapy versus waiting list control for paranoid ideation and social avoidance in patients with psychotic disorders: a single-blind randomised controlled trial. Lancet Psychiatry. 2018;5(3):217–26 Elsevier.
49. Fleming T, Dixon R, Frampton C, Merry S. A pragmatic randomized controlled trial of computerized CBT (SPARX) for symptoms of depression among adolescents excluded from mainstream education. Behav Cogn Psychother. 2012;40(5):529–41. https://doi.org/10.1017/s1352465811000695 12/06. Fleming, Theresa, University of Auckland, Department of Psychological Medicine, Private Bag 92019, Auckland, New Zealand, 1142.
50. An LC, Demers MRS, Kirch MA, Considine-Dunn S, Nair V, Dasgupta K, Narisetty N, Resnicow K, Ahluwalia J. A randomized trial of an avatar-hosted multiple behavior change intervention for young adult smokers. J Natl Cancer Inst Monogr. 2013;2013(47):209–15 PMID:24395994, Oxford University Press.
51. Fitzpatrick KK, Darcy A, Vierhile M. Delivering cognitive behavior therapy to young adults with symptoms of depression and anxiety using a fully automated conversational agent (Woebot): a randomized controlled trial. JMIR Ment Heal. 2017;4(2) JMIR Publications Inc.
52. Schlicker S, Ebert DD, Middendorf T, Titzler I, Berking M. Evaluation of a text-message-based maintenance intervention for Major Depressive Disorder after inpatient cognitive behavioral therapy. J Affect Disord. 2017;227:305–12 Elsevier.
53. Lin J, Ebert DD, Lehr D, Berking M, Baumeister H. Internetbasierte Gesundheitsinterventionen: State of the Art und Einsatzmöglichkeiten in der Rehabilitation. Rehabilitation. 2013;52(03):155–63. https://doi.org/10.1055/s-0033-1343491.
54. Backhaus A, Agha Z, Maglione ML, Repp A, Ross B, Zuest D, Rice-Thorp NM, Lohr J, Thorp SR. Videoconferencing psychotherapy: a systematic review. Psychol Serv. 2012;9(2):111–31 PMID:22662727.
55. Ebert DD, Lehr D, Smit F, Zarski A-C, Riper H, Heber E, Cuijpers P, Berking M. Efficacy and cost-effectiveness of minimal guided and unguided internet-based mobile supported stress-management in employees with occupational stress: a three-armed randomised controlled trial. BMC Public Health. 2014;14(1):807. https://doi.org/10.1186/1471-2458-14-807.
56. Schueller SM, Tomasino KN, Mohr DC. Integrating human support into behavioral intervention technologies: the efficiency model of support. Clin Psychol Sci Pract. 2016;24(1):27–45. https://doi.org/10.1111/cpsp.12173.
57. Zarski A-C, Lehr D, Berking M, Riper H, Cuijpers P, Ebert DD. Adherence to internet-based mobile-supported stress management: a pooled analysis of individual participant data from three randomized controlled trials. J Med Internet Res. 2016;18(6):e146. https://doi.org/10.2196/jmir.4493.
58. Buntrock C, Ebert DD, Lehr D, Cuijpers P, Riper H, Smit F, Berking M. Evaluating the efficacy and cost-effectiveness of web-based indicated prevention of major depression: design of a randomised controlled trial. BMC Psychiatry BioMed Cent. 2014;14:25. https://doi.org/10.1186/1471-244x-14-25.
59. Lal S, Adair CE. E-mental health: a rapid review of the literature. Psychiatr Serv Am Psychiatric Assoc. 2014;65(1):24–32.
60. Henderson C, Evans-Lacko S, Thornicroft G. Mental illness stigma, help seeking, and public health programs. Am J Public Health 2013 [cited 2017 May 10];103(5):777–80 PMID:23488489.
61. Muñoz RF, Bunge EL, Chen K, Schueller SM, Bravin JI, Shaughnessy EA, Pérez-Stable EJ. Massive open online interventions: a novel model for delivering behavioral-health services

worldwide. Clin Psychol Sci. 2016;4(2):194–205 Los Angeles, CA: Sage Publications Sage CA.
62. Fairburn CG, Patel V. The impact of digital technology on psychological treatments and their dissemination. Behav Res Ther. 2017;88:19–25 Elsevier.
63. Erbe D, Eichert H-C, Riper H, Ebert DD. Blending face-to-face and internet-based interventions for the treatment of mental disorders in adults: systematic review. J Med Internet Res. 2017;19(9):e306 PMID:28916506.
64. Ebert DD, Nobis S, Lehr D, Baumeister H, Riper HM, Auerbach RP, Snoek F, Cuijpers P, Berking M. The 6-month effectiveness of Internet-based guided self-help for depression in adults with Type 1 and 2 diabetes mellitus. Diabet Med. 2016; https://doi.org/10.1111/dme.13173.
65. Nobis S, Lehr D, Ebert DD, Baumeister H, Snoek F, Riper H, Berking M. Efficacy of a web-based intervention with mobile phone support in treating depressive symptoms in adults with type 1 and type 2 diabetes: a randomized controlled trial. Diabetes Care. 2015 PMID:25710923.
66. Sander L, Paganini S, Lin J, Schlicker S, Ebert DD, Buntrock C, Baumeister H. Effectiveness and cost-effectiveness of a guided Internet- and mobile-based intervention for the indicated prevention of major depression in patients with chronic back pain-study protocol of the PROD-BP multicenter pragmatic RCT. BMC Psychiatry. 2017;17(1). https://doi.org/10.1186/s12888-017-1193-6.
67. van Bastelaar KMP, Pouwer F, Cuijpers P, Riper H, Snoek FJ. Web-based depression treatment for type 1 and type 2 diabetic patients: a randomized, controlled trial. Diabetes Care. 2011;34(2):320–5 PMID:21216855.
68. van Bastelaar KMP, Pouwer F, Cuijpers P, Riper H, Twisk JWR, Snoek FJ. Is a severe clinical profile an effect modifier in a Web-based depression treatment for adults with type 1 or type 2 diabetes? Secondary analyses from a randomized controlled trial. J Med Internet Res. 2012;14(1):e2 PMID:22262728.
69. Buntrock C, Ebert DD, Lehr D, Smit F, Riper H, Berking M, Cuijpers P. Effect of a web-based guided self-help intervention for prevention of major depression in adults with subthreshold depression a randomized clinical trial. JAMA J Am Med Assoc. 2016;315(17). https://doi.org/10.1001/jama.2016.4326.
70. Ebert DD, Buntrock C, Cuijpers P, K van Z, P C, C B, C B, H B. Online intervention for prevention of major depression—reply. JAMA. 2016;316(8):881. https://doi.org/10.1001/jama.2016.9586 American Medical Association.
71. Bockting CLH, Kok GD, van der Kamp L, Smit F, van Valen E, Schoevers R, van Marwijk H, Cuijpers P, Riper H, Dekker J, Beck AT. Disrupting the rhythm of depression using Mobile Cognitive Therapy for recurrent depression: randomized controlled trial design and protocol. BMC Psychiatry. 2011;11:12 PMID:21235774.
72. Ebert DD, Gollwitzer M, Riper H, Cuijpers P, Baumeister H, Berking M. For whom does it work? moderators of outcome on the effect of a transdiagnostic internet-based maintenance treatment after inpatient psychotherapy: randomized controlled trial. J Med Internet Res. 2013;15(10):e191 PMID:24113764.
73. Ebert DD, Tarnowski T, Gollwitzer M, Sieland B, Berking M. A transdiagnostic internet-based maintenance treatment enhances the stability of outcome after inpatient cognitive behavioral therapy: a randomized controlled trial. Psychother Psychosom. 2013;82(4):246–56 PMID:23736751.
74. Golkaramnay V, Bauer S, Haug S, Wolf M, Kordy H. The exploration of the effectiveness of group therapy through an Internet chat as aftercare: a controlled naturalistic study. Psychother Psychosom. 2007;76(4):219–25 PMID:17570960.
75. Kok G, Bockting C, Burger H, Smit F, Riper H. Mobile cognitive therapy: adherence and acceptability of an online intervention in remitted recurrently depressed patients. Internet Interv. 2014;1(2):65–73. https://doi.org/10.1016/j.invent.2014.05.002.

76. Holländare F, Johnsson S, Randestad M, Tillfors M, Carlbring P, Andersson G, Engström I. Randomized trial of Internet-based relapse prevention for partially remitted depression. Acta Psychiatr Scand. 2011;124(4):285–94 Wiley Online Library.
77. Andrews G, Cuijpers P, Craske MG, McEvoy P, Titov N. Computer therapy for the anxiety and depressive disorders is effective, acceptable and practical health care: a meta-analysis. PLoS One. 2010;5(10):e13196 PMID:20967242.
78. Ebert DD, Zarski A-C, Christensen H, Stikkelbroek Y, Cuijpers P, Berking M, Riper H. Internet and computer-based cognitive behavioral therapy for anxiety and depression in youth: a meta-analysis of randomized controlled outcome trials. PLoS One 2015;10(3). https://doi.org/10.1371/journal.pone.0119895.
79. Mayo-Wilson E, Montgomery P. Media-delivered cognitive behavioural therapy and behavioural therapy (self-help) for anxiety disorders in adults. Cochrane database Syst Rev. 2013;9:CD005330 PMID:24018460.
80. Richards D, Richardson T. Computer-based psychological treatments for depression: a systematic review and meta-analysis. Clin Psychol Rev. 2012;32(4):329–42 PMID:22466510 Elsevier Ltd.
81. Richards D, Richardson T, Timulak L, McElvaney J. The efficacy of internet-delivered treatment for generalized anxiety disorder: a systematic review and meta-analysis. Internet Interv. 2015;2(3):272–82. https://doi.org/10.1016/j.invent.2015.07.003 Elsevier B.V.
82. Olthuis JV, Watt MC, Bailey K, Hayden JA, Stewart SH. Therapist-supported Internet cognitive behavioural therapy for anxiety disorders in adults. Cochrane database Syst Rev. 2015;3:CD011565 PMID:25742186.
83. Josephine K, Josefine L, Philipp D, David E, Harald B. Internet- and mobile-based depression interventions for people with diagnosed depression: a systematic review and meta-analysis. J Affect Disord. 2017;223:28–40 PMID:28715726 Elsevier.
84. Kuester A, Niemeyer H, Knaevelsrud C. Internet-based interventions for posttraumatic stress: a meta-analysis of randomized controlled trials. Clin Psychol Rev. 2016; 1–16 PMID:26655959 Elsevier Ltd.
85. Zachariae R, Lyby MS, Ritterband L, O'Toole MS. Efficacy of Internet-delivered cognitive-behavioral therapy for insomnia–a systematic review and meta-analysis of randomized controlled trials. Sleep Med Rev. 2015;30:1–10 PMID:26615572.
86. Melioli T, Bauer S, Franko DL, Moessner M, Ozer F, Chabrol H, Rodgers RF. Reducing eating disorder symptoms and risk factors using the internet: a meta-analytic review. Int J Eat Disord. 2016;49(1):19–31. https://doi.org/10.1002/eat.22477.
87. Buhrman M, Gordh T, Andersson G. Internet interventions for chronic pain including headache: a systematic review. Internet Interv. 2016;4:17–34. https://doi.org/10.1016/j.invent.2015.12.001.
88. Riper H, Blankers M, Hadiwijaya H, Cunningham J, Clarke S, Wiers R, Ebert D, Cuijpers P. Effectiveness of guided and unguided low-intensity internet interventions for adult alcohol misuse: a meta-analysis. PLoS One. 2014;9(6). https://doi.org/10.1371/journal.pone.0099912.
89. Andersson E, Enander J, Andrén P, Hedman E, Ljótsson B, Hursti T, Bergström J, Kaldo V, Lindefors N, Andersson G, Rück C. Internet-based cognitive behaviour therapy for obsessive–compulsive disorder: a randomized controlled trial. Psychol Med. 2012;42(10):2193–203 PMID:22348650.
90. Herbst N, Voderholzer U, Thiel N, Schaub R, Knaevelsrud C, Stracke S, Hertenstein E, Nissen C, Külz AK. No talking, just writing! Efficacy of an Internet-based cognitive behavioral therapy with exposure and response prevention in obsessive compulsive disorder. Psychother Psychosom. 2014;83(3):165–75 PMID:24732962.
91. Lenhard F, Andersson E, Mataix-Cols D, Rück C, Vigerland S, Högström J, Hillborg M, Brander G, Ljungström M, Ljótsson B, Serlachius E. Therapist-guided, internet-delivered cognitive-behavioral therapy for adolescents with obsessive-compulsive disorder:

a randomized controlled trial. J Am Acad Child Adolesc Psychiatry. 2017;56(1):10–19.e2. https://doi.org/10.1016/j.jaac.2016.09.515.
92. Storch EA, Caporino NE, Morgan JR, Lewin AB, Rojas A, Brauer L, Larson MJ, Murphy TK. Preliminary investigation of web-camera delivered cognitive-behavioral therapy for youth with obsessive-compulsive disorder. Psychiatry Res. 2011;189(3):407–12 PMID:21684018.
93. Gottlieb JD et al. Web-based cognitive-behavioral therapy for auditory hallucinations in persons with psychosis: a pilot study. Schizophr Res. 2013;145(1–3):82–7 PMID:23410709 Elsevier.
94. Harper KM. An Investigation of an Internet-based cognitive behavioral Therapy Program fpr auditory hallucinations. University of North Carolina at Chapel Hill; 2013.
95. Enander J, Andersson E, Mataix-Cols D, Lichtenstein L, Alstroem K, Andersson G, Ljotsson B, Rueck C. Therapist guided internet based cognitive behavioural therapy for body dysmorphic disorder: single blind randomised controlled trial. BMJ-BRITISH Med J. 2016;352. https://doi.org/10.1136/bmj.i241.
96. Hidalgo-Mazzei D, Mateu A, Reinares M, Matic A, Vieta E, Colom F. Internet-based psychological interventions for bipolar disorder: review of the present and insights into the future. J Affect Disord. 2015;188:1–13 PMID:26342885.
97. Andersson E, Walén C, Hallberg J, Paxling B, Dahlin M, Almlöv J, Källström R, Wijma K, Carlbring P, Andersson G. A randomized controlled trial of guided Internet-delivered cognitive behavioral therapy for erectile dysfunction. J Sex Med. 2011;8(10):2800–9 PMID:21797983.
98. Jones LM, Mccabe MP. The effectiveness of an internet-based psychological treatment program for female sexual dysfunction. J Sex Med. 2011;8(10):2781–92 PMID:21771279.
99. Schover LR, Canada AL, Yuan Y, Sui D, Neese L, Jenkins R, Rhodes MM. A randomized trial of internet-based versus traditional sexual counseling for couples after localized prostate cancer treatment. Cancer. 2012;118(2):500–9 PMID:21953578.
100. van Lankveld JJ, Leusink P, van Diest S, Gijs L, Slob AK. Internet-based brief sextherapy for heterosexual men with sexual dysfunctions: a randomized controlled pilot trial. J Sex Med. 2009;6(8):2224–36.
101. Zarski A-C, Berking M, Fackiner C, Rosenau C, Ebert DD. Internet-based guided self-help for vaginal penetration difficulties: results of a randomized controlled pilot trial. J Sex Med. 2017;14(2). https://doi.org/10.1016/j.jsxm.2016.12.232.
102. Abbott J-AM, Kaldo V, Klein B, Austin D, Hamilton C, Piterman L, Williams B, Andersson G. A Cluster Randomised Trial of an Internet-Based Intervention Program for Tinnitus Distress in an Industrial Setting. Cogn Behav Ther 2009;38(3):162–73 PMID:19675959.
103. Nyenhuis N, Zastrutzki S, Weise C, Jäger B, Kröner-Herwig B. The efficacy of minimal contact interventions for acute tinnitus: a randomised controlled study. Cogn Behav Ther. 2013;42(2):127–38 PMID:22413736.
104. Andersson G, Strömgren T, Ström L, Lyttkens L. Randomized controlled trial of internet-based cognitive behavior therapy for distress associated with tinnitus. Psychosom Med. 64(5):810–6 PMID:12271112.
105. Weise C, Kleinstäuber M, Andersson G. Internet-delivered cognitive-behavior therapy for tinnitus. Psychosom Med. 2016;78(4):501–10 PMID:26867083.
106. Kaldo V, Levin S, Widarsson J, Buhrman M, Larsen H-C, Andersson G. Internet versus group cognitive-behavioral treatment of distress associated with tinnitus: a randomized controlled trial. Behav Ther. 2008;39(4):348–59 PMID:19027431.
107. Nyenhuis N, Zastrutzki S, Jäger B, Kröner-Herwig B. An internet-based cognitive-behavioural training for acute tinnitus: secondary analysis of acceptance in terms of satisfaction, trial attrition and non-usage attrition. Cogn Behav Ther. 2013;42(2):139–45 PMID:23205617.
108. Hesser H, Westin VZ, Andersson G. Acceptance as a mediator in internet-delivered acceptance and commitment therapy and cognitive behavior therapy for tinnitus. J Behav Med. 2014;37(4):756–67 PMID:23881309.

109. Hesser H, Gustafsson T, Lundén C, Henrikson O, Fattahi K, Johnsson E, Westin VZ, Carlbring P, Mäki-Torkko E, Kaldo V, Andersson G. A randomized controlled trial of internet-delivered cognitive behavior therapy and acceptance and commitment therapy in the treatment of tinnitus. J Consult Clin Psychol. 2012;80(4):649–61 PMID:22250855.
110. Jasper K, Weise C, Conrad I, Andersson G, Hiller W, Kleinstäuber M. Internet-based guided self-help versus group cognitive behavioral therapy for chronic tinnitus: a randomized controlled trial. Psychother Psychosom. 2014;83(4):234–46 PMID:24970708.
111. Andersson G, Cuijpers P, Carlbring P, Riper H, Hedman E. Guided Internet-based vs. face-to-face cognitive behavior therapy for psychiatric and somatic disorders: a systematic review and meta-analysis. World Psychiatry. 2014;13(3):288–95 PMID:25273302.
112. Wagner B, Knaevelsrud C, Maercker A. Internet-based cognitive-behavioral therapy for complicated grief: a randomized controlled trial. Death Stud. 2006;30(5):429–53 PMID:16610157.
113. Wagner B, Maercker A. A 1.5-year follow-up of an internet-based intervention for complicated grief. J Trauma Stress. 2007;20(4):625–29. https://doi.org/10.1002/jts.20230 Wiley Subscription Services, Inc., A Wiley Company.
114. Eisma MC, Boelen PA, van den Bout J, Stroebe W, Schut HAW, Lancee J, Stroebe MS. Internet-based exposure and behavioral activation for complicated grief and rumination: a randomized controlled trial. Behav Ther. 2015;46(6):729–48 Elsevier.
115. Luquiens A, Tanguy M-L, Lagadec M, Benyamina A, Aubin H-J, Reynaud M. The efficacy of three modalities of internet-based psychotherapy for non-treatment-seeking online problem gamblers: a randomized controlled trial. J Med Internet Res. 2016;18(2):e36 PMID:26878894.
116. Carlbring P, Smit F. Randomized trial of internet-delivered self-help with telephone support for pathological gamblers. J Consult Clin Psychol. 2008;76(6):1090–4 PMID:19045977.
117. Nijhof SL, Priesterbach LP, Uiterwaal C, Bleijenberg G, Kimpen JLL, Putte EM. Internet-based therapy for adolescents with chronic fatigue syndrome: long-term follow-up. Pediatrics; 2013. p. e1788–95. https://doi.org/10.1542/peds.2012-2007.
118. Nijhof SL, et al. Effectiveness of internet-based cognitive behavioural treatment for adolescents with chronic fatigue syndrome (FITNET): a randomised controlled trial. Lancet (London, England). 2012;379(9824):1412–8 PMID:22385683 Elsevier.
119. Franklin JC, Fox KR, Franklin CR, Kleiman EM, Ribeiro JD, Jaroszewski AC, Hooley JM, Nock MK. A brief mobile app reduces nonsuicidal and suicidal self-injury: evidence from three randomized controlled trials. J Consult Clin Psychol. 2016;84(6):544–57 PMID:27018530.
120. Mohr DC, Ho J, Hart TL, Baron KG, Berendsen M, Beckner V, Cai X, Cuijpers P, Spring B, Kinsinger SW, Schroder KE, Duffecy J. Control condition design and implementation features in controlled trials: a meta-analysis of trials evaluating psychotherapy for depression. Transl Behav Med. 2014 [cited 2017 May 10];4(4):407–23 PMID:25584090 Springer.
121. Andersson G, Topooco N, Havik O, Nordgreen T. Internet-supported versus face-to-face cognitive behavior therapy for depression. Expert Rev Neurother. 2016;16(1):55–60 PMID:26610160.
122. Carlbring P, Andersson G, Cuijpers P, Riper H, Hedman-Lagerlöf E. Internet-based vs. face-to-face cognitive behavior therapy for psychiatric and somatic disorders: an updated systematic review and meta-analysis. Cogn Behav Ther. 2018;47(1):1–18 PMID:29215315.
123. Ebert DD, Berking M, Cuijpers P, Lehr D, P?rtner M, Baumeister H. Increasing the acceptance of internet based mental health interventions in primary care patients with depressive symptoms. A randomized controlled trial. J Affect Disord. 2015;176. https://doi.org/10.1016/j.jad.2015.01.056.
124. Baumeister H, Nowoczin L, Lin J, Seifferth H, Seufert J, Laubner K, Ebert DD. Impact of an acceptance facilitating intervention on diabetes patients' acceptance of Internet-based interventions for depression: a randomized controlled trial. Diabetes Res Clin Pract. 2014;105(1):30–9. https://doi.org/10.1016/j.diabres.2014.04.031.

125. Apolinário-Hagen J, Vehreschild V, Alkoudmani RM. Current views and perspectives on e-mental health: an exploratory survey study for understanding public attitudes toward internet-based psychotherapy in Germany. JMIR Ment Heal. 2017;4(1):e8 PMID:28232298.
126. Apolinário-Hagen J, Harrer M, Kählke F, Fritsche L, Salewski C, Ebert DD. Public attitudes toward guided internet-based therapies: web-based survey study. JMIR Ment Health. 2018;5(2):e10735 PMID:29764797, JMIR Publications Inc.
127. Dorow M, Löbner M, Pabst A, Stein J, Riedel-Heller SG. Preferences for depression treatment including internet-based interventions: results from a large sample of primary care patients. Front Psychiatry. 2018;9 Frontiers Media SA.
128. Sethi S, Campbell AJ, Ellis LA. The use of computerized self-help packages to treat adolescent depression and anxiety. J Technol Hum Serv. 2010, 31;28(3):144–60. https://doi.org/10.1080/15228835.2010.508317 Routledge.
129. Lindhiem O, Bennett CB, Rosen D, Silk J. Mobile technology boosts the effectiveness of psychotherapy and behavioral interventions: a meta-analysis. Behav Modif. 2015;39(6):785–804 PMID:26187164.
130. Kenwright M, Liness S, Marks I. Reducing demands on clinicians by offering computer-aided self-help for phobia/panic. Feasibility study. Br J Psychiatry. 2001;179:456–9 PMID:11689405.
131. Marks IM, Kenwright M, McDonough M, Whittaker M, Mataix-Cols D. Saving clinicians' time by delegating routine aspects of therapy to a computer: a randomized controlled trial in phobia/panic disorder. Psychol Med. 2004 PMID:14971623, Cambridge University Press.
132. Kenter RMF, van de Ven PM, Cuijpers P, Koole G, Niamat S, Gerrits RS, Willems M, van Straten A. Costs and effects of Internet cognitive behavioral treatment blended with face-to-face treatment: results from a naturalistic study. Internet Interv. 2015;2(1):77–83. https://doi.org/10.1016/j.invent.2015.01.001.
133. Kleiboer A, Smit J, Bosmans J, Ruwaard J, Andersson G, Topooco N, Berger T, Krieger T, Botella C, Ba?os R, Chevreul K, Araya R, Cerga-Pashoja A, Cie?lak R, Rogala A, Vis C, Draisma S, Schaik A, Kemmeren L, Ebert D, Berking M, Funk B, Cuijpers P, Riper H. European COMPARative Effectiveness research on blended Depression treatment versus treatment-as-usual (E-COMPARED): study protocol for a randomized controlled, non-inferiority trial in eight European countries. Trials 2016;17(1). https://doi.org/10.1186/s13063-016-1511-1.
134. Kooistra LC, Wiersma JE, Ruwaard J, van Oppen P, Smit F, Lokkerbol J, Cuijpers P, Riper H. Blended vs. face-to-face cognitive behavioural treatment for major depression in specialized mental health care: study protocol of a randomized controlled cost-effectiveness trial. BMC Psychiatry 2014;14(1):290 PMID:25326035.
135. Romijn G, Riper H, Kok R, Donker T, Goorden M, Roijen LH Van, Kooistra L, Balkom A Van, Koning J. Cost-effectiveness of blended vs. face-to- face cognitive behavioural therapy for severe anxiety disorders : study protocol of a randomized controlled trial. BMC Psychiatry BMC Psychiatry; 2015;1–10. https://doi.org/10.1186/s12888-015-0697-1.
136. Vara D, Herrero R, Etchemendy E, Espinoza M, Baños R, García-Palacios A, Lera G, Folch B, Palop-Larrea V, Vázquez P. Efficacy and cost-effectiveness of a blended cognitive behavioral therapy for depression in Spanish primary health care: study protocol for a randomised non-inferiority trial. BMC Psychiatry BioMed Central. 2018;18(1):74.
137. Ebert DD, Cuijpers P, Muñoz RF, Baumeister H. Prevention of mental health disorders using internet and mobile-based interventions: a narrative review and recommendations for future research. Front Psychiatry Front. 2017;8:116. https://doi.org/10.3389/fpsyt.2017.00116.
138. Buntrock C, Ebert DD, Lehr D, Smit F, Riper H, Berking M, Cuijpers P. Effect of a web-based guided self-help intervention for prevention of major depression in adults with subthreshold depression. JAMA. 2016;315(17):1854. https://doi.org/10.1001/jama.2016.4326 American Medical Association.
139. Buntrock C, Ebert D, Lehr D, Riper H, Smit F, Cuijpers P, Berking M. Effectiveness of a web-based cognitive behavioural intervention for subthreshold depression:

pragmatic randomised controlled trial. Psychother Psychosom. 2015;84(6). https://doi.org/10.1159/000438673.
140. Buntrock C, Berking M, Smit F, Lehr D, Nobis S, Riper H, Cuijpers P, Ebert D. Preventing depression in adults with subthreshold depression: health-economic evaluation alongside a pragmatic randomized controlled trial of a web-based intervention. J Med Internet Res. 2017;19(1):e5. https://doi.org/10.2196/jmir.6587.
141. Holländare F, Anthony SA, Randestad M, Tillfors M, Carlbring P, Andersson G, Engström I. Two-year outcome of internet-based relapse prevention for partially remitted depression. Behav Res Ther. 2013. https://doi.org/10.1016/j.brat.2013.08.002.
142. Klein NS, Kok GD, Burger H, van Valen E, Riper H, Cuijpers P, Dekker J, Smit F, van der Heiden C, Bockting CLH. No sustainable effects of an internet-based relapse prevention program over 24 months in recurrent depression: primary outcomes of a randomized controlled trial. Psychother Psychosom. 2018;87(1):55–57 Karger Publishers.
143. Riper H, Blankers M, Hadiwijaya H, Cunningham J, Clarke S, Wiers R, Ebert DD, Cuijpers P. Effectiveness of guided and unguided low-intensity internet interventions for adult alcohol misuse: a meta-analysis. PLoS One. 2014;9(6):e99912 PMID:24937483, Stewart R, editor, Public Library of Science.
144. Boß L, Lehr D, Schaub MP, Paz Castro R, Riper H, Berking M, Ebert DD. Efficacy of a web-based intervention with and without guidance for employees with risky drinking: results of a three-arm randomized controlled trial. Addiction Wiley Online Library; 2017.
145. Ebert DD, Berking M, Thiart H, Riper H, Laferton J, Lehr D. Restoring depleted resources: efficacy and mechanisms of change of an Internet-based unguided recovery training for better sleep and psychological detachment from work. Heal Psychol. 2015;Suppl(1). https://doi.org/10.1037/hea0000277.
146. Thiart H, Lehr D, Ebert DD, Berking M, Riper H. Log in and breathe out: internet-based recovery training for sleepless employees with work-related strain–results of a randomized controlled trial. Scand J Work Environ Health. 2015; https://doi.org/10.5271/sjweh.3478 Online Fir.
147. Lehr D, Heber E, Sieland B, Hillert A, Funk B, Ebert DD. Occupational eMental Health and teachers' health: a meta-analytic review on the efficacy of internet-based intervention for promoting mental health in teachers | "Occupational eMental Health" in der Lehrergesundheit: Ein metaanalytisches Review zur Wirksa. Pravention und Gesundheitsforderung 2016; https://doi.org/10.1007/s11553-016-0541-6.
148. Heber E, Lehr D, Ebert DD, Berking M, Riper H. Web-based and mobile stress management intervention for employees: results of a randomised controlled trial. J Med Internet Res. 2016;18(1):e21. https://doi.org/10.2196/jmir.5112.
149. Ebert DD, Heber E, Berking M, Riper H, Cuijpers P, Funk B, Lehr D. Self-guided internet-based and mobile-based stress management for employees: results of a randomised controlled trial. Occup Environ Med. 2016;73(5). https://doi.org/10.1136/oemed-2015-103269.
150. Ebert DD, Lehr D, Heber E, Riper H, Cuijpers P, Berking M. Internet- and mobile-based stress management for employees with adherence-focused guidance: efficacy and mechanism of change. Scand J Work Environ Health. 2016 Sept 1 [cited 2017 May 2];42(5):382–94 PMID:27249161.
151. Harrer M, Adam SH, Fleischmann RJ, Baumeister H, Auerbach RP, Bruffaerts R, Cuijpers P, Kessler RC, Berking M, Lehr D, Ebert DD. Effectiveness of an internet- and app-based intervention for college students with elevated stress: results of a randomized controlled trial (in press). J Med Internet Res. 2018.
152. Heber E, Ebert DD, Lehr D, Cuijpers P, Berking M, Nobis S, Riper H. The benefit of web- and computer-based interventions for stress: a systematic review and meta-analysis. J Med Internet Res. 2017;19(2):e32 PMID:28213341.
153. van Zoonen K, Buntrock C, Ebert DD, Smit F, Reynolds III CF, Beekman ATF, Cuijpers P. Preventing the onset of major depressive disorder: a meta-analytic review of psychological interventions. Int J Epidemiol. 2014;43(2). https://doi.org/10.1093/ije/dyt175.

154. Moreno-Peral P, Conejo-Cerón S, Rubio-Valera M, Fernández A, Navas-Campaña D, Rodríguez-Morejón A, Motrico E, Rigabert A, Luna J de D, Martín-Pérez C, Rodríguez-Bayón A, Ballesta-Rodríguez MI, Luciano JV, Bellón JÁ. Effectiveness of psychological and/or educational interventions in the prevention of anxiety. JAMA Psychiatry. 2017;56(9):1026–33. https://doi.org/10.1001/jamapsychiatry.2017.2509 Wiley, New York.
155. Üstün TB, Ayuso-Mateos JL, Chatterji S, Mathers C, Murray CJL. Global burden of depressive disorders in the year 2000. Br J psychiatry RCP. 2004;184(5):386–92.
156. Olesen J, Gustavsson A, Svensson M, Wittchen H, Jönsson B. The economic cost of brain disorders in Europe. Eur J Neurol. 2012;19(1):155–62 Wiley Online Library.
157. Sobocki P, Jönsson B, Angst J, Rehnberg C. Cost of depression in Europe. J Ment Health Policy Econ. 2006;9(2):87–98 PMID:17007486.
158. Drummond MF, Sculpher MJ, Torrance GW, O'Brien BJ, Stoddart GL. Methods for the economic evaluation of health care programmes. Oxford: Oxford University Press; 2005.
159. Donker T, Blankers M, Hedman E, Ljotsson B, Petrie K, Christensen H. Economic evaluations of Internet interventions for mental health: a systematic review. Psychol Med. 2015;45(16):3357–76 Cambridge University Press.
160. Paganini S, Teigelkötter W, Buntrock C, Baumeister H. Economic evaluations of internet- and mobile-based interventions for the treatment and prevention of depression: a systematic review. J Affect Disord. 2017;225:733–55. https://doi.org/10.1016/j.jad.2017.07.018 Elsevier.
161. Andrews G, Williams AD. Internet psychotherapy and the future of personalized treatment. Depress Anxiety. 2014;31(11):912–5 PMID:25407580.
162. Williams AD, O'Moore K, Mason E, Andrews G. The effectiveness of internet cognitive behaviour therapy (iCBT) for social anxiety disorder across two routine practice pathways. Internet Interv. 2014;1(4):225–9. https://doi.org/10.1016/j.invent.2014.11.001.
163. Hedman E, Ljótsson B, Kaldo V, Hesser H, El Alaoui S, Kraepelien M, Andersson E, Rück C, Svanborg C, Andersson G, Lindefors N. Effectiveness of Internet-based cognitive behaviour therapy for depression in routine psychiatric care. J Affect Disord. 2014;155(1):49–58 PMID:24238951.
164. El Alaoui S, Hedman E, Ljótsson B, Lindefors N. Long-term effectiveness and outcome predictors of therapist-guided internet-based cognitive-behavioural therapy for social anxiety disorder in routine psychiatric care. BMJ Open. 2015;5(6):e007902 PMID:26105031.
165. El Alaoui S, Hedman E, Kaldo V, Hesser H, Kraepelien M, Andersson E, Rück C, Andersson G, Ljótsson B, Lindefors N. Effectiveness of Internet-based cognitive-behavior therapy for social anxiety disorder in clinical psychiatry. J Consult Clin Psychol. 2015;83(5):902–14 PMID:26009780.
166. Nordgreen T, Gjestad R, Andersson G, Carlbring P, Havik OE. The implementation of guided Internet-based cognitive behaviour therapy for panic disorder in a routine-care setting: effectiveness and implementation efforts. Cogn Behav Ther. 2017;1–14 PMID:28714775.
167. Titov N, Dear BF, Staples LG, Bennett-Levy J, Klein B, Rapee RM, Andersson G, Purtell C, Bezuidenhout G, Nielssen OB. The first 30 months of the MindSpot Clinic: evaluation of a national e-mental health service against project objectives. Aust New Zeal J Psychiatry. 2016;000486741667159 PMID:27733709.
168. Titov N, Dear BF, Staples LG, Bennett-Levy J, Klein B, Rapee RM, Shann C, Richards D, Andersson G, Ritterband L, Purtell C, Bezuidenhout G, Johnston L, Nielssen OB. MindSpot Clinic: an accessible, efficient, and effective online treatment service for anxiety and depression. Psychiatr Serv. 2015;66(10):1043–50 PMID:26130001.
169. Titov N, Dear B, Nielssen O, Staples L, Hadjistavropoulos H, Nugent M, Adlam K, Nordgreen T, Bruvik KH, Hovland A. ICBT in routine care: a descriptive analysis of successful clinics in five countries. Internet Interv. 2018 Elsevier.
170. Karyotaki E, Riper H, Twisk J, Hoogendoorn A, Kleiboer A, Mira A, Mackinnon A, Meyer B, Botella C, Littlewood E, Andersson G, Christensen H, Klein JP, Schröder J,

Bretón-López J, Scheider J, Griffiths K, Farrer L, Huibers MJH, Phillips R, Gilbody S, Moritz S, Berger T, Pop V, Spek V, Cuijpers P. Efficacy of Self-guided Internet-Based Cognitive Behavioral Therapy in the Treatment of Depressive Symptoms. JAMA Psychiatry. 2017. https://doi.org/10.1001/jamapsychiatry.2017.0044.
171. Ebert DD, Baumeister H. Internet-based self-help interventions for depression in routine care. JAMA Psychiatry. 2017;1(4):205–15. https://doi.org/10.1001/jamapsychiatry.2017.1394.
172. Lin J, Faust B, Ebert DD, Krämer L, Baumeister H. A web-based acceptance-facilitating intervention for identifying patients' acceptance, uptake, and adherence of internet- and mobile-based pain interventions: randomized controlled trial. J Med Internet Res. 2018;20(8):e244 Toronto, Canada: JMIR Publications Inc.
173. Littlewood E, Duarte A, Hewitt C, Knowles S, Palmer S, Walker S, Andersen P, Araya R, Barkham M, Bower P, Brabyn S, Brierley G, Cooper C, Gask L, Kessler D, Lester H, Lovell K, Muhammad U, Parry G, Richards DA, Richardson R, Tallon D, Tharmanathan P, White D, Gilbody S. A randomised controlled trial of computerised cognitive behaviour therapy for the treatment of depression in primary care: the Randomised Evaluation of the Effectiveness and Acceptability of Computerised Therapy (REEACT) trial. Health Technol Assess. 2015;19(101):1–174 PMID:26685904.
174. Brabyn S, Araya R, Barkham M, Bower P, Cooper C, Duarte A, Kessler D, Knowles S, Lovell K, Littlewood E, Mattock R, Palmer S, Pervin J, Richards D, Tallon D, White D, Walker S, Worthy G, Gilbody S. The second Randomised Evaluation of the Effectiveness, cost-effectiveness and Acceptability of Computerised Therapy (REEACT-2) trial: does the provision of telephone support enhance the effectiveness of computer-delivered cognitive behaviour therapy? A ra. Health Technol Assess (Rockv). 2016;20(89):1–64 PMID:27922448.
175. Muñoz RF. The efficiency model of support and the creation of digital apothecaries. Clin Psychol Sci Pract. 2017;24(1):46–9. https://doi.org/10.1111/cpsp.12174.
176. Mohr DC, Weingardt KR, Reddy M, Schueller SM. Three problems with current digital mental health research and three things we can do about them. Psychiatr Serv. 2017;68(5):427–29. https://doi.org/10.1176/appi.ps.201600541 American Psychiatric Association Arlington, VA.
177. Kaltenthaler E, Parry G, Beverley C, Ferriter M. Computerised cognitive-behavioural therapy for depression: systematic review. Br J Psychiatry. 2008;193(3):181 84 Cambridge University Press.
178. Lillevoll K, Vangberg H, Griffiths K, Waterloo K, Eisemann M. Uptake and adherence of a self-directed internet-based mental health intervention with tailored e-mail reminders in senior high schools in Norway. BMC Psychiatry. 2014. https://doi.org/10.1186/1471-244x-14-14.
179. Whiteside U, Lungu A, Richards J, Simon GE, Clingan S, Siler J, Snyder L, Ludman E. Designing messaging to engage patients in an online suicide prevention intervention: survey results from patients with current suicidal ideation. J Med Internet Res. 2014;16(2) JMIR Publications Inc.
180. Woodford J, Farrand P, Bessant M, Williams C. Recruitment into a guided internet based CBT (iCBT) intervention for depression: lesson learnt from the failure of a prevalence recruitment strategy. Contemp Clin Trials. 2011;32(5):641–8. https://doi.org/10.1016/j.cct.2011.04.013.
181. Paul CL, Piterman L, Shaw JE, Kirby C, Forshaw KL, Robinson J, Thepwongsa I, Sanson-Fisher RW. Poor uptake of an online intervention in a cluster randomised controlled trial of online diabetes education for rural general practitioners. Trials Engl. 2017;18(1):137 PMID:28335809.
182. Meisel SF, Drury H, Perera-Delcourt RP. Therapists' attitudes to offering eCBT in an inner-city IAPT service: a survey study. Cogn Behav Ther. 2018;11 Cambridge University Press.

183. Hennemann S, Beutel ME, Zwerenz R. Ready for eHealth? Health professionals' acceptance and adoption of ehealth interventions in inpatient routine care. J Health Commun. 2017;22(3):274–84 Taylor & Francis.
184. Mitchell N, Gordon PK. Attitudes towards computerised CBT for depression amongst a student population. Behav Cogn Psychother. 2007 [cited 2017 Mar 22];35(04):421–30. https://doi.org/10.1017/s1352465807003700.
185. Cranen K, Veld RH in't, Ijzerman M, Vollenbroek-Hutten M. Change of patients' perceptions of telemedicine after brief use. Telemed J e-Health. 2011;17(7):530–35. https://doi.org/10.1089/tmj.2010.0208.
186. Baumeister H, Seifferth H, Lin J, Nowoczin L, L?king M, Ebert D. Impact of an acceptance facilitating intervention on patients' acceptance of internet-based pain interventions: a randomized controlled trial. Clin J Pain. 2015;31(6). https://doi.org/10.1097/ajp.0000000000000118.
187. Ebert DD, Berking M, Cuijpers P, Lehr D, Pörtner M, Baumeister H. Increasing the acceptance of internet-based mental health interventions in primary care patients with depressive symptoms. A randomized controlled trial. J Affect Disord. 2015; https://doi.org/10.1016/j.jad.2015.01.056.
188. Ebert DD, Donkin L, Andersson G, Andrews G, Berger T, Carlbring P, Rozenthal A, Choi I, Laferton JAC, Johansson R, Kleiboer A, Lange A, Lehr D, Reins JA, Funk B, Newby J, Perini S, Riper H, Ruwaard J, Sheeber L, Snoek FJ, Titov N, Ünlü Ince B, Van Bastelaar K, Vernmark K, Van Straten A, Warmerdam L, Salsman N, Cuijpers P. Does Internet-based guided-self-help for depression cause harm? An individual participant data meta-analysis on deterioration rates and its moderators in randomized controlled trials. Psychol Med. 2016;46(13). https://doi.org/10.1017/s0033291716001562.
189. Christensen H, Farrer L, Batterham PJ, Mackinnon A, Griffiths KM, Donker T. The effect of a web-based depression intervention on suicide ideation: secondary outcome from a randomised controlled trial in a helpline. BMJ Open. 2013;3(6):1–10 PMID:23811172.
190. Mewton L, Andrews G. Cognitive behaviour therapy via the internet for depression: a useful strategy to reduce suicidal ideation. J Affect Disord. 2014;170C:78–84 PMID:25233243.
191. Rozental A, Boettcher J, Andersson G, Schmidt B, Carlbring P. Negative effects of internet interventions: a qualitative content analysis of patients' experiences with treatments delivered online. Cogn Behav Ther. 2015;44(3):223–36 PMID:25705924.
192. Boettcher J, Rozental A, Andersson G, Carlbring P. Side effects in Internet-based interventions for Social Anxiety Disorder. Internet Interv. 2014;1(1):3–11. https://doi.org/10.1016/j.invent.2014.02.002.
193. Rozental A, Magnusson K, Boettcher J, Andersson G, Carlbring P. For better or worse: an individual patient data meta-analysis of deterioration among participants receiving Internet-based cognitive behavior therapy. J Consult Clin Psychol. 2017;85(2):160–77 PMID:27775414.
194. Wampold BE, Imel ZE. The Great psychotherapy debate. the evidence for what makes psychotherapy work. 2nd ed. New York: Routledge; 2015.
195. Domhardt M, Geßlein H, von Rezori RE, Baumeister H. Internet-and mobile-based interventions for anxiety disorders: a meta-analytic review of intervention components. Depress Anxiety. 2018 Wiley Online Library.
196. Ebert DD, Berking M, Thiart H, Riper H, Laferton JAC, Cuijpers P, Sieland B, Lehr D. Restoring depleted resources: efficacy and mechanisms of change of an internet-based unguided recovery training for better sleep and psychological detachment from work. Heal Psychol. 2015;34(Suppl):1240–51 PMID:26651465.
197. Warmerdam L, van Straten A, Jongsma J, Twisk J, Cuijpers P. Online cognitive behavioral therapy and problem-solving therapy for depressive symptoms: exploring mechanisms of change. J Behav Ther Exp Psychiatry. 2010;41(1):64–70 PMID:19913781.
198. Ebert DD, Lehr D, Heber E, Riper H, Cuijpers P, Berking M. Internet-and mobile-based stress management for employees with adherence-focused guidance: efficacy and mechanism of change. Scand J Environ Heal. 2016;42(5):382–94.

199. Boettcher J, Renneberg B, Berger T. Patient expectations in internet-based self-help for social anxiety. Cogn Behav Ther. 2013;42(3):203–14 PMID:23697570.
200. Lévesque A, Campbell ANC, Pavlicova M, Hu M-C, Walker R, McClure EA, Ghitza UE, Bailey G, Stitzer M, Nunes EV. Coping strategies as a mediator of internet-delivered psychosocial treatment: secondary analysis from a NIDA CTN multisite effectiveness trial. Addict Behav. 2017;65:74–80 PMID:27776269.
201. Mogoaşe C, Cobeanu O, David O, Giosan C, Szentagotai A. Internet-based psychotherapy for adult depression: what about the mechanisms of change? J Clin Psychol. 2017;73(1):5–64 PMID:27684405.
202. Hedman E, Andersson E, Andersson G, Lindefors N, Lekander M, Rück C, Ljótsson B. Mediators in internet-based cognitive behavior therapy for severe health anxiety. PLoS One. 2013;8(10):e77752. https://doi.org/10.1371/journal.pone.0077752 Thombs B, editor. Guilford Press.
203. Graham AL, Papandonatos GD, Cobb CO, Cobb NK, Niaura RS, Abrams DB, Tinkelman DG. Internet and telephone treatment for smoking cessation: mediators and moderators of short-term abstinence. Nicotine Tob Res. 2015;17(3):299–308 PMID:25156528, Oxford University Press.
204. Ljótsson B, Hesser H, Andersson E, Lindfors P, Hursti T, Rück C, Lindefors N, Andersson G, Hedman E. Mechanisms of change in an exposure-based treatment for irritable bowel syndrome. J Consult Clin Psychol. 2013;81(6):1113–26 PMID:23750460.
205. Morgan AJ, Mackinnon AJ, Jorm AF. Behavior change through automated e-mails: mediation analysis of self-help strategy use for depressive symptoms. Behav Res Ther. 2013;51(2):57–62 PMID:23262114.
206. Alpay L, van der Boog P, Dumaij A. An empowerment-based approach to developing innovative e-health tools for self-management. Health Inf J. 2011;17(4):247–55 Sage Publications Sage UK: London, England.
207. Crisp D, Griffiths K, Mackinnon A, Bennett K, Christensen H. An online intervention for reducing depressive symptoms: secondary benefits for self-esteem, empowerment and quality of life. Psychiatry Res. 2014;216(1):60–66 Elsevier.
208. Samoocha D, Bruinvels DJ, Elbers NA, Anema JR, van der Beek AJ. Effectiveness of web-based interventions on patient empowerment: a systematic review and meta-analysis. J Med Internet Res. 2010;12. https://doi.org/10.2196/jmir.1286.
209. Kuijpers W, Groen WG, Aaronson NK, van Harten WH. A systematic review of web-based interventions for patient empowerment and physical activity in chronic diseases: relevance for cancer survivors. J Med Internet Res. 2013;15(2) JMIR Publications Inc.
210. Cowpertwait L, Clarke D. Effectiveness of web-based psychological interventions for depression: a meta-analysis. Int J Ment Health Addict. 2013;11(2):247–68. https://doi.org/10.1007/s11469-012-9416-z Springer.
211. Baumeister H, Reichler L, Munzinger M, Lin J. The impact of guidance on Internet-based mental health interventions-a systematic review. Internet Interv. 2014:205–15. https://doi.org/10.1016/j.invent.2014.08.003.
212. Johansson R, Andersson G. Internet-based psychological treatments for depression. Expert Rev Neurother. 2012;12(7):861–9; quiz 870 PMID:22853793.
213. Andersson G, Carlbring P, Berger T, Almlov J, Cuijpers P. What makes internet therapy work? Cogn Behav Ther. 2009;38(sup1):55–60 PMID:19675956.
214. Andersson G, Paxling B, Wiwe M, Vernmark K, Felix CB, Lundborg L, Furmark T, Cuijpers P, Carlbring P. Therapeutic alliance in guided internet-delivered cognitive behavioural treatment of depression, generalized anxiety disorder and social anxiety disorder. Behav Res Ther. 2012;50(9):544–50 PMID:22728647.
215. Ebert DD, Hannig W, Tarnowski T, Sieland B, Götzky B, Berking M. Web-based rehabilitation aftercare following inpatient psychosomatic treatment. Rehabilitation (Stuttg). 2013;52(3):164–72 PMID:23761205.

216. Preschl B, Maercker A, Wagner B. The working alliance in a randomized controlled trial comparing online with face-to-face cognitive-behavioral therapy for depression. BMC Psychiatry. 2011;11(1):189 PMID:22145768.
217. Cook JE, Doyle C. Working alliance in online therapy as compared to face-to-face therapy: preliminary results. Cyberpsychol Behav. 2002;5(2):95–105 PMID:12025884.
218. Bengtsson J, Nordin S, Carlbring P. Therapists' Experiences of conducting cognitive behavioural therapy online vis-à-vis face-to-face. Cogn Behav Ther. 2015;1–10 PMID:26090947, Routledge.
219. Knaevelsrud C, Maercker A. Does the quality of the working alliance predict treatment outcome in online psychotherapy for traumatized patients? J Med Internet Res. 2006;8(4):e31 PMID:17213049.
220. Mohr DC, Cuijpers P, Lehman K. Supportive accountability: a model for providing human support to enhance adherence to eHealth interventions. J Med Internet Res. 2011 Mar 10 [cited 2017 May 15];13(1):e30 PMID:21393123.
221. Terhorst Y, Rathner EM, Baumeister H, Sander L. Help from the App-Store? a systematic review of apps for depression (in press). Verhaltenstherapie 2017.
222. Sucala M, Cuijpers P, Muench F, Cardoş R, Soflau R, Dobrean A, Achimas-Cadariu P, David D. Anxiety: there is an app for that. A systematic review of anxiety apps. Depress Anxiety. 2017;34(6):518–25 Wiley Online Library.
223. Neary M, Schueller SM. State of the field of mental health apps. Cogn Behav Pract. 2018 Elsevier.
224. Klein JP, Knaevelsrud C, Bohus M, Ebert DD, Gerlinger G, Günther K, Jacobi C, Löbner M, Riedel-Heller SG, Sander J. Internetbasierte Selbstmanagementinterventionen. Nervenarzt Springer; 2018;1–9.
225. Alpaydin E. Introduction to machine learning. MIT press; 2009. ISBN:0262303264.
226. Yarkoni T, Westfall J. Choosing prediction over explanation in psychology: lessons from machine learning. Perspect Psychol Sci. 2017;12(6):1100–22 SAGE Publications Sage CA: Los Angeles, CA.
227. Breiman L. Statistical modeling: the two cultures (with comments and a rejoinder by the author). Stat Sci. 2001;16(3):199–231 Institute of Mathematical Statistics.
228. Ward JS, Barker A. Undefined by data: a survey of big data definitions; 2013. arXiv:13095821.
229. Becker D, van Breda W, Funk B, Hoogendoorn M, Ruwaard J, Riper H. Predictive modeling in e-mental health: a common language framework. Internet Interv Neth. 2018;12:57–67 PMID:30135769.
230. Wongkoblap A, Vadillo MA, Curcin V. Researching mental health disorders in the era of social media: systematic review. J Med Internet Res. 2017;19(6):e228 PMID:28663166 Canada.
231. Kang K, Yoon C, Kim EY. Identifying depressive users in Twitter using multimodal analysis. In: 2016 International conference on big data smart comput (BigComp), IEEE;2016. p. 231–38.
232. Park S, Kim I, Lee SW, Yoo J, Jeong B, Cha M. Manifestation of depression and loneliness on social networks: a case study of young adults on Facebook. In: Proceedings of the 18th ACM conference on comput support cooperative work social computing, ACM; 2015. p. 557–70.
233. Hu Q, Li A, Heng F, Li J, Zhu T. Predicting depression of social media user on different observation windows. In: 2015 IEEE/WIC/ACM International conference on web intelligence agent technology (WI-IAT), IEEE; 2015. p. 361–64.
234. Tsugawa S, Kikuchi Y, Kishino F, Nakajima K, Itoh Y, Ohsaki H. Recognizing depression from twitter activity. In: Proceedings of the 33rd annual ACM conference on human factors in computing systems, ACM; 2015. p. 3187–96.
235. Park S, Lee SW, Kwak J, Cha M, Jeong B. Activities on Facebook reveal the depressive state of users. J Med Internet Res. 2013;15(10) JMIR Publications Inc.

236. Wang X, Zhang C, Sun L. An improved model for depression detection in micro-blog social network. In: 2013 IEEE 13th international conference on data mining workshops, IEEE; 2013. p. 80–87.
237. De Choudhury M, Counts S, Horvitz E. Social media as a measurement tool of depression in populations. In: Proceedings of the 5th annual ACM web science conference, ACM; 2013. p. 47–56.
238. Mowery D, Smith H, Cheney T, Stoddard G, Coppersmith G, Bryan C, Conway M. Understanding depressive symptoms and psychosocial stressors on Twitter: a corpus-based study. J Med Internet Res. 2017;19(2):e48 PMID:28246066 Canada.
239. Reece A. Chronology and patterns of psychiatric morbidity in substance-dependent and medical patients. Australas Psychiatry. 2009;17(2):170–71. https://doi.org/10.1177/10398562090170020З US: Informa Healthcare.
240. Reece AG, Danforth CM. Instagram photos reveal predictive markers of depression. EPJ Data Sci. 2017;6(1):15 SpringerOpen.
241. Lin H, Jia J, Guo Q, Xue Y, Li Q, Huang J, Cai L, Feng L. User-level psychological stress detection from social media using deep neural network. In: Proceedings of the 22nd ACM international conference on multimedia, ACM; 2014. p. 507–16.
242. Schwartz HA, Sap M, Kern ML, Eichstaedt JC, Kapelner A, Agrawal M, Blanco E, Dziurzynski L, Park G, Stillwell D. Predicting individual well-being through the language of social media. In: Biocomput 2016 proceedings of the pacific symposium world scientific; 2016. p. 516–27.
243. Liu P, Tov W, Kosinski M, Stillwell DJ, Qiu L. Do Facebook status updates reflect subjective well-being? Cyberpsychology, Behav Soc Netw. 2015;18(7):373–79 Mary Ann Liebert, Inc., 140 Huguenot Street, 3rd Floor New Rochelle, NY 10801 USA.
244. Settanni M, Azucar D, Marengo D. Predicting individual characteristics from digital traces on social media: a meta-analysis. Cyberpsychology Behav Soc Netw. 2018;21(4):217–28 Mary Ann Liebert, Inc. 140 Huguenot Street, 3rd Floor New Rochelle, NY 10801 USA.
245. Coppersmith G, Dredze M, Harman C. Quantifying mental health signals in Twitter. In: Proceedings of the workshop on computational linguistics and clinical psychology: from linguistic signal to clinical reality; 2014. p. 51–60.
246. Harman G, Dredze MH. Measuring post traumatic stress disorder in Twitter. In: ICWSM 2014.
247. Chancellor S, Lin Z, Goodman EL, Zerwas S, De Choudhury M. Quantifying and predicting mental illness severity in online pro-eating disorder communities. In: Proceedings of the 19th ACM conference on computer-supported cooperative work & social computing, ACM; 2016. p. 1171–84.
248. Moessner M, Feldhege J, Wolf M, Bauer S. Analyzing big data in social media: text and network analyses of an eating disorder forum. Int J Eat Disord. 2018 PMID:29746710 United States.
249. Song J, Song TM, Seo D-C, Jin JH. Data mining of web-based documents on social networking sites that included suicide-related words among korean adolescents. J Adolesc Health. 2016;59(6):668–73 PMID:27693129 United States.
250. Braithwaite SR, Giraud-Carrier C, West J, Barnes MD, Hanson CL. Validating machine learning algorithms for Twitter data against established measures of suicidality. JMIR Ment Heal. 2016;3(2) JMIR Publications Inc.
251. O'Dea B, Wan S, Batterham PJ, Calear AL, Paris C, Christensen H. Detecting suicidality on Twitter. Internet Interv. 2015;2(2):183–88 Elsevier.
252. Zhang L, Huang X, Liu T, Li A, Chen Z, Zhu T. Using linguistic features to estimate suicide probability of Chinese microblog users. In: International conference on human centered computing, Springer; 2014. p. 549–59.
253. Burnap P, Colombo G, Amery R, Hodorog A, Scourfield J. Multi-class machine classification of suicide-related communication on Twitter. Online Soc networks media. 2017;2:32–44 Elsevier.

254. Torous J, Larsen ME, Depp C, Cosco TD, Barnett I, Nock MK, Firth J. Smartphones, sensors, and machine learning to advance real-time prediction and interventions for suicide prevention: a review of current progress and next steps. Curr Psychiatry Rep. 2018;20(7):51 PMID:29956120 United States.
255. Kornfield R, Sarma PK, Shah DV, McTavish F, Landucci G, Pe-Romashko K, Gustafson DH. Detecting recovery problems just in time: application of automated linguistic analysis and supervised machine learning to an online substance abuse forum. J Med Internet Res. 2018;20(6):e10136 PMID:29895517 Canada.
256. Cepeda MS, Reps J, Fife D, Blacketer C, Stang P, Ryan P. Finding treatment-resistant depression in real-world data: how a data-driven approach compares with expert-based heuristics. Depress Anxiety. 2017. https://doi.org/10.1002/da.22705 US: Wiley.
257. Morales S, Barros J, Echavarri O, Garcia F, Osses A, Moya C, Maino MP, Fischman R, Nunez C, Szmulewicz T, Tomicic A. Acute mental discomfort associated with suicide behavior in a clinical sample of patients with affective disorders: ascertaining critical variables using artificial intelligence tools. Front Psychiatry. 2017;8:7 PMID:28210230 Switzerland.
258. Whelan R, Watts R, Orr CA, Althoff RR, Artiges E, Banaschewski T, Barker GJ, Bokde ALW, Buchel C, Carvalho FM, Conrod PJ, Flor H, Fauth-Buhler M, Frouin V, Gallinat J, Gan G, Gowland P, Heinz A, Ittermann B, Lawrence C, Mann K, Martinot J-L, Nees F, Ortiz N, Paillere-Martinot M-L, Paus T, Pausova Z, Rietschel M, Robbins TW, Smolka MN, Strohle A, Schumann G, Garavan H. Neuropsychosocial profiles of current and future adolescent alcohol misusers. Nature 2014;512(7513):185–89 PMID:25043041 England.
259. Chekroud AM, Foster D, Zheutlin AB, Gerhard DM, Roy B, Koutsouleris N, Chandra A, Esposti MD, Subramanyan G, Gueorguieva R, Paulus M, Krystal JH. Predicting barriers to treatment for depression in a U.S. National sample: a cross-sectional, proof-of-concept study. Psychiatr Serv. 2018;69(8):927–34 PMID:29962307 United States.
260. Jordan P, Shedden-Mora MC, Lowe B. Predicting suicidal ideation in primary care: an approach to identify easily assessable key variables. Gen Hosp Psychiatry. 2018;51:106–11 PMID:29428582 United States.
261. Tran T, Luo W, Phung D, Harvey R, Berk M, Kennedy RL, Venkatesh S. Risk stratification using data from electronic medical records better predicts suicide risks than clinician assessments. BMC Psychiatry. 2014;14(1):76 BioMed Central.
262. Hoexter MQ, Dougherty DD, Shavitt RG, D'Alcante CC, Duran FLS, Lopes AC, Diniz JB, Batistuzzo MC, Evans KC, Bressan RA, Busatto GF, Miguel EC. Differential prefrontal gray matter correlates of treatment response to fluoxetine or cognitive-behavioral therapy in obsessive–compulsive disorder. Eur Neuropsychopharmacol. 2013;23(7):569–80. https://doi.org/10.1016/j.euroneuro.2012.06.014 Hoexter, Marcelo Q., Department of Psychiatry, University of Sao Paulo Medical School, Rua Ovidio Pires de Campos 785–31 andar Ala Norte-sala 9 (PROTOC), CEP 05403-010, Sao Paulo, Brazil: Elsevier Science.
263. Lueken U, Straube B, Yang Y, Hahn T, Beesdo-Baum K, Wittchen H-U, Konrad C, Strohle A, Wittmann A, Gerlach AL, Pfleiderer B, Arolt V, Kircher T. Separating depressive comorbidity from panic disorder: a combined functional magnetic resonance imaging and machine learning approach. J Affect Disord Neth. 2015;184:182–92 PMID:26093832.
264. Jimenez-Serrano S, Tortajada S, Garcia-Gomez JM. A mobile health application to predict postpartum depression based on machine learning. Telemed J E Health. 2015;21(7):567–74 PMID:25734829 United States.
265. Marzano L, Bardill A, Fields B, Herd K, Veale D, Grey N, Moran P. The application of mHealth to mental health: opportunities and challenges. Lancet Psychiatry. 2015;2(10):942–48 Elsevier.
266. Pantelopoulos A, Bourbakis NG. A survey on wearable sensor-based systems for health monitoring and prognosis. IEEE Trans Syst Man, Cybern Part C (Appl Rev IEEE). 2010;40(1):1–12.

267. Schueller SM, Begale M, Penedo FJ, Mohr DC. Purple: a modular system for developing and deploying behavioral intervention technologies. J Med Internet Res. 2014;16(7) JMIR Publications Inc.
268. Torous J, Kiang M V, Lorme J, Onnela J-P. New tools for new research in psychiatry: a scalable and customizable platform to empower data driven smartphone research. JMIR Ment Health. 2016;3(2) JMIR Publications Inc.
269. Mohr DC, Zhang M, Schueller SM. Personal sensing: understanding mental health using ubiquitous sensors and machine learning. Annu Rev Clin Psychol. 2017;13:23–47 PMID:28375728 United States.
270. Bourla A, Mouchabac S, El Hage W, Ferreri F. e-PTSD: an overview on how new technologies can improve prediction and assessment of Posttraumatic Stress Disorder (PTSD). Eur J Psychotraumatol. 2018;9(sup1):1424448 PMID:29441154 United States.
271. Trull TJ, Solhan MB, Tragesser SL, Jahng S, Wood PK, Piasecki TM, Watson D. Affective instability: measuring a core feature of borderline personality disorder with ecological momentary assessment. J Abnorm Psychol. 2008;117(3):647 American Psychological Association.
272. Dennis ML, Scott CK, Funk RR. Ecological momentary assessment to predict the risk of relapse. Drug Alcohol Depend. 2015;146:e262. https://doi.org/10.1016/j.drugalcdep.2014.09.179.
273. Chih M-Y, Patton T, McTavish FM, Isham AJ, Judkins-Fisher CL, Atwood AK, Gustafson DH. Predictive modeling of addiction lapses in a mobile health application. J Subst Abuse Treat. 2014;46(1):29–35 PMID:24035143 United States.
274. Saeb S, Kording K, Mohr DC. Making activity recognition robust against deceptive behavior. PLoS One. 2015;10(12):e0144795 PMID:26659118 United States.
275. Pratap A, Atkins DC, Renn BN, Tanana MJ, Mooney SD, Anguera JA, Arean PA. The accuracy of passive phone sensors in predicting daily mood. Depress Anxiety. 2018 PMID:30129691 United States.
276. Hung GC-L, Yang P-C, Chang C-C, Chiang J-H, Chen Y-Y. Predicting negative emotions based on mobile phone usage patterns: an exploratory study. JMIR Res Protoc. 2016;5(3):e160 PMID:27511748 Canada.
277. Jaques N, Taylor S, Azaria A, Ghandeharioun A, Sano A, Picard R. Predicting students' happiness from physiology, phone, mobility, and behavioral data. In: International conference on affective computing and intelligent interaction and workshops [proceedings]. ACII United States; 2015:222–28 PMID:28515966.
278. Sano A. Measuring college students' sleep, stress, mental health and wellbeing with wearable sensors and mobile phones. Massachusetts Institute of Technology; 2016 [cited 2017 May 10]. https://dspace.mit.edu/handle/1721.1/106066.
279. Sano A, Taylor S, McHill AW, Phillips AJ, Barger LK, Klerman E, Picard R. Identifying objective physiological markers and modifiable behaviors for self-reported stress and mental health status using wearable sensors and mobile phones: observational study. J Med Internet Res. 2018;20(6).e210 PMID:29884610 Canada.
280. Lu H, Frauendorfer D, Rabbi M, Mast MS, Chittaranjan GT, Campbell AT, Gatica-Perez D, Choudhury T. StressSense: detecting stress in unconstrained acoustic environments using smartphones. In: Proceedings of the 2012 ACM conference on ubiquitous computing, ACM; 2012. p. 351–60.
281. Abbas H, Garberson F, Glover E, Wall DP. Machine learning approach for early detection of autism by combining questionnaire and home video screening. J Am Med Inform Assoc. 2018;25(8):1000–07 PMID:29741630, England.
282. Sathyanarayana A, Joty S, Fernandez-Luque L, Ofli F, Srivastava J, Elmagarmid A, Arora T, Taheri S. Sleep quality prediction from wearable data using deep learning. JMIR mHealth uHealth. 2016;4(4):e125 PMID:27815231 Canada.

283. Barnett I, Torous J, Staples P, Sandoval L, Keshavan M, Onnela J-P. Relapse prediction in schizophrenia through digital phenotyping: a pilot study. Neuropsychopharmacology. 2018;1 Nature Publishing Group.
284. Depp CA, Thompson WK, Frank E, Swartz HA. Prediction of near-term increases in suicidal ideation in recently depressed patients with bipolar II disorder using intensive longitudinal data. J Affect Disord. 2017;208:363–8 Elsevier.
285. Hofmann SG, Curtiss J, McNally RJ. A complex network perspective on clinical science. Perspect Psychol Sci. 2016;11(5):597–605 PMID:27694457.
286. van de Leemput IA, Wichers M, Cramer AOJ, Borsboom D, Tuerlinckx F, Kuppens P, van Nes EH, Viechtbauer W, Giltay EJ, Aggen SH. Critical slowing down as early warning for the onset and termination of depression. Proc Natl Acad Sci. 2014;111(1):87–92.
287. Fried EI, van Borkulo CD, Cramer AOJ, Boschloo L, Schoevers RA, Borsboom D. Mental disorders as networks of problems: a review of recent insights. Soc Psychiatry Psychiatr Epidemiol. 2017;52(1):1–10. https://doi.org/10.1007/s00127-016-1319-z.
288. Wampold BE, Imel ZE. The great psychotherapy debate: the evidence for what makes psychotherapy work. Routledge; 2015. ISBN:1136672605.
289. Loerinc AG, Meuret AE, Twohig MP, Rosenfield D, Bluett EJ, Craske MG. Response rates for CBT for anxiety disorders: need for standardized criteria. Clin Psychol Rev. 2015;42:72–82 Elsevier.
290. Hofmann SG, Asnaani A, Vonk IJJ, Sawyer AT, Fang A. The efficacy of cognitive behavioral therapy: a review of meta-analyses. Cognit Ther Res. 2012;36(5):427–40. https://doi.org/10.1007/s10608-012-9476-1.
291. Rush AJ, Trivedi MH, Wisniewski SR, Nierenberg AA, Stewart JW, Warden D, Niederehe G, Thase ME, Lavori PW, Lebowitz BD. Acute and longer-term outcomes in depressed outpatients requiring one or several treatment steps: a STAR* D report. Am J Psychiatry Am Psychiatric Assoc. 2006;163(11):1905–17.
292. Kessler RC. The potential of predictive analytics to provide clinical decision support in depression treatment planning. Curr Opin Psychiatry. 2018;31(1):32–39 Wolters Kluwer.
293. Riley RD, Ensor J, Snell KIE, Debray TPA, Altman DG, Moons KGM, Collins GS. External validation of clinical prediction models using big datasets from e-health records or IPD meta-analysis: opportunities and challenges. BMJ. 2016;353:i3140 PMID:27334381 England.
294. Gori A, Lauro-Grotto R, Giannini M, Schuldberg D. Predicting treatment outcome by combining different assessment tools: toward an integrative model of decision support in psychotherapy. J Psychother Integr. 2010;20(2):251–69. https://doi.org/10.1037/a0019768 Gori, Alessio, Department of Psychology, University of Florence, via di San Salvi, 12, 50135, Florence, Italy: Educational Publishing Foundation.
295. Hoogendoorn M, Berger T, Schulz A, Stolz T, Szolovits P. Predicting social anxiety treatment outcome based on therapeutic email conversations. IEEE J Biomed Health Inf. 2017;21(5):1449–59 IEEE.
296. Lenhard F, Sauer S, Andersson E, Mansson KN, Mataix-Cols D, Ruck C, Serlachius E. Prediction of outcome in internet-delivered cognitive behaviour therapy for paediatric obsessive-compulsive disorder: a machine learning approach. Int J Methods Psychiatr Res. 2018;27(1) PMID:28752937 United States.
297. Mansson KNT, Frick A, Boraxbekk C-J, Marquand AF, Williams SCR, Carlbring P, Andersson G, Furmark T. Predicting long-term outcome of Internet-delivered cognitive behavior therapy for social anxiety disorder using fMRI and support vector machine learning. Transl Psychiatry 2015;5:e530 PMID:25781229 United States.
298. Askland KD, Garnaat S, Sibrava NJ, Boisseau CL, Strong D, Mancebo M, Greenberg B, Rasmussen S, Eisen J. Prediction of remission in obsessive compulsive disorder using a novel machine learning strategy. Int J Methods Psychiatr Res. 2015;24(2):156–69 Wiley Online Library.

299. Sundermann B, Bode J, Lueken U, Westphal D, Gerlach AL, Straube B, Wittchen H-U, Strohle A, Wittmann A, Konrad C, Kircher T, Arolt V, Pfleiderer B. Support vector machine analysis of functional magnetic resonance imaging of interoception does not reliably predict individual outcomes of cognitive behavioral therapy in panic disorder with agoraphobia. Front Psychiatry. 2017;8:99 PMID:28649205, Switzerland.
300. Kessler RC, van Loo HM, Wardenaar KJ, Bossarte RM, Brenner LA, Cai T, Ebert DD, Hwang I, Li J, de Jonge P, Nierenberg AA, Petukhova MV, Rosellini AJ, Sampson NA, Schoevers RA, Wilcox MA, Zaslavsky AM. Testing a machine-learning algorithm to predict the persistence and severity of major depressive disorder from baseline self-reports. Mol Psychiatry. 2016;21(10):1366–71. https://doi.org/10.1038/mp.2015.198.
301. Kessler RC, Hwang I, Hoffmire CA, McCarthy JF, Petukhova M V, Rosellini AJ, Sampson NA, Schneider AL, Bradley PA, Katz IR. Developing a practical suicide risk prediction model for targeting high-risk patients in the Veterans health Administration. Int J Methods Psychiatr Res. 2017;26(3):e1575 Wiley Online Library.
302. Perlis RH. A clinical risk stratification tool for predicting treatment resistance in major depressive disorder. Biol Psychiatry. 2013;74(1):7–14 Elsevier.
303. Mikus A, Hoogendoorn M, Rocha A, Gama J, Ruwaard J, Riper H. Predicting short term mood developments among depressed patients using adherence and ecological momentary assessment data. Internet Interv Neth. 2018;12:105–10 PMID:30135774.
304. Chekroud AM, Zotti RJ, Shehzad Z, Gueorguieva R, Johnson MK, Trivedi MH, Cannon TD, Krystal JH, Corlett PR. Cross-trial prediction of treatment outcome in depression: a machine learning approach. Lancet Psychiatry. 2016;3(3):243–250 Elsevier.
305. Chekroud AM, Gueorguieva R, Krumholz HM, Trivedi MH, Krystal JH, McCarthy G. Reevaluating the efficacy and predictability of antidepressant treatments: a symptom clustering approach. JAMA Psychiatry. 2017;74(4):370–378 American Medical Association.
306. Spring Health. https://www.springhealth.com/about/.
307. Bremer V, Becker D, Kolovos S, Funk B, van Breda W, Hoogendoorn M, Riper H. Predicting therapy success and costs for personalized treatment recommendations using baseline characteristics: data-driven analysis. J Med Internet Res. 2018;20(8):e10275 PMID:30131318 Canada.
308. Ebert DD, Cuijpers P. It time to invest in the prevention of depression. JAMA Netw. 2018;1(2):e180335. https://doi.org/10.1001/jamanetworkopen.2018.0335 Open American Medical Association.
309. Walton A, Nahum-Shani I, Crosby L, Klasnja P, Murphy S. Optimizing digital integrated care via micro-randomized trials. Clin Pharmacol Ther. 2018 Wiley Online Library.
310. Collins LM. Optimization of behavioral, biobehavioral, and biomedical interventions: the multiphase optimization strategy (MOST). Springer; 2018. ISBN:3319722069.
311. Nahum-Shani I, Hekler EB, Spruijt-Metz D. Building health behavior models to guide the development of just-in-time adaptive interventions: a pragmatic framework. Heal Psychol. 2015;34(S):1209 American Psychological Association.
312. Nahum-Shani I, Smith SN, Spring BJ, Collins LM, Witkiewitz K, Tewari A, Murphy SA. Just-in-time adaptive interventions (JITAIs) in mobile health: key components and design principles for ongoing health behavior support. Ann Behav Med. 1–17. https://doi.org/10.1007/s12160-016-9830-8 Springer US.
313. Goldstein SP, Evans BC, Flack D, Juarascio A, Manasse S, Zhang F, Forman EM. Return of the JITAI: applying a just-in-time adaptive intervention framework to the development of m health solutions for addictive behaviors. Int J Behav Med. 2017;24(5):673–82 PMID:28083725 England.
314. Jaimes LG, Llofriu M, Raij A. A stress-free life: just-in-time interventions for stress via real-time forecasting and intervention adaptation. In: BODYNETS 2014-9th international conference on body area networks 2014;1:197–203. https://doi.org/10.4108/icst.bodynets.2014.258237.

315. Thomas JG, Bond DS. Behavioral response to a just-in-time adaptive intervention (JITAI) to reduce sedentary behavior in obese adults: implications for JITAI optimization. Health Psychol. 2015;34(Suppl):1261–7 PMID:26651467.
316. Spruijt-Metz D, Nilsen W. Dynamic models of behavior for just-in-time adaptive interventions. IEEE Pervasive Comput. 2014;13(3):13–7. https://doi.org/10.1109/MPRV.2014.46.
317. Juarascio AS, Parker MN, Lagacey MA, Godfrey KM. Just-in-time adaptive interventions: a novel approach for enhancing skill utilization and acquisition in cognitive behavioral therapy for eating disorders. Int J Eat Disord. 2018 PMID:30051495 United States.
318. Wahle F, Kowatsch T, Fleisch E, Rufer M, Weidt S. Mobile sensing and support for people with depression: a pilot trial in the wild. JMIR mHealth uHealth. 2016;4(3):e111 PMID:27655245 Canada.
319. Burns MN, Begale M, Duffecy J, Gergle D, Karr CJ, Giangrande E, Mohr DC. Harnessing context sensing to develop a mobile intervention for depression. J Med Internet Res. 2011;13(3):e55 PMID:21840837 Canada.
320. Ben-Zeev D, Kaiser SM, Brenner CJ, Begale M, Duffecy J, Mohr DC. Development and usability testing of FOCUS: a smartphone system for self-management of schizophrenia. Psychiatr Rehbil J. 2013;36(4):289–96 PMID:24015913.
321. Gustafson DH, McTavish FM, Chih M-Y, Atwood AK, Johnson RA, Boyle MG, Levy MS, Driscoll H, Chisholm SM, Dillenburg L, Isham A, Shah D. A smartphone application to support recovery from alcoholism: a randomized clinical trial. JAMA Psychiatry. 2014;71(5):566–72 PMID:24671165.
322. Pina L, Rowan K, Roseway a, Johns P, Hayes GR, Czerwinski M. In situ cues for ADHD parenting strategies using mobile technology. In: Proceedings of the-PERVASIVEHEALTH 2014 8th international conference on pervasive computing technologies for healthcare 2014. p. 17–24. https://doi.org/10.4108/icst.pervasivehealth.2014.254958.
323. Forman EM, Goldstein SP, Zhang F, Evans BC, Manasse SM, Butryn ML, Juarascio AS, Abichandani P, Martin GJ, Foster GD. OnTrack: development and feasibility of a smartphone app designed to predict and prevent dietary lapses. Transl Behav Med. 2018 PMID:29617911 England.
324. Luxton DD. Artificial intelligence in behavioral and mental health care. Academic Press; 2015. ISBN:0128007923.
325. DeVault D, Artstein R, Benn G, Dey T, Fast E, Gainer A, Georgila K, Gratch J, Hartholt A, Lhommet M. SimSensei Kiosk: a virtual human interviewer for healthcare decision support. In: Proceedings of the 2014 international conference on Autonomous agents and multiagent systems. International Foundation for Autonomous Agents and Multiagent Systems; 2014. p. 1061–68.
326. Bird T, Mansell W, Wright J, Gaffney H, Tai S. Manage your life online: a web-based randomized controlled trial evaluating the effectiveness of a problem-solving intervention in a student sample. Behav Cogn Psychother. 2018;1–13 PMID:29366432 United States.
327. Burton C, Tatar AS, McKinstry B, Matheson C, Matu S, Moldovan R, Macnab M, Farrow E, David D, Pagliari C, Blanco AS, Wolters M. Pilot randomised controlled trial of Help4Mood, an embodied virtual agent-based system to support treatment of depression. J Telemed Telecare. 2016;22(6):348–355. https://doi.org/10.1177/1357633x15609793 Burton, Christopher, University of Aberdeen Aberdeen, Aberdeen, United Kingdom: Sage Publications.
328. Crutzen R, Peters G-JY, Portugal SD, Fisser EM, Grolleman JJ. An artificially intelligent chat agent that answers adolescents' questions related to sex, drugs, and alcohol: an exploratory study. J Adolesc Health. 2011;48(5):514–19 PMID:21501812 United States.
329. Provoost S, Lau HM, Ruwaard J, Riper H. Embodied conversational agents in clinical psychology: a scoping review. J Med Internet Res. 2017;19(5) JMIR Publications Inc.
330. Khoury MJ, Ioannidis JPA. Big data meets public health. Science. 2014;346(6213):1054–55 (80-) American Association for the Advancement of Science.
331. Pearl J. Myth, confusion, and science in causal analysis. 2009.

332. Goldstein BA, Navar AM, Pencina MJ, Ioannidis J. Opportunities and challenges in developing risk prediction models with electronic health records data: a systematic review. J Am Med Inf Assoc. 2017;24(1):198–208 Oxford University Press.
333. Moons KGM, Altman DG, Reitsma JB, Ioannidis JPA, Macaskill P, Steyerberg EW, Vickers AJ, Ransohoff DF, Collins GS. Transparent Reporting of a multivariable prediction model for Individual Prognosis or Diagnosis (TRIPOD): explanation and elaboration. Ann Intern Med. 2015;162(1):W1–W73 American College of Physicians.
334. DeMasi O, Kording K, Recht B. Meaningless comparisons lead to false optimism in medical machine learning. PLoS One. 2017;12(9):e0184604 PMID:28949964 United States.
335. Pearl J, MacKenzie D. The book of why: the new science of cause and effect. 1st ed. New York: Basic Books; 2018.
336. Chen JH, Asch SM. Machine learning and prediction in medicine—beyond the peak of inflated expectations. N Engl J Med. 2017;376(26):2507 NIH Public Access.
337. Cadwalladr C, Graham-Harrison E. Revealed: 50 million Facebook profiles harvested for Cambridge Analytica in major data breach. Guard 2018;17.
338. Gianfrancesco MA, Tamang S, Yazdany J, Schmajuk G. Potential biases in machine learning algorithms using electronic health record data. JAMA Intern Med. 2018.
339. Nahum-Shani I, Smith SN, Spring BJ, Collins LM, Witkiewitz K, Tewari A, Murphy SA. Just-in-time adaptive interventions (JITAIs) in mobile health: key components and design principles for ongoing health behavior support. Ann Behav Med. 2017;52(6):446–62 Oxford University Press US.

Cognitive Behavioral Therapy for Insomnia in the Digital Age

Seung-Gul Kang and Yong-Ku Kim

Abstract

Cognitive behavioral therapy for insomnia (CBT-I) is the treatment of choice for insomnia; however, it is not widely used due to a lack of experienced therapists and its relatively high clinical cost. Recently, Internet and mobile CBT-I have been developed to replace face-to-face CBT-I, and research on this topic has been increasing. In addition, attempts have been made to use wearable devices for sleep–wake estimation. Studies on digital CBT-I thus far have shown favorable treatment effects in general, but the problem of a high dropout rate has not been sufficiently improved. In addition, more sophisticated technology is needed to develop fully automated digital CBT-I. As part of efforts to maximize the treatment effectiveness of future insomnia patients, research and development of mobile and Internet CBT-I and improvement of sleep tracker accuracy and validation studies are needed.

Keywords

Sleep · Insomnia · Cognitive behavioral therapy for insomnia · Internet · Mobile application · Wearable device

S.-G. Kang (✉)
Department of Psychiatry, Gil Medical Center, College of Medicine, Gachon University, 21, Namdong-daero 774 beon-gil, Namdong-gu, Incheon 21565, Republic of Korea
e-mail: kangsg@gachon.ac.kr

Y.-K. Kim
Department of Psychiatry, Korea University College of Medicine, Seoul, Republic of Korea
e-mail: yongku@korea.edu

© Springer Nature Singapore Pte Ltd. 2019

Y.-K. Kim (ed.), *Frontiers in Psychiatry*, Advances in Experimental Medicine and Biology 1192, https://doi.org/10.1007/978-981-32-9721-0_30

Introduction

Many people in modern society have sleep insufficiency and sleep–wake problems and disorders. In particular, insomnia is one of the most common distressing and clinically important sleep disturbances. Approximately, one-third of adults experience insomnia symptoms on a weekly basis [1–3], and 10–20% of the population reportedly have a chronic insomnia disorder [4]. This issue not only impairs one's quality of life and functioning but also affects many aspects of health, including cognitive function, mental health, physical health, and suicide risk [5]. An even more serious problem is that approximately 50% of insomniacs have chronic courses of the disorder [3]. The primary causes of the chronic condition include poor sleep hygiene and an irregular sleep–wake schedule, which lead to the perpetuation of insomnia symptoms and the cognitive distortion toward sleep and insomnia. The chronicity of insomnia symptoms will further worsen sleep hygiene; these symptoms include extended time in bed (TIB), frequent and long napping, irregular sleep–wake schedule, worry and complaints about poor sleep, and sequelae of insomnia.

Evidence-based treatments for insomnia include cognitive behavioral therapy for insomnia (CBT-I) and pharmacotherapy using hypnotics approved by the Food and Drug Administration. While pharmacotherapy is the most commonly used treatment method, hypnotics, including benzodiazepine receptor agonists, are related to side effects, tolerance, and dependence [6]. Pharmacotherapy is usually not curative and leads to the disappearance of efficacy after 1 year [7]. CBT-I is equally as effective as and more long-lasting than pharmacotherapy [8]; therefore, CBT-I is recommended for the primary treatment of chronic insomnia.

However, successful CBT-I (a) requires an experienced and well-trained therapist to be available for the patient, (b) requires substantial time and cost for many treatment sessions (weekly for 6 to 8 weeks), and (c) should be restricted to a specific time and/or place due to its nature and methodology. These factors often constitute obstacles to successful CBT-I for many patients.

For this reason, CBT-I has been developed and studied using methods other than classical face-to-face CBT-I. Due to the development of the Internet and personal computers, CBT-I through these means have been introduced. Recently, much interest and attention have been directed to mobile applications and wearable devices for the treatment of insomnia. The use of sleep telemedicine developed by the American Academy of Sleep Medicine (AASM) in the management of sleep disorders in clinical applications has recently been receiving attention, but few related systematic studies and reviews have been conducted. This chapter aims to describe CBT-I using the Internet, mobile applications, and wearable devices.

Internet CBT-I

As described in the previous section, to overcome the inefficiency and cost problems of classical CBT-I, many studies and developments have been performed on Internet CBT-I. Many Internet CBT-I programs have had patients keep a sleep diary, which resulted in a time-in-bed (TIB) readjustment and improved sleep hygiene.

The first randomized controlled trial (RCT) evaluating the efficacy of Internet CBT-I was reported in 2004 and compared a 5-week Internet-delivered sleep management program to a wait-list control intervention in a clinical sample with insomnia ($n = 109$) [9]. The 5-week intervention program consisted of sleep restriction, stimulus control, and cognitive restructuring [9]. The results showed significant improvements in the Internet treatment group on many sleep measures, including total sleep time (TST), total wake TIB, and sleep efficiency [9]. However, improvements were also found in the control group, and between-groups effect sizes were low [9].

Ritterband et al. randomized 45 adults with insomnia into an Internet intervention group ($n = 22$) and a wait-list control group ($n = 23$) [10]. In this study, the Internet intervention program was based on established face-to-face CBT-I incorporating sleep restriction, stimulus control, sleep hygiene, cognitive restructuring, and relapse prevention [10]. Intention-to-treat analyses showed that Insomnia Severity Index (ISI) scores improved significantly from 15.7 to 6.6 for the Internet group but did not change for the control group [10]. The Internet group maintained the treatment effects until 6 months after treatment [10]. Recent studies and meta-analyses have found that therapist-guided online CBT-I is as effective as face-to-face treatment [6, 11, 12]. According to the meta-analysis of 11 RCTs of 1460 subjects, the effects of Internet CBT-I were comparable to those in face-to-face CBT-I, and the effect lasted for 4–48 weeks [6]. The longer treatment duration and higher level of personal support were associated with larger effect sizes in the study [6]. In some of the studies included in this meta-analysis, Internet CBT-I did not show significant effects on insomnia severity or sleep efficiency [13–15], but most of the other studies showed a significant effect [6]. A second meta-analysis using data from 15 RCTs suggested similar effectiveness for Internet CBT-I [16]. The sleep efficiency improved by 7.2%, the ISI score increased by 4.3 points, and the TST increased by 20 min in Internet CBT-I compared to control group [16].

Several direct comparisons have been made between adults receiving Internet CBT-I and those receiving face-to-face CBT-I [17, 18]. A study comparing Internet CBT-I to face-to-face CBT-I suggested that face-to-face CBT-I outperforms Internet CBT-I [18]. However, another study did not find any difference in the treatment effectiveness between group CBT-I and Internet CBT-I [17].

Many studies on Internet CBT-I have noted problems with high attrition and dropout rates. Unlike face-to-face CBT-I, the patients with low motivation easily quit Internet CBT-I, and noncompletion rates have ranged from 43 to 99% in the previous literature [19]. There was an opinion that the treatment outcome might seem to be better than the actual outcome in previous Internet CBT-I studies, since the follow-up data of dropout patients were not included [6]. Therefore, more detailed data regarding dropout patients should be reported appropriately.

Internet CBT-I has been shown to be effective in patients other than those with primary insomnia. Studies have shown that the Internet CBT-I is also effective in patients with mental disorders, including depressive and anxiety disorders [13, 20–25]. A recent meta-analysis including 10 studies also found that the effect sizes of the digital CBT-I in persons with depression or anxiety were in the range of those with noncomorbid insomnia [26]. The findings of RCTs have also suggested that digital CBT-I can be used to effectively treat insomnia in patients with physical illnesses such as tinnitus [27, 28], elevated blood pressure [29], and cancer [30].

Interactive web design, animation techniques, automated media-rich web applications, and personal clinical support designed to maximize the effectiveness of Internet CBT-I can improve the motivation and treatment outcomes of patients [11, 31]. Anderson et al. developed an interactive video-based, online CBT-I program that required participants to complete screening for sleep disorders other than insomnia prior to CBT-I [11]. Espie et al. made the Internet CBT-I program, which comprised a fully automated media-rich web application [31]. An animated virtual therapist named "Prof" conducted a progress review with the participant and explored the sleep diary data in the program [31], and the CBT-I algorithms fed the delivery of information, support, and advice, including TIB window and cognitive techniques [31].

Long-term treatment effects of Internet CBT-I were also studied. Ritterband et al. studied the long-term effectiveness of the Int CBT-I program Sleep Healthy Using the Internet [SHUTi] in 303 chronic insomnia patients [32]. Their CBT-I intervention program, which was a 6-week web-based, fully automated, and tailored CBT-I intervention method, was investigated at baseline, in the short term (9 weeks), at 6 months, and in the long term in the RCT [32]. The results showed that the group × time interaction was significant for all three primary sleep outcomes (ISI score, sleep onset latency, and wake after sleep onset). Treatment effects were maintained at the 1-year follow-up for ISI score and sleep onset latency. All secondary sleep outcomes except TST also showed significant group × time interactions, favoring the Internet CBT-I group [32].

Mobile CBT-I and Wearable Devices

While Internet-based self-help CBT-I can be more convenient than relying on in-person visits, the lack of interaction or communication with therapists can easily lead to patients experiencing decreased motivation to participate and poor adherence [12, 33]. Functions such as reminders may increase patient adherence to therapy [34]; however, these ancillary functions may have limitations in solving the fundamental problems of Internet CBT-I. Recently, there has been strong interest in the development and research of mobile CBT-I using smartphones to maximize the effect of such interventions given the high prevalence of smartphone use.

In recent years, several mobile applications delivering CBT-I programs have been developed, and an initial stage of research has been published. Kuhn et al. published a paper on the description and clinician perceptions of a mobile application called CBT-I Coach [35]. CBT-I Coach is a smartphone app based on the CBT-I manual of the Veterans Affairs National Center for Posttraumatic Stress Disorder and was designed to enhance CBT-I [35]. The clinicians answered that the CBT-I Coach app would improve care for insomnia; the majority intended to use it in the future [35], and 59.9% of participants reported a favorable impression of its impact on outcomes and homework adherence two years after the CBT-I Coach release [35]. Babson et al. published a pilot study using CBT-I Coach ($n=2$) and placebo control (mood-tracking app) ($n=2$) for veterans with cannabis use disorders and sleep problems, and the participants found that mobile apps were user-friendly and helpful [36]. The authors reported improved sleep efficiency, increased sleep quality, and decreased cannabis use after treatment [36]. Koffel et al. investigated the feasibility, the impact on adherence and sleep outcomes, and the acceptability in the pilot RCT ($n=18$) [37]. They reported that the participants used the app consistently as intended and showed improved sleep outcomes [37]. In addition, it was highly acceptable to use and was shown not to compromise the benefits of CBT-I [37].

Chen et al. reported a case report demonstrating the utilization, advantages, and limitations of mobile CBT-I apps in female Taiwanese older adults [38]. They developed a mobile app called "Win-Win aSleep" (WWaS) for assisting six-session CBT-I [38, 39]. A 64 year-old Chinese woman diagnosed with psychophysiological insomnia successfully discontinued all hypnotics and maintained satisfactory sleep quality after CBT-I using this app [38].

Horsch et al. investigated the efficacy of CBT-I delivered via the Sleepcare mobile phone app in an RCT [40]. They recruited Dutch participants with mild insomnia disorder and randomly assigned them to the app ($n=74$) or to the wait-list control ($n=77$) [40]. The app was fully automated, adjusted itself to a participant's progress, and came with a sleep diary, a sleep restriction exercise, a relaxation exercise, and a sleep hygiene exercise [40]. The results showed significant interaction effects favoring the app on insomnia severity and sleep efficiency [40]. This finding is meaningful in that this study was performed on a relatively large number of subjects using fully automated mobile phone apps that patients can use on their own.

Recently, Kyle et al. reported a research plan of the DISCO trial (Defining the Impact of Improved Sleep Cognitive Function) for the impact of mobile CBT-I on cognitive functioning and sleep in 404 subjects with insomnia disorder [41]. In this study, the participants will be randomized to digitally automated CBT-I delivered by mobile and/or web or to a wait-list control [41]. The primary outcome in this study is self-reported cognitive function at posttreatment (10 weeks) [41].

The traditional CBT-I is effective; however, the treatment is apt to be interrupted due to the long period of therapy. The self-regulated learning (SRL) program is known to help students manage their learning plans and improve learning outcomes [42]. Chu et al. developed the SRL program for mobile phones and investigated the utility of this program [43]. The results showed that this program was useful and easy to use. In addition, the number of students with insomnia significantly decreased [43].

In addition to its good portability, mobile phones have the advantage of easy wireless synchronization with wearable devices, such as the Fitbit and smartwatch. Many commercial devices, such as the Fitbit, Beautyrest, Emfit, Withings, Beddit, S+, Zeo, Jawbone, and Plex, have been released, and they determine sleeping and waking based on accelerometry, electroencephalography, ballistocardiography, and heart rate. AASM also designed and has been promoting a sleep telemedicine (Sleep TM) program, a telemedicine platform that uses wearable devices for sleep measurement. The appropriate use of the telemedicine program as a treatment method in addition to face-to-face treatments and the assurance of its quality will ensure its helpfulness to patients and therapists in educating about, monitoring, and managing sleep disorders such as insomnia and sleep apnea [44–46]. Additionally, the use of such wearable devices would help overcome the inaccuracies, limitations, and inconveniences of sleep diaries (e.g., sleep state misperception, discrepancy between subjective and objective sleep duration, underestimation of sleeping time and TIB, underreport of nap, and difficulty in the calculation of sleep diary measures).

These merits of commercial sleep trackers have attracted attention from sleep-disorder patients as well as healthy individuals. However, before using a wearable device to measure sleeping and waking in clinical and research fields, one must consider the question of how accurate wearable devices are. Despite rapidly growing interest, the accuracy of wearable devices has not been verified. Previous validation studies of commercial sleep trackers compared the first Fitbit Classic [47], the Fitbit Ultra [48], the Fitbit Flex [49], and the Jawbone UP [50, 51] with polysomnography. The results reported in these studies differed according to the sleep trackers, the age and clinical characteristics of the participants, the study design, and the method of statistical analysis. It has been claimed that validation studies for the clinical populations with each sleep disorder are needed [47]. Because of this problem, we investigated the validity of the Fitbit Flex in adult insomnia disorder patients and good sleepers [52]. The intraclass correlation coefficients were excellent for the Fitbit Flex normal mode in comparison with polysomnography for TST in both groups, and the level of agreement was high in good sleepers [52]. However, the level of acceptable agreement was significantly lower in the insomnia disorder group than in the good-sleepers group [52].

In spite of the limitations due to differences in participants' characteristics and the research designs, recent Fitbit trackers tend to show improved accuracy compared to older models [52]. The first relevant study reported that the sleep efficiency and TST of healthy adult participants, as measured by the Fitbit Classic released in 2008, were overestimated by 14.5% and 67.1 min, respectively, relative to those in polysomnography [47]. The sleep efficiency and TST of healthy children measured using the Fitbit Ultra released in 2011 were overestimated by 8% and 41 min, respectively [48]. The sleep efficiency and TST for the good sleepers in our study, as measured by the Fitbit Flex released in 2013, were overestimated by 1.75% and 6.5 min, respectively, relative to those in polysomnography. A recent study that measured the sleep efficiency and TST of healthy adolescents using the Fitbit Charge HR released in 2015 found that they were overestimated by 1.8% and 8 min, respectively [53]. These findings suggest that newer models of Fitbit devices to measure the sleep efficiency and TST in normal populations tend to reduce the degree of overestimation of sleep measurement. However, despite the remarkable development in the new sleep trackers and actigraphs, most of the sleep–wake scoring algorithms have not been open to the public, and their standardization has not been completed, making it a major hindrance to using these devices in clinical research and patient care [52].

We created a task force team comprising psychiatrists with expertise in sleep medicine and two information technology companies and developed the CBT-I mobile app (CBT-I-MA), which is synchronizable with wearable devices, from the spring of 2013 to the summer of 2015 [46]. We developed only an Android version of the app because more than 90% of smartphone users were using the Android operating system (OS) in South Korea when we started the development of the CBT-I-MA in 2013 [54]. The app could be used with or without sleep trackers, depending on the user's preferences. The core components of the app focused on sleep restriction therapy and stimulus control therapy. The important features were as follows: (a) sleep diary maintained by the users, (b) sleep and activity data viewer of the sleep trackers, (c) prescription for TIB, (d) educational video clips about sleep and insomnia, (e) questionnaires evaluating the severity of insomnia and sleepiness, (f) list of suggestions to improve sleep quality, (g) relaxation of the mind and body, and (h) reminder to the users to complete the sleep diary to increase adherence and to stop consuming caffeine after a specific time [46].

The sleep diary of the CBT-I-MA comprises time to bed, time of waking up, global rating of the quality of the last night's sleep, sleep latency, number of awakenings, duration of time out of bed at night, duration of wake time after sleep onset, napping hours, and the use of caffeine and hypnotics. Once the user completes the sleep diary in the app, the sleep efficiency and TST are automatically calculated to show the subjective sleep state of the last night. In addition, the sleep diary menu conveniently calculates the average TIB, sleep efficiency, and TST over the last week. If a user feels uncomfortable writing in a sleep diary every day, he or she can wear the sleep tracker only and synchronize it with the CBT-I-MA. The user can use Bluetooth to synchronize the sleep–wake record to the smartphone with the Fitbit app. The sleep record is then downloaded from the Fitbit web server and eventually synchronized to the CBT-I-MA, and then the sleep efficiency and TST from the Fitbit web server are automatically displayed in the CBT-I-MA. The participants' information in the app was uploaded to the therapists' web server, which was built for the research project [46].

Our team also studied and reported the effectiveness of the CBT-I-MA [46]. The main finding in our study was that CBT-I using our app was related to significant improvements in sleep, with an intermediate effect size for sleep diary sleep efficiency and with a large effect size for sleep efficiency measured by the Pittsburgh Sleep Quality Index. The sleep efficiency from actigraphy also improved significantly, although the effect size was small. The secondary outcome measures (sleep latency in the sleep diary and total scores for the ISI, Pittsburgh Sleep Quality Index, Beck Depression Inventory, Beck Anxiety Inventory, and Dysfunctional Beliefs and Attitudes about Sleep Scale) also showed positive results, with effect sizes ranging from small to large. In addition to these findings, the response rate of the intervention was excellent, although the number of participants was small [46].

The study participants showed relatively good adherence to the CBT-I-MA. They entered information into the sleep diary on an average of 24.3 days out of 28 days. Of the participants, 89.5% answered that our app was convenient to use, 94.7% answered that they used our app frequently, 94.7% answered that our app was effective for coping with insomnia, and 100% answered that they would use our app again in the future [46].

We also compared the improvement in the sleep measurements between the group using the sleep tracker and the group not using the sleep tracker [46]. Due to the small effect size and different characteristics of the two groups, there was no difference in the change in sleep efficiency between the two groups after treatment [46]. Although the treatment results in the group using the sleep tracker were not superior to those in the group not using the sleep tracker, using a sleep tracker is still likely to provide several benefits to patients and therapists. Although the sleep–wake-measuring algorithm of the Fitbit tracker does not seem to be very accurate [47, 48, 51], its use provides an objective sleep–wake record and may overcome the problems related to the sleep state misperception of sleep efficiency and may also correct the cognitive distortion. Moreover, the automatic napping measurement function of sleep trackers can correct the common issue of underreporting excessive napping by insomnia patients with a long TIB. Some insomniacs become too meticulous or obsessive about completing the sleep diary, which is stressful and can finally result in insomniacs giving up keeping sleep diary; the sleep–wake record obtained by using a wearable device can also provide a solution to such cases. The sleep trackers can also help the elderly [38] and children [51], who are likely to experience difficulties in voluntarily keeping a sleep diary. Using a sleep tracker also allows for the possibility of treating sleep disorders other than insomnia.

Using an app to keep a sleep diary avoids the possibility of losing the sleep diary, unlike pencil-and-paper methods. Patients can enter information into the sleep diary anywhere and more easily than even an Internet-based sleep diary. Using an app also makes it easy for patients and clinicians to review the sleep diaries and calculate the average sleep efficiency and TST for the most recent week. Since the CBT-I-MA enables therapists to review a sleep diary by transferring it via an Internet server, therapists can know the sleep–wake status of patients and provide appropriate and timely advice without requiring face-to-face visits.

Considerations for the Digital CBT-I

The terms that are used to describe the technology using the Internet, computers, and mobile phones include the following: Internet CBT-I, computerized CBT-I, mobile CBT-I, online CBT-I, and electronic CBT-I [55]. To describe these terms in one category, the term digital CBT was proposed [55]. Digital CBT-I could be classified into digital CBT-I as support, guided digital CBT-I, and fully automated digital CBT-I, according to the level of automatization, scalability, and the amount of therapists' time needed [55]. The digital CBT-I as support is the least extensive form of digital involvement in CBT-I, and the clinicians or health care professionals use these digital elements when they provide face-to-face CBT-I. The guided digital CBT-I programs combine clinicians' support and automated programs. Compared to face-to-face CBT-I, this class of digital CBT-I can reduce the time investment of health professionals [55, 56]. Fully automated digital CBT-I can function without any form of support of human therapists and provide the treatment using text with interactive components, video, or animation [56].

Since fully automated Internet CBT-I does not require the support of a therapist, it can be a cost-effective means of treating many people if the program is ideally structured. Espie et al. investigated the effect of the fully automated digital CBT-I using web and/or mobile channels on sleep-related quality of life, functional health, psychological well-being, and improvement in insomnia, which was a mediating factor in a large population (n = 1711) [57]. Compared to the control group, the use of this mobile CBT-I program was associated with a large improvement in sleep-related quality of life and a small improvement in psychological well-being and functional health [57]. A large improvement in insomnia mediated these treatment outcomes [57]. These findings suggested that digital CBT-I improves both daytime and nighttime aspects of insomnia [57].

However, there is concern regarding whether insomnia patients would fully understand and follow the CBT-I protocol provided by the automated digital CBT-I. Many components of CBT-I, such as sleep time restriction therapy and stimulus control therapy, have characteristics that are the opposite of those related to sleep hygiene in patients with insomnia and have a concept that may not be easily understood or accepted by insomnia patients. Therefore, the process of persuading patients to understand the concepts of CBT-I and addressing their resistance is critical in face-to-face CBT-I. In automated digital CBT-I, in which there are no healers at all, early dropout can occur.

Since digital CBT-I programs also collect and treat people's personal health information, similar to any other health care app, security and privacy issues are very important. Health information privacy is the right to control the collection, use, and disclosure of an individual's identifiable health data, and security encompasses the technological, administrative, and physical tools or safeguards that protect identifiable health data from unwarranted access or disclosure [58]. The app designers and users should be aware of and cautious about these issues at the time of app design and ensure that an insecure application would not be released. In mobile health applications, privacy- and security-related laws should consider information requirements, cover data, consent requirements, security, authentication, confidentiality, access control, data retention, data transfer, and breach notification obligation [59].

Conclusions

Sleep and wake problems and sleep deprivation are common and serious challenges in modern society. Many individuals and clinicians do not cope effectively with sleep problems, and there is an unmet need for greater clinical expertise in addressing sleep problems. For this reason, the use of Internet CBT-I, mobile apps for CBT-I, and wearable devices have been increasing in the field of sleep, and studies are underway. However, research on mobile apps and wearable devices is still lacking.

The use of digital programs and wearable devices in sleep–wake disorders is expected to enhance the effectiveness of CBT-I and sleep habits. To use commercial sleep trackers for clinical and research purposes, it is necessary to improve and evaluate the accuracy of determining the sleep–wake period.

Acknowledgements This research was supported by Basic Science Research Program through the National Research Foundation of Korea (NRF) funded by the Ministry of Education (NRF-2017R1D1A1B03032431). This research was supported by Basic Science Research Program through the National Research Foundation of Korea funded by the Ministry of Education (NRF-2013R1A1A2059105).

References

1. Winkelman JW. Clinical practice. Insomnia Disorder. NEJM. 2015;373(15):1437–44.
2. Ohayon MM. Epidemiology of insomnia: what we know and what we still need to learn. Sleep Med Rev. 2002;6(2):97–111.
3. Buysse DJ. Insomnia. JAMA. 2013;309(7):706–16.
4. Chung KF, Yeung WF, Ho FY, Yung KP, Yu YM, Kwok CW. Cross-cultural and comparative epidemiology of insomnia: the Diagnostic and statistical manual (DSM), International classification of diseases (ICD) and International classification of sleep disorders (ICSD). Sleep Med. 2015;16(4):477–82.
5. Blank M, Zhang J, Lamers F, Taylor AD, Hickie IB, Merikangas KR. Health correlates of insomnia symptoms and comorbid mental disorders in a nationally representative sample of US adolescents. Sleep. 2015;38(2):197–204.
6. Zachariae R, Lyby MS, Ritterband LM, O'Toole MS. Efficacy of internet-delivered cognitive-behavioral therapy for insomnia-a systematic review and meta-analysis of randomized controlled trials. Sleep Med Rev. 2016;30:1–10.
7. Riemann D, Perlis ML. The treatments of chronic insomnia: a review of benzodiazepine receptor agonists and psychological and behavioral therapies. Sleep Med Rev. 2009;13(3):205–14.
8. Mitchell MD, Gehrman P, Perlis M, Umscheid CA. Comparative effectiveness of cognitive behavioral therapy for insomnia: a systematic review. BMC Fam Pract. 2012;13:40.
9. Strom L, Pettersson R, Andersson G. Internet-based treatment for insomnia: a controlled evaluation. J Consult Clin Psychol. 2004;72(1):113–20.
10. Ritterband LM, Thorndike FP, Gonder-Frederick LA, Magee JC, Bailey ET, Saylor DK, et al. Efficacy of an Internet-based behavioral intervention for adults with insomnia. Arch Gen Psychiatry. 2009;66(7):692–8.
11. Anderson KN, Goldsmith P, Gardiner A. A pilot evaluation of an online cognitive behavioral therapy for insomnia disorder-targeted screening and interactive Web design lead to improved sleep in a community population. Nat Sci Sleep. 2014;6:43–9.
12. Ho FY, Chung KF, Yeung WF, Ng TH, Kwan KS, Yung KP, et al. Self-help cognitive-behavioral therapy for insomnia: a meta-analysis of randomized controlled trials. Sleep Med Rev. 2015;19:17–28.
13. Pillai V, Anderson JR, Cheng P, Bazan L, Bostock S, Roth TA, et al. The anxiolytic effects of cognitive behavior therapy for insomnia: preliminary results from a web-delivered protocol. J Sleep Med Disord. 2015;2(2):a-7.

14. Suzuki E, Tsuchiya M, Hirokawa K, Taniguchi T, Mitsuhashi T, Kawakami N. Evaluation of an internet-based self-help program for better quality of sleep among Japanese workers: a randomized controlled trial. J Occup Health. 2008;50(5):387–99.
15. Vincent N, Penner S, Lewycky S. What predicts patients' perceptions of improvement in insomnia? J Sleep Res. 2006;15(3):301–8.
16. Seyffert M, Lagisetty P, Landgraf J, Chopra V, Pfeiffer PN, Conte ML, et al. Internet-delivered cognitive behavioral therapy to treat insomnia: a systematic review and meta-analysis. PLoS ONE. 2016;11(2):e0149139.
17. Blom K, Tarkian Tillgren H, Wiklund T, Danlycke E, Forssen M, Soderstrom A, et al. Internet-vs. group-delivered cognitive behavior therapy for insomnia: a randomized controlled non-inferiority trial. Behav Res Ther. 2015;70:47–55.
18. Lancee J, van Straten A, Morina N, Kaldo V, Kamphuis JH. Guided online or face-to-face cognitive behavioral treatment for insomnia: a randomized wait-list controlled trial. Sleep. 2016;39(1):183–91.
19. Price M, Gros DF, McCauley JL, Gros KS, Ruggiero KJ. Nonuse and dropout attrition for a web-based mental health intervention delivered in a post-disaster context. Psychiatry. 2012;75(3):267–84.
20. Luik AI, Bostock S, Chisnall L, Kyle SD, Lidbetter N, Baldwin N, et al. Treating depression and anxiety with digital cognitive behavioural therapy for insomnia: a real world NHS evaluation using standardized outcome measures. Behav Cogn Psychother. 2017;45(1):91–6.
21. Lancee J, Sorbi MJ, Eisma MC, van Straten A, van den Bout J. The effect of support on internet-delivered treatment for insomnia: does baseline depression severity matter? Behav Ther. 2014;45(4):507–16.
22. Lancee J, van den Bout J, van Straten A, Spoormaker VI. Baseline depression levels do not affect efficacy of cognitive-behavioral self-help treatment for insomnia. Depress Anxiety. 2013;30(2):149–56.
23. Christensen H, Batterham PJ, Gosling JA, Ritterband LM, Griffiths KM, Thorndike FP, et al. Effectiveness of an online insomnia program (SHUTi) for prevention of depressive episodes (the GoodNight Study): a randomised controlled trial. Lancet Psychiatry. 2016;3(4):333–41.
24. Blom K, Jernelov S, Kraepelien M, Bergdahl MO, Jungmarker K, Ankartjarn L, et al. Internet treatment addressing either insomnia or depression, for patients with both diagnoses: a randomized trial. Sleep. 2015;38(2):267–77.
25. Thorndike FP, Ritterband LM, Gonder-Frederick LA, Lord HR, Ingersoll KS, Morin CM. A randomized controlled trial of an internet intervention for adults with insomnia: effects on comorbid psychological and fatigue symptoms. J Clin Psychol. 2013;69(10):1078–93.
26. Ye YY, Zhang YF, Chen J, Liu J, Li XJ, Liu YZ, et al. Internet-based cognitive behavioral therapy for insomnia (ICBT-i) improves comorbid anxiety and depression-a meta-analysis of randomized controlled trials. PLoS ONE. 2015;10(11):e0142258.
27. Jasper K, Weise C, Conrad I, Andersson G, Hiller W, Kleinstauber M. Internet-based guided self-help versus group cognitive behavioral therapy for chronic tinnitus: a randomized controlled trial. Psychother Psychosom. 2014;83(4):234–46.
28. Weise C, Kleinstauber M, Andersson G. Internet-delivered cognitive-behavior therapy for tinnitus: a randomized controlled trial. Psychosom Med. 2016;78(4):501–10.
29. McGrath ER, Espie CA, Power A, Murphy AW, Newell J, Kelly C, et al. Sleep to lower elevated blood pressure: a randomized controlled trial (SLEPT). Am J Hypertens. 2017;30(3):319–27.
30. Ritterband LM, Bailey ET, Thorndike FP, Lord HR, Farrell-Carnahan L, Baum LD. Initial evaluation of an Internet intervention to improve the sleep of cancer survivors with insomnia. Psychooncology. 2012;21(7):695–705.
31. Espie CA, Kyle SD, Williams C, Ong JC, Douglas NJ, Hames P, et al. A randomized, placebo-controlled trial of online cognitive behavioral therapy for chronic insomnia disorder delivered via an automated media-rich web application. Sleep. 2012;35(6):769–81.

32. Ritterband LM, Thorndike FP, Ingersoll KS, Lord HR, Gonder-Frederick L, Frederick C, et al. Effect of a web-based cognitive behavior therapy for insomnia intervention with 1-year follow-up: a randomized clinical trial. JAMA Psychiatry. 2017;74(1):68–75.
33. Horsch C, Lancee J, Beun RJ, Neerincx MA, Brinkman WP. Adherence to technology-mediated insomnia treatment: a meta-analysis, interviews, and focus groups. J Med Internet Res. 2015;17(9):e214.
34. Horsch C, Spruit S, Lancee J, van Eijk R, Beun RJ, Neerincx M, et al. Reminders make people adhere better to a self-help sleep intervention. Health Technol. 2017;7(2):173–88.
35. Kuhn E, Weiss BJ, Taylor KL, Hoffman JE, Ramsey KM, Manber R, et al. CBT-I coach: a description and clinician perceptions of a mobile app for cognitive behavioral therapy for insomnia. J Clin Sleep Med. 2016;12(4):597–606.
36. Babson KA, Ramo DE. Baldini L, Vandrey R, Bonn-Miller MO. Mobile app-delivered cognitive behavioral therapy for insomnia: feasibility and initial efficacy among veterans with cannabis use disorders. JMIR Res Protoc. 2015;4(3):e87.
37. Koffel E, Kuhn E, Petsoulis N, Erbes CR, Anders S, Hoffman JE, et al. A randomized controlled pilot study of CBT-I Coach: feasibility, acceptability, and potential impact of a mobile phone application for patients in cognitive behavioral therapy for insomnia. Health Inf J. 2016.
38. Chen YX, Hung YP, Chen HC. Mobile application-assisted cognitive behavioral therapy for insomnia in an older adult. Telemed J E Health. 2016;22(4):332–4.
39. Chen Y-X, Chen H-C, Chen L-X, Hu J-W, Shie C-K, Lin Y-S, et al., editors. Enhancing adherence to cognitive behavioral therapy for insomnia through machine and social persuasion. In: Green Computing and Communications (GreenCom), 2013 (IEEE and Internet of Things (iThings/CPSCom), IEEE International Conference on and IEEE Cyber, Physical and Social Computing; 2013: IEEE).
40. Horsch CH, Lancee J, Griffioen-Both F, Spruit S, Fitrianie S, Neerincx MA, et al. Mobile phone-delivered cognitive behavioral therapy for insomnia: a randomized waitlist controlled trial. J Med Internet Res. 2017;19(4):e70.
41. Kyle SD, Hurry MED, Emsley R, Luik AI, Omlin X, Spiegelhalder K, et al. Effects of digital Cognitive Behavioural Therapy for Insomnia on cognitive function: study protocol for a randomised controlled trial. Trials. 2017;18(1):281.
42. Azevedo R, Cromley JG. Does training on self-regulated learning facilitate students' learning with hypermedia? J Educ Psychol. 2004;96(3):523.
43. Chu HC, Liu YM, Kuo FR. A mobile sleep-management learning system for improving students' sleeping habits by integrating a self-regulated learning strategy: randomized controlled trial. JMIR Mhealth Uhealth. 2018;6(10):e11557.
44. Baig MM, Antonescu-Turcu A, Ratarasarn K. Impact of sleep telemedicine protocol in management of sleep apnea: A 5-year VA experience. Telemed J E Health. 2016.
45. Singh J, Badr MS, Diebert W, Epstein L, Hwang D, Karres V, et al. American academy of sleep medicine (AASM) position paper for the use of telemedicine for the diagnosis and treatment of sleep disorders. J Clin Sleep Med. 2015;11(10):1187–98.
46. Kang SG, Kang JM, Cho SJ, Ko KP, Lee YJ, Lee HJ, et al. Cognitive behavioral therapy using a mobile application synchronizable with wearable devices for insomnia treatment: a pilot study. J Clin Sleep Med. 2017;13(4):633–40.
47. Montgomery-Downs HE, Insana SP, Bond JA. Movement toward a novel activity monitoring device. Sleep Breath. 2012;16(3):913–7.
48. Meltzer LJ, Hiruma LS, Avis K, Montgomery-Downs H, Valentin J. Comparison of a commercial accelerometer with polysomnography and actigraphy in children and adolescents. Sleep. 2015;38(8):1323–30.
49. Mantua J, Gravel N, Spencer RM. Reliability of sleep measures from four personal health monitoring devices compared to research-based actigraphy and polysomnography. Sensors (Basel). 2016;16(5).

50. de Zambotti M, Baker FC, Colrain IM. Validation of sleep-tracking technology compared with polysomnography in adolescents. Sleep. 2015;38(9):1461–8.
51. Toon E, Davey MJ, Hollis SL, Nixon GM, Horne RS, Biggs SN. Comparison of commercial wrist-based and smartphone accelerometers, actigraphy, and PSG in a clinical cohort of children and adolescents. J Clin Sleep Med. 2016;12(3):343–50.
52. Kang SG, Kang JM, Ko KP, Park SC, Mariani S, Weng J. Validity of a commercial wearable sleep tracker in adult insomnia disorder patients and good sleepers. J Psychosom Res. 2017;97:38–44.
53. de Zambotti M, Baker FC, Willoughby AR, Godino JG, Wing D, Patrick K, et al. Measures of sleep and cardiac functioning during sleep using a multi-sensory commercially-available wristband in adolescents. Physiol Behav. 2016;158:143–9.
54. Android rules S. Korean market in 2013: data: Yonhap News. http://english.yonhapnews.co.kr/business/2014/01/21/30/0501000000AEN20140121001000320F.html.
55. Luik AI, Kyle SD, Espie CA. Digital cognitive behavioral therapy (dCBT) for insomnia: a state-of-the-science review. Curr Sleep Med Rep. 2017;3(2):48–56.
56. Lancee J, van den Bout J, Sorbi MJ, van Straten A. Motivational support provided via email improves the effectiveness of internet-delivered self-help treatment for insomnia: a randomized trial. Behav Res Ther. 2013;51(12):797–805.
57. Espie CA, Emsley R, Kyle SD, Gordon C, Drake CL, Siriwardena AN, et al. Effect of digital cognitive behavioral therapy for insomnia on health, psychological well-being, and sleep-related quality of life: a randomized clinical trial. JAMA Psychiatry. 2018.
58. Cohn SP. Privacy and confidentiality in the nationwide health information network. http://www.ncvhshhsgov/060622lth.tm (2006).
59. Martinez-Perez B, de la Torre-Diez I, Lopez-Coronado M. Privacy and security in mobile health apps: a review and recommendations. J Med Syst. 2015;39(1):181.

MIX
Papier aus verantwortungsvollen Quellen
Paper from responsible sources
FSC® C105338

If you have any concerns about our products,
you can contact us on
ProductSafety@springernature.com

In case Publisher is established outside the EU,
the EU authorized representative is:
Springer Nature Customer Service Center GmbH
Europaplatz 3, 69115 Heidelberg, Germany

Printed by Libri Plureos GmbH
in Hamburg, Germany